ABSTRACT ALGEBRA
An Introduction

second edition

THOMAS W. HUNGERFORD
Cleveland State University

BROOKS/COLE

™

THOMSON LEARNING

Australia • Canada • Mexico • Singapore • Spain •
United Kingdom • United States

BROOKS/COLE

THOMSON LEARNING

Developmental Editor: Patrick Farace
Marketing Manager: Nick Agnew
Production Manager: Alicia Jackson
Project Editor: Bonnie Boehme
Text Designer: Kathryn Needle
Art Director: Joan Wendt
Art and Design Coordinator: Kathleen Flanagan
Illustraton Supervisor: Sue Kinney

Illustrator: Progressive Information Technologies
Copy Editor: Sara Black
Cover Designer: Chazz Bjanes
Cover Printer: Lehigh Press
Compositor: Progressive Information Technologies
Printer: Courier Westford
Cover Image: Ron Lowery/The Stock Market

Printed in the United States of America

9   10   11   12   05   04   03

For more information about our products, contact us at:
Thomson Learning Academic Resource Center
1-800-423-0563

For permission to use material from this text, contact us by:
Phone: 1-800-730-2214
Fax: 1-800-730-2215
Web: http://www.thomsonrights.com

Library of Congress Catalog Number: 95-072840
ABSTRACT ALGEBRA, AN INTRODUCTION
Second Edition
ISBN: 0-03-010559-5

Asia
Thomson Learning
60 Albert Street, #15-01
Albert Complex
Singapore 189969

Australia
Nelson Thomson Learning
102 Dodds Street
South Melbourne, Victoria 3205
Australia

Canada
Nelson Thomson Learning
1120 Birchmount Road
Toronto, Ontario M1K 5G4
Canada

Europe/Middle East/Africa
Thomson Learning
Berkshire House
168-173 High Holborn
London WC1 V7AA
United Kingdom

Latin America
Thomson Learning
Seneca, 53
Colonia Polanco
11560 Mexico D.F.
Mexico

Spain
Paraninfo Thomson Learning
Calle/Magallanes, 25
28015 Madrid, Spain

Dedicated to
Vincent O. McBrien
and to the memory of
Raymond J. Swords, S.J.

Preface

◆

This book is intended for a first undergraduate course in modern abstract algebra. Linear algebra is not a prerequisite. The flexible design of the text (which is fully explained in the TO THE INSTRUCTOR section) makes it suitable for courses of various lengths and different levels of mathematical sophistication, ranging from a traditional abstract algebra course to one with a more applied flavor. As in the first edition, the emphasis is on clarity of exposition and the goal has been to produce a book that an average student can read with minimal outside assistance.

The motto of this revision has been "if it ain't broke, don't fix it." So the general approach and the topics covered are the same as in the first edition. There are a few changes in the order of the presentation. Chapters 8 to 10 have been renumbered so that the two group theory chapters are now adjacent to each other, as are the two field theory chapters. By popular demand, Lagrange's Theorem has been moved from the end of Chapter 7 to the middle and the material on the structure of finite groups integrated into the remainder of the chapter. The two sections on the symmetric and alternating groups, which were separated in the first edition, are now in the same chapter.

The breadth of coverage is substantially the same as before, but a number of discussions (e.g., subrings, principal ideals, and simple groups) have been expanded and two new ones have been added (vector spaces and an advanced section on the structure of finite groups). Minor improvements in exposition have been made throughout the book, and some significant rewriting has been done in a few cases. The majority of the exercises are the same, but ones that "didn't work" as teaching devices have been deleted or altered and some new exercises have been added. Virtually all of these changes are in response to the experience of classroom instructors (myself and others) and are designed to improve clarity and student learning.

There are a few changes in layout and format. Major definitions are now in shaded boxes. In addition, notice boxes have been inserted at appropriate places in the first half of the text to call the reader's attention to applications and excursions (from Chapters 12 to 16) that can be covered at that particular point.

One thing that has not changed is the treatment of rings, fields, and polynomials before groups. I still believe that this is the best way to introduce students to modern algebra, and the reception given the first edition of this book indicates that many other mathematicians agree.

Also unchanged are the *thematic development* and *organizational overview* that set this book apart. Chapters 1–7 and 9 are organized around three themes: arithmetic, congruence, and abstract structures. Each theme is developed first for the integers, then for polynomials, and finally for rings and groups. This enables students to see where many abstract concepts come from, why they are important, and how they relate to one another. The congruence theme is strongly emphasized in the development of quotient rings. Consequently, students can see that ideals, cosets, and quotient rings are a natural extension of familiar concepts in the integers rather than an unmotivated mystery. Dealing with congruence in polynomial domains before arbitrary rings not only eases the transition but also provides some nontrivial results on field extensions much earlier than customary.

To assist the student in forming a coherent overview, the interconnections of the basic areas of algebra are frequently pointed out in the text and in the THEMATIC TABLE OF CONTENTS on page xviii. Each horizontal line of that table shows how a particular theme or subtheme is developed first for integers, then for polynomials, rings, and groups. Each vertical column exhibits the interplay of all the themes for a particular topic.

The new edition has benefited from the comments of many students and mathematicians who used the first edition or reviewed the manuscript for the revision. My warm thanks to:

Tony Evans, Wright State University
Rodney Forcade, Brigham Young University
Richard Greechie, Loyola University
Kent Harris, Western Illinois University
Haruzo Hida, University of California, Los Angeles
David Leep, University of Kentucky
Wayne Lewis, Texas Tech University
Joseph Liang, University of South Florida
Frank B. Miles, California State University at Dominguez Hills
William Staton, University of Mississippi
Theodore Sundstrom, Grand Valley State University
Bhushan Wadhwa, Cleveland State University
Marysia T. Weiss, Hofstra University

Special thanks are due to the accuracy reviewers who reviewed both the original and final versions of the manuscript and read galleys and page proofs:

Lenny Jones, Shippensburg University
James Madden, Louisiana State University

James Madden and Daniel B. Shapiro (who prepared the Instructor's Manual) also provided a variety of helpful suggestions that greatly improved the book.

Nevertheless, I am fully responsible for any errors or other infelicities in the text.

It is a pleasure to acknowledge the invaluable assistance of the Saunders staff, particularly Liz Widdicombe, Patrick Farace, and Bonnie Boehme. Their fine work has made a difficult undertaking seem easy.

Finally, a very special thank you to my wife, Mary Alice, for her patience, understanding, and support during the preparation of this revision.

T.W.H

Topical Table of Contents

* Sections in the Core Course marked * may be omitted or postponed. See the beginning of each such section for specifics.

To The Instructor

The Core Course (Chapters 1–7) and selections from the Other Topics (Chapters 8–11) can be used for a traditional abstract algebra course. Sections and chapters of the Core Course marked with an asterisk in the TOPICAL TABLE OF CONTENTS (page viii) may be postponed or omitted in a shorter course; see the beginning of each such section for specifics.

For a course with a more applied flavor, virtually all the Excursions and Applications (Chapters 12–16) may be integrated into Chapters 1 to 7, beginning as early as Chapter 2. See the TOPICAL TABLE OF CONTENTS and the Interdependence of Chapters Chart on page xv for specifics.

The following information should enable you to design a course of appropriate emphasis, length, and level of sophistication.

PREREQUISITES AND PRELIMINARIES The usual preliminary material is in Appendices A to D. Depending on your situation, these appendices may be used as Chapter 0, introduced later as needed, or omitted entirely. Here are some guidelines:

Appendix A (Logic and Proof) This is a prerequisite for *any* abstract algebra book but often is not a formal part of the course.

Appendix B (Sets and Functions) The first part (basic terminology) is used from the beginning—and is probably familiar to most students. The middle part (Cartesian products and binary operations) is first needed in Section 3.1, and the last five pages (injective and surjective functions) in Section 3.3.

Appendix C (Induction) The so-called Principle of Complete Induction is first used in Section 4.1; ordinary induction first appears in Section 4.4. The equivalence of induction and well ordering (Theorem C.4) is not needed in the body of the text.

Appendix D (Equivalence Relations) The text is designed according to my personal preference, which is to introduce important examples in Sections 2.1, 5.1, 6.1, and 7.5 *before* introducing the formal definition of equivalence relation. So Appendix D need not be covered until Section 9.4.

If you prefer to define equivalence relations first and treat the major examples as special cases, you may cover Appendix D prior to Section 2.1 and then skip an occasional proof in the sections listed above.

SUPPLEMENTARY MATERIAL   Some topics that did not fit conveniently into the main presentation are in Appendices E to G.

Appendix E   (The Binomial Theorem) Although it is used only in Section 9.6 and occasional exercises elsewhere, the theorem is relevant to the discussion of arithmetic in rings (Section 3.2, page 59).

Appendix F   (Matrix Algebra) This material is a prerequisite for Chapter 16 but may be omitted by students who have had linear algebra. It can also be used for additional examples of noncommutative rings in Chapter 3.

Appendix G   (Polynomials) This provides a formal definition of polynomials and indeterminates. It may be included in Section 4.1, as noted there, if desired.

EXERCISES   Exercises in Group A involve routine calculations or straightforward proofs. Those in Group B require a reasonable amount of thought, but the vast majority should be accessible to most students. Group C consists of difficult exercises.

Whenever possible, the exercises in each group are arranged in the order that the corresponding topics are treated in the text. Consequently, some relatively easy exercises in groups A or B may appear near the end of that group.

Answers (or hints) for approximately half of the odd-numbered exercises are given at the end of the book. Answers for the remaining exercises are in the Instructor's Manual, which was prepared by Daniel B. Shapiro and is available from Brooks/Cole.

CHAPTER AND SECTION INTERDEPENDENCE   See next page.

CHAPTER AND SECTION INTERDEPENDENCE Unless noted otherwise in the chapter introduction, each section in a chapter depends on the preceding ones. Chapter interdependence is given by the chart on the next page, in which A $\longrightarrow$ B means that Chapter A is a prerequisite for Chapter B. A broken arrow A $\dashrightarrow$ B indicates that Chapter B depends only on parts of Chapter A; see the TOPICAL TABLE OF CONTENTS on page viii for specifics. The chart is broken down into sections for Chapters 13 and 16.

Brooks/Cole may provide complimentary instructional aids and supplements or supplement packages to those adopters qualified under our adoption policy. Please contact your sales representative for more information. If as an adopter or potential user you receive supplements you do not need, please return them to your sales representative or send them to

Attn: Returns Department
Troy Warehouse
465 South Lincoln Drive
Troy, MO 63379

CHAPTER INTERDEPENDENCE

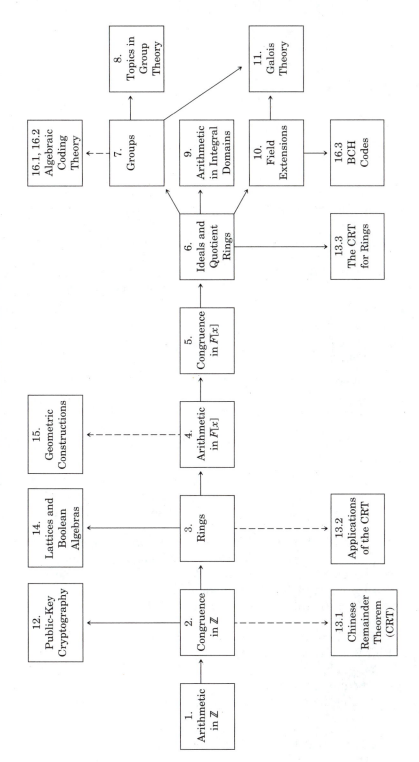

To The Student

OVERVIEW This book begins with things you've seen before: grade-school arithmetic and the algebra of polynomials from high school. We shall then see that these familiar objects fit into a larger framework—that they are just special cases of certain abstract algebraic structures.

To assist you in the essential task of tying the various topics together in a coherent fashion in your mind, the book is organized around several themes. Here is a brief description of the three most important themes in the first seven chapters:

Arithmetic You will see how the familiar properties of division, remainders, factorization, and primes in the integers carry over to polynomials and then to more general algebraic systems.

Congruence Congruence of integers may be familiar to you in the form of "clock arithmetic."* But the concept is deeper and more far-reaching than you might suspect. It leads to new miniature arithmetic systems that serve as a model for what can be done with polynomials and other systems. Congruence and the related concept of a quotient object are the keys to understanding abstract algebra.

Abstract Algebraic Structures The essential properties of familiar systems (integers, rational numbers, polynomials) are identified. Then we consider other mathematical systems that share some or all of these properties.

PROOFS The emphasis in this course, much more than in high-school algebra, is on the rigorous, logical development of the subject. If you are one of the many students who have had little experience with mathematical logic and reading or writing proofs, you should begin by reading Appendix A. It summarizes the basic rules of logic and the proof techniques that are used throughout this book.

Read the text with pencil and paper in hand *before* looking at the exercises. When you read the statement of a theorem, be sure you know the meaning of all the terms in the statement of the theorem. For example, if it says "every finite integral domain is a field," review the definitions of "integral domain" and "field"—if necessary, use the index to find the definitions.

---

* When the hour hand of a clock moves 3 hours or 15 hours from 12, it ends in the same position, so $3 = 15$ on the clock; similarly $0 = 12$. If the hour hand starts at 12 and moves 8 hours, then moves an additional 9 hours, it finishes at 5; so $8 + 9 = 5$ on the clock.

Once you understand what the theorem *claims* is true, then turn to the proof. Remember: There is a great deal of difference between *understanding* a proof in the text and *constructing* one yourself. Just as you can appreciate a new building without being an architect, you can verify the validity of proofs presented by others *even if you can't see how anyone ever thought of doing it this way in the first place.*

Begin by skimming through the proof to get an idea of its general outline before worrying about the details in each step. It's easier to understand an argument if you know approximately where it's headed. Then go back to the beginning and read the proof carefully, line by line. If it says "such and such is true by Theorem 5.18," check to see just what Theorem 5.18 says and be sure you understand why it applies here. It you get stuck, take that part on faith and finish the rest of the proof. Then go back and see if you can figure out the sticky point.

When you're really stuck, *ask your instructor.* He or she will welcome questions that arise from a serious effort on your part.

EXERCISES Mathematics is not a spectator sport. You can't expect to learn mathematics without *doing* mathematics, any more than you could learn to swim without getting in the water. That's why there are so many exercises in this book.

The exercises in groups A are usually straightforward. If you can't do almost all of them, you don't really understand the material. The exercises in group B often require a reasonable amount of thought—and for most of us, some trial and error as well. But the vast majority of them are within your grasp. The exercises in group C are usually difficult . . . a good test for strong students.

Many of the exercises will ask you to prove something. As you build up your skill in reading the proofs of others (as discussed above), you will find it easier to make proofs of your own. The proof techniques presented in Appendix A may also be helpful.

Answers (or hints) for approximately half of the odd-numbered exercises are given at the end of the book.

KEEPING IT ALL STRAIGHT The different branches of algebra are connected in a myriad of ways that are not always clear to either the beginner or the experienced practitioner. So it's no wonder that students sometimes have trouble seeing how the various topics tie together, or even *if* they do. Keeping in mind the three themes mentioned above will help, especially if you

regularly consult the THEMATIC TABLE OF CONTENTS

on the next page. Unlike the usual table of contents, which lists the topics in numerical order by chapter and section, the Thematic Table is arranged in logical order, so you can see how everything fits together.

Thematic Table of Contents

DIRECTIONS

Generally speaking, each block represents a generalization of the block immediately to its left. Reading from left to right shows how the theme or subtheme listed in the left-hand column is developed first in the integers, then in polynomials, and finally in rings and groups. Each vertical column shows how all the themes are carried out for the topic listed at the top of the column.

RINGS	GROUPS
9. Arithmetic in Integral Domains 9.1 Euclidean Domains	
{ 9.1 Euclidean Domains 9.2 Principal Ideal Domains and Unique Factorization Domains 9.3 Factorization of Quadratic Integers	
6. Ideals and Quotient Rings 6.1 Ideals and Congruence	**7. Groups** { 7.5 Congruence 7.6 Normal Subgroups
6.2 Quotient Rings and Homomorphisms	{ 7.7 Quotient Groups 7.8 Quotient Groups and Homomorphisms
6.3 The Structure of R/I When I Is Prime or Maximal	
3. Rings 3.1 Definition and Examples of Rings	**7. Groups** 7.1 Definition and Examples of Groups
3.2 Basic Properties of Rings	{ 7.2 Basic Properties of Groups 7.3 Subgroups
3.3 Isomorphism	7.4 Isomorphism
	7.5 Lagrange's Theorem 7.9 The Symmetric and Alternating Groups 7.10 The Simplicity of A_n

Part

1

THE CORE COURSE

CHAPTER **1**

Arithmetic in $\mathbb{Z}$ Revisited

$\blacklozenge$

$\mathbf{A}$lgebra grew out of arithmetic and depends heavily on it. So we begin our study of abstract algebra with a review of those facts from arithmetic that are used frequently in the rest of this book and provide a model for much of the work we do. We stress primarily the underlying pattern and properties rather than methods of computation. Nevertheless, the fundamental concepts are ones that you have seen before.

1.1 THE DIVISION ALGORITHM

Our starting point is the set of all integers $\mathbb{Z} = \{0, \pm 1, \pm 2, \ldots\}$. We assume that you are familiar with the arithmetic of integers and with the usual order relation ($<$) on the set $\mathbb{Z}$. We also assume the

WELL-ORDERING AXIOM *Every nonempty subset of the set of nonnegative integers contains a smallest element.*

If you think of the nonnegative integers laid out on the usual number line, it is intuitively plausible that each subset contains an element that lies to the left of all the other elements in the subset—that is the smallest element. On the other hand, the Well-Ordering Axiom does not hold in the set $\mathbb{Z}$ of all integers (there is no smallest negative integer). Nor does it hold in the set of all nonnegative rational numbers (the subset of all positive rationals does not contain a smallest element because, for any positive rational number r, there is always a smaller positive rational—for instance, $r/2$).

Consider the following grade school division problem:

$$
\begin{array}{r}
\text{Quotient} \longrightarrow \quad 11 \\
\text{Divisor} \longrightarrow 7\overline{)82} \\
\text{Dividend} \longrightarrow \quad 7 \\
\hline
12 \\
\quad 7 \\
\hline
\text{Remainder} \longrightarrow \quad 5
\end{array}
\qquad
\begin{array}{r}
\text{Check:} \quad 11 \longleftarrow \text{Quotient} \\
\times 7 \longleftarrow \text{Divisor} \\
\hline
77 \\
+5 \longleftarrow \text{Remainder} \\
\hline
82 \longleftarrow \text{Dividend}
\end{array}
$$

The division process stops when we reach a remainder that is less than the divisor. All the essential facts are contained in the checking procedure, which may be verbally summarized like this:

$$\text{dividend} = (\text{divisor})(\text{quotient}) + (\text{remainder}).$$

Here is a formal statement of this idea, in which the dividend is denoted by a, the divisor by b, the quotient by q, and the remainder by r:

THEOREM 1.1   (THE DIVISION ALGORITHM)   *Let a, b be integers with $b > 0$. Then there exist unique integers q and r such that*

$$a = bq + r \qquad and \qquad 0 \le r < b.$$

This theorem allows for the possibility that the dividend a may be negative. This is one reason the theorem contains the additional statement that the quotient q and the remainder r are *unique*. In grade school the idea that a division problem might have more than one correct answer never arises, but without some restrictions on the remainder, such a result is quite possible here. Simply requiring the remainder r to be less than the divisor b does *not* guarantee a unique quotient or remainder. For instance, if $a = -14$ and $b = 3$, then there are two possibilities for q and r:

$$q = -4, r = -2: \qquad -14 = 3(-4) - 2 \qquad \text{with} \qquad -2 < 3$$

and

$$q = -5, r = 1: \qquad -14 = 3(-5) + 1 \qquad \text{with} \qquad 1 < 3.$$

When the remainder r is also required to be nonnegative (as in the theorem), then we do have uniqueness, as will be shown in the proof.

The fundamental idea underlying the proof of Theorem 1.1 is that division is just repeated subtraction. For example, the division of 82 by 7 is just a shorthand method for repeatedly subtracting 7:

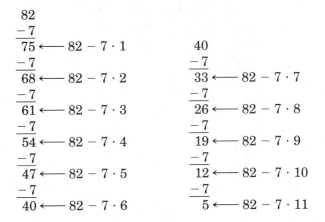

The subtractions continue until you reach a nonnegative number less than 7 (in this case 5). The number 5 is the remainder, and the *number* of multiples of 7 that were subtracted (namely, 11, as shown at the right of the subtractions) is the quotient.

The first part of the proof of Theorem 1.1 is essentially a formalization of this subtraction process. Since a is being divided by b, we will successively subtract b from a and consider numbers of the form $a - bx$, where x is any integer (just as we considered $82 - 7x$ with $x = 1, 2, \ldots, 11$ above). The smallest such *nonnegative* number is the remainder, and the corresponding value of x is the quotient (in the example, $5 = 82 - 7 \cdot 11$ was the remainder and $x = 11$ the quotient).

Proof of Theorem 1.1*   Let a, b be fixed integers with $b > 0$. Consider the set S of all integers of the form

$$a - bx, \qquad \text{where } x \text{ is an integer and } a - bx \geq 0.$$

We first show that the set S is nonempty by finding a value of x for which $a - bx \geq 0$. Since b is a positive integer, we must have $b \geq 1$. Multiplying both sides of this inequality by $|a|$, we see that

$$b|a| \geq |a| \geq -a$$

and hence that $a + b|a| \geq 0$. Now let $x = -|a|$ and consider $a - bx$:

$$a - bx = a - b(-|a|) = a + b|a| \geq 0.$$

This means that when $x = -|a|$, the number $a - bx$ is an element of S. Therefore S is nonempty.

---

* For an alternate proof by induction of part of the theorem, see Appendix C.

By the Well-Ordering Axiom, S contains a smallest element—call it r. Since $r \in S$, r is of the form $a - bx$ for some x, say $x = q$. Thus we have found integers q and r such that

$$r = a - bq \qquad \text{or, equivalently,} \qquad a = bq + r.$$

Since $r \in S$, we know that $r \geq 0$. We now show that $r < b$. Suppose, on the contrary, that $r \geq b$. Then $r - b \geq 0$, so that

$$0 \leq r - b = (a - bq) - b = a - b(q + 1).$$

Since $a - b(q + 1)$ is nonnegative, it is an element of S by definition. But since b is positive, it is certainly true that $r - b < r$. Thus

$$a - b(q + 1) = r - b < r.$$

The last inequality states that $a - b(q + 1)$—which is an element of S—is less than r, the *smallest* element of S. This is a contradiction. So we must have $r < b$. Hence we have found integers q and r such that $a = bq + r$ and $0 \leq r < b$.

To complete the proof, we must show that q and r are the *only* numbers with these properties (that's what "unique" means in the statement of the theorem). To do this, we suppose that for some integers q_1 and r_1 we also have $a = bq_1 + r_1$, with $0 \leq r_1 < b$. Then we prove that $q = q_1$ and $r = r_1$.

Since $a = bq + r$ and $a = bq_1 + r_1$, we have

$$bq + r = bq_1 + r_1$$

so that

(∗) $$b(q - q_1) = r_1 - r.$$

Furthermore, $0 \leq r < b$ and $0 \leq r_1 < b$. Multiplying the first inequality by -1 (and reversing the direction of the inequality), we see that

$$-b < -r \leq 0$$
$$0 \leq r_1 < b$$

Adding these two inequalities term by term and using Equation (∗) shows that

$$-b < r_1 - r < b$$
$$-b < b(q - q_1) < b$$
$$-1 < q - q_1 < 1.$$

But $q - q_1$ is an *integer* (because q and q_1 are integers) and the only integer strictly between -1 and 1 is 0. Therefore $q - q_1 = 0$ and $q = q_1$. Substituting $q - q_1 = 0$ in Equation (∗) shows that $r_1 - r = 0$ and hence $r = r_1$. Thus the quotient and remainder are unique, and the proof is complete. ◆*

---

* The symbol ◆ indicates the end of a proof.

A version of the Division Algorithm also holds when the divisor is negative.

COROLLARY 1.2   *Let a and c be integers with c ≠ 0. Then there exist unique integers q and r such that*

$$a = cq + r \qquad and \qquad 0 \leq r < |c|.$$

Proof   Exercise 4.   ◆

◆ **EXERCISES**

A.   **1.** Find the quotient and remainder when a is divided by b:

    (a) $a = 302, b = 19$ **(b)** $a = -302, b = 19$

    (c) $a = 0, b = 19$ **(d)** $a = 2000, b = 17$

    (e) $a = 2001, b = 17$ **(f)** $a = 2002, b = 17$

2. If you divide 59 by 7 on a calculator, the answer is displayed as 8.428571429. Using the Division Algorithm, we see that when 59 is divided by 7, the quotient is 8 and the remainder is 3. How can you use the calculator to obtain this integer quotient and remainder? More generally, develop an algorithm for use with your calculator that will produce the quotient and remainder for a given dividend and positive divisor. Be sure your algorithm works for both positive and negative dividends.

3. Use the calculator algorithm developed in Exercise 2 to find the quotient and remainder when a is divided by b:

    (a) $a = 517, b = 83$ **(b)** $a = -612, b = 74$

    (c) $a = 7,965,532; b = 127$ **(d)** $a = 8,126,493; b = 541$

    (e) $a = -9,217,645; b = 617$ **(f)** $a = 171,819,920; b = 4321$

B.   **4.** Prove Corollary 1.2. [*Hint:* First use Theorem 1.1 with $b = |c|$ to find q, r such that $a = |c| q + r$ and $0 \leq r < |c|$.]

5. Prove that the square of any integer a is either of the form $3k$ or of the form $3k + 1$ for some integer k. [*Hint:* By the Division Algorithm, a must be of the form $3q$ or $3q + 1$ or $3q + 2$.]

6. Use the Division Algorithm to prove that every odd integer is either of the form $4k + 1$ or of the form $4k + 3$ for some integer k.

7. Let n be a positive integer. Prove that a and c leave the same remainder when divided by n if and only if $a - c = nk$ for some integer k.

8. (a) Divide 5^2, 7^2, 11^2, 15^2, and 27^2 by 8 and note the remainder in each case.

   (b) Make a conjecture about the remainder when the square of an odd integer is divided by 8.

   (c) Prove your conjecture.

9. Prove that the cube of any integer has to be exactly one of these forms: $9k$ or $9k + 1$ or $9k + 8$ for some integer k.

1.2 DIVISIBILITY

An important case of division occurs when the remainder is 0, that is, when the divisor is a factor of the dividend. Here is a formal definition:

> **DEFINITION •** *Let a and b be integers with $b \neq 0$. We say that b **divides** a (or that b **is a divisor of** a, or that b **is a factor of** a) if $a = bc$ for some integer c. In symbols, "b divides a" is written $b \mid a$ and "b does not divide a" is written $b \nmid a$.*

EXAMPLE $3 \mid 24$ because $24 = 3 \cdot 8$, but $3 \nmid 17$. Negative divisors are allowed: $-6 \mid 54$ because $54 = (-6)(-9)$, but $-6 \nmid (-13)$.

EXAMPLE Every nonzero integer b divides 0 because $0 = b \cdot 0$. For every integer a, we have $1 \mid a$ because $a = 1 \cdot a$.

Remark If b divides a, then $a = bc$ for some c. Hence $-a = b(-c)$, so that $b \mid (-a)$. An analogous argument shows that every divisor of $-a$ is also a divisor of a. Therefore

a and $-a$ have the same divisors.

Remark Suppose $a \neq 0$ and $b \mid a$. Then $a = bc$, so that $|a| = |b||c|$. Consequently, $0 \leq |b| \leq |a|$. This last inequality is equivalent to $-|a| \leq b \leq |a|$. Therefore

(i) every divisor of the nonzero integer a is less than or equal to $|a|$;

(ii) a nonzero integer has only finitely many divisors.

All the divisors of the integer 12 are

$$1, -1, 2, -2, 3, -3, 4, -4, 6, -6, 12, -12.$$

Similarly, all the divisors of 30 are

$$1, -1, 2, -2, 3, -3, 5, -5, 6, -6, 10, -10, 15, -15, 30, -30.$$

The **common divisors** of 12 and 30 are the numbers that divide both 12 and 30, that is, the numbers that appear on both of the preceding lists:

$$1, -1, 2, -2, 3, -3, 6, -6.$$

The largest of these common divisors, namely 6, is called the "greatest common divisor" of 12 and 30. This is an example of the following definition.

> **DEFINITION•**   *Let a and b be integers, not both 0. The* **greatest common divisor (gcd)** *of a and b is the largest integer d that divides both a and b. In other words, d is the gcd of a and b provided that*
>
> *(i) $d \mid a$ and $d \mid b$;*
> *(ii) if $c \mid a$ and $c \mid b$, then $c \leq d$.*
>
> *The greatest common divisor of a and b is usually denoted (a,b).*

If a and b are not both 0, then their gcd exists and is unique. The reason is that a nonzero integer has only finitely many divisors, and so there are only a finite number of common divisors. Hence there must be a unique largest one. Furthermore, the greatest common divisor of a and b satisfies the inequality

$$(a,b) \geq 1$$

because 1 is a common divisor of a and b.

> **EXAMPLE** $(12,30) = 6$, as shown above. The only common divisors of 10 and 21 are 1 and -1. Hence $(10,21) = 1$. Two integers whose greatest common divisor is 1, such as 10 and 21, are said to be **relatively prime.**

> **EXAMPLE**   The common divisors of an integer a and 0 are just the divisors of a. If $a > 0$, then the largest divisor of a is clearly a itself. Hence, if $a > 0$, then $(a,0) = a$.

Listing all the divisors of two large integers in order to find their gcd can be quite time-consuming. A relatively quick method for finding gcd's in such cases is presented in Theorem 1.6.

We have seen that $6 = (12,30)$. A little arithmetic shows that something else is true here: 6 is a *linear combination* of 12 and 30. For instance,

$$6 = 12(-2) + 30(1) \quad \text{and} \quad 6 = 12(8) + 30(-3).$$

You can readily find other integers u and v such that $6 = 12u + 30v$. The following theorem shows that the same thing is possible for any greatest common divisor.

THEOREM 1.3 *Let a and b be integers, not both 0, and let d be their greatest common divisor. Then there exist (not necessarily unique) integers u and v such that $d = au + bv$. Furthermore, d is the smallest positive integer that can be written in the form $au + bv$.*

> **Warning**   Read the theorem carefully. The fact that $d = au + bv$ does *not* imply that $d = (a,b)$. See Exercise 27.

Proof of Theorem 1.3   Let S be the set of all linear combinations of a and b, that is

$$S = \{am + bn \mid m, n \in \mathbb{Z}\}.$$

We shall find a particular element of S and show that it is the gcd. First note that $a^2 + b^2 = aa + bb$ is in S and $a^2 + b^2 \geq 0$. Since a and b are not both 0, $a^2 + b^2$ must be positive. Therefore S contains positive integers and hence must contain a smallest positive integer by the Well-Ordering Axiom. Let t denote this smallest positive element of S. By the definition of S, we know that $t = au + bv$ for some integers u and v. We claim that t is the gcd of a and b, that is, $t = d$. To prove this, we first show that $t \mid a$. By the Division Algorithm, there are integers q and r such that $a = tq + r$, with $0 \leq r < t$. Consequently,

$$r = a - tq,$$
$$r = a - (au + bv)q = a - aqu - bvq,$$
$$r = a(1 - qu) + b(-vq).$$

Thus r is a linear combination of a and b, and hence $r \in S$. Since $r < t$ (the smallest positive element of S), we know that r is not positive. Since $r \geq 0$, the only possibility is that $r = 0$. Therefore, $a = tq + r = tq + 0 = tq$, so that $t \mid a$. A similar argument shows that $t \mid b$. Hence, t is a common divisor of a and b.

Let c be any other common divisor of a and b, so that $c \mid a$ and $c \mid b$. Then $a = cr$ and $b = cs$ for some integers r and s. Consequently,

$$t = au + bv = (cr)u + (cs)v = c(ru + sv).$$

The first and last parts of this equation show that $c \mid t$. Hence, $c \leq |t|$. But t is positive, so $|t| = t$. Thus $c \leq t$. This shows that t is the *greatest* common divisor d and completes the proof of the theorem.   ◆

COROLLARY 1.4 *Let a and b be integers, not both 0, and let d be a positive integer. Then d is the greatest common divisor of a and b if and only if d satisfies these conditions:*

 (i) $d \mid a$ and $d \mid b$;

 (ii) if $c \mid a$ and $c \mid b$, then $c \mid d$.

Proof Suppose $d = (a,b)$. Then d satisfies condition (i) by definition. To verify condition (ii), suppose $c \mid a$ and $c \mid b$, so that $a = cr$ and $b = cs$. By Theorem 1.3, there are integers u, v such that $d = au + bv$. Consequently, $d = au + bv = (cr)u + (cs)v = c(ru + sv)$, which says that $c \mid d$. Therefore d satisfies condition (ii).

 Conversely, suppose d is a positive integer that satisfies the two conditions. Then d is a common divisor of a and b by (i). If c is any other common divisor, then $c \mid d$ by (ii). Consequently, by Remark (i) on page 7, $c \le |d|$. But d is positive, so $|d| = d$. Hence, $c \le d$. Therefore d is the greatest common divisor. ◆

 The answer to the following question will be needed on several occasions. If $a \mid bc$, then under what conditions is it true that $a \mid b$ or $a \mid c$? It is certainly not always true, as this example shows:

$$6 \mid 3 \cdot 4, \quad \text{but} \quad 6 \nmid 3 \quad \text{and} \quad 6 \nmid 4.$$

Note that 6 has a nontrivial factor in common with 3 and another in common with 4. When a divisor of bc has no common factors (except ± 1) with either b or c, then there is a useful answer to the question.

THEOREM 1.5 *If $a \mid bc$ and $(a,b) = 1$, then $a \mid c$.*

Proof Since $(a,b) = 1$, Theorem 1.3 shows that $au + bv = 1$ for some integers u and v. Multiplying this equation by c shows that $acu + bcv = c$. But $a \mid bc$, so that $bc = ar$ for some r. Therefore

$$c = acu + bcv = acu + (ar)v = a(cu + rv).$$

The first and last parts of this equation show that $a \mid c$. ◆

The Euclidean Algorithm

 For any integer b, we know that b and $-b$ have the same divisors. Consequently, the common divisors of a and b are the same as the common divisors of a and $-b$. Therefore, the greatest common divisors must be the same, that is,

$(a,b) = (a,-b)$. Using similar arguments, we see that

$$(a,b) = (a,-b) = (-a,b) = (-a,-b).$$

So a method for finding the gcd of two positive integers can also be used to find the gcd of *any* two integers. Here is a reasonably efficient method:

THEOREM 1.6 (THE EUCLIDEAN ALGORITHM)* *Let a and b be positive integers with $a \geq b$. If $b \mid a$, then $(a,b) = b$. If $b \nmid a$, then apply the division algorithm repeatedly as follows:*

$$\begin{array}{ll} a = bq_0 + r_0 & 0 < r_0 < b \\ b = r_0 q_1 + r_1 & 0 \leq r_1 < r_0 \\ r_0 = r_1 q_2 + r_2 & 0 \leq r_2 < r_1 \\ r_1 = r_2 q_3 + r_3 & 0 \leq r_3 < r_2 \\ r_2 = r_3 q_4 + r_4 & 0 \leq r_4 < r_3 \\ \quad \vdots \end{array}$$

This process ends when a remainder of 0 is obtained. This must occur after a finite number of steps; that is, for some integer t:

$$\begin{array}{ll} r_{t-2} = r_{t-1} q_t + r_t & 0 < r_t < r_{t-1} \\ r_{t-1} = r_t q_{t+1} + 0 \end{array}$$

Then r_t, the last nonzero remainder, is the greatest common divisor of a and b.

Before proving the theorem, we consider a numerical example that should make the process clear. We use the Euclidean Algorithm to find $(324,148)$. We have $a = 324$, $b = 148$; using the Division Algorithm, we find that $q = 2$ and $r_0 = 28$:

(1) $\qquad 324 = 148 \cdot 2 + 28$

(2) $\qquad 148 = 28 \cdot 5 + 8$

(3) $\qquad 28 = 8 \cdot 3 + 4$

$\qquad\qquad 8 = 4 \cdot 2 + 0$

Note that the divisor in each line becomes the dividend in the next line and the remainder in each line becomes the divisor in the next line.

The last nonzero remainder is 4, and so $(324,148) = 4$. We now use backsubstitution in the Equations (1)–(3) to write 4 as a linear combination of 324

---

* This theorem will be used only in occasional exercises.

and 148:

$$4 = 28 - 8 \cdot 3 \qquad \text{[This is just Equation (3).]}$$

$$4 = 28 - (148 - 28 \cdot 5)3 \qquad \text{[Equation (2) has been used to rewrite the number 8.]}$$

$$4 = 28 - 148 \cdot 3 + 28 \cdot 15$$

$$4 = 28 \cdot 16 - 148 \cdot 3$$

$$4 = (324 - 148 \cdot 2)16 - 148 \cdot 3 \qquad \text{[Equation (1) has been used to rewrite the number 28.]}$$

$$4 = 324 \cdot 16 - 148 \cdot 32 - 148 \cdot 3$$

$$4 = 324 \cdot 16 + 148(-35)$$

The proof of Theorem 1.6 will be an immediate consequence of

LEMMA 1.7 *If $a, b, q, r \in \mathbb{Z}$ and $a = bq + r$, then $(a,b) = (b,r)$.*

Proof If c is a common divisor of a and b, then $a = cs$ and $b = ct$ for some integers s and t. Consequently,

$$r = a - bq = cs - (ct)q = c(s - tq).$$

Hence $c \mid r$, so that c is also a common divisor of b and r. Conversely, suppose e is a common divisor of b and r, so that $b = ex$ and $r = ey$. Then

$$a = bq + r = (ex)q + ey = e(xq + y).$$

Thus $e \mid a$, so that e is also a common divisor of a and b. Therefore the set S of all common divisors of a and b is the same as the set T of all common divisors of b and r. Hence the largest element in S, namely (a,b), is the same as the largest element in T, namely (b,r). ◆

Proof of Theorem 1.6 If $b \mid a$, then $a = bq + 0$, so that $(a,b) = (b,0) = b$ by Lemma 1.7. If $b \nmid a$, then repeated application of Lemma 1.7 (to each division listed in the statement of the theorem) shows that

$$(a,b) = (b,r_0) = (r_0,r_1) = (r_1,r_2) = \cdots$$
$$= (r_{t-2}, r_{t-1}) = (r_{t-1}, r_t) = (r_t, 0) = r_t. \quad ◆$$

◆ EXERCISES

A. **1.** Find the greatest common divisors:

    (a) $(56, 72)$ **(b)** $(24, 138)$

    (c) $(143, 227)$ **(d)** $(314, 159)$

(e) (306, 657)   **(f)** (272, 1479)

(g) (4144, 7696)   **(h)** (12378, 3054)

2. Prove that $b \mid a$ if and only if $(-b) \mid a$.

3. If $a \mid b$ and $b \mid c$, prove that $a \mid c$.

4. **(a)** If $a \mid b$ and $a \mid c$, prove that $a \mid (b + c)$.

   (b) If $a \mid b$ and $a \mid c$, prove that $a \mid (br + ct)$ for any $r, t \in \mathbb{Z}$.

5. If $a \mid b$ and $b \mid a$, is it true that $a = b$? If not, what is true?

6. If $a \mid b$ and $c \mid d$, prove that $ac \mid bd$.

7. Prove or disprove: If $a \mid (b + c)$, then $a \mid b$ or $a \mid c$.

8. If $r \in \mathbb{Z}$ and r is a nonzero solution of $x^2 + ax + b = 0$ (where $a, b \in \mathbb{Z}$), prove that $r \mid b$.

9. If $(a,0) = 1$, what can a possibly be?

10. Prove that $(n, n + 1) = 1$ for every integer n.

11. If $n \in \mathbb{Z}$, what are the possible values of

    (a) $(n, n + 2)$ **(b)** $(n, n + 6)$

12. Suppose that $(a,b) = 1$ and $(a,c) = 1$. Are any of the following statements false? Justify your answers.

    (a) $(ab, a) = 1$ **(b)** $(b, c) = 1$

    (c) $(bc, a) = 1$ **(d)** $(ab, c) = 1$

13. If $k = abc + 1$, then prove that $(k,a) = (k,b) = (k,c) = 1$.

14. Find the smallest positive integer in the given set:

    (a) $\{6u + 15v \mid u, v \in \mathbb{Z}\}$ **(b)** $\{12r + 17s \mid r, s \in \mathbb{Z}\}$

B. 15. Express each of the greatest common divisors (a,b) in Exercise 1 as a linear combination of a and b.

16. If $(a,b) = d$, prove that $\left(\dfrac{a}{d}, \dfrac{b}{d}\right) = 1$.

17. If $a \mid c$ and $b \mid c$, must ab divide c? What if $(a,b) = 1$?

18. If $c > 0$, prove that $(ca, cb) = c(a,b)$.

19. If $a \mid (b + c)$ and $(b,c) = 1$, prove that $(a,b) = 1 = (a,c)$.

20. Prove or disprove each of the following statements.

    (a) If $2 \nmid a$, then $4 \mid (a^2 - 1)$.

    (b) If $2 \nmid a$, then $8 \mid (a^2 - 1)$.

21. Prove that $(a, a + b) = d$ if and only if $(a,b) = d$.

22. Prove that $(a,b) = (a, b + at)$ for every $t \in \mathbb{Z}$.

23. Prove that $(a, (b,c)) = ((a,b), c)$.

24. If $(a,c) = 1$ and $(b,c) = 1$, prove that $(ab,c) = 1$.

25. Use induction to show that if $(a,b) = 1$, then $(a, b^n) = 1$ for all $n \geq 1$.*

26. Let a, b, $c \in \mathbb{Z}$. Prove that the equation $ax + by = c$ has integer solutions if and only if $(a,b) \mid c$.

27. **(a)** If a, b, u, $v \in \mathbb{Z}$ are such that $au + bv = 1$, prove that $(a,b) = 1$.

   (b) Show by example that if $au + bv = d > 1$, then (a,b) may not be d.

28. If $a \mid c$ and $b \mid c$ and $(a,b) = d$, prove that $ab \mid cd$.

29. If $c \mid ab$ and $(c,a) = d$, prove that $c \mid db$.

30. If $a_1, a_2, \ldots, a_n$ are integers, not all zero, then their **greatest common divisor** is the largest integer d such that $d \mid a_i$ for every i. Prove that there exist integers u_i such that $d = a_1 u_1 + a_2 u_2 + \cdots + a_n u_n$. [*Hint:* Adapt the proof of Theorem 1.3.]

31. The **least common multiple** of nonzero integers a and b is the smallest positive integer m such that $a \mid m$ and $b \mid m$; m is usually denoted $[a,b]$. Prove that

   (a) whenever $a \mid k$ and $b \mid k$, then $[a,b] \mid k$;

   (b) $[a,b] = ab/(a,b)$ if $a > 0$ and $b > 0$.

32. Prove that a positive integer is divisible by 3 if and only if the sum of its digits is divisible by 3. [*Hint:* $10^3 = 999 + 1$ and similarly for other powers of 10.]

33. Prove that a positive integer is divisible by 9 if and only if the sum of its digits is divisible by 9. [See Exercise 32.]

C. 34. Prove that

   (a) $(a,b) \mid (a + b, a - b)$;

   (b) if a is odd and b is even, then $(a,b) = (a + b, a - b)$;

   (c) if a and b are odd, then $2(a,b) = (a + b, a - b)$.

35. Prove or disprove each of the following statements. [*Hint:* See Exercise 20.]

   (a) If $2 \nmid a$, then $24 \mid (a^2 - 1)$.

   (b) If $2 \nmid a$ and $3 \nmid a$, then $24 \mid (a^2 - 1)$.

36. Prove that for every n, $(n + 1, n^2 - n + 1)$ is 1 or 3.

* Induction is discussed in Appendix C.

1.3 PRIMES AND UNIQUE FACTORIZATION

Every nonzero integer n except ± 1 has at least four distinct divisors, namely 1, -1, n, $-n$. Integers that have *only* these four divisors play a crucial role.

> **DEFINITION** • *An integer p is said to be **prime** if $p \neq 0, \pm 1$ and the only divisors of p are ± 1 and $\pm p$.*

EXAMPLE $3, -5, 7, -11, 13$, and -17 are prime, but 15 is not (because 15 has divisors other than ± 1 and ± 15, such as 3 and 5). The integer 4567 is prime; to prove this from the definition requires a tedious check of all its possible divisors.

It is not difficult to show that there are infinitely many distinct primes (Exercise 23). Because an integer p has the same divisors as $-p$, we see that

$$p \text{ is prime if and only if } -p \text{ is prime.}$$

If p and q are both prime and $p \mid q$, then p must be one of $1, -1, q, -q$. But since p is prime, $p \neq \pm 1$. Hence,

$$\text{if } p \text{ and } q \text{ are prime and } p \mid q, \text{ then } p = \pm q.$$

Under what conditions does a divisor of a product bc necessarily divide b or c? Theorem 1.5 gave one answer to this question. Here is another.

THEOREM 1.8 *Let p be an integer with $p \neq 0, \pm 1$. Then p is prime if and only if p has this property:*

$$\textit{whenever } p \mid bc, \textit{ then } p \mid b \textit{ or } p \mid c.$$

Proof Suppose p is prime and $p \mid bc$. Consider the gcd of p and b. Now (p,b) must be a positive divisor of the prime p. So the only possibilities are $(p,b) = 1$ and $(p,b) = \pm p$ (whichever is positive). If $(p,b) = \pm p$, then $p \mid b$. If $(p,b) = 1$, then $p \mid c$ by Theorem 1.5. In every case, therefore, $p \mid b$ or $p \mid c$. The converse is left to the reader (Exercise 4). ◆

COROLLARY 1.9 *If p is prime and $p \mid a_1 a_2 \cdots a_n$, then p divides at least one of the a_i.*

Proof If $p \mid a_1(a_2 a_3 \cdots a_n)$, then $p \mid a_1$ or $p \mid a_2 a_3 \cdots a_n$ by Theorem 1.8. If $p \mid a_1$, we are finished. If $p \mid a_2(a_3 a_4 \cdots a_n)$, then $p \mid a_2$ or $p \mid a_3 a_4 \cdots a_n$ by Theorem 1.8 again. If $p \mid a_2$, we are finished; if not, continue this process, using Theorem 1.8 repeatedly. After at most n steps, there must be an a_i that is divisible by p. ◆

Choose an integer other than 0, ± 1. If you factor it "as much as possible," you will find that it is a product of one or more primes. For example,

$$12 = 4 \cdot 3 = 2 \cdot 2 \cdot 3,$$
$$60 = 12 \cdot 5 = 2 \cdot 2 \cdot 3 \cdot 5,$$
$$113 = 113 \text{ (prime)}.$$

In this context, *we allow the possibility of a "product" with just one factor* in case the number we begin with is actually a prime. What was done in these examples can always be done:

THEOREM 1.10   *Every integer n except* 0, ± 1 *is the product of primes.*

Proof   First note that if n is a product of primes, say $n = p_1 p_2 \cdots p_k$, then $-n = (-p_1)p_2 \cdots p_k$ is also a product of primes. Consequently, we need prove the theorem only when $n > 1$. Let S be the set of all integers greater than 1 that are *not* the product of primes. We shall show that S is empty. Assume on the contrary that S is nonempty. Then S contains a smallest integer m by the Well-Ordering Axiom. Since $m \in S$, m is not itself prime. Hence m must have positive divisors other than 1 or m, say $m = ab$ with $1 < a < m$ and $1 < b < m$. Since both a and b are less than m (the smallest element of S), neither a nor b is in S. By the definition of S, both a and b *are* the product of primes, say

$$a = p_1 p_2 \cdots p_r \quad \text{and} \quad b = q_1 q_2 \cdots q_s,$$

with $r \geq 1$, $s \geq 1$, and each p_i, q_j prime. Therefore

$$m = ab = p_1 p_2 \cdots p_r q_1 q_2 \cdots q_s$$

is a product of primes, so that $m \notin S$. We have reached a contradiction: $m \in S$ and $m \notin S$. Therefore, S must be empty.   ◆

An integer other than 0, ± 1 that is not prime is called **composite.** Although a composite integer may have several different prime factorizations, such as

$$45 = 3 \cdot 3 \cdot 5,$$
$$45 = (-3) \cdot 5 \cdot (-3),$$
$$45 = 5 \cdot 3 \cdot 3,$$
$$45 = (-5) \cdot (-3) \cdot 3,$$

these factorizations are essentially the same. The only differences are the order of the factors and the insertion of minus signs. You can readily convince yourself that every prime factorization of 45 has exactly three prime factors, say $q_1 q_2 q_3$. Furthermore, by rearranging and relabeling the q's, you will always have $3 = \pm q_1$, $3 = \pm q_2$, and $5 = \pm q_3$. This is an example of

THEOREM 1.11 (THE FUNDAMENTAL THEOREM OF ARITHMETIC) *Every integer n except 0, ± 1 is a product of primes. This prime factorization is unique in the following sense: If*

$$n = p_1 p_2 \cdots p_r \quad and \quad n = q_1 q_2 \cdots q_s$$

with each p_i, q_j prime, then $r = s$ (that is, the number of factors is the same) and after reordering and relabeling the q's,

$$p_1 = \pm q_1, \qquad p_2 = \pm q_2, \qquad p_3 = \pm q_3, \ldots, p_r = \pm q_r.$$

Proof Every integer n except 0, ± 1 has at least one prime factorization by Theorem 1.10. Suppose that n has two prime factorizations, as listed in the statement of the theorem. Then

$$p_1(p_2 p_3 \cdots p_r) = q_1 q_2 q_3 \cdots q_s,$$

so that $p_1 \mid q_1 q_2 \cdots q_s$. By Corollary 1.9, p_1 must divide one of the q_j. By reordering and relabeling the q's if necessary, we may assume that $p_1 \mid q_1$. Since p_1 and q_1 are prime, we must have $p_1 = \pm q_1$. Consequently,

$$\pm q_1 p_2 p_3 \cdots p_r = q_1 q_2 q_3 \cdots q_s.$$

Dividing both sides by q_1 shows that

$$p_2(\pm p_3 p_4 \cdots p_r) = q_2 q_3 q_4 \cdots q_s,$$

so that $p_2 \mid q_2 q_3 \cdots q_s$. By Corollary 1.9, p_2 must divide one of the q_j; as before, we may assume $p_2 \mid q_2$. Hence, $p_2 = \pm q_2$ and

$$\pm q_2 p_3 p_4 \cdots p_r = q_2 q_3 q_4 \cdots q_s.$$

Dividing both sides by q_2 shows that

$$p_3(\pm p_4 \cdots p_r) = q_3 q_4 \cdots q_s.$$

We continue in this manner, repeatedly using Corollary 1.9 and eliminating one prime on each side at every step. If $r > s$, then after s steps all the q's will be eliminated and we will have $\pm p_{s+1} p_{s+2} \cdots p_r = 1$. Since the only divisors of 1 are ± 1 and $p_r \neq \pm 1$ (because p_r is prime), this conclusion is a contradiction. A similar argument shows that the assumption $s > r$ also leads to a contradiction. Therefore we must have $r = s$; and after r steps the elimination process ends with $p_1 = \pm q_1, p_2 = \pm q_2, \ldots, p_r = \pm q_r$. ◆

If consideration is restricted to positive integers, then there is a stronger version of unique factorization:

COROLLARY 1.12 *Every integer $n > 1$ can be written in one and only one way in the form $n = p_1 p_2 p_3 \cdots p_r$, where the p_i are positive primes such that $p_1 \leq p_2 \leq p_3 \leq \cdots \leq p_r$.*

Proof Exercise 5. ◆

◆ EXERCISES

A. **1.** Express each number as a product of primes:

    (a) 5040   **(b)** -2345

    (c) 45,670   **(d)** 2,042,040

2. Let p be an integer other than $0, \pm 1$. Prove that p is prime if and only if for each $a \in \mathbb{Z}$ either $(a,p) = 1$ or $p \mid a$.

3. Let p be an integer other than $0, \pm 1$. Prove that p is prime if and only if it has this property: Whenever r and s are integers such that $p = rs$, then $r = \pm 1$ or $s = \pm 1$.

4. Let p be an integer other than $0, \pm 1$ with this property: Whenever b and c are integers such that $p \mid bc$, then $p \mid b$ or $p \mid c$. Prove that p is prime. [*Hint:* If d is a divisor of p, say $p = dt$, then $p \mid d$ or $p \mid t$. Show that this implies $d = \pm p$ or $d = \pm 1$.]

5. Prove Corollary 1.12.

6. If p is prime and $p \mid a^n$, is it true that $p^n \mid a^n$? Justify your answer.

7. (a) List all the positive integer divisors of $3^s 5^t$, where $s, t \in \mathbb{Z}$ and $s, t > 0$.

    (b) If $r, s, t \in \mathbb{Z}$ are positive, how many positive divisors does $2^r 3^s 5^t$ have?

8. Prove that $(a,b) = 1$ if and only if there is no prime p such that $p \mid a$ and $p \mid b$.

9. If p is prime and $(a,b) = p$, then $(a^2, b^2) = ?$

10. Prove or disprove each of the following statements:

    (a) If p is prime and $p \mid (a^2 + b^2)$ and $p \mid (c^2 + d^2)$, then $p \mid (a^2 - c^2)$.

    (b) If p is prime and $p \mid (a^2 + b^2)$ and $p \mid (c^2 + d^2)$, then $p \mid (a^2 + c^2)$.

    (c) If p is prime and $p \mid a$ and $p \mid (a^2 + b^2)$, then $p \mid b$.

B. **11.** If $a = p_1^{r_1} p_2^{r_2} p_3^{r_3} \cdots p_k^{r_k}$ and $b = p_1^{s_1} p_2^{s_2} p_3^{s_3} \cdots p_k^{s_k}$, where $p_1, p_2, \ldots, p_k$ are distinct positive primes and each $r_i, s_i \geq 0$, then prove that

    (a) $(a,b) = p_1^{n_1} p_2^{n_2} p_3^{n_3} \cdots p_k^{n_k}$, where for each i, $n_i = $ minimum of r_i, s_i.

    (b) $[a,b] = p_1^{t_1} p_2^{t_2} p_3^{t_3} \cdots p_k^{t_k}$, where $t_i = $ maximum of r_i, s_i. [See Exercise 31 in Section 1.2.]

12. (a) If $3 \mid (a^2 + b^2)$, prove that $3 \mid a$ and $3 \mid b$. [*Hint:* If $3 \nmid a$, then $a = 3k + 1$ or $3k + 2$.]

    (b) If $5 \mid (a^2 + b^2 + c^2)$, prove that $5 \mid a$ or $5 \mid b$ or $5 \mid c$.

13. If $c^2 = ab$ and $(a,b) = 1$, prove that a and b are perfect squares.

14. Let $n = p_1^{r_1} p_2^{r_2} \cdots p_k^{r_k}$, where $p_1, p_2, \ldots, p_k$ are distinct primes and each $r_i \geq 0$. Prove that n is a perfect square if and only if each r_i is even.

15. Prove that $a \mid b$ if and only if $a^3 \mid b^3$.

16. If n is a positive integer, prove that there exist n consecutive composite integers. [*Hint:* Consider $(n + 1)! + 2, (n + 1)! + 3, (n + 1)! + 4, \ldots$.]

17. If $p \geq 5$ is prime, prove that $p^2 + 2$ is composite. [*Hint:* Consider the possible remainders when p is divided by 6.]

18. Prove or disprove: The sums

$$1 + 2 + 4, \qquad 1 + 2 + 4 + 8, \qquad 1 + 2 + 4 + 8 + 16, \ldots$$

are alternately prime and composite.

19. If $n \in \mathbb{Z}$ and $n \neq 0$, prove that n can be written uniquely in the form $n = 2^k m$, where $k \geq 0$ and m is odd.

20. (a) Prove that there are no nonzero integers a, b such that $a^2 = 2b^2$. [*Hint:* Use the Fundamental Theorem of Arithmetic or Theorem 1.8.]

(b) Prove that $\sqrt{2}$ is irrational. [*Hint:* Use proof by contradiction (Appendix A). Assume that $\sqrt{2} = a/b$ (with $a, b \in \mathbb{Z}$) and use part (a) to reach a contradiction.]

21. If p is a positive prime, prove that $\sqrt{p}$ is irrational. [See Exercise 20.]

22. (a) Prove that $\sqrt{30}$ is irrational.

(b) Prove that $\sqrt[3]{2}$ is irrational.

23. (Euclid) Prove that there are infinitely many primes. [*Hint:* Use proof by contradiction (Appendix A). Assume there are only finitely many primes p_1, $p_2, \ldots, p_k$, and reach a contradiction by showing that the number $p_1 p_2 \cdots p_k + 1$ is not divisible by any of $p_1, p_2, \ldots, p_k$.]

C. 24. Prove or disprove: If n is a positive integer, then $n = p + a^2$, where $a \in \mathbb{Z}$ and p is prime or $p = 1$.

25. Prove or disprove: If n is an integer and $n > 2$, then there exists a prime p such that $n < p < n!$.

26. Let $S = \{4n + 1 \mid n \in \mathbb{Z}, n \geq 0\}$. If $a, b \in S$, we say that a is an *S-divisor* of b provided that $b = ac$ for some $c \in S$. An *S-prime* is an element $q \neq 1$ of S whose only *S*-divisors are itself and 1.

(a) Prove that every element of S (except 1) is a product of *S*-primes. [*Hint:* Copy the proof of Theorem 1.10. Does anything have to be changed?]

(b) Use the number 441 to show that factorization as a product of *S*-primes need not be unique.

27. (a) Let a be a positive integer. If $\sqrt{a}$ is rational, prove that $\sqrt{a}$ is an integer.

(b) Let r be a rational number and a an integer such that $r^n = a$. Prove that r is an integer. [Part (a) is the case when $n = 2$.]

28. Let p, q be primes with $p \geq 5$, $q \geq 5$. Prove that $24 \mid (p^2 - q^2)$.

1.4 PRIMALITY TESTING*

The results of the preceding sections are difficult to use on large numbers (200 digits or more). In the past this made no difference. Recently, however, the need for secret codes in the electronic communication of business, military, and scientific information has given primality testing and factoring a new importance. One of the most promising new coding methods, the RSA cipher system, which is discussed in Chapter 12, makes use of large primes. Its security depends on the fact that factoring large numbers is very difficult.

In theory it is easy to determine if a given positive integer n is prime. You simply verify that n is not divisible by any of the integers from 2 through $n/2$ (the largest possible proper factor of n). Since any divisor of n is itself a product of primes, you need only check to see if n is divisible by the primes from 2 through $n/2$. Even with a computer this may be a formidable task. The following primality test reduces the amount of work considerably.

THEOREM 1.13 *Let $n > 1$. If n has no prime factor less than or equal to $\sqrt{n}$, then n is prime.*

By using this theorem, any number less than 1,000,000 can be tested for primality by performing at <u>most 168</u> divisions, because there are only 168 primes less than $1,000 = \sqrt{1,000,000}$.

Proof of Theorem 1.13   By Theorem 1.10, $n = p_1 p_2 \cdots p_k$ with each p_i prime. By hypothesis, $p_i > \sqrt{n}$ for every i. If there are two or more prime factors in this decomposition, then

$$n = p_1 p_2 p_3 \cdots p_k > \sqrt{n}\sqrt{n}p_3 \cdots p_k = np_3 \cdots p_k \geq n.$$

Thus $n > n$, which is a contradiction. Therefore, we must have $n = p_1$. ◆

The following process, known as the **Sieve of Erathosthenes,** can be used to determine all the positive primes less than or equal to a given positive integer n. List all the integers from 2 to n. The first number on the list is the

---

* This section is optional and may be omitted.

prime 2. Cross out all multiples of 2 that are larger than 2. The first number after 2 on the remaining list is the prime 3. Cross out all multiples of 3 that are larger than 3. The first number after 3 on the remaining list is the prime 5. Continue in this fashion. At each stage all multiples of a prime p (except p itself) are crossed out, and the first number after p on the remaining list is necessarily prime. Once you reach a prime greater than $\sqrt{n}$, all the integers that have not been crossed out must be prime by Theorem 1.13 (because they are not multiples of any prime less than $\sqrt{n}$).

The efficiency of the Sieve of Erathosthenes is obviously related to the *number* of primes less than or equal to n. This number can be approximated by the following theorem, which was conjectured by K. F. Gauss in 1793 (at age 16) and first proved by J. Hadamard and (independently) C. J. de la Vallee-Poussin in 1896. A proof of the theorem, which is beyond the scope of this book, is given in H. E. Rose [16].

THEOREM 1.14 (THE PRIME NUMBER THEOREM) *If n is a large positive integer, then the number $\pi(n)$ of positive primes less than or equal to n is approximately equal to $n / \ln n$, where $\ln n$ is the natural logarithm of n. More precisely,*

$$\lim_{n \to \infty} \left(\frac{\pi(n)}{n/\ln n} \right) = 1.$$

To test an integer n for primality by means of Theorem 1.13 requires dividing n by every prime less than or equal to $\sqrt{n}$. By Theorem 1.14 the number of primes less than or equal to $\sqrt{n}$ is approximately $\sqrt{n}/(\ln \sqrt{n}) = 2\sqrt{n}/\ln n)$. The fastest computers take at least $(\ln n)/10^6$ seconds to perform one such division. So it would take approximately

$$\frac{2\sqrt{n}}{\ln n} \cdot \frac{\ln n}{10^6} = \frac{2\sqrt{n}}{10^6} \text{ seconds}$$

to check that n was prime, assuming that you already knew all the primes less than or equal to $\sqrt{n}$. Using this direct approach, it would take more than 63 years to verify that a 30-digit number was prime.

Current computer-based primality tests for large numbers depend in part on probabilistic methods. It can be shown, for example, that every odd prime n has the property that $2^{n-1} - 1$ is divisible by n (Lemma 12.2). Unfortunately, the converse is not true: There are composite numbers n for which n divides $2^{n-1} - 1$. But there are relatively few of them. So if n does divide $2^{n-1} - 1$, then there is a very high probability that n is prime.

Although high probability is not the same as certainty, the probabilistic tests can be used to determine likely candidates. Then other tests can be used to prove primality. In 1995, the fastest such computer methods could determine if a 100-digit number is prime in 33 seconds. A 200-digit number took 8 minutes, and a 1000-digit number, one week.

There are a few much larger numbers that have been verified to be prime. They are the **Mersenne primes**—prime numbers of the form $2^p - 1$. Not all numbers of this form are prime, but some of them are, including $7 = 2^3 - 1$ and $2^{859,433} - 1$. This last number has 258,716 digits.

After these feats of computation in primality testing, it may come as a bit of a surprise that there are no similar results for factorization. Even if you know that a number is the product of just two large primes, factoring it may be virtually impossible. The following chart indicates the time needed to find the prime factorization of integers of various sizes, using a state-of-the-art computer and the most efficient known algorithm:*

Number of Decimal Digits	Approximate Time
100	less than a month
200	50,000 years
300	one billion years
500	5×10^{15} years

These times (accurate in 1995) will decrease in future years as better algorithms and more powerful supercomputers are developed. Nevertheless, the problem of factoring large numbers will undoubtedly remain formidable for years to come.

◆ EXERCISES

A. **1.** Which of the following numbers are prime:

    (a) 701   **(b)** 1009

    (c) 1949   **(d)** 1951

  2. Use the Sieve of Erathosthenes to find all positive primes less than 300.

  3. Primes p and q are said to be twin primes if $q = p + 2$. For example, 3 and 5 are twin primes; so are 11 and 13. Find all pairs of positive twin primes less than 200.

  4. (a) Verify that $2^5 - 1$ and $2^7 - 1$ are prime.

    (b) Show that $2^{11} - 1$ is not prime.

---

* These figures apply to the hardest cases—numbers that are the product of large primes of approximately the same digit length. A number with some small prime factors can often be factored much more quickly.

5. Prove that every composite three-digit positive integer has a prime factor less than or equal to 31.

B. **6.** Let $n > 1$. If n has no prime factor less than or equal to $\sqrt[3]{n}$, prove that n is either prime or the product of two primes.

7. Let $p > 1$. If $2^p - 1$ is prime, prove that p is prime. [*Hint:* Prove the contrapositive: If p is composite, so is $2^p - 1$. *Note:* The converse is false by Exercise 4(b).]

8. Let a, b be nonnegative integers not both 1. If k is odd and $k \geq 3$, prove that $a^k + b^k$ is not prime.

9. If $k \geq 1$ and $2^k + 1$ is prime, prove that $k = 2^t$ for some t. [*Hint:* Use Exercise 8 above and Exercise 19 in Section 1.3.]

10. If n is a positive integer of the form $4k + 3$, prove that n has a prime factor of the form $4k + 3$.

11. Show that there are infinitely many primes of the form $4k + 3$. [*Hint:* See Exercise 10.]

12. Find all positive integers n such that n, $n + 2$, and $n + 4$ are all prime.

13. Let p be prime and $1 \leq k < p$. Prove that p divides the binomial coefficient $\dbinom{p}{k}$. $\left[\text{Recall that } \dbinom{p}{k} = \dfrac{p!}{k!(p - k)!} . \right]$

14. Let p_n be the nth positive prime in the usual ordering. Use induction to prove that $p_n \leq 2^{2^{n-1}}$. [*Hint:* First verify that $2^n - 1 = 1 + 2 + 2^2 + \cdots + 2^{n-1}$; then use this fact in your proof.]

15. If you have a programmable calculator, use Theorem 1.13 to write a program that will produce all the prime factors of a positive integer n. [The idea is to divide n successively by $k = 2, 3, 5, 7, 9, \ldots, \sqrt{n}$. When you divide by k and get a nonzero remainder (that is, the fractional part of n/k is nonzero), then k is not a factor; so move on to $k + 2$ (except when going from 2 to 3). When you get remainder 0 (that is, the fractional part of n/k is 0), then k is a factor of n and the next division can be performed on n/k. At each stage have the calculator display the factors that have been found. Although such a program is not the most efficient possible, it will allow you to factor a ten-digit integer in a few minutes or less.]

CHAPTER **2**

Congruence in ℤ and Modular Arithmetic

This chapter is a bridge between the arithmetic studied in the preceding chapter and the concepts of abstract algebra introduced in the next chapter. Basic concepts of arithmetic are extended here to include the idea of "congruence modulo n." Congruence leads to the construction of the set $\mathbb{Z}_n$ of all congruence classes. This construction will serve as a model for many similar constructions in the rest of this book. It also provides our first example of a system of arithmetic that shares many fundamental properties with ordinary arithmetic and yet differs significantly from it.

2.1 CONGRUENCE AND CONGRUENCE CLASSES

The concept of "congruence" may be thought of as a generalization of the equality relation. Two integers a and b are equal if their difference is 0 or, equivalently, if their difference is a multiple of 0. If n is a positive integer, we say that two integers are *congruent modulo n* if their difference is a multiple of n. To say that $a - b = nk$ for some integer k means that n divides $a - b$. So we have this formal definition:

> **DEFINITION** • *Let a, b, n be integers with $n > 0$. Then **a is congruent to b modulo n** [written "$a \equiv b \pmod{n}$"], provided that n divides $a - b$.*

EXAMPLE $17 \equiv 5$ (mod 6) because 6 divides $17 - 5 = 12$. Similarly, $4 \equiv 25$ (mod 7) because 7 divides $4 - 25 = -21$, and $6 \equiv -4$ (mod 5) because 5 divides $6 - (-4) = 10$.

Remark   In the notation "$a \equiv b$ (mod n)," the symbols "$\equiv$" and "(mod n)" are really parts of a single symbol; "$a \equiv b$" by itself is meaningless. Some texts write "$a \equiv_n b$" instead of "$a \equiv b$ (mod n)." Although this single-symbol notation is advantageous, we shall stick with the traditional "(mod n)" notation here.

The symbol used to denote congruence looks very much like an equal sign. This is no accident since the relation of congruence has many of the same properties as the relation of equality. For example, we know that equality is

reflexive: $a = a$ for every integer a;

symmetric: if $a = b$, then $b = a$;

transitive: if $a = b$ and $b = c$, then $a = c$.

We now see that congruence modulo n is also reflexive, symmetric, and transitive.

THEOREM 2.1   *Let n be a positive integer. For all a, b, $c \in \mathbb{Z}$,*

   (1) $a \equiv a$ (mod n);

   (2) if $a \equiv b$ (mod n), then $b \equiv a$ (mod n);

   (3) if $a \equiv b$ (mod n) and $b \equiv c$ (mod n), then $a \equiv c$ (mod n).

Proof   (1) $a - a = 0$ and $n \mid 0$. Hence $a \equiv a$ (mod n).

   (2) $a \equiv b$ (mod n) means that $a - b = nk$ for some integer k. Therefore, $b - a = -(a - b) = -nk = n(-k)$. The first and last parts of this equation say that $n \mid (b - a)$. Hence, $b \equiv a$ (mod n).

   (3) If $a \equiv b$ (mod n) and $b \equiv c$ (mod n), then by the definition of congruence, there are integers k and t such that $a - b = nk$ and $b - c = nt$. Therefore,

$$(a - b) + (b - c) = nk + nt$$
$$a - c = n(k + t).$$

Thus $n \mid (a - c)$ and, hence, $a \equiv c$ (mod n).   ◆

Several essential arithmetic and algebraic manipulations depend on this key fact:

   If $a = b$ and $c = d$, then $a + c = b + d$ and $ac = bd$.

We now show that the same thing is true for congruence.

THEOREM 2.2 *If a ≡ b (mod n) and c ≡ d (mod n), then*

(1) *a + c ≡ b + d (mod n);*

(2) *ac ≡ bd (mod n).*

Proof (1) By the definition of congruence, there are integers k and t such that $a - b = nk$ and $c - d = nt$. Therefore,

$$(a - b) + (c - d) = nk + nt$$
$$a + c - b - d = n(k + t)$$
$$(a + c) - (b + d) = n(k + t).$$

Thus n divides $(a + c) - (b + d)$ so that $a + c ≡ b + d$ (mod n).

(2) Using the fact that $-bc + bc = 0$, we have

$$ac - bd = ac + 0 - bd = ac - bc + bc - bd = (a - b)c + b(c - d)$$
$$= (nk)c + b(nt) = n(kc + bt).$$

The first and last parts of this equation say that $n \mid (ac - bd)$. Therefore, $ac ≡ bd$ (mod n). ◆

With the equality relation, it's easy to see what numbers are equal to a given number a—just a itself. With congruence, however, the story is different and leads to some interesting consequences.

> **DEFINITION•** *Let a and n be integers with n > 0. The **congruence class of a modulo n** (denoted [a]) is the set of all those integers that are congruent to a modulo n, that is,*
>
> $$[a] = \{b \mid b \in ℤ \quad and \quad b ≡ a \;(mod\; n)\}.$$

To say that $b ≡ a$ (mod n) means that $b - a = kn$ for some integer k or, equivalently, that $b = a + kn$. Thus

$$[a] = \{b \mid b ≡ a \;(mod\; n)\} = \{b \mid b = a + kn \text{ with } k \in ℤ\}$$
$$= \{a + kn \mid k \in ℤ\}.$$

EXAMPLE In congruence modulo 5, we have

$$[9] = \{9 + 5k \mid k \in ℤ\} = \{9, 9 \pm 5, 9 \pm 10, 9 \pm 15, \ldots\}$$
$$= \{\ldots, -11, -6, -1, 4, 9, 14, 19, 24, \ldots\}.$$

EXAMPLE   The meaning of the symbol "[ ]" depends on the context. In congruence modulo 3, for instance,

$$[2] = \{2 + 3k \mid k \,\epsilon\, \mathbb{Z}\} = \{ \ldots , -7, -4, -1, 2, 5, 8, \ldots \},$$

but in congruence modulo 5 the congruence class [2] is the set

$$\{2 + 5k \mid k \,\epsilon\, \mathbb{Z}\} = \{ \ldots , -13, -8, -3, 2, 7, 12, \ldots \}.$$

This ambiguity will not cause any difficulty in what follows because only one modulus will be discussed at a time.

EXAMPLE   In congruence modulo 3, the congruence class

$$[2] = \{ \ldots , -7, -4, -1, 2, 5, 8, \ldots \}.$$

Notice, however, that $[-1]$ is the same class because

$$[-1] = \{-1 + 3k \mid k \,\epsilon\, \mathbb{Z}\} = \{ \ldots , -7, -4, -1, 2, 5, \ldots \}.$$

Furthermore, $2 \equiv -1 \pmod 3$. This is an example of the following theorem.

THEOREM 2.3 $a \equiv c$ *(mod n) if and only if* $[a] = [c]$.

Observe that the proof of this theorem does not use the definition of congruence. Instead, it uses only the fact that congruence is reflexive, symmetric, and transitive (Theorem 2.1).

Proof of Theorem 2.3   Assume $a \equiv c \pmod n$. To prove that $[a] = [c]$, we first show that $[a] \subseteq [c]$. To do this, let $b \,\epsilon\, [a]$. Then by definition $b \equiv a \pmod n$. Since $a \equiv c \pmod n$, we have $b \equiv c \pmod n$ by transitivity. Therefore, $b \,\epsilon\, [c]$ and $[a] \subseteq [c]$. Reversing the roles of a and c in this argument and using the fact that $c \equiv a$ by symmetry show that $[c] \subseteq [a]$. Therefore, $[a] = [c]$.

Conversely, assume that $[a] = [c]$. Since $a \equiv a \pmod n$ by reflexivity, we have $a \,\epsilon\, [a]$ and, hence, $a \,\epsilon\, [c]$. By the definition of $[c]$, we see that $a \equiv c \pmod n$. ◆

If A and C are two sets, there are usually three possibilities: Either A and C are disjoint, or $A = C$, or $A \cap C$ is nonempty but $A \neq C$. With congruence classes, however, there are only two possibilities:

COROLLARY 2.4   *Two congruence classes modulo n are either disjoint or identical.*

Proof If $[a]$ and $[c]$ are disjoint, there is nothing to prove. Suppose that $[a] \cap [c]$ is nonempty. Then there is an integer b with $b \in [a]$ and $b \in [c]$. By the definition of congruence class, $b \equiv a \pmod{n}$ and $b \equiv c \pmod{n}$. Therefore, by symmetry and transitivity, $a \equiv c \pmod{n}$. Hence, $[a] = [c]$ by Theorem 2.3. ◆

COROLLARY 2.5 *Let $n > 1$ be an integer and consider congruence modulo n.*

(1) If a is any integer and r is the remainder when a is divided by n, then $[a] = [r]$.

(2) There are exactly n distinct congruences classes, namely, $[0]$, $[1]$, $[2]$, . . . , $[n - 1]$.

Proof (1) Let $a \in \mathbb{Z}$. By the Division Algorithm, $a = nq + r$, with $0 \leq r < n$. Thus $a - r = qn$, so that $a \equiv r \pmod{n}$. By Theorem 2.3, $[a] = [r]$.

(2) If $[a]$ is any congruence class, then (1) shows that $[a] = [r]$ with $0 \leq r < n$. Hence, $[a]$ must be one of $[0], [1], [2], \ldots, [n - 1]$. To complete the proof, we must show that these n classes are all distinct. To do this, suppose that $0 \leq s < t < n$. Then $t - s$ is a positive integer and less than n. Thus n does not divide $t - s$ and, hence, $t \not\equiv s \pmod{n}$. Since no two of $0, 1, 2, \ldots, n - 1$ are congruent, the classes $[0], [1], [2], \ldots, [n - 1]$ are all distinct, by Theorem 2.3. ◆

> **DEFINITION**• *The set of all congruence classes modulo n is denoted $\mathbb{Z}_n$ (which is read "$\mathbb{Z}$ mod n").*

There are several points to be careful about here. The elements of $\mathbb{Z}_n$ are *classes,* not single integers. So the statement $[5] \in \mathbb{Z}_n$ is true, but the statement $5 \in \mathbb{Z}_n$ is not. Furthermore, every element of $\mathbb{Z}_n$ can be denoted in many different ways. For example, we know that

$$2 \equiv 5 \pmod{3} \qquad 2 \equiv -1 \pmod{3} \qquad 2 \equiv 14 \pmod{3}.$$

Therefore, by Theorem 2.3, $[2] = [5] = [-1] = [14]$ in $\mathbb{Z}_3$. Even though each element of $\mathbb{Z}_n$ (that is, each congruence class) has infinitely many different labels, there are only finitely many distinct classes by Corollary 2.5, which says in effect that

The set $\mathbb{Z}_n$ has exactly n elements.

For example, the set $\mathbb{Z}_3$ consists of the three elements $[0], [1], [2]$.

◆ **EXERCISES**

A.   **1.** If $a \equiv b \pmod{n}$ and $k \mid n$, is it true that $a \equiv b \pmod{k}$?

   2. (a) If $k \equiv 1 \pmod 4$, then what is $6k + 5$ congruent to modulo 4?

   (b) If $r \equiv 3 \pmod{10}$ and $s \equiv -7 \pmod{10}$, then what is $2r + 3s$ congruent to modulo 10?

   3. If $a \in \mathbb{Z}$, prove that a^2 is not congruent to 2 modulo 4 or to 3 modulo 4.

   4. Prove that every odd integer is congruent to 1 modulo 4 or to 3 modulo 4.

   5. Prove that $a \equiv b \pmod n$ if and only if $a^2 + b^2 \equiv 2ab \pmod{n^2}$.

   6. If a, b are integers such that $a \equiv b \pmod p$ for every positive prime p, prove that $a = b$.

   7. If $a, m, n \in \mathbb{Z}$ with $m > 0, n > 0$, prove that $[a^m] = [a^n]$ in $\mathbb{Z}_2$. [*Hint:* Theorem 2.3.]

   8. Show that $a^{p-1} \equiv 1 \pmod p$ for the given p and a:

   (a) $a = 2, p = 5$ **(b)** $a = 4, p = 7$ **(c)** $a = 3, p = 11$

   9. Prove that

   (a) $(n - a)^2 \equiv a^2 \pmod n$ **(b)** $(2n - a)^2 \equiv a^2 \pmod{4n}$

   10. If $p \geq 5$ and p is prime, prove that $[p] = [1]$ or $[p] = [5]$ in $\mathbb{Z}_6$. [*Hint:* Theorem 2.3 and Corollary 2.5.]

   11. Find all solutions of each congruence:

   (a) $2x \equiv 3 \pmod 5$ **(b)** $3x \equiv 1 \pmod 7$

   (c) $6x \equiv 9 \pmod{15}$ **(d)** $6x \equiv 10 \pmod{15}$

   12. Which of the following congruences have solutions:

   (a) $x^2 \equiv 1 \pmod 3$ **(b)** $x^2 \equiv 2 \pmod 7$ **(c)** $x^2 \equiv 3 \pmod{11}$

   13. (a) If $a \equiv b \pmod{2n}$, prove that $a^2 \equiv b^2 \pmod{4n}$.

   (b) If $a \equiv b \pmod{3n}$, prove that $a^3 \equiv b^3 \pmod{9n}$.

   14. If $a \equiv 2 \pmod 4$, prove that there are no integers c and d such that $a = c^2 - d^2$.

   15. If $a \equiv 3 \pmod 4$, prove that there are no integers c and d such that $a = c^2 + d^2$.

   16. (a) If a is a nonnegative integer, prove that a is congruent to its last digit mod 10 [for example, $27 \equiv 7 \pmod{10}$].

   (b) Show that no perfect square has 2, 3, 7, or 8 as its last digit.

17. Prove that $a \equiv b \pmod{n}$ if and only if a and b leave the same remainder when divided by n.

18. If $a \in \mathbb{Z}$, prove that the last digit of a^4 is 0, 1, 5, or 6. [*Hint:* See Exercise 16.]

B. 19. (a) Prove or disprove: If $a^2 \equiv b^2 \pmod{n}$, then $a \equiv b \pmod{n}$ or $a \equiv -b \pmod{n}$.

  (b) Do part (a) when n is prime.

20. If $[a] = [1]$ in $\mathbb{Z}_n$, prove that $(a,n) = 1$. Show by example that the converse may be false.

21. Prove or disprove: If $[a] = [b]$ in $\mathbb{Z}_n$, then $(a,n) = (b,n)$.

22. If $a^2 \equiv 1 \pmod{2}$, prove that $a^2 \equiv 1 \pmod{4}$.

23. (a) Show that $10^n \equiv 1 \pmod{9}$ for every positive n.

  (b) Prove that every positive integer is congruent to the sum of its digits mod 9 [for example, $38 \equiv 11 \pmod{9}$].

24. Use congruences (not a calculator) to show that $(125698)(23797) \neq 2891235306$. [*Hint:* See Exercise 23.]

25. Prove that $10^n \equiv (-1)^n \pmod{11}$ for every positive n.

26. (a) Give an example to show that the following statement is false: If $ab \equiv ac \pmod{n}$ and $a \not\equiv 0 \pmod{n}$, then $b \equiv c \pmod{n}$.

  (b) Prove that the statement in part (a) is true whenever $(a,n) = 1$.

27. (a) Prove or disprove: If $ab \equiv 0 \pmod{n}$, then $a \equiv 0 \pmod{n}$ or $b \equiv 0 \pmod{n}$.

  (b) Do part (a) when n is prime.

28. Prove or disprove: If $c \equiv a \pmod{r}$ and $c \equiv b \pmod{s}$, then $a \equiv b \pmod{(r,s)}$.

29. If $a^2 \equiv 6 \pmod{10}$, prove that $a^2 - 6 \equiv 10 \pmod{20}$.

30. Prove that $a^5 \equiv a \pmod{30}$ for every integer a.

31. If $(a,n) = 1$, prove that there is an integer b such that $ab \equiv 1 \pmod{n}$.

32. Let a, b, n be integers with $n > 0$. If (a,n) does not divide b, prove that the congruence $ax \equiv b \pmod{n}$ has no solution.

C. 33. If $(a,n) = 1$, prove that the congruence $ax \equiv b \pmod{n}$ has exactly one solution t such that $0 \le t < n$.

> ***Excursion***   The Chinese Remainder Theorem (Section 13.1) may be covered at this point if desired.

2.2 MODULAR ARITHMETIC

The finite set $\mathbb{Z}_n$ is closely related to the infinite set $\mathbb{Z}$. So it is natural to ask if it is possible to define addition and multiplication in $\mathbb{Z}_n$ and do some reasonable kind of arithmetic there. To define addition in $\mathbb{Z}_n$, we must have some way of taking two classes in $\mathbb{Z}_n$ and producing another class—their sum. Because addition of integers *is* defined, the following *tentative* definition seems worth investigating:

The sum of the classes $[a]$ and $[c]$ is the class containing $a + c$

or, in symbols,

$$[a] \oplus [c] = [a + c],$$

where addition of classes is denoted by $\oplus$ to distinguish it from ordinary addition of integers.

We can try a similar *tentative* definition for multiplication:

The product of $[a]$ and $[c]$ is the class containing ac:

$$[a] \odot [c] = [ac],$$

where $\odot$ denotes multiplication of classes.

> **EXAMPLE** In $\mathbb{Z}_5$ we have $[3] \oplus [4] = [3 + 4] = [7] = [2]$ and $[3] \odot [2] = [3 \cdot 2] = [6] = [1]$.

Everything seems to work so far, but there is a possible difficulty. Every element of $\mathbb{Z}_n$ can be written in many different ways. In $\mathbb{Z}_5$, for instance, $[3] = [13]$ and $[4] = [9]$. In the preceding example, we saw that $[3] \oplus [4] = [2]$ in $\mathbb{Z}_5$. Do we get the same answer if we use $[13]$ in place of $[3]$ and $[9]$ in place of $[4]$? In this case the answer is "yes" because

$$[13] \oplus [9] = [13 + 9] = [22] = [2].$$

But how do we know that the answer will be the same no matter which way we write the classes?

To get some idea of the kind of thing that might go wrong, consider these five classes of integers:

$$A = \{ \ldots, -14, -8, -2, 0, 6, 12, 18, \ldots \}$$
$$B = \{ \ldots, -11, -7, -3, 1, 5, 9, 13, \ldots \}$$
$$C = \{ \ldots, -9, -5, -1, 3, 7, 11, 15, \ldots \}$$
$$D = \{ \ldots, -16, -10, -4, 2, 8, 14, 20, \ldots \}$$
$$E = \{ \ldots, -18, -12, -6, 4, 10, 16, 22, \ldots \}.$$

These classes, like the classes in $\mathbb{Z}_5$, have the following basic properties: Every integer is in one of them, and any two of them are either disjoint or identical. Since 1 is in B and 7 is in C, we could define $B + C$ as the class containing $1 + 7 = 8$, that is, $B + C = D$. But B is also the class containing -3 and C the class containing 15, and so $B + C$ ought to be the class containing $-3 + 15 = 12$. But 12 is in A, so that $B + C = A$. Thus you get different answers, depending on which "representatives" you choose from the classes B and C. Obviously you can't have any meaningful concept of addition if the answer is one thing this time and something else another time.

In order to remove the word "tentative" from our definition of addition and multiplication in $\mathbb{Z}_n$, we must first prove that these operations do not depend on the choice of representatives from the various classes. Here is what's needed:

THEOREM 2.6 *If* $[a] = [b]$ *and* $[c] = [d]$ *in* $\mathbb{Z}_n$, *then*

$$[a + c] = [b + d] \qquad and \qquad [ac] = [bd].$$

Proof Since $[a] = [b]$, we know that $a \equiv b \pmod{n}$ by Theorem 2.3. Similarly, $[c] = [d]$ implies that $c \equiv d \pmod{n}$. Therefore, by Theorem 2.2,

$$a + c \equiv b + d \pmod{n} \qquad and \qquad ac \equiv bd \pmod{n}.$$

Hence, by Theorem 2.3 again,

$$[a + c] = [b + d] \qquad and \qquad [ac] = [bd]. \quad \blacklozenge$$

Because of Theorem 2.6, we know that the following formal definition of addition and multiplication of classes is independent of the choice of representatives from each class:

DEFINITION• *Addition and multiplication in* $\mathbb{Z}_n$ *are defined by*

$$[a] \oplus [c] = [a + c] \qquad and \qquad [a] \odot [c] = [ac].$$

EXAMPLE Here are the complete addition and multiplication tables for $\mathbb{Z}_5$ (verify that these calculations are correct):*

---

* These tables are read like this: If $[a]$ appears in the left-hand vertical column and $[c]$ in the top horizontal row of the addition table, for example, then the sum $[a] \oplus [c]$ appears at the intersection of the horizontal row containing $[a]$ and the vertical column containing $[c]$.

$\oplus$	[0]	[1]	[2]	[3]	[4]
[0]	[0]	[1]	[2]	[3]	[4]
[1]	[1]	[2]	[3]	[4]	[0]
[2]	[2]	[3]	[4]	[0]	[1]
[3]	[3]	[4]	[0]	[1]	[2]
[4]	[4]	[0]	[1]	[2]	[3]

$\odot$	[0]	[1]	[2]	[3]	[4]
[0]	[0]	[0]	[0]	[0]	[0]
[1]	[0]	[1]	[2]	[3]	[4]
[2]	[0]	[2]	[4]	[1]	[3]
[3]	[0]	[3]	[1]	[4]	[2]
[4]	[0]	[4]	[3]	[2]	[1]

And here are the tables for $\mathbb{Z}_6$:

$\oplus$	[0]	[1]	[2]	[3]	[4]	[5]
[0]	[0]	[1]	[2]	[3]	[4]	[5]
[1]	[1]	[2]	[3]	[4]	[5]	[0]
[2]	[2]	[3]	[4]	[5]	[0]	[1]
[3]	[3]	[4]	[5]	[0]	[1]	[2]
[4]	[4]	[5]	[0]	[1]	[2]	[3]
[5]	[5]	[0]	[1]	[2]	[3]	[4]

$\odot$	[0]	[1]	[2]	[3]	[4]	[5]
[0]	[0]	[0]	[0]	[0]	[0]	[0]
[1]	[0]	[1]	[2]	[3]	[4]	[5]
[2]	[0]	[2]	[4]	[0]	[2]	[4]
[3]	[0]	[3]	[0]	[3]	[0]	[3]
[4]	[0]	[4]	[2]	[0]	[4]	[2]
[5]	[0]	[5]	[4]	[3]	[2]	[1]

Now that addition and multiplication are defined in $\mathbb{Z}_n$, we want to compare the properties of these "miniature arithmetics" with the well-known properties of $\mathbb{Z}$. The key facts about arithmetic in $\mathbb{Z}$ (and the usual titles for these properties) are as follows. For all $a, b, c \in \mathbb{Z}$:

1. If $a, b \in \mathbb{Z}$, then $a + b \in \mathbb{Z}$. [closure for addition]

2. $a + (b + c) = (a + b) + c$. [associative addition]

3. $a + b = b + a$. [commutative addition]

4. $a + 0 = a = 0 + a$. [additive identity]

5. For each $a \in \mathbb{Z}$, the equation $a + x = 0$ has a solution in $\mathbb{Z}$.

6. If $a, b \in \mathbb{Z}$, then $ab \in \mathbb{Z}$. [closure for multiplication]

7. $a(bc) = (ab)c$. [associative multiplication]

8. $a(b + c) = ab + ac$ and
   $(a + b)c = ac + bc$.   [distributive laws]

9. $ab = ba$ [commutative multiplication]

10. $a \cdot 1 = a = 1 \cdot a$ [multiplicative identity]

11. If $ab = 0$, then $a = 0$ or $b = 0$.

By using the tables in the preceding example, you can verify that the first ten of these properties hold in $\mathbb{Z}_5$ and $\mathbb{Z}_6$ and that property 11 holds in $\mathbb{Z}_5$ and fails in $\mathbb{Z}_6$. But using tables is not a very efficient method of proof (especially for verifying associativity or distributivity). So the proof that properties $1-10$ hold for any $\mathbb{Z}_n$ is based on the definition of the operations in $\mathbb{Z}_n$ and on the fact that these properties are known to be valid in $\mathbb{Z}$.

THEOREM 2.7 *For any classes* $[a]$, $[b]$, $[c]$ *in* $\mathbb{Z}_n$,

1. *If* $[a] \in \mathbb{Z}_n$ *and* $[b] \in \mathbb{Z}_n$, *then* $[a] \oplus [b] \in \mathbb{Z}_n$.
2. $[a] \oplus ([b] \oplus [c]) = ([a] \oplus [b]) \oplus [c]$.
3. $[a] \oplus [b] = [b] \oplus [a]$.
4. $[a] \oplus [0] = [a] = [0] \oplus [a]$.
5. *For each* $[a]$ *in* $\mathbb{Z}_n$, *the equation* $[a] \oplus X = [0]$ *has a solution in* $\mathbb{Z}_n$.
6. *If* $[a] \in \mathbb{Z}_n$ *and* $[b] \in \mathbb{Z}_n$, *then* $[a] \odot [b] \in \mathbb{Z}_n$.
7. $[a] \odot ([b] \odot [c]) = ([a] \odot [b]) \odot [c]$.
8. $[a] \odot ([b] \oplus [c]) = [a] \odot [b] \oplus [a] \odot [c]$ *and*
   $([a] \oplus [b]) \odot [c] = [a] \odot [c] \oplus [b] \odot [c]$.
9. $[a] \odot [b] = [b] \odot [a]$.
10. $[a] \odot [1] = [a] = [1] \odot [a]$.

Proof Properties 1 and 6 are an immediate consequence of the definition of $\oplus$ and $\odot$ in $\mathbb{Z}_n$. To prove 2, note that by the definition of addition,

$$[a] \oplus ([b] \oplus [c]) = [a] \oplus [b + c] = [a + (b + c)].$$

In $\mathbb{Z}$ we know that $a + (b + c) = (a + b) + c$. So the classes of these integers must be the same in $\mathbb{Z}_n$; that is, $[a + (b + c)] = [(a + b) + c]$. By the definition of addition in $\mathbb{Z}_n$, we have

$$[(a + b) + c] = [a + b] \oplus [c] = ([a] \oplus [b]) \oplus [c].$$

This proves 2. The proofs of 3, 7, 8, and 9 are analogous (Exercise 4). Properties 4 and 10 are proved by a direct calculation; for instance, $[a] \odot [1] = [a \cdot 1] = [a]$. For property 5, it is easy to see that $X = [-a]$ is a solution of the equation since $[a] \oplus [-a] = [a + (-a)] = [0]$. ◆

New Notation

We have been very careful to distinguish integers in $\mathbb{Z}$ and classes in $\mathbb{Z}_n$ and have even used different symbols for the operations in the two systems. By now, however, you should be reasonably comfortable with the fundamental ideas and familiar with arithmetic in $\mathbb{Z}_n$. So we shall adopt a new notation that is widely used in mathematics, even though it has the flaw that the same symbol represents two totally different entities.

Whenever the context makes clear that we are dealing with $\mathbb{Z}_n$, we shall abbreviate the class notation "[a]" and write simply "a." In $\mathbb{Z}_6$, for instance, we might say $6 = 0$, which is certainly true for classes in $\mathbb{Z}_6$ even though it is nonsense if 6 and 0 are ordinary integers. We shall use an ordinary plus sign for addition in $\mathbb{Z}_n$ and either a small dot or juxtaposition for multiplication. For example, in $\mathbb{Z}_5$ we may write things like

$$4 + 1 = 0 \quad \text{or} \quad 3 \cdot 4 = 2 \quad \text{or} \quad 4 + 4 = 3.$$

On those few occasions where this usage might cause confusion, we will return to the brackets notation for classes.

> **EXAMPLE** In this new notation, the addition and multiplication tables for $\mathbb{Z}_3$ are:
>
> | + | 0 | 1 | 2 |
> |---|---|---|---|
> | 0 | 0 | 1 | 2 |
> | 1 | 1 | 2 | 0 |
> | 2 | 2 | 0 | 1 |
>
> | · | 0 | 1 | 2 |
> |---|---|---|---|
> | 0 | 0 | 0 | 0 |
> | 1 | 0 | 1 | 2 |
> | 2 | 0 | 2 | 1 |

The same exponent notation used in ordinary arithmetic is also convenient in $\mathbb{Z}_n$. If $a \in \mathbb{Z}_n$ and k is a positive integer, then a^k denotes the product $aaa \cdots a$ (k factors) in $\mathbb{Z}_n$.

> **EXAMPLE** In $\mathbb{Z}_5$, $3^2 = 3 \cdot 3 = 4$ and $3^4 = 3 \cdot 3 \cdot 3 \cdot 3 = 1$.

Warning *Exponents are ordinary integers*—not elements of $\mathbb{Z}_n$. In $\mathbb{Z}_3$, for instance, $2^4 = 2 \cdot 2 \cdot 2 \cdot 2 = 1$ and $2^1 = 2$, so that $2^4 \neq 2^1$ even though $4 = 1$ in $\mathbb{Z}_3$.

◆ EXERCISES

A. **1.** Write out the addition and multiplication tables for

    (a) $\mathbb{Z}_2$ **(b)** $\mathbb{Z}_4$ **(c)** $\mathbb{Z}_7$ **(d)** $\mathbb{Z}_{12}$

2. The set $\mathbb{Z}_n$ contains only n elements. To solve an equation in $\mathbb{Z}_n$ you need only substitute these n elements in the equation to see which ones are solutions. Solve these equations:

(a) $x^2 = 1$ in $\mathbb{Z}_8$ **(b)** $x^4 = 1$ in $\mathbb{Z}_5$

(c) $x^2 + 3x + 2 = 0$ in $\mathbb{Z}_6$ **(d)** $x^2 + 1 = 0$ in $\mathbb{Z}_{12}$

3. (a) Find an element a in $\mathbb{Z}_7$ such that every nonzero element of $\mathbb{Z}_7$ is a power of a.

   (b) Do part (a) in $\mathbb{Z}_5$.   **(c)** Can you do part (a) in $\mathbb{Z}_6$?

4. Prove parts 3, 7, 8, 9 of Theorem 2.7.

5. (a) Solve the equation $x + x + x + x + x = 0$ in $\mathbb{Z}_5$.

   (b) Solve the equation $x + x + x = 0$ in $\mathbb{Z}_3$.

   (c) Solve the equation $x + x + x + x = 0$ in $\mathbb{Z}_4$.

6. Prove or disprove: If $ab = 0$ in $\mathbb{Z}_n$, then $a = 0$ or $b = 0$.

7. Prove or disprove: If $ab = ac$ and $a \neq 0$ in $\mathbb{Z}_n$, then $b = c$.

B. **8. (a)** Solve the equation $x^2 + x = 0$ in $\mathbb{Z}_5$.

   (b) Solve the equation $x^2 + x = 0$ in $\mathbb{Z}_6$.

   (c) If p is prime, prove that the only solutions of $x^2 + x = 0$ in $\mathbb{Z}_p$ are 0 and $p - 1$.

9. (a) In $\mathbb{Z}_5$ compute $(a + b)^5$. [*Hint:* Exercise 5(a) may be helpful.]

   (b) In $\mathbb{Z}_3$ compute $(a + b)^3$. [*Hint:* See Exercise 5(b).]

   (c) In $\mathbb{Z}_2$ compute $(a + b)^2$.

   (d) Based on the results of parts (a)–(c), what do you think $(a + b)^7$ is equal to in $\mathbb{Z}_7$?

10. (a) Find all a in $\mathbb{Z}_5$ for which the equation $ax = 1$ has a solution. Then do the same thing for

   (b) $\mathbb{Z}_4$ **(c)** $\mathbb{Z}_3$ **(d)** $\mathbb{Z}_6$

11. The usual ordering of $\mathbb{Z}$ by $<$ is transitive and behaves nicely with respect to addition. Show that there is *no* ordering of $\mathbb{Z}_n$ such that

     (i) if $a < b$ and $b < c$, then $a < c$;
     (ii) if $a < b$, then $a + c < b + c$ for every c in $\mathbb{Z}_n$.

[*Hint:* If there is such an ordering with $0 < 1$, then adding 1 repeatedly to both sides shows that $0 < 1 < 2 < \cdots < n - 1$ by (ii). Thus $0 < n - 1$ by (i). Add 1 to each side and get a contradiction. Make a similar argument when $1 < 0$.]

2.3 THE STRUCTURE OF $\mathbb{Z}_p$ WHEN p IS PRIME

Some of the $\mathbb{Z}_n$ do not share all the nice properties of $\mathbb{Z}$. For instance, the product of nonzero integers in $\mathbb{Z}$ is always nonzero, but in $\mathbb{Z}_6$ we have $2 \cdot 3 = 0$ even though $2 \neq 0$ and $3 \neq 0$. On the other hand, the multiplication table on page 33 shows that the product of nonzero elements in $\mathbb{Z}_5$ is always nonzero. Indeed, $\mathbb{Z}_5$ has a much stronger property than $\mathbb{Z}$. When $a \neq 0$, the equation $ax = 1$ has a solution in $\mathbb{Z}$ if and only if $a = \pm 1$. But the multiplication table for $\mathbb{Z}_5$ shows that, for any $a \neq 0$, the equation $ax = 1$ has a solution in $\mathbb{Z}_5$; for example,

$$x = 3 \text{ is a solution of } 2x = 1$$
$$x = 4 \text{ is a solution of } 4x = 1.$$

More generally, whenever n is prime, $\mathbb{Z}_n$ has special properties:

THEOREM 2.8 *If $p > 1$ is an integer, then the following conditions are equivalent:**

(1) p is prime.

(2) For any $a \neq 0$ in $\mathbb{Z}_p$, the equation $ax = 1$ has a solution in $\mathbb{Z}_p$.

(3) Whenever $ab = 0$ in $\mathbb{Z}_p$, then $a = 0$ or $b = 0$.

The proof of this theorem illustrates the two basic techniques for proving statements that involve $\mathbb{Z}_n$:

- (i) Translate equations in $\mathbb{Z}_n$ into equivalent congruence statements in $\mathbb{Z}$. Then the properties of congruence and arithmetic in $\mathbb{Z}$ can be used. The brackets notation for elements of $\mathbb{Z}_n$ may be necessary to avoid confusion.

 (ii) Use the arithmetic properties of $\mathbb{Z}_n$ directly, without involving arithmetic in $\mathbb{Z}$. In this case, the brackets notation in $\mathbb{Z}_n$ isn't needed.

Proof of Theorem 2.8 (1) $\Rightarrow$ (2) We use the first technique. Suppose p is prime and $[a] \neq [0]$ in $\mathbb{Z}_p$. Then in $\mathbb{Z}$, $a \not\equiv 0 \pmod{p}$ by Theorem 2.3. Hence, $p \nmid a$ by the definition of congruence. Now the gcd of a and p is a positive divisor of p and thus must be either p or 1. Since (a,p) also divides a and $p \nmid a$, we must have $(a,p) = 1$. By Theorem 1.3, $au + pv = 1$ for some integers u and v. Hence, $au - 1 = p(-v)$, so that $au \equiv 1 \pmod{p}$. Therefore $[au] = [1]$ in $\mathbb{Z}_p$ by Theorem 2.3. Thus $[a][u] = [au] = [1]$, so that $x = [u]$ is a solution of $[a]x = [1]$.

---

* See Appendix A for the meaning of "the following conditions are equivalent" and what must be done to prove such a statement.

$(2) \Rightarrow (3)$ We use the second technique. Suppose $ab = 0$ in $\mathbb{Z}_p$. If $a = 0$, there is nothing to prove. If $a \neq 0$, then by (2) there exists $u \in \mathbb{Z}_p$ such that $au = 1$. Then

$$0 = u \cdot 0 = u(ab) = (ua)b = (au)b = 1 \cdot b = b.$$

In every case, therefore, we have $a = 0$ or $b = 0$.

$(3) \Rightarrow (1)$ Back to the first technique. Let a be any divisor of p, say $p = ab$. To prove p is prime, we must show that $a = \pm 1$ or $\pm p$. Now, $p = ab$ implies $ab \equiv 0 \pmod{p}$, so that $[a][b] = [ab] = [0]$ in $\mathbb{Z}_p$ by Theorem 2.3. Then by (3), either $[a] = [0]$ or $[b] = [0]$. Now, $[a] = [0]$ implies that $a \equiv 0 \pmod{p}$, so that $p \mid a$, say $a = pw$. Thus $p = ab = pwb$. Dividing both ends by p shows that $wb = 1$. Since w and b are integers, the only possibilities are $w = \pm 1$ and $b = \pm 1$. Hence, $b = \pm 1$ and $a = pw = p(\pm 1) = \pm p$. On the other hand, a similar argument shows that $[b] = [0]$ implies that $a = \pm 1$. Therefore, p is prime. ◆

COROLLARY 2.9 *Let p be a positive prime. For any $a \neq 0$ and any b in $\mathbb{Z}_p$, the equation $ax = b$ has a unique solution in $\mathbb{Z}_p$.*

Proof We must prove two things: (1) a solution exists; and (2) it is unique.

Existence By (2) of Theorem 2.8 the equation $ax = 1$ has a solution in $\mathbb{Z}_p$, say $x = u$. Hence, $au = 1$. Multiplying both sides of this equation by b shows that $aub = b$. Hence, $x = ub$ is a solution of $ax = b$.

Uniqueness To show that this is the only solution, suppose $x = w$ is also a solution of $ax = b$ so that $aw = b$. Then

$$a(w - ub) = aw - aub = b - b = 0.$$

Multiplying both ends of this equation by u and using the fact that $ua = au = 1$ shows that $w - ub = 0$. Hence, $w = ub$ and $x = ub$ is the unique solution of $ax = b$. ◆

According to Corollary 2.9, every equation of the form $ax = b$ (with $a \neq 0$) has a solution in $\mathbb{Z}_n$ when n is prime. When n is not prime, *some* such equations may have solutions. In $\mathbb{Z}_{10}$, for example, $x = 6$ is a solution of $3x = 8$.

COROLLARY 2.10 *Let a, b, n be integers with $n > 1$ and $(a, n) = 1$. Then the equation $[a]x = [b]$ has a unique solution in $\mathbb{Z}_n$.*

Proof First, we claim that the equation $[a]x = [1]$ has a solution in $\mathbb{Z}_n$. In the proof of $(1) \Rightarrow (2)$ of Theorem 2.8, the primeness of p is used only to establish the fact that $(a,p) = 1$. From that point on, with n in place of p, the proof applies equally well to the present situation and shows that $[a]x = [1]$ has the solution

$x = [u]$, where $au + nv = 1$. The fact that $[a][u] = [1]$ can now be used, exactly as it was in the proof of Corollary 2.9, to show that $x = [ub]$ is the unique solution of $[a]x = [b]$.   ◆

The proofs of Corollaries 2.9 and 2.10 provide a practical method for solving the equation $[a]x = [b]$ in $\mathbb{Z}_n$ when $(a,n) = 1$: Find integers u, v such that $au + nv = 1$ (this can be done by trial and error or by using the Euclidean Algorithm as in the example on pages 11 and 12). Then $x = [ub]$ is a solution.

> **EXAMPLE**   To solve $24x = 5$ in $\mathbb{Z}_{95}$, we note that $24(4) + 95(-1) = 1$. Here $a = 24$, $n = 95$, $b = 5$, and $u = 4$; so a solution is $x = 4 \cdot 5 = 20$.

When a and n are not relatively prime, the equation $ax = b$ may have no solutions or several solutions in $\mathbb{Z}_n$. In $\mathbb{Z}_{10}$, for example, you can easily verify that $5x = 1$ has no solution and that $6x = 4$ has both $x = 4$ and $x = 9$ as solutions. The reasons for this are given in the following theorem, which includes Corollaries 2.9 and 2.10 as special cases.

THEOREM 2.11   *Let a, b, n be integers with $n > 1$ and let $d = (a,n)$. Then*

(1) *The equation $[a]x = [b]$ has solutions in $\mathbb{Z}_n$ if and only if $d \mid b$.*

(2) *If $d \mid b$, then the equation $[a]x = [b]$ has d distinct solutions in $\mathbb{Z}_n$.*

Proof   Exercises 8–10.   ◆

◆ EXERCISES

A.   **1.** For which a does the equation $ax = 1$ have a solution:

   (a) in $\mathbb{Z}_7$?   **(b)** in $\mathbb{Z}_8$?   **(c)** in $\mathbb{Z}_9$?   **(d)** in $\mathbb{Z}_{10}$?

2. How many solutions does the equation $6x = 4$ have in

   (a) $\mathbb{Z}_7$?   **(b)** $\mathbb{Z}_8$?   **(c)** $\mathbb{Z}_9$?   **(d)** $\mathbb{Z}_{10}$?

3. Without using Theorem 2.8, prove that if p is prime and $ab = 0$ in $\mathbb{Z}_p$, then $a = 0$ or $b = 0$. [*Hint:* Theorem 1.8.]

4. If n is composite, prove that there exist a, $b \in \mathbb{Z}_n$ such that $a \neq 0$ and $b \neq 0$ but $ab = 0$.

5. (a) Give three examples of equations of the form $ax = b$ in $\mathbb{Z}_{12}$ that have no nonzero solutions.

   (b) For each of the equations in part (a), does the equation $ax = 0$ have a nonzero solution?

B.   **6.** Let a and n be integers with $n > 1$. Prove that $(a,n) = 1$ in $\mathbb{Z}$ if and only if the equation $[a]x = [1]$ in $\mathbb{Z}_n$ has a solution.

7. Solve each of the following equations.

    (a) $12x = 2$ in $\mathbb{Z}_{19}$ **(b)** $7x = 2$ in $\mathbb{Z}_{24}$

    (c) $31x = 1$ in $\mathbb{Z}_{50}$ **(d)** $34x = 1$ in $\mathbb{Z}_{97}$

    (e) $27x = 2$ in $\mathbb{Z}_{40}$ **(f)** $15x = 5$ in $\mathbb{Z}_{63}$

8. Let a, b, n be integers with $n > 1$ and let $d = (a,n)$. If the equation $[a]x = [b]$ has a solution in $\mathbb{Z}_n$, prove that $d \mid b$. [*Hint:* If $x = [r]$ is a solution, then $[ar] = [b]$ so that $ar - b = kn$ for some integer k.]

9. Let a, b, n be integers with $n > 1$. Let $d = (a,n)$ and assume $d \mid b$. Prove that the equation $[a]x = [b]$ has a solution in $\mathbb{Z}_n$ as follows.

    (a) Explain why there are integers u, v, a_1, b_1, n_1 such that $au + nv = d$, $a = da_1$, $b = db_1$, $n = dn_1$.

    (b) Show that each of

$$[ub_1], [ub_1 + n_1], [ub_1 + 2n_1], [ub_1 + 3n_1], \ldots , [ub_1 + (d-1)n_1]$$

    is a solution of $[a]x = [b]$.

10. Let a, b, n be integers with $n > 1$. Let $d = (a,n)$ and assume $d \mid b$. Prove that the equation $[a]x = [b]$ has d distinct solutions in $\mathbb{Z}_n$ as follows.

    (a) Show that the solutions listed in Exercise 9(b) are all distinct. [*Hint:* $[r] = [s]$ if and only if $n \mid (r - s)$.]

    (b) If $x = [r]$ is any solution of $[a]x = [b]$, show that $[r] = [ub_1 + kn_1]$ for some integer k with $0 \le k \le d - 1$. [*Hint:* $[ar] - [aub_1] = [0]$ (why?), so that $n \mid (a(r - ub_1))$. Show that $n_1 \mid (a_1(r - ub_1))$ and use Theorem 1.5 to show that $n_1 \mid (r - ub_1)$.]

11. Use Exercise 9 to solve the following equations.

    (a) $15x = 9$ in $\mathbb{Z}_{18}$ **(b)** $25x = 10$ in $\mathbb{Z}_{65}$.

12. Let a, b, n be integers with $n > 1$. Describe the solutions in $\mathbb{Z}$ of the congruence $ax \equiv b \pmod{n}$.

C.   **13.** Let $a \ne 0$ in $\mathbb{Z}_n$. Prove that $ax = 0$ has a nonzero solution in $\mathbb{Z}_n$ if and only if $ax = 1$ has no solution.

Application   Public Key Cryptography (Chapter 12) may be covered at this point if desired.

CHAPTER **3**

Rings

We have seen that many rules of ordinary arithmetic hold not only in $\mathbb{Z}$ but also in the miniature arithmetics $\mathbb{Z}_n$. You know other mathematical systems, such as the real numbers, in which many of these same rules hold. Your high-school algebra courses dealt with the arithmetic of polynomials.

The fact that similar rules of arithmetic hold in different systems suggests that it might be worthwhile to consider the common features of such systems. In the long run, this might save a lot of work: If we can prove a theorem about one system using only the properties that it has in common with a second system, then the theorem is also valid in the second system. By "abstracting" the common core of essential features, we can develop a general theory that includes as special cases $\mathbb{Z}$, $\mathbb{Z}_n$, and the other familiar systems. Results proved for this general theory will apply simultaneously to all the systems covered by the theory. This process of abstraction will allow us to discover the real reasons a particular statement is true (or false, for that matter) without getting bogged down in nonessential details. In this way a deeper understanding of all the systems involved should result.

So we now begin the development of *abstract* algebra. This chapter is just the first step and consists primarily of definitions, examples, and terminology. Systems that share a minimal number of fundamental properties with $\mathbb{Z}$ and $\mathbb{Z}_n$ are called *rings*. Other names are applied to rings that may have additional properties, as you will see in Section 3.1. The elementary facts about arithmetic and algebra in arbitrary rings are developed in Section 3.2. In Section 3.3 we consider rings that appear to be different from one another but actually are "essentially the same" except for the labels on their elements. A simple example of this occurs when you write the integers in roman numerals instead of arabic ones.

3.1 Definition and Examples of Rings

We begin the process of abstracting the common features of familiar systems with this definition:

DEFINITION• *A **ring** is a nonempty set R equipped with two operations* (usually written as addition and multiplication) that satisfy the following axioms. For all a, b, c ∈ R:*

1. *If a ∈ R and b ∈ R, then a + b ∈ R.*   [closure for addition]
2. *a + (b + c) = (a + b) + c.*   [associative addition]
3. *a + b = b + a.*   [commutative addition]
4. *There is an element 0_R in R such that $a + 0_R = a = 0_R + a$ for every a ∈ R.*   [additive identity or zero element]
5. *For each a ∈ R, the equation $a + x = 0_R$ has a solution in R.*
6. *If a ∈ R and b ∈ R, then ab ∈ R.*   [closure for multiplication]
7. *a(bc) = (ab)c.*   [associative multiplication]
8. *a(b + c) = ab + ac and (a + b)c = ac + bc.*   [distributive laws]

These axioms are the bare minimum needed for a system to resemble $\mathbb{Z}$ and $\mathbb{Z}_n$. But $\mathbb{Z}$ and $\mathbb{Z}_n$ have several additional properties that are worth special mention:

DEFINITION• *A **commutative ring** is a ring R that satisfies this axiom:*

9. *ab = ba for all a, b ∈ R.*   [commutative multiplication]

---

* "Operation" and "closure" are defined in Appendix B.

> **DEFINITION•** *A **ring with identity** is a ring R that contains an element 1_R satisfying this axiom:*
>
> 10. $a1_R = a = 1_Ra$ for all $a \in R$.   [*multiplicative identity*]

In the following examples, the verification of most of the axioms is left to the reader.

EXAMPLE   The set of integers $\mathbb{Z}$, with the usual addition and multiplication, is a commutative ring with identity (see page 33).

EXAMPLE   The set $\mathbb{Z}_n$, with the usual addition and multiplication of classes, is a commutative ring with identity by Theorem 2.7.

EXAMPLE   Let E be the set of even integers with the usual addition and multiplication. Since the sum or product of two even integers is also even, the closure axioms (1 and 6) hold. Since 0 is an even integer, E has an additive identity element (Axiom 4). If a is even, then the solution of $a + x = 0$ (namely $-a$) is also even, and so Axiom 5 holds. The other axioms (associativity, commutativity, distributivity) hold for *all* integers and, therefore, are true whenever a, b, c are even. Consequently, E is a commutative ring. E does *not* have an identity, however, because no *even* integer e has the property that $ae = a = ea$ for every even integer a.

EXAMPLE   The set of odd integers with the usual addition and multiplication is *not* a ring. Among other things, Axiom 1 fails: The sum of two odd integers is not odd.

Although the definition of ring was constructed with $\mathbb{Z}$ and $\mathbb{Z}_n$ as models, there are many rings that aren't at all like these models. In these rings, the elements may not be numbers or classes of numbers, and their operations may have nothing to do with "ordinary" addition and multiplication.

EXAMPLE   The set $T = \{r, s, t, z\}$ equipped with the addition and multiplication defined by the following tables is a ring:

+	z	r	s	t
z	z	r	s	t
r	r	z	t	s
s	s	t	z	r
t	t	s	r	z

·	z	r	s	t
z	z	z	z	z
r	z	z	r	r
s	z	z	s	s
t	z	z	t	t

You may take our word for it that Axioms 2, 7, and 8 hold. The element z is the additive identity—the element denoted 0_R in Axiom 4. It behaves in the same way the number 0 does in $\mathbb{Z}$ (that's why the notation 0_R is used in the axiom), but z is not the integer 0—in fact, it's not any kind of number. Nevertheless, we shall call z the "zero element" of the ring T. In order to verify Axiom 5, you must show that each of the equations

$$r + x = z \qquad s + x = z \qquad t + x = z \qquad z + x = z$$

has a solution in T. This is easily seen to be the case from the addition table; for example, $x = r$ is the solution of $r + x = z$ because $r + r = z$. Note that T is *not* a commutative ring; for instance, $rs = r$ and $sr = z$, so that $rs \neq sr$.

EXAMPLE Let $M(\mathbb{R})$ be the set of all 2×2 matrices over the real numbers, that is, $M(\mathbb{R})$ consists of all arrays

$$\begin{pmatrix} a & b \\ c & d \end{pmatrix}, \qquad \text{where } a, b, c, d \text{ are real numbers.}$$

Two matrices are equal provided that the entries in corresponding positions are equal; that is,

$$\begin{pmatrix} a & b \\ c & d \end{pmatrix} = \begin{pmatrix} r & s \\ t & u \end{pmatrix} \qquad \text{if and only if} \qquad a = r, b = s, c = t, d = u.$$

For example,

$$\begin{pmatrix} 4 & 0 \\ -3 & 1 \end{pmatrix} = \begin{pmatrix} 2 + 2 & 0 \\ 1 - 4 & 1 \end{pmatrix} \qquad \text{but} \qquad \begin{pmatrix} 1 & 3 \\ 5 & 2 \end{pmatrix} \neq \begin{pmatrix} 3 & 5 \\ 1 & 2 \end{pmatrix}.$$

Addition of matrices is defined by

$$\begin{pmatrix} a & b \\ c & d \end{pmatrix} + \begin{pmatrix} a' & b' \\ c' & d' \end{pmatrix} = \begin{pmatrix} a + a' & b + b' \\ c + c' & d + d' \end{pmatrix}.$$

For example,

$$\begin{pmatrix} 3 & -2 \\ 5 & 1 \end{pmatrix} + \begin{pmatrix} 4 & 7 \\ 6 & 0 \end{pmatrix} = \begin{pmatrix} 3 + 4 & -2 + 7 \\ 5 + 6 & 1 + 0 \end{pmatrix} = \begin{pmatrix} 7 & 5 \\ 11 & 1 \end{pmatrix}.$$

Multiplication of matrices is defined by

$$\begin{pmatrix} a & b \\ c & d \end{pmatrix}\begin{pmatrix} w & x \\ y & z \end{pmatrix} = \begin{pmatrix} aw + by & ax + bz \\ cw + dy & cx + dz \end{pmatrix}.$$

For example,

$$\begin{pmatrix} 2 & 3 \\ 0 & -4 \end{pmatrix}\begin{pmatrix} 1 & -5 \\ 6 & 7 \end{pmatrix} = \begin{pmatrix} 2 \cdot 1 + 3 \cdot 6 & 2(-5) + 3 \cdot 7 \\ 0 \cdot 1 + (-4)6 & 0(-5) + (-4)7 \end{pmatrix}$$

$$= \begin{pmatrix} 20 & 11 \\ -24 & -28 \end{pmatrix}.$$

Reversing the order of the factors in matrix multiplication *may* produce a different answer, as is the case here:

$$\begin{pmatrix} 1 & -5 \\ 6 & 7 \end{pmatrix}\begin{pmatrix} 2 & 3 \\ 0 & -4 \end{pmatrix} = \begin{pmatrix} 1\cdot 2 + (-5)0 & 1\cdot 3 + (-5)(-4) \\ 6\cdot 2 + 7\cdot 0 & 6\cdot 3 + 7(-4) \end{pmatrix}$$

$$= \begin{pmatrix} 2 & 23 \\ 12 & -10 \end{pmatrix}.$$

So this multiplication is not commutative. With a bit of work, you can verify that $M(\mathbb{R})$ is a ring with identity. The zero element is the matrix

$$\begin{pmatrix} 0 & 0 \\ 0 & 0 \end{pmatrix}$$

and

$$X = \begin{pmatrix} -a & -b \\ -c & -d \end{pmatrix}$$

is a solution of

$$\begin{pmatrix} a & b \\ c & d \end{pmatrix} + X = \begin{pmatrix} 0 & 0 \\ 0 & 0 \end{pmatrix}.$$

The multiplicative identity element (Axiom 10) is the matrix

$$I = \begin{pmatrix} 1 & 0 \\ 0 & 1 \end{pmatrix};$$

for instance,

$$\begin{pmatrix} a & b \\ c & d \end{pmatrix}\begin{pmatrix} 1 & 0 \\ 0 & 1 \end{pmatrix} = \begin{pmatrix} a\cdot 1 + b\cdot 0 & a\cdot 0 + b\cdot 1 \\ c\cdot 1 + d\cdot 0 & c\cdot 0 + d\cdot 1 \end{pmatrix}$$

$$= \begin{pmatrix} a & b \\ c & d \end{pmatrix}.$$

Note that the product of nonzero elements of $M(\mathbb{R})$ may be the zero element; for example,

$$\begin{pmatrix} 4 & 6 \\ 2 & 3 \end{pmatrix}\begin{pmatrix} -3 & -9 \\ 2 & 6 \end{pmatrix} = \begin{pmatrix} 4(-3) + 6\cdot 2 & 4(-9) + 6\cdot 6 \\ 2(-3) + 3\cdot 2 & 2(-9) + 3\cdot 6 \end{pmatrix}$$

$$= \begin{pmatrix} 0 & 0 \\ 0 & 0 \end{pmatrix}.$$

EXAMPLE $M(\mathbb{Z}), M(\mathbb{Q}), M(\mathbb{C})$, and $M(\mathbb{Z}_n)$ denote, respectively, the 2×2 matrices with entries in the integers $\mathbb{Z}$, the rational numbers $\mathbb{Q}$, the complex numbers $\mathbb{C}$, and the ring $\mathbb{Z}_n$.* With addition and multiplication defined as in the preceding example, $M(\mathbb{Z}), M(\mathbb{Q}), M(\mathbb{C})$, and $M(\mathbb{Z}_n)$ (with $n \geq 2$) are all noncommutative rings with identity.

EXAMPLE   Let T be the set of all continuous functions from $\mathbb{R}$ to $\mathbb{R}$, where $\mathbb{R}$ is the set of real numbers. As in calculus, $f + g$ and fg are the functions defined by

$$(f + g)(x) = f(x) + g(x) \qquad \text{and} \qquad (fg)(x) = f(x)g(x).$$

It is proved in calculus that the sum and product of continuous functions are also continuous, and so T is closed under addition and multiplication (Axioms 1 and 6). You can readily verify that T is a commutative ring with identity. The zero element is the function h given by $h(x) = 0$ for all $x \in \mathbb{R}$. The identity element is the function e given by $e(x) = 1$ for all $x \in \mathbb{R}$. Once again the product of nonzero elements of T may turn out to be the zero element; see Exercise 30.

We have seen that some rings do *not* have the property that the product of two nonzero elements is always nonzero. But some of the rings that *do* have this property, such as $\mathbb{Z}$, occur frequently enough to merit a title.

DEFINITION• *An **integral domain** is a commutative ring R with identity $1_R \neq 0_R$ that satisfies this axiom:*

11. Whenever $a, b \in R$ and $ab = 0_R$, then $a = 0_R$ or $b = 0_R$.

The condition $1_R \neq 0_R$ is needed to exclude the zero ring (that is, the single-element ring $\{0_R\}$) from the class of integral domains. Note that Axiom 11 is logically equivalent to its contrapositive.**

Whenever $a \neq 0_R$ and $b \neq 0_R$, then $ab \neq 0_R$.

EXAMPLE   The ring $\mathbb{Z}$ of integers is an integral domain. If p is prime, then $\mathbb{Z}_p$ is an integral domain by Theorem 2.8. On the other hand, $\mathbb{Z}_6$ is not an integral domain because $4 \cdot 3 = 0$, even though $4 \neq 0$ and $3 \neq 0$.

You should be familiar with the set $\mathbb{Q}$ of rational numbers, which consists of all fractions a/b with $a, b \in \mathbb{Z}$ and $b \neq 0$. Equality of fractions, addition, and

---

* Throughout this book, $\mathbb{R}$ always denotes the real numbers, $\mathbb{Q}$ the rational numbers, and $\mathbb{C}$ the complex numbers.

** See Appendix A for a discussion of contrapositives.

multiplication are given by the usual rules:

$$\frac{a}{b} = \frac{r}{s} \qquad \text{if and only if} \qquad as = br$$

$$\frac{a}{b} + \frac{c}{d} = \frac{ad + bc}{bd} \qquad\qquad \frac{a}{b} \cdot \frac{c}{d} = \frac{ac}{bd}$$

It is easy to verify that $\mathbb{Q}$ is an integral domain. But $\mathbb{Q}$ has an additional property that does not hold in $\mathbb{Z}$: Every equation of the form $ax = 1$ (with $a \neq 0$) has a solution in $\mathbb{Q}$. Therefore, $\mathbb{Q}$ is an example of the next definition.

> **DEFINITION•** *A **field** is a commutative ring R with identity $1_R \neq 0_R$ that satisfies this axiom:*
>
> 12. *For each $a \neq 0_R$ in R, the equation $ax = 1_R$ has a solution in R.*

Once again the condition $1_R \neq 0_R$ is needed to exclude the zero ring. Note that Axiom 11 is not mentioned explicitly in the definition of a field. However, Axiom 11 does hold in fields, as we shall see in Theorem 3.9 below.

EXAMPLE The set $\mathbb{R}$ of real numbers, with the usual addition and multiplication, is a field. If p is a prime, then $\mathbb{Z}_p$ is a field by Theorem 2.8.

EXAMPLE The set $\mathbb{C}$ of complex numbers consists of all numbers of the form $a + bi$, where $a, b \in \mathbb{R}$ and $i^2 = -1$. Equality in $\mathbb{C}$ is defined by:

$$a + bi = r + si \qquad \text{if and only if} \qquad a = r \text{ and } b = s.$$

The set $\mathbb{C}$ is a field with addition and multiplication given by:

$$(a + bi) + (c + di) = (a + c) + (b + d)i$$
$$(a + bi)(c + di) = (ac - bd) + (ad + bc)i.$$

The field $\mathbb{R}$ of real numbers is contained in $\mathbb{C}$ because $\mathbb{R}$ consists of all complex numbers of the form $a + 0i$. If $a + bi \neq 0$ in $\mathbb{C}$, then the solution of the equation $(a + bi)x = 1$ is $x = c + di$, where

$$c = a/(a^2 + b^2) \in \mathbb{R} \qquad \text{and} \qquad d = -b/(a^2 + b^2) \in \mathbb{R} \text{ (verify!)}.$$

EXAMPLE Let K be the set of all 2×2 matrices of the form

$$\begin{pmatrix} a & b \\ -b & a \end{pmatrix},$$

where a and b are real numbers. We claim that K is a field. For any two matrices in K,

$$\begin{pmatrix} a & b \\ -b & a \end{pmatrix} + \begin{pmatrix} c & d \\ -d & c \end{pmatrix} = \begin{pmatrix} a+c & b+d \\ -b-d & a+c \end{pmatrix}$$

$$\begin{pmatrix} a & b \\ -b & a \end{pmatrix} \cdot \begin{pmatrix} c & d \\ -d & c \end{pmatrix} = \begin{pmatrix} ac-bd & ad+bc \\ -ad-bc & ac-bd \end{pmatrix}.$$

In each case the matrix on the right is in K because the entries along the **main diagonal** (upper left to lower right) are the same and the entries on the opposite diagonal (upper right to lower left) are negatives of each other. Therefore, K is closed under addition and multiplication. K is commutative because

$$\begin{pmatrix} c & d \\ -d & c \end{pmatrix}\begin{pmatrix} a & b \\ -b & a \end{pmatrix} = \begin{pmatrix} ac-bd & ad+bc \\ -ad-bc & ac-bd \end{pmatrix} = \begin{pmatrix} a & b \\ -b & a \end{pmatrix}\begin{pmatrix} c & d \\ -d & c \end{pmatrix}.$$

Clearly, the zero matrix and the identity matrix I are in K. If

$$A = \begin{pmatrix} a & b \\ -b & a \end{pmatrix}$$

is not the zero matrix, then verify that the solution of $AX = I$ is

$$X = \begin{pmatrix} a/d & -b/d \\ b/d & a/d \end{pmatrix} \epsilon K, \qquad \text{where } d = a^2 + b^2.$$

Whenever the rings in the preceding examples are mentioned, you may assume that addition and multiplication are the operations defined above, unless there is some specific statement to the contrary. You should be aware, however, that a given set (such as $\mathbb{Z}$) may be made into a ring in many different ways by defining different addition and multiplication operations on it. See Exercises 14 and 18–22 for examples.

Now that we know a variety of different kinds of rings, we can use them to produce new rings in the following way.

EXAMPLE Let T be the Cartesian product $\mathbb{Z}_6 \times \mathbb{Z}$, as defined in Appendix B. Define addition in T by the rule

$$(a,z) + (a',z') = (a + a', z + z').$$

The plus sign is being used in three ways here: In the first coordinate on the right-hand side of the equal sign, $+$ denotes addition in $\mathbb{Z}_6$; in the second coordinate, $+$ denotes addition in $\mathbb{Z}$; the $+$ on the left of the equal sign is the addition in T that is being defined. Since $\mathbb{Z}_6$ is a ring and a, $a' \epsilon \mathbb{Z}_6$, the first coordinate on the right, $a + a'$, is in $\mathbb{Z}_6$. Similarly

$z + z' \in \mathbb{Z}$. Therefore, addition in T is closed. Multiplication is defined similarly:

$$(a,z)(a',z') = (aa', zz').$$

For example, $(3,5) + (4,9) = (3 + 4,\ 5 + 9) = (1,14)$ and $(3,5)(4,9) = (3 \cdot 4, 5 \cdot 9) = (0,45)$. You can readily verify that T is a commutative ring with identity. The zero element is $(0,0)$, and the multiplicative identity is $(1,1)$. What was done here can be done for any two rings.

THEOREM 3.1 *Let R and S be rings. Define addition and multiplication on the Cartesian product $R \times S$ by*

$$(r,s) + (r',s') = (r + r', s + s') \text{ and}$$
$$(r,s)(r',s') = (rr', ss').$$

Then $R \times S$ is a ring. If R and S are both commutative, then so is $R \times S$. If both R and S have an identity, then so does $R \times S$.

Proof Exercise 27. ◆

Subrings

If R is a ring and S is a subset of R, then S may or may not itself be a ring under the operations in R. In the ring $\mathbb{Z}$ of integers, for example, the subset E of even integers is a ring, but the subset O of odd integers is not, as we saw on page 43. When a subset S of a ring R is itself a ring under the addition and multiplication in R, then we say that S is a **subring** of R.

EXAMPLE $\mathbb{Z}$ is a subring of the ring $\mathbb{Q}$ of rational numbers and $\mathbb{Q}$ is a subring of the field $\mathbb{R}$ of all real numbers. Since $\mathbb{Q}$ is itself a field, we say that $\mathbb{Q}$ is a **subfield** of $\mathbb{R}$.

Proving that a subset S of a ring R is actually a subring is easier than proving directly that S is a ring. For instance, since $a + b = b + a$ for all elements of R, this fact is also true when a, b happen to be in the subset S. Thus Axiom 3 (commutative addition) automatically holds in any subset S of a ring. In fact, to prove that a subset of a ring is actually a subring, you need only verify a few of the axioms for a ring, as the next theorem shows.

THEOREM 3.2 *Suppose that R is a ring and that S is a subset of R such that*

(i) S is closed under addition (if a, $b \in S$, then $a + b \in S$);

(ii) S is closed under multiplication (if a, $b \in S$, then $ab \in S$);

(iii) $0_R \in S$;

(iv) If $a \in S$, then the solution of the equation $a + x = 0_R$ is in S.

Then S is a subring of R.

Note condition (iv) carefully. To verify it, you need not show that the equation $a + x = 0_R$ *has* a solution—we already know that it does because R is a ring. You need only show that this solution is an element of S (which implies that Axiom 5 holds for S).

Proof of Theorem 3.2   As noted before the theorem, Axioms 2, 3, 7, and 8 hold for *all* elements of R, and so they necessarily hold for the elements of the subset S. Axioms 1, 6, 4, and 5 hold by (i)–(iv). ◆

EXAMPLE   The subset $S = \{0,3\}$ of $\mathbb{Z}_6$ is closed under addition and multiplication ($0 + 0 = 0$; $0 + 3 = 3$; $3 + 3 = 0$; similarly, $0 \cdot 0 = 0 = 0 \cdot 3$; $3 \cdot 3 = 3$). By the definition of S we have $0 \in S$. Finally, the equation $0 + x = 0$ has solution $x = 0 \in S$, and the equation $3 + x = 0$ has solution $x = 3 \in S$. Therefore, S is a subring of $\mathbb{Z}_6$ by Theorem 3.2.

◆ **EXERCISES**

A. **1.** The following subsets of $\mathbb{Z}$ (with ordinary addition and multiplication) satisfy all but one of the axioms for a ring. In each case, which axiom fails?

(a) The set S of all odd integers and 0.

(b) The set of nonnegative integers.

2. Let $R = \{0, e, b, c\}$ with addition and multiplication defined by the following tables. Assume associativity and distributivity and show that R is a ring with identity. Is R commutative? Is R a field?

+	0	e	b	c
0	0	e	b	c
e	e	0	c	b
b	b	c	0	e
c	c	b	e	0

·	0	e	b	c
0	0	0	0	0
e	0	e	b	c
b	0	b	b	0
c	0	c	0	c

3. Let $F = \{0, e, a, b\}$ with operations given by the following tables. Assume associativity and distributivity and show that F is a field.

+	0	e	a	b
0	0	e	a	b
e	e	0	b	a
a	a	b	0	e
b	b	a	e	0

·	0	e	a	b
0	0	0	0	0
e	0	e	a	b
a	0	a	b	e
b	0	b	e	a

4. **(a)** Show that the set R of all multiples of 3 is a subring of $\mathbb{Z}$.

   (b) Let k be a fixed integer. Show that the set of all multiples of k is a subring of $\mathbb{Z}$.

5. Which of the following six sets are subrings of $M(\mathbb{R})$? Which ones have an identity?

   (a) All matrices of the form $\begin{pmatrix} 0 & r \\ 0 & 0 \end{pmatrix}$ with $r \in \mathbb{Q}$.

   (b) All matrices of the form $\begin{pmatrix} a & b \\ 0 & c \end{pmatrix}$ with $a, b, c \in \mathbb{Z}$.

   (c) All matrices of the form $\begin{pmatrix} a & a \\ b & b \end{pmatrix}$ with $a, b \in \mathbb{R}$.

   (d) All matrices of the form $\begin{pmatrix} a & 0 \\ a & 0 \end{pmatrix}$ with $a \in \mathbb{R}$.

   (e) All matrices of the form $\begin{pmatrix} a & 0 \\ 0 & a \end{pmatrix}$ with $a \in \mathbb{R}$.

   (f) All matrices of the form $\begin{pmatrix} a & 0 \\ 0 & 0 \end{pmatrix}$ with $a \in \mathbb{R}$.

6. Is the subset $\{1, -1, i, -i\}$ a subring of $\mathbb{C}$?

7. Let R be a ring and consider the subset R^* of $R \times R$ defined by $R^* = \{(r,r) \mid r \in R\}$.

   (a) If $R = \mathbb{Z}_6$, list the elements of R^*.

   (b) For any ring R, show that R^* is a subring of $R \times R$.

8. Let R and S be rings and consider these subsets of $R \times S$:
$$\overline{R} = \{(r,0_S) \mid r \in R\}, \qquad \overline{S} = \{(0_R,s) \mid s \in S\}.$$

   (a) If $R = \mathbb{Z}_3$ and $S = \mathbb{Z}_5$, what are the sets $\overline{R}$ and $\overline{S}$?

   (b) For any rings R and S, show that $\overline{R}$ is a subring of $R \times S$.

   (c) For any rings R and S, show that $\overline{S}$ is a subring of $R \times S$.

9. Let $\mathbb{Z}[\sqrt{2}]$ denote the set $\{a + b\sqrt{2} \mid a,b \in \mathbb{Z}\}$. Show that $\mathbb{Z}[\sqrt{2}]$ is a subring of $\mathbb{R}$.

10. Let $\mathbb{Z}[i]$ denote the set $\{a + bi \mid a,b \in \mathbb{Z}\}$. Show that $\mathbb{Z}[i]$ is a subring of $\mathbb{C}$.

11. Let $S = \{a,b,c\}$ and let $P(S)$ be the set of all subsets of S; denote the elements of $P(S)$ as follows:
$$S = \{a, b, c\}; \quad D = \{a, b\}; \quad E = \{a, c\}; \quad F = \{b, c\};$$
$$A = \{a\}; \quad B = \{b\}; \quad C = \{c\}; \quad 0 = \varnothing.$$

Define addition and multiplication in $P(S)$ by these rules:

$$M + N = (M - N) \cup (N - M) \quad \text{and} \quad MN = M \cap N.$$

Write out the addition and multiplication tables for $P(S)$. Also, see Exercise 36.

12. Let T be the ring of all continuous functions from $\mathbb{R}$ to $\mathbb{R}$. Let $S = \{f \in T \mid f(2) = 0\}$. Is S a subring of T?

13. Write out the addition and multiplication tables for

    (a) $\mathbb{Z}_2 \times \mathbb{Z}_3$ **(b)** $\mathbb{Z}_2 \times \mathbb{Z}_2$ **(c)** $\mathbb{Z}_3 \times \mathbb{Z}_3$

14. Define a new multiplication in $\mathbb{Z}$ by the rule "$ab = 0$ for all $a, b \in \mathbb{Z}$." Show that, with ordinary addition and this new multiplication, $\mathbb{Z}$ is a commutative ring.

15. Define a new multiplication in $\mathbb{Z}$ by the rule "$ab = 1$ for all $a, b \in \mathbb{Z}$." With ordinary addition and this new multiplication, is $\mathbb{Z}$ a ring?

B. 16. Show that the subset $R = \{0, 3, 6, 9, 12, 15\}$ of $\mathbb{Z}_{18}$ is a subring. Does R have an identity?

17. Show that the subset $S = \{0, 2, 4, 6, 8\}$ of $\mathbb{Z}_{10}$ is a subring. Does S have an identity?

18. Define a new addition $\oplus$ and multiplication $\odot$ on $\mathbb{Z}$ by

$$a \oplus b = a + b - 1 \quad \text{and} \quad a \odot b = a + b - ab,$$

where the operations on the right-hand side of the equal signs are ordinary addition, subtraction, and multiplication. Prove that, with the new operations $\oplus$ and $\odot$, $\mathbb{Z}$ is an integral domain.

19. Let E be the set of even integers with ordinary addition. Define a new multiplication $*$ on E by the rule "$a * b = ab/2$" (where the product on the right is ordinary multiplication). Prove that with these operations E is a commutative ring with identity.

20. Define a new addition and multiplication on $\mathbb{Z}$ by

$$a \oplus b = a + b - 1 \quad \text{and} \quad a \odot b = ab - (a + b) + 2.$$

Prove that with these new operations $\mathbb{Z}$ is an integral domain.

21. Define a new addition and multiplication on $\mathbb{Q}$ by

$$r \oplus s = r + s + 1 \quad \text{and} \quad r \odot s = rs + r + s.$$

Prove that with these new operations $\mathbb{Q}$ is a commutative ring with identity. Is it an integral domain?

22. Let L be the set of positive real numbers. Define a new addition and multiplication on L by

$$a \oplus b = ab \qquad \text{and} \qquad a \otimes b = a^{\log b}.$$

(a) Is L a ring under these operations?

(b) Is L a commutative ring?

(c) Is L a field?

23. Let p be a positive prime and let R be the set of all rational numbers that can be written in the form r/p^i with $r, i \in \mathbb{Z}$, and $i \geq 0$. Note that $\mathbb{Z} \subseteq R$ because each $n \in \mathbb{Z}$ can be written as n/p^0. Show that R is a subring of $\mathbb{Q}$.

24. The addition table and part of the multiplication table for a three-element ring are given below. Use the distributive laws to complete the multiplication table.

+	r	s	t
r	r	s	t
s	s	t	r
t	t	r	s

$\cdot$	r	s	t
r	r	r	r
s	r	t	
t	r		

25. Do Exercise 24 for this four-element ring:

+	w	x	y	z
w	w	x	y	z
x	x	y	z	w
y	y	z	w	x
z	z	w	x	y

$\cdot$	w	x	y	z
w	w	w	w	w
x	w	y		
y	w		w	
z	w		w	y

26. Show that $M(\mathbb{Z}_2)$ (all 2×2 matrices with entries in $\mathbb{Z}_2$) is a 16-element non-commutative ring with identity.

27. Prove Theorem 3.1.

28. Let $J = \{0, 2\} \subseteq \mathbb{Z}_4$ and $K = \{0, 3, 6, 9\} \subseteq \mathbb{Z}_{12}$. Show that $J \times K$ is a subring of $\mathbb{Z}_4 \times \mathbb{Z}_{12}$.

29. Prove or disprove:

(a) If R and S are integral domains, then $R \times S$ is an integral domain.

(b) If R and S are fields, then $R \times S$ is a field.

30. Let T be the ring of continuous functions from $\mathbb{R}$ to $\mathbb{R}$ and let f, g be given by

$$f(x) = \begin{cases} 0 & \text{if } x \le 2 \\ x - 2 & \text{if } x > 2 \end{cases} \qquad g(x) = \begin{cases} 2 - x & \text{if } x \le 2 \\ 0 & \text{if } x > 2. \end{cases}$$

Show that $f, g \in T$ and that $fg = 0_T$. Therefore T is not an integral domain.

31. Let $\mathbb{Q}(\sqrt{2}) = \{r + s\sqrt{2} \mid r, s \in \mathbb{Q}\}$. Show that $\mathbb{Q}(\sqrt{2})$ is a subfield of $\mathbb{R}$. [*Hint:* To show that the solution of $(r + s\sqrt{2})x = 1$ is actually in $\mathbb{Q}(\sqrt{2})$, multiply $1/(r + s\sqrt{2})$ by $(r - s\sqrt{2})/(r - s\sqrt{2})$.]

32. Let d be an integer that is not a perfect square. Show that $\mathbb{Q}(\sqrt{d}) = \{a + b\sqrt{d} \mid a, b \in \mathbb{Q}\}$ is a subfield of $\mathbb{C}$. [*Hint:* See Exercise 31.]

33. Let S be the set of all elements $a + b\sqrt[3]{2} + c\sqrt[3]{4}$ with $a, b, c \in \mathbb{Q}$. Show that S is a subring of $\mathbb{R}$. (S is actually a subfield, but this is more difficult to prove.)

34. A **division ring** is a ring R with identity $1_R \ne 0_R$ that satisfies Axiom 12 (for each $a \ne 0_R$ in R, the equation $ax = 1_R$ has a solution in R). Thus a field is a commutative division ring. See Exercise 35 for a noncommutative example. Suppose R is a division ring and a, b are nonzero elements of R.

(a) If $bb = b$, prove that $b = 1_R$. [*Hint:* Let v be the solution of $bx = 1_R$ and note that $bv = b^2v$.]

(b) If u is the solution of the equation $ax = 1_R$, prove that u is also a solution of the equation $xa = 1_R$. (Remember that R may not be commutative.) [*Hint:* Use part (a) with $b = ua$.]

35. In the ring $M(\mathbb{C})$, let

$$\mathbf{1} = \begin{pmatrix} 1 & 0 \\ 0 & 1 \end{pmatrix} \qquad \mathbf{i} = \begin{pmatrix} i & 0 \\ 0 & -i \end{pmatrix} \qquad \mathbf{j} = \begin{pmatrix} 0 & 1 \\ -1 & 0 \end{pmatrix} \qquad \mathbf{k} = \begin{pmatrix} 0 & i \\ i & 0 \end{pmatrix}$$

The product of a real number and a matrix is the matrix given by this rule:

$$r \begin{pmatrix} t & u \\ v & w \end{pmatrix} = \begin{pmatrix} rt & ru \\ rv & rw \end{pmatrix}$$

The set H of **real quaternions** consists of all matrices of the form

$$a\mathbf{1} + b\mathbf{i} + c\mathbf{j} + d\mathbf{k} = a \begin{pmatrix} 1 & 0 \\ 0 & 1 \end{pmatrix} + b \begin{pmatrix} i & 0 \\ 0 & -i \end{pmatrix} + c \begin{pmatrix} 0 & 1 \\ -1 & 0 \end{pmatrix} + d \begin{pmatrix} 0 & i \\ i & 0 \end{pmatrix}$$

$$= \begin{pmatrix} a & 0 \\ 0 & a \end{pmatrix} + \begin{pmatrix} bi & 0 \\ 0 & -bi \end{pmatrix} + \begin{pmatrix} 0 & c \\ -c & 0 \end{pmatrix} + \begin{pmatrix} 0 & di \\ di & 0 \end{pmatrix}$$

$$= \begin{pmatrix} a + bi & c + di \\ -c + di & a - bi \end{pmatrix},$$

where $a, b, c,$ and d are real numbers.

(a) Prove that

$$i^2 = j^2 = k^2 = -1 \qquad ij = -ji = k$$
$$jk = -kj = i \qquad ki = -ik = j.$$

(b) Show that H is a noncommutative ring with identity.

(c) Show that H is a division ring (defined in Exercise 34). [*Hint:* If $M = a\mathbf{1} + b\mathbf{i} + c\mathbf{j} + d\mathbf{k}$, then verify that the solution of the equation $Mx = \mathbf{1}$ is the matrix $ta\mathbf{1} - tb\mathbf{i} - tc\mathbf{j} - td\mathbf{k}$, where $t = 1/(a^2 + b^2 + c^2 + d^2)$.]

(d) Show that the equation $x^2 = -1$ has infinitely many solutions in H. [*Hint:* Consider quaternions of the form $0\mathbf{1} + b\mathbf{i} + c\mathbf{j} + d\mathbf{k}$, where $b^2 + c^2 + d^2 = 1$.]

36. Let S be a set and let $P(S)$ be the set of all subsets of S. Define addition and multiplication in $P(S)$ by the rules

$$M + N = (M - N) \cup (N - M) \qquad \text{and} \qquad MN = M \cap N.$$

(a) Prove that $P(S)$ is a commutative ring with identity. [The verification of additive associativity and distributivity is a bit messy, but an informal discussion using Venn diagrams is adequate for appreciating this example. See Exercise 11 for a special case.]

(b) Show that every element of $P(S)$ satisfies the equations $x^2 = x$ and $x + x = 0_{P(S)}$.

C. 37. Consider $\mathbb{R} \times \mathbb{R}$ with the usual coordinatewise addition (as in Theorem 3.1) and a new multiplication given by

$$(a,b)(c,d) = (ac - bd, ad + bc).$$

Show that with these operations $\mathbb{R} \times \mathbb{R}$ is a field.

38. (a) Find all matrices A in $M(\mathbb{R})$ with the property $AB = BA$ for every matrix B in $M(\mathbb{R})$.

(b) Show that the set of matrices found in part (a) is a subring of $M(\mathbb{R})$. It is called the **center** of $M(\mathbb{R})$.

39. Let r and s be positive integers such that r divides $ks + 1$ for some k with $1 \le k < r$. Prove that the subset $\{0, r, 2r, 3r, \ldots, (s - 1)r\}$ of $\mathbb{Z}_{rs}$ is a ring with identity $ks + 1$ under the usual addition and multiplication in $\mathbb{Z}_{rs}$. Exercise 17 is a special case of this result.

Application Applications of the Chinese Remainder Theorem (Section 13.2) may be covered at this point if desired.

3.2 BASIC PROPERTIES OF RINGS

When you do arithmetic in $\mathbb{Z}$, you often use far more than the axioms for an integral domain. For instance, subtraction appears regularly, as do cancellation, the various rules for multiplying negative numbers, and the fact that $a \cdot 0 = 0$. We now show that many of these same properties hold in every ring.

Subtraction is not mentioned in the axioms for a ring, and we cannot just assume that such an operation exists in an arbitrary ring. If we want to define a subtraction operation in a ring, we must do so in terms of addition, multiplication, and the ring axioms. The first step is

THEOREM 3.3 *For any element a in a ring R, the equation $a + x = 0_R$ has a unique solution.*

Proof We know that $a + x = 0_R$ has at least one solution u by Axiom 5. If v is also a solution, then $a + u = 0_R$ and $a + v = 0_R$, so that

$$v = 0_R + v = (a + u) + v = (u + a) + v = u + (a + v) = u + 0_R = u.$$

Therefore, u is the only solution. ◆

We can now define negatives and subtraction in any ring by copying what happens in familiar rings such as $\mathbb{Z}$. Let R be a ring and $a \in R$. By Theorem 3.3 the equation $a + x = 0_R$ has a unique solution. Using notation adapted from $\mathbb{Z}$, we denote this unique solution by the symbol "$-a$." Since addition is commutative,

$-a$ **is the unique element of R such that**

$a + (-a) = 0_R = (-a) + a.$

In familiar rings, this definition coincides with the known concept of the negative of an element. More importantly, it provides a meaning for "negative" in *any* ring.

> **EXAMPLE** In the ring $\mathbb{Z}_6$, the solution of the equation $2 + x = 0$ is 4, and so in this ring $-2 = 4$. Similarly, $-9 = 5$ in $\mathbb{Z}_{14}$ because 5 is the solution of $9 + x = 0$.

Subtraction in a ring is now defined by the rule

$b - a$ **means** $b + (-a)$.

In $\mathbb{Z}$ and other familiar rings, this is just ordinary subtraction. In other rings we have a new operation.

⊏ **EXAMPLE** In $\mathbb{Z}_6$ we have $1 - 2 = 1 + (-2) = 1 + 4 = 5$.

In junior high school you learned many computational and algebraic rules for dealing with negatives and subtraction. The next two theorems show that these rules are valid in any ring. Although these facts are not particularly interesting in themselves, it is essential to establish their validity so that we may do arithmetic in arbitrary rings.

THEOREM 3.4 *If $a + b = a + c$ in a ring R, then $b = c$.*

Proof Adding $-a$ to both sides of $a + b = a + c$ and then using associativity and negatives show that

$$-a + (a + b) = -a + (a + c)$$
$$(-a + a) + b = (-a + a) + c$$
$$0_R + b = 0_R + c$$
$$b = c. \quad ◆$$

THEOREM 3.5 *For any elements a and b of a ring R,*

(1) $a \cdot 0_R = 0_R = 0_R \cdot a$.
(2) $a(-b) = -(ab) = (-a)b$.
(3) $-(-a) = a$.
(4) $-(a + b) = (-a) + (-b)$.
(5) $-(a - b) = -a + b$.
(6) $(-a)(-b) = ab$.

If R has an identity, then

(7) $(-1_R)a = -a$.

Proof (1) Since $0_R + 0_R = 0_R$, the distributive law shows that

$$a \cdot 0_R + a \cdot 0_R = a(0_R + 0_R) = a \cdot 0_R = a \cdot 0_R + 0_R.$$

Applying Theorem 3.4 to the first and last parts of this equation shows that $a \cdot 0_R = 0_R$. The proof that $0_R \cdot a = 0_R$ is similar.

(2) By definition, $-(ab)$ is the *unique* solution of the equation $ab + x = 0_R$, and so any other solution of this equation must be equal to $-(ab)$. But $x = a(-b)$ is a solution because, by the distribution law and (1),

$$ab + a(-b) = a[b + (-b)] = a[0_R] = 0_R.$$

Therefore, $a(-b) = -(ab)$. The other parts are proved similarly.

(3) By definition, $-(-a)$ is the unique solution of $(-a) + x = 0_R$. But a is a solution of this equation since $(-a) + a = 0_R$. Hence, $-(-a) = a$ by uniqueness.

(4) By definition, $-(a + b)$ is the unique solution of $(a + b) + x = 0_R$, but $(-a) + (-b)$ is also a solution:

$$(a + b) + [(-a) + (-b)] = b + [a + (-a)] + (-b)$$
$$= (b + 0_R) + (-b)$$
$$= b + (-b) = 0_R.$$

Therefore, $-(a + b) = (-a) + (-b)$ by uniqueness.

(5) By the definition of subtraction and (4) and (3),

$$-(a - b) = -(a + (-b)) = (-a) + (-(-b)) = -a + b.$$

(6) By (3) and repeated use of (2),

$$(-a)(-b) = -[a(-b)] = -[-(ab)] = ab.$$

(7) By (2),

$$(-1_R)a = -(1_R a) = -(a) = -a. \quad \blacklozenge$$

Subtraction provides a faster method than Theorem 3.2 for showing that a subset of a ring is actually a subring.

THEOREM 3.6 *Let S be a nonempty subset of a ring R such that*

(1) S is closed under subtraction (if a, b ∈ S, then a − b ∈ S);

(2) S is closed under multiplication (if a, b ∈ S, then ab ∈ S).

Then S is a subring of R.

Proof We shall show that S satisfies the four conditions of Theorem 3.2. Certainly S satisfies condition (ii) of Theorem 3.2, which is the same as (2) here. Since S is nonempty, there is some element c with $c \in S$. Applying (1) with $a = c = b$, we see that $c - c = 0_R$ is in S. Thus S satisfies condition (iii) of Theorem 3.2. Now if a is any element of S, then by (1), $0_R - a = -a$ is also in S; in other words, the solution of the equation $a + x = 0_R$ is in S. Hence condition (iv) of Theorem 3.2 is satisfied. Finally, if $a, b \in S$, then $-b$ is also in S so that $a - (-b) = a + b$ is in S by (1). So condition (i) of Theorem 3.2 is satisfied. Therefore, S is a subring of R by Theorem 3.2. $\blacklozenge$

When doing ordinary arithmetic, exponent notation is a definite convenience, as is its additive analogue (for instance, $a + a + a = 3a$). We now carry these concepts over to arbitrary rings. If R is a ring, $a \in R$, and n is a positive integer, then we *define*

$$a^n = aaa \cdots a \qquad (n \text{ factors}).$$

It is easy to verify that for any $a \in R$ and positive integers m and n,

$$a^m a^n = a^{m+n} \qquad \text{and} \qquad (a^m)^n = a^{mn}.$$

If R has an identity and $a \neq 0_R$, then we define a^0 to be the element 1_R. In this case, the exponent rules are valid for all $m, n \geq 0$.

If R is a ring, $a \in R$, and n is a positive integer, then we define

$$na = a + a + a + \cdots + a \qquad (n \text{ summands}).$$

$$-na = (-a) + (-a) + (-a) + \cdots + (-a) \qquad (n \text{ summands}).$$

Finally, we define $0a = 0_R$. In familiar rings this is nothing new, but in other rings it gives a meaning to the "product" of an integer n and a ring element a.

> **EXAMPLE** Let R be a ring and $a, b \in R$. Then
> $$(a + b)^2 = (a + b)(a + b) = a(a + b) + b(a + b)$$
> $$= aa + ab + ba + bb = a^2 + ab + ba + b^2.$$
>
> Be careful here. If $ab \neq ba$, then you *can't* combine the middle terms. If R is a commutative ring, however, then $ab = ba$ and we have the familiar pattern
> $$(a + b)^2 = a^2 + ab + ba + b^2 = a^2 + ab + ab + b^2 = a^2 + 2ab + b^2.$$
>
> For a calculation of $(a + b)^n$ in a commutative ring, with $n > 2$, see the Binomial Theorem in Appendix E.

Having established the basic facts about arithmetic in rings, we turn now to algebra and the solution of equations.

THEOREM 3.7 *Let R be a ring and let $a, b \in R$. Then the equation $a + x = b$ has the unique solution $x = b - a$.*

Proof $x = b - a$ is a solution because

$$a + [b - a] = a + [b + (-a)] = a + [-a + b]$$
$$= [a + (-a)] + b = 0_R + b = b.$$

If w is any other solution, then $a + w = b = a + (b - a)$. Hence, $w = b - a$ by Theorem 3.4. Therefore, $x = b - a$ is the only solution. ◆

Suppose that a is a nonzero element of a ring R with identity (possibly noncommutative) and that each of the equations $ax = 1_R$ and $xa = 1_R$ has a solution, say u is a solution of $ax = 1_R$ and v a solution of $xa = 1_R$. Then $au = 1_R$ and $va = 1_R$ so that

$$u = 1_R u = (va)u = v(au) = v1_R = v.$$

Thus u is an element such that $au = 1_R = ua$. Some special terminology is used to describe this situation.

> **DEFINITION•** *An element a in a ring R with identity is called a **unit** if there exists $u \in R$ such that $au = 1_R = ua$. In this case the element u is called the **(multiplicative) inverse** of a and is denoted a^{-1}.*

The notation for the inverse of a unit is modeled on the usual exponent notation in the real numbers, where $a^{-1} = 1/a$. Every nonzero element of $\mathbb{R}$ is a unit. In fact, *every nonzero element of a field is a unit* by Axiom 12. There are units in other rings as well.

> **EXAMPLE**   In $\mathbb{Z}_{10}$ the element 7 is a unit because $7 \cdot 3 = 1 = 3 \cdot 7$. In this case $7^{-1} = 3$ and $3^{-1} = 7$. More generally, Corollary 2.10 shows that whenever $(a,n) = 1$, then $[a]$ is a unit in $\mathbb{Z}_n$.

> **EXAMPLE**   The only units in $\mathbb{Z}$ are 1 and -1 (why?). But in the noncommutative matrix ring $M(\mathbb{Z})$ there are many units; for instance,
>
> $$\begin{pmatrix} 3 & 2 \\ 7 & 5 \end{pmatrix}$$
>
> is a unit because
>
> $$\begin{pmatrix} 3 & 2 \\ 7 & 5 \end{pmatrix} \begin{pmatrix} 5 & -2 \\ -7 & 3 \end{pmatrix} = \begin{pmatrix} 1 & 0 \\ 0 & 1 \end{pmatrix} = \begin{pmatrix} 5 & -2 \\ -7 & 3 \end{pmatrix} \begin{pmatrix} 3 & 2 \\ 7 & 5 \end{pmatrix}.$$
>
> Units in a matrix ring are called **invertible matrices.**

In Corollary 2.10 we saw that the equation $[a]x = [b]$ in $\mathbb{Z}_n$ has a unique solution whenever $(a,n) = 1$, that is, whenever $[a]$ is a unit in $\mathbb{Z}_n$. A similar result holds in the general case.

THEOREM 3.8   *Let R be a ring with identity and $a,b \in R$. If a is a unit, then each of the equations $ax = b$ and $ya = b$ has a unique solution in R.*

Each equation has a unique solution, but in the noncommutative case the solution of $ax = b$ may not be the same as the solution of $ya = b$; see Exercise 4 for an example.

Proof of Theorem 3.8 Since a is a unit, it has an inverse $a^{-1} \in R$ such that $aa^{-1} = 1_R = a^{-1}a$. Then $x = a^{-1}b$ is a solution of $ax = b$ because $a(a^{-1}b) = (aa^{-1})b = 1_Rb = b$. If $x = c$ is another solution, then $ac = b$ and $c = 1_Rc = (a^{-1}a)c = a^{-1}(ac) = a^{-1}b$. Therefore, $x = a^{-1}b$ is the only solution. A similar argument shows that $y = ba^{-1}$ is the unique solution of $ya = b$. ◆

The fact that every nonzero element in a field is a unit leads to the following result.

THEOREM 3.9 *Every field F is an integral domain.*

Proof Since a field is a commutative ring with identity $1_F \neq 0_F$, we need only verify Axiom 10. Suppose $ab = 0_F$. We must show that $a = 0_F$ or $b = 0_F$. If $a = 0_F$, there is nothing to prove. If $a \neq 0_F$, then a is a unit and multiplying both sides of $ab = 0_F$ by a^{-1} shows that

$$a^{-1}ab = a^{-1}0_F$$
$$1_Fb = 0_F$$
$$b = 0_F. \quad ◆$$

Integral domains have several useful properties not shared by other rings, such as the following.

THEOREM 3.10 *Cancellation is valid in any integral domain R: If $a \neq 0_R$ and $ab = ac$ in R, then $b = c$.*

Cancellation may fail in rings that are not integral domains. In $\mathbb{Z}_{12}$, for instance, $2 \cdot 4 = 2 \cdot 10$, but $4 \neq 10$.

Proof of Theorem 3.10 If $ab = ac$, then $ab - ac = 0_R$, so that $a(b - c) = 0_R$. Since $a \neq 0_R$, Axiom 11 implies that $b - c = 0_R$, or equivalently, $b = c$. ◆

The converse of Theorem 3.9 is false in general ($\mathbb{Z}$ is an integral domain that is not a field) but true in the finite case:

THEOREM 3.11 *Every finite integral domain R is a field.*

Proof Since R is a commutative ring with identity, we need only show that for each $a \neq 0_R$, the equation $ax = 1_R$ has a solution. Let $a_1, a_2, \ldots, a_n$ be the distinct elements of R and suppose $a_t \neq 0_R$. To show that $a_tx = 1_R$ has a solution, consider the products $a_ta_1, a_ta_2, a_ta_3, \ldots, a_ta_n$. If $a_i \neq a_j$, then we must have $a_ta_i \neq a_ta_j$ (because $a_ta_i = a_ta_j$ would imply that $a_i = a_j$ by cancellation). Therefore, $a_ta_1, a_ta_2, \ldots, a_ta_n$ are n distinct elements of R. However,

R has exactly n elements all together, and so these must be all the elements of R in some order. In particular, for some j, $a_t a_j = 1_R$. Therefore, the equation $a_t x = 1_R$ has a solution and R is a field.   ◆

We close this section with another bit of convenient terminology:

> **DEFINITION•** *An element a in a ring R is a **zero divisor** if*
>
> *(1) $a \neq 0_R$*
>
> *(2) there exists a nonzero element b of R such that either $ab = 0_R$ or $ba = 0_R$.*

For example, 3 is zero divisor in $\mathbb{Z}_6$ because $3 \cdot 2 = 0$. Note that 0_R is *not* a zero divisor. An integral domain may be described as a nonzero commutative ring with identity that has no zero divisors. No unit is a zero divisor (Exercise 13), but an element that is not a zero divisor need not be a unit (for instance, 2 is neither a zero divisor nor a unit in $\mathbb{Z}$).

◆ **EXERCISES**

A.   **1.** Let R be a ring and $a, b \in R$.

    (a) $(a + b)(a - b) = ?$ **(b)** $(a + b)^3 = ?$

    (c) What are the answers in parts (a) and (b) if R is commutative?

2. An element e of a ring R is said to be **idempotent** if $e^2 = e$.

    (a) Find four idempotent elements in the ring $M(\mathbb{R})$.

    (b) Find all idempotents in $\mathbb{Z}_{12}$.

    (c) Prove that the only idempotents in an integral domain R are 0_R and 1_R.

3. (a) Can a ring R have more than one zero element? [*Hint:* If there were more than one, how many solutions would the equation $0_R + x = 0_R$ have?]

    (b) Can a ring R with identity have more than one identity element?

4. In the ring $M(\mathbb{R})$, let

$$A = \begin{pmatrix} 1 & 2 \\ 0 & 1 \end{pmatrix} \quad B = \begin{pmatrix} 4 & 1 \\ 1 & -1 \end{pmatrix} \quad C = \begin{pmatrix} 2 & 3 \\ 1 & -1 \end{pmatrix} \quad D = \begin{pmatrix} 4 & -7 \\ 1 & -3 \end{pmatrix}.$$

Show that the equations $AX = B$ and $XA = B$ have different solutions [*Hint:* Check to see if C or D is a solution of either equation.]

5. Let S and T be subrings of a ring R.

   (a) Is $S \cap T$ a subring of R?

   (b) Is $S \cup T$ a subring of R?

6. Let R be a ring and b a fixed element of R. Let $T = \{rb \mid r \in R\}$. Prove that T is a subring of R.

7. **(a)** Let S be the subset of $\mathbb{Z}_{10}$ defined by $S = \{r \in \mathbb{Z}_{10} \mid 5r = 0\}$. List the elements of S. Is S a subring of $\mathbb{Z}_{10}$?

   (b) Let $B = \begin{pmatrix} 0 & 0 \\ 1 & 0 \end{pmatrix}$ and let S be the subset of $M(\mathbb{R})$ defined by $S = \left\{ A \in M(\mathbb{R}) \;\middle|\; AB = \begin{pmatrix} 0 & 0 \\ 0 & 0 \end{pmatrix} \right\}$. What does a typical element of S look like? Is S a subring of $M(\mathbb{R})$?

   (c) Let R be any ring and b a fixed element of R. Let $S = \{r \in R \mid rb = 0_R\}$. Prove that S is a subring of R. (Parts (a) and (b) are special cases of this result.)

8. Let R be a ring and let $C = \{a \in R \mid ra = ar \text{ for every } r \in R\}$.

   (a) When $R = M(\mathbb{R})$, find at least one element of $M(\mathbb{R})$ that is *not* in C and at least three elements of $M(\mathbb{R})$ that are in C.

   (b) If R is any ring, prove that C is a subring of R.

9. Let R be a ring with identity and let $S = \{n1_R \mid n \in Z\}$. Prove that S is a subring of R. [The definition of na with $n \in Z$, $a \in R$ is on page 59. Also see Exercise 27.]

10. Prove or disprove: The set of units in a ring R with identity is a subring of R.

11. **(a)** If a and b are units in a ring R with identity, prove that ab is a unit whose inverse is $(ab)^{-1} = b^{-1}a^{-1}$.

   (b) Give an example to show that if a and b are units, then $a^{-1}b^{-1}$ need not be the multiplicative inverse of ab.

12. **(a)** Prove that $[a]$ is a unit in $\mathbb{Z}_n$ if and only if $(a,n) = 1$ in $\mathbb{Z}$.

   (b) Prove that $[a]$ is a nonunit in $\mathbb{Z}_n$ if and only if $[a]$ is a zero divisor.

13. Is it possible for a unit in a ring with identity to be a zero divisor?

14. Let R and S be nonzero rings (meaning that each of them contains at least one nonzero element). Show that $R \times S$ contains zero divisors.

15. Let R be a ring and $a \in R$. Assume that $a \neq 0_R$ and that a is not a zero divisor. Prove that whenever $ab = ac$ in R, then $b = c$.

16. (a) If ab is a zero divisor in a ring R, prove that a or b is a zero divisor.

    (b) If a or b is a zero divisor in a commutative ring R and $ab \neq 0_R$, prove that ab is a zero divisor.

17. Let R be a ring and a, $b \in R$. Let m and n be positive integers.

    (a) Show that $a^m a^n = a^{m+n}$ and $(a^m)^n = a^{mn}$.

    (b) Under what conditions is it true that $(ab)^n = a^n b^n$?

B. 18. Let S be a subring of a ring R. Prove that $0_S = 0_R$. [*Hint:* For $a \in S$, consider the equation $a + x = a$.]

19. Let S be a subring of a ring R with identity.

    (a) If S has an identity, show by example that 1_S may not be the same as 1_R.

    (b) If both R and S are fields, prove that $1_S = 1_R$.

    (c) If S has an identity, show that either $1_S = 1_R$ or 1_S is a zero divisor in R. An example of the latter possibility is given in Exercise 17 of Section 3.1.

20. Let R be a set equipped with an addition and multiplication satisfying Axioms 1, 2, and 4–10 (pages 42–43).

    (a) Show that the solution of $a + x = 0_R$ is also a solution of $y + a = 0_R$.

    (b) Prove that R is a ring. [*Hint:* To show that $a + b = b + a$, expand $(a + b)(1_R + 1_R)$ in two ways.]

21. (a) Let R be a ring and a, $b \in R$. Let m and n be nonnegative integers and prove that

        (i) $(m + n)a = ma + na$.

        (ii) $m(a + b) = ma + mb$.

        (iii) $m(ab) = (ma)b = a(mb)$.

        (iv) $(ma)(nb) = mn(ab)$.

    (b) Do part (a) when m and n are any integers.

22. Let R and S be rings with identity. What are the units in the ring $R \times S$?

23. Let R be a commutative ring with identity. Prove that R is an integral domain if and only if cancellation holds in R (that is, $a \neq 0_R$ and $ab = ac$ in R imply $b = c$).

24. Let R be a commutative ring with identity and $b \in R$. Let T be the subring of all multiples of b (see Exercise 6). If u is a unit in R and $u \in T$, prove that $T = R$.

25. A **Boolean ring** is a ring R with identity in which $x^2 = x$ for every $x \in R$. For examples, see Exercises 11 and 36 in Section 3.1. If R is a Boolean ring, prove that

(a) $a + a = 0_R$ for every $a \in R$. [*Hint:* Expand $(a + a)^2$.]

(b) R is commutative. [*Hint:* Expand $(a + b)^2$.]

26. Assume that $R = \{0_R, 1_R, a, b\}$ is a ring and that a and b are units. Write out the multiplication table of R.

27. Let R be a ring with identity and b a fixed element of R and let $S = \{nb \mid n \in \mathbb{Z}\}$. Is S necessarily a subring of R? [Exercise 9 is the case when $b = 1_R$.]

28. Let R be a ring with identity. If ab and a are units in R, prove that b is a unit.

29. Let R be a ring with identity and no zero divisors. If ab is a unit in R, prove that a and b are units.

30. An element a of a ring is **nilpotent** if $a^n = 0_R$ for some positive integer n. Prove that R has no nonzero nilpotent elements if and only if 0_R is the unique solution of the equation $x^2 = 0_R$.

The following definition is needed for Exercises 31–33. Let R be a ring with identity. If there is a smallest positive integer n such that $n1_R = 0_R$, then R is said to have **characteristic n.** If no such n exists, R is said to have **characteristic zero.**

31. (a) Show that $\mathbb{Z}$ has characteristic zero and $\mathbb{Z}_n$ has characteristic n.

(b) What is the characteristic of $\mathbb{Z}_4 \times \mathbb{Z}_6$?

32. Prove that a finite ring with identity has characteristic n for some $n > 0$.

33. Let R be a ring with identity of characteristic $n > 0$.

(a) Prove that $na = 0_R$ for every $a \in R$.

(b) If R is an integral domain, prove that n is prime.

34. Prove that

$$\begin{pmatrix} a & b \\ c & d \end{pmatrix}$$

is a unit in the ring $M(\mathbb{R})$ if and only if $ad - bc \neq 0$. In this case, verify that its inverse is

$$\begin{pmatrix} d/t & -b/t \\ -c/t & a/t \end{pmatrix},$$

where $t = ad - bc$.

C. 35. Let R be a ring without identity. Let T be the set $R \times \mathbb{Z}$. Define addition and multiplication in T by these rules:

$$(r,m) + (s,n) = (r + s, n + m).$$

$$(r,m)(s,n) = (rs + ms + nr, mn).$$

(a) Prove that T is a ring with identity.

(b) Let R^* consist of all elements of the form $(r,0)$ in T. Prove that R^* is a subring of T.

36. (a) Let a and b be nilpotent elements in a commutative ring R (see Exercise 30). Prove that $a + b$ and ab are also nilpotent. [You will need the Binomial Theorem from Appendix E.]

(b) Let N be the set of all nilpotent elements of R. Show that N is a subring of R.

37. Let R be a ring such that $x^3 = x$ for every $x \in R$. Prove that R is commutative.

38. Let R be a nonzero finite commutative ring with no zero divisors. Prove that R is a field.

3.3 ISOMORPHISMS AND HOMOMORPHISMS

If you were unfamiliar with roman numerals and came across a discussion of integer arithmetic written solely with roman numerals, it might take you some time to realize that this arithmetic was essentially the same as the familiar arithmetic in $\mathbb{Z}$ except for the labels on the elements. Here is a less trivial example.

EXAMPLE   Consider the subset $S = \{0, 2, 4, 6, 8\}$ of $\mathbb{Z}_{10}$. With the addition and multiplication of $\mathbb{Z}_{10}$, S is actually a commutative ring, as can be seen from these tables:*

+	0	6	2	8	4
0	0	6	2	8	4
6	6	2	8	4	0
2	2	8	4	0	6
8	8	4	0	6	2
4	4	0	6	2	8

·	0	6	2	8	4
0	0	0	0	0	0
6	0	6	2	8	4
2	0	2	4	6	8
8	0	8	6	4	2
4	0	4	8	2	6

* The reason the elements of S are listed in this order will become clear in a moment.

A careful examination of the tables shows that S is a field with five elements and that the multiplicative identity of this field is the element 6.

We claim that S is "essentially the same" as the field $\mathbb{Z}_5$ except for the labels on the elements. You can see this as follows. Write out addition and multiplication tables for $\mathbb{Z}_5$. To avoid any possible confusion with elements of S, denote the elements of $\mathbb{Z}_5$ by $\bar{0}$, $\bar{1}$, $\bar{2}$, $\bar{3}$, $\bar{4}$. Then relabel the entries in the $\mathbb{Z}_5$ tables according to this scheme:

$$\text{Relabel } \bar{0} \text{ as } 0, \quad \text{relabel } \bar{1} \text{ as } 6, \quad \text{relabel } \bar{2} \text{ as } 2,$$
$$\text{relabel } \bar{3} \text{ as } 8, \quad \text{relabel } \bar{4} \text{ as } 4.$$

Look what happens:

$+$	$\bar{0}$ (0)	$\bar{1}$ (6)	$\bar{2}$ (2)	$\bar{3}$ (8)	$\bar{4}$ (4)
$\bar{0}$ (0)	$\bar{0}$ (0)	$\bar{1}$ (6)	$\bar{2}$ (2)	$\bar{3}$ (8)	$\bar{4}$ (4)
$\bar{1}$ (6)	$\bar{1}$ (6)	$\bar{2}$ (2)	$\bar{3}$ (8)	$\bar{4}$ (4)	$\bar{0}$ (0)
$\bar{2}$ (2)	$\bar{2}$ (2)	$\bar{3}$ (8)	$\bar{4}$ (4)	$\bar{0}$ (0)	$\bar{1}$ (6)
$\bar{3}$ (8)	$\bar{3}$ (8)	$\bar{4}$ (4)	$\bar{0}$ (0)	$\bar{1}$ (6)	$\bar{2}$ (2)
$\bar{4}$ (4)	$\bar{4}$ (4)	$\bar{0}$ (0)	$\bar{1}$ (6)	$\bar{2}$ (2)	$\bar{3}$ (8)

$\cdot$	$\bar{0}$ (0)	$\bar{1}$ (6)	$\bar{2}$ (2)	$\bar{3}$ (8)	$\bar{4}$ (4)
$\bar{0}$ (0)	$\bar{0}$ (0)	$\bar{0}$ (0)	$\bar{0}$ (0)	$\bar{0}$ (0)	$\bar{0}$ (0)
$\bar{1}$ (6)	$\bar{0}$ (0)	$\bar{1}$ (6)	$\bar{2}$ (2)	$\bar{3}$ (8)	$\bar{4}$ (4)
$\bar{2}$ (2)	$\bar{0}$ (0)	$\bar{2}$ (2)	$\bar{4}$ (4)	$\bar{1}$ (6)	$\bar{3}$ (8)
$\bar{3}$ (8)	$\bar{0}$ (0)	$\bar{3}$ (8)	$\bar{1}$ (6)	$\bar{4}$ (4)	$\bar{2}$ (2)
$\bar{4}$ (4)	$\bar{0}$ (0)	$\bar{4}$ (4)	$\bar{3}$ (8)	$\bar{2}$ (2)	$\bar{1}$ (6)

By relabeling the elements of $\mathbb{Z}_5$, you obtain the addition and multiplication tables for S. Thus the operations in $\mathbb{Z}_5$ and S work in exactly the same way— the only difference is the way the elements are labeled. As far as ring structure goes, S is just the ring $\mathbb{Z}_5$ with new labels on the elements. In more technical terms, $\mathbb{Z}_5$ and S are said to be *isomorphic*.

In general, *isomorphic rings are rings that have the same structure, in the sense that the addition and multiplication tables of one are the tables of the other with the elements suitably relabeled.* Although this intuitive idea is adequate for small finite systems, we need a rigorous mathematical definition of isomorphism that agrees with this intuitive idea *and* is readily applicable to large rings as well.

There are two aspects to the intuitive idea that "R and S are isomorphic": relabeling and comparing the tables. Relabeling means that every element of R

is paired with a unique element of S (its new label). In other words, there is a function $f: R \to S$ that assigns to each $r \in R$ its new label $f(r) \in S$. In the preceding example, we used the relabeling function $f: \mathbb{Z}_5 \to S$, given by:

$$f(\bar{0}) = 0 \quad f(\bar{1}) = 6 \quad f(\bar{2}) = 2 \quad f(\bar{3}) = 8 \quad f(\bar{4}) = 4.$$

Such a function must have these additional properties:

(i) Distinct elements of R must get distinct new labels:

If $r \neq r'$ in R, then $f(r) \neq f(r')$ in S.

(ii) Every element of S must be the label of some element in R:*

For each $s \in S$, there is an $r \in R$ such that $f(r) = s$.

Statements (i) and (ii) simply say that the function f must be both injective and surjective,** that is, *f must be a bijection.*

However, a bijection (relabeling scheme) f won't be an isomorphism unless the tables of R become the tables of S when f is applied. In this case, if $a + b = c$ in R, then the tables of R and S must look like this:

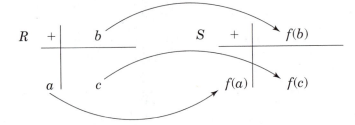

The S-table shows that $f(a) + f(b) = f(c)$. However, $a + b$ and c are the *same element* of R, and so $f(a + b) = f(c)$, and therefore

$$f(a + b) = f(a) + f(b).$$

This is the condition that f must satisfy in order for f to change the addition tables of R into those of S. The analogous condition on f for the multiplication tables is $f(ab) = f(a)f(b)$. We now can state a formal definition of isomorphism:

---

* Otherwise, we couldn't possibly get the complete tables of S from those of R.
** Injective, surjective, and bijective functions are discussed in Appendix B.

> **DEFINITION•** *A ring R is **isomorphic** to a ring S (in symbols, $R \cong S$) if there is a function $f: R \to S$ such that*
>
> *(i) f is injective;*
>
> *(ii) f is surjective;*
>
> *(iii) $f(a + b) = f(a) + f(b)$ and $f(ab) = f(a) f(b)$ for all $a, b \in R$.*
>
> *In this case the function f is called an **isomorphism.***

Warning   In order to be an isomorphism, a function must satisfy *all three* of the conditions in the definition. It is quite possible for a function to satisfy any two of these conditions but not the third; see Exercises 4, 23 and 30.

EXAMPLE   In the last example on page 47, we considered the field K of all 2×2 matrices of the form

$$\begin{pmatrix} a & b \\ -b & a \end{pmatrix},$$

where a and b are real numbers. We claim that K is isomorphic to the field $\mathbb{C}$ of complex numbers. To prove this, define a function $f: K \to \mathbb{C}$ by the rule

$$f\begin{pmatrix} a & b \\ -b & a \end{pmatrix} = a + bi.$$

To show that f is injective, suppose

$$f\begin{pmatrix} a & b \\ -b & a \end{pmatrix} = f\begin{pmatrix} r & s \\ -s & r \end{pmatrix}.$$

Then by the definition of f, $a + bi = r + si$ in $\mathbb{C}$. By the rules of equality in $\mathbb{C}$, we must have $a = r$ and $b = s$. Hence, in K

$$\begin{pmatrix} a & b \\ -b & a \end{pmatrix} = \begin{pmatrix} r & s \\ -s & r \end{pmatrix},$$

so that f is injective. The function f is surjective because any complex number $a + bi$ is the image under f of the matrix

$$\begin{pmatrix} a & b \\ -b & a \end{pmatrix}$$

in K. Finally, for any matrices A and B in K, we must show that $f(A + B) = f(A) + f(B)$ and $f(AB) = f(A)f(B)$. We have

$$f\left[\begin{pmatrix} a & b \\ -b & a \end{pmatrix} + \begin{pmatrix} c & d \\ -d & c \end{pmatrix}\right] = f\begin{pmatrix} a + c & b + d \\ -b - d & a + c \end{pmatrix}$$
$$= (a + c) + (b + d)i$$
$$= (a + bi) + (c + di)$$
$$= f\begin{pmatrix} a & b \\ -b & a \end{pmatrix} + f\begin{pmatrix} c & d \\ -d & c \end{pmatrix}$$

and

$$f\left[\begin{pmatrix} a & b \\ -b & a \end{pmatrix}\begin{pmatrix} c & d \\ -d & c \end{pmatrix}\right] = f\begin{pmatrix} ac - bd & ad + bc \\ -ad - bc & ac - bd \end{pmatrix}$$
$$= (ac - bd) + (ad + bc)i$$
$$= (a + bi)(c + di)$$
$$= f\begin{pmatrix} a & b \\ -b & a \end{pmatrix}f\begin{pmatrix} c & d \\ -d & c \end{pmatrix}.$$

Therefore, f is an isomorphism.

It is quite possible to relabel the elements of a single ring in such a way that the ring is isomorphic to itself.

EXAMPLE   Let $f : \mathbb{C} \to \mathbb{C}$ be the complex conjugation map given by $f(a + bi) = a - bi$.* The function f satisfies

$$f[(a + bi) + (c + di)] = f[(a + c) + (b + d)i]$$
$$= (a + c) - (b + d)i = (a - bi) + (c - di)$$
$$= f(a + bi) + f(c + di)$$

and

$$f[(a + bi)(c + di)] = f[(ac - bd) + (ad + bc)i]$$
$$= (ac - bd) - (ad + bc)i = (a - bi)(c - di)$$
$$= f(a + bi)f(c + di).$$

You can readily verify that f is both injective and surjective (Exercise 15). Therefore f is an isomorphism.

---

* The function f has a geometric interpretation in the complex plane, where $a + bi$ is identified with the point (a,b): It reflects the plane in the x-axis.

> **EXAMPLE** If R is any ring and $\iota_R : R \to R$ is the identity map given by $\iota_R(r) = r$, then for any $a, b \in R$
>
> $$\iota_R(a + b) = a + b = \iota_R(a) + \iota_R(b) \quad \text{and} \quad \iota_R(ab) = ab = \iota_R(a)\iota_R(b).$$
>
> Since ι_R is obviously bijective, it is an isomorphism.

Our intuitive notion of isomorphism is symmetric: "R is isomorphic to S" means the same thing as "S is isomorphic to R." The formal definition of isomorphism is not symmetric, however, since it requires a function from R onto S but no function from S onto R. This apparent asymmetry is easily remedied. If $f : R \to S$ is an isomorphism, then f is a bijective function of sets. Therefore, f has an inverse function $g : S \to R$ such that $g \circ f = \iota_R$ (the identity function on R) and $f \circ g = \iota_S$.* It is not hard to verify that the function g is actually an isomorphism (Exercise 27). Thus $R \cong S$ implies that $S \cong R$, and symmetry is restored.

Homomorphisms

Many functions that are not injective or surjective satisfy condition (iii) of the definition of isomorphism. Such functions are given a special name.

> **DEFINITION•** *Let R and S be rings. A function $f : R \to S$ is said to be a **homomorphism** if*
>
> $$f(a + b) = f(a) + f(b) \quad and \quad f(ab) = f(a)f(b) \quad for\ all\ a, b \in R.$$

Thus every isomorphism is a homomorphism, but as the following examples show, a homomorphism need not be an isomorphism because a homomorphism may fail to be injective or surjective.

> **EXAMPLE** For any rings R and S the **zero map** $z : R \to S$ given by $z(r) = 0_S$ for every $r \in R$ is a homomorphism because for any $a, b \in R$
>
> $$z(a + b) = 0_S = 0_S + 0_S = z(a) + z(b)$$
>
> and
>
> $$z(ab) = 0_S = 0_S \cdot 0_S = z(a)z(b).$$
>
> When both R and S contain nonzero elements, then the zero map is neither injective nor surjective.

* See Appendix B for details.

EXAMPLE The function $f:\mathbb{Z} \to \mathbb{Z}_6$ given by $f(a) = [a]$ is a homomorphism because of the way that addition and subtraction are defined in $\mathbb{Z}_6$: for any $a, b \in \mathbb{Z}$

$$f(a + b) = [a + b] = [a] + [b] = f(a) + f(b)$$

and

$$f(ab) = [ab] = [a][b] = f(a)f(b).$$

The homomorphism f is surjective, but not injective (why?).

EXAMPLE The map $g:\mathbb{R} \to M(\mathbb{R})$ given by

$$g(r) = \begin{pmatrix} 0 & 0 \\ -r & r \end{pmatrix}$$

is a homomorphism because for any $r, s \in \mathbb{R}$

$$g(r) + g(s) = \begin{pmatrix} 0 & 0 \\ -r & r \end{pmatrix} + \begin{pmatrix} 0 & 0 \\ -s & s \end{pmatrix} = \begin{pmatrix} 0 & 0 \\ -r - s & r + s \end{pmatrix}$$

$$= \begin{pmatrix} 0 & 0 \\ -(r + s) & r + s \end{pmatrix} = g(r + s)$$

and

$$g(r)g(s) = \begin{pmatrix} 0 & 0 \\ -r & r \end{pmatrix}\begin{pmatrix} 0 & 0 \\ -s & s \end{pmatrix} = \begin{pmatrix} 0 & 0 \\ -rs & rs \end{pmatrix} = g(rs).$$

The homomorphism g is injective but not surjective (Exercise 24).

Warning Not all functions are homomorphisms. The properties

$$f(a + b) = f(a) + f(b) \qquad \text{and} \qquad f(ab) = f(a)f(b)$$

fail for many functions. For example, if $f:\mathbb{R} \to \mathbb{R}$ given by $f(x) = x + 2$, then

$$f(3 + 4) = f(7) = 9 \qquad \text{but} \qquad f(3) + f(4) = 5 + 6 = 11$$

so that $f(3 + 4) \neq f(3) + f(4)$. Similarly, $f(3 \cdot 4) \neq f(3)f(4)$ because

$$f(3 \cdot 4) = f(12) = 14, \qquad \text{but} \qquad f(3)f(4) = 5 \cdot 6 = 30.$$

THEOREM 3.12 *Let $f:R \to S$ be a homomorphism of rings. Then*

(1) $f(0_R) = 0_S$.

(2) $f(-a) = -f(a)$ for every $a \in R$.

(3) $f(a - b) = f(a) - f(b)$ for all $a, b \in R$.

If R is a ring with identity and f is surjective, then

(4) *S is a ring with identity and* $f(1_R) = 1_S$.

(5) *Whenever u is a unit in R, then f(u) is a unit in S and* $f(u)^{-1} = f(u^{-1})$.

Proof (1) Since f is a homomorphism and $0_R + 0_R = 0_R$ in R,

$$f(0_R) + f(0_R) = f(0_R + 0_R) = f(0_R) \text{ in } S.$$

Subtracting $f(0_R)$ from each end shows that $f(0_R) = 0_S$.

(2) $f(a) + f(-a) = f(a + (-a)) = f(0_R) = 0_S$. So $f(-a)$ is a solution of $f(a) + x = 0_S$. But the unique solution of this equation is $-f(a)$ by Theorem 3.3. Hence, $f(-a) = -f(a)$ by uniqueness.

(3) By the definition of subtraction, the fact that f is a homomorphism, and (2)

$$f(a - b) = f(a + (-b)) = f(a) + f(-b) = f(a) + (-f(b)) = f(a) - f(b).$$

(4) We shall show that $f(1_R) \epsilon S$ is the identity element of S. Let s be any element of S. Then since f is surjective, $s = f(r)$ for some $r \epsilon R$. Hence,

$$s \cdot f(1_R) = f(r)f(1_R) = f(r \cdot 1_R) = f(r) = s$$

and, similarly, $f(1_R) \cdot s = s$. Therefore, S has $f(1_R)$ as its identity element, that is, $f(1_R) = 1_S$ (see Exercise 3(b) in Section 3.2).

(5) Since u is a unit in R, there is an element v in R such that $uv = 1_R = vu$. Hence, by (4)

$$f(u)f(v) = f(uv) = f(1_R) = 1_S.$$

Similarly, $vu = 1_R$ implies that $f(v)f(u) = 1_S$. Therefore, $f(u)$ is a unit in S, with inverse $f(v)$. In other words, $f(u)^{-1} = f(v)$. Since $v = u^{-1}$, we see that $f(u)^{-1} = f(u^{-1})$. ◆

If $f : R \to S$ is a function, then the **image** of f is this subset of S:

$$I = \{s \epsilon S \mid s = f(r) \text{ for some } r \epsilon R\} = \{f(r) \mid r \epsilon R\}.$$

If f is surjective, then $I = S$ by the definition of surjective. In any case we have:

COROLLARY 3.13 *If $f : R \to S$ is a homomorphism of rings, then the image of f is a subring of S.*

Proof The image I is nonempty because $0_S = f(0_R) \epsilon I$ by (1) of Theorem 3.12. The definition of homomorphism shows that I is closed under multiplication: If $f(a)$, $f(b) \epsilon I$, then $f(a)f(b) = f(ab) \epsilon I$. Similarly, I is closed under subtraction because $f(a) - f(b) = f(a - b) \epsilon I$ by Theorem 3.12. Therefore, I is a subring of S by Theorem 3.6. ◆

Existence of Isomorphisms

If you suspect that two rings are isomorphic, there are no hard and fast rules for finding a function that is an isomorphism between them. However, the properties of homomorphisms in Theorem 3.12 can sometimes be helpful.

EXAMPLE If there is an isomorphism f from $\mathbb{Z}_{12}$ to the ring $\mathbb{Z}_3 \times \mathbb{Z}_4$, then $f(1) = (1,1)$ by part (4) of Theorem 3.12. Since f is a homomorphism, it has to satisfy

$$f(2) = f(1 + 1) = f(1) + f(1) = (1,1) + (1,1) = (2,2)$$
$$f(3) = f(2 + 1) = f(2) + f(1) = (2,2) + (1,1) = (0,3)$$
$$f(4) = f(3 + 1) = f(3) + f(1) = (0,3) + (1,1) = (1,0).$$

Continuing in this fashion shows that *if f is an isomorphism, then it must be this bijective function:*

$f(1) = (1,1)$	$f(4) = (1,0)$	$f(7) = (1,3)$	$f(10) = (1,2)$
$f(2) = (2,2)$	$f(5) = (2,1)$	$f(8) = (2,0)$	$f(11) = (2,3)$
$f(3) = (0,3)$	$f(6) = (0,2)$	$f(9) = (0,1)$	$f(0) = (0,0).$

All we have shown up to here is that this bijective function f is the only *possible* isomorphism. To show that this f actually *is* an isomorphism, we must verify that it is a homomorphism. This can be done either by writing out the tables (tedious) or by observing that the rule of f can be described this way:

$$f([a]_{12}) = ([a]_3, [a]_4),$$

where $[a]_{12}$ denotes the congruence class of the integer a in $\mathbb{Z}_{12}$, $[a]_3$ denotes the class of a in $\mathbb{Z}_3$, and $[a]_4$ the class of a in $\mathbb{Z}_4$. (*Verify that this last statement is correct.*) Then

$f([a]_{12} + [b]_{12}) = f([a + b]_{12})$	[definition of addition in $\mathbb{Z}_{12}$]
$\qquad = ([a + b]_3, [a + b]_4)$	[definition of f]
$\qquad = ([a]_3 + [b]_3, [a]_4 + [b]_4)$	[definition of addition in $\mathbb{Z}_3$ and $\mathbb{Z}_4$]
$\qquad = ([a]_3, [a]_4) + ([b]_3, [b]_4)$	[definition of addition in $\mathbb{Z}_3 \times \mathbb{Z}_4$]
$\qquad = f([a]_{12}) + f([b]_{12})$	[definition of f].

An identical argument using multiplication in place of addition shows that $f([a]_{12}[b]_{12}) = f([a]_{12})f([b]_{12})$. Therefore, f is an isomorphism and $\mathbb{Z}_{12} \cong \mathbb{Z}_3 \times \mathbb{Z}_4$.

Up to now we have concentrated on showing that various rings are isomorphic, but sometimes it is equally important to demonstrate that two rings are *not* isomorphic. To do this, you must show that there is *no* possible function from one to the other satisfying the three conditions of the definition.

EXAMPLE $\mathbb{Z}_6$ is not isomorphic to $\mathbb{Z}_{12}$ or to $\mathbb{Z}$ because it is not possible to have a surjective function from a six-element set to a larger set (or an injective one from a larger set to $\mathbb{Z}_6$).

To show that two infinite rings or two finite rings with the same number of elements are not isomorphic, it is usually best to proceed indirectly.

EXAMPLE The rings $\mathbb{Z}_4$ and $\mathbb{Z}_2 \times \mathbb{Z}_2$ are not isomorphic. To show this, suppose on the contrary that $f : \mathbb{Z}_4 \to \mathbb{Z}_2 \times \mathbb{Z}_2$ is an isomorphism. Then $f(0) = (0,0)$ and $f(1) = (1,1)$ by Theorem 3.12. Consequently, $f(2) = f(1 + 1) = f(1) + f(1) = (1,1) + (1,1) = (0,0)$. Since f is injective and $f(0) = f(2)$, we have a contradiction. Therefore, no isomorphism is possible.

Suppose that $f : R \to S$ is an isomorphism and the elements $a, b, c, \ldots$ of R have a particular property. If the elements $f(a), f(b), f(c), \ldots$ of S have the same property, then we say that the property is **preserved by isomorphism**. According to parts (1), (4), and (5) of Theorem 3.12, for example, the property of being the zero element or the identity element or a unit is preserved by isomorphism. A property that is preserved by isomorphism can sometimes be used to prove that two rings are *not* isomorphic, as in the following examples.

EXAMPLE In the ring $\mathbb{Z}_8$ the elements 1, 3, 5, and 7 are units by Corollary 2.10 (with $b = 1$). Since being a unit is preserved by isomorphism, any isomorphism from $\mathbb{Z}_8$ to another ring with identity will map these four units to four units in the other ring. Consequently, $\mathbb{Z}_8$ is not isomorphic to any ring with less than four units. In particular, $\mathbb{Z}_8$ is not isomorphic to $\mathbb{Z}_4 \times \mathbb{Z}_2$ because there are only two units in this latter ring, namely $(1,1)$ and $(3,1)$ as you can readily verify.

EXAMPLE None of $\mathbb{Q}$, $\mathbb{R}$, or $\mathbb{C}$ is isomorphic to $\mathbb{Z}$ because every nonzero element in the fields $\mathbb{Q}$, $\mathbb{R}$, and $\mathbb{C}$ is a unit, whereas $\mathbb{Z}$ has only two units (1 and -1).

EXAMPLE Suppose R is a commutative ring and $f : R \to S$ is an isomorphism. Then for any $a, b \in R$, we have $ab = ba$ in R. Therefore, in S

$$f(a)f(b) = f(ab) = f(ba) = f(b)f(a).$$

Hence, S is also commutative because any two elements of S are of the form $f(a), f(b)$ (since f is surjective). In other words, the property of being a commutative ring is preserved by isomorphism. Therefore, no commutative ring can be isomorphic to a noncommutative ring.

◆ **EXERCISES**

A. **1.** Write out the addition and multiplication tables for $\mathbb{Z}_6$ and for $\mathbb{Z}_2 \times \mathbb{Z}_3$. Use them to show that $\mathbb{Z}_6 \cong \mathbb{Z}_2 \times \mathbb{Z}_3$.

2. Use tables to show that $\mathbb{Z}_2 \times \mathbb{Z}_2$ is isomorphic to the ring R of Exercise 2 in Section 3.1.

3. Let R be a ring and let R^* be the subring of $R \times R$ consisting of all elements of the form (a,a). Show that the function $f : R \to R^*$ given by $f(a) = (a,a)$ is an isomorphism.

4. Let S be the subring $\{0, 2, 4, 6, 8\}$ of $\mathbb{Z}_{10}$ and let $\mathbb{Z}_5 = \{\bar{0}, \bar{1}, \bar{2}, \bar{3}, \bar{4},\}$ (notation as in the Example on page 43). Use tables to show that the following bijection from $\mathbb{Z}_5$ to S is *not* an isomorphism:

$$\bar{0} \longrightarrow 0 \qquad \bar{1} \longrightarrow 2 \qquad \bar{2} \longrightarrow 4 \qquad \bar{3} \longrightarrow 6 \qquad \bar{4} \longrightarrow 8.$$

5. Let R and S be rings and let $\bar{R}$ be the subring of $R \times S$ consisting of all elements of the form $(a,0_S)$. Show that the function $f : R \to \bar{R}$ given by $f(a) = (a,0_S)$ is an isomorphism.

6. Prove that the field $\mathbb{R}$ of real numbers is isomorphic to the ring of all 2×2 matrices of the form $\begin{pmatrix} 0 & 0 \\ 0 & a \end{pmatrix}$, with $a \in \mathbb{R}$. [*Hint:* Consider the function f given by $f(a) = \begin{pmatrix} 0 & 0 \\ 0 & a \end{pmatrix}$.]

7. Let $\mathbb{Q}(\sqrt{2})$ be as in Exercise 31 of Section 3.1. Prove that the function $f : \mathbb{Q}(\sqrt{2}) \to \mathbb{Q}(\sqrt{2})$ given by $f(a + b\sqrt{2}) = a - b\sqrt{2}$ is an isomorphism.

8. Prove that $\mathbb{R}$ is isomorphic to the ring of all 2×2 matrices of the form $\begin{pmatrix} a & 0 \\ 0 & a \end{pmatrix}$, where $a \in \mathbb{R}$.

9. If $f : \mathbb{Z} \to \mathbb{Z}$ is an isomorphism, prove that f is the identity map. [*Hint:* What are $f(1), f(1 + 1), \ldots$?]

10. Which of the following functions are homomorphisms?

(a) $f : \mathbb{Z} \to \mathbb{Z}$, defined by $f(x) = -x$.

(b) $f : \mathbb{Z}_2 \to \mathbb{Z}_2$, defined by $f(x) = -x$.

(c) $g : \mathbb{Q} \to \mathbb{Q}$, defined by $g(x) = \dfrac{1}{x^2 + 1}$.

(d) $h : \mathbb{R} \to M(\mathbb{R})$, defined by $h(a) = \begin{pmatrix} -a & 0 \\ a & 0 \end{pmatrix}$.

(e) $f : \mathbb{Z}_{12} \to \mathbb{Z}_4$, defined by $f([x]_{12}) = [x]_4$, where $[u]_n$ denotes the class of the integer u in $\mathbb{Z}_n$.

11. (a) Give an example of a ring R and a function $f : R \to R$ such that $f(a + b) = f(a) + f(b)$ for all $a, b \in R$, but $f(ab) \neq f(a)f(b)$ for some $a, b \in R$.

(b) Give an example of a ring R and a function $f : R \to R$ such that $f(ab) = f(a)f(b)$ for all $a, b \in R$, but $f(a + b) \neq f(a) + f(b)$ for some $a, b \in R$.

12. Give an example of a ring R and a function $f : R \to R$ such that $f(a + b) = f(a)f(b)$ for all $a, b \in R$ and $f(a) \neq 0_R$ for all $a \in R$. Is your function f a homomorphism?

13. Let $f : R \to S$ be a homomorphism of rings. If r is a zero divisor in R, is $f(r)$ a zero divisor in S?

B. 14. Let T, R, and F be the four-element rings whose tables are given in the example on page 43 and in Exercises 2 and 3 of Section 3.1. Show that no two of these rings are isomorphic.

15. Show that the complex conjugation function $f : \mathbb{C} \to \mathbb{C}$ (whose rule is $f(a + bi) = a - bi$) is a bijection.

16. Show that the isomorphism of $\mathbb{Z}_5$ and S in the first example of this section is given by the function whose rule is $f([x]_5) = [6x]_{10}$. Give a direct proof (without using tables) that this map is a homomorphism.

17. Show that $S = \{0, 4, 8, 12, 16, 20, 24\}$ is a subring of $\mathbb{Z}_{28}$. Then prove that the map $f : \mathbb{Z}_7 \to S$ given by $f([x]_7) = [8x]_{28}$ is an isomorphism.

18. Let E be the ring of even integers with the $*$ multiplication defined in Exercise 19 of Section 3.1. Show that the map $f : E \to \mathbb{Z}$ given by $f(x) = x/2$ is an isomorphism.

19. Let $\mathbb{Z}^*$ denote the ring of integers with the $\oplus$ and $\odot$ operations defined in Exercise 18 of Section 3.1. Prove that $\mathbb{Z}$ is isomorphic to $\mathbb{Z}^*$.

20. Let $\overline{\mathbb{Z}}$ denote the ring of integers with the $\oplus$ and $\odot$ operations defined in Exercise 20 of Section 3.1. Prove that $\overline{\mathbb{Z}}$ is isomorphic to $\mathbb{Z}$.

21. Let $\mathbb{R} \times \mathbb{R}$ be the field of Exercise 37 of Section 3.1. Show that $\mathbb{R} \times \mathbb{R}$ is isomorphic to the field $\mathbb{C}$ of complex numbers.

22. (a) Show that $\mathbb{R} \times \mathbb{R}$ is a ring with the usual coordinatewise addition (as in Theorem 3.1) and multiplication given by the rule $(a,b)(c,d) = (ac,bc)$.

(b) Show that the ring of part (a) is isomorphic to the ring of all matrices in $M(\mathbb{R})$ of the form $\begin{pmatrix} a & 0 \\ b & 0 \end{pmatrix}$.

23. Let L be the ring of all matrices in $M(\mathbb{Z})$ of the form $\begin{pmatrix} a & 0 \\ b & c \end{pmatrix}$. Show that the function $f:L \to \mathbb{Z}$ given by $f\begin{pmatrix} a & 0 \\ b & c \end{pmatrix} = a$ is a surjective homomorphism but not an isomorphism.

24. Show that the homomorphism g in the second Example on page 72 is injective but not surjective.

25. **(a)** If $g:R \to S$ and $f:S \to T$ are homomorphisms, show that $f \circ g:R \to T$ is a homomorphism.

(b) If f and g are isomorphisms, show that $f \circ g$ is also an isomorphism.

26. **(a)** Give an example of a homomorphism $f:R \to S$ such that R has an identity but S does not. Does this contradict part (4) of Theorem 3.12?

(b) Give an example of a homomorphism $f:R \to S$ such that S has an identity but R does not.

27. Let $f:R \to S$ be an isomorphism of rings and let $g:S \to R$ be the inverse function of f (as defined in Appendix B). Show that g is also an isomorphism. [*Hint:* To show $g(a + b) = g(a) + g(b)$, consider the images of the left- and right-hand side under f and use the facts that f is a homomorphism and $f \circ g$ is the identity map.]

28. Let $f:R \to S$ be a homomorphism of rings and let $K = \{r \in R \mid f(r) = 0_S\}$. Prove that K is a subring of R.

29. Let $f:R \to S$ be a homomorphism of rings and T a subring of S. Let $P = \{r \in R \mid f(r) \in T\}$. Prove that P is a subring of R.

30. Assume $n \equiv 1 \pmod{m}$. Show that the function $f:\mathbb{Z}_m \to \mathbb{Z}_{mn}$ given by $f([x]_m) = [nx]_{mn}$ is an injective homomorphism but not an isomorphism when $n \geq 2$ (notation as in Exercise 10(e)).

31. **(a)** Let T be the ring of continuous functions from $\mathbb{R}$ to $\mathbb{R}$, as defined on page 46. Let $\theta:T \to \mathbb{R}$ be the function defined by $\theta(f) = f(5)$. Prove that θ is a surjective homomorphism. Is θ an isomorphism?

(b) Is part (a) true if 5 is replaced by any constant $c \in \mathbb{R}$?

32. If $f:R \to S$ is an isomomorphism of rings, which of the following properties are preserved by this isomorphism? Justify your answers.

(a) $a \in R$ is a zero divisor.

(b) $a \in R$ is idempotent (see Exercise 2 in Section 3.2).

(c) R is an integral domain.

33. Show that the first ring is not isomorphic to the second.

   (a) E and $\mathbb{Z}$ **(b)** $\mathbb{R} \times \mathbb{R} \times \mathbb{R} \times \mathbb{R}$ and $M(\mathbb{R})$

   (c) $\mathbb{Z}_4 \times \mathbb{Z}_{14}$ and $\mathbb{Z}_{16}$ **(d)** $\mathbb{Q}$ and $\mathbb{R}$

   (e) $\mathbb{Z} \times \mathbb{Z}_2$ and $\mathbb{Z}$ **(f)** $\mathbb{Z}_4 \times \mathbb{Z}_4$ and $\mathbb{Z}_{16}$

34. (a) If $f : R \to S$ is a homomorphism of rings, show that for any $r \in R$ and $n \in \mathbb{Z}$, $f(nr) = nf(r)$.

   (b) Prove that isomorphic rings with identity have the same characteristic. [See Exercises 31–33 of Section 3.2.]

   (c) If $f : R \to S$ is a homomorphism of rings with identity, is it true that R and S have the same characteristic?

35. Let T be the ring of continuous functions from $\mathbb{R}$ to $\mathbb{R}$. If $f \in T$ and $f^2 = f$, show that f is either the constant function $f(x) = 0$ or the constant function $f(x) = 1$. [*Hint:* For any $a \in \mathbb{R}$, $f(a)$ must be a solution of $x^2 = x$, and f is continuous.]

36. Prove that the ring T of Exercise 35 is not isomorphic to $\mathbb{R} \times \mathbb{R}$. [*Hint:* Find four solutions of $x^2 = x$ in $\mathbb{R} \times \mathbb{R}$; use Exercises 32(b) and 35.]

37. Let R be a ring without identity. Let T be the ring with identity of Exercise 35 in Section 3.2. Show that R is isomorphic to the subring R^* of T. Thus, if R is identified with R^*, then R is a subring of a ring with identity.

C. 38. For each positive integer k, let $k\mathbb{Z}$ denote the ring of all integer multiples of k (see Exercise 4 of Section 3.1). Prove that if $m \neq n$, then $m\mathbb{Z}$ is not isomorphic to $n\mathbb{Z}$.

39. Let $m, n \in \mathbb{Z}$ with $(m,n) = 1$ and let $f : \mathbb{Z}_{mn} \to \mathbb{Z}_m \times \mathbb{Z}_n$ be the function given by $f([a]_{mn}) = ([a]_m, [a]_n)$. (Notation as in Exercise 10(e). The Example on page 74 is the case $m = 3$, $n = 4$.)

   (a) Show that the map f is well-defined, that is, show that if $[a]_{mn} = [b]_{mn}$ in $\mathbb{Z}_{mn}$, then $[a]_m = [b]_m$ in $\mathbb{Z}_m$ and $[a]_n = [b]_n$ in $\mathbb{Z}_n$.

   (b) Prove that f is an isomorphism. [*Hint:* Adapt the proof in the Example on page 74; the difference is that proving f is a bijection takes more work here.]

40. If $(m,n) \neq 1$, prove that $\mathbb{Z}_{mn}$ is not isomorphic to $\mathbb{Z}_m \times \mathbb{Z}_n$.

Excursion   Lattices and Boolean Algebras (Chapter 14) may be covered at this point if desired.

CHAPTER 4

Arithmetic in F[x]

In Chapter 1 we examined grade-school arithmetic from an advanced standpoint and developed some important properties of the ring $\mathbb{Z}$ of integers. In this chapter we follow a parallel path, but the starting point here is high-school algebra—specifically, polynomials with coefficients in the field $\mathbb{R}$ of real numbers, such as

$$x^2 - 3x - 5, \qquad 6x^3 - 3x^2 + 7x + 4, \qquad x^{12} - 1.$$

Dealing with polynomials means dealing with the mysterious symbol "x", which is used in three different ways in high-school algebra. First, x often "stands for" a number, as in the equation $12x - 8 = 0$, where x is the number $\frac{2}{3}$. Second, x sometimes doesn't seem to stand for any particular number but is treated as if it were a number in simplification exercises such as this one:

$$\frac{x^3 + x}{x^2 + 1} = \frac{x(x^2 + 1)}{x^2 + 1} = x.$$

Third, x is also used as the variable in the rules of functions such as $f(x) = 3x + 5$.

Now that you know what rings and fields are, we shall consider polynomials with coefficients in any ring and attempt to clear up some of the mystery about the nature of x. In Sections 4.1–4.3, we shall see that when x is given a meaning similar to the second way it is used in high school, then the polynomials with coefficients in a field F form a ring (denoted $F[x]$) whose structure is remarkably similar to that of the ring $\mathbb{Z}$ of integers. In many cases the proofs for $\mathbb{Z}$ given in Chapter 1 carry over almost verbatim to $F[x]$.

In Sections 4.4–4.6 we consider tests to determine whether a polynomial is irreducible (the analogue of testing an integer for primality). Here the development is not an exact copy of what was done in the integers. The reason is that the polynomial ring $F[x]$ has features that have no analogues in the ring of integers, namely, the concepts of the root of a polynomial and of a polynomial function (which correspond to the first and third uses of x in high school).

4.1 POLYNOMIAL ARITHMETIC AND THE DIVISION ALGORITHM

The underlying idea here is to define "polynomial" in a way that is the obvious extension of polynomials with real-number coefficients. Let R be any ring. A *polynomial with coefficients in R* is an expression of the form

$$a_0 + a_1 x + a_2 x^2 + \cdots + a_n x^n,$$

where n is a nonnegative integer and $a_i \in R$.

This informal definition raises several questions: What is x? Is it an element of R? If not, what does it mean to multiply x by a ring element? In order to answer these questions, note that an expression of the form $a_0 + a_1 x + a_2 x^2 + \cdots + a_n x^n$ makes sense, provided that the a_i and x are all elements of some larger ring. An analogy might be helpful here. The number π is not in the ring $\mathbb{Z}$ of integers, but expressions such as $3 - 4\pi + 12\pi^2 + \pi^3$ and $8 - \pi^2 + 6\pi^5$ make sense in the real numbers. Furthermore, it is not difficult to verify that the set of all numbers of the form

$$a_0 + a_1 \pi + a_2 \pi^2 + \cdots + a_n \pi^n, \quad \text{with } n \geq 0 \text{ and } a_i \in \mathbb{Z}$$

is a subring of $\mathbb{R}$ that contains both $\mathbb{Z}$ and π (Exercise 2).

For the present we shall think of polynomials with coefficients in a ring R in much the same way, as elements of a larger ring that contains both R and a special element x that is not in R. This is analogous to the situation in the preceding paragraph with R in place of $\mathbb{Z}$ and x in place of π, except that here we don't know anything about the element x or even if such a larger ring exists. The following theorem provides the answer, as well as a definition of "polynomial."

THEOREM 4.1   *If R is a ring, then there exists a ring P that contains an element x that is not in R and has these properties:*

   (i) R is a subring of P.

   (ii) $xa = ax$ for every $a \in R$.

(iii) Every element of P can be written in the form

$$a_0 + a_1 x + a_2 x^2 + \cdots + a_n x^n, \qquad \text{for some } n \geq 0 \text{ and } a_i \in R.$$

(iv) The representation of elements in P in (iii) is unique in this sense: If $n \leq m$ and

$$a_0 + a_1 x + a_2 x^2 + \cdots + a_n x^n = b_0 + b_1 x + b_2 x^2 + \cdots + b_m x^m,$$

then $a_i = b_i$ for $i \leq n$ and $b_i = 0_R$ for each $i > n$.

(v) $a_0 + a_1 x + a_2 x^2 + \cdots + a_n x^n = 0_R$ if and only if $a_i = 0_R$ for every i.

The elements of the ring P in Theorem 4.1 are called **polynomials with coefficients in R,** the elements a_i are called **coefficients,** and the special element x is called an **indeterminate.*** A proof of Theorem 4.1, showing just how the ring P and the element x are constructed, is given in Appendix G, which may be covered at this time if desired. We shall assume Theorem 4.1 here.

Property (ii) of Theorem 4.1 does not imply that the ring P is commutative, but only that the special element x commutes with each element of the subring R (whose elements may not commute with each other). Note that property (v) is the special case of property (iv) obtained by letting each $b_i = 0_R$. The expression in property (v) is *not* an equation to be solved for x. In this context, asking what value of x makes $a_0 + a_1 x + a_2 x^2 + \cdots + a_n x^n = 0_R$ is as meaningless as asking what value of π makes $3 + 5\pi - 7\pi^2 = 0$ because x is a specific element of the ring P, not a variable that can be assigned values.**

Hereafter, we use the standard notation:

> **The ring P of polynomials with coefficients in R will be denoted by R[x].**

Since the product of any ring element with the zero element 0_R is just 0_R, terms with zero coefficients in a polynomial may be omitted or inserted as needed.

> **EXAMPLE** The polynomial $2 + 0x - x^2 + 0x^3 + 5x^4$ in $\mathbb{R}[x]$ is usually written $2 - x^2 + 5x^4$. If we want the same powers of x to appear in the polynomials $1 + x^2$ and $4 + x + 2x^3$, we write:
>
> $$1 + 0x + x^2 + 0x^3 \qquad \text{and} \qquad 4 + x + 0x^2 + 2x^3.$$

---

* Although in common use, the term "indeterminate" is misleading. As shown in Appendix G, there is nothing undetermined or ambiguous about x. It is a specific element of the larger ring P and is not an element of R.

** Variables and equations will be dealt with in Section 4.4.

The rules for adding and multiplying polynomials follow directly from the fact that $R[x]$ is a ring.

> **EXAMPLE**   If $f(x) = 1 + 5x - x^2 + 4x^3 + 2x^4$ and $g(x) = 4 + 2x + 3x^2 + x^3$ in $\mathbb{Z}_7[x]$, then the commutative, associative, and distributive laws show that
>
> $$\begin{aligned} f(x) + g(x) &= (1 + 5x - x^2 + 4x^3 + 2x^4) + (4 + 2x + 3x^2 + x^3 + 0x^4) \\ &= (1 + 4) + (5 + 2)x + (-1 + 3)x^2 + (4 + 1)x^3 + (2 + 0)x^4 \\ &= 5 + 0x + 2x^2 + 5x^3 + 2x^4 = 5 + 2x^2 + 5x^3 + 2x^4. \end{aligned}$$

> **EXAMPLE**   The product of $1 - 7x + x^2$ and $2 + 3x$ in $\mathbb{Q}[x]$ is found by using the distributive law repeatedly:
>
> $$\begin{aligned} (1 - 7x + x^2)(2 + 3x) &= 1(2 + 3x) - 7x(2 + 3x) + x^2(2 + 3x) \\ &= 1(2) + 1(3x) - 7x(2) - 7x(3x) + x^2(2) + x^2(3x) \\ &= 2 + 3x - 14x - 21x^2 + 2x^2 + 3x^3 \\ &= 2 - 11x - 19x^2 + 3x^3. \end{aligned}$$

The preceding examples are typical of the general case. You add polynomials by adding the corresponding coefficients, and you multiply polynomials by using the distributive laws and collecting like powers of x. Thus **polynomial addition** is given by the rule:*

$$(a_0 + a_1x + a_2x^2 + \cdots + a_nx^n) + (b_0 + b_1x + b_2x^2 + \cdots + b_nx^n)$$
$$= (a_0 + b_0) + (a_1 + b_1)x + (a_2 + b_2)x^2 + \cdots + (a_n + b_n)x^n$$

and **polynomial multiplication** is given by the rule:

$$(a_0 + a_1x + a_2x^2 + \cdots + a_nx^n)(b_0 + b_1x + b_2x^2 + \cdots + b_mx^m)$$
$$= a_0b_0 + (a_0b_1 + a_1b_0)x + (a_0b_2 + a_1b_1 + a_2b_0)x^2 + \cdots + a_nb_mx^{n+m}.$$

For each $k \geq 0$, the coefficient of x^k in the product is

$$a_0b_k + a_1b_{k-1} + a_2b_{k-2} + \cdots + a_{k-2}b_2 + a_{k-1}b_1 + a_kb_0 = \sum_{i=0}^{k} a_ib_{k-i},$$

where $a_i = 0_R$ if $i > n$ and $b_j = 0_R$ if $j > m$.

It follows readily from this description of multiplication in $R[x]$ that if R is commutative, then so is $R[x]$ (Exercise 7). Furthermore, if R has a multiplicative identity 1_R, then 1_R is also the multiplicative identity of $R[x]$ (Exercise 8).

---

* We may assume that the same powers of x appear by inserting zero coefficients where necessary.

DEFINITION• *Let $f(x) = a_0 + a_1 x + a_2 x^2 + \cdots + a_n x^n$ be a polynomial in $R[x]$ with $a_n \neq 0_R$. Then a_n is called the **leading coefficient** of $f(x)$. The **degree** of $f(x)$ is the integer n; it is denoted "deg $f(x)$." In other words, deg $f(x)$ is the largest exponent of x that appears with a nonzero coefficient, and this coefficient is the leading coefficient.*

EXAMPLE The degree of $3 - x + 4x^2 - 7x^3 \in \mathbb{R}[x]$ is 3, and its leading coefficient is -7. Similarly, deg $(3 + 5x) = 1$ and deg $(x^{12}) = 12$. The degree of $2 + x + 4x^2 - 0x^3 + 0x^5$ is 2 (the largest exponent of x with a *nonzero* coefficient); its leading coefficient is 4.

The ring R that we start with is a subring of the polynomial ring $R[x]$. The elements of R, considered as polynomials in $R[x]$, are called **constant polynomials.** The polynomials of degree 0 in $R[x]$ are precisely the nonzero constant polynomials. Note that

the constant polynomial 0_R does not have a degree

(because no power of x appears with *nonzero* coefficient).

THEOREM 4.2 *If R is an integral domain and $f(x)$, $g(x)$ are nonzero polynomials in $R[x]$, then*

$$deg[\,f(x)g(x)] = deg\ f(x) + deg\ g(x).$$

Proof Suppose $f(x) = a_0 + a_1 x + a_2 x^2 + \cdots + a_n x^n$ and $g(x) = b_0 + b_1 x + b_2 x^2 + \cdots + b_m x^m$ with $a_n \neq 0_R$ and $b_m \neq 0_R$, so that deg $f(x) = n$ and deg $g(x) = m$. Then

$$f(x)g(x) = a_0 b_0 + (a_0 b_1 + a_1 b_0)x + (a_2 b_0 + a_1 b_1 + a_0 b_2)x^2 + \cdots + a_n b_m x^{n+m}.$$

The largest exponent of x that can *possibly* have a nonzero coefficient is $n + m$. But $a_n b_m \neq 0_R$ because R is an integral domain and $a_n \neq 0_R$ and $b_m \neq 0_R$. Therefore, $f(x)g(x)$ is nonzero and deg$[\,f(x)g(x)] = n + m = $ deg $f(x) + $ deg $g(x)$. ◆

COROLLARY 4.3 *If R is an integral domain, then so is $R[x]$.*

Proof Since R is a commutative ring with identity, so is $R[x]$ (Exercises 7 and 8). The proof of Theorem 4.2 shows that the product of nonzero polynomials in $R[x]$ is nonzero. Therefore, $R[x]$ is an integral domain. ◆

Observe that the first part of the proof of Theorem 4.2 is valid for any ring R and shows that

if $f(x)$, $g(x)$ and $f(x)g(x)$ are nonzero, then

deg$[\,f(x)g(x)] \leq$ deg $f(x) +$ deg $g(x)$.

Strict inequality can occur if the ring has zero divisors. For instance, in $\mathbb{Z}_6[x]$, $2x^4$ has degree 4, and $1 + 3x^2$ has degree 2, but their product does *not* have degree $4 + 2 = 6$:

$$2x^4(1 + 3x^2) = 2x^4 + 2 \cdot 3x^6 = 2x^4 + 0x^6 = 2x^4.$$

For information on the degree of the *sum* of polynomials, see Exercise 4 and 10.

The Division Algorithm in $F[x]$

Our principal interest in the rest of this chapter will be polynomials with coefficients in a field F (such as $\mathbb{Q}$ or $\mathbb{R}$ or $\mathbb{Z}_5$). As noted in the chapter introduction, the domain $F[x]$ has many of the same properties as the domain $\mathbb{Z}$ of integers, including:

THEOREM 4.4 (THE DIVISION ALGORITHM IN $F[x]$) *Let F be a field and $f(x)$, $g(x) \in F[x]$ with $g(x) \neq 0_F$. Then there exist unique polynomials $q(x)$ and $r(x)$ such that*

$$f(x) = g(x)q(x) + r(x)$$
and either $\quad r(x) = 0_F \quad$ *or* $\quad deg\ r(x) < deg\ g(x).$

Compare this with the Division Algorithm in $\mathbb{Z}$ (Theorem 1.1), which says that for any integers a and b with b positive, there exist unique integers q and r such that

$$a = bq + r \quad \text{and} \quad 0 \le r < b.$$

The only change here is to replace statements such as "$r < b$" in $\mathbb{Z}$ by statements involving degree in $F[x]$.

You have probably used the Division Algorithm to check division problems such as this one, where you can readily verify that $f(x) = g(x)q(x) + r(x)$:

$$
\begin{array}{r}
\frac{3}{2}x^2 + x + 1 \qquad\qquad \longleftarrow \text{quotient } q(x) \\
2x^3 + 1 \overline{\smash{\big)}\ 3x^5 + 2x^4 + 2x^3 + 4x^2 + x - 2} \quad \longleftarrow \text{dividend } f(x) \\
\underline{3x^5 + \tfrac{3}{2}x^2 } \quad \longleftarrow (\tfrac{3}{2}x^2)g(x) \\
2x^4 + 2x^3 + \tfrac{5}{2}x^2 + x - 2 \quad \longleftarrow f(x) - (\tfrac{3}{2}x^2)g(x) \\
\underline{2x^4 \phantom{+ 2x^3 + \tfrac{5}{2}x^2} + x } \\
2x^3 + \tfrac{5}{2}x^2 - 2 \\
\underline{2x^3 \phantom{+ \tfrac{5}{2}x^2 + x} + 1} \\
\tfrac{5}{2}x^2 - 3 \quad \longleftarrow \text{remainder } r(x)
\end{array}
$$

By examining this division problem, we can get a clue as to how to prove Theorem 4.4. The first term of the quotient, $\frac{3}{2}x^2$, is obtained by dividing the leading term of the dividend, $3x^5$, by the leading term of the divisor, $2x^3$: $3x^5/2x^3 = \frac{3}{2}x^2$. The product of this term and the divisor, $(\frac{3}{2}x^2)g(x)$, is then subtracted from the dividend. Then the same process is repeated with the divisor and $2x^4 + 2x^3 - \frac{5}{2}x^2 + x - 2$ as the new dividend (note that its degree is one less than that of the original dividend): The second term of the quotient is $2x^4/2x^3 = x$, and so on. This repetitive process suggests a proof by mathematical induction.

Proof of Theorem 4.4   We first prove the existence of the polynomials $q(x)$ and $r(x)$.

   Case 1: If $f(x) = 0_F$ or if deg $f(x) <$ deg $g(x)$, then the theorem is true with $q(x) = 0_F$ and $r(x) = f(x)$ because $f(x) = g(x)0_F + f(x)$.

   Case 2: If $f(x) \neq 0_F$ and deg $g(x) \leq$ deg $f(x)$, then the proof of existence is by induction on the degree of the dividend $f(x)$.* If deg $f(x) = 0$, then deg $g(x) = 0$ also. Hence, $f(x) = a$ and $g(x) = b$ for some nonzero $a, b \in F$. Since F is a field, b is a unit and $a = b(b^{-1}a) + 0_F$. Thus the theorem is true with $q(x) = b^{-1}a$ and $r(x) = 0_F$.

   Assume inductively that the theorem is true whenever the dividend has degree less than n. This part of the proof is presented in two columns. The left-hand column is the formal proof, while the right-hand column refers to the long division example on page 85. The example will help you understand what's being done in the proof.

PROOF	EXAMPLE

We must show that the theorem is true whenever the dividend $f(x)$ has degree n, say

$$f(x) = a_n x^n + \cdots + a_1 x + a_0$$

with $a_n \neq 0_F$. The divisor $g(x)$ must have the form

$$g(x) = b_m x^m + \cdots + b_1 x + b_0$$

with $b_m \neq 0_F$ and $m \leq n$. We begin as we would in the long division of $g(x)$ into $f(x)$. Since F is a field and $b_m \neq 0_F$, b_m is a unit. Multiply the divisor $g(x)$ by $a_n b_m^{-1} x^{n-m}$ to obtain

$a_n b_m^{-1} x^{n-m} g(x)$
$\quad = a_n b_m^{-1} x^{n-m}(b_m x^m + \cdots + b_1 x + b_0)$
$\quad = a_n x^n + a_n b_m^{-1} b_{m-1} x^{n-1} + \cdots + a_n b_m^{-1} b_0 x^{n-m}$

EXAMPLE column:

$$f(x) = 3x^5 + 2x^4 + 2x^3 + 4x^2 + x - 2$$

$\overbrace{\qquad}$
$a_n x^n$

$$g(x) = 2x^3 + 1$$

$\overbrace{\quad}$
$b_m x^m$

$$a_n b_m^{-1} x^{n-m} = 3 \cdot 2^{-1} x^{5-3} = \frac{3}{2}x^2$$

$\overbrace{\qquad}$
first term of the quotient

$$\frac{3}{2}x^2 g(x) = \frac{3}{2}x^2(2x^3 + 1)$$
$$= 3x^5 + \frac{3}{2}x^2$$

* We use the Principle of Complete Induction; see Appendix C.

Since $a_n b_m^{-1} x^{n-m} g(x)$ and $f(x)$ have the same degree and the same leading coefficient, the difference

$$f(x) - a_n b_m^{-1} x^{n-m} g(x)$$

$$f(x) - \tfrac{3}{2} x^2 g(x)$$

is a polynomial of degree *less than* n (or possibly the zero polynomial). Now apply the induction hypothesis with $g(x)$ as divisor and the polynomial $f(x) - a_n b_m^{-1} x^{n-m} g(x)$ as dividend (or use Case 1 if this dividend is zero). By induction there exist polynomials $q_1(x)$ and $r(x)$ such that

$$= f(x) - (3x^5 + \tfrac{3}{2} x^2)$$
$$= 2x^4 + 2x^3 + \tfrac{5}{2} x^2 + x - 2$$

$$\underbrace{\phantom{= 2x^4 + 2x^3 + \tfrac{5}{2} x^2 + x - 2}}$$

fourth line of long division

$$f(x) - a_n b_m^{-1} x^{n-m} g(x) = g(x) q_1(x) + r(x) \qquad \text{and}$$
$$r(x) = 0_F \quad \text{or} \quad \deg r(x) < \deg g(x).$$

$$q_1(x) = x + 1 \qquad r(x) = \tfrac{5}{2} x^2 - 3$$

$$\underbrace{} \qquad \underbrace{\phantom{r(x) = \tfrac{5}{2} x^2 - 3}}$$

last part of   *remainder*
the quotient

Therefore,

$$f(x) = g(x)[a_n b_m^{-1} x^{n-m} + q_1(x)] + r(x) \qquad \text{and}$$
$$r(x) = 0_F \quad \text{or} \quad \deg r(x) < \deg g(x).$$

Thus the theorem is true with $q(x) = a_n b_m^{-1} x^{n-m} + q_1(x)$ when $\deg f(x) = n$. This completes the induction and shows that $q(x)$ and $r(x)$ always exist for any divisor and dividend.

To prove that $q(x)$ and $r(x)$ are unique, suppose that $q_2(x)$ and $r_2(x)$ are polynomials such that

$$f(x) = g(x) q_2(x) + r_2(x) \qquad \text{and} \qquad r_2(x) = 0_F \text{ or } \deg r_2(x) < \deg g(x).$$

Then

$$g(x) q(x) + r(x) = f(x) = g(x) q_2(x) + r_2(x),$$

so that

$$g(x)[q(x) - q_2(x)] = r_2(x) - r(x).$$

If $q(x) - q_2(x)$ is nonzero, then by Theorem 4.2 the degree of the left side is $\deg g(x) + \deg[q(x) - q_2(x)]$, a number greater than or equal to $\deg g(x)$. However, both $r_2(x)$ and $r(x)$ have degree strictly less than $\deg g(x)$, and so the right-hand side of the equation must also have degree strictly less than $\deg g(x)$ (Exercise 10). This is a contradiction. Therefore $q(x) - q_2(x) = 0_F$, or, equivalently, $q(x) = q_2(x)$. Since the left side is zero, we must have $r_2(x) - r(x) = 0_F$, so that $r_2(x) = r(x)$. Thus the polynomials $q(x)$ and $r(x)$ are unique. ◆

◆ EXERCISES

NOTE: *R denotes a ring and F a field.*

A. **1.** Perform the indicated operation and simplify your answer:

   (a) $(3x^4 + 2x^3 - 4x^2 + x + 4) + (4x^3 + x^2 + 4x + 3)$ in $\mathbb{Z}_5[x]$

   (b) $(x + 1)^3$ in $\mathbb{Z}_3[x]$

   (c) $(x - 1)^5$ in $\mathbb{Z}_5[x]$

   (d) $(x^2 - 3x + 2)(2x^3 - 4x + 1)$ in $\mathbb{Z}_7[x]$

2. Show that the set of all real numbers of the form

$$a_0 + a_1\pi + a_2\pi^2 + \cdots + a_n\pi^n, \qquad \text{with } n \geq 0 \text{ and } a_i \in \mathbb{Z}$$

is a subring of $\mathbb{R}$ that contains both $\mathbb{Z}$ and π.

3. **(a)** List all polynomials of degree 3 in $\mathbb{Z}_2[x]$.

   (b) List all polynomials of degree less than 3 in $\mathbb{Z}_3[x]$.

4. In each part, give an example of polynomials $f(x), g(x) \in \mathbb{Q}[x]$ that satisfy the given condition:

   (a) The deg of $f(x) + g(x)$ is less than the maximum of deg $f(x)$ and deg $g(x)$.

   (b) Deg $[f(x) + g(x)]$ = max {deg $f(x)$, deg $g(x)$}.

5. Find polynomials $q(x)$ and $r(x)$ such that $f(x) = g(x)q(x) + r(x)$, and $r(x) = 0$ or deg $r(x) <$ deg $g(x)$:

   (a) $f(x) = 3x^4 - 2x^3 + 6x^2 - x + 2$ and $g(x) = x^2 + x + 1$ in $\mathbb{Q}[x]$.

   (b) $f(x) = x^4 - 7x + 1$ and $g(x) = 2x^2 + 1$ in $\mathbb{Q}[x]$.

   (c) $f(x) = 2x^4 + x^2 - x + 1$ and $g(x) = 2x - 1$ in $\mathbb{Z}_5[x]$.

   (d) $f(x) = 4x^4 + 2x^3 + 6x^2 + 4x + 5$ and $g(x) = 3x^2 + 2$ in $\mathbb{Z}_7[x]$.

6. Which of the following subsets of $R[x]$ are subrings of $R[x]$? Justify your answer:

   (a) All polynomials with constant term 0_R.

   (b) All polynomials of degree 2.

   (c) All polynomials of degree $\leq k$, where k is a fixed positive integer.

   (d) All polynomials in which the odd powers of x have zero coefficients.

   (e) All polynomials in which the even powers of x have zero coefficients.

7. If R is commutative, show that $R[x]$ is also commutative.

8. If R has multiplicative identity 1_R, show that 1_R is also the multiplicative identity of $R[x]$.

9. If $c \in R$ is a zero divisor in a commutative ring R, then is c also a zero divisor in $R[x]$?

B. 10. If $f(x), g(x) \in R[x]$ and $f(x) + g(x) \neq 0_R$, show that

$$\deg[f(x) + g(x)] \leq \max \{\deg f(x), \deg g(x)\}.$$

11. If F is a field, show that $F[x]$ is not a field. [*Hint:* Is x a unit in $F[x]$?]

12. Let F be a field and let $f(x)$ be a nonzero polynomial in $F[x]$. Show that $f(x)$ is a unit in $F[x]$ if and only if $\deg f(x) = 0$.

13. Let R be a commutative ring. If $a_n \neq 0_R$ and $a_0 + a_1x + a_2x^2 + \cdots + a_nx^n$ is a zero divisor in $R[x]$, prove that a_n is a zero divisor in R.

14. (a) Let R be an integral domain and $f(x), g(x) \in R[x]$. Assume that the leading coefficient of $g(x)$ is a unit in R. Verify that the Division Algorithm holds for $f(x)$ as dividend and $g(x)$ as divisor. [*Hint:* Adapt the proof of Theorem 4.4. Where is the hypothesis that F is a field used there?]

(b) Give an example in $\mathbb{Z}[x]$ to show that part (a) may be false if the leading coefficient of $g(x)$ is not a unit. [*Hint:* Exercise 5(b).]

15. Let R be an integral domain. Assume that the Division Algorithm always holds in $R[x]$. Prove that R is a field.

16. Let $\varphi : R[x] \to R$ be the function that maps each polynomial in $R[x]$ onto its constant term (an element of R). Show that φ is a surjective homomorphism of rings.

17. Let $\varphi : \mathbb{Z}[x] \to \mathbb{Z}_n[x]$ be the function that maps the polynomial $a_0 + a_1x + \cdots + a_kx^k$ in $\mathbb{Z}[x]$ onto the polynomial $[a_0] + [a_1]x + \cdots + [a_k]x^k$, where $[a]$ denotes the class of the integer a in $\mathbb{Z}_n$. Show that φ is a surjective homomorphism of rings.

18. Let $D : \mathbb{R}[x] \to \mathbb{R}[x]$ be the derivative map defined by

$$D(a_0 + a_1x + a_2x^2 + \cdots + a_nx^n) = a_1 + 2a_2x + 3a_3x^2 + \cdots + na_nx^{n-1}.$$

Is D a homomorphism of rings? An isomorphism?

C. 19. Let $h : R \to S$ be a homomorphism of rings and define a function $\overline{h} : R[x] \to S[x]$ by the rule

$$\overline{h}(a_0 + a_1x + \cdots + a_nx^n) = h(a_0) + h(a_1)x + h(a_2)x^2 + \cdots + h(a_n)x^n.$$

Prove that

(a) $\overline{h}$ is a homomorphism of rings.

(b) $\overline{h}$ is injective if and only if h is injective.

(c) $\overline{h}$ is surjective if and only if h is surjective.

(d) If $R \cong S$, then $R[x] \cong S[x]$.

B. 20. Let R be a commutative ring and let $k(x)$ be a fixed polynomial in $R[x]$. Prove that there exists a unique homomorphism $\varphi : R[x] \to R[x]$ such that

$$\varphi(r) = r \text{ for all } r \in R \qquad \text{and} \qquad \varphi(x) = k(x).$$

21. Explain why the proof of Theorem 1.1 cannot be directly carried over to $F[x]$. Why is induction necessary to prove Theorem 4.4?

4.2 DIVISIBILITY IN $F[x]$

All the results of Section 1.2 on divisibility and greatest common divisors in $\mathbb{Z}$ now carry over, with only minor modifications, to the ring of polynomials over a field. In fact, most of the proofs in this section simply tell you to adapt the proof of the corresponding theorem for $\mathbb{Z}$ to $F[x]$. It is important that you actually *write out* these adaptations because the ability to do so is a guarantee that you really understand the analogy between $\mathbb{Z}$ and polynomials. *Throughout this section, F always denotes a field.*

> **DEFINITION•** *Let F be a field and $f(x)$, $g(x) \in F[x]$ with $f(x)$ nonzero. We say that $f(x)$ **divides** $g(x)$ [or $f(x)$ is a **factor** of $g(x)$], and write $f(x) \mid g(x)$, if $g(x) = f(x)h(x)$ for some $h(x) \in F[x]$.*

EXAMPLE $(2x + 1) \mid (6x^2 - x - 2)$ in $\mathbb{Q}[x]$ because $6x^2 - x - 2 = (2x + 1)(3x - 2)$. Furthermore, every constant multiple of $2x + 1$ also divides $6x^2 - x - 2$. For instance, $5(2x + 1) = 10x + 5$ divides $6x^2 - x - 2$ because $6x^2 - x - 2 = 5(2x + 1)[\frac{1}{5}(3x - 2)]$. A similar argument in the general case shows that

if $f(x)$ divides $g(x)$, then $cf(x)$ also divides $g(x)$ for each nonzero $c \in F$.

This example shows that a nonzero polynomial may have infinitely many divisors. In contrast, a nonzero integer has only a finite number of divisors. The example also illustrates this fact: If $g(x)$ is nonzero, then

every divisor of $g(x)$ has degree less than or equal to deg $g(x)$.

To prove this, suppose $f(x) \mid g(x)$, say $g(x) = f(x)h(x)$. By Theorem 4.2, $\deg g(x) = \deg f(x) + \deg h(x)$. Hence, $0 \le \deg f(x) \le \deg g(x)$.

As we learned earlier, the greatest common divisor of two integers is the largest integer that divides both of them. By analogy, the greatest common divisor of two polynomials $f(x), g(x) \in F[x]$ ought to be the polynomial of highest degree that divides both of them. But such a greatest common divisor would not be unique because each constant multiple of it would have the same degree and would also divide both $f(x)$ and $g(x)$. In order to guarantee a unique gcd, we modify this definition slightly by introducing a new concept. A polynomial in $F[x]$ is said to be **monic** if its leading coefficient is 1_F. For instance, $x^3 + x + 2$ is monic in $\mathbb{Q}[x]$, but $2x + 1$ is not.

> **DEFINITION •** *Let F be a field and f(x), g(x) $\in$ F[x], not both zero. The **greatest common divisor (gcd)** of f(x) and g(x) is the monic polynomial d(x) of highest degree that divides both f(x) and g(x). In other words, d(x) is the gcd of f(x) and g(x), provided that d(x) is monic and*
>
> *(i) d(x) | f(x) and d(x) | g(x);*
> *(ii) if c(x) | f(x) and c(x) | g(x), then deg c(x) $\leq$ deg d(x).*

Since the degree of a common divisor of $f(x)$ and $g(x)$ cannot exceed either $\deg f(x)$ or $\deg g(x)$, it is clear that there is at least one gcd for $f(x)$ and $g(x)$. In Theorem 4.5 we show that there is only one, thus justifying the definition's reference to *the* greatest common divisor.

EXAMPLE   To find the gcd of $3x^2 + x + 6$ and 0 in $\mathbb{Q}[x]$, we note that the common divisors of highest degree are just the divisors of $3x^2 + x + 6$ of degree 2. These include $3x^2 + x + 6$ itself and *all nonzero constant multiples* of this polynomial—in particular, the monic polynomial

$$\frac{1}{3}(3x^2 + x + 6) = x^2 + \frac{1}{3}x + 2.$$

Hence, $x^2 + \frac{1}{3}x + 2$ is a gcd of $3x^2 + x + 6$ and 0.

EXAMPLE   You can easily verify these factorizations in $\mathbb{Q}[x]$:

$$f(x) = 2x^4 + 5x^3 - 5x - 2 = (2x + 1)(x + 2)(x + 1)(x - 1),$$
$$g(x) = 2x^3 - 3x^2 - 2x = (2x + 1)(x - 2)x.$$

It appears that $2x + 1$ is a common divisor of highest degree of $f(x)$ and $g(x)$. In this case, the constant multiple $\frac{1}{2}(2x + 1) = x + \frac{1}{2}$ is a *monic* common divisor of highest degree. A proof that $x + \frac{1}{2}$ actually is the greatest common divisor is given at the end of this section. You can also verify that this gcd can be written in the form $f(x)u(x) + g(x)v(x)$:

$$x + \frac{1}{2} = (2x^4 + 5x^3 - 5x - 2)\left(\frac{7}{48}x - \frac{1}{4}\right)$$

$$+ (2x^3 - 3x^2 - 2x)\left(-\frac{7}{48}x^2 - \frac{1}{3}x - \frac{1}{48}\right).$$

THEOREM 4.5 *Let F be a field and $f(x)$, $g(x) \in F[x]$, not both zero. Then there is a unique greatest common divisor $d(x)$ of $f(x)$ and $g(x)$. Furthermore, there exist (not necessarily unique) polynomials $u(x)$ and $v(x)$ such that $d(x) = f(x)u(x) + g(x)v(x)$.*

Proof Let $t(x)$ be a monic polynomial of smallest degree in the set $S = \{f(x)m(x) + g(x)n(x) \mid m(x), \; n(x) \in F[x]\}$. By the definition of S, $t(x) = f(x)u(x) + g(x)v(x)$ for some $u(x)$, $v(x) \in F[x]$. Now adapt the proof of Theorem 1.3 to $F[x]$ as follows: Replace integer inequalities with inequalities involving the degrees of polynomials and show that $t(x)$ is a gcd of $f(x)$ and $g(x)$.

Next, suppose $d(x)$ is any gcd of $f(x)$ and $g(x)$. To prove uniqueness, we must show that $d(x) = t(x)$. Since $d(x)$ is a common divisor, we have $f(x) = d(x)a(x)$ and $g(x) = d(x)b(x)$ for some $a(x)$, $b(x) \in F[x]$. Therefore,

$$t(x) = f(x)u(x) + g(x)v(x) = [d(x)a(x)]u(x) + [d(x)b(x)]v(x)$$
$$= d(x)[a(x)u(x) + b(x)v(x)].$$

By Theorem 4.2,

$$\deg t(x) = \deg d(x) + \deg [a(x)u(x) + b(x)v(x)].$$

However, $t(x)$ and $d(x)$ have the same degree because they are gcd's. Hence, $\deg [a(x)u(x) + b(x)v(x)] = 0$, so that $a(x)u(x) + b(x)v(x) = c$ for some nonzero $c \in F$. Therefore, $t(x) = d(x)c$. Now the gcd $d(x)$ is monic, say $d(x) = x^n + \cdots + a_1x + a_0$. Hence, $cd(x) = cx^n + \cdots + ca_1x + ca_0$. But $cd(x)$ is the monic polynomial $t(x)$. Therefore, $c = 1_F$ and $d(x) = t(x)$. Thus the gcd is unique and $d(x) = f(x)u(x) + g(x)v(x)$. ◆

COROLLARY 4.6 *Let F be a field and $f(x)$, $g(x) \in F[x]$, not both zero. A monic polynomial $d(x) \in F[x]$ is the greatest common divisor of $f(x)$ and $g(x)$ if and only if $d(x)$ satisfies these conditions:*

(i) $d(x) \mid f(x)$ and $d(x) \mid g(x)$;

(ii) if $c(x) \mid f(x)$ and $c(x) \mid g(x)$, then $c(x) \mid d(x)$.

Proof Adapt the proof of Corollary 1.4 to $F[x]$. ◆

Polynomials $f(x)$ and $g(x)$ are said to be **relatively prime** if their greatest common divisor is 1_F.

THEOREM 4.7   *Let F be a field and $f(x), g(x), h(x) \in F[x]$. If $f(x) \mid g(x)h(x)$ and $f(x)$ and $g(x)$ are relatively prime, then $f(x) \mid h(x)$.*

Proof   Adapt the proof of Theorem 1.5 to $F[x]$.   ◆

The Euclidean Algorithm (Theorem 1.6) also carries over to $F[x]$ and provides an efficient method of calculating greatest common divisors. The only change from $\mathbb{Z}$ is that the last nonzero remainder in the process is a common divisor of highest degree but not necessarily monic. It must be multiplied by an appropriate constant to produce the gcd.

> **EXAMPLE**   To find the gcd of $f(x) = 2x^4 + 5x^3 - 5x - 2$ and $g(x) = 2x^3 - 3x^2 - 2x$ in $\mathbb{Q}[x]$, we use the Division Algorithm repeatedly until we reach a remainder of zero. The divisor and remainder in each step become the dividend and divisor in the next step:
>
> $$2x^4 + 5x^3 - 5x - 2 = (2x^3 - 3x^2 - 2x)(x + 4) + (14x^2 + 3x - 2)$$
>
> $$2x^3 - 3x^2 - 2x = (14x^2 + 3x - 2)\left(\frac{1}{7}x - \frac{12}{49}\right) + \left(-\frac{48}{49}x - \frac{24}{49}\right)$$
>
> $$14x^2 + 3x - 2 = \left(-\frac{48}{49}x - \frac{24}{49}\right)\left(-\frac{343}{24}x + \frac{49}{12}\right) + 0.$$
>
> Therefore, the last nonzero remainder, $-\dfrac{48}{49}x - \dfrac{24}{49}$, is a common divisor of highest degree, and so the gcd is the monic polynomial
>
> $$\left(-\frac{49}{48}\right)\left(-\frac{48}{49}x - \frac{24}{49}\right) = x + \frac{1}{2}.$$

◆ EXERCISES

NOTE: *F denotes a field.*

A.   **1.** If $f(x) \in F[x]$, show that every nonzero constant polynomial divides $f(x)$.

2. If $f(x) = c_n x^n + \cdots + c_0$ with $c_n \neq 0_F$, what is the gcd of $f(x)$ and 0_F?

3. If $a, b \in F$ and $a \neq b$, show that $x + a$ and $x + b$ are relatively prime in $F[x]$.

4. **(a)** Let $f(x), g(x) \in F[x]$. If $f(x) \mid g(x)$ and $g(x) \mid f(x)$, show that $f(x) = cg(x)$ for some nonzero $c \in F$.

  (b) If $f(x)$ and $g(x)$ in part (a) are monic, show that $f(x) = g(x)$.

5. Use the Euclidean Algorithm to find the gcd of the given polynomials:

   (a) $x^4 - x^3 - x^2 + 1$ and $x^3 - 1$ in $\mathbb{Q}[x]$

   (b) $x^5 + x^4 + 2x^3 - x^2 - x - 2$ and $x^4 + 2x^3 + 5x^2 + 4x + 4$ in $\mathbb{Q}[x]$

   (c) $x^4 + 3x^3 + 2x + 4$ and $x^2 - 1$ in $\mathbb{Z}_5[x]$

   (d) $4x^4 + 2x^3 + 6x^2 + 4x + 5$ and $3x^3 + 5x^2 + 6x$ in $\mathbb{Z}_7[x]$

   (e) $x^3 - ix^2 + 4x - 4i$ and $x^2 + 1$ in $\mathbb{C}[x]$

   (f) $x^4 + x + 1$ and $x^2 + x + 1$ in $\mathbb{Z}_2[x]$

6. Express each of the gcd's in Exercise 5 as a linear combination of the two polynomials.

B. 7. Let $f(x) \in F[x]$ and assume that $f(x) \mid g(x)$ for every nonconstant $g(x) \in F[x]$. Show that $f(x)$ is a constant polynomial. [*Hint:* See Exercise 3.]

8. Let $f(x), g(x) \in F[x]$, not both zero, and let $d(x)$ be their gcd. If $h(x)$ is a common divisor of $f(x)$ and $g(x)$ of highest possible degree, then prove that $h(x) = cd(x)$ for some nonzero $c \in F$.

9. If $f(x)$ is relatively prime to 0_F, what can be said about $f(x)$?

10. Find the gcd of $x + a + b$ and $x^3 - 3abx + a^3 + b^3$ in $\mathbb{Q}[x]$.

11. Fill in the details of the proof of Theorem 4.5.

12. Prove Corollary 4.6.

13. Prove Theorem 4.7.

14. Let $f(x), g(x), h(x) \in F[x]$, with $f(x)$ and $g(x)$ relatively prime. If $f(x) \mid h(x)$ and $g(x) \mid h(x)$, prove that $f(x)g(x) \mid h(x)$.

15. Let $f(x), g(x), h(x) \in F[x]$, with $f(x)$ and $g(x)$ relatively prime. If $h(x) \mid f(x)$, prove that $h(x)$ and $g(x)$ are relatively prime.

16. Let $f(x), g(x), h(x) \in F[x]$, with $f(x)$ and $g(x)$ relatively prime. Prove that the gcd of $f(x)h(x)$ and $g(x)$ is the same as the gcd of $h(x)$ and $g(x)$.

4.3 IRREDUCIBLES AND UNIQUE FACTORIZATION

Throughout this section F always denotes a field. Before carrying over the results of Section 1.3 on unique factorization in $\mathbb{Z}$ to the ring $F[x]$, we must first examine an area in which $\mathbb{Z}$ differs significantly from $F[x]$. In $\mathbb{Z}$ there are only two units,* namely ± 1, but a polynomial ring may have many more units.

* "Unit" is defined on page 60.

THEOREM 4.8 *Let R be an integral domain. Then $f(x)$ is a unit in $R[x]$ if and only if $f(x)$ is a constant polynomial that is a unit in R.*

Proof Suppose $f(x)$ is a unit in $R[x]$, so that $f(x)g(x) = 1_F$ for some $g(x)$. By Theorem 4.2,

$$\deg f(x) + \deg g(x) = \deg[f(x)g(x)] = \deg[1_R] = 0.$$

Since all degrees are nonnegative integers, $\deg f(x) = 0$ and $\deg g(x) = 0$. Hence, $f(x)$ and $g(x)$ are constant polynomials, that is, elements of R. Consequently, the fact that $f(x)g(x) = 1_R$ shows that $f(x)$ is a unit in R. Conversely, if $f(x) = b$, with b a unit in R, let $g(x) = b^{-1}$. Then $f(x)g(x) = bb^{-1} = 1_R$. Therefore, $f(x)$ is a unit in $R[x]$. ◆

Theorem 4.8 may be false if R is not an integral domain. For example, $5x + 1$ is a unit in $\mathbb{Z}_{25}[x]$ because $(5x + 1)(20x + 1) = 1$ (verify!). Since every nonzero element in a field is a unit, Theorem 4.8 immediately implies:

COROLLARY 4.9 *Let F be a field. Then $f(x)$ is a unit in $F[x]$ if and only if $f(x)$ is a nonzero constant polynomial.*

Corollary 4.9 shows that each of $\mathbb{Q}[x]$, $\mathbb{R}[x]$, and $\mathbb{C}[x]$ has infinitely many units, whereas the ring $\mathbb{Z}[x]$ has just two units (± 1) by Theorem 4.8.

An element a in a commutative ring with identity R is said to be an **associate** of an element b of R if $a = bu$ for some unit u. In this case b is also an associate of a because u^{-1} is a unit and $b = au^{-1}$. In the ring $\mathbb{Z}$, the only associates of an integer n are n and $-n$ because ± 1 are the only units. If F is a field, then Corollary 4.9 shows that $f(x)$ is an associate of $g(x)$ in $F[x]$ if and only if $f(x) = cg(x)$ for some nonzero $c \in F$.

Recall that a nonzero integer p is prime in $\mathbb{Z}$ if it is not ± 1 (that is, p is not a unit in $\mathbb{Z}$) and its only divisors are ± 1 (the units) and $\pm p$ (the associates of p). In $F[x]$ the units are the nonzero constants, which suggests the following definition.

> **DEFINITION** • *Let F be a field. A nonconstant polynomial $p(x) \in F[x]$ is said to be **irreducible*** if its only divisors are its associates and the nonzero constant polynomials (units). A nonconstant polynomial that is not irreducible is said to be **reducible.***

---

* You could just as well call such as polynomial "prime," but "irreducible" is the customary term with polynomials.

EXAMPLE The polynomial $x + 2$ is irreducible in $\mathbb{Q}[x]$ because, by Theorem 4.2, all its divisors must have degree 0 or 1. Divisors of degree 0 are nonzero constants. If $f(x) \mid (x + 2)$, say $x + 2 = f(x)g(x)$, and if $\deg f(x) = 1$, then $g(x)$ has degree 0, so that $g(x) = c$. Thus $c^{-1}(x + 2) = f(x)$, and $f(x)$ is an associate of $x + 2$. A similar argument in the general case shows that

every polynomial of degree 1 in $F[x]$ is irreducible in $F[x]$.

The definition of irreducibility is a natural generalization of the concept of primality in $\mathbb{Z}$. In most high-school texts, however, a polynomial is defined to be irreducible if it is *not* the product of polynomials of lower degree. The next theorem shows that these two definitions are equivalent.

THEOREM 4.10 *Let F be a field. A nonzero polynomial $f(x)$ is reducible in $F[x]$ if and only if $f(x)$ can be written as the product of two polynomials of lower degree.*

Proof If $f(x)$ is reducible, then it must have a divisor $g(x)$ that is neither an associate nor a nonzero constant, say $f(x) = g(x)h(x)$. If either $g(x)$ or $h(x)$ has the same degree as $f(x)$, then the other must have degree 0 by Theorem 4.2. Since a polynomial of degree 0 is a nonzero constant in F, this means that either $g(x)$ is a constant or an associate of $f(x)$, contrary to hypothesis. Therefore, both $g(x)$ and $h(x)$ have lower degree than $f(x)$. The converse is Exercise 8. ◆

Various other tests for irreducibility are presented in Sections 4.4 to 4.6. For now, we note that the concept of irreducibility is not an absolute one. For instance, $x^2 + 1$ is reducible in $\mathbb{C}[x]$ because $x^2 + 1 = (x + i)(x - i)$ and neither factor is a constant or an associate of $x^2 + 1$. But $x^2 + 1$ is irreducible in $\mathbb{Q}[x]$ (Exercise 6).

The following theorem shows that irreducibles in $F[x]$ have essentially the same divisibility properties as do primes in $\mathbb{Z}$. Condition (3) in the theorem is often used to prove that a polynomial is irreducible; in many books, (3) is given as the definition of "irreducible."

THEOREM 4.11 *Let F be a field and $p(x)$ a nonconstant polynomial in $F[x]$. Then the following conditions are equivalent:*

  (1) $p(x)$ is irreducible.

  (2) If $b(x)$ and $c(x)$ are any polynomials such that $p(x) \mid b(x)c(x)$, then $p(x) \mid b(x)$ or $p(x) \mid c(x)$.

  (3) If $r(x)$ and $s(x)$ are any polynomials such that $p(x) = r(x)s(x)$, then $r(x)$ or $s(x)$ is a nonzero constant polynomial.

Proof $(1) \Rightarrow (2)$ Adapt the proof of Theorem 1.8 to $F[x]$. Replace statements about $\pm p$ by statements about the associates of $p(x)$; replace statements about ± 1 by statements about units (nonzero constant polynomials) in $F[x]$; use Theorem 4.7 in place of Theorem 1.5.

$(2) \Rightarrow (3)$ If $p(x) = r(x)s(x)$, then $p(x) \mid r(x)$ or $p(x) \mid s(x)$, by (2). If $p(x) \mid r(x)$, say $r(x) = p(x)v(x)$, then $p(x) = r(x)s(x) = p(x)v(x)s(x)$. Since $F[x]$ is an integral domain, we can cancel $p(x)$ by Theorem 3.10 and conclude that $1_F = v(x)s(x)$. Thus $s(x)$ is a unit, and hence by Corollary 4.9, $s(x)$ is a nonzero constant. A similar argument shows that if $p(x) \mid s(x)$, then $r(x)$ is a nonzero constant.

$(3) \Rightarrow (1)$ Let $c(x)$ be any divisor of $p(x)$, say $p(x) = c(x)d(x)$. Then by (3), either $c(x)$ or $d(x)$ is a nonzero constant. If $d(x) = d \neq 0_F$, then multiplying both sides of $p(x) = c(x)d(x) = dc(x)$ by d^{-1} shows that $c(x) = d^{-1}p(x)$. Thus in every case, $c(x)$ is a nonzero constant or an associate of $p(x)$. Therefore, $p(x)$ is irreducible. ◆

COROLLARY 4.12 *Let F be a field and $p(x)$ an irreducible polynomial in $F[x]$. If $p(x) \mid a_1(x)a_2(x) \cdots a_n(x)$, then $p(x)$ divides at least one of the $a_i(x)$.*

Proof Adapt the proof of Corollary 1.9 to $F[x]$. ◆

THEOREM 4.13 *Let F be a field. Every nonconstant polynomial $f(x)$ in $F[x]$ is a product of irreducible polynomials in $F[x]$.* * *This factorization is unique in the following sense: If*

$$f(x) = p_1(x)p_2(x) \cdots p_r(x) \qquad and \qquad f(x) = q_1(x)q_2(x) \cdots q_s(x)$$

with each $p_i(x)$ and $q_j(x)$ irreducible, then $r = s$ (that is, the number of irreducible factors is the same). After the $q_j(x)$ are reordered and relabeled, if necessary,

$$p_i(x) \text{ is an associate of } q_i(x) \qquad (i = 1, 2, 3, \ldots, r).$$

Proof To show that $f(x)$ is a product of irreducibles, adapt the proof of Theorem 1.10 to $F[x]$: Let S be the set of all nonconstant polynomials that are *not* the product of irreducibles, and use a proof by contradiction to show that S is empty. To prove that this factorization is unique up to associates, suppose $f(x) = p_1(x)p_2(x) \cdots p_r(x) = q_1(x)q_2(x) \cdots q_s(x)$ with each $p_i(x)$ and $q_j(x)$ irreducible. Then $p_1(x)[p_2(x) \cdots p_r(x)] = q_1(x)q_2(x) \cdots q_s(x)$, so that $p_1(x)$ divides $q_1(x)q_2(x) \cdots q_s(x)$. Corollary 4.12 shows that $p_1(x) \mid q_j(x)$ for some j. After rearranging and relabeling the $q(x)$'s if necessary, we may assume that $p_1(x) \mid q_1(x)$. Since $q_1(x)$ is irreducible, $p_1(x)$ must be either a constant or an associate of $q_1(x)$. However, $p_1(x)$ is irreducible, and so it is not a constant.

---

* We allow the possibility of a product with just one factor in case $f(x)$ is itself irreducible.

Therefore, $p_1(x)$ is an associate of $q_1(x)$, with $p_1(x) = c_1 q_1(x)$ for some constant c_1. Thus

$$q_1(x)[c_1 p_2(x) p_3(x) \cdots p_r(x)] = p_1(x) p_2(x) \cdots p_r(x) = q_1(x) q_2(x) \cdots q_s(x).$$

Cancelling $q_1(x)$ on each end, we have

$$p_2(x)[c_1 p_3(x) \cdots p_r(x)] = q_2(x) q_3(x) \cdots q_s(x).$$

Complete the argument by adapting the proof of Theorem 1.11 to $F[x]$, replacing statements about $\pm q_j$ with statements about associates of $q_j(x)$.   ◆

◆ EXERCISES

NOTE: *F denotes a field and p a positive prime integer.*

A.   **1.** Find a monic associate of

(a) $3x^3 + 2x^2 + x + 5$ in $\mathbb{Q}[x]$ (b) $3x^5 - 4x^2 + 1$ in $\mathbb{Z}_5[x]$

(c) $ix^3 + x - 1$ in $\mathbb{C}[x]$

2. Prove that every nonzero $f(x) \in F[x]$ has a unique monic associate in $F[x]$.

3. List all associates of

(a) $x^2 + x + 1$ in $\mathbb{Z}_5[x]$ (b) $3x + 2$ in $\mathbb{Z}_7[x]$

4. Show that a nonzero polynomial in $\mathbb{Z}_p[x]$ has exactly $p - 1$ associates.

5. Prove that $f(x)$ and $g(x)$ are associates in $F[x]$ if and only if $f(x) \mid g(x)$ and $g(x) \mid f(x)$.

6. Show that $x^2 + 1$ is irreducible in $\mathbb{Q}[x]$. [*Hint:* If not, it must factor as $(ax + b)(cx + d)$ with $a, b, c, d \in \mathbb{Q}$; show that this is impossible.]

7. Prove that $f(x)$ is irreducible in $F[x]$ if and only if each of its associates is irreducible.

8. Complete the proof of Theorem 4.10.

9. Find all irreducible polynomials of

(a) degree 2 in $\mathbb{Z}_2[x]$ (b) degree 3 in $\mathbb{Z}_2[x]$

(c) degree 2 in $\mathbb{Z}_3[x]$

10. Is the given polynomial irreducible:

(a) $x^2 - 3$ in $\mathbb{Q}[x]$? In $\mathbb{R}[x]$?

(b) $x^2 + x - 2$ in $\mathbb{Z}_3[x]$? In $\mathbb{Z}_7[x]$?

11. Show that $x^3 - 3$ is irreducible in $\mathbb{Z}_7[x]$.

12. Express $x^4 - 4$ as a product of irreducibles in $\mathbb{Q}[x]$, in $\mathbb{R}[x]$, and in $\mathbb{C}[x]$.

13. (a) Find a polynomial of positive degree in $\mathbb{Z}_9[x]$ that is a unit.

   (b) Show that every polynomial (except the constant polynomials 3 and 6) in $\mathbb{Z}_9[x]$ can be written as the product of two polynomials of positive degree.

14. Show that $x^2 + x$ can be factored in two ways in $\mathbb{Z}_6[x]$ as the product of non-constant polynomials that are not units.

15. Use unique factorization to find the gcd in $\mathbb{C}[x]$ of $(x - 3)^3(x - 4)^4(x - i)^2$ and $(x - 1)(x - 3)(x - 4)^3$.

B. 16. Prove that $p(x)$ is irreducible in $F[x]$ if and only if for every $g(x) \in F[x]$, either $p(x) \mid g(x)$ or $p(x)$ is relatively prime to $g(x)$.

17. Prove Theorem 4.11.

18. Without using statement (2), prove directly that statement (1) is equivalent to statement (3) in Theorem 4.11.

19. Prove Corollary 4.12.

20. If $p(x)$ and $q(x)$ are nonassociate irreducibles in $F[x]$, prove that $p(x)$ and $q(x)$ are relatively prime.

21. (a) By counting products of the form $(x + a)(x + b)$, show that there are exactly $(p^2 + p)/2$ monic polynomials of degree 2 that are *not* irreducible in $\mathbb{Z}_p[x]$.

   (b) Show that there are exactly $(p^2 - p)/2$ monic irreducible polynomials of degree 2 in $\mathbb{Z}_p[x]$.

22. (a) Show that $x^3 + a$ is reducible in $\mathbb{Z}_3[x]$ for each $a \in \mathbb{Z}_3$.

   (b) Show that $x^5 + a$ is reducible in $\mathbb{Z}_5[x]$ for each $a \in \mathbb{Z}_5$.

23. (a) Show that $x^2 + 2$ is irreducible in $\mathbb{Z}_5[x]$.

   (b) Factor $x^4 - 4$ as a product of irreducibles in $\mathbb{Z}_5[x]$.

24. Prove Theorem 4.13.

25. Prove that every nonconstant $f(x) \in F[x]$ can be written in the form $cp_1(x)p_2(x) \cdots p_n(x)$, with $c \in F$ and each $p_i(x)$ monic irreducible in $F[x]$. Show further that if $f(x) = dq_1(x)q_2(x) \cdots q_m(x)$ with $d \in F$ and each $q_j(x)$ monic irreducible in $F[x]$, then $m = n$, $c = d$, and after reordering and relabeling if necessary, $p_i(x) = q_i(x)$ for each i.

26. Prove that there is no irreducible polynomial of degree 2 in $\mathbb{C}[x]$.

4.4 POLYNOMIAL FUNCTIONS, ROOTS, AND REDUCIBILITY

In the parallel development of $F[x]$ and $\mathbb{Z}$, the next step is to consider criteria for irreducibility of polynomials (the analogue of primality testing for integers). Unlike the situation in the integers, there are a number of such criteria for polynomials whose implementation does not depend on a computer. Most of them are based on the fact that every polynomial in $F[x]$ induces a function from F to F. The properties of this function (in particular, the places where it is zero) are closely related to the reducibility or irreducibility of the polynomial.

Throughout this section, R is a commutative ring. Associated with each polynomial $a_n x^n + \cdots + a_2 x^2 + a_1 x + a_0$ in $R[x]$ is a function $f : R \to R$ whose rule is

$$\text{for each } r \in R, \quad f(r) = a_n r^n + \cdots + a_2 r^2 + a_1 r + a_0.$$

The function f induced by a polynomial in this way is called a **polynomial function.**

EXAMPLE The polynomial $x^2 + 5x + 3 \in \mathbb{R}[x]$ induces the function $f : \mathbb{R} \to \mathbb{R}$ whose rule is $f(r) = r^2 + 5r + 3$ for each $r \in \mathbb{R}$.

EXAMPLE The polynomial $x^4 + x + 1 \in \mathbb{Z}_3[x]$ induces the function $f : \mathbb{Z}_3 \to \mathbb{Z}_3$ whose rule is $f(r) = r^4 + r + 1$. Thus

$$f(0) = 0^4 + 0 + 1 = 1, \quad f(1) = 1^4 + 1 + 1 = 0,$$
$$f(2) = 2^4 + 2 + 1 = 1.$$

The polynomial $x^3 + x^2 + 1 \in \mathbb{Z}_3[x]$ induces the function $g : \mathbb{Z}_3 \to \mathbb{Z}_3$ given by

$$g(0) = 0^3 + 0^2 + 1 = 1, \quad g(1) = 1^3 + 1^2 + 1 = 0,$$
$$g(2) = 2^3 + 2^2 + 1 = 1.$$

Thus f and g are the *same* function on $\mathbb{Z}_3$, even though they are induced by *different* polynomials in $\mathbb{Z}_3[x]$.*

Although the distinction between a polynomial and the polynomial function it induces is clear, the customary notation is quite ambiguous. For example, you will see a statement such as $f(x) = x^2 - 3x + 2$. Depending on the context, $f(x)$ denotes either the polynomial $x^2 - 3x + 2 \in \mathbb{R}[x]$ or the rule of its induced function $f : \mathbb{R} \to \mathbb{R}$. The symbol x is being used in two different ways

---

* Remember that functions f and g are equal if $f(r) = g(r)$ for every r in the domain.

here. In the polynomial $x^2 - 3x + 2$, x is an indeterminate (transcendental element) of the ring $R[x]$.* But in the polynomial function $f : \mathbb{R} \to \mathbb{R}$, the symbol x is used as a variable to describe the rule of the function. It might be better to use one symbol for an indeterminate and another for a variable, but the practice of using x for both is so widespread you may as well get used to it.

The use of the same notation for both the polynomial and its induced function also affects the language that is used. For instance, one says "evaluate the polynomial $3x^2 - 5x + 4$ at $x = 2$" or "substitute $x = 2$ in $3x^2 - 5x + 4$" when what is really meant is "find $f(2)$ when f is the function induced by the polynomial $3x^2 - 5x + 4$."

The truth or falsity of certain statements depends on whether x is treated as an indeterminate or a variable. For instance, in the ring $\mathbb{R}[x]$, where x is an indeterminate (special element of the ring), the statement $x^2 - 3x + 2 = 0$ is *false* because, by Theorem 4.1, a polynomial is zero if and only if all its coefficients are zero. When x is a variable, however, as in the rule of the polynomial function $f(x) = x^2 - 3x + 2$, things are different. Here it is perfectly reasonable to ask which elements of $\mathbb{R}$ are mapped to 0 by the function f, that is, for which values of the variable x is it true that $x^2 - 3x + 2 = 0$. It may help to remember that statements about the variable x occur in the ring R, whereas statements about the indeterminate x occur in the polynomial ring $R[x]$.

Roots of Polynomials

Questions about the reducibility of a polynomial can sometimes be answered by considering its induced polynomial function. The key to this analysis is the concept of a root.

> **DEFINITION** • *Let R be a commutative ring and $f(x) \in R[x]$. An element a of R is said to be a **root** of the polynomial $f(x)$ if $f(a) = 0_R$, that is, if the induced function $f : R \to R$ maps a to 0_R.*

EXAMPLE The roots of the polynomial $f(x) = x^2 - 3x + 2 \in \mathbb{R}[x]$ are the values of the variable x for which $f(x) = 0$, that is, the solutions of the equation $x^2 - 3x + 2 = 0$. It is easy to see that the roots are 1 and 2.

EXAMPLE The polynomial $x^2 + 1 \in \mathbb{R}[x]$ has no roots in $\mathbb{R}$ because there are no real-number solutions of the equation $x^2 + 1 = 0$. However, if $x^2 + 1$ is considered as a polynomial in $\mathbb{C}[x]$, then it has i and $-i$ as roots because these are the solutions in $\mathbb{C}$ of $x^2 + 1 = 0$.

---

* See page 547 in Appendix G for more information.

As illustrated in the preceding examples, questions about roots of polynomials are simply questions about the solutions of polynomial equations. Such questions are a major component of both classical and modern algebra, and we shall return to them repeatedly. For now, however, we concentrate on the issues related to reducibility in $F[x]$, where F is a field.

THEOREM 4.14   (THE REMAINDER THEOREM)   *Let F be a field, $f(x) \in F[x]$, and $a \in F$. The remainder when $f(x)$ is divided by the polynomial $x - a$ is $f(a)$.*

This theorem says, for example, that the remainder when the polynomial $f(x) = x^3 - 4x^2 + 3x + 5$ is divided by $x - 2$ is $f(2) = 2^3 - 4 \cdot 2^2 + 3 \cdot 2 + 5 = 3$. You can check this by long division. To find the remainder when $f(x)$ is divided by $x + 1$, note that $x + 1 = x - (-1)$. Hence, the remainder is $f(-1) = (-1)^3 - 4(-1)^2 + 3(-1) + 5 = -3$.

Proof of Theorem 4.14   By the Division Algorithm, $f(x) = (x - a)q(x) + r(x)$, where the remainder $r(x)$ either is 0_F or has smaller degree than the divisor $x - a$. Thus $\deg r(x) = 0$ or $r(x) = 0_F$. In either case, $r(x) = c$ for some $c \in F$. Hence, $f(x) = (x - a)q(x) + c$, so that $f(a) = (a - a)q(a) + c = 0_F + c = c$.   ◆

Let $F, f(x)$, and a be as in the Remainder Theorem. Then $x - a$ is a factor of the polynominal $f(x)$ if and only if the remainder when $f(x)$ is divided by $x - a$ is zero, that is, if and only if $f(a) = 0_F$. We have proved

THEOREM 4.15   (THE FACTOR THEOREM)   *Let F be a field, $f(x) \in F[x]$, and $a \in F$. Then a is a root of the polynomial $f(x)$ if and only if $x - a$ is a factor of $f(x)$ in $F[x]$.*

EXAMPLE   To show that $x^7 - x^5 + 2x^4 - 3x^2 - x + 2$ is reducible in $\mathbb{Q}[x]$, note that 1 is a root of this polynomial. Therefore, $x - 1$ is a factor.

COROLLARY 4.16   *Let F be a field and $f(x)$ a nonzero polynomial of degree n in $F[x]$. Then $f(x)$ has at most n roots in F.*

This corollary is also true for integral domains (Exercise 21), but it may be false for other rings, even division rings (see Exercise 17 in this section and also Exercise 35(d) in Section 3.1).

Proof of Corollary 4.16   The proof is by induction on the degree n. If $n = 0$, then $f(x)$ is a nonzero constant polynomial and therefore has no roots, and so the theorem is true for $n = 0$.

Assume inductively that the corollary is true for all polynomials of degree $k - 1$, and suppose $\deg f(x) = k$. If $f(x)$ has no roots in F, then the theorem is true. If $f(x)$ has a root $a \in F$, then by the Factor Theorem $f(x) = (x - a)g(x)$. If $c \in F$ is any root of $f(x)$ other than a, then $0_F = f(c) = (c - a)g(c)$. Since $c - a \neq 0_F$ and $F[x]$ is an integral domain, we must have $g(c) = 0_F$. Therefore, the only roots of $f(x)$ in F are a and the roots of $g(x)$. By Theorem 4.2, $k = \deg f(x) = \deg(x - a) + \deg g(x) = 1 + \deg g(x)$, so that $\deg g(x) = k - 1$. By the induction hypothesis, $g(x)$ has at most $k - 1$ roots in F. Therefore, $f(x)$ has a total of at most $1 + (k - 1) = k$ roots in F. Thus the corollary is true for $n = k$ and, hence by induction, for all n. $\blacklozenge$

COROLLARY 4.17 *Let F be a field and $f(x) \in F[x]$, with $\deg f(x) \geq 2$. If $f(x)$ is irreducible in $F[x]$, then $f(x)$ has no roots in F.*

Proof If $f(x)$ is irreducible, then it has no factor of the form $x - a$ in $F[x]$. Therefore, $f(x)$ has no roots in F by the Factor Theorem. $\blacklozenge$

The converse of Corollary 4.17 is false in general. For example, $x^4 + 2x^2 + 1 = (x^2 + 1)(x^2 + 1)$ has no roots in $\mathbb{Q}$ but is reducible in $\mathbb{Q}[x]$. However, the converse is true for degrees 2 and 3.

COROLLARY 4.18 *Let F be a field and let $f(x) \in F[x]$ be a polynomial of degree 2 or 3. Then $f(x)$ is irreducible in $F[x]$ if and only if $f(x)$ has no roots in F.*

Proof One half of the statement is true by Corollary 4.17. To prove the other half, suppose that $f(x)$ has no roots in F. Then $f(x)$ has no first-degree factor in $F[x]$ because every first-degree polynomial $cx + d$ in $F[x]$ has a root in F, namely $-c^{-1}d$. Therefore, if $f(x) = r(x)s(x)$, neither $r(x)$ nor $s(x)$ has degree 1. Since $f(x)$ has degree 2 or 3, either $r(x)$ or $s(x)$ must have degree 0 by Theorem 4.2, that is, either $r(x)$ or $s(x)$ is a nonzero constant. Hence, $f(x)$ is irreducible by Theorem 4.11. $\blacklozenge$

EXAMPLE To show that $x^3 + x + 1$ is irreducible in $\mathbb{Z}_5[x]$, you need only verify that none of 0, 1, 2, 3, 4 $\in \mathbb{Z}_5$ is a root.

We close this section by returning to its starting point, polynomial functions. The last example on page 100 shows that two different polynomials in $F[x]$ may induce the same function from F to F. We now see that this cannot occur if F is infinite.

COROLLARY 4.19 *Let F be an infinite field and $f(x), g(x) \in F[x]$. Then $f(x)$ and $g(x)$ induce the same function from F to F if and only if $f(x) = g(x)$ in $F[x]$.*

Proof Suppose that $f(x)$ and $g(x)$ induce the same function from F to F. Then $f(a) = g(a)$, so that $f(a) - g(a) = 0_F$ for every $a \in F$. This means that every element of F is a root of the polynomial $f(x) - g(x)$. Since F is infinite, this is impossible by Corollary 4.16 unless $f(x) - g(x)$ is the zero polynomial, that is, $f(x) = g(x)$. The converse is obvious. ◆

◆ **EXERCISES**

NOTE: *F denotes a field.*

A. 1. **(a)** Find a nonzero polynomial in $\mathbb{Z}_2[x]$ that induces the zero function on $\mathbb{Z}_2$.

   (b) Do the same in $\mathbb{Z}_3[x]$.

2. Find the remainder when $f(x)$ is divided by $g(x)$:

   (a) $f(x) = x^{10} + x^8$ and $g(x) = x - 1$ in $\mathbb{Q}[x]$

   (b) $f(x) = 2x^5 - 3x^4 + x^3 - 2x^2 + x - 8$ and $g(x) = x - 10$ in $\mathbb{Q}[x]$

   (c) $f(x) = 10x^{75} - 8x^{65} + 6x^{45} + 4x^{37} - 2x^{15} + 5$ and $g(x) = x + 1$ in $\mathbb{Q}[x]$

   (d) $f(x) = 2x^5 - 3x^4 + x^3 + 2x + 3$ and $g(x) = x - 3$ in $\mathbb{Z}_5[x]$

3. Determine if $h(x)$ is a factor of $f(x)$:

   (a) $h(x) = x + 2$ and $f(x) = x^3 - 3x^2 - 4x - 12$ in $\mathbb{R}[x]$

   (b) $h(x) = x - \frac{1}{2}$ and $f(x) = 2x^4 + x^3 + x - \frac{3}{4}$ in $\mathbb{Q}[x]$

   (c) $h(x) = x + 2$ and $f(x) = 3x^5 + 4x^4 + 2x^3 - x^2 + 2x + 1$ in $\mathbb{Z}_5[x]$

   (d) $h(x) = x - 3$ and $f(x) = x^6 - x^3 + x - 5$ in $\mathbb{Z}_7[x]$

4. **(a)** For what value of k is $x - 2$ a factor of $x^4 - 5x^3 + 5x^2 + 3x + k$ in $\mathbb{Q}[x]$?

   (b) For what value of k is $x + 1$ a factor of $x^4 + 2x^3 - 3x^2 + kx + 1$ in $\mathbb{Z}_5[x]$?

5. Show that $x - 1_F$ divides $a_n x^n + \cdots + a_2 x^2 + a_1 x + a_0$ in $F[x]$ if and only if $a_0 + a_1 + a_2 + \cdots + a_n = 0_F$.

6. **(a)** Verify that every element of $\mathbb{Z}_3$ is a root of $x^3 - x \in \mathbb{Z}_3[x]$.

   (b) Verify that every element of $\mathbb{Z}_5$ is a root of $x^5 - x \in \mathbb{Z}_5[x]$.

   (c) Make a conjecture about the roots of $x^p - x \in \mathbb{Z}_p[x]$ (p prime).

7. Use the Factor Theorem to show that $x^7 - x$ factors in $\mathbb{Z}_7[x]$ as $x(x - 1)(x - 2)(x - 3)(x - 4)(x - 5)(x - 6)$, without doing any polynomial multiplication.

8. Determine if the given polynomial is irreducible:

   (a) $x^2 - 7$ in $\mathbb{R}[x]$ (b) $x^2 - 7$ in $\mathbb{Q}[x]$

   (c) $x^2 + 7$ in $\mathbb{C}[x]$ (d) $2x^3 + x^2 + 2x + 2$ in $\mathbb{Z}_5[x]$

   (e) $x^3 - 9$ in $\mathbb{Z}_{11}[x]$ (f) $x^4 + x^2 + 1$ in $\mathbb{Z}_3[x]$

9. List all monic irreducible polynomials of degree 2 in $\mathbb{Z}_3[x]$. Do the same in $\mathbb{Z}_5[x]$.

10. Find a prime $p > 5$ such that $x^2 + 1$ is reducible in $\mathbb{Z}_p[x]$.

11. Find an odd prime p for which $x - 2$ is a divisor of $x^4 + x^3 + 3x^2 + x + 1$ in $\mathbb{Z}_p[x]$.

B. 12. If $a \in F$ is a nonzero root of $c_n x^n + c_{n-1}x^{n-1} + \cdots + c_1 x + c_0 \in F[x]$, show that a^{-1} is a root of $c_0 x^n + c_1 x^{n-1} + \cdots + c_{n-1}x + c_n$.

13. (a) If $f(x)$ and $g(x)$ are associates in $F[x]$, show that they have the same roots in F.

   (b) If $f(x), g(x) \in F[x]$ have the same roots in F, are they associates in $F[x]$?

14. (a) Suppose $r, s \in F$ are roots of $ax^2 + bx + c \in F[x]$ (with $a \neq 0_F$). Use the Factor Theorem to show that $r + s = -a^{-1}b$ and $rs = a^{-1}c$.

   (b) Suppose $r, s, t \in F$ are roots of $ax^3 + bx^2 + cx + d \in F[x]$ (with $a \neq 0_F$). Show that $r + s + t = -a^{-1}b$ and $rs + st + rt = a^{-1}c$ and $rst = -a^{-1}d$.

15. Prove that $x^2 + 1$ is reducible in $\mathbb{Z}_p[x]$ if and only if there exist integers a and b such that $p = a + b$ and $ab \equiv 1 \pmod{p}$.

16. Let $f(x), g(x) \in F[x]$ have degree $\leq n$ and let $c_0, c_1, \ldots, c_n$ be distinct elements of F. If $f(c_i) = g(c_i)$ for $i = 0, 1, \ldots, n$, prove that $f(x) = g(x)$ in $F[x]$.

17. Find a polynomial of degree 2 in $\mathbb{Z}_6[x]$ that has four roots in $\mathbb{Z}_6$. Does this contradict Corollary 4.16?

18. Let $\varphi : \mathbb{C} \to \mathbb{C}$ be an isomorphism of rings such that $\varphi(a) = a$ for each $a \in \mathbb{Q}$. Suppose $r \in \mathbb{C}$ is a root of $f(x) \in \mathbb{Q}[x]$. Prove that $\varphi(r)$ is also a root of $f(x)$.

19. We say that $a \in F$ is a **multiple root** of $f(x) \in F[x]$ if $(x - a)^k$ is a factor of $f(x)$ for some $k \geq 2$.

   (a) Prove that $a \in \mathbb{R}$ is a multiple root of $f(x) \in \mathbb{R}[x]$ if and only if a is a root of both $f(x)$ and $f'(x)$, where $f'(x)$ is the derivative of $f(x)$.

   (b) If $f(x) \in \mathbb{R}[x]$ and if $f(x)$ is relatively prime to $f'(x)$, prove that $f(x)$ has no multiple root in $\mathbb{R}$.

20. Let R be an integral domain. Then the Division Algorithm holds in $R[x]$ whenever the divisor is monic, by Exercise 14 in Section 4.1. Use this fact to show that the Remainder and Factor Theorems hold in $R[x]$.

21. If R is an integral domain and $f(x)$ is a nonzero polynomial of degree n in $R[x]$, prove that $f(x)$ has at most n roots in R. [*Hint:* Exercise 20.]

22. Show that Corollary 4.19 holds if F is an infinite integral domain. [*Hint:* See Exercise 21.]

23. Let $f(x), g(x), h(x) \in F[x]$ and $a \in F$.

    (a) If $f(x) = g(x) + h(x)$ in $F[x]$, show that $f(a) = g(a) + h(a)$ in F.

    (b) If $f(x) = g(x)h(x)$ in $F[x]$, show that $f(a) = g(a)h(a)$ in F.

    Where were these facts used in this section?

24. Let a be a fixed element of F and define a map $\varphi_a : F[x] \to F$ by $\varphi_a[f(x)] = f(a)$. Prove that φ_a is a surjective homomorphism of rings. The map φ_a is called an **evaluation homomorphism**; there is one for each $a \in F$.

25. Let $\mathbb{Q}[\pi]$ be the set of all real numbers of the form

$$r_0 + r_1\pi + r_2\pi^2 + \cdots + a_n\pi^n, \qquad \text{with } n \geq 0 \text{ and } r_i \in \mathbb{Q}.$$

    (a) Show that $\mathbb{Q}[\pi]$ is a subring of $\mathbb{R}$.

    (b) Show that the function $\theta : \mathbb{Q}[x] \to \mathbb{Q}[\pi]$ defined by $\theta(f(x)) = f(\pi)$ is an isomorphism. You may assume the following nontrivial fact: π is not the root of any nonzero polynomial with rational coefficients. Therefore, Theorem 4.1 is true with $R = \mathbb{Q}$ and π in place of x. However, see Exercise 26.

26. Let $\mathbb{Q}[\sqrt{2}]$ be the set of all real numbers of the form

$$r_0 + r_1\sqrt{2} + r_2(\sqrt{2})^2 + \cdots + r_n(\sqrt{2})^n, \qquad \text{with } n \geq 0 \text{ and } r_i \in \mathbb{Q}.$$

    (a) Show that $\mathbb{Q}[\sqrt{2}]$ is a subring of $\mathbb{R}$.

    (b) Show that the function $\theta : \mathbb{Q}[x] \to \mathbb{Q}[\sqrt{2}]$ defined by $\theta(f(x)) = f(\sqrt{2})$ is a surjective homomorphism, but not an isomorphism. Thus Theorem 4.1 is *not* true with $R = \mathbb{Q}$ and $\sqrt{2}$ in place of x. Compare this with Exercise 25.

27. Let T be the set of all polynomial functions from F to F. Show that T is a commutative ring with identity, with operations defined as in calculus: For each $r \in F$,

$$(f + g)(r) = f(r) + g(r) \qquad \text{and} \qquad (fg)(r) = f(r)g(r).$$

[*Hint:* To show that T is closed under addition and multiplication, use Exercise 23 to verify that $f + g$ and fg are the polynomial functions induced by the sum and product polynomials $f(x) + g(x)$ and $f(x)g(x)$, respectively.]

28. Let T be the ring of all polynomial functions from $\mathbb{Z}_3$ to $\mathbb{Z}_3$ (see Exercise 27).

    (a) Show that T is a finite ring with zero divisors. [*Hint:* Consider $f(x) = x + 1$ and $g(x) = x^2 + 2x$.]

    (b) Show that T cannot possibly be isomorphic to $\mathbb{Z}_3[x]$. Then see Exercise 29.

C. 29. If F is an infinite field, prove that the polynomial ring $F[x]$ is isomorphic to the ring T of all polynomial functions from F to F (Exercise 27). [*Hint:* Define a map $\varphi : F[x] \to T$ by assigning to each polynomial $f(x) \in F[x]$ its induced function in T; φ is injective by Corollary 4.19.]

30. Let $\varphi : F[x] \to F[x]$ be an isomorphism such that $\varphi(a) = a$ for every $a \in F$. Prove that $f(x)$ is irreducible in $F[x]$ if and only if $\varphi(f(x))$ is.

31. (a) Show that the map $\varphi : F[x] \to F[x]$ given by $\varphi(f(x)) = f(x + 1_F)$ is an isomorphism such that $\varphi(a) = a$ for every $a \in F$.

  (b) Use Exercise 30 to show that $f(x)$ is irreducible in $F[x]$ if and only if $f(x + 1_F)$ is.

4.5 IRREDUCIBILITY IN $\mathbb{Q}[x]$*

The central theme of this section is that factoring in $\mathbb{Q}[x]$ can be reduced to factoring in $\mathbb{Z}[x]$. Then elementary number theory can be used to check polynomials with integer coefficients for irreducibility. We begin by noting a fact that will be used frequently:

> **If $f(x) \in \mathbb{Q}[x]$, then $cf(x)$ has integer coefficients for some nonzero integer c.**

For example, consider

$$f(x) = x^5 + \tfrac{2}{3}x^4 + \tfrac{3}{4}x^3 - \tfrac{1}{6}.$$

The least common denominator of the coefficients of $f(x)$ is 12, and $12f(x)$ has integer coefficients:

$$12f(x) = 12[x^5 + \tfrac{2}{3}x^4 + \tfrac{3}{4}x^3 - \tfrac{1}{6}] = 12x^5 + 8x^4 + 9x^3 - 2.$$

According to the Factor Theorem, finding first-degree factors of a polynomial $g(x) \in \mathbb{Q}[x]$ is equivalent to finding the roots of $g(x)$ in $\mathbb{Q}$. Now, $g(x)$ has the same roots as $cg(x)$ for any nonzero constant c. When c is chosen so that $cg(x)$ has integer coefficients, we can find the roots of $g(x)$ by using

THEOREM 4.20 (RATIONAL ROOT TEST)   *Let* $f(x) = a_n x^n + a_{n-1}x^{n-1} + \cdots + a_1 x + a_0$ *be a polynomial with integer coefficients. If* $r \neq 0$ *and the rational number* r/s *(in lowest terms) is a root of* $f(x)$, *then* $r \mid a_0$ *and* $s \mid a_n$.

---

* This section is used only in Chapters 10, 11, and 15. It may be omitted until then, if desired. Section 4.6 is independent of this section.

Proof First consider the case when $s = 1$, that is, the case when the integer r is a root of $f(x)$. Now $f(r) = 0$ means that $a_n r^n + a_{n-1} r^{n-1} + \cdots + a_1 r + a_0 = 0$. Hence

$$a_0 = -a_n r^n - a_{n-1} r^{n-1} - \cdots - a_1 r$$
$$a_0 = r(-a_n r^{n-1} - a_{n-1} r^{n-2} - \cdots - a_1),$$

which says that r divides a_0.

In the general case, we use essentially the same strategy. Since r/s is a root of $f(x)$, we have

$$a_n \left(\frac{r^n}{s^n}\right) + a_{n-1}\left(\frac{r^{n-1}}{s^{n-1}}\right) + \cdots + a_1\left(\frac{r}{s}\right) + a_0 = 0.$$

We need an equation involving only integers (as in the case when $s = 1$). So multiply both sides by s^n, rearrange, and factor as before:

$$a_n r^n + a_{n-1} s r^{n-1} + \cdots + a_1 s^{n-1} r + a_0 s^n = 0$$

(*)
$$a_0 s^n = -a_n r^n - a_{n-1} s r^{n-1} - \cdots - a_1 s^{n-1} r$$
$$a_0 s^n = r[-a_n r^{n-1} - a_{n-1} s r^{n-2} - \cdots - a_1 s^{n-1}].$$

This last equation says that r divides $a_0 s^n$, which is not quite what we want. However, since r/s is in lowest terms, we have $(r,s) = 1$. It follows that $(r,s^n) = 1$ (a prime that divides s^n also divides s, by Corollary 1.9). Since $r \mid a_0 s^n$ and $(r,s^n) = 1$, Theorem 1.5 shows that $r \mid a_0$. A similar argument proves that $s \mid a_n$ (just rearrange Equation (*) so that $a_n r^n$ is on one side and everything else is on the other side). ◆

EXAMPLE The possible roots in $\mathbb{Q}$ of $f(x) = 2x^4 + x^3 - 21x^2 - 14x + 12$ are of the form r/s, where r is one of $\pm 1, \pm 2, \pm 3, \pm 4, \pm 6,$ or ± 12 (the divisors of the constant term, 12) and s is ± 1 or ± 2 (the divisors of the leading coefficient, 2). Hence, the Rational Root Test reduces the search for roots of $f(x)$ to this finite list of possibilities:

$$1, -1, 2, -2, 3, -3, 4, -4, 6, -6, 12, -12, \frac{1}{2}, -\frac{1}{2}, \frac{3}{2}, -\frac{3}{2}.$$

It is tedious but straightforward to substitute each of these in $f(x)$ to find that -3 and $\frac{1}{2}$ are the only roots of $f(x)$ in $\mathbb{Q}$.* By the Factor Theorem, both $x - (-3) = x + 3$ and $x - \frac{1}{2}$ are factors of $f(x)$. Long division shows that

$$f(x) = (x + 3)(x - \tfrac{1}{2})(2x^2 - 4x - 8).$$

---

* A graphing calculator will reduce the amount of computation significantly. Since the x-intercepts of the graph of $y = f(x)$ are the roots of $f(x)$, you can eliminate any numbers from the list that aren't near an intercept. In this case, the graph indicates that you need only check $-3, \frac{1}{2}$, and $-\frac{3}{2}$.

The quadratic formula shows that the roots of $2x^2 - 4x - 8$ are $1 \pm \sqrt{5}$, neither of which is in $\mathbb{Q}$. Therefore, $2x^2 - 4x - 8$ is irreducible in $\mathbb{Q}[x]$ by Corollary 4.18. Hence, we have factored $f(x)$ as a product of irreducible polynomials in $\mathbb{Q}[x]$.

EXAMPLE   The only possible roots of $g(x) = x^3 + 4x^2 + x - 1$ in $\mathbb{Q}$ are 1 and -1 (why?). Verify that neither 1 nor -1 is a root of $g(x)$. Hence $g(x)$ is irreducible in $\mathbb{Q}[x]$ by Corollary 4.18.

If $f(x) \in \mathbb{Q}[x]$, then $cf(x)$ has integer coefficients for some nonzero integer c. Any factorization of $cf(x)$ in $\mathbb{Z}[x]$ leads to factorization of $f(x)$ in $\mathbb{Q}[x]$. So it appears that tests for irreducibility in $\mathbb{Q}[x]$ can be restricted to polynomials with integer coefficients. However, we must first rule out the possibility that a polynomial with integer coefficients could factor in $\mathbb{Q}[x]$ but not in $\mathbb{Z}[x]$. In order to do this, we need

LEMMA 4.21   *Let* $f(x)$, $g(x)$, $h(x) \in \mathbb{Z}[x]$ *with* $f(x) = g(x)h(x)$. *If* p *is a prime that divides every coefficient of* $f(x)$, *then either* p *divides every coefficient of* $g(x)$ *or* p *divides every coefficient of* $h(x)$.

Proof   Let $f(x) = a_0 + a_1x + \cdots + a_kx^k$, $g(x) = b_0 + b_1x + \cdots + b_mx^m$, and $h(x) = c_0 + c_1x + \cdots + c_nx^n$. We use a proof by contradiction. If the lemma is false, then p does not divide some coefficient of $g(x)$ and some coefficient of $h(x)$. Let b_r be the *first* coefficient of $g(x)$ that is *not* divisible by p, and let c_t be the *first* coefficient of $h(x)$ that is *not* divisible by p. Then $p \mid b_i$ for $i < r$ and $p \mid c_j$ for $j < t$. Consider the coefficient a_{r+t} of $f(x)$. Since $f(x) = g(x)h(x)$,

$$a_{r+t} = b_0c_{r+t} + \cdots + b_{r-1}c_{t+1} + b_rc_t + b_{r+1}c_{t-1} + \cdots + b_{r+t}c_0.$$

Consequently,

$$b_rc_t = a_{r+t} - [b_0c_{r+t} + \cdots + b_{r-1}c_{t+1}] - [b_{r+1}c_{t-1} + \cdots + b_{r+t}c_0].$$

Now, $p \mid a_{r+t}$ by hypothesis. Also, p divides each term in the first pair of brackets because r was chosen so that $p \mid b_i$ for each $i < r$. Similarly, p divides each term in the second pair of brackets because $p \mid c_j$ for each $j < t$. Since p divides every term on the right side, we see that $p \mid b_rc_t$. Therefore, $p \mid b_r$ or $p \mid c_t$ by Theorem 1.8. This contradicts the fact that neither b_r nor c_t is divisible by p.   ◆

THEOREM 4.22   *Let* $f(x)$ *be a polynomial with integer coefficients. Then* $f(x)$ *factors as a product of polynomials of degrees* m *and* n *in* $\mathbb{Q}[x]$ *if and only if* $f(x)$ *factors as a product of polynomials of degrees* m *and* n *in* $\mathbb{Z}[x]$.

Proof   Obviously, if $f(x)$ factors in $\mathbb{Z}[x]$, it factors in $\mathbb{Q}[x]$. Conversely, suppose $f(x) = g(x)h(x)$ in $\mathbb{Q}[x]$. Let c and d be nonzero integers such that $cg(x)$ and $dh(x)$ have integer coefficients. Then $cdf(x) = [cg(x)][dh(x)]$ in $\mathbb{Z}[x]$ with deg $cg(x) =$

deg $g(x)$ and deg $dh(x) = $ deg $h(x)$. Let p be any prime divisor of cd, say $cd = pt$. Then p divides every coefficient of the polynomial $cdf(x)$. By Lemma 4.21, p divides either every coefficient of $cg(x)$ or every coefficient of $dh(x)$, say the former. Then $cg(x) = pk(x)$ with $k(x) \in \mathbb{Z}[x]$ and deg $k(x) = $ deg $g(x)$. Therefore, $ptf(x) = cdf(x) = [cg(x)][dh(x)] = [pk(x)][dh(x)]$. Cancelling p on each end, we have $tf(x) = k(x)[dh(x)]$ in $\mathbb{Z}[x]$.

Now repeat the same argument with any prime divisor of t and cancel that prime from both sides of the equation. Continue until every prime factor of cd has been cancelled. Then the left side of the equation will be $\pm f(x)$, and the right side will be a product of two polynomials in $\mathbb{Z}[x]$, one with the same degree as $g(x)$ and one with the same degree as $h(x)$. ◆

EXAMPLE We claim that $f(x) = x^4 - 5x^2 + 1$ is irreducible in $\mathbb{Q}[x]$. The proof is by contradiction. If $f(x)$ is reducible, it can be factored as the product of two nonconstant polynomials in $\mathbb{Q}[x]$. If either of these factors has degree 1, then $f(x)$ has a root in $\mathbb{Q}$. But the Rational Root Test shows that $f(x)$ has no roots in $\mathbb{Q}$. (The only possibilities are ± 1, and neither is a root.) Thus if $f(x)$ is reducible, the only possible factorization is as a product of two quadratics, by Theorem 4.2. In this case Theorem 4.22 shows that there is such a factorization in $\mathbb{Z}[x]$. Furthermore, there is a factorization as a product of *monic* quadratics in $\mathbb{Z}[x]$ by Exercise 10, say

$$(x^2 + ax + b)(x^2 + cx + d) = x^4 - 5x^2 + 1$$

with $a, b, c, d \in \mathbb{Z}$. Multiplying out the left-hand side, we have

$$x^4 + (a + c)x^3 + (ac + b + d)x^2 + (bc + ad)x + bd$$
$$= x^4 + 0x^3 - 5x^2 + 0x + 1.$$

Equal polynomials have equal coefficients; hence,

$$a + c = 0 \qquad ac + b + d = -5 \qquad bc + ad = 0 \qquad bd = 1.$$

Since $a + c = 0$, we have $a = -c$, so that

$$-5 = ac + b + d = -c^2 + b + d,$$

or, equivalently,

$$5 = c^2 - b - d.$$

However, $bd = 1$ in $\mathbb{Z}$ implies that $b = d = 1$ or $b = d = -1$, and so there are only these two possibilities:

$$5 = c^2 - 1 - 1 \qquad \text{or} \qquad 5 = c^2 + 1 + 1$$
$$7 = c^2 \qquad\qquad\qquad 3 = c^2.$$

There is no integer whose square is 3 or 7, and so a factorization of $f(x)$ as a product of quadratics in $\mathbb{Z}[x]$, and, hence in $\mathbb{Q}[x]$, is impossible. Therefore, $f(x)$ is irreducible in $\mathbb{Q}[x]$.

The brute-force methods of the preceding example are less effective for polynomials of high degree because the system of equations that must be solved is complicated and difficult to handle in a systematic way. However, the irreducibility of certain polynomials of high degree is easily established by

THEOREM   4.23 (EISENSTEIN'S   CRITERION) *Let $f(x) = a_n x^n + \cdots + a_1 x + a_0$ be a nonconstant polynomial with integer coefficients. If there is a prime p such that p divides each of $a_0, a_1, \ldots, a_{n-1}$ but p does not divide a_n and p^2 does not divide a_0, then $f(x)$ is irreducible in $\mathbb{Q}[x]$.*

Proof   The proof is by contradiction. If $f(x)$ is reducible, then by Theorem 4.22 it can be factored in $\mathbb{Z}[x]$, say

$$f(x) = (b_0 + b_1 x + \cdots + b_r x^r)(c_0 + c_1 x + \cdots + c_s x^s),$$

where each $b_i, c_j \in \mathbb{Z}$, $r \geq 1$, and $s \geq 1$. Note that $a_0 = b_0 c_0$. By hypothesis, $p \mid a_0$ and, hence, $p \mid b_0$ or $p \mid c_0$ by Theorem 1.8, say $p \mid b_0$. Since p^2 does not divide a_0, we see that c_0 is not divisible by p. We also have $a_n = b_r c_s$. Consequently, p does not divide b_r (otherwise a_n would be divisible by p, contrary to hypothesis). There may be other b_i not divisible by p as well. Let b_k be the first of the b_i not divisible by p; then $0 < k \leq r < n$ and

$$p \mid b_i \text{ for } i < k \qquad \text{and} \qquad p \nmid b_k.$$

By the rules of polynomial multiplication,

$$a_k = b_0 c_k + b_1 c_{k-1} + \cdots + b_{k-1} c_1 + b_k c_0,$$

so that

$$b_k c_0 = a_k - b_0 c_k - b_1 c_{k-1} - \cdots - b_{k-1} c_1.$$

Since $p \mid a_k$ and $p \mid b_i$ for $i < k$, we see that p divides every term on the right-hand side of this equation. Hence, $p \mid b_k c_0$. By Theorem 1.8, p must divide b_k or c_0. This contradicts the fact that neither b_k nor c_0 is divisible by p. Therefore, $f(x)$ is irreducible in $\mathbb{Q}[x]$.   ◆

> **EXAMPLE**   The polynomial $x^{17} + 6x^{13} - 15x^4 + 3x^2 - 9x + 12$ is irreducible in $\mathbb{Q}[x]$ by Eisenstein's Criterion with $p = 3$.

> **EXAMPLE**   The polynomial $x^9 + 5$ is irreducible in $\mathbb{Q}[x]$ by Eisenstein's Criterion with $p = 5$. Similarly, $x^n + 5$ is irreducible in $\mathbb{Q}[x]$ for each $n \geq 1$. Thus

there are irreducible polynomials of every degree in $\mathbb{Q}[x]$.

Although Eisenstein's Criterion is very efficient, there are many polynomials to which it cannot be applied. In such cases other techniques are necessary. One such method involves reducing a polynomial mod p, in the following sense. Let p be a positive prime. For each integer a, let $[a]$ denote the congruence class of a in $\mathbb{Z}_p$. If $f(x) = a_k x^k + \cdots + a_1 x + a_0$ is a polynomial with integer coefficients, let $\bar{f}(x)$ denote the polynomial $[a_k]x^k + \cdots + [a_1]x + [a_0]$ in $\mathbb{Z}_p[x]$. For instance, if $f(x) = 2x^4 - 3x^2 + 5x + 7$ in $\mathbb{Z}[x]$, then in $\mathbb{Z}_3[x]$,

$$\bar{f}(x) = [2]x^4 - [3]x^2 + [5]x + [7]$$
$$= [2]x^4 - [0]x^2 + [2]x + [1] = [2]x^4 + [2]x + [1].$$

Notice that $f(x)$ and $\bar{f}(x)$ have the same degree. This will always be the case when the leading coefficient of $f(x)$ is not divisible by p (so that the leading coefficient of $\bar{f}(x)$ will not be the zero class in $\mathbb{Z}_p$).

THEOREM 4.24 *Let $f(x) = a_k x^k + \cdots + a_1 x + a_0$ be a polynomial with integer coefficients, and let p be a positive prime that does not divide a_k. If $\bar{f}(x)$ is irreducible in $\mathbb{Z}_p[x]$, then $f(x)$ is irreducible in $\mathbb{Q}[x]$.*

Proof Suppose, on the contrary, that $f(x)$ is reducible in $\mathbb{Q}[x]$. Then by Theorem 4.22, $f(x) = g(x)h(x)$ with $g(x), h(x)$ nonconstant polynomials in $\mathbb{Z}[x]$. Since p does not divide a_k, the leading coefficient of $f(x)$, it cannot divide the leading coefficients of $g(x)$ or $h(x)$ (whose product is a_k). Consequently, deg $\bar{g}(x) =$ deg $g(x)$ and deg $\bar{h}(x) =$ deg $h(x)$. In particular, neither $\bar{g}(x)$ nor $\bar{h}(x)$ is a constant polynomial in $\mathbb{Z}_p[x]$.

Verify that $f(x) = g(x)h(x)$ in $\mathbb{Z}[x]$ implies that $\bar{f}(x) = \bar{g}(x)\bar{h}(x)$ in $\mathbb{Z}_p[x]$ (Exercise 19). This contradicts the irreducibility of $\bar{f}(x)$ in $\mathbb{Z}_p[x]$. Therefore, $f(x)$ must be irreducible in $\mathbb{Q}[x]$. ◆

The usefulness of Theorem 4.24 depends on this fact: For each nonnegative integer k, there are only finitely many polynomials of degree k in $\mathbb{Z}_p[x]$ (Exercise 17). Therefore, it is always possible, in theory, to determine whether a given polynomial in $\mathbb{Z}_p[x]$ is irreducible by checking the finite number of possible factors. Depending on the size of p and on the degree of $f(x)$, this can often be done in a reasonable amount of time.

EXAMPLE To show that $f(x) = x^5 + 8x^4 + 3x^2 + 4x + 7$ is irreducible in $\mathbb{Q}[x]$, we reduce mod 2. In $\mathbb{Z}_2[x]$, $\bar{f}(x) = x^5 + x^2 + 1$.* It is easy to see that $\bar{f}(x)$ has no roots in $\mathbb{Z}_2$ and hence no first-degree factors in $\mathbb{Z}_2[x]$. The only quadratic polynomials in $\mathbb{Z}_2[x]$ are $x^2, x^2 + x, x^2 + 1$, and $x^2 + x + 1$. However, if $x^2, x^2 + x = x(x + 1)$, or $x^2 + 1 = (x + 1)(x + 1)$ were a factor,

---

* When no confusion is likely, we omit the brackets for elements of $\mathbb{Z}_2$.

then $\bar{f}(x)$ would have a first-degree factor, which it doesn't. You can use long division to show that the remaining quadratic, $x^2 + x + 1$, is not a factor of $\bar{f}(x)$. Finally, $\bar{f}(x)$ cannot have a factor of degree 3 or 4 (if it did, the other factor would have degree 2 or 1, which is impossible). Therefore, $\bar{f}(x)$ is irreducible in $\mathbb{Z}_2[x]$. Hence, $f(x)$ is irreducible in $\mathbb{Q}[x]$.

Warning   If a polynomial in $\mathbb{Z}[x]$ reduces mod p to a polynomial that is *reducible* in $\mathbb{Z}_p[x]$, then no conclusion can be drawn from Theorem 4.24. Unfortunately, there may be many p for which the reduction of $f(x)$ is reducible in $\mathbb{Z}_p[x]$, even when $f(x)$ is actually irreducible in $\mathbb{Q}[x]$. Consequently, it may take more time to apply Theorem 4.24 than is first apparent.

◆ EXERCISES

A. **1.** Use the Rational Root Test to write each polynomial as a product of irreducible polynomials in $\mathbb{Q}[x]$:

   (a) $-x^4 + x^3 + x^2 + x + 2$ **(b)** $x^5 + 4x^4 + x^3 - x^2$

   (c) $3x^5 + 2x^4 - 7x^3 + 2x^2$ **(d)** $2x^4 - 5x^3 + 3x^2 + 4x - 6$

   (e) $2x^4 + 7x^3 + 5x^2 + 7x + 3$ **(f)** $6x^4 - 31x^3 + 25x^2 + 33x + 7$

2. Show that $\sqrt{p}$ is irrational for every positive prime integer p. [*Hint:* What are the roots of $x^2 - p$? Do you prefer this proof to the one in Exercises 20 and 21 of Section 1.3?]

3. If a monic polynomial with integer coefficients has a root in $\mathbb{Q}$, show that this root must be an integer.

4. Show that each polynomial is irreducible in $\mathbb{Q}[x]$, as in the example that follows Theorem 4.22:

   (a) $x^4 + 2x^3 + x + 1$ **(b)** $x^4 - 2x^2 + 8x + 1$

5. Use Eisenstein's Criterion to show that each polynomial is irreducible in $\mathbb{Q}[x]$:

   (a) $x^5 - 4x + 22$ **(b)** $10 - 15x + 25x^2 - 7x^4$

   (c) $5x^{11} - 6x^4 + 12x^3 + 36x - 6$

6. Show that there are infinitely many integers k such that $x^9 + 12x^5 - 21x + k$ is irreducible in $\mathbb{Q}[x]$.

7. Show that each polynomial $f(x)$ is irreducible in $\mathbb{Q}[x]$ by finding a prime p such that $f(x)$ is irreducible in $\mathbb{Z}_p[x]$:

   (a) $7x^3 + 6x^2 + 4x + 6$ **(b)** $9x^4 + 4x^3 - 3x + 7$

8. Give an example of a polynomial $f(x) \epsilon \mathbb{Z}[x]$ and a prime p such that $f(x)$ is reducible in $\mathbb{Q}[x]$ but $\bar{f}(x)$ is irreducible in $\mathbb{Z}_p[x]$. Does this contradict Theorem 4.24?

9. Give an example of a polynomial in $\mathbb{Z}[x]$ that is irreducible in $\mathbb{Q}[x]$ but factors when reduced mod 2, 3, 4, and 5.

10. If a monic polynomial with integer coefficients factors in $\mathbb{Z}[x]$ as a product of polynomials of degrees m and n, prove that it can be factored as a product of monic polynomials of degrees m and n in $\mathbb{Z}[x]$.

B. 11. Prove that $30x^n - 91$ (where $n \epsilon \mathbb{Z}, n > 1$) has no roots in $\mathbb{Q}$.

12. Let F be a field and $f(x) \epsilon F[x]$. If $c \epsilon F$ and $f(x + c)$ is irreducible in $F[x]$, prove that $f(x)$ is irreducible in $F[x]$. [*Hint:* Prove the contrapositive.]

13. Prove that $f(x) = x^4 + 4x + 1$ is irreducible in $\mathbb{Q}[x]$ by using Eisenstein's Criterion to show that $f(x + 1)$ is irreducible and applying Exercise 12.

14. Prove that $f(x) = x^4 + x^3 + x^2 + x + 1$ is irreducible in $\mathbb{Q}[x]$. [*Hint:* Use the hint for Exercise 20 with $p = 5$.]

15. Let $f(x) = a_n x^n + a_{n-1} x^{n-1} + \cdots + a_1 x + a_0$ be a polynomial with integer coefficients. If p is a prime such that $p \mid a_1, p \mid a_2, \ldots, p \mid a_n$ but $p \nmid a_0$ and $p^2 \nmid a_n$, prove that $f(x)$ is irreducible in $\mathbb{Q}[x]$. [*Hint:* Let $y = 1/x$ in $f(x)/x^n$; the resulting polynomial is irreducible, by Theorem 4.23.]

16. Show by example that this statement is false: If $f(x) \epsilon \mathbb{Z}[x]$ and there is no prime p satisfying the hypotheses of Theorem 4.23, then $f(x)$ is reducible in $\mathbb{Q}[x]$.

17. Show that there are $n^{k+1} - n^k$ polynomials of degree k in $\mathbb{Z}_n[x]$.

18. Which of these polynomials are irreducible in $\mathbb{Q}[x]$:

    (a) $x^4 - x^2 + 1$ **(b)** $x^4 + x + 1$

    (c) $x^5 + 4x^4 + 2x^3 + 3x^2 - x + 5$ **(d)** $x^5 + 5x^2 + 4x + 7$

19. If $f(x) = a_n x^n + \cdots + a_1 x + a_0, g(x) = b_r x^r + \cdots + b_1 x + b_0$, and $h(x) = c_s x^s + \cdots + c_1 x + c_0$ are polynomials in $\mathbb{Z}[x]$ such that $f(x) = g(x)h(x)$, show that in $\mathbb{Z}_n[x], \bar{f}(x) = \bar{g}(x)\bar{h}(x)$. Also, see Exercise 17 in Section 4.1.

C. 20. Prove that for p prime, $f(x) = x^{p-1} + x^{p-2} + \cdots + x^2 + x + 1$ is irreducible in $\mathbb{Q}[x]$. [*Hint:* $(x - 1)f(x) = x^p - 1$, so that $f(x) = (x^p - 1)/(x - 1)$ and $f(x + 1) = [(x + 1)^p - 1]/x$. Expand $(x + 1)^p$ by the Binomial Theorem (Appendix E) and note that p divides $\binom{p}{k}$ when $k > 0$. Use Eisenstein's Criterion to show that $f(x + 1)$ is irreducible; apply Exercise 12.]

Excursion   Geometric Constructions (Chapter 15) may be covered at this point if desired.

4.6 IRREDUCIBILITY IN $\mathbb{R}[x]$ AND $\mathbb{C}[x]$*

Unlike the situation in $\mathbb{Q}[x]$, it is possible to give an explicit description of all the irreducible polynomials in $\mathbb{R}[x]$ and $\mathbb{C}[x]$. Consequently, you can immediately tell if a polynomial in $\mathbb{R}[x]$ or $\mathbb{C}[x]$ is irreducible without any elaborate tests or criteria. These facts are a consequence of the following theorem, which was first proved by Gauss in 1799:

THEOREM 4.25 (THE FUNDAMENTAL THEOREM OF ALGEBRA) *Every nonconstant polynomial in* $\mathbb{C}[x]$ *has a root in* $\mathbb{C}$.

This theorem is sometimes expressed in other terminology by saying that the field $\mathbb{C}$ is **algebraically closed.** Every known proof of the theorem depends significantly on facts from analysis and/or the theory of functions of a complex variable. For this reason, we shall consider only some of the implications of the Fundamental Theorem on irreducibility in $\mathbb{C}[x]$ and $\mathbb{R}[x]$. For a proof, see Hungerford [7].

COROLLARY 4.26 *A polynomial is irreducible in* $\mathbb{C}[x]$ *if and only if it has degree* 1.

Proof A polynomial $f(x)$ of degree ≥ 2 in $\mathbb{C}[x]$ has a root in $\mathbb{C}$ by Theorem 4.25 and hence a first-degree factor by the Factor Theorem. Therefore $f(x)$ is reducible in $\mathbb{C}[x]$, and every irreducible polynomial in $\mathbb{C}[x]$ must have degree 1. Conversely, every first-degree polynomial is irreducible (page 96). ◆

COROLLARY 4.27 *Every nonconstant polynomial* $f(x)$ *of degree n in* $\mathbb{C}[x]$ *can be written in the form* $c(x - a_1)(x - a_2) \cdots (x - a_n)$ *for some c, $a_1, a_2, \ldots, a_n \in$* $\mathbb{C}$. *This factorization is unique except for the order of the factors.*

Proof By Theorem 4.13, $f(x)$ is a product of irreducible polynomials in $\mathbb{C}[x]$. Each of them has degree 1 by Corollary 4.26, and there are exactly n of them by Theorem 4.2. Therefore,

$$f(x) = (r_1 x + s_1)(r_2 x + s_2) \cdots (r_n x + s_n)$$
$$= r_1(x - (-r_1^{-1} s_1)) r_2 (x - (-r_2^{-1} s_2)) \cdots r_n (x - (-r_n^{-1} s_n))$$
$$= c(x - a_1)(x - a_2) \cdots (x - a_n),$$

where $c = r_1 r_2 \cdots r_n$ and $a_i = r_i^{-1} s_i$. Uniqueness follows from Theorem 4.13; see Exercise 25 in Section 4.3. ◆

---

* This section is used only in Chapters 10 and 11. It may be omitted until then, if desired.

To obtain a description of all the irreducible polynomials in $\mathbb{R}[x]$, we need

LEMMA 4.28 *If $f(x)$ is a polynomial in $\mathbb{R}[x]$ and $a + bi$ is a root of $f(x)$ in $\mathbb{C}$, then $a - bi$ is also a root of $f(x)$.*

Proof If $c = a + bi \in \mathbb{C}$ (with $a, b \in \mathbb{R}$), let $\bar{c}$ denote $a - bi$. Verify that for any c, $d \in \mathbb{C}$,

$$\overline{(c + d)} = \bar{c} + \bar{d} \qquad \text{and} \qquad \overline{cd} = \bar{c}\bar{d}.$$

Also note that $\bar{c} = c$ *if and only if c is a real number*. Now, if $f(x) = a_n x^n + \cdots + a_1 x + a_0$ and c is a root of $f(x)$, then $f(c) = 0$, so that

$$\begin{aligned}
0 = \bar{0} = \overline{f(c)} &= \overline{a_n c^n + \cdots + a_1 c + a_0} \\
&= \overline{a_n}\,\overline{c}^n + \cdots + \overline{a_1}\overline{c} + \overline{a_0} \\
&= a_n \overline{c}^n + \cdots + a_1 \overline{c} + a_0 \qquad \text{(because each } a_i \in \mathbb{R}). \\
&= f(\bar{c})
\end{aligned}$$

Therefore $\bar{c} = a - bi$ is also a root of $f(x)$. $\blacklozenge$

THEOREM 4.29 *A polynomial $f(x)$ is irreducible in $\mathbb{R}[x]$ if and only if $f(x)$ is a first-degree polynomial or*

$$f(x) = ax^2 + bx + c \qquad \text{with } b^2 - 4ac < 0.$$

Proof The proof that the two kinds of polynomials mentioned in the theorem are in fact irreducible is left to the reader (Exercise 7). Conversely, suppose $f(x)$ has degree ≥ 2 and is irreducible in $\mathbb{R}[x]$. Then $f(x)$ has a root w in $\mathbb{C}$ by Theorem 4.25. Lemma 4.28 shows that $\overline{w}$ is also a root of $f(x)$. Furthermore, $w \neq \overline{w}$ (otherwise w would be a real root of $f(x)$, contradicting the irreducibility of $f(x)$). Consequently, by the Factor Theorem, $x - w$ and $x - \overline{w}$ are factors of $f(x)$ in $\mathbb{C}[x]$; that is, $f(x) = (x - w)(x - \overline{w})h(x)$ for some $h(x)$ in $\mathbb{C}[x]$. Let $g(x) = (x - w)(x - \overline{w})$; then $f(x) = g(x)h(x)$ in $\mathbb{C}[x]$. Furthermore, if $w = r + si$ (with r, $s \in \mathbb{R}$), then

$$\begin{aligned}
g(x) = (x - w)(x - \overline{w}) &= (x - (r + si))(x - (r - si)) \\
&= x^2 - 2rx + (r^2 + s^2).
\end{aligned}$$

Hence, the coefficients of $g(x)$ are real numbers.

We now show that $h(x)$ also has real coefficients. The Division Algorithm in $\mathbb{R}[x]$ shows that there are polynomials $q(x)$, $r(x)$ in $\mathbb{R}[x]$ such that $f(x) = g(x)q(x) + r(x)$, with $r(x) = 0$ or $\deg r(x) < \deg g(x)$. In $\mathbb{C}[x]$, however, we have $f(x) = g(x)h(x) + 0$. Since $q(x)$ and $r(x)$ can be considered as polynomials in $\mathbb{C}[x]$, the uniqueness part of the Division Algorithm in $\mathbb{C}[x]$ shows that $q(x) = h(x)$ and $r(x) = 0$. Thus $h(x) = q(x) \in \mathbb{R}[x]$. Since $f(x) = g(x)h(x)$ and $f(x)$ is irreducible in $\mathbb{R}[x]$ and deg $g(x) = 2$, $h(x)$ must be a constant $d \ \epsilon \ \mathbb{R}$. Consequently, $f(x) = dg(x)$ is a quadratic polynomial in $\mathbb{R}[x]$ and hence has the form $ax^2 + bx + c$ for some $a, b, c \ \epsilon \ \mathbb{R}$. Since $f(x)$ has no roots in $\mathbb{R}$, the quadratic formula (Exercise 6) shows that $b^2 - 4ac < 0$. ◆

COROLLARY 4.30   *Every polynomial* $f(x)$ *of odd degree in* $\mathbb{R}[x]$ *has a root in* $\mathbb{R}$.

Proof   By Theorem 4.13, $f(x) = p_1(x)p_2(x) \cdots p_k(x)$ with each $p_i(x)$ irreducible in $\mathbb{R}[x]$. Each $p_i(x)$ has degree 1 or 2 by Theorem 4.29. Theorem 4.2 shows that

$$\deg f(x) = \deg p_1(x) + \deg p_2(x) + \cdots + \deg p_k(x).$$

Since $f(x)$ has odd degree, at least one of the $p_i(x)$ must have degree 1. Therefore, $f(x)$ has a first-degree factor in $\mathbb{R}[x]$ and, hence, a root in $\mathbb{R}$. ◆

It may seem that the Fundamental Theorem and its corollaries settle all the basic questions about polynomial equations. Unfortunately, things aren't quite that simple. None of the known proofs of the Fundamental Theorem provides a constructive way to find the roots of a specific polynomial.* Therefore, even though we know that every polynomial equation has a solution in $\mathbb{C}$, we may not be able to solve a particular equation.

Polynomial equations of degree less than 5 are no problem. The quadratic formula shows that the solutions of any second-degree polynomial equation can be obtained from the coefficients of the polynomials by taking sums, differences, products, quotients, and square roots. There are analogous, but more complicated, formulas involving cube and fourth roots for third- and fourth-degree polynomial equations (see page 387 for one version of the cubic formula). However, there are no such formulas for finding the roots of all fifth-degree or higher-degree polynomials. This remarkable fact, which was proved nearly two centuries ago, is discussed in Section 11.3.

◆ **EXERCISES**

A.   **1.** Find all the roots in $\mathbb{C}$ of each polynomial (one root is already given):

   (a) $x^4 - 3x^3 + x^2 + 7x - 30$; root $1 - 2i$

   (b) $x^4 - 2x^3 - x^2 + 6x - 6$; root $1 + i$

   (c) $x^4 - 4x^3 + 3x^2 + 14x + 26$; root $3 + 2i$

---

* It may seem strange that it is possible to prove that a root exists without actually exhibiting one, but such "existence theorems" are quite common in mathematics. A very rough analogy is the situation that occurs when a person is killed by a sniper's bullet. The police know that there *is* a killer, but actually *finding* the killer may be difficult or impossible.

2. Find a polynomial in $\mathbb{R}[x]$ that satisfies the given conditions:

   (a) Monic of degree 3 with 2 and $3 + i$ as roots

   (b) Monic of least possible degree with $1 - i$ and $2i$ as roots

   (c) Monic of least possible degree with 3 and $4i - 1$ as roots

3. Factor each polynomial as a product of irreducible polynomials in $\mathbb{Q}[x]$, in $\mathbb{R}[x]$, and in $\mathbb{C}[x]$:

   (a) $x^4 - 2$ (b) $x^3 + 1$ (c) $x^3 - x^2 - 5x + 5$

4. Factor $x^2 + x + 1 + i$ in $\mathbb{C}[x]$.

B. 5. Show that a polynomial of odd degree in $\mathbb{R}[x]$ with no multiple roots must have an odd number of real roots.

6. Let $f(x) = ax^2 + bx + c \in \mathbb{R}[x]$ with $a \neq 0$. Prove that the roots of $f(x)$ in $\mathbb{C}$ are

$$\frac{-b + \sqrt{b^2 - 4ac}}{2a} \quad \text{and} \quad \frac{-b - \sqrt{b^2 - 4ac}}{2a}.$$

   [*Hint:* Show that $ax^2 + bx + c = 0$ is equivalent to $x^2 + (b/a)x = -c/a$; then complete the square to find x.]

7. Prove that every $ax^2 + bx + c \in \mathbb{R}[x]$ with $b^2 - 4ac < 0$ is irreducible in $\mathbb{R}[x]$. [*Hint:* See Exercise 6].

8. If $a + bi$ is a root of $x^3 - 3x^2 + 2ix + i - 1 \in \mathbb{C}[x]$, then is it true that $a - bi$ is also a root?

CHAPTER 5

Congruence in $F[x]$ and Congruence-Class Arithmetic

$\blacklozenge$

In this chapter we continue to explore the analogy between the ring $\mathbb{Z}$ of integers and the ring $F[x]$ of polynomials with coefficients in a field F. We shall see that the concepts of congruence and congruence-class arithmetic carry over from $\mathbb{Z}$ to $F[x]$ with practically no changes. Because of the additional features of the polynomial ring $F[x]$ (polynomial functions and roots), these new congruence-class rings have a much richer structure than do the rings $\mathbb{Z}_n$. This additional structure leads to a striking result: Given any polynomial over any field, we can find a root of that polynomial in some larger field.

5.1 CONGRUENCE IN $F[x]$ AND CONGRUENCE CLASSES

The concept of congruence of integers depends only on some basic facts about divisibility in $\mathbb{Z}$. If F is a field, then the polynomial ring $F[x]$ has essentially the same divisibility properties as does $\mathbb{Z}$. So it is not surprising that the concept of congruence in $\mathbb{Z}$ and its basic properties (Section 2.1) can be carried over to $F[x]$ almost verbatim.

> **DEFINITION•** *Let F be a field and $f(x)$, $g(x)$, $p(x) \in F[x]$ with $p(x)$ nonzero. Then $f(x)$ **is congruent to $g(x)$ modulo $p(x)$**—written $f(x) \equiv g(x)$ (mod $p(x)$)—provided that $p(x)$ divides $f(x) - g(x)$.*

EXAMPLE   In $\mathbb{Q}[x]$, $x^2 + x + 1 \equiv x + 2 \pmod{x + 1}$ because

$$(x^2 + x + 1) - (x + 2) = x^2 - 1 = (x + 1)(x - 1).$$

EXAMPLE   In $\mathbb{R}[x]$, $3x^4 + 4x^2 + 2x + 2 \equiv x^3 + 3x^2 + 3x + 4 \pmod{x^2 + 1}$ because long division shows that

$$(3x^4 + 4x^2 + 2x + 2) - (x^3 + 3x^2 + 3x + 4) = 3x^4 - x^3 + x^2 - x - 2$$
$$= (x^2 + 1)(3x^2 - x - 2).$$

THEOREM 5.1   *Let F be a field and p(x) a nonzero polynomial in F[x]. Then the relation of congruence modulo p(x) is*

(1) *reflexive:* $f(x) \equiv f(x) \pmod{p(x)}$ *for all* $f(x) \in F[x]$;

(2) *symmetric: if* $f(x) \equiv g(x) \pmod{p(x)}$, *then* $g(x) \equiv f(x) \pmod{p(x)}$;

(3) *transitive: if* $f(x) \equiv g(x) \pmod{p(x)}$ *and* $g(x) \equiv h(x) \pmod{p(x)}$, *then* $f(x) \equiv h(x) \pmod{p(x)}$.

Proof   Adapt the proof of Theorem 2.1 with $p(x)$, $f(x)$, $g(x)$, $h(x)$ in place of n, a, b, c.   ◆

THEOREM 5.2   *Let F be a field and p(x) a nonzero polynomial in F[x]. If* $f(x) \equiv g(x) \pmod{p(x)}$ *and* $h(x) \equiv k(x) \pmod{p(x)}$, *then*

(1) $f(x) + h(x) \equiv g(x) + k(x) \pmod{p(x)}$,

(2) $f(x)h(x) \equiv g(x)k(x) \pmod{p(x)}$.

Proof   Adapt the proof of Theorem 2.2 with $p(x)$, $f(x)$, $g(x)$, $h(x)$, $k(x)$ in place of n, a, b, c, d.   ◆

DEFINITION• *Let F be a field and* $f(x)$, $p(x) \in F[x]$ *with* $p(x)$ *nonzero. The **congruence class** (or **residue class**) of* $f(x)$ *modulo* $p(x)$ *is denoted* $[f(x)]$ *and consists of all polynomials in F[x] that are congruent to* $f(x)$ *modulo* $p(x)$, *that is,*

$$[f(x)] = \{g(x) \mid g(x) \in F[x] \text{ and } g(x) \equiv f(x) \pmod{p(x)}\}.$$

Since $g(x) \equiv f(x) \pmod{p(x)}$ means that $g(x) - f(x) = k(x)p(x)$ for some $k(x) \in F[x]$ or, equivalently, that $g(x) = f(x) + k(x)p(x)$, we see that

$$[f(x)] = \{g(x) \mid g(x) \equiv f(x) \pmod{p(x)}\}$$
$$= \{f(x) + k(x)p(x) \mid k(x) \in F[x]\}.$$

EXAMPLE   Consider congruence modulo $x^2 + 1$ in $\mathbb{R}[x]$. The congruence class of $2x + 1$ is the set

$$\{(2x + 1) + k(x)(x^2 + 1) \mid k(x) \in \mathbb{R}[x]\}.$$

The Division Algorithm shows that the elements of this set are the polynomials in $\mathbb{R}[x]$ that leave remainder $2x + 1$ when divided by $x^2 + 1$.

EXAMPLE Consider congruence modulo $x^2 + x + 1$ in $\mathbb{Z}_2[x]$. To find the congruence class of x^2, we note that $x^2 \equiv x + 1 \pmod{x^2 + x + 1}$ because $x^2 - (x + 1) = x^2 - x - 1 = (x^2 + x + 1)1$ (remember that $1 + 1 = 0$ in $\mathbb{Z}_2$, so that $1 = -1$). Therefore, $x + 1$ is a member of the congruence class $[x^2]$. In fact, the next theorem shows that $[x + 1] = [x^2]$.

THEOREM 5.3 $f(x) \equiv g(x) \pmod{p(x)}$ *if and only if* $[f(x)] = [g(x)]$.

Proof Adapt the proof of Theorem 2.3 with $f(x), g(x), p(x)$, and Theorem 5.1 in place of a, c, n, and Theorem 2.1. ◆

COROLLARY 5.4 *Two congruence classes modulo p(x) are either disjoint or identical.*

Proof Adapt the proof of Corollary 2.4. ◆

Under congruence modulo n in $\mathbb{Z}$, there are exactly n distinct congruence classes (Corollary 2.5). These classes are $[0], [1], \ldots, [n - 1]$. Note that there is a class for each possible remainder under division by n. In $F[x]$ the possible remainders under division by a polynomial of degree n are all the polynomials of degree less than n (and, of course, 0). So the analogue of Corollary 2.5 is

COROLLARY 5.5 *Let F be a field and p(x) a polynomial of degree n in F[x]. Let S be the set consisting of the zero polynomial and all the polynomials of degree less than n in F[x]. Then every congruence class modulo p(x) is the class of some polynomial in S, and the congruence classes of different polynomials in S are distinct.*

Proof Use the Division Algorithm as in the first part of the proof of Corollary 2.5 to show that every polynomial in $F[x]$ is congruent modulo $p(x)$ to a polynomial in S. Hence, by Theorem 5.3 every congruence class is equal to the class of some polynomial in S. Two different polynomials in S cannot be congruent modulo $p(x)$ because their difference has degree less than n and hence is not divisible by $p(x)$. Therefore, different polynomials in S must be in distinct congruence classes by Theorem 5.3. ◆

The set of all congruence classes modulo $p(x)$ is denoted

$$F[x]/(p(x)),$$

which is the notational analogue of $\mathbb{Z}_n$.

EXAMPLE Consider congruence modulo $x^2 + 1$ in $\mathbb{R}[x]$. There is a congruence class for each possible remainder on division by $x^2 + 1$. Now, the possible remainders are polynomials of the form $rx + s$ (with $r, s \in \mathbb{R}$; one or both of r, s may possibly be 0). Therefore, $\mathbb{R}[x]/(x^2 + 1)$ consists of infinitely many distinct congruence classes, including

$$[0], [x], [x + 1], [5x + 3], \left[\frac{7}{9}x + 2\right], [x - 7], \ldots .$$

Corollary 5.5 states that $[rx + s] = [cx + d]$ if and only if $rx + s$ is *equal* (not just congruent) to $cx + d$. By the definition of polynomial equality, $rx + s = cx + d$ if and only if $r = c$ and $s = d$. Therefore, every element of $\mathbb{R}[x]/(x^2 + 1)$ can be written *uniquely* in the form $[rx + s]$.

EXAMPLE Consider congruence modulo $x^2 + x + 1$ in $\mathbb{Z}_2[x]$. The possible remainders on division by $x^2 + x + 1$ are the polynomials of the form $ax + b$ with $a, b \in \mathbb{Z}_2$. Thus there are only four possible remainders: $0, 1, x$, and $x + 1$. Therefore, $\mathbb{Z}_2[x]/(x^2 + x + 1)$ consists of four congruence classes: $[0], [1], [x]$, and $[x + 1]$.

EXAMPLE The pattern in the previous example works in the general case. If $p(x) \in \mathbb{Z}_n[x]$ has degree k, then the possible remainders on division by $p(x)$ are of the form $a_0 + a_1x + \cdots + a_{k-1}x^{k-1}$, with $a_i \in \mathbb{Z}_n$. There are n possibilities for each of the k coefficients $a_0, \ldots, a_{k-1}$, and so there are n^k different polynomials of this form. Consequently, by Corollary 5.5, there are exactly n^k distinct congruence classes in $\mathbb{Z}_n[x]/(p(x))$.

◆ **EXERCISES**

NOTE: *F denotes a field and p(x) a nonzero polynomial in F[x].*

A. **1.** Let $f(x), g(x), p(x) \in \mathbb{Q}[x]$. Determine whether $f(x) \equiv g(x) \pmod{p(x)}$.

(a) $f(x) = x^5 - 2x^4 + 4x^3 - 3x + 1; g(x) = 3x^4 + 2x^3 - 5x^2 + 2;$
$p(x) = x^2 + 1.$

(b) $f(x) = x^4 + 2x^3 - 3x^2 + x - 5; g(x) = x^4 + x^3 - 5x^2 + 12x - 25;$
$p(x) = x^2 - 3x + 4.$

(c) $f(x) = 3x^5 + 4x^4 + 5x^3 - 6x^2 + 5x - 7; g(x) = 2x^5 + 6x^4 + x^3 + 2x^2 +$
$2x - 5; p(x) = x^3 - x^2 + x - 1.$

2. If $p(x)$ is a nonzero constant polynomial in $F[x]$, show that any two polynomials in $F[x]$ are congruent modulo $p(x)$.

3. How many distinct congruence classes are there modulo $x^3 + x + 1$ in $\mathbb{Z}_2[x]$? List them.

4. Show that, under congruence modulo $x^3 + 2x + 1$ in $\mathbb{Z}_3[x]$, there are exactly 27 distinct congruence classes.

5. Show that there are infinitely many distinct congruence classes modulo $x^2 - 2$ in $\mathbb{Q}[x]$. Describe them.

6. Let $a \in F$. Describe the congruence classes in $F[x]$ modulo the polynomial $x - a$.

7. If $p(x)$ has degree k in $\mathbb{Z}_n[x]$, how many distinct congruence classes are there modulo $p(x)$?

B. 8. Prove or disprove: If $p(x)$ is relatively prime to $k(x)$ and $f(x)k(x) \equiv g(x)k(x)$ (mod $p(x)$), then $f(x) \equiv g(x)$ (mod $p(x)$).

9. Prove that $f(x) \equiv g(x)$ (mod $p(x)$) if and only if $f(x)$ and $g(x)$ leave the same remainder when divided by $p(x)$.

10. Prove or disprove: If $p(x)$ is irreducible in $F[x]$ and $f(x)g(x) \equiv 0_F$ (mod $p(x)$), then $f(x) \equiv 0_F$ (mod $p(x)$) or $g(x) \equiv 0_F$ (mod $p(x)$).

11. If $p(x)$ is not irreducible in $F[x]$, prove that there exist $f(x), g(x) \in F[x]$ such that $f(x) \not\equiv 0_F$ (mod $p(x)$) and $g(x) \not\equiv 0_F$ (mod $p(x)$) but $f(x)g(x) \equiv 0_F$ (mod $p(x)$).

12. If $f(x)$ is relatively prime to $p(x)$, prove that there is a polynomial $g(x) \in F[x]$ such that $f(x)g(x) \equiv 1_F$ (mod $p(x)$).

13. Suppose $f(x), g(x) \in \mathbb{R}[x]$ and $f(x) \equiv g(x)$ (mod x). What can be said about the graphs of $y = f(x)$ and $y = g(x)$?

5.2 CONGRUENCE-CLASS ARITHMETIC

Congruence in the integers led to the rings $\mathbb{Z}_n$. Similarly, congruence in $F[x]$ also produces new rings and fields. These turn out to be much richer in structure than the rings $\mathbb{Z}_n$. The development here closely parallels Section 2.2.

THEOREM 5.6 *Let F be a field and p(x) a nonconstant polynomial in F[x]. If* $[f(x)] = [g(x)]$ *and* $[h(x)] = [k(x)]$ *in* $F[x]/(p(x))$, *then,*

$$[f(x) + h(x)] = [g(x) + k(x)] \qquad and \qquad [f(x)h(x)] = [g(x)k(x)].$$

Proof Copy the proof of Theorem 2.6. ◆

Because of Theorem 5.6 we can now define addition and multiplication of congruence classes just as we did in the integers and be certain that these operations are independent of the choice of representatives in each congruence class.

> **DEFINITION•** *Let F be a field and p(x) a nonconstant polynomial in F[x]. Addition and multiplication in F[x]/(p(x)) are defined by*
>
> $$[f(x)] + [g(x)] = [f(x) + g(x)],$$
>
> $$[f(x)][g(x)] = [f(x)g(x)].$$

EXAMPLE Consider congruence modulo $x^2 + 1$ in $\mathbb{R}[x]$. The sum of the classes $[2x + 1]$ and $[3x + 5]$ is the class

$$[(2x + 1) + (3x + 5)] = [5x + 6].$$

The product is

$$[2x + 1][3x + 5] = [(2x + 1)(3x + 5)] = [6x^2 + 13x + 5].$$

As noted in the first example on page 122, every congruence class in $\mathbb{R}[x]/(x^2 + 1)$ can be written in the form $[ax + b]$. To express the class $[6x^2 + 13x + 5]$ in this form, we divide $6x^2 + 13x + 5$ by $x^2 + 1$ and find that

$$6x^2 + 13x + 5 = 6(x^2 + 1) + (13x - 1).$$

It follows that $6x^2 + 13x + 5 \equiv 13x - 1 \pmod{x^2 + 1}$, and hence $[6x^2 + 13x + 5] = [13x - 1]$.

EXAMPLE In an example on page 122, we saw that $\mathbb{Z}_2[x]/(x^2 + x + 1)$ consists of four classes: $[0]$, $[1]$, $[x]$, and $[x + 1]$. Using the definition of addition of classes, we see that $[x + 1] + [1] = [x + 1 + 1] = [x]$ (remember that $1 + 1 = 0$ in $\mathbb{Z}_2$). Similar calculations produce the following addition table for $\mathbb{Z}_2[x]/(x^2 + x + 1)$:

+	[0]	[1]	[x]	[x + 1]
[0]	[0]	[1]	[x]	[x + 1]
[1]	[1]	[0]	[x + 1]	[x]
[x]	[x]	[x + 1]	[0]	[1]
[x + 1]	[x + 1]	[x]	[1]	[0]

Most of the multiplication table for $\mathbb{Z}_2[x]/(x^2 + x + 1)$ is easily obtained from the definition:

·	[0]	[1]	[x]	[x + 1]
[0]	[0]	[0]	[0]	[0]
[1]	[0]	[1]	[x]	[x + 1]
[x]	[0]	[x]		
[x + 1]	[0]	[x + 1]		

To fill in the rest of the table, note, for example, that

$$[x] \cdot [x + 1] = [x(x + 1)] = [x^2 + x].$$

Now long division or simple addition in $\mathbb{Z}_2[x]$ shows that $x^2 + x = (x^2 + x + 1) + 1$. Therefore, $x^2 + x \equiv 1 \pmod{x^2 + x + 1}$, so that $[x^2 + x] = [1]$. A similar calculation shows that $[x] \cdot [x] = [x^2] = [x + 1]$ (because $x^2 = (x^2 + x + 1) + (x + 1)$ in $\mathbb{Z}_2[x]$).

If you examine the tables in the preceding example, you will see that $\mathbb{Z}_2[x]/(x^2 + x + 1)$ is a commutative ring with identity (in fact, a field). In view of our experience with $\mathbb{Z}$ and $\mathbb{Z}_n$, this is not too surprising. What is unexpected is the upper left-hand corners of the two tables (the sums and products of [0] and [1]). It is easy to see that the subset $F^* = \{[0], [1]\}$ is actually a *subring* of $\mathbb{Z}_2[x]/(x^2 + x + 1)$ and that F^* is isomorphic to $\mathbb{Z}_2$ (the tables for the two systems are identical except for the brackets in F^*). These facts illustrate the next theorem.

THEOREM 5.7 *Let F be a field and p(x) a nonconstant polynomial in F[x]. Then the set F[x] / (p(x)) of congruence classes modulo p(x) is a commutative ring with identity. Furthermore, F[x] / (p(x)) contains a subring F* that is isomorphic to F.*

Proof To prove that $F[x]/(p(x))$ is a commutative ring with identity, adapt the proof of Theorem 2.7 to the present case. Let F^* be the subset of $F[x]/(p(x))$ consisting of the congruence classes of all the constant polynomials; that is, $F^* = \{[a] \mid a \in F\}$. Verify that F^* is a subring of $F[x]/(p(x))$ (Exercise 10). Define a map $\varphi: F \to F^*$ by $\varphi(a) = [a]$. This definition shows that φ is surjective. The definitions of addition and multiplication in $F[x]/(p(x))$ show that

$$\varphi(a + b) = [a + b] = [a] + [b] = \varphi(a) + \varphi(b) \quad \text{and}$$
$$\varphi(ab) = [ab] = [a] \cdot [b] = \varphi(a) \cdot \varphi(b).$$

Therefore, φ is a homomorphism.

To see that φ is injective, suppose $\varphi(a) = \varphi(b)$. Then $[a] = [b]$, so that $a \equiv b \pmod{p(x)}$. Hence, $p(x)$ divides $a - b$. However, $p(x)$ has degree ≥ 1, and $a - b \in F$. This is impossible unless $a - b = 0$. Therefore, $a = b$ and φ is injective. Thus $\varphi: F \to F^*$ is an isomorphism. ◆

We began with a field F and a polynomial $p(x)$ in $F[x]$. We have now constructed a ring $F[x]/(p(x))$ that contains an isomorphic *copy* of F. What we would really like is a ring that contains the field F *itself*. There are two possible ways to accomplish this, as illustrated in the following example.

EXAMPLE In the last example on page 124, we used the polynomial $x^2 + x + 1$ in $\mathbb{Z}_2[x]$ to construct the ring $\mathbb{Z}_2[x]/(x^2 + x + 1)$, which contains a subset $F^* = \{[0],[1]\}$ that is isomorphic to $\mathbb{Z}_2$. Suppose we *identify* $\mathbb{Z}_2$ with its isomorphic copy F^* inside $\mathbb{Z}_2[x]/(x^2 + x + 1)$ and write the elements of F^* as if they were in $\mathbb{Z}_2$. Then the tables on pages 124–125 become

+	0	1	[x]	[x + 1]
0	0	1	[x]	[x + 1]
1	1	0	[x + 1]	[x]
[x]	[x]	[x + 1]	0	1
[x + 1]	[x + 1]	[x]	1	0

·	0	1	[x]	[x + 1]
0	0	0	0	0
1	0	1	[x]	[x + 1]
[x]	0	[x]	[x + 1]	1
[x + 1]	0	[x + 1]	1	[x]

We now have a ring that has $\mathbb{Z}_2$ as a subset. If this procedure makes you a bit uneasy (is $\mathbb{Z}_2$ really a subset?), you can use the following alternate route to the same end. Let E be any four-element set that actually contains $\mathbb{Z}_2$ as a subset, say $E = \{0, 1, r, s\}$. Define addition and multiplication in E by

+	0	1	r	s
0	0	1	r	s
1	1	0	s	r
r	r	s	0	1
s	s	r	1	0

·	0	1	r	s
0	0	0	0	0
1	0	1	r	s
r	0	r	s	1
s	0	s	1	r

A comparison of the tables for $\mathbb{Z}_2[x]/(x^2 + x + 1)$ and those for E shows that these two rings are isomorphic (replacing $[x]$ by r and $[x + 1]$ by s changes one set of tables into the other). Therefore, E is essentially the same ring we obtained before. However, E does contain $\mathbb{Z}_2$ as an honest-to-goodness subset, without any identification.

What was done in the preceding example can be done in the general case. Given a field F and a polynomial $p(x)$ in $F[x]$, we can construct a ring that contains F as a subset. The customary way to do this is to identify F with its isomorphic copy F^* inside $F[x]/(p(x))$ and to consider F to be a subset of $F[x]/(p(x))$. If doing this makes you uncomfortable, keep in mind that you can always build a ring isomorphic to $F[x]/(p(x))$ that genuinely contains F as a subset, as in the preceding example. Because this latter approach tends to get cumbersome, we shall follow the usual custom and identify F with F^* hereafter. Consequently, when a, $b \in F$, we shall write $b[x]$ instead of $[b][x]$ and $a + b[x]$ instead of $[a] + [b][x] = [a + bx]$. Then Theorem 5.7 can be reworded:

THEOREM 5.8   *Let F be a field and $p(x)$ a nonconstant polynomial in $F[x]$. Then $F[x]/(p(x))$ is a commutative ring with identity that contains F.*

If a and n are integers such that $(a,n) = 1$, then according to Corollary 2.10, the equation $[a]x = [1]$ has a solution in the ring $\mathbb{Z}_n$, that is, $[a]$ is a unit in $\mathbb{Z}_n$. Here is the analogue for polynomials.

THEOREM 5.9   *Let F be a field and $p(x)$ a nonconstant polynomial in $F[x]$. If $f(x) \in F[x]$ and $f(x)$ is relatively prime to $p(x)$, then $[f(x)]$ is a unit in $F[x]/(p(x))$.*

Proof   By Theorem 4.5 there are polynomials $u(x)$ and $v(x)$ such that $f(x)u(x) + p(x)v(x) = 1$. Hence, $f(x)u(x) - 1 = p(x)v(x)$, which implies that $[f(x)u(x)] = [1]$ by Theorem 5.3. Therefore, $[f(x)][u(x)] = [f(x)u(x)] = [1]$, so that $[f(x)]$ is a unit in $F[x]/(p(x))$.   ◆

EXAMPLE   Since $x^2 - 2$ is irreducible in $\mathbb{Q}[x]$, $2x + 5$ and $x^2 - 2$ are relatively prime in $\mathbb{Q}[x]$. (Why?) Hence, $[2x + 5]$ is a unit in the ring $\mathbb{Q}[x]/(x^2 - 2)$. The proof of Theorem 5.9 shows that its inverse is $[u(x)]$, where $(2x + 5)u(x) + (x^2 - 2)v(x) = 1$. Using the Euclidean Algorithm as in the integer example on pages 11–12, we find that

$$(2x + 5)\left(-\frac{2}{17}x + \frac{5}{17}\right) + (x^2 - 2)\left(\frac{4}{17}\right) = 1.$$

Therefore, $[-\frac{2}{17}x + \frac{5}{17}]$ is the inverse of $[2x + 5]$ in $\mathbb{Q}[x]/(x^2 - 2)$.

◆ **EXERCISES**

A. In Exercises 1–4, write out the addition and multiplication tables for the congruence-class ring $F[x]/(p(x))$. In each case, is $F[x]/(p(x))$ a field?

    1. $F = \mathbb{Z}_2; p(x) = x^3 + x + 1$ **2.** $F = \mathbb{Z}_3; p(x) = x^2 + 1$

    3. $F = \mathbb{Z}_2; p(x) = x^2 + 1$ **4.** $F = \mathbb{Z}_5; p(x) = x^2 + 1$

B. In Exercises 5–8, each element of the given congruence-class ring can be written in the form $[ax + b]$ (Why?). Determine the rules for addition and multiplication of congruence classes. (In other words, if the product $[ax + b][cx + d]$ is the class $[rx + s]$, describe how to find r and s from a, b, c, d, and similarly for addition.)

    5. $\mathbb{R}[x]/(x^2 + 1)$ [*Hint:* See the first example on page 124.]

    6. $\mathbb{Q}[x]/(x^2 - 2)$ **7.** $\mathbb{Q}[x]/(x^2 - 3)$ **8.** $\mathbb{Q}[x]/(x^2)$

    9. Show that $\mathbb{R}[x]/(x^2 + 1)$ is a field by verifying that every nonzero congruence class $[ax + b]$ is a unit. [*Hint:* Show that the inverse of $[ax + b]$ is $[cx + d]$, where $c = -a/(a^2 + b^2)$ and $d = b/(a^2 + b^2)$.]

    10. Let F be a field and $p(x) \in F[x]$. Prove that $F^* = \{[a] \mid a \in F\}$ is a subring of $F[x]/(p(x))$.

    11. Show that the ring in Exercise 8 is *not* a field.

    12. Write out a complete proof of Theorem 5.6 (that is, carry over to $F[x]$ the proof of the analogous facts for $\mathbb{Z}$).

    13. Prove the first statement of Theorem 5.7.

    14. In each part explain why $[f(x)]$ is a unit in $F[x]/(p(x))$ and find its inverse.

        (a) $[f(x)] = [2x - 3] \in \mathbb{Q}[x]/(x^2 - 2)$

        (b) $[f(x)] = [x^2 + x + 1] \in \mathbb{Z}_3[x]/(x^2 + 1)$

        (c) $[f(x) = [x^2 + x + 1] \in \mathbb{Z}_2[x]/(x^3 + x + 1)$

C. 15. Find a fourth-degree polynomial in $\mathbb{Z}_2[x]$ whose roots are the four elements of the field $\mathbb{Z}_2[x]/(x^2 + x + 1)$, whose tables are given on page 126. [*Hint:* The Factor Theorem may be helpful.]

    16. Show that $\mathbb{Q}[x]/(x^2 - 2)$ is a field.

5.3 THE STRUCTURE OF $F[x]/(p(x))$ WHEN $p(x)$ IS IRREDUCIBLE

When p is a prime integer, $\mathbb{Z}_p$ is a field (and, of course, an integral domain). You can probably guess the analogous result for $F[x]$ (Theorem 5.10). But there is much more going on here than just a parallel of what was done in $\mathbb{Z}$. We have already seen that $F[x]/(p(x))$ contains the field F. In this section we see that it also contains a *root* of $p(x)$.

THEOREM 5.10   *Let F be a field and $p(x)$ a nonconstant polynomial in $F[x]$. Then the following statements are equivalent:*

(1) $p(x)$ is irreducible in $F[x]$.

(2) $F[x]/(p(x))$ is a field.

(3) $F[x]/(p(x))$ is an integral domain.

Proof   Adapt the proof of Theorem 2.8 by replacing $\mathbb{Z}$ by $F[x]$, $\mathbb{Z}_p$ by $F[x]/(p(x))$, Theorem 1.3 by Theorem 4.5, and Theorem 2.3 by Theorem 5.3. The following observations should be helpful.

(1) $\Rightarrow$ (2) Note that when $p(x)$ is irreducible, the gcd of $a(x)$ and $p(x)$ is either 1_F or a monic associate of $p(x)$. If $a(x)$ is not divisible by $p(x)$, then it is not divisible by any associate of $p(x)$. Therefore, if $p(x) \nmid a(x)$, then their gcd must be 1_F.

(2) $\Rightarrow$ (3) Never mind Theorem 2.8 here; this is an immediate consequence of Theorem 3.9.

(3) $\Rightarrow$ (1) Note that if $p(x) = a(x)b(x)$ and $w(x)b(x) = 1_F$, then $b(x)$ is a constant, so that $a(x)$ is an associate of $p(x)$. This is the analogue of saying $a = \pm p$ in $\mathbb{Z}$. $\blacklozenge$

Theorem 5.10 can be used to construct finite fields. If p is prime and $f(x)$ is irreducible in $\mathbb{Z}_p[x]$ of degree k, then $\mathbb{Z}_p[x]/(f(x))$ is a field by Theorem 5.10. The last example on page 122 shows that this field has p^k elements. Finite fields are discussed further in Section 10.6, where it is shown that there are irreducible polynomials of every positive degree in $\mathbb{Z}_p[x]$ and, hence, finite fields of all possible prime power orders. See Exercise 9 for an example.

Let F be a field and $p(x)$ an irreducible polynomial in $F[x]$. Let K denote the field of congruence classes $F[x]/(p(x))$. By Theorems 5.8 and 5.10, F is a subfield of the field K. One also says that K is an **extension field** of F. Polynomials in $F[x]$ can be considered to have coefficients in the larger field K, and we can ask about the roots of such polynomials in K. In particular, what can be said about the roots of the polynomial $p(x)$ that we started with? Even though $p(x)$ is irreducible in $F[x]$, it may have roots in the extension field K.

> **EXAMPLE**   The polynomial $p(x) = x^2 + x + 1$ has no roots in $\mathbb{Z}_2$ and is, therefore, irreducible in $\mathbb{Z}_2[x]$ by Corollary 4.18. Consequently, $K = \mathbb{Z}_2[x]/(x^2 + x + 1)$ is an extension field of $\mathbb{Z}_2$ by Theorem 5.10. Using the tables for K on page 126, we see that
>
> $$[x]^2 + [x] + 1 = [x + 1] + [x] + 1 = 1 + 1 = 0.$$

This result may be a little easier to absorb if we use a different notation. Let $\alpha = [x]$. Then the calculation above says that $\alpha^2 + \alpha + 1 = 0$; that is, α is a root in K of $p(x) = x^2 + x + 1$. It's important to note here that you don't really need the tables for K to prove that α is a root of $p(x)$ because we know that $x^2 + x + 1 \equiv 0 \pmod{x^2 + x + 1}$. Consequently, $[x^2 + x + 1] = 0$ in K, and by the definition of congruence-class arithmetic,

$$\alpha^2 + \alpha + 1 = [x]^2 + [x] + 1 = [x^2 + x + 1] = 0.$$

For the general case we have

THEOREM 5.11   *Let F be a field and p(x) an irreducible polynomial in F[x]. Then F[x]/(p(x)) is an extension field of F that contains a root of p(x).*

Proof   Let $K = F[x]/(p(x))$. Then K is an extension field of F by Theorems 5.8 and 5.10. Let $p(x) = a_n x^n + \cdots + a_1 x + a_0$, where each a_i is in F and, hence, in K. Let $\alpha = [x]$ in K. We shall show that α is a root of $p(x)$. By the definition of congruence-class arithmetic in K,

$$a_n \alpha^n + \cdots + a_1 \alpha + a_0 = a_n [x]^n + \cdots + a_1 [x] + a_0$$
$$= [a_n x^n + \cdots + a_1 x + a_0]$$
$$= [p(x)] = 0_F \qquad (\text{because } p(x) \equiv 0_F \pmod{p(x)}).$$

Therefore, $\alpha \in K$ is a root of $p(x)$.   ◆

COROLLARY 5.12   *Let F be a field and f(x) a nonconstant polynomial in F[x]. Then there is an extension field K of F that contains a root of f(x).*

Proof   By Theorem 4.13, $f(x)$ has an irreducible factor $p(x)$ in $F[x]$. By Theorem 5.11, $K = F[x]/(p(x))$ is an extension field of F that contains a root of $p(x)$. Since every root of $p(x)$ is a root of $f(x)$, K contains a root of $f(x)$.   ◆

The implications of Theorem 5.11 run much deeper than might first appear. Throughout the history of mathematics, the passage from a known number system to a new, larger system has often been greeted with doubt and distrust. In the Middle Ages, some mathematicians refused to acknowledge the existence of negative numbers. When complex numbers were introduced in the seventeenth century, there was uneasiness—which extended into the last century—because some mathematicians would not accept the idea that there could be a number whose square is -1, that is, a root of $x^2 + 1$. One cause for these difficulties was the lack of a suitable framework in which to view the situation. Abstract algebra provides such a framework. Theorem 5.11 and its corollary, then, take care of the doubt and uncertainty.

It is instructive to consider the complex numbers from this point of view. Instead of asking about a *number* whose square is -1, we ask, "Is there a field containing $\mathbb{R}$ in which the polynomial $x^2 + 1$ has a root?" Since $x^2 + 1$ is irreducible in $\mathbb{R}[x]$, Theorem 5.11 tells us that the answer is yes: $K = \mathbb{R}[x]/(x^2 + 1)$ is an extension field of $\mathbb{R}$ that contains a root of $x^2 + 1$, namely $\alpha = [x]$. In the field K, α is an element whose square is -1. But how is the field K related to the field of complex numbers introduced earlier in the book?

As is noted in the first example on page 122, every element of $K = \mathbb{R}[x]/(x^2 + 1)$ can be written uniquely in the form $[ax + b]$ with $a, b \in \mathbb{R}$. Since we are identifying each element $r \in \mathbb{R}$ with the element $[r]$ in K, we see that every element of K can be written uniquely in the form

$$[a + bx] = [a] + [b][x] = a + b\alpha.$$

Addition in K is given by the rule

$$(a + b\alpha) + (c + d\alpha) = [a + bx] + [c + dx] = [(a + bx) + (c + dx)]$$
$$= [(a + c) + (b + d)x] = [a + c] + [b + d][x].$$

so that

$$(a + b\alpha) + (c + d\alpha) = (a + c) + (b + d)\alpha.$$

Multiplication in K is given by the rule

$$(a + b\alpha)(c + d\alpha) = [a + bx][c + dx] = [(a + bx)(c + dx)]$$
$$= [ac + (ad + bc)x + bdx^2]$$
$$= ac + (ad + bc)\alpha + bd\alpha^2.$$

However, α is a root of $x^2 + 1$, and so $\alpha^2 = -1$. Therefore, the rule for multiplication in K becomes

$$(a + b\alpha)(c + d\alpha) = (ac - bd) + (ad + bc)\alpha.$$

If the symbol α is replaced by the symbol i, then these rules become the usual rules for adding and multiplying complex numbers. In formal language, the field K is isomorphic to the field $\mathbb{C}$, with the isomorphism f being given by $f(a + b\alpha) = a + bi$.

Up to now we have taken the position that the field $\mathbb{C}$ of complex numbers was already known. The field K constructed above then turns out to be isomorphic to the known field $\mathbb{C}$. A good case can be made, however, for not assuming any previous knowledge of the complex numbers and using the preceding example as a *definition* instead. In other words, we can define $\mathbb{C}$ to be the field $\mathbb{R}[x]/(x^2 + 1)$. Such a definition is obviously too sophisticated to use on high-school students, but for mature students it has the definite advantage of removing any lingering doubts about the validity of the complex numbers and their arithmetic.* Had this definition been available several centuries ago, the introduction of the complex numbers might have caused no stir whatsoever.

◆ **EXERCISES**

NOTE: *F always denotes a field.*

A. **1.** Determine whether the given congruence-class ring is a field.

   (a) $\mathbb{Z}_3[x]/(x^3 + 2x^2 + x + 1)$

   (b) $\mathbb{Z}_5[x]/(2x^3 - 4x^2 + 2x + 1)$

   (c) $\mathbb{Z}_2[x]/(x^4 + x^2 + 1)$

B. **2. (a)** Verify that $\mathbb{Q}(\sqrt{2}) = \{r + s\sqrt{2} \mid r, s \in \mathbb{Q}\}$ is a subfield of $\mathbb{R}$.

   (b) Show that $\mathbb{Q}(\sqrt{2})$ is isomorphic to $\mathbb{Q}[x]/(x^2 - 2)$. [*Hint:* Exercise 6 in Section 5.2 may be helpful.]

3. If $a \in F$, describe the field $F[x]/(x - a)$.

4. Let $p(x)$ be irreducible in $F[x]$. Without using Theorem 5.10, prove that if $[f(x)][g(x)] = [0_F]$ in $F[x]/(p(x))$, then $[f(x)] = [0_F]$ or $[g(x)] = [0_F]$. [*Hint:* Exercise 10 in Section 5.1.]

5. (a) Verify that $\mathbb{Q}(\sqrt{3}) = \{r + s\sqrt{3} \mid r, s \in \mathbb{Q}\}$ is a subfield of $\mathbb{R}$.

   (b) Show that $\mathbb{Q}(\sqrt{3})$ is isomorphic to $\mathbb{Q}[x]/(x^2 - 3)$.

6. Let $p(x)$ be irreducible in $F[x]$. If $[f(x)] \neq [0_F]$ in $F[x]/(p(x))$ and $h(x) \in F[x]$, prove that there exists $g(x) \in F[x]$ such that $[f(x)][g(x)] = [h(x)]$ in $F[x]/(p(x))$. [*Hint:* Corollary 2.9.]

7. If $f(x) \in F[x]$ has degree n, prove that there exists an extension field E of F such that $f(x) = c_0(x - c_1)(x - c_2) \cdots (x - c_n)$ for some (not necessarily distinct) $c_i \in E$. In other words, E contains all the roots of $f(x)$.

8. If $p(x)$ is an irreducible quadratic polynomial in $F[x]$, show that $F[x]/(p(x))$ contains all the roots of $p(x)$.

9. (a) Show that $\mathbb{Z}_2[x]/(x^3 + x + 1)$ is a field.

   (b) Show that the field $\mathbb{Z}_2[x]/(x^3 + x + 1)$ contains all three roots of $x^3 + x + 1$.

---

* Only a minor rearrangement of this book is needed to accommodate such a definition. A few examples in Chapter 3 would have to be omitted, and the discussion of irreducibility in $\mathbb{C}[x]$ and $\mathbb{R}[x]$ (Section 4.6) would have to be postponed. All the intervening material in Chapter 5 is independent of any formal knowledge of the complex numbers.

10. Show that $\mathbb{Q}[x]/(x^2 - 2)$ is not isomorphic to $\mathbb{Q}[x]/(x^2 - 3)$. [*Hint:* Exercises 2 and 5 may be helpful.]

11. Let K be a ring that contains $\mathbb{Z}_6$ as a subring. Show that $p(x) = 3x^2 + 1 \in \mathbb{Z}_6[x]$ has no roots in K. Thus, Corollary 5.12 may be false if F is not a field. [*Hint:* If u were a root, then $0 = 2 \cdot 0$ and $3u^2 + 1 = 0$. Derive a contradiction.]

12. Show that $2x^3 + 4x^2 + 8x + 3 \in \mathbb{Z}_{16}[x]$ has no roots in any ring K that contains $\mathbb{Z}_{16}$ as a subring. [See Exercise 11.]

C. 13. Show that every polynomial of degree 1, 2, or 4 in $\mathbb{Z}_2[x]$ has a root in $\mathbb{Z}_2[x]/(x^4 + x + 1)$.

CHAPTER 6

Ideals and Quotient Rings

◆

Congruence in the integers led us to the finite arithmetics $\mathbb{Z}_n$ and helped motivate the definition of a ring. Congruence in the polynomial ring $F[x]$ resulted in a new class of rings consisting of the various $F[x]/(p(x))$. These rings enabled us to construct extension fields of F that contained roots of the polynomial $p(x)$. In this chapter the concept of congruence is extended to arbitrary rings, producing additional rings and a deeper understanding of algebraic structure.

You will see that much of the discussion is an exact parallel of the development of congruence in $\mathbb{Z}$ (Chapter 2) and in $F[x]$ (Chapter 5). Nevertheless, the results here are considerably broader than the earlier ones.

6.1 IDEALS AND CONGRUENCE

Our goal is to develop a notion of congruence in arbitrary rings that includes as special cases congruence modulo n in $\mathbb{Z}$ and congruence modulo $p(x)$ in $F[x]$. We begin by taking a second look at some examples of congruence in $\mathbb{Z}$ and $F[x]$ from a somewhat different viewpoint than before.

> **EXAMPLE** In the ring $\mathbb{Z}$, $a \equiv b \pmod 3$ means that $a - b$ is a multiple of 3. Let I be the set of all multiples of 3, so that
>
> $$I = \{0, \pm 3, \pm 6, \ldots \}.$$
>
> Then congruence modulo 3 may be characterized like this:
>
> $$a \equiv b \pmod 3 \quad \text{means} \quad a - b \in I.$$

Observe that the subset I is actually a *subring* of $\mathbb{Z}$ (sums and products of multiples of 3 are also multiples of 3). Furthermore, the product of any integer and a multiple of 3 is itself a multiple of 3. Thus the subring I has this property:

Whenever $k \in \mathbb{Z}$ and $i \in I$, then $ki \in I$.

EXAMPLE   The notation $f(x) \equiv g(x) \pmod{x^2 - 2}$ in the polynomial ring $\mathbb{Q}[x]$ means that $f(x) - g(x)$ is a multiple of $x^2 - 2$. Let I be the set of all multiples of $x^2 - 2$ in $\mathbb{Q}[x]$, that is, $I = \{h(x)(x^2 - 2) \mid h(x) \in \mathbb{Q}[x]\}$. Once again, it is not difficult to check that I is a subring of $\mathbb{Q}[x]$ with this property:

Whenever $k(x) \in \mathbb{Q}[x]$ and $t(x) \in I$, then $k(x)t(x) \in I$

(the product of any polynomial with a multiple of $x^2 - 2$ is itself a multiple of $x^2 - 2$). Congruence modulo $x^2 - 2$ may be described in terms of I:

$$f(x) \equiv g(x) \pmod{x^2 - 2} \qquad \text{means} \qquad f(x) - g(x) \in I.$$

These examples suggest that congruence in a ring R might be defined in terms of certain subrings. If I were such a subring, we might define $a \equiv b \pmod{I}$ to mean $a - b \in I$. The subring I might consist of all multiples of a fixed element, as in the preceding examples, but there is no reason for restricting to this situation. The examples indicate that the key property for such a subring I is that it "absorb products": Whenever you multiply an element of I by *any* element of the ring (either inside or outside I), the resulting product is an element of I. The set of all multiples of a fixed element has this absorption property. We shall see that many other subrings have it as well. Because such subrings play a crucial role in what follows, we pause to give them a name and to consider their basic properties.

DEFINITION• *A subring I of a ring R is an **ideal** provided:*

Whenever $r \in R$ and $a \in I$, then $ra \in I$ and $ar \in I$.

The double absorption condition that $ra \in I$ and $ar \in I$ is necessary for noncommutative rings. When R is commutative, as in the preceding examples, this condition reduces to $ra \in I$.

EXAMPLE   The zero ideal in a ring R consists of the single element 0_R. This is a subring that absorbs all products since $r0_R = 0_R = 0_R r$ for every $r \in R$. The entire ring R is also an ideal.

EXAMPLE   In the ring $\mathbb{Z}[x]$ of all polynomials with integer coefficients, let I be the set of polynomials whose constant terms are even integers. Thus $x^3 + x + 6$ is in I, but $4x^2 + 3$ is not. Verify that I is an ideal in $\mathbb{Z}[x]$ (Exercise 2).

EXAMPLE   Let T be the ring of all continuous functions from $\mathbb{R}$ to $\mathbb{R}$, as described on page 46. Let I be the subset consisting of those functions g such that $g(2) = 0$. Then I is a subring of T (Exercise 12 of Section 3.1). If f is any function in T and if $g \in I$, then

$$(fg)(2) = f(2)g(2) = f(2) \cdot 0 = 0.$$

Therefore, $fg \in I$. Similarly, $gf \in I$, so that I is an ideal in T.

EXAMPLE   The subring $\mathbb{Z}$ of the rational numbers is *not* an ideal in $\mathbb{Q}$ because $\mathbb{Z}$ fails to have the absorption property. For instance, $\frac{1}{2} \in \mathbb{Q}$ and $5 \in \mathbb{Z}$, but their product, $\frac{5}{2}$, is not in $\mathbb{Z}$.

EXAMPLE   Verify that the set I of all matrices of the form $\begin{pmatrix} a & 0 \\ b & 0 \end{pmatrix}$ with $a, b \in \mathbb{R}$ forms a subring of the ring $M(\mathbb{R})$ of all 2×2 matrices over the reals. It is easy to see that I absorbs products on the *left*:

$$\begin{pmatrix} r & s \\ t & u \end{pmatrix}\begin{pmatrix} a & 0 \\ b & 0 \end{pmatrix} = \begin{pmatrix} ra + sb & 0 \\ ta + ub & 0 \end{pmatrix} \in I.$$

But I is not an ideal in $M(\mathbb{R})$ because it may not absorb products on the right—for instance,

$$\begin{pmatrix} 1 & 0 \\ 2 & 0 \end{pmatrix}\begin{pmatrix} 3 & 4 \\ 5 & 6 \end{pmatrix} = \begin{pmatrix} 3 & 4 \\ 6 & 8 \end{pmatrix} \notin I.$$

One sometimes says that I is a **left ideal,** but not a two-sided ideal, in $M(\mathbb{R})$.

The following generalization of Theorem 3.6 often simplifies the verification that a particular subset of a ring is an ideal.

THEOREM 6.1   *A nonempty subset I of a ring R is an ideal if and only if it has these properties:*

(i) *if $a, b \in I$, then $a - b \in I$;*

(ii) *if $r \in R$ and $a \in I$, then $ra \in I$ and $ar \in I$.*

Proof   Every ideal certainly has these two properties. Conversely, suppose I has properties (i) and (ii). Then I absorbs products by (ii), so we need only verify that I is a subring. Property (i) states that I is closed under subtraction. Since I is a subset of R, the product of any two elements of I must be in I by (ii). In other words, I is closed under multiplication. Therefore, I is a subring of R by Theorem 3.6. ◆

Finitely Generated Ideals

In the first example of this section we saw that the set I of all multiples of 3 is an ideal in $\mathbb{Z}$. This fact is a special case of

THEOREM 6.2 *Let R be a commutative ring with identity, $c \in R$, and I the set of all multiples of c in R, that is, $I = \{rc \mid r \in R\}$. Then I is an ideal.*

Proof If $r_1, r_2, r \in R$ and $r_1c, r_2c \in I$, then

$$r_1c - r_2c = (r_1 - r_2)c \in I \qquad \text{and} \qquad r(r_1c) = (rr_1)c \in I$$

because $r_1 - r_2$ and rr_1 are elements of R. Similarly, since R is commutative, $(r_1c)r = (rr_1)c \in I$. Therefore, I is an ideal by Theorem 6.1. ◆

The ideal I in Theorem 6.2 is called the **principal ideal generated by c** and hereafter will be denoted by (c). In the ring $\mathbb{Z}$, for example, (3) indicates the ideal of all multiples of 3. In any commutative ring R with identity, the principal ideal (1_R) is the entire ring R because $r = r1_R$ for every $r \in R$. It can be shown that every ideal in $\mathbb{Z}$ is a principal ideal (Exercise 38). However, there are ideals in other rings that are not principal, that is, ideals that do not consist of all the multiples of a particular element of the ring.

> **EXAMPLE** We have seen that the set I of all polynomials with even constant terms is an ideal in the ring $\mathbb{Z}[x]$. We claim that I is *not* a principal ideal. To prove this, suppose, on the contrary, that I consists of all multiples of some polynomial $p(x)$. Since the constant polynomial 2 is in I, 2 must be a multiple of $p(x)$. By Theorem 4.2, this is possible only if $p(x)$ has degree 0, that is, if $p(x)$ is a constant, say $p(x) = c$. Since $p(x) \in I$, the constant c must be an even integer. Since 2 is a multiple of $p(x) = c$, the only possibility is $c = \pm 2$. On the other hand, $x \in I$ because it has even constant term 0. Therefore, x must be a multiple of $p(x) = \pm 2$. This is impossible since all the polynomials involved have integer coefficients. Therefore, I does not consist of all multiples of $p(x)$ and is not a principal ideal.

In a commutative ring with identity, a principal ideal consists of all multiples of a fixed element. Here is a generalization of that idea.

THEOREM 6.3 *Let R be a commutative ring with identity and c_1, c_2, . . . , $c_n \in R$. Then the set $I = \{r_1c_1 + r_2c_2 + \cdots + r_nc_n \mid r_1,\, r_2,\, \ldots,\, r_n \in R\}$ is an ideal in R.*

Proof Exercise 8.   ◆

The ideal I in Theorem 6.3 is called the **ideal generated by c_1, c_2, . . . , c_n** and is sometimes denoted by $(c_1, c_2,\, \ldots,\, c_n)$. Such an ideal is said to be **finitely generated.** A principal ideal is the special case $n = 1$, that is, an ideal generated by a single element.* The generators of a finitely generated ideal need not be unique, that is, the ideal generated by $c_1, c_2,\, \ldots, c_n$ might be the same set as the ideal generated by $d_1, d_2,\, \ldots, d_k$, even though no c_i is equal to any d_j (Exercise 12).

> **EXAMPLE** In the ring $\mathbb{Z}[x]$, the ideal generated by the polynomial x and the constant polynomial 2 consists of all polynomials of the form
>
> $$f(x)x + g(x)2, \qquad \text{with } f(x), g(x) \in Z[x].$$
>
> It can be shown that this ideal is the ideal I of all polynomials with even constant term, which was discussed in the preceding example (Exercise 11).

Congruence

Now that you are familiar with ideals, we can define congruence in an arbitrary ring:

> **DEFINITION•** *Let I be an ideal in a ring R and let a, $b \in R$. Then **a is congruent to b modulo I** [written $a \equiv b \;(mod\; I)$] provided that $a - b \in I$.*

The example on page 134 shows that congruence modulo 3 in the integers is the same thing as congruence modulo the ideal I, where I is the principal ideal (3) of all multiples of 3. Similarly, the example on page 135 shows that congruence modulo $x^2 - 2$ in $\mathbb{Q}[x]$ is the same as congruence modulo the principal ideal $(x^2 - 2)$. Thus congruence modulo an ideal includes as a special case the concepts of congruence in $\mathbb{Z}$ and $F[x]$ used earlier in this book.

---

* When a commutative ring does not have an identity, the ideal generated by $c_1, c_2,\, \ldots, c_n$ is defined somewhat differently (see Exercise 31).

> **EXAMPLE**   Let T be the ring of all continuous functions from $\mathbb{R}$ to $\mathbb{R}$ and let I be the ideal of all functions g such that $g(2) = 0$. If $f(x) = x^2 + 6$ and $h(x) = 5x$, then the function $f - h$ is in I because
>
> $$(f - h)(2) = f(2) - h(2) = (2^2 + 6) - (5 \cdot 2) = 0.$$
>
> Therefore, $f \equiv h \pmod{I}$.

THEOREM 6.4   *Let I be an ideal in a ring R. Then the relation of congruence modulo I is*

(1) *reflexive: $a \equiv a \pmod{I}$ for every $a \in R$;*

(2) *symmetric: if $a \equiv b \pmod{I}$, then $b \equiv a \pmod{I}$;*

(3) *transitive: if $a \equiv b \pmod{I}$ and $b \equiv c \pmod{I}$, then $a \equiv c \pmod{I}$.*

This theorem generalizes Theorems 2.1 and 5.1. Observe that the proof is virtually identical to that of Theorem 2.1—just replace "divisible by n" with "is an element of I."

Proof of Theorem 6.4   (1) $a - a = 0_R \in I$; hence, $a \equiv a \pmod{I}$.

(2) $a \equiv b \pmod{I}$ means that $a - b = i$ for some $i \in I$. Therefore, $b - a = -(a - b) = -i$. Since I is an ideal, the negative of an element of I is also in I, and so $b - a = -i \in I$. Hence, $b \equiv a \pmod{I}$.

(3) If $a \equiv b \pmod{I}$ and $b \equiv c \pmod{I}$, then by the definition of congruence, there are elements i and j in I such that $a - b = i$ and $b - c = j$. Therefore, $a - c = (a - b) + (b - c) = i + j$. Since the ideal I is closed under addition, $i + j \in I$ and, hence, $a \equiv c \pmod{I}$.   ◆

THEOREM 6.5   *Let I be an ideal in a ring R. If $a \equiv b \pmod{I}$ and $c \equiv d \pmod{I}$, then*

(1) $a + c \equiv b + d \pmod{I}$;

(2) $ac \equiv bd \pmod{I}$.

This theorem generalizes Theorems 2.2 and 5.2. Its proof is quite similar to theirs once you make the change to the language of ideals.

Proof of Theorem 6.5   (1) By the definition of congruence, there are $i, j \in I$ such that $a - b = i$ and $c - d = j$. Therefore, $(a + c) - (b + d) = (a - b) + (c - d) = i + j \in I$. Hence, $a + c \equiv b + d \pmod{I}$.

(2) $ac - bd = ac - bc + bc - bd = (a - b)c + b(c - d) = ic + bj$. Since the ideal I absorbs products on both left and right, $ic \in I$ and $bj \in I$. Hence, $ac - bd = ic + bj \in I$. Therefore, $ac \equiv bd \pmod{I}$.   ◆

If I is an ideal in a ring R and $a \in R$, then the **congruence class of a modulo I** is the set of all elements of R that are congruent to a modulo I, that is, the set

$$\{b \in R \mid b \equiv a \;(\text{mod } I)\} = \{b \in R \mid b - a \in I\}$$
$$= \{b \in R \mid b - a = i, \text{ with } i \in I\}$$
$$= \{b \in R \mid b = a + i, \text{ with } i \in I\}$$
$$= \{a + i \mid i \in I\}.$$

Consequently, we shall denote the congruence class of a modulo I by the symbol $a + I$ rather than the symbol $[a]$ that was used in $\mathbb{Z}$ and $F[x]$. The plus sign in $a + I$ is just a formal symbol; we have not defined the sum of an element and an ideal. In this context, the congruence class $a + I$ is usually called a (left) **coset** of I in R.

THEOREM 6.6   *Let I be an ideal in a ring R and let a, c $\in$ R. Then a $\equiv$ c (mod I) if and only if a + I = c + I.*

Proof With only minor notational changes, the proof of Theorem 2.3 carries over almost verbatim to the present case. Simply replace "mod n" by "mod I" and "$[a]$" by "$a + I$"; use Theorem 6.4 in place of Theorem 2.1. ◆

COROLLARY 6.7   *Let I be an ideal in a ring R. Then two cosets of I are either disjoint or identical.*

Proof Copy the proof of Corollary 2.4 with the obvious notational changes. ◆

If I is an ideal in a ring R, then the set of all cosets of I (congruence classes modulo I) is denoted R/I.

EXAMPLE Let I be the principal ideal (3) in the ring $\mathbb{Z}$. Then the cosets of I are just the congruence classes modulo 3, and so there are three distinct cosets: $0 + I = [0]$, $1 + I = [1]$, and $2 + I = [2]$. The set $\mathbb{Z}/I$ of all cosets is precisely the set $\mathbb{Z}_3$ in our previous notation.

EXAMPLE Let I be the ideal in $\mathbb{Z}[x]$ consisting of all polynomials with even constant terms. We claim that $\mathbb{Z}[x]/I$ consists of exactly two distinct cosets, namely, $0 + I$ and $1 + I$. To see this, consider any coset $f(x) + I$. The constant term of $f(x)$ is either even or odd. If it is even, then $f(x) \in I$, so that $f(x) \equiv 0 \;(\text{mod } I)$. Therefore, $f(x) + I = 0 + I$ by Theorem 6.6. If $f(x)$ has odd constant term, then $f(x) - 1$ has even constant term, so that $f(x) \equiv 1 \;(\text{mod } I)$. Thus $f(x) + I = 1 + I$ by Theorem 6.6.

> **EXAMPLE**   Let T be the ring of continuous functions from $\mathbb{R}$ to $\mathbb{R}$ and let I be the ideal of all functions g such that $g(2) = 0$. Note that for each real number r, the constant function f_r (whose rule is $f_r(x) = r$) is an element of T. Let $h(x)$ be any element of T. Then $h(2)$ is some real number, say $h(2) = c$, and
>
> $$(h - f_c)(2) = h(2) - f_c(2) = c - c = 0.$$
>
> Thus $h - f_c \in I$, so that $h \equiv f_c \pmod{I}$ and, hence, $h + I = f_c + I$. Consequently, every coset of I can be written in the form $f_r + I$ for some real number r. Furthermore, if $c \neq d$, then $f_c(2) \neq f_d(2)$, so that $[f_c - f_d](2) \neq 0$ and $f_c - f_d \notin I$. Hence, $f_c \not\equiv f_d \pmod{I}$ and $f_c + I \neq f_d + I$. Therefore, there are infinitely many distinct cosets of I, one for each real number r.

◆ EXERCISES

NOTE: *R denotes a ring.*

A.   **1.** Show that the set K of all constant polynomials in $\mathbb{Z}[x]$ is a subring but not an ideal in $\mathbb{Z}[x]$.

   2. Show that the set I of all polynomials with even constant terms is an ideal in $\mathbb{Z}[x]$.

   3. (a) Show that the set $I = \{(k,0) \mid k \in \mathbb{Z}\}$ is an ideal in the ring $\mathbb{Z} \times \mathbb{Z}$.

   (b) Show that the set $T = \{(k,k) \mid k \in \mathbb{Z}\}$ is not an ideal in $\mathbb{Z} \times \mathbb{Z}$.

   4. Is the set $J = \left\{ \begin{pmatrix} 0 & 0 \\ 0 & r \end{pmatrix} \mid r \in \mathbb{R} \right\}$ an ideal in the ring $M(\mathbb{R})$ of 2×2 matrices over $\mathbb{R}$?

   5. Show that the set $K = \{\begin{pmatrix} a & b \\ 0 & 0 \end{pmatrix} \mid a, b \in \mathbb{R}\}$ is a subring of $M(\mathbb{R})$ that absorbs products on the right. Show that K is not an ideal because it may fail to absorb products on the left. Such a set K is sometimes called a **right ideal.**

   6. (a) Show that the set of nonunits in $\mathbb{Z}_8$ is an ideal.

   (b) Do part (a) for $\mathbb{Z}_9$. [Also, see Exercise 22.]

   7. List the distinct principal ideals in each ring:

   (a) $\mathbb{Z}_5$ **(b)** $\mathbb{Z}_9$ **(c)** $\mathbb{Z}_{12}$

   8. Prove Theorem 6.3.

   9. Let $c \in R$ and let $I = \{rc \mid r \in R\}$.

   (a) If R is commutative, prove that I is an ideal (that is, Theorem 6.2 is true even when R does not have an identity).

(b) If R is commutative but has no identity, is c an element of the ideal I? [*Hint*: Consider the ideal $\{2k \mid k \in E\}$ in the ring E of even integers. Also see Exercise 31.]

(c) Give an example to show that if R is not commutative, then I need not be an ideal.

10. If I is an ideal in R and J is an ideal in the ring S, prove that $I \times J$ is an ideal in the ring $R \times S$.

11. Show that the ideal generated by x and 2 in the ring $\mathbb{Z}[x]$ is the ideal I of all polynomials with even constant terms (see the Example after Theorem 6.3).

12. (a) Find principal ideals (c) and (d) in the ring $\mathbb{Z}$ such that $(c) = (d)$, but $c \neq d$.

(b) Show that $(4,6) = (2)$ in $\mathbb{Z}$, where $(4,6)$ is the ideal generated by 4 and 6 and (2) is the principal ideal generated by 2.

(c) Show that $(6,9,15) = (3)$ in $\mathbb{Z}$.

13. Let R be a ring with identity and let I be an ideal in R.

(a) If $1_R \in I$, prove that $I = R$.

(b) If I contains a unit, prove that $I = R$.

14. If I is an ideal in a field F, prove that $I = (0_F)$ or $I = F$. [*Hint*: Exercise 13.]

15. (a) If I and J are ideals in R, prove that $I \cap J$ is an ideal.

(b) If $\{I_k\}$ is a (possibly infinite) family of ideals in R, prove that the intersection of all the I_k is an ideal.

16. Give an example in $\mathbb{Z}$ to show that the set theoretic union of two ideals may not be an ideal (in fact, it may not even be a subring).

17. If I is an ideal in R and S is a subring of R, prove that $I \cap S$ is an ideal in S.

18. Let I and J be ideals in R. Prove that the set $K = \{a + b \mid a \in I, b \in J\}$ is an ideal in R that contains both I and J. K is called the **sum** of I and J and is denoted $I + J$.

19. If d is the greatest common divisor of a and b in $\mathbb{Z}$, show that $(a) + (b) = (d)$. (The sum of ideals is defined in Exercise 18.)

20. Let I and J be ideals in R. Is the set $K = \{ab \mid a \in I, b \in J\}$ an ideal in R? Compare Exercise 18.

21. (a) Verify that $I = \{0, 3\}$ is an ideal in $\mathbb{Z}_6$ and list all its distinct cosets.

(b) Verify that $I = \{0, 3, 6, 9, 12\}$ is an ideal in $\mathbb{Z}_{15}$ and list all its distinct cosets.

B. 22. Let R be a commutative ring with identity, and let N be the set of nonunits in R. Give an example to show that N need not be an ideal.

23. Let J be an ideal in R. Prove that I is an ideal, where

$$I = \{r \in R \mid rt = 0_R \text{ for every } t \in J\}.$$

24. Let I be an ideal in R. Prove that K is an ideal, where

$$K = \{a \in R \mid ra \in I \text{ for every } r \in R\}.$$

25. Let $f : R \to S$ be a homomorphism of rings and let

$$K = \{r \in R \mid f(r) = 0_S\}.$$

Prove that K is an ideal in R.

26. If I is an ideal in R, prove that $I[x]$ (polynomials with coefficients in I) is an ideal in the polynomial ring $R[x]$.

27. If $(m, n) = 1$ in $\mathbb{Z}$, prove that $(m) \cap (n)$ is the ideal (mn).

28. Prove that the set of nilpotent elements in a commutative ring R is an ideal. [*Hint:* See Exercise 36 in Section 3.2.]

29. Let R be an integral domain and $a, b \in R$. Show that $(a) = (b)$ if and only if $a = bu$ for some unit $u \in R$.

30. **(a)** Prove that the set J of all polynomials in $\mathbb{Z}[x]$ whose constant terms are divisible by 3 is an ideal.

(b) Show that J is not a principal ideal.

31. Let R be a commutative ring without identity and let $a \in R$. Show that $A = \{ra + na \mid r \in R, n \in \mathbb{Z}\}$ is an ideal containing a and that every ideal containing a also contains A. A is called the **principal ideal generated by a.**

32. If M is an ideal in a commutative ring R with identity and if $a \in R$ with $a \notin M$, prove that the set

$$J = \{m + ra \mid r \in R \text{ and } m \in M\}$$

is an ideal such that $M \subsetneq J$.

33. Let I be an ideal in $\mathbb{Z}$ such that $(3) \subseteq I \subseteq \mathbb{Z}$. Prove that either $I = (3)$ or $I = \mathbb{Z}$.

34. Let I and J be ideals in R. Let IJ denote the set of all possible finite sums of elements of the form ab (with $a \in I$, $b \in J$). Prove that IJ is an ideal.

35. Let R be a commutative ring with identity $1_R \neq 0_R$ whose only ideals are (0_R) and R. Prove that R is a field. [*Hint:* If $a \neq 0_R$, use the ideal (a) to find a multiplicative inverse for a.]

36. Let I be an ideal in a commutative ring R and let

$$J = \{r \in R \mid r^n \in I \text{ for some positive integer } n\}.$$

Prove that J is an ideal that contains I. [*Hint:* You will need the Binomial Theorem from Appendix E. Exercise 28 is the case when $I = (0_R)$.]

37. (a) Show that the ring $M(\mathbb{R})$ is not a division ring by exhibiting a matrix that has no multiplicative inverse. (Division rings are defined in Exercise 34 of Section 3.1.)

(b) Show that $M(\mathbb{R})$ has no ideals except the zero ideal and $M(\mathbb{R})$ itself. [*Hint:* If J is a nonzero ideal, show that J contains a matrix A with a nonzero entry c in the upper left-hand corner. Verify that

$$\begin{pmatrix} 1 & 0 \\ 0 & 0 \end{pmatrix} \cdot A \cdot \begin{pmatrix} c^{-1} & 0 \\ 0 & 0 \end{pmatrix} = \begin{pmatrix} 1 & 0 \\ 0 & 0 \end{pmatrix}$$ and that this matrix is in J. Simi-

larly, show that $\begin{pmatrix} 0 & 0 \\ 0 & 1 \end{pmatrix}$ is in J. What is their sum? See Exercise 13.]

38. Prove that every ideal in $\mathbb{Z}$ is principal. [*Hint:* If I is a nonzero ideal, show that I must contain positive elements and, hence, must contain a smallest positive element c (Why?). Since $c \in I$, every multiple of c is also in I; hence, $(c) \subseteq I$. To show that $I \subseteq (c)$, let a be any element of I. Then $a = cq + r$ with $0 \le r < c$ (Why?). Show that $r = 0$ so that $a = cq \in (c)$.]

39. (a) Prove that the set S of rational numbers (in lowest terms) with odd denominators is a subring of $\mathbb{Q}$.

(b) Let I be the set of elements of S with even numerators. Prove that I is an ideal in S.

(c) Show that S/I consists of exactly two distinct cosets.

40. (a) Let p be a prime integer and let T be the set of rational numbers (in lowest terms) whose denominators are not divisible by p. Prove that T is a ring.

(b) Let I be the set of elements of T whose numerators are divisible by p. Prove that I is an ideal in T.

(c) Show that T/I consists of exactly p distinct cosets.

41. Let J be the set of all polynomials with zero constant term in $\mathbb{Z}[x]$.

(a) Show that J is the principal ideal (x) in $\mathbb{Z}[x]$.

(b) Show that $\mathbb{Z}[x]/J$ consists of an infinite number of distinct cosets, one for each $n \in \mathbb{Z}$.

42. (a) Prove that the set T of matrices of the form $\begin{pmatrix} a & b \\ 0 & a \end{pmatrix}$ with $a, b \in \mathbb{R}$ is a subring of $M(\mathbb{R})$.

(b) Prove that the set I of matrices of the form $\begin{pmatrix} 0 & b \\ 0 & 0 \end{pmatrix}$ with $b \in \mathbb{R}$ is an ideal in the ring T.

(c) Show that every coset in T/I can be written in the form $\begin{pmatrix} a & 0 \\ 0 & a \end{pmatrix} + I$.

43. (a) Prove that the set S of matrices of the form $\begin{pmatrix} a & b \\ 0 & c \end{pmatrix}$ with $a, b, c \in \mathbb{R}$ is a subring of $M(\mathbb{R})$.

    (b) Prove that the set I of matrices of the form $\begin{pmatrix} 0 & b \\ 0 & 0 \end{pmatrix}$ with $b \in \mathbb{R}$ is an ideal in the ring S.

    (c) Show that there are infinitely many distinct cosets in S/I, one for each pair in $\mathbb{R} \times \mathbb{R}$.

C. 44. Let F be a field. Prove that every ideal in $F[x]$ is principal. [*Hint:* Use the Division Algorithm to show that the nonzero ideal I in $F[x]$ is $(p(x))$, where $p(x)$ is a polynomial of smallest possible degree in I.]

45. Prove that a subring S of $\mathbb{Z}_n$ has an identity if and only if there is an element u in S such that $u^2 = u$ and S is the ideal (u).

6.2 QUOTIENT RINGS AND HOMOMORPHISMS

We now show that the set of congruence classes modulo an ideal is itself a ring. As you might expect, this is a straightforward generalization of what we did with congruence classes in $\mathbb{Z}$ and $F[x]$. However, you may not have expected these rings of congruence classes to have close connections with some topics studied in Chapter 3, isomorphisms and homomorphisms. These connections are explored in detail and provide new insight into the structure of rings.

Let I be an ideal in a ring R. The elements of the set R/I are the cosets of I (congruence classes modulo I), that is, all sets of the form $a + I = \{a + i \mid i \in I\}$. In order to define addition and multiplication of cosets as we did with congruence classes in $\mathbb{Z}$ and $F[x]$, we need

THEOREM 6.8   *Let I be an ideal in a ring R. If $a + I = b + I$ and $c + I = d + I$ in R/I, then*

$$(a + c) + I = (b + d) + I \quad \text{and} \quad ac + I = bd + I.$$

Proof   This is a generalization of Theorem 2.6, in slightly different notation. Replace "$[a]$" by "$a + I$" and copy the proof of Theorem 2.6, using Theorems 6.5 and 6.6 in place of Theorems 2.2 and 2.3.   ◆

We can now define addition and multiplication in R/I just as we did in $\mathbb{Z}_n$ and $F[x]/(p(x))$: The sum of the coset $a + I$ (congruence class of a) and the coset $c + I$ (congruence class of c) is the coset $(a + c) + I$ (congruence class of $a + c$). In symbols,

$$(a + I) + (c + I) = (a + c) + I.$$

This statement may be a bit confusing because the plus sign is used with three entirely different meanings:

> as a formal symbol to denote a coset: $a + I$;
>
> as an operation on elements of R: $a + c$;
>
> as the addition operation on cosets that is being defined.*

The important thing is that, because of Theorem 6.8, coset addition is independent of the choice of representative elements in each coset. Even if we replace $a + I$ by an equal coset $b + I$ and replace $c + I$ by an equal coset $d + I$, the resulting coset sum, namely $(b + d) + I$, is the same as $(a + b) + I$.

Multiplication of cosets is defined similarly and is independent of the choice of representatives by Theorem 6.8:

$$(a + I)(c + I) = ac + I.$$

EXAMPLE If I is the principal ideal (3) in $\mathbb{Z}$, then addition and multiplication of cosets is the same as addition and multiplication of congruence classes in Section 2.2. Thus $\mathbb{Z}/I$ is just the ring $\mathbb{Z}_3$.

EXAMPLE If F is a field, $p(x)$ is a polynomial in $F[x]$, and I is the principal ideal $(p(x))$, then cosets of I are precisely congruence classes modulo $p(x)$, so that addition and multiplication of cosets is done exactly as it was in Section 5.2. Thus $F[x]/I$ is the congruence class ring $F[x]/(p(x))$.

EXAMPLE Let I be the ideal of polynomials with even constant terms in $\mathbb{Z}[x]$. As we saw on page 140, $\mathbb{Z}[x]/I$ consists of just two distinct cosets, $0 + I$ and $1 + I$. We have $(1 + I) + (1 + I) = (1 + 1) + I = 2 + I$, but $2 \in I$, so that $2 \equiv 0 \pmod{I}$ and, hence, $2 + I = 0 + I$. Similar calculations produce the following tables for $\mathbb{Z}[x]/I$. It is easy to see that $\mathbb{Z}[x]/I$ is a ring (in fact, a field) isomorphic to $\mathbb{Z}_2$:

$+$	$0 + I$	$1 + I$
$0 + I$	$0 + I$	$1 + I$
$1 + I$	$1 + I$	$0 + I$

$\cdot$	$0 + I$	$1 + I$
$0 + I$	$0 + I$	$0 + I$
$1 + I$	$0 + I$	$1 + I$

These examples illustrate the following theorem, which should not be very surprising in view of your previous experience with $\mathbb{Z}$ and $F[x]$.

---

* This ambiguity can be avoided by using a different notation for cosets, such as $[a]$, and a different symbol for coset addition, such as $\oplus$. The notation above is customary, however, and once you're used to it, there should be no confusion.

THEOREM 6.9 *Let I be an ideal in a ring R. Then*

(1) *R/I is a ring, with addition and multiplication of cosets as defined previously.*

(2) *If R is commutative, then R/I is a commutative ring.*

(3) *If R has an identity, then so does the ring R/I.*

Proof (1) With the usual change of notation ("$a + I$" instead of "$[a]$"), the proof of Theorem 2.7 carries over to the present situation since that proof depends only on the fact that $\mathbb{Z}$ is a ring. Don't take our word for it, though; write out the proof in detail for yourself.

(2) If R is commutative and $a, c \in R$, then $ac = ca$. Consequently, in R/I we have $(a + I)(c + I) = ac + I = ca + I = (c + I)(a + I)$. Hence, R/I is commutative.

(3) The identity in R/I is the coset $1_R + I$ because $(a + I)(1_R + I) = a1_R + I = a + I$ and similarly $(1_R + I)(a + I) = a + I$. ◆

The ring R/I is called the **quotient ring** (or **factor ring**) of R by I. One sometimes speaks of factoring out the ideal I to obtain the quotient ring R/I.

Homomorphisms

Quotient rings are the natural generalization of congruence-class arithmetic in $\mathbb{Z}$ and $F[x]$. As is often the case in mathematics, however, a concept developed with one idea in mind may have unexpected linkages with other important mathematical concepts. That is precisely the situation here. We shall now see that the concept of homomorphism that arose in our study of isomorphism of rings in Chapter 3 is closely related to ideals and quotient rings.

THEOREM 6.10 *Let $f: R \to S$ be a homomorphism of rings and let*

$$K = \{r \in R \mid f(r) = 0_S\}.$$

Then K is an ideal in the ring R.

Proof According to Theorem 6.1 a nonempty subset of R is an ideal if it is closed under subtraction and absorbs products. In this case, K is nonempty because $0_R \in K$ (since $f(0_R) = 0_S$ by Theorem 3.12). To prove that K is closed under subtraction, we must show that for $a, b \in K$, the element $a - b$ is also in K. To show $a - b \in K$, we must show that $f(a - b) = 0_S$. This follows from the fact that f is a homomorphism and that $f(a) = 0_S$ and $f(b) = 0_S$ (because $a, b \in K$):

$$f(a - b) = f(a) - f(b) = 0_S - 0_S = 0_S.$$

To prove that K absorbs products we must first verify that $ra \in K$ for any $r \in R$ and $a \in K$, that is, that $f(ra) = 0_S$; here's the proof:

$$f(ra) = f(r)f(a) = f(r)0_S = 0_S.$$

A similar argument shows that $ar \in K$. Therefore K is an ideal by Theorem 6.1. ◆

The ideal K in Theorem 6.10 is called the **kernel** of the homomorphism f. When the kernel is a "large" ideal in R, then many nonzero elements of R are mapped by f to 0_S. The other extreme is to have a kernel that is as small as possible, namely, the zero ideal. In this case, we have

THEOREM 6.11 *Let $f:R \rightarrow S$ be a homomorphism of rings with kernel K. Then $K = (0_R)$ if and only if f is injective.*

Proof Suppose $K = (0_R)$ and $f(a) = f(b)$. Then because f is a homomorphism, $f(a - b) = f(a) - f(b) = 0_S$. Hence, $a - b$ is in the kernel K, so that $a - b = 0_R$ and $a = b$. Therefore, f is injective. Conversely, if f is injective and $f(c) = 0_S$, then $f(c) = f(0_R)$ by Theorem 3.12. Therefore, $c = 0_R$ by injectivity. Hence, the kernel K consists of the single element 0_R. ◆

Theorem 6.10 states that every kernel is an ideal. Conversely, every ideal is the kernel of a homomorphism:

THEOREM 6.12 *Let I be an ideal in a ring R. Then the map $\pi:R \rightarrow R/I$ given by $\pi(r) = r + I$ is a surjective homomorphism with kernel I.*

The map π is called the **natural homomorphism** from R to R/I.

Proof of Theorem 6.12 The map π is surjective because given any coset $r + I$ in R/I, $\pi(r) = r + I$. The definition of addition and multiplication in R/I shows that π is a homomorphism:

$$\pi(r + s) = (r + s) + I = (r + I) + (s + I) = \pi(r) + \pi(s);$$
$$\pi(rs) = rs + I = (r + I)(s + I) = \pi(r)\pi(s).$$

The kernel of π is the set of elements $r \in R$ such that $\pi(r) = 0_R + I$ (the zero element in R/I). However, $\pi(r) = 0_R + I$ if and only if $r + I = 0_R + I$, which occurs if and only if $r \equiv 0_R \pmod{I}$, that is, if and only if $r \in I$. Therefore, I is the kernel of π. ◆

The natural homomorphism π in Theorem 6.12 is a special case of a more general situation. If $f:R \rightarrow S$ is a surjective homomorphism of rings, we say that S is a **homomorphic image** of R. If f is actually an isomorphism (so that S is an isomorphic image of R), then we know that R and S have identical structure. Whenever one of them has a particular algebraic property, the other one has it too. If f is not an isomorphism, then properties of one ring may not hold in the other. However, the properties of S and the homomorphism f often give us some useful information about R. An analogy with sculpture and photography may be helpful: If $f:R \to S$ is an isomorphism, then S is an exact, three-dimensional replica of R. If f is only a surjective homomorphism, then S is a two-dimensional photographic image of R in which some features of R are accurately reflected but others are distorted or missing. The next theorem tells us precisely how R, S, and the kernel of f are related in these circumstances.

THEOREM 6.13   (FIRST ISOMORPHISM THEOREM)   *Let $f:R \to S$ be a surjective homomorphism of rings with kernel K. Then the quotient ring R/K is isomorphic to S.*

The theorem states that every homomorphic image of a ring R is isomorphic to a quotient ring R/K for some ideal K. Thus if you know all the quotient rings of R, then you know all the possible homomorphic images of R. The ideal K measures how much information is lost in passing from the ring R to the homomorphic image R/K. When $K = (0_R)$, then f is an isomorphism by Theorem 6.11, and no information is lost. But when K is large, quite a bit may be lost.

Proof of Theorem 6.13   We shall define a function φ from R/K to S and then show that it is an isomorphism. To define φ, we must associate with each coset $r + K$ of R/K an element of S. A natural choice for such an element would be $f(r) \in S$; in other words, we would like to define $\varphi:R/K \to S$ by the rule $\varphi(r + K) = f(r)$. The only possible problem is that a coset can be labeled by many different elements of R. So we must show that the value of φ depends only on the coset and not on the particular representative r chosen to name it. If $r + K = t + K$, then $r - t \in K$ by Theorem 6.6. Consequently, since f is a homomorphism, $f(r) - f(t) = f(r - t) = 0_S$. Therefore, $r + K = t + K$ implies that $f(r) = f(t)$. It follows that the map $\varphi:R/K \to S$ given by the rule $\varphi(r + K) = f(r)$ is a well-defined function, independent of how the coset is written.

If $s \in S$, then $s = f(r)$ for some $r \in R$ because f is surjective. Thus $s = f(r) = \varphi(r + K)$, and φ is surjective. If $\varphi(r + K) = \varphi(c + K)$, then $f(r) = f(c)$, so that $0_S = f(r) - f(c) = f(r - c)$. Hence, $r - c \in K$, which implies that $r + K = c + K$ by Theorem 6.6. Therefore, φ is injective. Finally, φ is a homomorphism because f is:

$$\varphi[(c + K)(d + K)] = \varphi(cd + K) = f(cd) = f(c)f(d)$$
$$= \varphi(c + K)\varphi(d + K)$$

and

$$\varphi[(c + K) + (d + K)] = \varphi[(c + d) + K] = f(c + d) = f(c) + f(d)$$
$$= \varphi(c + K) + \varphi(d + K).$$

Therefore, $\varphi:R/K \to S$ is an isomorphism.   ◆

The First Isomorphism Theorem is a useful tool for determining the structure of quotient rings, as illustrated in the following examples.

> **EXAMPLE** In the ring $\mathbb{Z}[x]$, the principal ideal (x) consists of all multiples of x, that is, all polynomials with constant term 0. What does the quotient ring $\mathbb{Z}[x]/(x)$ look like? We can answer the question by using the function $\theta : \mathbb{Z}[x] \to \mathbb{Z}$, which maps each polynomial to its constant term. The function θ is certainly surjective because each $k \in \mathbb{Z}$ is the image of the polynomial $x + k$ in $\mathbb{Z}[x]$. Furthermore, θ is a homomorphism of rings (Exercise 1). The kernel of θ consists of all those polynomials that are mapped to 0, that is, all polynomials with constant term 0. Thus the kernel of θ is the ideal (x). By Theorem 6.13 the quotient ring $\mathbb{Z}[x]/(x)$ is isomorphic to $\mathbb{Z}$.

> **EXAMPLE** Let T be the ring of continuous functions from $\mathbb{R}$ to $\mathbb{R}$ and I the ideal of all functions g such that $g(2) = 0$. On page 141 we saw that T/I consists of the cosets $f_r + I$, one for each real number r, where $f_r : \mathbb{R} \to \mathbb{R}$ is the constant function given by $f_r(x) = r$ for every x. This suggests the possibility that the quotient ring T/I might be isomorphic to the field $\mathbb{R}$. We shall use Theorem 6.13 to show that this is indeed the case by constructing a surjective homomorphism from T to $\mathbb{R}$ whose kernel is the ideal I. Let $\varphi : T \to \mathbb{R}$ be the function defined by $\varphi(f) = f(2)$. Then φ is surjective because for every real number r, $r = f_r(2) = \varphi(f_r)$. Furthermore, φ is a homomorphism of rings:
>
> $$\varphi(f + h) = (f + h)(2) = f(2) + h(2) = \varphi(f) + \varphi(h)$$
> $$\varphi(fh) = (fh)(2) = f(2)h(2) = \varphi(f)\varphi(h).$$
>
> By definition, the kernel of φ is the set
>
> $$\{g \in T \mid \varphi(g) = 0\} = \{g \in T \mid g(2) = 0\}.$$
>
> Thus the kernel is precisely the ideal I. By Theorem 6.13, T/I is isomorphic to $\mathbb{R}$.

> **EXAMPLE** What do the homomorphic images of the ring $\mathbb{Z}$ look like? To answer this question, suppose that $f : \mathbb{Z} \to S$ is a surjective homomorphism. If f is actually an isomorphism, then S looks exactly like $\mathbb{Z}$, of course (in terms of algebraic structure). If f is surjective, but not an isomorphism (that is, not injective), then the kernel K of f is a nonzero ideal in $\mathbb{Z}$ by Theorem 6.11. Since K is an ideal in $\mathbb{Z}$, K must be a principal ideal, say $K = (n)$ for some $n \neq 0$, by Exercise 38 in Section 6.1. By Theorem 6.13, S is isomorphic to $\mathbb{Z}/K = \mathbb{Z}/(n) = \mathbb{Z}_n$. Thus every homomorphic image of $\mathbb{Z}$ is isomorphic either to $\mathbb{Z}$ or to $\mathbb{Z}_n$ for some n.

◆ **EXERCISES**

A. **1.** Show that the map $\theta:\mathbb{Z}[x] \rightarrow \mathbb{Z}$ that sends each polynomial $f(x)$ to its constant term is a surjective homomorphism.

2. Show that every homomorphic image of a field F is isomorphic either to F itself or to the zero ring. [*Hint:* See Exercise 14 in Section 6.1 and Exercise 9 below.]

3. If F is a field, R a nonzero ring, and $f:F \rightarrow R$ a surjective homomorphism, prove that f is an isomorphism.

4. Let $[a]_n$ denote the congruence class of the integer a modulo n.

  (a) Show that the map $f:\mathbb{Z}_{12} \rightarrow \mathbb{Z}_4$ that sends $[a]_{12}$ to $[a]_4$ is a well-defined, surjective homomorphism.

  (b) Find the kernel of f.

5. Let I be an ideal in an integral domain R. Is it true that R/I is also an integral domain?

6. (a) List all principal ideals in $\mathbb{Z}_{12}$.

  (b) For each ideal I in part (a), write out the addition and multiplication tables of $\mathbb{Z}_{12}/I$.

  (c) Show that every homomorphic image of $\mathbb{Z}_{12}$ is isomorphic to one of these rings: $0, \mathbb{Z}_2, \mathbb{Z}_3, \mathbb{Z}_4, \mathbb{Z}_6, \mathbb{Z}_{12}$.

7. Let T be the ring of all multiples of 3 in $\mathbb{Z}$. Let $I = \{0, \pm 6, \pm 12, \pm 18, \ldots \}$.

  (a) Show that I is an ideal in T.

  (b) Write out the addition and multiplication tables of T/I and verify that T/I is a field. Note that T/I has a multiplicative identity even though T does not.

8. (a) Let $I = \{0, 3\}$ in $\mathbb{Z}_6$. Verify that I is an ideal and show that $\mathbb{Z}_6/I \cong \mathbb{Z}_3$.

  (b) Let $J = \{0, 5\}$ in $\mathbb{Z}_{10}$. Verify that J is an ideal and show that $\mathbb{Z}_{10}/J \cong \mathbb{Z}_5$.

9. If R is a ring, show that $R/(0_R) \cong R$.

B. **10.** Let R and S be rings. Show that $\pi:R \times S \rightarrow R$ given by $\pi(r, s) = r$ is a surjective homomorphism whose kernel is isomorphic to S.

11. Let I and K be ideals in a ring R, with $K \subseteq I$. Prove that $I/K = \{a + K \mid a \in I\}$ is an ideal in the quotient ring R/K.

12. (a) Let $f:R \rightarrow S$ be a surjective homomorphism of rings and let I be an ideal in R. Prove that $f(I)$ is an ideal in S, where $f(I) = \{s \in S \mid s = f(a)$ for some $a \in I\}$.

(b) Show by example that part (a) may be false if f is not surjective.

13. If R is a commutative ring with identity and (x) is the principal ideal generated by x in $R[x]$, prove that $R[x]/(x) \cong R$.

14. Let I be an ideal in a noncommutative ring R such that $ab - ba \in I$ for all $a, b \in R$. Prove that R/I is commutative.

15. Let I be an ideal in a ring R. Prove that every element in R/I has a square root if and only if for every $a \in R$, there exists $b \in R$ such that $a - b^2 \in I$.

16. Let I be an ideal in a ring R. Prove that every element in R/I is a solution of $x^2 = x$ if and only if for every $a \in R$, $a^2 - a \in I$.

17. Let I be an ideal in a commutative ring R. Prove that R/I has an identity if and only if there exists $e \in R$ such that $ea - a \in I$ for every $a \in R$. [See Exercise 7 for an example.]

18. Let $I \neq R$ be an ideal in a commutative ring R with identity. Prove that R/I is an integral domain if and only if whenever $ab \in I$, either $a \in I$ or $b \in I$.

19. Suppose I and J are ideals in a ring R and let $f: R \rightarrow R/I \times R/J$ be the function defined by $f(a) = (a + I, a + J)$.

    (a) Prove that f is a homomorphism of rings.

    (b) Is f surjective? [*Hint*: Consider the case when $R = \mathbb{Z}, I = (2), J = (4)$.]

    (c) What is the kernel of f?

20. Let R be a commutative ring with identity with the property that every ideal in R is principal. Prove that every homomorphic image of R has the same property.

21. **(a)** Show that $R = \left\{ \begin{pmatrix} a & 0 \\ b & c \end{pmatrix} \mid a, b, c \in \mathbb{Z} \right\}$ is a ring with identity.

    (b) Show that the map $f: R \rightarrow \mathbb{Z}$ given by $f \begin{pmatrix} a & 0 \\ b & c \end{pmatrix} = a$ is a surjective homomorphism.

    (c) What is the kernel of f?

22. Let $f: R \rightarrow S$ be a homomorphism of rings with kernel K. Let I be an ideal in R such that $I \subseteq K$. Show that $\bar{f}: R/I \rightarrow S$ given by $\bar{f}(r + I) = f(r)$ is a well-defined homomorphism.

23. Use the First Isomorphism Theorem to show that $\mathbb{Z}_{20}/(5) \cong \mathbb{Z}_5$.

24. Let $f: R \rightarrow S$ be a homomorphism of rings. If J is an ideal in S and $I = \{r \in R \mid f(r) \in J\}$, prove that I is an ideal in R that contains the kernel of f.

25. (a) Let R be a ring with identity. Show that the map $f: \mathbb{Z} \to R$ given by $f(k) = k1_R$ is a homomorphism.

   (b) Show that the kernel of f is the ideal (n), where n is the characteristic of R. [*Hint:* See Exercise 31 in Section 3.2 and Exercise 38 in Section 6.1.]

26. Find at least three idempotents in the quotient ring $\mathbb{Q}[x]/(x^4 + x^2)$. [See Exercise 2 in Section 3.2.]

27. Let R be a commutative ring and J the ideal of all nilpotent elements of R (as in Exercise 28 in Section 6.1). Prove that the quotient ring R/J has no non-zero nilpotent elements.

28. Let S and I be as in Exercise 39 of Section 6.1. Prove that $S/I \cong \mathbb{Z}_2$.

29. Let T and I be as in Exercise 40 of Section 6.1. Prove that $T/I \cong \mathbb{Z}_p$.

30. Let T and I be as in Exercise 42 of Section 6.1. Prove that $T/I \cong \mathbb{R}$.

31. Let S and I be as in Exercise 43 of Section 6.1. Prove that $S/I \cong \mathbb{R} \times \mathbb{R}$.

C. 32. (The Second Isomorphism Theorem) Let I and J be ideals in a ring R. Then $I \cap J$ is an ideal in I, and J is an ideal in $I + J$ by Exercises 17 and 18 of Section 6.1. Prove that $\dfrac{I}{I \cap J} \cong \dfrac{I + J}{J}$. [*Hint:* Show that $f: I \to (I + J)/J$ given by $f(a) = a + J$ is a surjective homomorphism with kernel $I \cap J$.]

33. (The Third Isomorphism Theorem) Let I and K be ideals in a ring R such that $K \subseteq I$. Then I/K is an ideal in R/K by Exercise 11. Prove that $(R/K)/(I/K) \cong R/I$. [*Hint:* Show that the map $f: R/K \to R/I$ given by $f(r + K) = r + I$ is a well-defined surjective homomorphism with kernel I/K.]

34. (a) Let K be an ideal in a ring R. Prove that every ideal in the quotient ring R/K is of the form I/K for some ideal I in R. [*Hint:* Exercises 11 and 24.]

   (b) If $f: R \to S$ is a surjective homomorphism of rings with kernel K, prove that there is a bijective function from the set of all ideals of S to the set of all ideals of R that contain K. [*Hint:* Part (a) and Exercise 12.]

> **Excursion**   The Chinese Remainder Theorem for Rings (Section 13.3) may be covered at this point if desired.

6.3 THE STRUCTURE OF *R/I* WHEN *I* IS PRIME OR MAXIMAL*

Quotient rings were developed as a natural generalization of the rings $\mathbb{Z}_p$ and $F[x]/(p(x))$. When p is prime and $p(x)$ irreducible, then $\mathbb{Z}_p$ and $F[x]/(p(x))$ are

---

* This section is not used in the sequel and may be omitted if desired.

fields. In this section we explore the analogue of this situation for quotient rings of commutative rings. We shall determine the conditions necessary for a quotient ring to be either an integral domain or a field.

Primes in $\mathbb{Z}$ and irreducibles in $F[x]$ play essentially the same role in the structure of the congruence class rings. Our first task in arbitrary commutative rings is to find some reasonable way of describing this role in terms of ideals. According to Theorem 1.8, a nonzero integer p (other than ± 1) is prime if and only if p has this property: Whenever $p \mid bc$, then $p \mid b$ or $p \mid c$. To say that $p \mid a$ means that a is a multiple of p, that is, a is an element of the principal ideal (p) of all multiples of p. Thus this property of primes can be rephrased in terms of ideals:

> If $p \neq 0, \pm 1$, then p is prime if and only if
> whenever $bc \in (p)$, then $b \in (p)$ or $c \in (p)$.

The condition $p \neq \pm 1$ guarantees that 1 is not a multiple of p and, hence, that the ideal (p) is not all of $\mathbb{Z}$. Using this situation as a model, we have this

DEFINITION• *An ideal P in a commutative ring R is said to be **prime** if $P \neq R$ and whenever $bc \in P$, then $b \in P$ or $c \in P$.*

EXAMPLE As shown above, the principal ideal (p) is prime in $\mathbb{Z}$ whenever p is a prime integer. On the other hand, the ideal $P = (6)$ is not prime in $\mathbb{Z}$ because $2 \cdot 3 \in P$ but $2 \notin P$ and $3 \notin P$.

EXAMPLE The zero ideal in any integral domain R is prime because $ab = 0_R$ implies $a = 0_R$ or $b = 0_R$.

EXAMPLE The implication $(1) \Rightarrow (2)$ of Theorem 4.11 shows that if F is a field and $p(x)$ is irreducible in $F[x]$, then the principal ideal $(p(x))$ is prime in $F[x]$.

EXAMPLE Let I be the ideal of polynomials with even constant terms in $\mathbb{Z}[x]$. Then I is not principal (page 137) and clearly $I \neq \mathbb{Z}[x]$. Let $f(x) = a_n x^n + \cdots + a_0$ and $g(x) = b_m x^m + \cdots + b_0$ be polynomials in $\mathbb{Z}[x]$ such that $f(x)g(x) \in I$. Then the constant term of $f(x)g(x)$, namely $a_0 b_0$, must be even. Since the product of two odd integers is odd, we conclude that either a_0 is even (that is, $f(x) \in I$) or b_0 is even (that is, $g(x) \in I$). Therefore, I is a prime ideal.

The ideal I in the preceding example is prime, and the quotient ring $\mathbb{Z}[x]/I$ is a field (see the last example on page 146). Similarly, $\mathbb{Z}/(p) = \mathbb{Z}_p$ is a field when p is prime. However, the next example shows that R/P may *not* always be a field when P is prime.

> **EXAMPLE**   The principal ideal (x) in the ring $\mathbb{Z}[x]$ consists of polynomials that are multiples of x, that is, polynomials with zero constant terms. Hence, $(x) \neq \mathbb{Z}[x]$. If $f(x) = a_n x^n + \cdots + a_0$ and $g(x) = b_m x^m + \cdots + b_0$ and $f(x)g(x) \in I$, then the constant term of $f(x)g(x)$, namely $a_0 b_0$, must be 0. This can happen only if $a_0 = 0$ or $b_0 = 0$, that is, only if $f(x) \in (x)$ or $g(x) \in (x)$. Therefore, (x) is a prime ideal. However, the example on page 150 shows that the quotient ring $\mathbb{Z}[x]/(x)$ is isomorphic to $\mathbb{Z}$. Therefore, $\mathbb{Z}[x]/(x)$ is an integral domain but not a field.

In light of the preceding example, the next theorem is the best we can do with prime ideals.

THEOREM 6.14   *Let P be an ideal in a commutative ring R with identity. Then P is a prime ideal if and only if the quotient ring R/P is an integral domain.*

Proof   We shall frequently use the following fact, which follows from Theorem 6.6. For any ideal P in R

$$(*) \qquad a + P = 0_R + P \text{ in } R/P \qquad \text{if and only if} \qquad a \in P.$$

Suppose P is prime. By Theorem 6.9, R/P is a commutative ring with identity. In order to prove that R/P is an integral domain, we must show that its identity is not the zero element and that it has no zero divisors. Since P is prime, $P \neq R$. Consequently, $1_R \notin P$ because any ideal containing 1_R must be the whole ring. However, $1_R \notin P$ implies that $1_R + P \neq 0_R + P$ in R/P by $(*)$. Now we show that R/P has no zero divisors. If $(b + P)(c + P) = 0_R + P$, then $bc + P = 0_R + P$ and $bc \in P$ by $(*)$. Hence $b \in P$ or $c \in P$. Thus $b + P = 0_R + P$ or $c + P = 0_R + P$, so that R/P has no zero divisors. Therefore R/P is an integral domain.

Now assume that R/P is an integral domain. Then by definition $1_R + P \neq 0_R + P$ and hence $1_R \notin P$ by $(*)$. Therefore $P \neq R$. To complete the proof that P is prime we assume that $bc \in P$ and show that $b \in P$ or $c \in P$. Now if $bc \in P$, then in R/P we have $(b + P)(c + P) = bc + P = 0_R + P$ by $(*)$. Thus $b + P = 0_R + P$ or $c + P = 0_R + P$ because R/P has no zero divisors. Hence $b \in P$ or $c \in P$ by $(*)$. Therefore P is prime.   ◆

Since the quotient ring modulo a prime ideal is not necessarily a field, it is natural to ask what conditions an ideal must satisfy in order for the quotient ring to be a field.

> **EXAMPLE**   Consider the ideal (3) in $\mathbb{Z}$. We know that $\mathbb{Z}/(3) = \mathbb{Z}_3$ is a field. Now consider the ideal (3). Suppose J is an ideal such that $(3) \subseteq J \subseteq \mathbb{Z}$. If $J \neq (3)$, then there exists $a \in J$ with $a \notin (3)$. In particular, $3 \nmid a$, so that 3 and a are relatively prime. Hence, there are integers u and v such that $3u + av = 1$. Since 3 and a are in the ideal J, it follows that $1 \in J$. There-

fore $J = \mathbb{Z}$ by Exercise 13 of Section 6.1, and so *there are no ideals strictly between (3) and* $\mathbb{Z}$.

EXAMPLE   The quotient ring $\mathbb{Z}[x]/(x)$ is not a field. Furthermore, the ideal I of polynomials with even constant terms lies strictly between (x) and $\mathbb{Z}[x]$, that is, $(x) \subsetneq I \subsetneq \mathbb{Z}[x]$.

Here is a formal definition of the property suggested by these examples:

DEFINITION• *An ideal M in a ring R is said to be **maximal** if $M \neq R$ and whenever J is an ideal such that $M \subseteq J \subseteq R$, then $M = J$ or $J = R$.*

For example, the ideal (3) is maximal in $\mathbb{Z}$ and the ideal (x) is not maximal in $\mathbb{Z}[x]$. Note that a ring may have more than one maximal ideal. The ideal $\{0, 2, 4\}$ is maximal in $\mathbb{Z}_6$, and so is the ideal $\{0, 3\}$. There are infinitely many maximal ideals in $\mathbb{Z}$ (Exercise 3). Maximal ideals provide the following answer to the question posed above:

THEOREM 6.15   *Let M be an ideal in a commutative ring R with identity. Then M is a maximal ideal if and only if the quotient ring R/M is a field.*

Proof   We shall use the same consequence of Theorem 6.6 that was used in the proof of Theorem 6.14:

(∗) $a + M = 0_R + M$ in R/M if and only if $a \in M$.

Suppose R/M is a field. Then by definition $1_R + M \neq 0_R + M$ and hence $1_R \notin M$ *by* (∗). Therefore $M \neq R$. To show that M is maximal, we assume that J is an ideal with $M \subseteq J \subseteq R$ and show that $M = J$ or $J = R$. If $M = J$, there is nothing to prove. If $M \neq J$, then there exists $a \in J$ with $a \notin M$. Hence $a + M \neq 0_R + M$ in the field R/M, and $a + M$ has an inverse $b + M$ such that $(a + M)(b + M) = ab + M = 1_R + M$. Then $ab \equiv 1_R \pmod M$ by Theorem 6.6, so that $ab - 1_R = m$ for some $m \in M$. Thus $1_R = ab - m$. Since a and m are in the ideal J, it follows that $1_R \in J$ and $J = R$. Therefore M is a maximal ideal.

Now assume M is a maximal ideal in R. By Theorem 6.9, R/M is a commutative ring with identity. In order to prove that R/M is a field, we must show that its identity is not the zero element and that every nonzero element has a multiplicative inverse. Since M is maximal, $M \neq R$. Consequently, $1_R \notin M$ because any ideal containing 1_R must be the whole ring. However, $1_R \notin M$ implies that $1_R + M \neq 0_R + M$ in R/M by (∗). Now we show that every nonzero element of R/M has a multiplicative inverse. If $a + M$ is a nonzero element of R/M, then $a \notin M$ (otherwise $a + M$ would be the zero coset). The set

$$J = \{m + ra \mid r \in R \text{ and } m \in M\}$$

is an ideal in R that contains M by Exercise 32 of Section 6.1. Furthermore, $a = 0_R + 1_R a$ is in J, so that $M \neq J$. By maximality we must have $J = R$. Hence $1_R \in J$, which implies that $1_R = m + ca$ for some $m \in M$ and $c \in R$. Note that $ca - 1_R = -m \in M$, so that $ca \equiv 1_R \pmod{M}$, and hence $ca + M = 1_R + M$ by Theorem 6.6. Consequently, the coset $c + M$ is the inverse of $a + M$ in R/M:

$$(c + M)(a + M) = ca + M = 1_R + M.$$

Therefore R/M is a field.   ◆

COROLLARY 6.16   *In a commutative ring R with identity, every maximal ideal is prime.*

Proof   If M is a maximal ideal, then R/M is a field by Theorem 6.15. Hence, R/M is an integral domain by Theorem 3.9. Therefore, M is prime by Theorem 6.14.   ◆

Theorem 6.15 can be used to show that several familiar ideals are maximal.

EXAMPLE   The ideal I of polynomials with even constant terms in $\mathbb{Z}[x]$ is maximal because $\mathbb{Z}[x]/I$ is a field (see the last example on page 146).

EXAMPLE   Let T be the ring of continuous functions from $\mathbb{R}$ to $\mathbb{R}$ and let I be the ideal of all functions g such that $g(2) = 0$. On page 150 we saw that T/I is a field isomorphic to $\mathbb{R}$. Therefore, I is a maximal ideal in T.

◆ **EXERCISES**

A.   **1.** If n is a composite integer, prove that (n) is not a prime ideal in $\mathbb{Z}$.

2. Give an example to show that the intersection of two prime ideals need not be prime.

3. **(a)** Prove that a nonzero integer p is prime if and only if the ideal (p) is maximal in $\mathbb{Z}$.

(b) Let F be a field and $p(x) \in F[x]$. Prove that $p(x)$ is irreducible if and only if the ideal $(p(x))$ is maximal in $F[x]$.

4. Let R be a commutative ring with identity. Prove that R is an integral domain if and only if (0_R) is a prime ideal.

5. List all maximal ideals in $\mathbb{Z}_6$. Do the same in $\mathbb{Z}_{12}$.

6. (a) Show that there is exactly one maximal ideal in $\mathbb{Z}_8$. Do the same for $\mathbb{Z}_9$.

  (b) Show that $\mathbb{Z}_{10}$ and $\mathbb{Z}_{15}$ have more than one maximal ideal.

7. Let R be a commutative ring with identity. Prove that R is a field if and only if (0_R) is a maximal ideal.

8. If R is a finite commutative ring with identity, prove that every prime ideal in R is maximal. [*Hint:* Theorem 3.11.]

9. Let R be an integral domain in which every ideal is principal. If (p) is a nonzero prime ideal in R, prove that p has this property: Whenever p factors, $p = cd$, then c or d is a unit in R.

B. 10. Let p be a fixed prime and let J be the set of polynomials in $\mathbb{Z}[x]$ whose constant terms are divisible by p. Prove that J is a maximal ideal in $\mathbb{Z}[x]$.

11. Show that the principal ideal $(x - 1)$ in $\mathbb{Z}[x]$ is prime but not maximal.

12. If p is a prime integer, prove that M is a maximal ideal in $\mathbb{Z} \times \mathbb{Z}$, where $M = \{(pa, b) \,|\, a, b \,\epsilon\, \mathbb{Z}\}$.

13. Find an ideal in $\mathbb{Z} \times \mathbb{Z}$ that is prime but not maximal.

14. If P is a prime ideal in a commutative ring R, is the ideal $P \times P$ a prime ideal in $R \times R$?

15. (a) Let R be the set of integers equipped with the usual addition and multiplication given by $ab = 0$ for all $a, b \,\epsilon\, R$. Show that R is a commutative ring.

  (b) Show that $M = \{0, \pm 2, \pm 4, \pm 6, \ldots \}$ is a maximal ideal in R that is *not* prime. Explain why this result does not contradict Corollary 6.16.

16. Show that $M = \{0, \pm 4, \pm 8, \ldots \}$ is a maximal ideal in the ring E of even integers but E/M is *not* a field. Explain why this result does not contradict Theorem 6.15.

17. Let $f : R \rightarrow S$ be a surjective homomorphism of commutative rings. If J is a prime ideal in S, and $I = \{r \,\epsilon\, R \,|\, f(r) \,\epsilon\, J\}$, prove that I is a prime ideal in R.

18. Let P be an ideal in a commutative ring R with $P \neq R$. Prove that P is prime if and only if it has this property: Whenever A and B are ideals in R such that $AB \subseteq P$, then $A \subseteq P$ or $B \subseteq P$. [AB is defined in Exercise 34 of Section 6.1. This property is used as a definition of prime ideal in noncommutative rings.]

19. Assume that when R is a nonzero ring with identity, then every ideal of R except R itself is contained in a maximal ideal (the proof of this fact is beyond the scope of this book). Prove that a commutative ring R with identity has a unique maximal ideal if and only if the set of nonunits in R is an ideal. (Such a ring is called a **local ring**.)

C. 20. (a) Prove that $R = \{a + bi \mid a, b \in \mathbb{Z}\}$ is a subring of $\mathbb{C}$ and that

$$M = \{a + bi \mid 3 \mid a \text{ and } 3 \mid b\}$$

is a maximal ideal in R. [*Hint:* If $r + si \notin M$, then $3 \nmid r$ or $3 \nmid s$. Show that 3 does not divide $r^2 + s^2 = (r + si)(r - si)$. Then show that any ideal containing $r + si$ and M also contains 1.]

(b) Show that R/M is a field with nine elements.

21. Let R be as in Exercise 20. Show that J is not a maximal ideal in R, where $J = \{a + bi \mid 5 \mid a \text{ and } 5 \mid b\}$. [*Hint:* Consider the principal ideal $K = (2 + i)$ in R.]

22. If R and J are as in Exercise 21, show that $R/J \cong \mathbb{Z}_5 \times \mathbb{Z}_5$.

23. If R and K are as in Exercise 21, show that $R/K \cong \mathbb{Z}_5$.

24. Prove that $T = \{a + b\sqrt{2} \mid a, b \in \mathbb{Z}\}$ is a subring of $\mathbb{R}$ and $M = \{a + b\sqrt{2} \mid 5 \mid a \text{ and } 5 \mid b\}$ is a maximal ideal in T.

Alternative Routes   At this point there are three possibilities. You may explore a new algebraic concept, groups (Chapter 7), or continue further with either integral domains (Chapter 9) or fields (Chapter 10).

CHAPTER **7**

Groups

$\mathbf{R}$ings were a natural place to begin the study of abstract algebra because you already had a great deal of experience with particular rings (such as the integers, real numbers, and polynomials). Now it is time to consider some less familiar algebraic systems that arise in science and mathematics. The most important of these is the concept of a group, which is an algebraic system with a single operation. Groups arise naturally in the study of symmetry, geometric transformations and algebraic coding theory, and in the analysis of the roots of polynomial equations.

In the first eight sections of this chapter, we develop the group analogues of the concepts that have proved useful in the study of rings: subgroups, isomorphism, congruence, quotient groups, and homomorphisms. Along the way, we investigate the structure of finite groups, a topic with a much different flavor than ring theory. The last two sections deal with some important finite groups, the symmetric and alternating groups.

7.1 DEFINITION AND EXAMPLES OF GROUPS

A group is an algebraic system with one operation. Some groups, as we shall see, arise from rings by ignoring one of the ring operations and concentrating on the other. But many groups have no such direct connection with rings. The most important of these latter groups (the ones that were the historical starting point of group theory) developed from the study of permutations.* Consequently, we begin with a consideration of permutations.

Informally, a permutation of a set T is just a rearrangement of its elements. For example, there are six possible permutations of $T = \{1, 2, 3\}$:

$$1\,2\,3 \quad 1\,3\,2 \quad 2\,1\,3 \quad 2\,3\,1 \quad 3\,1\,2 \quad 3\,2\,1.$$

Each such ordering determines a bijective function from T to T: map 1 to the first element of the ordering, 2 to the second, and 3 to the third. For instance, 2 3 1 determines the function $f : T \rightarrow T$ whose rule is $f(1) = 2; f(2) = 3; f(3) = 1$. Conversely, every bijective function from T to T defines a rearrangement of the elements from 1, 2, 3 to $f(1), f(2), f(3)$. Consequently, we *define* a **permutation of a set T** to be a bijective function from T to T. This definition preserves the informal idea of rearrangement and has the advantage of being applicable to infinite sets. For now, however, we shall concentrate on finite sets and develop a convenient notation for dealing with their permutations.

> **EXAMPLE** Let $T = \{1, 2, 3\}$. The permutation f whose rule is $f(1) = 2$, $f(2) = 3$, $f(3) = 1$ may be represented by the array $\begin{pmatrix} 1\,2\,3 \\ 2\,3\,1 \end{pmatrix}$, in which the image under f of an element in the first row is listed immediately below it in the second row. Using this notation, the six permutations of T are
>
> $$\begin{pmatrix} 1 & 2 & 3 \\ 1 & 2 & 3 \end{pmatrix} \quad \begin{pmatrix} 1 & 2 & 3 \\ 1 & 3 & 2 \end{pmatrix} \quad \begin{pmatrix} 1 & 2 & 3 \\ 2 & 1 & 3 \end{pmatrix}$$
> $$\begin{pmatrix} 1 & 2 & 3 \\ 2 & 3 & 1 \end{pmatrix} \quad \begin{pmatrix} 1 & 2 & 3 \\ 3 & 1 & 2 \end{pmatrix} \quad \begin{pmatrix} 1 & 2 & 3 \\ 3 & 2 & 1 \end{pmatrix}.$$
>
> Since the composition of two bijective functions is itself bijective,** the composition of any two of these permutations is one of the six permutations on the list above. For instance, if $f = \begin{pmatrix} 1\,2\,3 \\ 3\,2\,1 \end{pmatrix}$ and $g = \begin{pmatrix} 1\,2\,3 \\ 2\,1\,3 \end{pmatrix}$, then $f \circ g$ is the function given by
>
> $$(f \circ g)(1) = f(g(1)) = f(2) = 2$$
> $$(f \circ g)(2) = f(g(2)) = f(1) = 3$$
> $$(f \circ g)(3) = f(g(3)) = f(3) = 1$$

---

* In the early nineteenth century, permutations played a key role in the attempt to find formulas for solving higher-degree polynomial equations similar to the quadratic formula. For more information, see Chapter 11.

** See Appendix B.

Thus $f \circ g = \begin{pmatrix} 1\,2\,3 \\ 2\,3\,1 \end{pmatrix}$. It is usually easier to make computations like this by visually tracing an element's progress as we first apply g and then f; for example,

$$\begin{pmatrix} 1 & 2 & 3 \\ 3 & 2 & 1 \end{pmatrix} \circ \begin{pmatrix} 1 & 2 & 3 \\ 2 & 1 & 3 \end{pmatrix} = \begin{pmatrix} 1 & 2 & 3 \\ 2 & 3 & 1 \end{pmatrix}$$

If we denote the set of permutations of T by S_3, then composition of functions ($\circ$) is an operation on the set S_3 with this property:

If $f \in S_3$ and $g \in S_3$, then $f \circ g \in S_3$.

Since composition of functions is associative,* we see that

$$(f \circ g) \circ h = f \circ (g \circ h) \qquad \text{for all } f, g, h \in S_3.$$

Verify that the identity permutation $I = \begin{pmatrix} 1\,2\,3 \\ 1\,2\,3 \end{pmatrix}$ has this property:

$$I \circ f = f \qquad \text{and} \qquad f \circ I = f \qquad \text{for every } f \in S_3.$$

Every bijection has an inverse function;* consequently,

if $f \in S_3$, then there exists $g \in S_3$ such that

$$f \circ g = I \qquad \text{and} \qquad g \circ f = I.$$

For instance, if $f = \begin{pmatrix} 1\,2\,3 \\ 3\,1\,2 \end{pmatrix}$, then $g = \begin{pmatrix} 1\,2\,3 \\ 2\,3\,1 \end{pmatrix}$ because

$$\begin{pmatrix} 1 & 2 & 3 \\ 3 & 1 & 2 \end{pmatrix} \circ \begin{pmatrix} 1 & 2 & 3 \\ 2 & 3 & 1 \end{pmatrix} = \begin{pmatrix} 1 & 2 & 3 \\ 1 & 2 & 3 \end{pmatrix}$$

and

$$\begin{pmatrix} 1 & 2 & 3 \\ 2 & 3 & 1 \end{pmatrix} \circ \begin{pmatrix} 1 & 2 & 3 \\ 3 & 1 & 2 \end{pmatrix} = \begin{pmatrix} 1 & 2 & 3 \\ 1 & 2 & 3 \end{pmatrix}.$$

You should determine the inverses of the other permutations in S_3 (Exercise 1). Finally, note that $f \circ g$ may not be equal to $g \circ f$; for instance,

$$\begin{pmatrix} 1 & 2 & 3 \\ 3 & 2 & 1 \end{pmatrix} \circ \begin{pmatrix} 1 & 2 & 3 \\ 2 & 1 & 3 \end{pmatrix} = \begin{pmatrix} 1 & 2 & 3 \\ 2 & 3 & 1 \end{pmatrix}$$

* See Appendix B.

but

$$\begin{pmatrix} 1 & 2 & 3 \\ 2 & 1 & 3 \end{pmatrix} \circ \begin{pmatrix} 1 & 2 & 3 \\ 3 & 2 & 1 \end{pmatrix} = \begin{pmatrix} 1 & 2 & 3 \\ 3 & 1 & 2 \end{pmatrix}.$$

By abstracting the key properties of S_3 under the operation $\circ$, we obtain this

DEFINITION • *A **group** is a nonempty set G equipped with a binary operation $*$ that satisfies the following axioms†:*

1. *Closure: If $a \in G$ and $b \in G$, then $a * b \in G$.*
2. *Associativity: $a * (b * c) = (a * b) * c$ for all $a, b, c \in G$.*
3. *There is an element $e \in G$ (called the **identity element**) such that $a * e = a = e * a$ for every $a \in G$.*
4. *For each $a \in G$, there is an element $d \in G$ (called the **inverse** of a) such that $a * d = e$ and $d * a = e$.*

*A group is said to be **abelian**‡ if it also satisfies this axiom:*

5. *Commutativity: $a * b = b * a$ for all $a, b \in G$.*

A group G is said to be **finite** (or of **finite order**) if it has a finite number of elements. In this case, the number of elements in G is called the **order of G** and is denoted $|G|$. A group with infinitely many elements is said to have **infinite order.**

EXAMPLE The discussion preceding the definition shows that S_3 is a nonabelian group of order 6, with the operation $*$ being composition of functions.

EXAMPLE The permutation group S_3 is just a special case of a more general situation. Let n be a fixed positive integer and let T be the set $\{1, 2, 3, \ldots, n\}$. Let S_n be the set of all permutations of T (that is, all bijections $T \to T$). We shall use the same notation for such functions as we did in S_3. In S_6, for instance, $\begin{pmatrix} 1 & 2 & 3 & 4 & 5 & 6 \\ 4 & 6 & 2 & 3 & 5 & 1 \end{pmatrix}$ denotes the permutation that takes 1 to 4, 2 to 6, 3 to 2, 4 to 3, 5 to 5, and 6 to 1. Since the composite of

---

† Binary operations are defined in Appendix B.
‡ In honor of the Norwegian mathematician N. H. Abel (1802–1829).

two bijective functions is bijective,* S_n is closed under the operation of composition. For example, in S_6

$$\begin{pmatrix} 1 & 2 & 3 & 4 & 5 & 6 \\ 3 & 5 & 2 & 4 & 1 & 6 \end{pmatrix} \circ \begin{pmatrix} 1 & 2 & 3 & 4 & 5 & 6 \\ 6 & 4 & 2 & 3 & 5 & 1 \end{pmatrix} = \begin{pmatrix} 1 & 2 & 3 & 4 & 5 & 6 \\ 6 & 4 & 5 & 2 & 1 & 3 \end{pmatrix}$$

(Remember that in composition of functions, we apply the right-hand function first and then the left-hand one. In this case, for instance, $4 \rightarrow 3 \rightarrow 2$, as shown by the arrows.) We claim that S_n is a group under this operation. Composition of functions is known to be associative, and every bijection has an inverse function under composition.* It is easy to verify that the identity permutation $\begin{pmatrix} 1 & 2 & 3 & \dots & n \\ 1 & 2 & 3 & & n \end{pmatrix}$ is the identity element of S_n. S_n is called the **symmetric group** on n symbols. The order of S_n is $n! = n(n-1)(n-2) \dots 2.1$ (Exercise 18).

EXAMPLE The preceding example is easily generalized. Let T be any nonempty set, possibly infinite. Let $A(T)$ be the set of all permutations of T (all bijective functions $T \rightarrow T$). The arguments given above for S_n carry over to $A(T)$ and show that $A(T)$ is a group under the operation of composition of functions (Exercise 10).

EXAMPLE Think of the plane as a sheet of thin, rigid plastic. Suppose you cut out a square, pick it up, and move it around,** then replace it so that it fits exactly in the cut-out space. Eight ways of doing this are shown below (where the square is centered at the origin and its corners numbered for easy reference). We claim that any motion of the square that ends with the square fitting exactly in the cut-out space has the same result as one of these eight (Exercise 12). For instance, a rotation of $540° (= 180° + 360°)$ puts the square in the same position as a rotation of $180°$.

---

* See Appendix B for details.
** Flip it, rotate it, turn it over, spin it, do whatever you want, as long as you don't bend, break, or distort it.

All Rotations Are Taken Counterclockwise Around the Center:

r_0 = rotation of $0°$

r_1 = rotation of $90°$

r_2 = rotation of $180°$

r_3 = rotation of $270°$

d = reflection in the x-axis

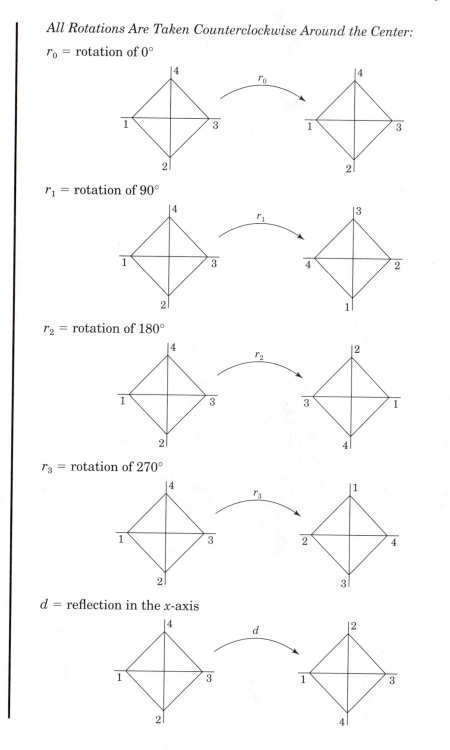

t = reflection in the y-axis

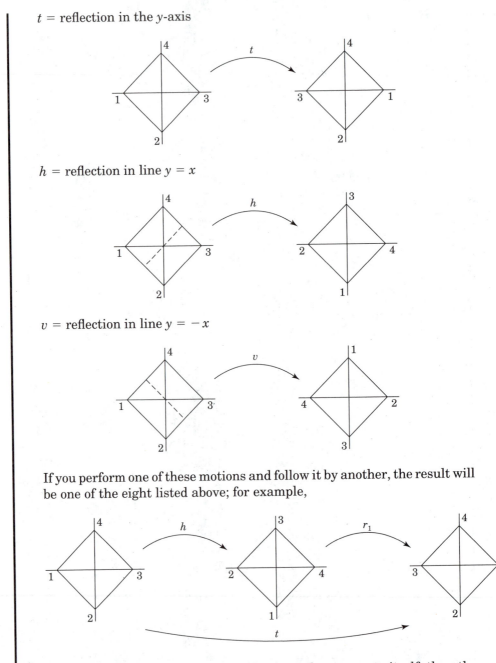

h = reflection in line $y = x$

v = reflection in line $y = -x$

If you perform one of these motions and follow it by another, the result will be one of the eight listed above; for example,

If you think of a motion as a function from the square to itself, then the idea of following one motion by another is just composition of functions. In the illustration above (h followed by r_1 is t), we can write $r_1 \circ h = t$ (remember $r_1 \circ h$ means first apply h, then apply r_1). Verify that the set

$$D_4 = \{r_0, r_1, r_2, r_3, h, v, d, t\}$$

equipped with the composition operation has this table:

$\circ$	r_0	r_1	r_2	r_3	d	h	t	v
r_0	r_0	r_1	r_2	r_3	d	h	t	v
r_1	r_1	r_2	r_3	r_0	h	t	v	d
r_2	r_2	r_3	r_0	r_1	t	v	d	h
r_3	r_3	r_0	r_1	r_2	v	d	h	t
d	d	v	t	h	r_0	r_3	r_2	r_1
h	h	d	v	t	r_1	r_0	r_3	r_2
t	t	h	d	v	r_2	r_1	r_0	r_3
v	v	t	h	d	r_3	r_2	r_1	r_0

Clearly D_4 is closed under $\circ$, and composition of functions is known to be associative. The table shows that r_0 is the identity element and that every element of D_4 has an inverse. For instance, $r_3 \circ r_1 = r_0 = r_1 \circ r_3$. Therefore, D_4 is a group. It is not abelian because, for example, $h \circ d \neq d \circ h$. D_4 is called the **dihedral group of degree 4** or the group of **symmetries of the square.**

EXAMPLE The group of symmetries of the square is just one of many symmetry groups. An analogous procedure can be carried out with any regular polygon of n sides. The resulting group D_n is called the **dihedral group of degree n**. The group D_3, for example, consists of the six symmetries of an equilateral triangle (counterclockwise rotations about the center of $0°$, $120°$, and $240°$; and the three reflections shown here and on the next page), with composition of functions as the operation:

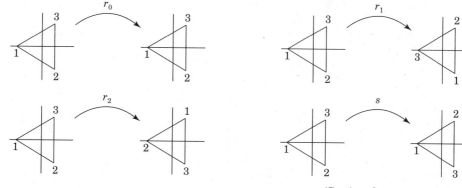

(Continued on next page)

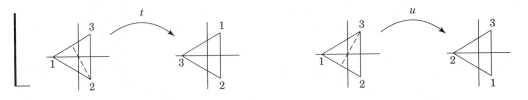

EXAMPLE The group of symmetries of the following figure consists of four elements (rotations about the center of 0° and 180° and reflections in the *x*- and *y*-axes):

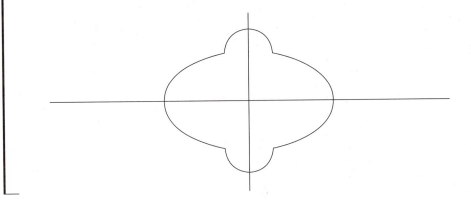

Symmetry groups arise frequently in art, architecture, and science. Crystallography and crystal physics use groups of symmetries of various three-dimensional shapes. The first accurate model of DNA (which led to the Nobel Prize for its creators) could not have been constructed without a recognition of the symmetry of the DNA molecule. Symmetry groups have been used by physicists to predict the existence of certain elementary particles that were later found experimentally.

Groups and Rings

A ring R has two associative operations, and it is natural to ask if R is a group under either one. For addition the answer is yes:

THEOREM 7.1 *Every ring is an abelian group under addition.*

Proof An examination of the first five axioms for a ring (page 42) shows that they are identical to the five axioms for an abelian group, with the operation $*$ being $+$, the identity element e being 0_R, and the inverse of a being $-a$. ◆

Hereafter when we use the word "group" without any qualification in referring to $\mathbb{Z}$ or $\mathbb{Z}_n$ or $\mathbb{R}$ or other well-known rings, it is understood that the group operation is addition.

Under multiplication, however, a nonzero ring is *never* a group (Exercise 5). But certain subsets of a ring may be groups under multiplication.

> **EXAMPLE** The subset $\{1, -1, i, -i\}$ of the complex numbers forms an abelian group of order 4 under multiplication. You can easily verify closure, and 1 is the identity element. Since $i(-i) = 1$, i and $-i$ are inverses of each other; -1 is its own inverse since $(-1)(-1) = 1$. Hence, Axiom 4 holds.

> **EXAMPLE** The positive rational numbers $\mathbb{Q}^{**}$ form an infinite abelian group under multiplication, because the product of positive numbers is positive, 1 is the identity element, and the inverse of a is $1/a$. Similarly, the positive reals form an abelian group under multiplication. The positive integers do not, however, since the equation $ax = 1$ (with $a \geq 2$) has no solution in $\mathbb{Z}$, so that a has no inverse under multiplication (Axiom 4 fails).

A ring R with identity always has at least one subset that is a group under multiplication. Recall that a *unit* in R is an element a that has a multiplicative inverse, that is, an element u such that $au = 1_R = ua$.

THEOREM 7.2 *If R is a ring with identity, then the set U of all units in R is a group under multiplication.*

Proof The product of units is a unit (Exercise 11 in Section 3.2), so U is closed under multiplication (Axiom 1). Multiplication in R is associative, so Axiom 2 holds. Since 1_R is obviously a unit, U has an identity element (Axiom 3). Axiom 4 holds in U by the definition of unit. Therefore, U is a group. ◆

> **EXAMPLE** Denote the multiplicative group of units in $\mathbb{Z}_n$ by U_n. According to Corollary 2.10 and Theorem 2.11, U_n consists of all $a \in \mathbb{Z}_n$ such that $(a, n) = 1$ (when a is considered as an ordinary integer). Thus the group of units in $\mathbb{Z}_8$ is $U_8 = \{1, 3, 5, 7\}$, and the group of units in $\mathbb{Z}_{15}$ is $U_{15} = \{1, 2, 4, 7, 8, 11, 13, 14\}$. Here is the operation table for U_8:
>
> | · | 1 | 3 | 5 | 7 |
> |---|---|---|---|---|
> | 1 | 1 | 3 | 5 | 7 |
> | 3 | 3 | 1 | 7 | 5 |
> | 5 | 5 | 7 | 1 | 3 |
> | 7 | 7 | 5 | 3 | 1 |

> **EXAMPLE** The group of units in the ring $M(\mathbb{R})$ of all 2×2 matrices over the real numbers is denoted $GL(2, \mathbb{R})$ and called the **general linear group.** According to Exercise 34 in Section 3.2, a matrix $\begin{pmatrix} a & b \\ c & d \end{pmatrix}$ is in $GL(2, \mathbb{R})$ if and only if $ad - bc \neq 0$. Consequently, both $\begin{pmatrix} 1 & -1 \\ 2 & 3 \end{pmatrix}$ and $\begin{pmatrix} 1 & 1 \\ 2 & 0 \end{pmatrix}$ are in $GL(2, \mathbb{R})$. By taking their product in both orders, you can verify that $GL(2, \mathbb{R})$ is a nonabelian group. The same notation is used in other situations: $GL(2, K)$ denotes the multiplicative group of units in the ring of 2×2 matrices with entries in the ring K.

If R is a field, then multiplication is commutative and every nonzero element is a unit (see page 60). These facts and Theorem 7.2 prove

COROLLARY 7.3 *The nonzero elements of a field form an abelian group under multiplication.*

New Groups from Old

The Cartesian product, with operations defined coordinatewise, allowed us to construct new rings from known ones. The same is true for groups.

THEOREM 7.4 *Let G (with operation $*$) and H (with operation $\diamond$) be groups. Define an operation $\cdot$ on $G \times H$ by*

$$(g, h) \cdot (g', h') = (g * g', h \diamond h').$$

Then $G \times H$ is a group. If G and H are abelian, then so is $G \times H$. If G and H are finite, then so is $G \times H$ and $|G \times H| = |G||H|$.

Proof Exercise 24. ◆

> **EXAMPLE** Both $\mathbb{Z}$ and $\mathbb{Z}_6$ are groups under addition. In $\mathbb{Z} \times \mathbb{Z}_6$ we have $(3, 5) \cdot (7, 4) = (3 + 7, 5 + 4) = (10, 3)$. The identity is $(0, 0)$, and the inverse of $(7, 4)$ is $(-7, 2)$.

> **EXAMPLE** Consider $\mathbb{R}^* \times D_4$, where $\mathbb{R}^*$ is the multiplicative group of nonzero real numbers. The table on page 167 shows that
>
> $$(2, r_1) \cdot (9, v) = (2 \cdot 9, r_1 \circ v) = (18, d).$$
>
> The identity element is $(1, r_0)$, and the inverse of $(8, r_3)$ is $(1/8, r_1)$.

◆ **EXERCISES**

A. **1.** Find the inverse of each permutation in S_3.

2. Find all elements f in S_3 such that $f \circ f \circ f = I$.

3. What is the order of each group:

(a) $\mathbb{Z}_{18}$ (b) D_4 (c) S_4 (d) S_5 (e) U_{18}

4. Determine whether the set G is a group under the operation $*$.

(a) $G = \{2, 4, 6, 8\}$ in $\mathbb{Z}_{10}$; $a * b = ab$

(b) $G = \{2^x \mid x \in \mathbb{Q}\}$; $a * b = ab$

(c) $G = \{x \in \mathbb{Z} \mid x \equiv 1 \pmod 5\}$; $a * b = ab$

(d) $G = \{x \in \mathbb{Q} \mid x \neq 1\}$; $a * b = a + b - ab$

(e) $G = \{x \in \mathbb{R} \mid x \neq -1\}$; $a * b = ab + a + b$

(f) $G = \{c + di \in \mathbb{C} \mid cd = 0 \text{ and } c + d \neq 0\}$; $a * b = ab$

5. Prove that a nonzero ring R is not a group under multiplication. [*Hint:* What is the inverse of 0_R?]

6. Use Theorem 2.11 to list the elements of each of these groups: U_4, U_6, U_{10}, U_{20}, U_{30}.

7. Write out the operation table for the group D_3 described in the example on pages 167–168.

8. Show that $G = \left\{ \begin{pmatrix} a & b \\ -b & a \end{pmatrix} \middle| a, b \in \mathbb{R}, \text{not both } 0 \right\}$ is an abelian group under matrix multiplication.

9. Consider the additive group $\mathbb{Z}_2$ and the multiplicative group $G = \{\pm 1, \pm i\}$ of complex numbers. Write out the operation table for the group $\mathbb{Z}_2 \times G$.

10. Let T be a nonempty set and $A(T)$ the set of all permutations of T. Show that $A(T)$ is a group under the operation of composition of functions.

11. (a) Give examples of nonabelian groups of orders 12, 16, 30, and 48. [*Hint:* Theorem 7.4 may be helpful.]

(b) Give an example of an abelian group of order 4 in which every non-identity element satisfies $x * x = e$.

B. **12.** Show that every rigid motion of the square (as described in the footnote on page 164) has the same result as an element of D_4. [*Hint:* The position of the square after any motion is completely determined by the location of corner 1 and by the orientation of the square—face up or face down.]

13. Write out the operation table for the symmetry groups of the following figures:

(a)    (b)   (c)

14. Let **1, i, j, k** be the following matrices with complex entries:

$$\mathbf{1} = \begin{pmatrix} 1 & 0 \\ 0 & 1 \end{pmatrix}, \quad \mathbf{i} = \begin{pmatrix} i & 0 \\ 0 & -i \end{pmatrix}, \quad \mathbf{j} = \begin{pmatrix} 0 & 1 \\ -1 & 0 \end{pmatrix}, \quad \mathbf{k} = \begin{pmatrix} 0 & i \\ i & 0 \end{pmatrix}.$$

Show that set $Q = \{1, i, -1, -i, j, k, -j, -k\}$ is a group under matrix multiplication by writing out its multiplication table. Q is called the **quaternion group.** [Exercise 35 in Section 3.1 may be helpful.]

15. If G is a group under the stated operation, prove it; if not, give a counterexample:

(a) $G = \mathbb{Q}; a * b = a + b + 3$

(b) $G = \{r \in \mathbb{Q} \mid r \neq 0\}; a * b = ab/3$

16. Let $K = \{r \in \mathbb{R} \mid r \neq 0, r \neq 1\}$. Let G consist of these six functions from K to K:

$$f(x) = \frac{1}{1-x} \qquad g(x) = \frac{x-1}{x} \qquad h(x) = \frac{1}{x}$$

$$i(x) = x \qquad j(x) = 1 - x \qquad k(x) = \frac{x}{x-1}$$

Is G a group under the operation of function composition?

17. Do the nonzero real numbers form a group under the operation given by $a * b = |a|b$, where $|a|$ is the absolute value of a?

18. Prove that S_n has order $n!$. [*Hint:* There are n possible images for 1; after one has been chosen, there are $n - 1$ possible images for 2; etc.]

19. Suppose G is a group with operation $*$. Define a new operation $\#$ on G by $a \# b = b * a$. Prove that G is a group under $\#$.

20. List the elements of the group D_5 (the symmetries of a regular pentagon). [*Hint:* The group has order 10.]

21. Let $SL(2, \mathbb{R})$ be the set of all 2×2 matrices $\begin{pmatrix} a & b \\ c & d \end{pmatrix}$ such that $a, b, c, d \in \mathbb{R}$ and $ad - bc = 1$. Prove that $SL(2, \mathbb{R})$ is a group under matrix multiplication. It is called the **special linear group.**

22. Prove that the set of nonzero real numbers is a group under the operation $*$ defined by

$$a * b = \begin{cases} ab & \text{if } a > 0 \\ a/b & \text{if } a < 0. \end{cases}$$

23. Prove that $\mathbb{R}^* \times \mathbb{R}$ is a group under the operation $*$ defined by
$(a, b) * (c, d) = (ac, bc + d)$.

24. Prove Theorem 7.4.

25. If $ab = ac$ in a group G, prove that $b = c$.

26. Prove that each element of a finite group G appears exactly once in each row
and exactly once in each column of the operation table. [*Hint:* Exercise 25.]

27. Here is part of the operation table for a group G whose elements are a, b, c, d.
Fill in the rest of the table. [*Hint:* Exercises 25 and 26.]

	a	b	c	d
a	a	b	c	d
b	b	a		
c	c		a	
d	d			

28. A partial operation table for a group $G = \{e, a, b, c, d, f\}$ is shown below.
Complete the table. [*Hint:* Exercises 25 and 26.]

	e	a	b	c	d	f
e	e	a	b	c	d	f
a	a	b	e	d		
b	b					
c	c	f				a
d	d					
f	f					

29. Let T be a set with at least three elements. Show that the permutation group
$A(T)$ (Exercise 10) is nonabelian.

30. Let T be an infinite set and let $A(T)$ be the group of permutations of T (Exercise 10). Let $M = \{f \in A(T) \mid f(t) \neq t$ for only a finite number of $t \in T\}$. Prove
that M is a group.

31. If $a, b \in \mathbb{R}$ with $a \neq 0$, let $T_{a,b} : \mathbb{R} \to \mathbb{R}$ be the function given by $T_{a,b}(x) = ax + b$. Prove that the set $G = \{T_{a,b} \mid a, b \in \mathbb{R}$ with $a \neq 0\}$ forms a nonabelian
group under composition of functions.

32. Let $H = \{T_{1,b} \mid b \in \mathbb{R}\}$ (notation as in Exercise 31). Prove that H is an abelian
group under composition of functions.

C. 33. If $f \in S_n$, prove that $f^k = I$ for some positive integer k, where f^k means
$f \circ f \circ f \circ \cdots \circ f$ (k times) and I is the identity permutation.

34. Let $G = \{0, 1, 2, 3, 4, 5, 6, 7\}$ and assume G is a group under an operation $*$ with these properties:

(i) $a * b \leq a + b$ for all $a, b \in G$;

(ii) $a * a = 0$ for all $a \in G$.

Write out the operation table for G. [*Hint:* Exercises 25 and 26 may help.]

7.2 BASIC PROPERTIES OF GROUPS

Before exploring the deeper concepts of group theory, we must develop some additional terminology and establish some elementary facts. We begin with a change in notation.

The operation in an arbitrary group was denoted by $*$ in the preceding section to avoid any confusion with rings. Now that you are comfortable with groups, we can switch to the **standard multiplicative notation.** Instead of $a * b$, we shall write ab when discussing abstract groups. However, particular groups in which the operation is addition (such as $\mathbb{Z}$) will still be written additively.

Although we have spoken of *the* inverse of an element or *the* identity element of a group, the definition of a group says nothing about inverses or identities being unique. Our first theorem settles the question, however.

THEOREM 7.5 *Let G be a group and let a, b, c ∈ G. Then*

(1) G has a unique identity element.

(2) Cancellation holds in G:

$$If\ ab = ac,\ then\ b = c; \qquad if\ ba = ca,\ then\ b = c.$$

(3) Each element of G has a unique inverse.

Proof (1) The group G has at least one identity by the definition of a group. If e and e' are both identity elements of G, then $ea = a = ae$ and $e'a = a = ae'$ for every $a \in G$. The first of these equations, with $a = e'$, says that $ee' = e'$. The last equation, with $a = e$, says that $e = ee'$. Therefore, $e' = ee' = e$, so that there is exactly one identity element.

(2) By the definition of a group, the element a has at least one inverse d such that $da = e = ad$. If $ab = ac$, then $d(ab) = d(ac)$. By associativity and the properties of inverses and identities,

$$(da)b = (da)c$$
$$eb = ec$$
$$b = c.$$

The second statement is proved similarly.

(3) Suppose that d and d' are both inverses of $a \in G$. Then $ad = e = ad'$, so that $d = d'$ by (2). Therefore a has exactly one inverse. ◆

Hereafter the unique inverse of an element a in a group will be denoted a^{-1}. The uniqueness of a^{-1} means that

whenever $ay = e = ya$, then $y = a^{-1}$.

COROLLARY 7.6   *If G is a group and a, b $\in$ G, then*

(1) $(ab)^{-1} = b^{-1}a^{-1}$;

(2) $(a^{-1})^{-1} = a$.

Note the order of the elements in statement (1). A common mistake is to write the inverse of ab as $a^{-1}b^{-1}$, which may not be true in nonabelian groups. See Exercise 2 for an example.

Proof of Corollary 7.6   (1) We have

$$(ab)(b^{-1}a^{-1}) = a(bb^{-1})a^{-1} = aea^{-1} = aa^{-1} = e$$

and, similarly, $(b^{-1}a^{-1})(ab) = e$. Since the inverse of ab is unique by Theorem 7.5, $b^{-1}a^{-1}$ must be this inverse, that is, $(ab)^{-1} = b^{-1}a^{-1}$.

(2) By definition, $a^{-1}a = e$ and $(a^{-1})(a^{-1})^{-1} = e$, so that $a^{-1}a = a^{-1}(a^{-1})^{-1}$. Cancelling a^{-1} by Theorem 7.5 shows that $a = (a^{-1})^{-1}$. ◆

Let G be a group and let $a \in G$. We define $a^2 = aa$, $a^3 = aaa$, and for any positive integer n,

$$a^n = aaa \cdots a \qquad \textbf{(\textit{n} factors).}$$

We also define $a^0 = e$ and

$$a^{-n} = a^{-1}a^{-1}a^{-1} \cdots a^{-1} \qquad \textbf{(\textit{n} factors).}$$

These definitions are obviously motivated by the usual exponent notation in $\mathbb{R}$ and other familiar rings. But be careful in the nonabelian case when, for instance, $(ab)^n$ may not be equal to a^nb^n. Some exponent rules, however, *do* hold in groups:

THEOREM 7.7   *Let G be a group and let a $\in$ G. Then for all m, n in $\mathbb{Z}$,*

$$a^ma^n = a^{m+n} \qquad and \qquad (a^m)^n = a^{mn}.$$

Proof The proof consists of a verification of each statement in each possible case ($m \geq 0$, $n \geq 0$; $m \geq 0$, $n < 0$; etc.) and is left to the reader (Exercise 19). ◆

NOTE ON ADDITIVE NOTATION To avoid confusion, the operation in certain groups must be written as addition (for example, the additive group of real numbers since multiplication here has a completely different meaning). When the group operation is written as addition, then the identity element is denoted by 0 instead of e and the inverse of a is denoted by $-a$ instead of a^{-1}. Exponentiation and Theorem 7.7 then take on a somewhat different form. The additive analogue of $aaa \cdots a$ (n factors) is $a + a + \cdots + a$ (n summands), and we write $na = a + a + \cdots + a$ instead of $a^n = aa \cdots a$. Similarly, the additive notation for a^{-n} is $(-n)a$, and the conclusion of Theorem 7.7 reads

$$ma + na = (m + n)a \qquad \text{and} \qquad n(ma) = (mn)a.$$

This notation is consistent with what was done with addition in rings.

Order of an Element

We return now to multiplicative notation for abstract groups. An element a in a group is said to have **finite order** if $a^k = e$ for some positive integer k.* In this case, the **order of the element a** is the *smallest* positive integer n such that $a^n = e$. The order of a is denoted $|a|$. An element a is said to have **infinite order** if $a^k \neq e$ for every positive integer k.

> **EXAMPLE** In the multiplicative group of nonzero real numbers, 2 has infinite order because $2^k \neq 1$ for all $k \geq 1$. In the group $G = \{\pm 1, \pm i\}$ under multiplication of complex numbers, the order of i is 4 because $i^2 = -1$, $i^3 = -i$, and $i^4 = 1$. Similarly, $|-i| = 4$. The element $\begin{pmatrix} 1 & 2 & 3 \\ 3 & 1 & 2 \end{pmatrix}$ in S_3 has order 3 because
>
> $$\begin{pmatrix} 1 & 2 & 3 \\ 3 & 1 & 2 \end{pmatrix}^2 = \begin{pmatrix} 1 & 2 & 3 \\ 2 & 3 & 1 \end{pmatrix} \qquad \text{and} \qquad \begin{pmatrix} 1 & 2 & 3 \\ 3 & 1 & 2 \end{pmatrix}^3 = \begin{pmatrix} 1 & 2 & 3 \\ 1 & 2 & 3 \end{pmatrix}.$$
>
> The identity element in a group has order 1.

> **EXAMPLE** In the additive group $\mathbb{Z}_{12}$, the element 8 has order 3 because $8 + 8 = 4$ and $8 + 8 + 8 = 0$.

In the multiplicative group of nonzero real numbers, the element 2 has infinite order and all the powers of 2 (2^{-3}, 2^0, 2^5, etc.) are distinct. On the other

---

* In additive notation, the condition is $ka = 0$.

hand, in the multiplicative group $G = \{\pm 1, \pm i\}$, the element i has order 4 and its powers are not distinct; for instance,

$$i^4 = 1 = i^0 \qquad \text{and} \qquad i^{10} = (i^4)^2 i^2 = i^2.$$

Observe that $i^{10} = i^2$ and $10 \equiv 2 \pmod 4$. These examples are illustrations of

THEOREM 7.8 *Let G be a group and let $a \in G$.*

(1) If a has infinite order, then the elements a^k, with $k \in \mathbb{Z}$, are all distinct.

(2) If $a^i = a^j$ with $i \neq j$, then a has finite order.

(3) If a has finite order n, then

$$a^k = e \text{ if and only if } n \mid k$$

and

$$a^i = a^j \text{ if and only if } i \equiv j \pmod n.$$

(4) If a has order n and $n = td$ with $d > 0$, then a^t has order d.

Part (2) of the theorem illustrates one of the ways that exponents behave differently in groups than in rings. For example, in the ring $\mathbb{Z}_6$ (whose nonzero elements do *not* form a group under multiplication), $2^3 = 2^1$, but $2^k \neq 1$ for any positive integer k.

Proof of Theorem 7.8 (1) and (2) Statement (1) is true if and only if statement (2) is true, because each statement is the contrapositive of the other, as explained in Appendix A. So we need only prove (2). Suppose that $a^i = a^j$ with $i > j$. Then multiplying both sides by a^{-j} shows that $a^{i-j} = a^{j-j} = a^0 = e$. Since $i - j > 0$, this says that a has finite order.

(3) If n divides k, say $k = nt$, then $a^k = a^{nt} = (a^n)^t = e^t = e$. Conversely, suppose that $a^k = e$. By the Division Algorithm, $k = nq + r$ with $0 \leq r < n$. Consequently,

$$e = a^k = a^{nq+r} = a^{nq}a^r = (a^n)^q a^r = e^q a^r = ea^r = a^r.$$

By the definition of order, n is the smallest positive integer with $a^n = e$. Since $r < n$, $a^r = e$ can occur only when $r = 0$. Thus, $k = nq + 0$ and n divides k. Finally, note that $a^i = a^j$ if and only if $a^{i-j} = a^0 = e$. By what was just proved, $a^{i-j} = e$ if and only if n divides $i - j$, that is, if and only if $i \equiv j \pmod n$.

(4) Since $|a| = n$, we have $(a^t)^d = a^{td} = a^n = e$. We must show that d is the smallest positive integer with this property. If k is any positive integer such that $(a^t)^k = e$, then $a^{tk} = e$. Therefore, $n \mid tk$ by part (3), say $tk = nr = (td)r$. Hence, $k = dr$. Since k and d are positive and $d \mid k$, we have $d \leq k$. ◆

COROLLARY 7.9 *Let G be an abelian group in which every element has finite order. If c ∈ G is an element of largest order in G (that is, |a| ≤ |c| for all a ∈ G), then the order of every element of G divides |c|.*

For example, $(1,0)$ has order 4 in the additive abelian group $\mathbb{Z}_4 \times \mathbb{Z}_2$ and every other element has order 1, 2, or 4 (Exercise 10(b)). Thus $(1,0)$ is an element of largest possible order, and the order of every element of the group divides 4, the order of $(1,0)$.

Proof of Corollary 7.9 Suppose, on the contrary, that $a \in G$ and $|a|$ does not divide $|c|$. Then there must be a prime p in the prime factorization of the integer $|a|$ that appears to a higher power than it does in the prime factorization of $|c|$. By prime factorization we can write $|a|$ as the product of a power of p and an integer that is not divisible by p and similarly for c. Thus there are integers m, n, r, s such that $|a| = p^r m$ and $|c| = p^s n$, with $(p, m) = 1 = (p, n)$ and $r > s$. By part (4) of Theorem 7.8, the element a^m has order p^r and c^{p^s} has order n. Exercise 31 shows that $a^m c^{p^s}$ has order $p^r n$. Hence, $|a^m c^{p^s}| = p^r n > p^s n = |c|$, contradicting the fact that c is an element of largest order. Therefore, $|a|$ divides $|c|$. ◆

◆ **EXERCISES**

NOTE: *Unless stated otherwise, G is a group with identity element e.*

A. **1.** If $c^2 = c$ in a group, prove that $c = e$.

2. Let $a = \begin{pmatrix} 1\ 2\ 3 \\ 3\ 1\ 2 \end{pmatrix}$ and $b = \begin{pmatrix} 1\ 2\ 3 \\ 1\ 3\ 2 \end{pmatrix}$ in S_3. Verify that $(ab)^{-1} \neq a^{-1}b^{-1}$.

3. If $a, b, c, d \in G$, then $(abcd)^{-1} = ?$

4. If $a, b \in G$ and $ab = e$, prove that $ba = e$.

5. Let $f: G \to G$ be given by $f(a) = a^{-1}$. Prove that f is a bijection.

6. Give an example of a group in which the equation $x^2 = e$ has more than two solutions.

7. Find the order of the given element.

(a) 5 in U_8

(b) $\begin{pmatrix} 1\ 2\ 3\ 4\ 5\ 6\ 7 \\ 2\ 3\ 7\ 5\ 1\ 4\ 6 \end{pmatrix}$ in S_7

(c) $\begin{pmatrix} 0 & -1 \\ 1 & 1 \end{pmatrix}$ in $GL(2, \mathbb{R})$

(d) $\begin{pmatrix} -\frac{1}{2} & \frac{1}{2} \\ -\frac{3}{2} & -\frac{1}{2} \end{pmatrix}$ in $GL(2, \mathbb{R})$

8. Give an example of a group that contains nonidentity elements of finite order and of infinite order.

9. **(a)** Find the order of the groups U_{10}, U_{12}, and U_{24}.

   (b) List the order of each element of the group U_{20}.

10. Find the order of every element in each group:

    (a) $\mathbb{Z}_4$ **(b)** $\mathbb{Z}_4 \times \mathbb{Z}_2$ **(c)** S_3 **(d)** D_4 **(e)** $\mathbb{Z}$

11. If G is a finite group of order n and $a \in G$, prove that $|a| \leq n$. [*Hint:* Consider the $n + 1$ elements $e = a^0 \, a, a^2, a^3, \ldots, a^n$. Are they all distinct?] Thus every element in a finite group has finite order. The converse, however, is false; see Exercise 15 in Section 7.7 for an infinite group in which every element has finite order.

12. True or false: A group of order n contains an element of order n. Justify your answer.

13. **(a)** If $a \in G$ and $a^{12} = e$, what order can a possibly have?

    (b) If $e \neq b \in G$ and $b^p = e$ for some prime p, what is $|b|$?

14. **(a)** If $a \in G$ and $|a| = 12$, find the orders of each of the elements $a, a^2, a^3, \ldots, a^{11}$.

    (b) Based on the evidence in part (a), make a conjecture about the order of a^k when $|a| = n$.

15. **(a)** Let $a, b \in G$. Prove that the equations $ax = b$ and $ya = b$ each have a unique solution in G. [*Hint:* Two things must be done for each equation: First find a solution and then show that it is the only solution.]

    (b) Show by example that the solution of $ax = b$ may not be the same as the solution of $ya = b$. [*Hint:* Consider S_3.]

16. Let $G = \{a_1, a_2, \ldots, a_n\}$ be a finite abelian group of order n. Let $x = a_1 a_2 \cdots a_n$. Prove that $x^2 = e$.

17. If $a, b \in G$, prove that $|bab^{-1}| = |a|$.

18. **(a)** Show that $a = \begin{pmatrix} 0 & 1 \\ -1 & -1 \end{pmatrix}$ has order 3 in $GL(2, \mathbb{R})$ and $b = \begin{pmatrix} 0 & -1 \\ 1 & 0 \end{pmatrix}$ has order 4.

    (b) Show that ab has infinite order.

B. 19. Prove Theorem 7.7.

20. Let $G = \{e, a, b,\}$ be a group of order 3. Write out the operation table for G. [*Hint:* Exercise 26 in Section 7.1.]

21. Let G be a group with this property: If $a, b, c \in G$ and $ab = ca$, then $b = c$. Prove that G is abelian.

22. If $(ab)^2 = a^2b^2$ for all $a, b, \in G$, prove that G is abelian.

23. Prove that G is abelian if and only if $(ab)^{-1} = a^{-1}b^{-1}$ for all $a, b \in G$.

24. Prove that every nonabelian group G has order at least 6; hence, every group of order 2, 3, 4, or 5 is abelian. [*Hint:* If $a, b \in G$ and $ab \neq ba$, show that the elements of the subset $H = \{e, a, b, ab, ba\}$ are all distinct. Show that either $a^2 \notin H$ or $a^2 = e$; in the latter case, verify that $aba \notin H$.]

25. If every nonidentity element of G has order 2, prove that G is abelian. [*Hint:* $|a| = 2$ if and only if $a \neq e$ and $a = a^{-1}$. Why?]

26. If $a \in G$, prove that $|a| = |a^{-1}|$.

27. If $a, b, c \in G$, prove that there is a unique element $x \in G$ such that $axb = c$.

28. If $a, b \in G$, prove that $|ab| = |ba|$.

29. **(a)** If $a, b \in G$ and $ab = ba$, prove that $(ab)^{|a||b|} = e$.

   (b) Show that part (a) may be false if $ab \neq ba$.

30. If $|G|$ is even, prove that G contains an element of order 2. [*Hint:* The identity element is its own inverse. See the hint for Exercise 25.]

31. Assume that $a, b \in G$ and $ab = ba$. If $|a|$ and $|b|$ are relatively prime, prove that ab has order $|a||b|$. [*Hint:* See Exercise 29.]

32. Suppose G has order 4, but contains no element of order 4.

   (a) Prove that no element of G has order 3. [*Hint:* If $|g| = 3$, then G consists of four distinct elements $g, g^2, g^3 = e, d$. Now gd must be one of these four elements. Show that each possibility leads to a contradiction.]

   (b) Explain why every nonidentity element of G has order 2.

   (c) Denote the elements of G by e, a, b, c and write out the operation table for G.

33. If $a, b \in G$, $b^6 = e$, and $ab = b^4a$, prove that $b^3 = e$ and $ab = ba$.

34. Suppose $a, b \in G$ with $|a| = 5$, $b \neq e$, and $aba^{-1} = b^2$. Find $|b|$.

35. If $(ab)^3 = a^3b^3$ and $(ab)^5 = a^5b^5$ for all $a, b \in G$, prove that G is abelian.

C. 36. If $(ab)^i = a^i b^i$ for three consecutive integers i and all $a, b \in G$, prove that G is abelian.

37. (a) Let G be a nonempty finite set equipped with an associative operation such that for all $a, b, c, d \in G$:

if $ab = ac$, then $b = c$ and if $bd = cd$, then $b = c$.

Prove that G is a group.

(b) Show that part (a) may be false if G is infinite.

38. Let G be a nonempty set equipped with an associative operation with these properties:

(i) There is an element $e \in G$ such that $ea = a$ for every $a \in G$.

(ii) For each $a \in G$, there exists $d \in G$ such that $da = e$.

Prove that G is a group.

39. Let G be a nonempty set equipped with an associative operation such that, for all $a, b \in G$, the equations $ax = b$ and $ya = b$ have solutions. Prove that G is a group.

7.3 SUBGROUPS

We continue our discussion of the basic properties of groups, with special attention to subgroups.

> **DEFINITION•** *A subset H of a group G is a **subgroup** of G if H is itself a group under the operation in G.*

Every group G has two subgroups: G itself and the one-element group $\{e\}$, which is called the **trivial subgroup.** All other subgroups are said to be **proper subgroups.**

EXAMPLE The set $\mathbb{R}^*$ of nonzero real numbers is a group under multiplication. The group $\mathbb{R}^{**}$ of positive real numbers is a proper subgroup of $\mathbb{R}^*$.

EXAMPLE Any ring R is a group under addition. If S is a subring of R, then S (considered as an additive group) is automatically a subgroup of R. In particular, every ideal in R is an additive subgroup of R.

> **EXAMPLE** The operation table for D_4 on page 167 shows that $H = \{r_0, r_1, r_2, r_3\}$ is a subgroup of D_4.

> **EXAMPLE** In the additive group $\mathbb{Z}_6 \times \mathbb{Z}_4$, let $H = \{(0, 0), (3, 0), (0, 2), (3, 2)\}$. Verify that H is a subgroup by writing out its addition table.

When proving that a subset of a group is a subgroup, it is never necessary to check associativity. Since the associative law holds for *all* elements of the group, it automatically holds when the elements are in some subset H. In fact, you need only verify two group axioms:

THEOREM 7.10 *A nonempty subset H of a group G is a subgroup of G provided that*

  (i) if a, $b \in H$, then $ab \in H$; and

  (ii) if $a \in H$, then $a^{-1} \in H$.

Proof Properties (i) and (ii) are the closure and inverse axioms for a group. Associativity holds in H, as noted above. Thus we need only verify that $e \in H$. Since H is nonempty, there exists an element $c \in H$. By (ii), $c^{-1} \in H$, and by (i) $cc^{-1} = e$ is in H. Therefore H is a group. ◆

> **EXAMPLE** Let H consist of all 2×2 matrices of the form $b = \begin{pmatrix} 1 & b \\ 0 & 1 \end{pmatrix}$ with $b \in \mathbb{R}$. Since $1 \cdot 1 - b \cdot 0 = 1$, H is a nonempty subset of the group $GL(2, \mathbb{R})$, which was defined on page 170. The product of two matrices in H is in H because
> $$\begin{pmatrix} 1 & b \\ 0 & 1 \end{pmatrix} \begin{pmatrix} 1 & c \\ 0 & 1 \end{pmatrix} = \begin{pmatrix} 1 & b+c \\ 0 & 1 \end{pmatrix}$$
> The inverse of $\begin{pmatrix} 1 & b \\ 0 & 1 \end{pmatrix}$ is $\begin{pmatrix} 1 & -b \\ 0 & 1 \end{pmatrix}$, which is also in H. Therefore, H is a subgroup of $GL(2, \mathbb{R})$ by Theorem 7.10.

When H is finite, just one axiom is sufficient to guarantee that H is a subgroup.

THEOREM 7.11 *Let H be a nonempty finite subset of a group G. If H is closed under the operation in G, then H is a subgroup of G.*

Proof By Theorem 7.10, we need only verify that the inverse of each element of H is also in H. If $a \in H$, then closure implies that $a^k \in H$ for every positive integer k. Since H is finite, these powers cannot all be distinct. So a has finite order n by Theorem 7.8 and $a^n = e$. Since $n - 1 \equiv -1 \pmod{n}$, we have $a^{n-1} = a^{-1}$ by Theorem 7.8. If $n > 1$, then $n - 1$ is positive and $a^{-1} = a^{n-1}$ is in H. If $n = 1$, then $a = e$ and $a^{-1} = e = a$, so that a^{-1} is in H. ◆

> **EXAMPLE** Let H consist of all permutations in S_5 that fix the element 1. In other words, $H = \{f \epsilon S_5 \mid f(1) = 1\}$. H is a finite set since S_5 is a finite group. If g, $h \epsilon H$, then $g(1) = 1$ and $h(1) = 1$. Hence, $(g \circ h)(1) = g(h(1)) = g(1) = 1$. Thus $g \circ h \epsilon H$ and H is closed. Therefore, H is a subgroup of S_5 by Theorem 7.11.

The Center of a Group

If G is a group, then the **center** of G is the subset denoted $Z(G)$ and defined by

$$Z(G) = \{a \epsilon G \mid ag = ga \text{ for every } g \epsilon G\}.$$

In other words, an element of G is in $Z(G)$ if and only if it commutes with every element of G. If G is an abelian group, then $Z(G) = G$ because all elements commute with each other. When G is nonabelian, however, $Z(G)$ is not all of G.

> **EXAMPLE** The center of S_3 consists of the identity element alone because this is the only element that commutes with every element of S_3 (Exercise 8).

> **EXAMPLE** The operation table for D_4 on page 167 shows that r_1 commutes with some elements of D_4 (for instance, $r_1 \circ r_3 = r_3 \circ r_1$). However, it does not commute with *every* element of D_4 because $r_1 \circ d \neq d \circ r_1$. Hence, r_1 is not in $Z(D_4)$ nor is d. Careful examination of the table shows that $Z(D_4) = \{r_0, r_2\}$ since these are the only elements that commute with every element of D_4. It is easy to verify that $\{r_0, r_2\}$ is a subgroup of D_4. This is an example of the following result.

THEOREM 7.12 *The center $Z(G)$ of a group G is a subgroup of G.*

Proof For every $g \epsilon G$, we have $eg = g = ge$. Hence, $e \epsilon Z(G)$ and $Z(G)$ is nonempty. If a, $b \epsilon Z(G)$, then for any $g \epsilon G$ we have $ag = ga$ and $bg = gb$, so that

$$(ab)g = a(bg) = a(gb) = (ag)b = (ga)b = g(ab).$$

Therefore, $ab \epsilon Z(G)$. Finally, if $a \epsilon Z(G)$ and $g \epsilon G$, then $ag = ga$. Multiplying both sides of this equation on the left and right by a^{-1} shows that

$$a^{-1}(ag)a^{-1} = a^{-1}(ga)a^{-1}$$
$$ga^{-1} = a^{-1}g$$

Therefore, $a^{-1} \epsilon Z(G)$ and $Z(G)$ is a subgroup by Theorem 7.10. ◆

Cyclic Groups

An important type of subgroup can be constructed as follows. If G is a group and $a \in G$, let $\langle a \rangle$ denote the set of all powers of a:

$$\langle a \rangle = \{ \ldots, a^{-3}, a^{-2}, a^{-1}, a^0, a^1, a^2, \ldots \} = \{a^n \mid n \in \mathbb{Z}\}.$$

The product of any two elements of $\langle a \rangle$ is also in $\langle a \rangle$ because $a^i a^j = a^{i+j}$. The inverse of a^k is a^{-k}, which is also in $\langle a \rangle$. By Theorem 7.10, $\langle a \rangle$ is a subgroup of G. We have proved

THEOREM 7.13 *If G is a group and $a \in G$, then $\langle a \rangle = \{a^n \mid n \in \mathbb{Z}\}$ is a subgroup of G.*

The group $\langle a \rangle$ is called the **cyclic subgroup generated by a.** If the subgroup $\langle a \rangle$ is the entire group G, we say that G is a **cyclic group.** Note that every cyclic group is abelian since $a^i a^j = a^{i+j} = a^j a^i$.

> **EXAMPLE** The multiplicative group of units in the ring $\mathbb{Z}_{15}$ is $U_{15} = \{1, 2, 4, 7, 8, 11, 13, 14\}$ by Theorem 2.11. In order to determine the cyclic subgroup generated by 7, we compute
>
> $$7^1 = 7 \qquad 7^2 = 4 \qquad 7^3 = 13 \qquad 7^4 = 1 = 7^0.$$
>
> Therefore, the element 7 has order 4 in U_{15}. Let 7^i be any element of $\langle 7 \rangle$. The integer i must be congruent modulo 4 to one of 0, 1, 2, or 3. Hence, by Theorem 7.8, 7^i is equal to one of 7^0, 7^1, 7^2, or 7^3. Thus the cyclic subgroup $\langle 7 \rangle$ is $\{7, 4, 13, 1\}$. This cyclic subgroup has the same order (namely 4) as the element that generates it. Note that the cyclic subgroup $\langle 13 \rangle$ generated by 13 is the same as the cyclic subgroup $\langle 7 \rangle$ since $13^2 = 4$, $13^3 = 7$, and $13^4 = 1$. Thus, the same cyclic subgroup may be generated by different elements.

The argument used in the preceding example works in general and provides the connection between the two uses of the word "order." It states, in effect, that the order of an element a is the same as the order of the cyclic subgroup generated by a.

THEOREM 7.14 *Let G be a group and let $a \in G$.*

(1) If a has infinite order, then $\langle a \rangle$ is an infinite subgroup consisting of the distinct elements a^k, with $k \in \mathbb{Z}$.

(2) If a has finite order n, then $\langle a \rangle$ is a subgroup of order n and $\langle a \rangle = \{e = a^0, a^1, a^2, a^3, \ldots, a^{n-1}\}$.

Proof (1) This is an immediate consequence of part (1) of Theorem 7.8.

(2) Let a^i be any element of $\langle a \rangle$. Then i is congruent modulo n to one of 0, 1, 2, . . . , $n - 1$. Consequently, by part (3) of Theorem 7.8, a^i must be equal to one of $a^0, a^1, a^2, \ldots, a^{n-1}$. Furthermore, no two of these powers of a are equal since no two of the integers 0, 1, 2, . . . , $n - 1$ are congruent modulo n. Therefore, $\langle a \rangle = \{a^0, a^1, a^2, \ldots, a^{n-1}\}$ is a group of order n. ◆

When the group operation is written as addition, then the "powers" of an element a are

$$a, \qquad a + a = 2a, \qquad a + a + a = 3a, \ldots$$
$$-a, \qquad -a - a = -2a, \qquad -a - a - a = -3a, \ldots$$

and $0a = 0$. Hence, $\langle a \rangle = \{na \mid n \in \mathbb{Z}\}$. The set E of even integers, for instance, is a cyclic subgroup of the additive group $\mathbb{Z}$ because $E = \{n2 \mid n \in \mathbb{Z}\} = \langle 2 \rangle$.

EXAMPLE Since $\mathbb{Z} = \{n1 \mid n \in \mathbb{Z}\}$, we see that $\mathbb{Z}$ is an infinite cyclic group with generator 1, that is, $\mathbb{Z} = \langle 1 \rangle$. Similarly, each of the additive groups $\mathbb{Z}_n$ is a cyclic group because $\mathbb{Z}_n$ consists of the "powers" of $1 : 1$, $2 = 1 + 1$, $3 = 1 + 1 + 1$, etc.

The subgroup $\{1, -1, i, -i\}$ of the multiplicative group of nonzero elements of $\mathbb{C}$ is the cyclic subgroup $\langle i \rangle$ because $i^2 = -1, i^3 = -i$, and $i^4 = 1$. Similarly, the multiplicative group of nonzero elements of the field $\mathbb{Z}_7$ is the cyclic group $\langle 3 \rangle$, as you can easily verify. These examples are special cases of the following theorem.

THEOREM 7.15 *If G is a finite subgroup of the multiplicative group of nonzero elements of a field F, then G is cyclic.*

Proof Let $c \in G$ be an element of largest order (there must be one since G is finite), say $|c| = m$. If $a \in G$, then $|a|$ divides m by Corollary 7.9, so that $a^m = 1_F$ by part (3) of Theorem 7.8. Thus every element of G is a root of the polynomial $x^m - 1_F$. Since this polynomial has at most m roots by Corollary 4.16, we must have $|G| \le m$. But $\langle c \rangle$ is a subgroup of G of order m by Theorem 7.14. Therefore, $\langle c \rangle$ must be all of G, that is, G is cyclic. ◆

Now that we know what cyclic groups look like, the next step is to examine the possible subgroups of a cyclic group.

THEOREM 7.16 *Every subgroup of a cyclic group is itself cyclic.*

Proof Suppose $G = \langle a \rangle$ and H is a subgroup of G. If $H = \langle e \rangle$, then H is the cyclic subgroup generated by e (all of whose powers are just e). If $H \ne \langle e \rangle$, then H contains a nonidentity element of G, say a^i with $i \ne 0$. Since H is a subgroup, the inverse element a^{-i} is also in H. One of i or $-i$ is positive, and so H contains positive powers of a. Let k be the smallest positive integer such that $a^k \in H$. We claim that H is the cyclic subgroup generated by a^k. To prove this, we must show that every element of H is a power of a^k. If $h \in H$, then $h \in G$, so that $h = a^m$ for some m. By the Division Algorithm, $m = kq + r$ with $0 \le r < k$. Consequently, $r = m - kq$ and

$$a^r = a^{m-kq} = a^m a^{-kq} = a^m (a^k)^{-q}.$$

Both a^m and a^k are in H. Therefore, $a^r \in H$ by closure. Since a^k is the *smallest* positive power of a in H and since $r < k$, we must have $r = 0$. Therefore, $m = kq$ and $h = a^m = a^{kq} = (a^k)^q \in \langle a^k \rangle$. Hence, $H = \langle a^k \rangle$. ◆

For additional information on the structure of cyclic groups and their subgroups, see Exercises 38–40.

Generators of a Group

Suppose G is a group and $a \in G$. Think of the cyclic subgroup $\langle a \rangle$ as being constructed from the one-element set $S = \{a\}$ in this way: Form all possible products of a and a^{-1} in every possible order. Of course, each such product reduces to a single element of the form a^n. We want to generalize this procedure by beginning with a set S that may contain more than one element.

THEOREM 7.17 *Let S be a nonempty subset of a group G. Let $\langle S \rangle$ be the set of all possible products, in every order, of elements of S and their inverses.* Then*

(1) *$\langle S \rangle$ is a subgroup of G that contains set S.*

(2) *If H is a subgroup of G that contains the set S, then H contains the entire subgroup $\langle S \rangle$.*

This theorem shows that $\langle S \rangle$ is the smallest subgroup of G that contains the set S. In the special case when $S = \{a\}$, the group $\langle S \rangle$ is just the cyclic subgroup $\langle a \rangle$, which is the smallest subgroup of G that contains a. The group $\langle S \rangle$ is called the **subgroup generated by S.** If $\langle S \rangle$ is the entire group G, we say that S **generates** G and refer to the elements of S as the **generators** of the group.

Proof of Theorem 7.17   (1) $\langle S \rangle$ is nonempty because the set S is nonempty and every element of S (considered as a one-element product) is an element of $\langle S \rangle$. If $a, b \in \langle S \rangle$, then a is of the form $a_1 a_2 \cdots a_k$, where $k \ge 1$ and each a_i is either an element of S or the inverse of an element of S. Similarly, $b = b_1 b_2 \cdots b_t$, with $t \ge 1$ and each b_i either an element of S or the inverse of an element of S.

---

* We allow the possibility of a product with one element so that elements of S will be in $\langle S \rangle$.

Therefore, the product $ab = a_1 a_2 \cdots a_k b_1 b_2 \cdots b_t$ consists of elements of S or inverses of elements of S. Hence, $ab \in \langle S \rangle$, and $\langle S \rangle$ is closed. The inverse of the element $a = a_1 a_2 \cdots a_k$ of $\langle S \rangle$ is $a^{-1} = a_k^{-1} \cdots a_2^{-1} a_1^{-1}$ by Corollary 7.6. Since each a_i is either an element of S or the inverse of an element of S, the same is true of a_i^{-1}. Therefore, $a^{-1} \in \langle S \rangle$. Hence, $\langle S \rangle$ is a subgroup of G by Theorem 7.10.

(2) Any subgroup that contains the set S must include the inverse of every element of S. By closure, this subgroup must also contain all possible products, in every order, of elements of S and their inverses. Therefore, every subgroup that contains S must also contain the entire group $\langle S \rangle$. ◆

EXAMPLE The group $U_{15} = \{1, 2, 4, 7, 8, 11, 13, 14\}$ is generated by the set $S = \{7, 11\}$ since

$$7^1 = 7 \qquad 7^2 = 4 \qquad 7^3 = 13 \qquad 7^4 = 1$$
$$11^1 = 11 \qquad 7 \cdot 11 = 2 \qquad 7^2 \cdot 11 = 14 \qquad 7^3 \cdot 11 = 8.$$

Different sets of elements may generate the same group. For instance, you can readily verify that U_{15} is also generated by the set $\{2, 13\}$ (Exercise 17).

EXAMPLE Using the operation table on page 167, we see that in the group D_4,

$$(r_1)^1 = r_1 \qquad (r_1)^2 = r_2 \qquad (r_1)^3 = r_3 \qquad (r_1)^4 = r_0$$
$$h^1 = h \qquad r_1 \circ h = t \qquad (r_1)^2 \circ h = v \qquad (r_1)^3 \circ h = d.$$

Therefore, D_4 is generated by $\{r_1, h\}$. Note that the representation of group elements in terms of the generators is not unique; for instance,

$$(r_1)^3 \circ h = d \qquad \text{and} \qquad r_1 \circ h \circ (r_1)^2 = d.$$

◆ EXERCISES

A. **1.** Let H be a subgroup of a group G. If e_G is the identity element of G and e_H is the identity element of H, prove that $e_G = e_H$.

2. Show that the analogue of Exercise 1 for rings may be false: If G is a ring with multiplicative identity e_G and H is a subring with identity e_H, then e_H may not be equal to e_G. [*Hint:* Consider the subring $\{(n, 0) \mid n \in \mathbb{Z}\}$ of $\mathbb{Z} \times \mathbb{Z}$.]

3. (a) Let H and K be subgroups of a group G. Prove that $H \cap K$ is a subgroup of G.

(b) Let $\{H_i\}$ be any collection of subgroups of G. Prove that $\bigcap H_i$ is a subgroup of G.

4. Let H and K be subgroups of a group G.

  (a) Show by example that $H \cup K$ need not be a subgroup of G.

  (b) Prove that $H \cup K$ is a subgroup of G if and only if $H \subseteq K$ or $K \subseteq H$.

5. Let G_1 be a subgroup of a group G and H_1 a subgroup of a group H. Prove that $G_1 \times H_1$ is a subgroup of $G \times H$.

6. Let G be an abelian group, k a fixed positive integer, and $H = \{a \in G| \ |a| \text{ divides } k\}$. Prove that H is a subgroup of G.

7. Let G be an abelian group and let T be the set of elements of G with finite order. Prove that T is a subgroup of G; it is called the **torsion subgroup**. (This result may not hold if G is nonabelian; see Exercise 18 of Section 7.2.)

8. Show that the center of S_3 is the identity subgroup.

9. **(a)** If G is a group and $ab \in Z(G)$, is it true that a and b are in $Z(G)$? [*Hint*: D_4.]

  (b) If G is a group and $ab \in Z(G)$, prove that $ab = ba$.

10. If a is the only element of order 2 in a group G, prove that $a \in Z(G)$.

11. List all the cyclic subgroups of

  (a) U_{15} **(b)** U_{30}

12. **(a)** List all the cyclic subgroups of D_4.

  (b) List at least one subgroup of D_4 that is not cyclic.

13. List the elements of the subgroup $\langle a \rangle$ of S_7, where

$$a = \begin{pmatrix} 1\ 2\ 3\ 4\ 5\ 6\ 7 \\ 3\ 2\ 7\ 6\ 5\ 1\ 4 \end{pmatrix}.$$

14. Let G be a group and let $a \in G$. Prove that $\langle a \rangle = \langle a^{-1} \rangle$.

15. True or false: If every proper subgroup of a group G is cyclic, then G is cyclic. Justify your answer.

16. Show that $\mathbb{Q}^{**}$, the multiplicative group of positive rational numbers, is not a cyclic group. [*Hint*: if $1 \neq r \in \mathbb{Q}^{**}$, then there must be a rational between r and r^2.]

17. Show that U_{15} is generated by the set $\{2, 13\}$.

18. Show that $(1, 0)$ and $(0, 2)$ generate the additive group $\mathbb{Z} \times \mathbb{Z}_7$.

19. Show that the only generators of the additive cyclic group $\mathbb{Z}$ are 1 and -1.

20. Show that $(3, 1), (-2, -1)$, and $(4, 3)$ generate the additive group $\mathbb{Z} \times \mathbb{Z}$.

21. **(a)** Show that the additive group $\mathbb{Z}_2 \times \mathbb{Z}_3$ is cyclic.

(b) Show that the additive group $\mathbb{Z}_2 \times \mathbb{Z}_4$ is not cyclic but is generated by two elements.

B. 22. (a) Let p be prime and let b be a nonzero element of $\mathbb{Z}_p$. Show that $b^{p-1} = 1$. [*Hint:* Corollary 7.3 and Theorem 7.15.]

(b) Prove **Fermat's Little Theorem:** If p is a prime and a is any integer, then $a^p \equiv a \pmod{p}$. [*Hint:* Let b be the congruence class of a in $\mathbb{Z}_p$ and use part (a).]

23. Prove that a nonempty subset H of a group G is a subgroup of G if and only if whenever $a, b \in H$, then $ab^{-1} \in H$.

24. Let G be an abelian group and n a fixed positive integer.

(a) Prove that $H = \{a \in G \mid a^n = e\}$ is a subgroup of G.

(b) Show by example that part (a) may be false if G is nonabelian. [*Hint:* S_3.]

25. Let G be a group and $a \in G$. The **centralizer of a** is the set $C(a) = \{g \in G \mid ga = ag\}$. Prove that $C(a)$ is a subgroup of G.

26. If G is a group, prove that $Z(G) = \bigcap_{a \in G} C(a)$ (notation as in Exercise 25).

27. Prove that an element a is in the center of a group G if and only if $C(a) = G$ (notation as in Exercise 25).

28. Let $A(T)$ be the group of permutations of the set T and let T_1 be a nonempty subset of T. Prove that $H = \{f \in A(T) \mid f(t) = t \text{ for every } t \in T_1\}$ is a subgroup of $A(T)$.

29. Let T and T_1 be as in Exercise 28. Prove that $K = \{f \in A(T) \mid f(T_1) = T_1\}$ is a subgroup of $A(T)$ that contains the subgroup H of Exercise 28. Verify that if T_1 has more than one element, then $K \neq H$.

30. (a) Let H and K be subgroups of an abelian group G and let $HK = \{ab \mid a \in H, b \in K\}$. Prove that HK is a subgroup of G.

(b) Show that part (a) may be false if G is not abelian.

31. Let H be a subgroup of a group G and, for $x \in G$, let $x^{-1}Hx$ denote the set $\{x^{-1}ax \mid a \in H\}$. Prove that $x^{-1}Hx$ is a subgroup of G.

32. Let H be a subgroup of a group G and assume that $x^{-1}Hx \subseteq H$ for every $x \in G$ (notation as in Exercise 31). Prove that $x^{-1}Hx = H$ for each $x \in G$.

33. If H is a subgroup of a group G, then the **normalizer** of H is the set $N_H = \{x \in G \mid x^{-1}Hx = H\}$ (notation as in Exercise 31). Prove that N_H is a subgroup of G that contains H.

34. Prove that $H = \left\{ \begin{pmatrix} a & b \\ 0 & 1 \end{pmatrix} \middle| a = 1 \text{ or } -1, b \in \mathbb{Z} \right\}$ is a subgroup of $GL(2, \mathbb{Q})$.

35. Let G be an abelian group and n a fixed positive integer. Prove that $H = \{a^n \mid a \in G\}$ is a subgroup of G.

36. Let k be a positive divisor of the positive integer n. Prove that $H_k = \{a \in U_n \mid a \equiv 1 \pmod{k}\}$ is a subgroup of U_n.

37. List all the subgroups of $\mathbb{Z}_{12}$. Do the same for $\mathbb{Z}_{20}$.

38. Let $G = \langle a \rangle$ be a cyclic group of order n.

   (a) Prove that the cyclic subgroup generated by a^m is the same as the cyclic subgroup generated by a^d, where $d = (m, n)$. [*Hint:* It suffices to show that a^d is a power of a^m and vice versa. (Why?) Note that by Theorem 1.3, there are integers u and v such that $d = mu + nv$.]

   (b) Prove that a^m is a generator of G if and only if $(m, n) = 1$.

39. Let $G = \langle a \rangle$ be a cyclic group of order n. If H is a subgroup of G, show that $|H|$ is a divisor of n. [*Hint:* Exercise 38 and Theorem 7.16.]

40. Let $G = \langle a \rangle$ be a cyclic group of order n. If k is a positive divisor of n, prove that G has a unique subgroup of order k. [*Hint:* Consider the subgroup generated by $a^{n/k}$.]

41. Let G be an abelian group of order mn where $(m, n) = 1$. Assume that G contains an element a of order m and an element b of order n. Prove that G is cyclic with generator ab.

42. Show that the multiplicative group $\mathbb{R}^*$ of nonzero real numbers is not cyclic.

43. Show that the additive group $\mathbb{Q}$ is not cyclic.

44. Let G and H be groups. If $G \times H$ is a cyclic group, prove that G and H are both cyclic. (Exercise 21(b) shows that the converse is false.)

45. Prove that $\mathbb{Z}_m \times \mathbb{Z}_n$ is cyclic if and only if $(m, n) = 1$.

46. Prove that $\left\{ \begin{pmatrix} 1 & n \\ 0 & 1 \end{pmatrix} \Big| n \in \mathbb{Z} \right\}$ is a cyclic subgroup of $GL(2, \mathbb{R})$.

47. If $G \neq \langle e \rangle$ is a group that has no proper subgroups, prove that G is a cyclic group of prime order.

48. Is the additive group $G = \{a + b\sqrt{2} \mid a, b \in \mathbb{Z}\}$ cyclic?

49. **(a)** Show that the group U_{18} of units in $\mathbb{Z}_{18}$ is cyclic.

   (b) Show that the group U_{20} of units in $\mathbb{Z}_{20}$ is not cyclic.

50. If S is a nonempty subset of a group G, show that $\langle S \rangle$ is the intersection of the family of all subgroups H such that $S \subseteq H$.

7.4 ISOMORPHISMS AND HOMOMORPHISMS

Informally, two groups are said to be isomorphic if they have the same structure, differing only in the way their elements are labeled. In the case of finite groups, this means that the operation table of one group can be obtained from the operation table of the other by a suitable relabeling of the elements. Such a relabeling scheme is equivalent to a bijective function, and so we have the following formal definition (which is virtually identical to that given for rings):

> **DEFINITION•** *Let G and H be groups with the group operations denoted by $*$. G is **isomorphic** to H (written $G \cong H$) if there is a function $f : G \to H$ such that*
>
> > *(i) f is injective;*
> >
> > *(ii) f is surjective;*
> >
> > *(iii) $f(a * b) = f(a) * f(b)$ for all a, $b \in G$.*
>
> *In this case, the function f is called an **isomorphism**. A function that satisfies condition (iii) but is not necessarily injective or surjective is called a **homomorphism**.*

Just as in the case of rings, the definition of isomorphism is symmetric. If $f : G \to H$ is an isomorphism, then f is a bijective map of sets and, therefore, has a bijective inverse function $g : H \to G$ such that $f \circ g$ is the identity map on H and $g \circ f$ is the identity map on G.* A straightforward verification (Exercise 22) shows that g is also an isomorphism, that is, $g(c * d) = g(c) * g(d)$ for all c, $d \in H$. Thus $G \cong H$ if and only if $H \cong G$.

We have temporarily reverted to the $*$ notation for group operations to remind you that in a specific group the operation might be addition, or multiplication, or composition, or something else. For example, when G is an additive group and H a multiplicative one, then condition (iii) of the definition reads

* See Appendix B.

$f(a + b) = f(a)f(b)$. When G is multiplicative and H is additive, then (iii) becomes $f(ab) = f(a) + f(b)$.

EXAMPLE   The multiplicative group $U_8 = \{1, 3, 5, 7\}$ of units in $\mathbb{Z}_8$ is isomorphic to the additive group $\mathbb{Z}_2 \times \mathbb{Z}_2$. To prove this, let $f: U_8 \to \mathbb{Z}_2 \times \mathbb{Z}_2$ be defined by

$$f(1) = (0, 0) \qquad f(3) = (1, 0) \qquad f(5) = (0, 1) \qquad f(7) = (1, 1).$$

Clearly f is a bijection. Showing that $f(ab) = f(a) + f(b)$ for $a, b \in U_8$ is equivalent to showing that the operation table for $\mathbb{Z}_2 \times \mathbb{Z}_2$ can be obtained from that of U_8 simply by replacing each $a \in U_8$ by $f(a) \in \mathbb{Z}_2 \times \mathbb{Z}_2$.* Use the tables below to verify that this is indeed the case. Therefore, f is an isomorphism:

U_8						$\mathbb{Z}_2 \times \mathbb{Z}_2$			
$\circ$	1	3	5	7	$+$	(0,0)	(1,0)	(0,1)	(1,1)
1	1	3	5	7	(0,0)	(0,0)	(1,0)	(0,1)	(1,1)
3	3	1	7	5	(1,0)	(1,0)	(0,0)	(1,1)	(0,1)
5	5	7	1	3	(0,1)	(0,1)	(1,1)	(0,0)	(1,0)
7	7	5	3	1	(1,1)	(1,1)	(0,1)	(1,0)	(0,0)

EXAMPLE   The additive group $\mathbb{R}$ of real numbers is isomorphic to the multiplicative group $\mathbb{R}^{**}$ of positive real numbers. To prove this, let $f: \mathbb{R} \to \mathbb{R}^{**}$ be given by $f(r) = 10^r$. If $f(r) = f(s)$, then $10^r = 10^s$, so that $r = \log 10^r = \log 10^s = s$. Hence, f is injective. If $k \in \mathbb{R}^{**}$, then $r = \log k$ is a well-defined real number and $k = 10^{\log k} = 10^r = f(r)$. Thus f is surjective. Finally, f is a homomorphism because $f(r + s) = 10^{r+s} = 10^r 10^s = f(r)f(s)$. Therefore, f is an isomorphism and $\mathbb{R} \cong \mathbb{R}^{**}$.

EXAMPLE   Every isomorphism of rings $f: R \to S$ is also an isomorphism of additive abelian groups because f is bijective and $f(a + b) = f(a) + f(b)$ for all $a, b \in R$. The converse of this statement may be false, as we see next.

EXAMPLE   Let E be the additive group of even integers and let $f: \mathbb{Z} \to E$ be given by $f(a) = 2a$. You can readily verify that f is a bijection. Furthermore, for all $a, b \in \mathbb{Z}$, we have $f(a + b) = 2(a + b) = 2a + 2b = f(a) + f(b)$. Hence, f is an isomorphism of additive groups. However, f is *not* an isomorphism of rings; indeed, no ring isomorphism exists because $\mathbb{Z}$ is a ring with identity and E is not.

---

* See the discussion of ring isomorphism on pages 66–68 for an explanation.

EXAMPLE The groups S_3 and $\mathbb{Z}_6$ both have order 6, but they are *not* isomorphic. In fact, no abelian group (such as $\mathbb{Z}_6$) can be isomorphic to a nonabelian group (such as S_3). The proof of this fact is essentially the same as in the case of rings; see the last example on page 75.

EXAMPLE An isomorphism $G \to G$ is called an **automorphism** of the group G. Here is one way to construct an automorphism of G. Let c be a fixed element of G and define $f: G \to G$ by $f(a) = c^{-1}ac$.* Then f is a homomorphism because

$$f(a)f(b) = (c^{-1}ac)(c^{-1}bc) = c^{-1}a(cc^{-1})bc = c^{-1}abc = f(ab).$$

If $g \in G$, then $g = c^{-1}(cgc^{-1})c = f(cgc^{-1})$, so that f is surjective. If $f(a) = f(b)$, then $c^{-1}ac = c^{-1}bc$. Cancelling c on the right and c^{-1} on the left by Theorem 7.5, we have $a = b$; hence, f is injective. Therefore, f is an isomorphism, called the **inner automorphism of G** induced by c. For more about automorphisms, see Exercises 24, 25, 40, and 41.

The next theorem completely characterizes all cyclic groups.

THEOREM 7.18   *Every infinite cyclic group is isomorphic to $\mathbb{Z}$. Every finite cyclic group of order n is isomorphic to $\mathbb{Z}_n$.*

This theorem tells us, for example, that the cyclic subgroup $\langle r_1 \rangle$ of D_4 (a group of order 4) is isomorphic to the additive group $\mathbb{Z}_4$. Similarly, the infinite cyclic subgroup $\langle 2 \rangle$ of the multiplicative group $\mathbb{R}^*$ of nonzero real numbers is isomorphic to the additive group $\mathbb{Z}$.

Proof of Theorem 7.18   Let $G = \langle a \rangle$ be an infinite cyclic group and define $f: \mathbb{Z} \to G$ by $f(i) = a^i$. The map f is surjective by the definition of a cyclic group and is injective by Theorem 7.14. Since $f(i + j) = a^{i+j} = a^i a^j = f(i)f(j)$, f is an isomorphism. Therefore, $\mathbb{Z} \cong G$.

Now suppose $G = \langle b \rangle$ and b has order n. Then $\mathbb{Z}_n$ is the set $\{0, 1, 2, \ldots, n - 1\}$ and, by Theorem 7.14, $G = \{b^0, b^1\, b^2, \ldots, b^{n-1}\}$, so that the function $f: \mathbb{Z}_n \to G$ given by $f(i) = b^i$ is a bijection. To show that f is also a homomorphism, recall how addition in $\mathbb{Z}_n$ works when bracket notation isn't used: $i + j$ is equal to k, where $i + j \equiv k \pmod{n}$ and $0 \le k \le n - 1$. Therefore, by Theorem 7.8, $b^{i+j} = b^k$ in G. Consequently,

$$f(i + j) = f(k) = b^k = b^{i+j} = b^i b^j = f(i)f(j).$$

Therefore, f is an isomorphism and $\mathbb{Z}_n \cong G$.   ◆

---

* If G is abelian, then $f(a) = c^{-1}ac = c^{-1}ca = a$, so that f is the identity map. If G is nonabelian, however, then f may be a nontrivial function.

Recall that the **image** of a function $f:G \to H$ is a subset of H, namely Im $f = \{h \in H \mid h = f(a)$ for some $a \in G\}$. The function f can be considered as a surjective map from G to Im f.

THEOREM 7.19   *Let G and H be groups with identity elements e_G and e_H, respectively. If $f:G \to H$ is a homomorphism, then*

(1) $f(e_G) = e_H$.

(2) $f(a^{-1}) = f(a)^{-1}$ *for every $a \in G$.*

(3) Im f *is a subgroup of H.*

(4) *If f is injective, then $G \cong$ Im f.*

Proof   (1) Since f is a homomorphism, e_G is the identity in G, and e_H is the identity in H, we know that

$$f(e_G)f(e_G) = f(e_G e_G) = f(e_G) = e_H f(e_G).$$

Hence, $f(e_G) = e_H$ by cancellation (Theorem 7.5).

(2) By (1) we have

$$f(a^{-1})f(a) = f(a^{-1}a) = f(e_G) = e_H = f(a)^{-1}f(a).$$

Cancelling $f(a)$ on each end shows that $f(a^{-1}) = f(a)^{-1}$.

(3) The identity $e_H \in$ Im f by (1), and so Im f is nonempty. Since $f(a)f(b) = f(ab)$, Im f is closed. The inverse of each $f(a) \in$ Im f is also in Im f because $f(a)^{-1} = f(a^{-1})$ by (2). Therefore, Im f is a subgroup of H by Theorem 7.10.

(4) As noted before the theorem, f can be considered as a surjective function from G to Im f. If f is also an injective homomorphism, then f is an isomorphism.   ◆

Group theory began with the study of permutations and groups of permutations. The abstract definition of a group came later and may appear to be far more general than the concept of a group of permutations. The next theorem shows that this is not the case, however.

THEOREM 7.20   (CAYLEY'S THEOREM)   *Every group G is isomorphic to a group of permutations.*

Proof   Consider the group $A(G)$ of all permutations of the *set* G. Recall that $A(G)$ consists of all bijective *functions* from G to G with composition as the group operation. These functions need not be homomorphisms. To prove the theorem, we find a subgroup of $A(G)$ that is isomorphic to G.* We do this by

---

* The group $A(G)$ itself is usually far too large to be isomorphic to G. For instance, if G has order n, then $A(G)$ has order $n!$ by Exercise 18 of Section 7.1.

constructing an injective homomorphism of groups $f: G \to A(G)$; then G is isomorphic to the subgroup Im f of $A(G)$ by Theorem 7.19.

If $a \in G$, then we claim that the map $\varphi_a : G \to G$ defined by $\varphi_a(x) = ax$ is a bijection of sets (that is, an element of $A(G)$). This follows from the fact that if $b \in G$, then $\varphi_a(a^{-1}b) = a(a^{-1}b) = b$; hence, φ_a is surjective. If $\varphi_a(b) = \varphi_a(c)$, then $ab = ac$. Cancelling a by Theorem 7.5, we conclude that $b = c$. Therefore, φ_a is injective and, hence, a bijection. Thus $\varphi_a \in A(G)$.

Now define $f: G \to A(G)$ by $f(a) = \varphi_a$. For any $a, b \in G$, $f(ab) = \varphi_{ab}$ is the map from G to G given by $\varphi_{ab}(x) = abx$. On the other hand, $f(a) \circ f(b) = \varphi_a \circ \varphi_b$ is the map given by $(\varphi_a \circ \varphi_b)(x) = \varphi_a(\varphi_b(x)) = \varphi_a(bx) = abx$. Therefore, $f(ab) = f(a) \circ f(b)$ and f is a homomorphism of groups. Finally, suppose $f(a) = f(c)$, so that $\varphi_a(x) = \varphi_c(x)$ for all $x \in G$. Then $a = ae = \varphi_a(e) = \varphi_c(e) = ce = c$. Hence, f is injective. Therefore, $G \cong \text{Im } f$ by Theorem 7.19. ◆

COROLLARY 7.21 *Every finite group G of order n is isomorphic to a subgroup of the symmetric group S_n.*

Proof The group G is isomorphic to a subgroup H of $A(G)$ by Theorem 7.20. Since G is a set of n elements, $A(G)$ is isomorphic to S_n by Exercise 20. Consequently, H is isomorphic to a subgroup K of S_n by Exercise 12. By Exercise 9, $G \cong H$ and $H \cong K$ imply that $G \cong K$. ◆

Any homomorphism from a group G to a group of permutations is called a **representation** of G, and G is said to be **represented by a group of permutations**. The homomorphism $G \to A(G)$ in the proof of Theorem 7.20 is called the **left regular representation** of G. By the use of such representations, group theory can be reduced to the study of permutation groups. This approach is sometimes very advantageous because permutations are concrete objects that are readily visualized. Calculations with permutations are straightforward, which is not always the case in some groups. In certain situations, group representations are a very effective tool.

On the other hand, representation by permutations has some drawbacks. For one thing, a given group can be represented as a group of permutations in many ways—the homomorphism $G \to A(G)$ of Theorem 7.20 is just one of the possibilities (see Exercises 26, 33, and 35 for others). And many of these representations may be quite inefficient. According to Corollary 7.21, for example, every group of order 12 is isomorphic to a subgroup of S_{12}, but S_{12} has order $12! = 479,001,600$. Determining useful information about a subgroup of order 12 in a group that size is likely to be difficult at best.

Except for some special situations, then, the study of elementary group theory via the abstract definition (as we have been doing) rather than via concrete permutation representations is likely to be more effective. The abstract approach has the advantage of eliminating nonessential features and concentrating on the basic underlying structure. In the long run, this usually results in simpler proofs and better understanding.

◆ EXERCISES

A. **1. (a)** Show that the function $f: \mathbb{R} \to \mathbb{R}$ given by $f(x) = 3x$ is an isomorphism of additive groups.

   (b) Let $\mathbb{R}^{**}$ be the multiplicative group of positive real numbers. Show that $f: \mathbb{R}^{**} \to \mathbb{R}^{**}$ given by $f(x) = 3x$ is not a homomorphism of groups.

2. Show that the function $g: \mathbb{R}^{**} \to \mathbb{R}^{**}$ given by $g(x) = \sqrt{x}$ is an isomorphism.

3. (a) List the elements of the group $GL(2, \mathbb{Z}_2)$.

   (b) Show that $GL(2, \mathbb{Z}_2)$ is isomorphic to S_3 by writing out the operation tables for each group.

4. Show that U_5 is isomorphic to the additive group $\mathbb{Z}_4$.

5. Show that U_5 is isomorphic to U_{10}.

6. Prove that the additive group $\mathbb{Z}_6$ is isomorphic to the multiplicative group of nonzero elements in $\mathbb{Z}_7$.

7. Let $\mathbb{Q}^*$ be the multiplicative group of nonzero rational numbers. Prove that $\mathbb{Q}^*$ is isomorphic to the group G of Exercise 4(d) in Section 7.1

8. Let $\mathbb{R}^*$ be the multiplicative group of nonzero real numbers. Prove that $\mathbb{R}^*$ is isomorphic to the group G of Exercise 4(e) in Section 7.1.

9. If $f: G \to H$ and $g: H \to K$ are isomorphisms of groups, prove that the composite function $g \circ f: G \to K$ is also an isomorphism.

10. If $f: G \to H$ is a surjective homomorphism of groups and G is abelian, prove that H is abelian.

11. Let $f: G \to H$ be a homomorphism of groups. Prove that for each $a \in G$ and each integer n, $f(a^n) = f(a)^n$.

12. If $f: G \to H$ is an isomorphism of groups and if T is a subgroup of G, prove that T is isomorphic to the subgroup $f(T) = \{f(a) \mid a \in T\}$ of H.

13. (a) Let H be a subgroup of a group G and let $a \in G$. Prove that $a^{-1}Ha = \{a^{-1}ha \mid h \in H\}$ is a subgroup of G that is isomorphic to H. [*Hint:* Apply Exercise 12 when f is the inner automorphism induced by a.]

   (b) If H is finite, prove that $|H| = |a^{-1}Ha|$.

14. Let G, H, G_1, H_1 be groups such that $G \cong G_1$ and $H \cong H_1$. Prove that $G \times H \cong G_1 \times H_1$.

15. Prove that a group G is abelian if and only if the function $f: G \to G$ given by $f(x) = x^{-1}$ is a homomorphism of groups. In this case, show that f is an isomorphism.

B. 16. Let G be a multiplicative group and c a fixed element of G. Let H be the *set G* equipped with a new operation $*$ defined by $a * b = acb$.

    (a) Prove that H is a group.

    (b) Prove that the map $f: G \to H$ given by $f(x) = c^{-1}x$ is an isomorphism.

17. Let G be a multiplicative group. Let G^{op} be the set G equipped with a new operation $*$ defined by $a * b = ba$.

    (a) Prove that G^{op} is a group.

    (b) Prove that $G \cong G^{op}$. [*Hint:* Corollary 7.6 may be helpful.]

18. If $G = \langle a \rangle$ is a cyclic group and $f: G \to H$ is a surjective homomorphism of groups, show that $f(a)$ is a generator of H, that is, H is the cyclic group $\langle f(a) \rangle$. [*Hint:* Exercise 11.]

19. Assume that a and b are both generators of the cyclic group G, so that $G = \langle a \rangle$ and $G = \langle b \rangle$. Prove that the function $f: G \to G$ given by $f(a^i) = b^i$ is an automorphism of G.

20. Let T be a set of n elements and let $A(T)$ be the group of permutations of T. Prove that $A(T) \cong S_n$. [*Hint:* If the elements of T in some order are relabeled as $1, 2, \ldots, n$, then every permutation of T becomes a permutation of $1, 2, \ldots, n$.]

21. If $f: G \to H$ is an injective homomorphism of groups and $a \in G$, prove that $|f(a)| = |a|$.

22. Let $f: G \to H$ be an isomorphism of groups. Let $g: H \to G$ be the inverse function of f as defined in Appendix B. Prove that g is also an isomorphism of groups. [*Hint:* Exercise 27 in Section 3.3.]

23. Let $f: G \to H$ be a homomorphism of groups and let K be a subgroup of H. Prove that the set $\{a \in G \mid f(a) \in K\}$ is a subgroup of G.

24. Let G be a group and let **Aut G** be the set of all automorphisms of G. Prove that Aut G is a group under the operation of composition of functions. [*Hint:* Exercise 9 may help.]

25. Let G be a group and let Aut G be as in Exercise 24. Let **Inn G** be the set of all inner automorphisms of G (that is, isomorphisms of the form $f(a) = c^{-1}ac$ for some $c \in G$, as in the second example on page 193). Prove that Inn G is a subgroup of Aut G. [Note: Two different elements of G may induce the same inner automorphism, that is, we may have $c^{-1}ac = d^{-1}ad$ for all $a \in G$. Hence, $|\text{Inn } G| \le |G|$.]

26. (a) Show that $D_3 \cong S_3$. [*Hint:* D_3 is described on pages 167–168. Each motion in D_3 permutes the vertices; use this to define a function from D_3 to S_3.]

(b) Show that D_4 is isomorphic to a subgroup of S_4. [*Hint:* See the hint for part (a). This isomorphism represents D_4, a group of order 8, as a subgroup of a permutation group of order 4! = 24, whereas the left regular representation of Corollary 7.21 represents G as a subgroup of S_8, a group of order 8! = 40,320.]

27. Show that the additive groups $\mathbb{Z}$ and $\mathbb{Q}$ are not isomorphic.

28. Explain why the two groups are *not* isomorphic:

   (a) $\mathbb{Z}_6$ and S_3 **(b)** $\mathbb{Z}_4 \times \mathbb{Z}_2$ and D_4

   (c) $\mathbb{Z}_4 \times \mathbb{Z}_2$ and $\mathbb{Z}_2 \times \mathbb{Z}_2 \times \mathbb{Z}_2$ **(d)** $\mathbb{Z}$ and $\mathbb{R}$

29. (a) Show that U_8 is not isomorphic to U_{10}.

   (b) Show that U_{10} is not isomorphic to U_{12}.

   (c) Is U_8 isomorphic to U_{12}?

30. Prove that the additive group $\mathbb{R}$ of all real numbers is not isomorphic to the multiplicative group $\mathbb{R}^*$ of nonzero real numbers. [*Hint:* If there were an isomorphism $f : \mathbb{R} \to \mathbb{R}^*$, then $f(k) = -1$ for some k; use this fact to arrive at a contradiction.]

31. Show that D_4 is not isomorphic to the quaternion group of Exercise 14 of Section 7.1.

32. Prove that the additive group $\mathbb{Q}$ is *not* isomorphic to the multiplicative group $\mathbb{Q}^{**}$ of positive rational numbers, even though $\mathbb{R}$ and $\mathbb{R}^{**}$ *are* isomorphic.

33. Let G be a group and let $A(G)$ be the group of permutations of the set G. Define a function g from G to $A(G)$ by assigning to each $d \in G$ the inner automorphism induced by d^{-1} (as in the second example on page 193 with $c = d^{-1}$). Prove that g is a homomorphism of groups.

34. Let G be a group and $h \in A(G)$. Assume that $h \circ \varphi_a = \varphi_a \circ h$ for all $a \in G$ (where φ_a is as in the proof of Theorem 7.20). Prove that there exists $b \in G$ such that $h(x) = xb^{-1}$ for all $x \in G$.

35. (a) Let G be a group and $c \in G$. Prove that the map $\theta_c : G \to G$ given by $\theta_c(x) = xc^{-1}$ is an element of $A(G)$.

   (b) Prove that $h : G \to A(G)$ given by $h(c) = \theta_c$ is an injective homomorphism of groups. Thus G is isomorphic to the subgroup Im h of $A(G)$. This is the **right regular representation of G.**

36. Find the left regular representation of each group (that is, express each group as a permutation group as in the proof of Theorem 7.20):

(a) $\mathbb{Z}_3$ (b) $\mathbb{Z}_4$ (c) S_3

37. (a) Prove that $H = \left\{ \left(\begin{matrix} 1 - n & -n \\ n & 1 + n \end{matrix} \right) \middle| \ n \in \mathbb{Z} \right\}$ is a group under matrix multiplication.

(b) Prove that $H \cong \mathbb{Z}$.

38. (a) Prove that $K = \left\{ \left(\begin{matrix} 1 - 2n & n \\ -4n & 1 + 2n \end{matrix} \right) \middle| \ n \in \mathbb{Z} \right\}$ is a group under matrix multiplication.

(b) Is K isomorphic to $\mathbb{Z}$?

39. Prove that the additive group $\mathbb{Z}[x]$ is isomorphic to the multiplicative group $\mathbb{Q}^{**}$ of positive rationals. [*Hint:* Let $p_0, p_1, p_2, \ldots$ be the distinct positive primes in their usual order. Define $\varphi : \mathbb{Z}[x] \to \mathbb{Q}^{**}$ by

$$\varphi(a_0 + a_1 x + a_2 x^2 + \cdots + a_n x^n) = p_0^{a_0} p_1^{a_1} \cdots p_n^{a_n}.]$$

40. Prove that G is an abelian group if and only if Inn G consists of a single element. [*Hint:* See Exercise 25.]

41. (a) Verify that the group Inn D_4 has order 4. [*Hint:* See Exercise 25.]

(b) Prove that Inn $D_4 \cong \mathbb{Z}_2 \times \mathbb{Z}_2$.

42. Prove that Aut $\mathbb{Z} \cong \mathbb{Z}_2$. [*Hint:* What are the possible generators of the cyclic group $\mathbb{Z}$? See Exercises 18 and 19.]

43. Prove that Aut $\mathbb{Z}_n \cong U_n$. [*Hint:* See Exercise 19 above and Exercise 38 of Section 7.3.]

44. Prove that Aut $(\mathbb{Z}_2 \times \mathbb{Z}_2) \cong S_3$.

> ***Application***   Linear Codes (Section 16.1) may be covered at this point if desired.

7.5 CONGRUENCE AND LAGRANGE'S THEOREM

In this section we present the analogue for groups of the concept of congruence. Except for necessary notational changes, the first three results are virtually identical to those proved earlier for integers and for rings. The following chart shows this parallel development.

INTEGERS	RINGS	GROUPS
Theorem 2.1	Theorem 6.4	Theorem 7.22
Theorem 2.3	Theorem 6.6	Theorem 7.23
Corollary 2.4	Corollary 6.7	Corollary 7.24

The remainder of the section deals with concepts that have no analogue in ring theory but that provide useful tools for the study of groups.

Congruence modulo an ideal I in a ring R was defined purely in terms of the additive group of the ring: $a \equiv c \pmod{I}$ means $a - c \in I$. Consequently, the definition and basic properties can be carried over directly to groups. The only difference is that a ring is an additive group, and we usually use multiplicative notation for groups. Therefore, we must translate statements in additive notation into equivalent statements in multiplicative notation.* The following dictionary may be helpful for this translation.

ADDITIVE NOTATION	MULTIPLICATIVE NOTATION
$a + b$	ab
0	e
$-c$	c^{-1}
$a - b$	ab^{-1}

If we translate the definition of congruence for ideals from additive to multiplicative notation, we obtain

> **DEFINITION** • *Let K be a subgroup of a group G and let a, $b \in G$. Then a is **congruent to b modulo K** [written $a \equiv b \pmod{K}$] provided that $ab^{-1} \in K$.*

EXAMPLE   Let K be the subgroup $\{r_0, r_1, r_2, r_3\}$ of D_4. Then the operation table on page 167 shows that $d^{-1} = d$ and $h \circ d^{-1} = h \circ d = r_1 \in K$. Therefore, $h \equiv d \pmod{K}$.

THEOREM 7.22 *Let K be a subgroup of a group G. Then the relation of congruence modulo K is*

(1) reflexive: $a \equiv a \pmod{K}$ for all $a \in G$;

(2) symmetric: if $a \equiv b \pmod{K}$, then $b \equiv a \pmod{K}$;

(3) transitive: if $a \equiv b \pmod{K}$ and $b \equiv c \pmod{K}$, then $a \equiv c \pmod{K}$.

---

* There is a possibility of confusion here since rings also have a multiplication. In carrying over congruence results from rings to groups, we consider *only* the additive structure of rings and ignore their multiplication completely.

Proof Translate the proof of Theorem 6.4 from additive to multiplicative notation. Use K and G in place of I and R, and the dictionary above. ◆

If K is a subgroup of a group G and if $a \in G$, then the **congruence class** of a modulo K is the set of all elements of G that are congruent to a modulo K, that is, the set

$$\{b \in G \mid b \equiv a \ (\mathrm{mod}\ K)\} = \{b \in G \mid ba^{-1} \in K\}$$
$$= \{b \in G \mid ba^{-1} = k, \text{ with } k \in K\}.$$

Right multiplication by a shows that the statement $ba^{-1} = k$ is equivalent to $b = ka$. Therefore, the congruence class of a modulo K is the set

$$\{b \in G \mid b = ka, \text{ with } k \in K\} = \{ka \mid k \in K\}.$$

Consequently, the congruence class of a modulo K is denoted Ka. The set Ka is called a **right coset** of K in G. When the operation in the group G is addition, then right cosets are written additively: $K + a$.*

THEOREM 7.23 *Let K be a subgroup of a group G and let $a, c \in G$. Then $a \equiv c \ (\mathrm{mod}\ K)$ if and only if $Ka = Kc$.*

Proof With minor notational changes, the proof is essentially the same as that of Theorem 2.3 or Theorem 6.6. Just replace "mod n" with "mod K" and "$[a]$" with "Ka" and use Theorem 7.22 in place of Theorem 2.1. ◆

COROLLARY 7.24 *Let K be a subgroup of a group G. Then two right cosets of K are either disjoint or identical.*

Proof Copy the proof of Corollary 2.4 or Corollary 6.7 with the obvious notational changes. ◆

Lagrange's Theorem

At this point we temporarily leave the parallel treatment of congruence in groups and rings and use right cosets to develop some facts about finite groups that have no counterpart in ring theory. The discussion of congruence and the group theory analogues of ideals and quotient rings will be resumed in the next section.

---

* Cosets of an ideal I in a ring were denoted $a + I$ instead of $I + a$ because the additive group of a ring is abelian so that $a + i = i + a$ for every $i \in I$.

THEOREM 7.25   *Let K be a subgroup of a group G. Then*

    (1) G is the union of the right cosets of K: $G = \bigcup\limits_{a \in G} Ka$.

    (2) For each $a \in G$, there is a bijection $f : K \rightarrow Ka$. Consequently, if K is finite, any two right cosets of K contain the same number of elements.

Proof   (1) Since every right coset consists of elements of G, we have $\bigcup\limits_{a \in G} Ka \subseteq G$. If $b \in G$, then $b = eb \in Kb \subseteq \bigcup\limits_{a \in G} Ka$, so that $G \subseteq \bigcup\limits_{a \in G} Ka$. Hence, $G = \bigcup\limits_{a \in G} Ka$.

    (2) Define $f : K \rightarrow Ka$ by $f(x) = xa$. Then by the definition of Ka, f is surjective. If $f(x) = f(y)$, than $xa = ya$, so that $x = y$ by Theorem 7.5. Therefore, f is injective and, hence, a bijection. Consequently, if K is finite, every coset Ka has the same number of elements as K, namely $|K|$. ◆

    If H is a subgroup of a group G, then the number of distinct right cosets of H in G is called the **index of H in G** and is denoted $[G:H]$. If G is a finite group, then there can be only a finite number of distinct right cosets of H; hence, the index $[G:H]$ is finite. If G is an infinite group, then the index may be either finite or infinite.

> **EXAMPLE**   Let H be the cyclic subgroup $\langle 3 \rangle$ of the additive group $\mathbb{Z}$. Then H consists of all multiples of 3, and the cosets of H are just the congruence classes modulo 3; for instance,
>
> $$H + 2 = \{h + 2 \mid h \in H\} = \{3z + 2 \mid z \in \mathbb{Z}\} = [2].$$
>
> Since there are exactly three distinct congruence classes modulo 3 (cosets of H), we have $[\mathbb{Z}:H] = 3$.

> **EXAMPLE**   Under addition the group $\mathbb{Z}$ of integers is a subgroup of the group $\mathbb{Q}$ of rational numbers. By the definition of congruence and Theorem 7.23,
>
> $$\mathbb{Z} + a = \mathbb{Z} + c \qquad \text{if and only if} \qquad a - c \in \mathbb{Z}.$$
>
> Consequently, if $0 < c < a < 1$, then $\mathbb{Z} + a$ and $\mathbb{Z} + c$ are distinct cosets because $0 < a - c < 1$, which means that $a - c$ cannot be in $\mathbb{Z}$. Since there are infinitely many rationals between 0 and 1, there are an infinite number of distinct cosets of $\mathbb{Z}$ in $\mathbb{Q}$. Hence, $[\mathbb{Q}:\mathbb{Z}]$ is infinite.

THEOREM 7.26   (LAGRANGE'S THEOREM)   *If K is a subgroup of a finite group G, then the order of K divides the order of G. In particular, $|G| = |K| [G:K]$.*

Proof   It is convenient to adopt the following notation. If A is a finite set, then $|A|$ denotes the number of elements in A. Observe that if A and B are disjoint finite sets, then $|A \cup B| = |A| + |B|$. Now suppose that $[G:K] = n$ and denote the n distinct cosets of K in G by $Kc_1, Kc_2, \ldots, Kc_n$. By Theorem 7.25

$$G = Kc_1 \cup Kc_2 \cup \cdots \cup Kc_n.$$

Since these cosets are all distinct, they are mutually disjoint by Corollary 7.24. Consequently,

$$|G| = |Kc_1| + |Kc_2| + \cdots + |Kc_n|.$$

For each c_i, however, $|Kc_i| = |K|$ by Theorem 7.25. Therefore,

$$|G| = \underbrace{|K| + |K| + \cdots + |K|}_{n \text{ summands}} = |K|n = |K|[G\!:\!K].\quad \blacklozenge$$

Lagrange's Theorem shows that there are a limited number of possibilities for the subgroups of a finite group. For instance, a subgroup of a group of order 12 must have one of these orders: 1, 2, 3, 4, 6, or 12 (the only divisors of 12). Be careful, however, for these are only the *possible* orders of subgroups. Lagrange's Theorem does *not* say that a group G *must* have a subgroup of order k for every k that divides $|G|$. In fact, we shall see an example of a group of order 12 that has no subgroup of order 6 (Exercise 28 in Section 7.9). Lagrange's Theorem also puts limitations on the possible orders of elements in a group:

COROLLARY 7.27 *Let G be a finite group.*

(1) If $a \in G$, then the order of a divides the order of G.

(2) If $|G| = k$, then $a^k = e$ for every $a \in G$.

Proof (1) If $a \in G$ has order n, then the cyclic subgroup $\langle a \rangle$ of G has order n by Theorem 7.14. Consequently, n divides $|G|$ by Lagrange's Theorem.

(2) If $a \in G$ has order n, then $n \mid k$ by part (1), say $k = nt$. Therefore, $a^k = a^{nt} = (a^n)^t = e^t = e.\quad \blacklozenge$

The Structure of Finite Groups

A major goal of group theory is the classification of all finite groups up to isomorphism; that is, we would like to produce a list of groups such that every finite group is isomorphic to exactly one group on the list. This is a problem of immense difficulty, but a number of partial results have already been obtained. Theorem 7.18, for example, provides a classification of all cyclic groups; it says, in effect, that every nontrivial finite cyclic group is isomorphic to exactly one group on this list: $\mathbb{Z}_2, \mathbb{Z}_3, \mathbb{Z}_4, \ldots$. All finite abelian groups will be classified in Section 8.2.

We now use Lagrange's Theorem and its corollary to classify all groups of prime order and all groups of order less than 8. In the proofs below enough of the necessary calculations are included to show you how the argument goes, but you should take pencil and paper and supply all the missing computations.

THEOREM 7.28 *Let p be a positive prime integer. Every group of order p is cyclic and isomorphic to $\mathbb{Z}_p$.*

Proof If G is a group of order p and a is any nonidentity element of G, then the cyclic subgroup $\langle a \rangle$ is a group of order greater than 1. Since the order of the group $\langle a \rangle$ must divide p and since p is prime, $\langle a \rangle$ must be a group of order p. Thus $\langle a \rangle$ is all of G, and G is a cyclic group of order p. Therefore, $G \cong \mathbb{Z}_p$ by Theorem 7.18. ◆

THEOREM 7.29 *Every group of order 4 is isomorphic to either $\mathbb{Z}_4$ or $\mathbb{Z}_2 \times \mathbb{Z}_2$.*

Proof Let G be a group of order 4. Either G contains an element of order 4 or it does not. If it does, then the cyclic subgroup generated by this element has order 4 by Theorem 7.14 and, hence, must be all of G. Therefore, G is a cyclic group of order 4, and $G \cong \mathbb{Z}_4$ by Theorem 7.18. Now suppose that G does not contain an element of order 4. Let e, a, b, c be the distinct elements of G, with e the identity element. Since every element of G must have order dividing 4 by Corollary 7.27 and since e is the only element of order 1, each of a, b, c must have order 2. Thus the operation table of G must look like this:

	e	a	b	c
e	e	a	b	c
a	a	e		
b	b		e	
c	c			e

In order to fill in the missing entries, we first consider the product ab. If $ab = e$, then $ab = aa$ and, hence, $a = b$ by cancellation. This is a contradiction, and so $ab \neq e$. If $ab = a$, then $ab = ae$ and $b = e$ by cancellation, another contradiction. Similarly, $ab = b$ implies the contradiction $a = e$. Therefore, the only possibility is $ab = c$. Similar arguments show that there is only one possible operation table for G, namely,

	e	a	b	c
e	e	a	b	c
a	a	e	c	b
b	b	c	e	a
c	c	b	a	e

Let $f \colon G \to \mathbb{Z}_2 \times \mathbb{Z}_2$ be given by $f(e) = (0, 0)$, $f(a) = (1, 0)$, $f(b) = (0, 1)$, and $f(c) = (1, 1)$. Show that f is an isomorphism by comparing the operation tables of the two groups. ◆

THEOREM 7.30 *Every group G of order 6 is isomorphic to either $\mathbb{Z}_6$ or S_3.*

Proof If G contains an element of order 6, then G is a cyclic group of order 6 and, hence, is isomorphic to $\mathbb{Z}_6$ by Theorem 7.18. So suppose G contains no element of order 6. Then every nonidentity element of G has order 2 or 3 by Corollary 7.27. If every nonidentity element of G has order 2, then G is an abelian group by Exercise 25 of Section 7.2. If c and d are nonidentity elements of G, then the set $H = \{e, c, d, cd\}$ is closed under multiplication (because $c^2 = e = d^2$ and $cd = dc$). Hence, H is a subgroup of G by Theorem 7.11. This is a contradiction since no group of order 6 can have a subgroup of order 4 by Lagrange's Theorem. Therefore, the nonidentity elements of G cannot all have order 2, and G must contain an element a of order 3. Let N be the cyclic subgroup $\langle a \rangle = \{e, a, a^2\}$ and let b be any element of G that is not in N. The cosets $Ne = \{e, a, a^2\}$ and $Nb = \{b, ab, a^2b\}$ are not identical since $b \notin N = Ne$ and, hence, must be disjoint (Corollary 7.24). Therefore, G consists of the six elements e, a, a^2, b, ab, a^2b.

We now show that there is only one possible operation table for G. What are the possibilities for b^2? We claim that b^2 cannot be any of a, a^2, b, ab, or a^2b. For instance, if $b^2 = a$, then $b^4 = a^2$. However, b either has order 2 (in which case $a^2 = b^4 = b^2b^2 = ee = e$, a contradiction) or order 3 (in which case $a^2 = b^4 = b^3b = eb = b$, another contradiction since $b \notin N$). Similar arguments show that the only possibility is $b^2 = e$.

Next we determine the product ba. It is easy to see that ba cannot be any of b, e, a, or a^2 (for instance, $ba = a$ implies $b = e$). So the only possibilities are $ba = ab$ or $ba = a^2b$. If $ba = ab$, then verify that ba has order 6 by computing its powers. This contradicts our assumption that G has no element of order 6. Therefore, we must have $ba = a^2b$. Using these two facts:

$$b^2 = e \qquad \text{and} \qquad ba = a^2b,$$

we can now compute every product in G. For example, $ba^2 = (ba)a = (a^2b)a = a^2(ba) = a^2a^2b = a^4b = ab$.

Verify that the operation table for G must look like this:

	e	a	a^2	b	ab	a^2b
e	e	a	a^2	b	ab	a^2b
a	a	a^2	e	ab	a^2b	b
a^2	a^2	e	a	a^2b	b	ab
b	b	a^2b	ab	e	a^2	a
ab	ab	b	a^2b	a	e	a^2
a^2b	a^2b	ab	b	a^2	a	e

By comparing tables, show that G is isomorphic to S_3 under the correspondence

$$
\overset{e}{\underset{\downarrow}{}} \qquad \overset{a}{\underset{\downarrow}{}} \qquad \overset{a^2}{\underset{\downarrow}{}} \qquad \overset{b}{\underset{\downarrow}{}} \qquad \overset{ab}{\underset{\downarrow}{}} \qquad \overset{a^2b}{\underset{\downarrow}{}}
$$

$$
\begin{pmatrix} 1 & 2 & 3 \\ 1 & 2 & 3 \end{pmatrix}\begin{pmatrix} 1 & 2 & 3 \\ 2 & 3 & 1 \end{pmatrix}\begin{pmatrix} 1 & 2 & 3 \\ 3 & 1 & 2 \end{pmatrix}\begin{pmatrix} 1 & 2 & 3 \\ 2 & 1 & 3 \end{pmatrix}\begin{pmatrix} 1 & 2 & 3 \\ 3 & 2 & 1 \end{pmatrix}\begin{pmatrix} 1 & 2 & 3 \\ 1 & 3 & 2 \end{pmatrix}. \quad \blacklozenge
$$

The last three theorems provide a complete classification of all groups of order less than 8, as summarized in this table:

If G has order	then G is isomorphic to
2	$\mathbb{Z}_2$
3	$\mathbb{Z}_3$
4	$\mathbb{Z}_4$ or $\mathbb{Z}_2 \times \mathbb{Z}_2$
5	$\mathbb{Z}_5$
6	$\mathbb{Z}_6$ or S_3
7	$\mathbb{Z}_7$

The classification of groups is discussed further in Chapter 8, particularly in Section 8.5 where the preceding chart is extended to order 15.

◆ **EXERCISES**

A.   **1.** Let K be a subgroup of a group G and let $a \in G$. Prove that $Ka = K$ if and only if $a \in K$.

2. List the distinct right cosets of K in G.

   (a) $K = \{r_0, v\}; G = D_4$

   (b) $K = \{r_0, r_1, r_2, r_3\}; G = D_4$

   (c) $K = \left\{ \begin{pmatrix} 1\,2\,3 \\ 1\,2\,3 \end{pmatrix}, \begin{pmatrix} 1\,2\,3 \\ 1\,3\,2 \end{pmatrix} \right\}; G = S_3$

   (d) $K = \{1, 17\}; G = U_{32}$

3. Find the index $[G{:}H]$ when

   (a) $H = \{r_0, r_2\}$ and $G = D_4$

   (b) $H = \langle 3 \rangle$ and $G = \mathbb{Z}_{12}$

   (c) $H = \langle 3 \rangle$ and $G = \mathbb{Z}_{20}$

   (d) H is the subgroup generated by 12 and 20 and $G = \mathbb{Z}_{40}$

   (e) H is the subgroup generated by $\begin{pmatrix} 1\,2\,3\,4 \\ 2\,3\,4\,1 \end{pmatrix}$ and $G = S_4$

4. Suppose G is the cyclic group $\langle a \rangle$ and $|a| = 15$. If $K = \langle a^3 \rangle$, list all the distinct cosets of K in G.

5. What are the possible orders of the subgroups of G when G is

   (a) $\mathbb{Z}_{24}$ **(b)** S_4 **(c)** $D_4 \times \mathbb{Z}_{10}$

6. Give examples, other than those in the text, of infinite groups G and H such that

   (a) $[G{:}H]$ is finite   **(b)** $[G{:}H]$ is infinite

7. Let G be a finite group that has elements of every order from 1 through 12. What is the smallest possible value of $|G|$?

8. A group G has fewer than 100 elements and subgroups of orders 10 and 25. What is the order of G?

9. Let H and K, each of prime order p, be subgroups of a group G. If $H \neq K$, prove that $H \cap K = \langle e \rangle$.

10. If H and K are subgroups of a finite group G, prove that $|H \cap K|$ is a common divisor of $|H|$ and $|K|$.

B. 11. If G is a group with more than one element and G has no proper subgroups, prove that G is isomorphic to $\mathbb{Z}_p$ for some prime p.

12. If G is a group of order 25, prove that either G is cyclic or else every nonidentity element of G has order 5.

13. Let a be an element of order 30 in a group G. What is the index of $\langle a^4 \rangle$ in the group $\langle a \rangle$?

14. Let H and K be subgroups of a finite group G such that $[G{:}H] = p$ and $[G{:}K] = q$, with p and q distinct primes. Prove that pq divides $[G{:}H \cap K]$.

15. **(a)** If $n > 2$, prove that U_n contains an element of order 2. [*Hint:* What are the solutions of $x^2 = 1$ in $\mathbb{Z}_n$?]

    (b) If $n > 2$, prove that the order of the group U_n is even.

16. If p is prime, prove that U_{p^2} contains an element of order p. [*Hint:* Show that the subgroup H_p in Exercise 36 of Section 7.3 has order p.]

17. If $f{:}G \to H$ is a homomorphism of groups and if $a \in G$ has finite order, prove that $f(a)$ has finite order in H and $|f(a)|$ divides $|a|$.

18. Let H and K be subgroups of a group G such that $K \subseteq H$ and $[G{:}H]$ and $[H{:}K]$ are finite. Prove that $[G{:}K]$ is finite and $[G{:}K] = [G{:}H][H{:}K]$.

19. If p and q are primes, show that every proper subgroup of a group of order pq is cyclic.

20. (a) If G is an abelian group of order $2n$, with n odd, prove that G contains exactly one element of order 2. [*Hint:* Exercise 30 in Section 7.2.]

(b) Show by example that part (a) may be false if G is nonabelian.

21. Let G be an abelian group of order n and let k be a positive integer. If $(k, n) = 1$, prove that the function $f \colon G \to G$ given by $f(a) = a^k$ is an isomorphism.

22. If G is a group of order n and G has 2^{n-1} subgroups, prove that $G = \langle e \rangle$ or $G \cong \mathbb{Z}_2$.

23. Let G be a group of order 10.

(a) Prove that G contains an element of order 2.

(b) If G is nonabelian, prove that G contains five elements of order 2. [*Hint:* Use techniques similar to those in the proof of Theorem 7.30.]

24. Prove that a group of order 33 contains an element of order 3.

25. If a prime p divides the order of a finite group G, prove that the number of elements of order p in G is a multiple of $p - 1$.

C. 26. Let G be a group generated by elements a and b such that $|a| = 4$, $|b| = 2$, and $ba = a^3b$. Show that G is a group of order 8 and that G is isomorphic to D_4.

27. Let G be a group generated by elements a and b such that $|a| = 4$, $b^2 = a^2$, and $ba = a^3b$. Show that G is a group of order 8 and that G is isomorphic to the quaternion group of Exercise 14 in Section 7.1.

7.6 NORMAL SUBGROUPS

In the first part of Section 7.5 we carried over to groups all but one of the congruence results proved in Chapter 6 for rings. The remaining fact, which was essential for defining quotient rings (Theorem 6.5), states that for any ideal I, if $a \equiv b \pmod{I}$ and $c \equiv d \pmod{I}$, then $a + c \equiv b + d \pmod{I}$. Translating this into multiplicative notation (see the dictionary on page 200), the corresponding result for groups would be: For a subgroup K of a group G,

$$\text{if } a \equiv b \pmod{K} \text{ and } c \equiv d \pmod{K}, \text{ then } ac \equiv bd \pmod{K}.$$

If G is an abelian group, then this statement is true, as you can readily show by translating the proof of part (1) of Theorem 6.5 from additive to multiplicative notation. Unfortunately, however, *this statement may be false* for some nonabelian groups.

EXAMPLE   Let K be the subgroup $\{r_0, v\}$ of D_4. Then the operation table below shows that

$$r_1 \equiv t \;(\mathrm{mod}\;K) \text{ because } r_1 \circ t^{-1} = r_1 \circ t = v \;\epsilon\; K,$$
$$r_2 \equiv h \;(\mathrm{mod}\;K) \text{ because } r_2 \circ h^{-1} = r_2 \circ h = v \;\epsilon\; K,$$

but $r_1 \circ r_2 \not\equiv t \circ h \;(\mathrm{mod}\;K)$ because

$$(r_1 \circ r_2) \circ (t \circ h)^{-1} = (r_3) \circ (r_1)^{-1} = r_3 \circ r_3 = r_2 \;\not\epsilon\; K.$$

D_4	$\circ$	r_0	r_1	r_2	r_3	d	h	t	v
	r_0	r_0	r_1	r_2	r_3	d	h	t	v
	r_1	r_1	r_2	r_3	r_0	h	t	v	d
	r_2	r_2	r_3	r_0	r_1	t	v	d	h
	r_3	r_3	r_0	r_1	r_2	v	d	h	t
	d	d	v	t	h	r_0	r_3	r_2	r_1
	h	h	d	v	t	r_1	r_0	r_3	r_2
	t	t	h	d	v	r_2	r_1	r_0	r_3
	v	v	t	h	d	r_3	r_2	r_1	r_0

It really shouldn't come as a surprise that the group-theory analogue of Theorem 6.5 for ideals does not hold for arbitrary subgroups. After all, an ideal is a special kind of subring, and congruence facts for ideals don't hold for every subring. Presumably, an analogue of Theorem 6.5 might hold for "special" subgroups of a group.

As a first step toward identifying these special subgroups, we note a crucial difference between congruence in groups and in rings. If I is an ideal in a ring R, then because an ideal contains the negative of each of its elements, we have

$$a - b \;\epsilon\; I \qquad \text{if and only if} \qquad -(a - b) \;\epsilon\; I.$$

Since $-(a - b) = -a + b$ in a ring,

$$a - b \;\epsilon\; I \qquad \text{if and only if} \qquad -a + b \;\epsilon\; I.$$

Thus there are two equivalent ways to define congruence modulo I in a ring (namely, $a - b \;\epsilon\; I$ or $-a + b \;\epsilon\; I$). For groups, however, the situation is different because the multiplicative translation of "$-(a - b) = -a + b$" is "$(ab^{-1})^{-1} = a^{-1}b$", which may be *false* in nonabelian groups, as shown in Corollary 7.6 and Exercise 2 of Section 7.2. Consequently, when K is a subgroup of a nonabelian group G and $ab^{-1} \;\epsilon\; K$, it may not be true that $a^{-1}b \;\epsilon\; K$. Consequently, we define a second congruence relation in groups.

DEFINITION• *Let K be a subgroup of a group G and let a, b ∈ G. Then a is **left congruent to b modulo K** [written a ≈ b (mod K)] provided that a⁻¹b ∈ K.*

To avoid confusion hereafter, we shall refer to the congruence relation ≡ defined in Section 7.5 as **right congruence.** It is easy to verify that left congruence has the same basic properties as right congruence.

THEOREM 7.31   *Let K be a subgroup of G. Then the relation of left congruence modulo K is*

(1) *reflexive: a ≈ a (mod K) for every a ∈ K;*

(2) *symmetric: if a ≈ b (mod K), then b ≈ a (mod K);*

(3) *transitive: if a ≈ b (mod K) and b ≈ c (mod K), then a ≈ c (mod K).*

Proof   Exercise. Adapt the proof of Theorem 7.22.   ◆

The **left congruence class** of *a* modulo *K* is the set

$$\{b \in G \mid b \approx a \ (\text{mod } K)\} = \{b \in G \mid a \approx b(\text{mod } K)\}$$
$$= \{b \in G \mid a^{-1}b \in K\}$$
$$= \{b \in G \mid a^{-1}b = k, \text{ with } k \in K\}$$
$$= \{b \in G \mid b = ak, \text{ with } k \in K\}$$
$$= \{ak \mid k \in K\}$$

The left congruence class of *a* modulo *K* is denoted *aK* and is called a **left coset** of *K* in *G*. In additive notation, left cosets are written *a + K*.

THEOREM 7.32   *Let K be a subgroup of a group G and let a, c ∈ G. Then*

(1) *a ≈ c (mod K) if and only if aK = cK.*

(2) *Any two left cosets of K are either disjoint or identical.*

Proof   Adapt the proofs of Theorem 7.23 and Corollary 7.24.   ◆

Left and right congruence are the same for abelian groups because *ab⁻¹ ∈ K* if and only if $(ab^{-1})^{-1} \in K$ and for *abelian* groups, $(ab^{-1})^{-1} = (b^{-1})^{-1}a^{-1} = ba^{-1} = a^{-1}b$. But left and right congruence may be different relations in non-abelian groups.

EXAMPLE   Let *K* be the subgroup $\{r_0, v\}$ of D_4. The operation table on page 209 shows that $t \equiv r_1 \ (\text{mod } K)$ because

$$t \circ (r_1)^{-1} = t \circ r_3 = v \, \epsilon \, K,$$

but $t \not\equiv r_1 \,(\mathrm{mod}\, K)$ because

$$t^{-1} \circ r_1 = t \circ r_1 = h \notin K.$$

Thus t is right congruent, but not left congruent, to r_1 modulo K.

We know that the analogue of Theorem 6.5 holds for any subgroup of an abelian group and that right and left congruence coincide in abelian groups. Conversely, the two preceding examples show that the analogue of Theorem 6.5 fails for the subgroup $K = \{r_0, v\}$ of D_4, and that right and left congruence are not the same for this subgroup K. This suggests that Theorem 6.5 may carry over to subgroups for which right and left congruence are the same relation. Saying that right and left congruence modulo K are the same amounts to saying that the congruence class of each element a under right congruence (that is, the right coset Ka) is identical to its congruence class under left congruence (that is, the left coset aK). Subgroups with this property have a special name.

DEFINITION• *A subgroup N of a group G is said to be* ***normal*** *if $Na = aN$ for every $a \, \epsilon \, G$.*

EXAMPLE If N is a subgroup of an abelian group G and $a \, \epsilon \, G$, then $na = an$ for every $n \, \epsilon \, N$, so that the right coset Na is the same as the left coset aN. Hence, every subgroup of an abelian group is normal.

EXAMPLE Let K be the subgroup $\{r_0, v\}$ of D_4. Then the right coset Kd is the set $\{r_0 \circ d, v \circ d\} = \{d, r_3\}$ and the left coset dK is the set $\{d \circ r_0, d \circ v\} = \{d, r_1\}$. Thus $Kd \neq dK$ and, therefore, K is not a normal subgroup of D_4.

EXAMPLE Let M be the subgroup $\{r_0, r_2\}$ of D_4. Then the operation table for D_4 shows that $r_0 \circ a = a \circ r_0$ and $r_2 \circ a = a \circ r_2$ for every $a \, \epsilon \, D_4$. So it is certainly true that $Ma = aM$ for every $a \, \epsilon \, D_4$. Hence, M is a normal subgroup of D_4.

EXAMPLE Let G be any group and N the center of G. By the definition of "center" (page 183), $na = an$ for every $n \, \epsilon \, N$ and $a \, \epsilon \, G$, so that $Na = aN$ for every $a \, \epsilon \, G$. Hence, the center is a normal subgroup of G. [The previous example is a special case of this one since $\{r_0, r_2\}$ is the center of D_4.]

The preceding examples, though important, are misleading in that the elements of the normal subgroup N commute with all other elements of the group in each case. In the general case, however,

the condition $Na = aN$ does *not* imply that $na = an$ for every $n \in N$.

All you can conclude from $Na = aN$ is that if $na \in Na$, then na is also an element of the set aN—that is, there is some $t \in N$ such that $na = at \in aN$.

> **EXAMPLE**   Let N be the subgroup $\{r_0, r_1, r_2, r_3\}$ of D_4. Then the right coset Nv is the set
>
> $$\{r_0 \circ v, r_1 \circ v, r_2 \circ v, r_3 \circ v\} = \{v, d, h, t\}$$
>
> and the left coset vN is the set
>
> $$\{v \circ r_0, v \circ r_1, v \circ r_2, v \circ r_3\} = \{v, t, h, d\}.$$
>
> Thus $Nv = vN$, even though $nv \ne vn$ for every nonidentity element n in N (for instance, $r_1 \circ v = d$, but $v \circ r_1 = t$). Similar computations show that $Na = aN$ for every $a \in D_4$. Therefore, N is normal in D_4.

Other examples of normal subgroups appear in Exercises 3–8, 14, and 23. Normal subgroups are just what's needed to carry over Theorem 6.5 to groups.

THEOREM 7.33   *Let N be a normal subgroup of a group G. If $a \equiv b \pmod N$, and $c \equiv d \pmod N$, then $ac \equiv bd \pmod N$.*

Proof   By the definition of congruence, there are elements $m, n \in N$ such that $ab^{-1} = m$ and $cd^{-1} = n$. By Corollary 7.6 $(ac)(bd)^{-1} = acd^{-1}b^{-1} = anb^{-1}$. The element an is in the left coset aN. Since N is normal, $aN = Na$. Hence, $an = n_1 a$ for some $n_1 \in N$. Consequently, $(ac)(bd)^{-1} = anb^{-1} = n_1 ab^{-1} = n_1 m \in N$. Therefore, $ac \equiv bd \pmod N$.   ◆

Ideals were the appropriate concept for carrying over the facts about congruence in $\mathbb{Z}$ to arbitrary rings. We have now seen that normal subgroups play an analogous role in group theory. In the next section, we shall see that normal subgroups lead to quotient groups, just as ideals lead to quotient rings. We close this section with a theorem that provides alternate descriptions of normality. Verifying condition (2) or (3) in the theorem is often the easiest way to prove that a given subgroup is normal.

THEOREM 7.34   *The following conditions on a subgroup N of a group G are equivalent:*

(1) N is a normal subgroup of G.

(2) $a^{-1}Na \subseteq N$ for every $a \in G$, where $a^{-1}Na = \{a^{-1}na \mid n \in N\}$.

(3) $aNa^{-1} \subseteq N$ for every $a \in G$, where $aNa^{-1} = \{ana^{-1} \mid n \in N\}$.

(4) $a^{-1}Na = N$ for every $a \in G$.

(5) $aNa^{-1} = N$ for every $a \in G$.

Note that in (4), $a^{-1}Na = N$ does *not* mean that $a^{-1}na = n$ for each $n \in N$; all it means is that $a^{-1}na = n_1$ for some $n_1 \in N$. Analogous remarks apply to (2), (3), and (5).

Proof of Theorem 7.34 (1) $\Rightarrow$ (2) Suppose $n \in N$ and $a^{-1}na \in a^{-1}Na$. We must show that $a^{-1}na \in N$. Note that na is an element of the right coset Na. Since N is normal by (1), $Na = aN$. Hence, $na = an_1$ for some $n_1 \in N$. Thus $a^{-1}na = a^{-1}an_1 = en_1 = n_1 \in N$. Therefore, $a^{-1}Na \subseteq N$.

(2) $\Leftrightarrow$ (3) If (2) holds for *every* element of G, then it holds with a^{-1} in place of a, that is,

$$(*) \qquad\qquad (a^{-1})^{-1}Na^{-1} \subseteq N.$$

But $(a^{-1})^{-1} = a$, so that $(*)$ is statement (3): $aNa^{-1} \subseteq N$. Similarly, if (3) holds for every element of G, then it holds with a^{-1} in place of a, which implies statement (2).

(3) $\Rightarrow$ (4) Since (3) implies (2), we have $a^{-1}Na \subseteq N$. To prove $N \subseteq a^{-1}Na$, suppose $n \in N$. Then $n = a^{-1}(ana^{-1})a$. By (3) $ana^{-1} = n_2$ for some $n_2 \in N$. Thus $n = a^{-1}n_2a \in a^{-1}Na$, which proves that $N \subseteq a^{-1}Na$. Therefore, $a^{-1}Na = N$.

(4) $\Leftrightarrow$ (5) If (4) holds for *every* element of G, then it holds with a^{-1} in place of a, that is,

$$N = (a^{-1})^{-1}Na^{-1} = aNa^{-1}.$$

Similarly, if (5) holds for every element of G, then it holds with a^{-1} in place of a, which implies statement (4).

(5) $\Rightarrow$ (1) Suppose $n \in N$ and $an \in aN$. Then $ana^{-1} \in aNa^{-1} = N$ by (5), so that $ana^{-1} = n_3$ for some $n_3 \in N$. Multiplying this last equation on the right by a shows that $an = n_3a \in Na$. Therefore, $aN \subseteq Na$. Conversely, if $na \in Na$, then $a^{-1}na \in a^{-1}Na = N$ because (5) implies (4). Hence, $a^{-1}na = n_4$ for some $n_4 \in N$. Multiplying on the left by a shows that $na = an_4 \in aN$. Thus $Na \subseteq aN$. Therefore, $Na = aN$ for every $a \in G$ and N is a normal subgroup of G. ◆

◆ EXERCISES

A. **1.** Let K be a subgroup of a group G and let $a \in G$. Prove that $aK = K$ if and only if $a \in K$.

2. **(a)** Verify that Theorem 7.25 is valid when "right coset" is replaced by "left coset."

(b) Verify that the proof of Lagrange's Theorem is valid when "right coset" is replaced by "left coset."

(c) If K is a subgroup of a finite group G, prove that the number of left cosets of K is the same as the number of right cosets of K.

3. Prove that N is a normal subgroup of G by listing all its left and right cosets.

   (a) $G = S_3; N = \left\{ \begin{pmatrix} 1\ 2\ 3 \\ 2\ 3\ 1 \end{pmatrix}, \begin{pmatrix} 1\ 2\ 3 \\ 3\ 1\ 2 \end{pmatrix}, \begin{pmatrix} 1\ 2\ 3 \\ 1\ 2\ 3 \end{pmatrix} \right\}.$

   (b) $G = D_4; N = \{r_0, r_2, h, v\}.$

4. If G is a group, show that $\langle e \rangle$ and G are normal subgroups.

5. (a) Prove that $G = \left\{ \begin{pmatrix} a & b \\ 0 & d \end{pmatrix} \middle| a, b, d \in \mathbb{R} \text{ and } ad \neq 0 \right\}$ is a group under ma-

   trix multiplication and that $N = \left\{ \begin{pmatrix} 1 & b \\ 0 & 1 \end{pmatrix} \middle| b \in \mathbb{R} \right\}$ is a subgroup of G.

   (b) Use Theorem 7.34 to show that N is normal in G.

6. Prove that $\left\{ \begin{pmatrix} 1\ 2\ 3 \\ 2\ 1\ 3 \end{pmatrix}, \begin{pmatrix} 1\ 2\ 3 \\ 1\ 2\ 3 \end{pmatrix} \right\}$ is a subgroup of S_3 but not normal.

7. Let G and H be groups. Prove that $G^* = \{(a, e) \mid a \in G\}$ is a normal subgroup of $G \times H$.

8. (a) List all the cyclic subgroups of the quaternion group (Exercise 14 of Section 7.1).

   (b) Show that each of the subgroups in part (a) is normal.

9. Let N be a subgroup of a group G. Suppose that, for each $a \in G$, there exists $b \in G$ such that $Na = bN$. Prove that N is a normal subgroup.

10. If G is a group, prove that every subgroup of $Z(G)$ is normal in G. [Compare with Exercise 14.]

11. A subgroup N of a group G is said to be **characteristic** if $f(N) \subseteq N$ for *every* automorphism f of G. Prove that every characteristic subgroup is normal. (The converse is false, but this is harder to prove.)

12. Prove that for any group G, the center $Z(G)$ is a characteristic subgroup.

13. Let N be a subgroup of a group G. Prove that N is normal if and only if $f(N) = N$ for every inner automorphism f of G.

14. Show by example that if M is a normal subgroup of N and if N is a normal subgroup of a group G, then M need *not* be a normal subgroup of G; in other words, normality isn't transitive. [*Hint:* Consider $M = \{v, r_0\}$ and $N = \{h, v, r_2, r_0\}$ in D_4.]

B. 15. Let $f : G \to H$ be a homomorphism of groups and let $K = \{a \in G \mid f(a) = e_H\}$. Prove that K is a normal subgroup of G.

16. If K and N are normal subgroups of a group G, prove that $K \cap N$ is a normal subgroup of G.

17. Let N and K be subgroups of a group G. If N is normal in G, prove that $N \cap K$ is a normal subgroup of K.

18. **(a)** Let N and K be subgroups of a group G. If N is normal in G, prove that $NK = \{nk \mid n \in N, k \in K\}$ is a subgroup of G. [Compare Exercise 30(b) of Section 7.3.]

(b) If both N and K are normal subgroups of G, prove that NK is normal.

19. If K and N are normal subgroups of a group G such that $K \cap N = \langle e \rangle$, prove that $nk = kn$ for every $n \in N$, $k \in K$.

20. Prove that a subgroup of index 2 in a group G is normal. [*Hint:* For each $a \in G$, show that $a^{-1}Na \subseteq N$. When $a \notin N$, use the fact that every element of G is in N or Na (Theorem 7.25) and rule out the possibility that $a^{-1}na \in Na$ for any $n \in N$.]

21. If $f : G \to H$ is a surjective homomorphism of groups and if N is a normal subgroup of G, prove that $f(N)$ is a normal subgroup of H.

22. Prove that the function $f : GL(2, \mathbb{R}) \to \mathbb{R}^*$ given by $f\begin{pmatrix} a & b \\ c & d \end{pmatrix} = ad - bc$ is a homomorphism of multiplicative groups.

23. Prove that $SL(2, \mathbb{R})$ is a normal subgroup of $GL(2, \mathbb{R})$. [*Hint:* $SL(2, \mathbb{R})$ is defined in Exercise 21 of Section 7.1 Use Exercises 15 and 22 above.]

24. Let H be a subgroup of order n in a group G. If H is the only subgroup of order n, prove that H is normal. [*Hint:* Theorem 7.34 and Exercise 13 in Section 7.4.]

25. Prove that a subgroup N of a group G is normal if and only if it has this property: $ab \in N$ if and only if $ba \in N$, for all $a, b \in G$.

26. Prove that the cyclic subgroup $\langle a \rangle$ of a group G is normal if and only if for each $g \in G$, $ga = a^k g$ for some $k \in \mathbb{Z}$.

27. Let G be a finite group, N a cyclic normal subgroup of G, and H any subgroup of N. Prove that H is a normal subgroup of G. [Compare Exercise 14.]

28. Let A and B be normal subgroups of a group G such that $A \cap B = \langle e \rangle$ and $AB = G$ (see Exercise 18). Prove that $A \times B \cong G$. [*Hint:* Define $f : A \times B \to G$ by $f(a, b) = ab$ and use Exercise 19.]

29. Let H be a subgroup of a group G and let N_H be its normalizer (see Exercise 33 in Section 7.3). Prove that

(a) H is a normal subgroup of N_H.

(b) If H is a normal subgroup of a subgroup K of G, then $K \subseteq N_H$.

30. Prove that Inn G is a normal subgroup of Aut G. [See Exercise 25 of Section 7.4.]

31. Let T be a set with three or more elements and let $A(T)$ be the group of all permutations of T. If $a \in T$, let $H_a = \{f \in A(T) \mid f(a) = a\}$. Prove that H_a is a subgroup of $A(T)$ that is not normal.

32. If N is a normal subgroup of order 2 in a group G, prove that $N \subseteq Z(G)$.

33. Let H be a subgroup of a group G and let $N = \bigcap_{a \in G} a^{-1}Ha$. Prove that N is a normal subgroup of G.

34. Let G be a group that contains at least one subgroup of order n. Let $N = \bigcap K$, where the intersection is taken over all subgroups K of order n. Prove that N is a normal subgroup of G. [*Hint:* For each $a \in G$, verify that $a^{-1}Na = \bigcap a^{-1}Ka$, where the intersection is over all subgroups K of order n; use Exercise 13 of Section 7.4.]

35. Let G be a group all of whose subgroups are normal. If $a, b \in G$, prove that there is an integer k such that $ab = ba^k$.

36. If M is a characteristic subgroup of N and N is a normal subgroup of a group G, prove that M is a normal subgroup of G. [See Exercise 11.]

7.7 QUOTIENT GROUPS

Let N be a normal subgroup of a group G. Then

G/N denotes the set of all right cosets of N in G.

Our first goal is to define an operation on right cosets so that G/N becomes a group. Since right cosets are congruence classes, our experience with $\mathbb{Z}$ and other rings suggests that it would be reasonable to define such an operation as follows: The product of the coset Na (the congruence class of a) and the coset Nb (the congruence class of b) is the coset Nab (the congruence class of ab). In symbols, this definition reads

$$(Na)(Nb) = Nab.$$

As in the past, we must verify that the definition does not depend on the elements chosen to represent the various cosets, and so we must prove

THEOREM 7.35 *Let N be a normal subgroup of a group G. If $Na = Nc$ and $Nb = Nd$ in G/N, then $Nab = Ncd$.*

Proof $Na = Nc$ implies that $a \equiv c \pmod{N}$ by Theorem 7.23; similarly, $Nb = Nd$ implies that $b \equiv d \pmod{N}$. Therefore, $ab \equiv cd \pmod{N}$ by Theorem 7.33. Hence, $Nab = Ncd$ by Theorem 7.23. ◆

THEOREM 7.36 *Let N be a normal subgroup of a group G. Then*

(1) *G/N is a group under the operation defined by (Na)(Nc) = Nac.*

(2) *If G is finite, then the order of G/N is $|G| / |N|$.*

(3) *If G is an abelian group, then so is G/N.*

Proof (1) The operation in G/N is well defined by Theorem 7.35. The coset $N = Ne$ is the identity element in G/N since $(Na)(Ne) = Nae = Na$ and $(Ne)(Na) = Nea = Na$ for every Na in G/N. The inverse of Na is the coset Na^{-1} since $(Na)(Na^{-1}) = Naa^{-1} = Ne$ and, similarly, $(Na^{-1})(Na) = Ne$. Associativity in G/N follows from that in G:

$$[(Na)(Nb)](Nc) = (Nab)(Nc) = N(ab)c = Na(bc) = (Na)(Nbc) = (Na)[(Nb)(Nc)].$$

Therefore, G/N is a group.

(2) The order of G/N is the number of distinct right cosets of N, that is, the index $[G:N]$. By Lagrange's Theorem, $[G:N] = |G|/|N|$.

(3) Exercise 9. ◆

The group G/N is called the **quotient group** or **factor group** of G by N. A special case of this situation occurs when I is an ideal in a ring R. Under addition, I is a normal subgroup of R since addition in a ring is commutative. The quotient group R/I is just the additive group of the quotient ring R/I. Here are some examples that are not a consequence of ring theory.

EXAMPLE In the example on page 212 we saw that $N = \{r_0, r_1, r_2, r_3\}$ is a normal subgroup of D_4. The operation table for D_4 on page 209 shows that

$$Nr_0 = \{r_0 \circ r_0, r_1 \circ r_0, r_2 \circ r_0, r_3 \circ r_0\} = \{r_0, r_1, r_2, r_3\}$$
$$Nv = \{r_0 \circ v, r_1 \circ v, r_2 \circ v, r_3 \circ v\} = \{v, d, h, t\}.$$

Since every element of D_4 is in either Nr_0 or Nv and since any two cosets of N are either disjoint or identical (Corollary 7.24), every coset of N must be equal to Nr_0 or Nv. In other words, $D_4/N = \{Nr_0, Nv\}$. Since $r_0 \circ v = v = v \circ r_0$ and $v \circ v = r_0$, the operation table for the quotient group D_4/N is

	Nr_0	Nv
Nr_0	Nr_0	Nv
Nv	Nv	Nr_0

It is easy to see that D_4/N is isomorphic to the additive group $\mathbb{Z}_2$ (let Nr_0 correspond to 0 and Nv to 1; then compare tables).

EXAMPLE   On page 211 we saw that $M = \{r_0, r_2\}$ is a normal subgroup of D_4. Using the operation table for D_4, we find that D_4/M consists of these four cosets:

$$Mr_0 = \{r_0, r_2\} = Mr_2 \qquad Mr_1 = \{r_1, r_3\} = Mr_3$$
$$Mh = \{h, v\} = Mv \qquad Md = \{d, t\} = Mt.$$

We shall choose one way of representing each coset and list the elements of D_4/M as $Mr_0, Mr_1, Mh,$ and Md. When we compute products in D_4/M, we express the answers in terms of these four cosets. For instance, since $d \circ r_1 = v$ in D_4, we have $(Md)(Mr_1) = M(d \circ r_1) = Mv$; but $Mv = Mh$, so we write $(Md)(Mr_1) = Mh$ in the table below. You should fill in the missing entries:

	Mr_0	Mr_1	Mh	Md
Mr_0	Mr_0	Mr_1	Mh	Md
Mr_1	Mr_1	Mr_0		
Mh	Mh		Mr_0	
Md	Md	Mh		

The completed table shows that D_4/M is an abelian group in which every nonidentity element has order 2. So D_4/M is not cyclic. Hence, D_4/M is isomorphic to $\mathbb{Z}_2 \times \mathbb{Z}_2$ by Theorem 7.29.

EXAMPLE   Let N be the cyclic subgroup $\langle (1, 2) \rangle$ of the additive group $G = \mathbb{Z}_2 \times \mathbb{Z}_4$. Since $(1, 2) + (1, 2) = (0, 0)$, we see that $N = \{(0, 0), (1, 2)\}$. Consequently, G/N consists of these four cosets

$$N + (0, 0) = \{(0, 0), (1, 2)\} = N + (1, 2)$$
$$N + (1, 0) = \{(1, 0), (0, 2)\} = N + (0, 2)$$
$$N + (0, 1) = \{(0, 1), (1, 3)\} = N + (1, 3)$$
$$N + (1, 1) = \{(1, 1), (0, 3)\} = N + (0, 3)$$

and has the following addition table:

	$N + (0, 0)$	$N + (1, 0)$	$N + (0, 1)$	$N + (1, 1)$
$N + (0, 0)$	$N + (0, 0)$	$N + (1, 0)$	$N + (0, 1)$	$N + (1, 1)$
$N + (1, 0)$	$N + (1, 0)$	$N + (0, 0)$	$N + (1, 1)$	$N + (0, 1)$
$N + (0, 1)$	$N + (0, 1)$	$N + (1, 1)$	$N + (1, 0)$	$N + (0, 0)$
$N + (1, 1)$	$N + (1, 1)$	$N + (0, 1)$	$N + (0, 0)$	$N + (1, 0)$

Use the table to verify that G/N is a cyclic group of order 4 generated by $N + (0, 1)$. Therefore, $G/N \cong \mathbb{Z}_4$ by Theorem 7.18.

> **EXAMPLE** The subgroup $\mathbb{Z}$ of integers in the additive group $\mathbb{Q}$ of rational numbers is normal since $\mathbb{Q}$ is abelian. The second example on page 202 shows that there are infinitely many distinct cosets of $\mathbb{Z}$ in $\mathbb{Q}$. Consequently, the quotient group $\mathbb{Q}/\mathbb{Z}$ is an infinite abelian group (but *not* a quotient ring because $\mathbb{Z}$ is not an ideal in the ring $\mathbb{Q}$). Nevertheless, every element of $\mathbb{Q}/\mathbb{Z}$ has *finite* order (Exercise 15).

The Structure of Groups

If N is a normal subgroup of a group G, then the structure of each of the groups N, G, and G/N is related to the structure of the others. If we know enough information about two of these groups, we can often determine useful information about the third, as illustrated in the following theorems.

THEOREM 7.37 *Let N be a normal subgroup of a group G. Then G/N is abelian if and only if $aba^{-1}b^{-1} \epsilon N$ for all a, $b \epsilon G$.*

Proof G/N is abelian if and only if

$$Nab = NaNb = NbNa = Nba \qquad \text{for all } a, b \epsilon G.$$

But $Nab = Nba$ if and only if $(ab)(ba)^{-1} \epsilon N$ by Theorem 7.23; and $(ab)(ba)^{-1} = aba^{-1}b^{-1}$ by Corollary 7.6. Therefore, G/N is abelian if and only if $aba^{-1}b^{-1} \epsilon N$ for all a, $b \epsilon G$. ◆

The center $Z(G)$ of a group G was defined on page 183. In an example on page 211 we saw that $Z(G)$ is a normal subgroup of G.

THEOREM 7.38 *If G is a group such that the quotient group $G/Z(G)$ is cyclic, then G is abelian.*

Proof For notational convenience, denote $Z(G)$ by C. Since G/C is cyclic, it has a generator Cd, and every coset in G/C is of the form $(Cd)^k = Cd^k$ for some integer k. Let a and b be any elements of G. Since $a = ea$ is in the coset Ca and since $Ca = Cd^i$ for some i, we have $a = c_1 d^i$ for some $c_1 \epsilon C$. Similarly, $b = c_2 d^j$ for some $c_2 \epsilon C$ and integer j. Now $d^i d^j = d^{i+j} = d^{j+i} = d^j d^i$, and c_1 and c_2 commute with every element of G by the definition of the center. Consequently,

$$ab = (c_1 d^i)(c_2 d^j) = c_1 c_2 d^i d^j = c_2 c_1 d^j d^i = (c_2 d^j)(c_1 d^i) = ba.$$

Therefore, G is abelian. ◆

◆ **EXERCISES**

A.   **1.** Verify that $N = \left\{ \begin{pmatrix} 1\ 2\ 3 \\ 1\ 2\ 3 \end{pmatrix}, \begin{pmatrix} 1\ 2\ 3 \\ 2\ 3\ 1 \end{pmatrix}, \begin{pmatrix} 1\ 2\ 3 \\ 3\ 1\ 2 \end{pmatrix} \right\}$ is a normal subgroup of S_3 and show that $S_3/N \cong \mathbb{Z}_2$.

   2. Show that $\mathbb{Z}_6/N \cong \mathbb{Z}_3$, where N is the subgroup $\{0, 3\}$.

   3. Show that $\mathbb{Z}_{18}/M \cong \mathbb{Z}_6$, where M is the cyclic subgroup $\langle 6 \rangle$.

   4. Let $G = \mathbb{Z}_4 \times \mathbb{Z}_4$ and let N be the cyclic subgroup generated by $(3, 2)$. Show that $G/N \cong \mathbb{Z}_4$.

   5. Let $G = \mathbb{Z}_6 \times \mathbb{Z}_2$ and let N be the cyclic subgroup $\langle (1, 1) \rangle$. Describe the quotient group G/N.

   6. Show that $U_{32}/N \cong U_{16}$, where N is the subgroup $\{1, 17\}$.

   7. (a) Let M be the cyclic subgroup $\langle (0, 2) \rangle$ of the additive group $G = \mathbb{Z}_2 \times \mathbb{Z}_4$ and let N be the cyclic subgroup $\langle (1, 2) \rangle$, as in the second example on page 218. Verify that M is isomorphic N.

   (b) Write out the operation table of G/M, using the four cosets $M + (0, 0)$, $M + (1, 0)$, $M + (0, 1)$, $M + (1, 1)$.

   (c) Show that G/M is not isomorphic to G/N (the operation table for G/N is on page 218). Thus for normal subgroups M and N, the fact that $M \cong N$ does not imply that G/M is isomorphic to G/N.

   8. If N is a normal subgroup of a group G and if $x^2 \in N$ for every $x \in G$, prove that every nonidentity element of the quotient group G/N has order 2.

   9. If N is a subgroup of an abelian group G, prove that G/N is abelian.

   10. (a) Give an example of a group G such that $G/Z(G)$ is abelian.

   (b) Give an example of a group G such that $G/Z(G)$ is not abelian.

   11. (a) Show that $V = \left\{ \begin{pmatrix} 1\ 2\ 3\ 4 \\ 1\ 2\ 3\ 4 \end{pmatrix}, \begin{pmatrix} 1\ 2\ 3\ 4 \\ 2\ 1\ 4\ 3 \end{pmatrix}, \begin{pmatrix} 1\ 2\ 3\ 4 \\ 3\ 4\ 1\ 2 \end{pmatrix}, \begin{pmatrix} 1\ 2\ 3\ 4 \\ 4\ 3\ 2\ 1 \end{pmatrix} \right\}$ is a normal subgroup of S_4.

   (b) Write out the operation table for the group S_4/V.

B. **12.** Let $\mathbb{R}^*$ be the multiplicative group of nonzero real numbers and let N be the subgroup $\{1, -1\}$. Prove that $\mathbb{R}^*/N$ is isomorphic to the multiplicative group $\mathbb{R}^{**}$ of positive real numbers.

   13. Describe the quotient group $\mathbb{R}^*/\mathbb{R}^{**}$, where $\mathbb{R}^*$ and $\mathbb{R}^{**}$ are as in Exercise 12.

   14. If G is a cyclic group, prove that G/N is cyclic, where N is any subgroup of G.

15. (a) Prove that every element of $\mathbb{Q}/\mathbb{Z}$ has finite order.

(b) Prove that $\mathbb{Q}/\mathbb{Z}$ contains elements of every possible finite order.

16. Prove that the set of elements of finite order in the group $\mathbb{R}/\mathbb{Z}$ is the subgroup $\mathbb{Q}/\mathbb{Z}$.

17. Let G and H be groups and let G^* be the subset of $G \times H$ consisting of all (a, e) with $a \in G$.

(a) Show that G^* is isomorphic to G.

(b) Show that G^* is a normal subgroup of $G \times H$.

(c) Show that $(G \times H)/G^* \cong H$.

18. Let M and N be normal subgroups of a group G such that $M \cap N = \langle e \rangle$. Prove that G is isomorphic to a subgroup of $G/M \times G/N$.

19. If N is a normal subgroup of a group G and if every element of N and of G/N has finite order, prove that every element of G has finite order.

20. If N is a finite normal subgroup of a group G and if G/N contains an element of order n, prove that G contains an element of order n.

21. Let G be a group of order pq, with p and q (not necessarily distinct) primes. Prove that the center $Z(G)$ is either $\langle e \rangle$ or G.

22. A group H is said to be **finitely generated** if there is a finite subset S of H such that $H = \langle S \rangle$ (see Theorem 7.17). If N is a normal subgroup of a group G such that the groups N and G/N are finitely generated, prove that G is finitely generated.

23. Let G be a group and let S be the set of all elements of the form $aba^{-1}b^{-1}$ with $a, b \in G$. The subgroup G' generated by the set S (as in Theorem 7.17) is called the **commutator subgroup** of G. Prove

(a) G' is normal in G. [*Hint:* For any $g, a, b \in G$, show that $g^{-1}(aba^{-1}b^{-1})g = (g^{-1}ag)(g^{-1}bg)(g^{-1}a^{-1}g)(g^{-1}b^{-1}g)$ is in S.]

(b) G/G' is abelian.

24. Let G be the additive group $\mathbb{R} \times \mathbb{R}$.

(a) Show that $N = \{(x, y) \mid y = -x\}$ is a subgroup of G.

(b) Describe the quotient group G/N.

25. Let N be a normal subgroup of a group G and let G' be the commutator subgroup defined in Exercise 23. If $N \cap G' = \langle e \rangle$, prove that

(a) $N \subseteq Z(G)$ **(b)** The center of G/N is $Z(G)/N$.

26. If G is a group, prove that $G/Z(G)$ is isomorphic to the group Inn G of all inner automorphisms of G (see Exercise 25 in Section 7.4).

C. 27. Let A, B, N be normal subgroups of a group G such that $N \subseteq A, N \subseteq B$. If $G = AB$ and $A \cap B = N$, prove that $G/N \cong A/N \times B/N$. (The special case $N = \langle e \rangle$ is Exercise 28 in Section 7.6.)

7.8 QUOTIENT GROUPS AND HOMOMORPHISMS

In view of our experience with rings, it should come as no surprise that there is a close connection between normal subgroups, quotient groups, and homomorphisms. The first step in developing this connection is to carry over to groups the concept of a kernel. Recall that the kernel of a ring homomorphism is the set of elements that are mapped to the zero element (the identity element of the additive group). Translating this idea to multiplicative groups, we have

> **DEFINITION•** *Let $f: G \to H$ be a homomorphism of groups. Then the **kernel** of f is the set $\{a \in G \mid f(a) = e_H\}$.*

The kernel of a ring homomorphism is an ideal. Normal subgroups are the group-theory analogue of ideals, and we have

THEOREM 7.39 *Let $f: G \to H$ be a homomorphism of groups with kernel K. Then K is a normal subgroup of G.*

Proof If $c, d \in K$, then $f(c) = e_H$ and $f(d) = e_H$ by the definition of kernel. Hence, $f(cd) = f(c)f(d) = e_H e_H = e_H$, so that $cd \in K$. If $c \in K$, then by Theorem 7.19 $f(c^{-1}) = f(c)^{-1} = (e_H)^{-1} = e_H$. Thus $c^{-1} \in K$. Therefore, K is a subgroup of G by Theorem 7.10. To show that K is normal, we must verify that for any $a \in G$ and $c \in K$, $a^{-1}ca \in K$ (Theorem 7.34). However,

$$f(a^{-1}ac) = f(a^{-1})f(c)f(a) = f(a)^{-1}e_H f(a) = f(a)^{-1}f(a) = e_H.$$

Therefore, $a^{-1}ca \in K$ and K is normal. ◆

As in the case of rings, the kernel measures how far the homomorphism f is from being injective.

THEOREM 7.40 *Let $f: G \to H$ be a homomorphism of groups with kernel K. Then $K = \langle e_G \rangle$ if and only if f is injective.*

Proof Translate the proof of Theorem 6.11 from additive to multiplicative notation,* replacing R by G, S by H, 0_R by e_G, and Theorem 3.12 by Theorem 7.19. ◆

---

* Here and below, "translate" means change the parts of the ring-theory proof dealing with the additive group of the ring to multiplicative-group notation and ignore the parts of the proof dealing with ring multiplication.

Theorem 7.39 states that every kernel is a normal subgroup. Conversely, every normal subgroup is a kernel:

THEOREM 7.41 *If N is a normal subgroup of a group G, then the map* $\pi: G \to G/N$ *given by* $\pi(a) = Na$ *is a surjective homomorphism with kernel N.*

Proof Translate the proof of Theorem 6.12 from additive to multiplicative notation, replacing R by G, S by H, I by N, and $r + I$ by Nr. ◆

THEOREM 7.42 (FIRST ISOMORPHISM THEOREM FOR GROUPS) *Let* $f: G \to H$ *be a surjective homomorphism of groups with kernel K. Then the quotient group G/K is isomorphic to H.*

Proof Define $\varphi: G/K \to H$ by $\varphi(Ka) = f(a)$. To show that φ is a well-defined isomorphism, translate the proof of Theorem 6.13 from additive to multiplicative notation, replacing R by G, S by H, $r + K$ by Kr, $r - t$ by rt^{-1}, and 0_S by e_H. ◆

The First Isomorphism Theorem makes it easier to identify certain quotient groups.

EXAMPLE If G and H are groups, then we claim that the direct product $G \times H$ contains a normal subgroup G^* such that G^* is isomorphic to G and the quotient group $(G \times H)/G^*$ is isomorphic to H. To prove this, let $f: G \times H \to H$ be the function given by $f(a, b) = b$. Verify that f is a surjective homomorphism (Exercise 2). Let G^* be the kernel of f, so that $G^* = \{(a, b) \mid f(a, b) = e\} = \{(a, b) \mid b = e\} = \{(a, e) \mid a \in G\}$. By the First Isomorphism Theorem, $(G \times H)/G^* \cong H$, and it is easy to show that G is isomorphic to G^* (Exercise 3).

EXAMPLE Consider the multiplicative group $\mathbb{C}^*$ of nonzero complex numbers. Let N be the set of all complex numbers of absolute value 1, that is, $N = \{a + bi \mid a^2 + b^2 = 1\}$. We claim that N is a normal subgroup of $\mathbb{C}^*$ and that the quotient group $\mathbb{C}^*/N$ is isomorphic to the multiplicative group $\mathbb{R}^{**}$ of all positive real numbers. To prove this, define a function $f: \mathbb{C}^* \to \mathbb{R}^{**}$ by $f(a + bi) = a^2 + b^2$. Verify that f is a surjective homomorphism of multiplicative groups (Exercise 4). Since 1 is the identity element of $\mathbb{R}^{**}$, the kernel of f is the set of complex numbers of absolute value 1, that is, the set N. By Theorems 7.39 and 7.42, N is a normal subgroup of $\mathbb{C}^*$ and the quotient group $\mathbb{C}^*/N$ is isomorphic to $\mathbb{R}^{**}$.

Subgroups of Quotient Groups

Let N be a normal subgroup of a group G. What do the subgroups of G/N look like, and how are they related to the subgroups of G? We can give a partial answer to this question by demonstrating one way of constructing subgroups of G/N. Let K be a subgroup of G that contains N. Then N is certainly a subgroup of K. Since $Na = aN$ for every $a \in G$, then in particular $Na = aN$ for every $a \in K$. Hence, N is a normal subgroup of K, so that K/N is a group by Theorem 7.36. The elements of K/N are the cosets Na with $a \in K$. Clearly, every such coset is also a coset in $G/N = \{Nc \mid c \in G\}$. Therefore, K/N is a subgroup of G/N. If the group K that we start with is normal in G, then we have

THEOREM 7.43 (THIRD ISOMORPHISM THEOREM FOR GROUPS)* *Let K and N be normal subgroups of a group G with $N \subseteq K \subseteq G$. Then K/N is a normal subgroup of G/N, and the quotient group $(G/N)/(K/N)$ is isomorphic to G/K.*

Proof The basic idea of the proof is to define a surjective homomorphism from G/N to G/K whose kernel is K/N. Then the conclusion of the theorem will follow immediately from the First Isomorphism Theorem. First note that, if $Na = Nc$ in G/N, then $ac^{-1} \in N$ by Theorem 7.23 and the definition of congruence modulo N. Since $N \subseteq K$, this means that $ac^{-1} \in K$. Consequently, $Ka = Kc$ in G/K by Theorem 7.23 again. Therefore, the map $f: G/N \to G/K$ given by $f(Na) = Ka$ is a well-defined function, that is, independent of the coset representatives in G/N. Clearly f is surjective since any Ka in G/K is the image of Na in G/N. The definition of coset operation shows that

$$f(NaNb) = f(Nab) = Kab = KaKb = f(Na)f(Nb).$$

Hence, f is a homomorphism. Since the identity element of G/K is Ke, a coset Na is in the kernel of f if and only if $f(Na) = Ke$, that is, if and only if $Ka = Ke$. However, $Ka = Ke$ if and only if $a \in K$ by Theorem 7.23. Thus the kernel of f consists of all cosets Na with $a \in K$; in other words, K/N is the kernel of f. Therefore, K/N is a normal subgroup of G/N (Theorem 7.39), and by the First Isomorphism Theorem, $(G/N)/(K/N) = (G/N)/\text{kernel } f \cong G/K$. ◆

We now have complete information on those subgroups of G/N that arise from subgroups of G. Are these all the subgroups of G/N? The next theorem summarizes what we have done and answers this question in the affirmative.

THEOREM 7.44 *Let N be a normal subgroup of a group G and let K be any subgroup of G that contains N. Then*

  (1) K/N is a subgroup of G/N.

  (2) K/N is normal in G/N if and only if K is normal in G.

  (3) if T is any subgroup of G/N, then there is a subgroup H of G such that $N \subseteq H$ and $T = H/N$.

---

* Yes, Virginia, there *is* a Second Isomorphism Theorem; see Exercise 24.

Proof Theorem 7.43 and the paragraph preceding it prove statement (1) and half of (2). The other implication in (2) is Exercise 13.

(3) Let $H = \{a \in G \mid Na \in T\}$. If $a, b \in H$, then $Na \in T$ and $Nb \in T$. Since T is a subgroup, $Nab = NaNb \in T$. Hence, $ab \in H$ and H is closed. If $a \in H$, then $Na \in T$. Since T is a subgroup, $Na^{-1} = (Na)^{-1}$ is also in T. Hence, $a^{-1} \in H$ and H is a subgroup by Theorem 7.10. Note that if $a \in N$, then $Na = Ne$ by Theorem 7.23 because $a \equiv e \pmod{N}$. Thus $Na = Ne \in T$, so that $a \in H$. Therefore, $N \subseteq H$. The quotient group H/N consists of all cosets Na with $a \in H$; by the definition of H, this is the set of all Na in T. Thus $H/N = T$. ◆

Simple Groups

In Section 7.5 we considered the classification problem for finite groups — the attempt to produce a list of groups such that every finite group is isomorphic to exactly one group on the list. We now introduce the groups that apparently are the key to solving the classification problem. Recall that a group G always has two normal subgroups, the trivial group $\langle e \rangle$ and G itself (Exercise 4 in Section 7.6). A group G is said to be **simple** if its only normal subgroups are $\langle e \rangle$ and G.

> **EXAMPLE** If p is prime, then any (normal) subgroup H of the additive group $\mathbb{Z}_p$ must have order dividing p by Lagrange's Theorem. So H must have order 1 or p, so that $H = \langle 0 \rangle$ or $H = \mathbb{Z}_p$. Therefore, $\mathbb{Z}_p$ is simple.

THEOREM 7.45 *G is a simple abelian group if and only if G is isomorphic to the additive group $\mathbb{Z}_p$ for some prime p.*

Proof The preceding example shows that any group isomorphic to $\mathbb{Z}_p$ is simple. Conversely, suppose G is simple. Since every subgroup of an abelian group is normal, G has no subgroups at all, except $\langle e \rangle$ and G. So if a is any nonidentity element of G, then the cyclic subgroup $\langle a \rangle$ must be G itself. Since every infinite cyclic group is isomorphic to $\mathbb{Z}$ by Theorem 7.18 and $\mathbb{Z}$ has many proper subgroups, $G = \langle a \rangle$ must be a cyclic group of finite order n. We claim that n is prime. If n were composite, say $n = td$ with $1 < d < n$, then $\langle a^t \rangle$ would be a subgroup of G of order d by part (4) of Theorem 7.8, which is impossible since G is simple. Therefore, G is cyclic of prime order and, hence, is isomorphic to some $\mathbb{Z}_p$ by Theorem 7.18. ◆

Nonabelian simple groups are relatively rare. There are only five of order less than 1000 and only 56 of order less than 1,000,000. A large class of nonabelian simple groups, the alternating groups, is considered in Section 7.10.

We now show why simple groups are the basic building blocks for all groups. If G is a finite group, then it has only finitely many normal subgroups other than itself (and there is at least one such subgroup since $\langle e \rangle$ is normal).

Let G_1 be a normal subgroup (other than G) that has the largest possible order. We claim that G/G_1 is simple. If G/G_1 had a proper normal subgroup, then by Theorem 7.44 this subgroup would be of the form M/G_1, where M is a normal subgroup of G such that $G_1 \subsetneqq M \subsetneqq G$. In this case, M would be a normal subgroup other than G with order larger than $|G_1|$, a contradiction. Hence, G/G_1 is simple.

If $G_1 \neq \langle e \rangle$, let G_2 be a normal subgroup of G_1 (other than G_1) of largest possible order. (G_2 is normal in G_1, but need not be normal in G.) The argument in the preceding paragraph, with G_1 in place of G and G_2 in place of G_1, shows that G_1/G_2 is simple. Similarly, if $G_2 \neq \langle e \rangle$, there is a normal subgroup G_3 of G_2 such that $G_3 \neq G_2$ and G_2/G_3 is simple. This process can be continued until we reach some G_n which is the identity subgroup (and this must occur since the order of G_i gets smaller at each stage). Then we have a sequence of groups

$$G = G_0 \supsetneqq G_1 \supsetneqq G_2 \supsetneqq G_3 \supsetneqq \cdots \supsetneqq G_{n-1} \supsetneqq G_n = \langle e \rangle$$

such that each G_i is a normal subgroup of its predecessor and each quotient group G_i/G_{i+1} is simple. The simple groups $G_0/G_1, G_1/G_2, \ldots, G_{n-1}/G_n$ are called the **composition factors** of G.

It can be shown that the composition factors of a finite group G are independent of the choice of the subgroups G_i. In other words, if you made different choices of the G_i, the simple quotient groups you would obtain would be isomorphic to the ones obtained in the previous paragraph. This means that the composition factors of G are completely determined by the structure of G and suggests a strategy for solving the classification problem. If we could first classify all simple groups and then show how the composition factors of an arbitrary group determine the structure of the group, it would be possible to classify all groups.

The good news is that the first half of this plan has already succeeded. Over a 25-year period ending in 1981, a number of group theorists around the world worked on various aspects of the problem and eventually obtained a list of simple groups such that every finite simple group is isomorphic to exactly one group on the list. The complete proof of this spectacular result runs some 10,000 pages! For a brief history of the search for simple groups, see Gallian [24] or Steen [27].

◆ **EXERCISES**

A. **1.** Verify that each map is a group homomorphism and find its kernel:

(a) $f : \mathbb{Z}_{12} \to \mathbb{Z}_{12}$, where $f(x) = 3x$

(b) $f : \mathbb{Z} \to \mathbb{Z}_2 \times \mathbb{Z}_4$, where $[k]_n$ denotes the congruence class of k in $\mathbb{Z}_n$ and $f(k) = ([k]_2, [k]_4)$

(c) $g: \mathbb{Z}_8 \to \mathbb{Z}_2$, where $f([k]_8) = [k]_2$

(d) $\varphi: S_n \to S_{n+1}$, where for each $f \in S_n$, $\varphi(f) \in S_{n+1}$ is given by

$$\varphi(f)(k) = \begin{cases} f(k) & \text{if } 1 \le k \le n \\ n+1 & \text{if } k = n+1 \end{cases}$$

(e) $h: \mathbb{Z}_{18} \to \mathbb{Z}_3$, where $h([x]_{18}) = [2x]_3$

2. Let G and H be groups and let $f: G \times H \to H$ be given by $f(a, b) = b$. Show that f is a surjective homomorphism.

3. Let G and H be groups and let $G^* = \{(a, e_H) \mid a \in G\}$ and $H^* = \{(e_G, b) \mid b \in H\}$.

   (a) Show that G^* and H^* are normal subgroups of $G \times H$.

   (b) Show that $G \cong G^*$ and $H \cong H^*$.

   (c) Show that $(G \times H)/G^* \cong H$ and $(G \times H)/H^* \cong G$.

4. Prove that the function $f: \mathbb{C}^* \to \mathbb{R}^{**}$ given by $f(a + bi) = a^2 + b^2$ is a surjective homomorphism of groups.

5. (a) Produce a list of groups such that every homomorphic image of $\mathbb{Z}_{12}$ is isomorphic to exactly one group on the list. [*Hint:* See Exercise 18 in Section 7.4.]

   (b) Do the same for $\mathbb{Z}_{20}$.

6. (a) List all subgroups of $\mathbb{Z}_{12}/H$, where $H = \{0, 6\}$.

   (b) List all subgroups of $\mathbb{Z}_{20}/K$, where $K = \{0, 4, 8, 12, 16\}$.

7. Suppose that G is a simple group and $f: G \to H$ is a surjective homomorphism of groups. Prove that either f is an isomorphism or $H = \langle e \rangle$.

B. 8. If $k \mid n$ and $f: U_n \to U_k$ is given by $f([x]_n) = [x]_k$, show that f is a homomorphism and find its kernel (notation as in Exercise 1).

9. Let $\varphi: \mathbb{Z}[x] \to \mathbb{Z}$ be the function given by $\varphi(f(x)) = f(3)$. Show that φ is a homomorphism of additive groups and find its kernel.

10. Let G be an abelian group.

    (a) Show that $K = \{a \in G \mid |a| \le 2\}$ is a subgroup of G.

    (b) Show that $H = \{x^2 \mid x \in G\}$ is a subgroup of G.

    (c) Prove that $G/K \cong H$. [*Hint:* Define a surjective homomorphism from G to H with kernel K.]

11. Use the First Isomorphism Theorem to prove that $\mathbb{R}^*/\langle 1, -1 \rangle \cong \mathbb{R}^{**}$.

12. Let G and H be the groups in Exercises 31 and 32 of Section 7.1. Use the First Isomorphism Theorem to prove that H is normal in G and that G/H is isomorphic to the multiplicative group $\mathbb{R}^*$ of nonzero real numbers. [*Hint:* Consider the map $f: G \to \mathbb{R}^*$ given by $f(T_{a,b}) = a$.]

13. Let N be a normal subgroup of a group G and let K be any subgroup of G that contains N. Then K/N is a subgroup of G/N by part (1) of Theorem 7.44. If K/N is normal in G/N, prove that K is normal in G.

14. Let M be a normal subgroup of a group G and let N be a normal subgroup of a group H. Use the First Isomorphism Theorem to prove that $M \times N$ is a normal subgroup of $G \times H$ and that $(G \times H)/(M \times N) \cong G/M \times H/N$.

15. $SL(2, \mathbb{R})$ is a normal subgroup of $GL(2, \mathbb{R})$ by Exercise 23 of Section 7.6. Prove that $GL(2, \mathbb{R})/SL(2, \mathbb{R})$ is isomorphic to the multiplicative group $\mathbb{R}^*$ of nonzero real numbers.

16. Let R be a ring and let $r \in R$. Let $f: R \to R$ be the function given by $f(x) = rx$.

    (a) Show that f is a homomorphism of additive groups.

    (b) Show by example that f need not be a ring homomorphism.

    (c) What conditions on R and/or r would guarantee that f is a ring homomorphism?

17. (An exercise for those who know how to multiply 3×3 matrices.) Let G be the set of all matrices of the form

$$\begin{pmatrix} 1 & a & b \\ 0 & 1 & c \\ 0 & 0 & 1 \end{pmatrix}$$

    where $a, b, c \in \mathbb{Q}$.

    (a) Show that G is a group under matrix multiplication.

    (b) Find the center C of G and show that C is isomorphic to the additive group $\mathbb{Q}$.

    (c) Show that G/C is isomorphic to the additive group $\mathbb{Q} \times \mathbb{Q}$.

18. Let $f: G \to H$ be a surjective homomorphism of groups with kernel K and let M be a subgroup of H.

    (a) Prove that there is a subgroup N of G such that $K \subseteq N \subseteq G$ and $N/K \cong M$.

    (b) If M is normal in H, prove that N is normal in G and $G/N \cong H/M$.

19. Let $f: G \to H$ be a surjective homomorphism of groups with kernel K. Prove that there is a bijection between the set of all subgroups of H and the set of subgroups of G that contain K.

20. Let N be a normal subgroup of a group G and let $f: G \to H$ be a homomorphism of groups such that the restriction of f to N is an isomorphism $N \cong H$. Prove that $G \cong N \times K$, where K is the kernel of f. [*Hint:* Exercise 28 in Section 7.6.]

21. A group G is said to be **metabelian** if it has a subgroup N such that N is abelian, N is normal in G, and G/N is abelian.

   (a) Show that S_3 is metabelian.

   (b) Prove that every homomorphic image of a metabelian group is metabelian.

   (c) Prove that every subgroup of a metabelian group is metabelian.

22. Let N be a normal subgroup of a group G. Prove that G/N is simple if and only if there is no normal subgroup K such that $N \subsetneq K \subsetneq G$. [*Hint:* Theorem 7.44.]

23. The additive group $\mathbb{Z}[x]$ contains $\mathbb{Z}$ (the set of constant polynomials) as a normal subgroup. Show that $\mathbb{Z}[x]/\mathbb{Z}$ is isomorphic to $\mathbb{Z}[x]$. This example shows that $G/N \cong G$ does not necessarily imply that $N = \langle e \rangle$. [*Hint:* Consider the map $T: \mathbb{Z}[x] \to \mathbb{Z}[x]/\mathbb{Z}$ given by $T(f(x)) = \mathbb{Z} + xf(x)$.]

C. 24. (Second Isomorphism Theorem) Let K and N be subgroups of a group G, with N normal in G. Then $NK = \{nk \mid n \in N, k \in K\}$ is a subgroup of G that contains both K and N by Exercise 18 of Section 7.6.

   (a) Prove that N is a normal subgroup of NK.

   (b) Prove that the function $f: K \to NK/N$ given by $f(k) = Nk$ is a surjective homomorphism with kernel $K \cap N$.

   (c) Conclude that $K/(N \cap K) \cong NK/N$.

25. Cayley's Theorem 7.20 represents a group G as a subgroup of the permutation group $A(G)$. A more efficient way of representing G as a permutation group arises from the following generalized Cayley's Theorem. Let K be a subgroup of G and let T be the set of all distinct right cosets of K.

   (a) If $a \in G$, show that the map $f_a: T \to T$ given by $f_a(Kb) = Kba$ is a permutation of the set T.

   (b) Prove that the function $\varphi: G \to A(T)$ given by $\varphi(a) = f_{a^{-1}}$, is a homomorphism of groups whose kernel is contained in K.

   (c) If K is normal in G, prove that $K = \text{kernel } \varphi$.

   (d) Prove Cayley's Theorem by applying parts (b) and (c) with $K = \langle e \rangle$.

Application   Decoding Techniques (Section 16.2) may be covered at this point if desired.

7.9 THE SYMMETRIC AND ALTERNATING GROUPS

The finite symmetric groups S_n are important because, as we saw in Corollary 7.21, every finite group is isomorphic to a subgroup of some S_n. Certain subgroups of the S_n to be introduced here are also of interest since they are non-abelian simple groups. As we saw in Section 7.8, such groups are the basic building blocks for all groups.

The notation used up to now for elements of the symmetric groups is rather awkward and obscures some useful facts that are quite apparent in another notation. For this reason, we first develop this new notation.

Consider the permutation $\begin{pmatrix} 1\ 2\ 3\ 4\ 5\ 6 \\ 1\ 4\ 3\ 6\ 2\ 5 \end{pmatrix}$ in S_6. Note that 2 is mapped to 4, 4 is mapped to 6, 6 is mapped to 5, 5 is mapped back to 2, and the other two elements, 1 and 3, are mapped to themselves. All the essential information can be summarized by this diagram:

It isn't necessary to include the arrows here as long as we keep things in the same order. A complete description of this permutation is given by the symbol (2465), with the understanding that

each element is mapped to the element listed immediately to the right;

the last element in the string is mapped to the first;

elements not listed are mapped to themselves.

This is an example of *cycle notation*. Here is a formal definition.

DEFINITION• *Let $a_1, a_2, a_3, \ldots, a_k$ (with $k \geq 1$) be distinct elements of the set $\{1, 2, 3, \ldots, n\}$. Then $(a_1a_2a_3 \cdots a_k)$ denotes the permutation in S_n that maps a_1 to a_2, a_2 to a_3, $\ldots$, a_{k-1} to a_k, and a_k to a_1 and maps every other element of $\{1, 2, 3, \ldots, n\}$ to itself. $(a_1a_2a_3 \cdots a_k)$ is called a* **cycle of length k** *or a* **k-cycle.**

EXAMPLE In S_4, (143) is the 3-cycle that maps 1 to 4, 4 to 3, 3 to 1, and 2 to itself; it was written $\begin{pmatrix} 1\ 2\ 3\ 4 \\ 4\ 2\ 1\ 3 \end{pmatrix}$ in the old notation. Note that (143) may also be denoted by (431) or (314) since each of these indicates the function that maps 1 to 4, 4 to 3, 3 to 1, and 2 to 2.

EXAMPLE According to the definition above, the 1-cycle (3) in S_n is the permutation that maps 3 to 3 and maps every other element of $\{1, 2, \ldots, n\}$ to itself; in other words, (3) is the identity permutation. Similarly, for any k in $\{1, 2, \ldots, n\}$, the 1-cycle (k) is the identity permutation.

Strictly speaking, cycle notation is ambiguous since, for example, (163) might denote a permutation in S_6, in S_7, or in any S_n with $n \geq 6$. In context, however, this won't cause any problems because it will always be made clear which group S_n is under discussion.

Products in cycle notation can be visually calculated just as in the old notation. For example, we know that

$$\begin{pmatrix} 1 & 2 & 3 & 4 \\ 1 & 4 & 2 & 3 \end{pmatrix} \circ \begin{pmatrix} 1 & 2 & 3 & 4 \\ 2 & 4 & 1 & 3 \end{pmatrix} = \begin{pmatrix} 1 & 2 & 3 & 4 \\ 4 & 3 & 1 & 2 \end{pmatrix}.$$

(Remember that the product in S_n is composition of functions, and so the right-hand permutation is performed first.) In cycle notation, this product* becomes

$$(2 \ \ 4 \ \ 3)(1 \ \ 2 \ \ 4 \ \ 3) = (1 \ \ 4 \ \ 2 \ \ 3).$$

The arrows indicate the process: 1 is mapped to 2 and 2 is mapped to 4, so that the product maps 1 to 4. Similarly, 4 is mapped to 3 and 3 is mapped to 2, so that the product maps 4 to 2.

Two cycles are said to be **disjoint** if they have no elements in common. For instance, (13) and (2546) are disjoint cycles in S_6, but (13) and (345) are not since 3 appears in both cycles.

EXAMPLE As shown above, (243)(1243) = (1423). Reversing the order shows that

$$(1243)(243) = (2341).$$

Hence, the cycles (243) and (1234) do not commute with each other. On the other hand, you can easily verify that the disjoint cycles (13) and (2546) *do* commute:

$$(13)(2546) = \begin{pmatrix} 1 & 2 & 3 & 4 & 5 & 6 \\ 3 & 5 & 1 & 6 & 4 & 2 \end{pmatrix} = (2546)(13).$$

This is an illustration of the following theorem.

---

* Hereafter we shall omit the composition symbol ∘ and write the group operation in S_n multiplicatively.

THEOREM 7.46 *If* $\sigma = (a_1 a_2 \cdots a_k)$ *and* $\tau = (b_1 b_2 \cdots b_r)$ *are disjoint cycles in* S_n, *then* $\sigma\tau = \tau\sigma$.

Proof Exercise 12. ◆

It is not true that every permutation is a cycle, but every permutation can be expressed as the product of disjoint cycles. Consider, for example, the permutation $\begin{pmatrix} 1\,2\,3\,4\,5\,6\,7 \\ 5\,1\,7\,2\,4\,6\,3 \end{pmatrix}$ in S_7. Find an element that is not mapped to itself, say 1, and trace where it is sent by the permutation:

1 is mapped to 5,   5 is mapped to 4,   4 is mapped to 2,   and
2 is mapped to 1 (the element with which we started).

Thus the given permutation has the same action as the cycle (1542) on these four elements. Now look at any element other than 1, 5, 4, 2 that is not mapped onto itself, say 3. Note that

3 is mapped to 7,   and   7 is mapped to 3.

Thus the 2-cycle (37) has the same action on 7 and 3 as the given permutation. The only element now unaccounted for is 6, which is mapped to itself. You can now easily verify that the original permutation is the product of the two cycles we have found, that is,

$$\begin{pmatrix} 1 & 2 & 3 & 4 & 5 & 6 & 7 \\ 5 & 1 & 7 & 2 & 4 & 6 & 3 \end{pmatrix} = (1542)(37).$$

Although some care must be used and the notation is more cumbersome, essentially the same procedure works in the general case.

THEOREM 7.47 *Every permutation in* S_n *is the product of disjoint cycles.*[*]

Proof Adapt the procedure in the preceding example; see Exercise 36. ◆

Note that the identity permutation can be written as (1) = (12)(12). Similarly, the cycle (1234) is also a product of (nondisjoint) 2-cycles since (1234) = (14)(13)(12). A 2-cycle is usually called a **transposition.**

COROLLARY 7.48 *Every permutation in* S_n *is a product of transpositions.*

Proof Since every permutation is a product of cycles by Theorem 7.47, you need only verify that every cycle $(a_1 a_2 \cdots a_k)$ is a product of transpositions:

$$(a_1 a_2 \cdots a_k) = (a_1 a_k)(a_1 a_{k-1}) \cdots (a_1 a_3)(a_1 a_2)$$

---

[*] As usual, we allow the possibility of a product with just one cycle in it.

or alternatively,

$$(a_1a_2 \cdots a_k) = (a_1a_2)(a_2a_3) \cdots (a_{k-1}a_k). \quad \blacklozenge$$

This corollary can also be proved directly by induction, without using Theorem 7.47 (Exercise 21).

The Alternating Groups

The factorization of a permutation as a product of disjoint cycles is unique except for the order of the cycles (Exercise 20), but the factorization as a product of transformations is far from unique. For example,

$(123) = (13)(12)$ $(1235) = (15)(24)(24)(13)(23)(23)(12)$

$(123) = (13)(23)(12)(13)$ $(1235) = (13)(24)(35)(14)(24)$

$(123) = (23)(13)(12)(13)(12)(23)$ $(1235) = (15)(13)(12)$

Notice that the factorizations of (123) all have an even number of transpositions, whereas the factorizations of (1235) all have an odd number of transpositions. This suggests that the number of transpositions in the factorization of a permutation is always even or always odd. This is indeed the case, as we now prove.

LEMMA 7.49 *The identity permutation in S_n is not the product of an odd number of transpositions.*

Proof Suppose, on the contrary, that $(1) = \tau_k \cdots \tau_2\tau_1$ with each τ_i a transposition and k odd. Let c be a symbol that appears in at least one of these transpositions. Let τ_r be the first transposition (reading from *right to left*) in which c appears, say $\tau_r = (cd)$. Then c does not appear in $\tau_{r-1}, \cdots \tau_1$ and is, therefore, left fixed by these transpositions. If $r = k$, then c is left fixed by all the τ's except τ_k, so that the product—the identity permutation—maps c to d, a contradiction. Hence, $r < k$, and we can consider the transposition τ_{r+1}. It must have one of the following forms (where x, y, c, d denote distinct elements of $\{1, 2, \cdots n\}$):

       I. (xy) II. (xd) III. (cy) IV. (cd).

Consequently, there are four possibilities for the product $\tau_{r+1}\tau_r$:

     I. $(xy)(cd)$ II. $(xd)(cd)$ III. $(cy)(cd)$ IV. $(cd)(cd)$.

In case I, verify that $(xy)(cd) = (cd)(xy)$. Replace $(xy)(cd)$ by $(cd)(xy)$ in the product; this moves the first appearance of c one transposition to the left. In case II, verify that $(xd)(cd) = (xc)(xd)$; if we replace $(xd)(cd)$ by $(xc)(xd)$, then once again the first appearance of c is one transposition farther left. Show that a similar conclusion holds in case III by verifying that $(cy)(cd) = (cd)(dy)$.

Each repetition of the procedure in cases I–III moves the first appearance of c one transposition farther left. Eventually case IV must occur; otherwise, we could keep moving c until it first appears in the last permutation at the left, τ_k, which is impossible, as we saw in the first paragraph. In case IV, however, we have $\tau_{r+1}\tau_r = (cd)(cd) = (1)$. So we can delete these two transpositions and write (1) as a product of two fewer transpositions than before. Obviously, we can carry out the same argument for any symbol that appears in a transposition in the product. If the original product contains an odd number of transpositions, eliminating two at a time eventually reduces it to a single transposition $(1) = (ab)$, which is a contradiction. Therefore, (1) cannot be written as the product of an odd number of transpositions. ◆

THEOREM 7.50 *No permutation in S_n can be written as the product of an even number of transpositions and also as the product of an odd number of transpositions.*

Proof Suppose $\alpha \in S_n$ can be written as $\sigma_1\sigma_2 \cdots \sigma_k$ and as $\tau_1\tau_2 \cdots \tau_r$ with each σ_i, τ_j a transposition, k odd, and r even. Since every transposition is its own inverse [for instance, $(34)(34) = (1)$], Corollary 7.6 shows that

$$(1) = \alpha\alpha^{-1} = (\sigma_1 \cdots \sigma_k)(\tau_1 \cdots \tau_r)^{-1}$$
$$= \sigma_1 \cdots \sigma_k \tau_r^{-1} \cdots \tau_1^{-1}$$
$$= \sigma_1 \cdots \sigma_k \tau_r \cdots \tau_1.$$

Since k is odd and r is even, $k + r$ is odd, and we have written (1) as the product of an odd number of transpositions. This contradicts Lemma 7.49, and completes the proof of the theorem. ◆

A permutation in S_n is said to be **even** if it can be written as the product of an even number of transpositions and **odd** if it can be written as the product of an odd number of transpositions. For example, the identity permutation (1) is even and (134)(25) is odd because

$$(1) = (12)(12) \quad \text{and} \quad (134)(25) = (14)(13)(25).$$

Every permutation is either even or odd by Corollary 7.48, and no permutation is both even and odd by Theorem 7.50. The set of all even permutations in S_n is denoted A_n and is called the **alternating group of degree n;** the word "group" is justified by the following theorem.

THEOREM 7.51 *A_n is a normal subgroup of order $n!/2$ and index 2 in S_n.*

Proof Define a function $f: S_n \to \mathbb{Z}_2$ by $f(\sigma) = 0$ if σ is even and $f(\sigma) = 1$ if σ is odd. Since no permutation is both even and odd, f is a well-defined function. It is easy to see that the product of two even or two odd permutations is even and that the product of an odd and an even permutation is odd. Use these facts to verify that f is a surjective group homomorphism with kernel A_n. Therefore, A_n is a normal subgroup by Theorem 7.39 and $S_n/A_n \cong \mathbb{Z}_2$ by the First Isomorphism Theorem. Consequently, statement (2) of Theorem 7.36 shows that

$$2 = |\mathbb{Z}_2| = |S_n/A_n| = \frac{|S_n|}{|A_n|} = \frac{n!}{|A_n|}$$

which imples that $|A_n| = n!/2$. Finally, by Lagrange's Theorem

$$[S_n : A_n] = \frac{|S_n|}{|A_n|} = \frac{n!}{n!/2} = 2. \quad \blacklozenge$$

> **EXAMPLE**   The permutations (1), (123), and (132) in S_3 are even because
>
> $$(1) = (12)(12), \qquad (123) = (13)(12), \qquad \text{and} \qquad (132) = (12)(13).$$
>
> Since A_3 is a group of order $3!/2 = 3$, we see that $A_3 = \{(1), (123), (132)\}$.

> **EXAMPLE** A_4 is a group of order $4!/2 = 12$, which has no subgroup of order 6 (Exercise 28). Thus, the converse of Lagrange's Theorem is false.

◆ EXERCISES

A.   **1.** Write each permutation in cycle notation:

(a) $\begin{pmatrix} 1\,2\,3\,4\,5\,6\,7\,8\,9 \\ 7\,2\,1\,4\,5\,6\,3\,8\,9 \end{pmatrix}$ (b) $\begin{pmatrix} 1\,2\,3\,4\,5\,6\,7\,8\,9 \\ 2\,4\,3\,5\,7\,6\,8\,9\,1 \end{pmatrix}$

(c) $\begin{pmatrix} 1\,2\,3\,4\,5\,6\,7\,8\,9 \\ 4\,8\,1\,7\,5\,2\,6\,3\,9 \end{pmatrix}$ (d) $\begin{pmatrix} 1\,2\,3\,4\,5\,6\,7\,8\,9 \\ 1\,2\,5\,4\,7\,6\,9\,3\,8 \end{pmatrix}$

2. Compute each product:

(a) (12)(23)(34)   (b) (246)(147)(135)

(c) (12)(53214)(23)   (d) (1234)(2345)

3. Express as a product of disjoint cycles:

(a) $\begin{pmatrix} 1\,2\,3\,4\,5\,6\,7\,8\,9 \\ 2\,1\,3\,5\,4\,7\,9\,8\,6 \end{pmatrix}$ (b) $\begin{pmatrix} 1\,2\,3\,4\,5\,6\,7\,8\,9 \\ 3\,5\,1\,2\,4\,6\,8\,9\,7 \end{pmatrix}$

(c) $\begin{pmatrix} 1\,2\,3\,4\,5\,6\,7\,8\,9 \\ 3\,5\,1\,2\,4\,9\,8\,7\,6 \end{pmatrix}$ (d) (14)(27)(523)(34)(1472)

(e) (7236)(85)(571)(1537)(486)

4. Write each permutation in Exercise 3 as a product of transpositions.

5. Which of these permutations are even:

(a) (2468)   (b) (246)(134)   (c) (12)(123)(1234)

6. List the elements in each group:

(a) A_2 (b) A_3 (c) A_4

7. What is the order of each group:

(a) A_3 (b) A_4 (c) A_5 (d) A_{10}

8. Let $\sigma = (a_1 a_2 a_3 a_4 a_5 a_6)$. Find σ^i for $i = 1, 2, \ldots, 6$.

B. **9.** Prove that a k-cycle in the group S_n has order k.

10. Show that the inverse of $(a_1 a_2 \cdots a_k)$ in S_n is $(a_1 a_k a_{k-1} \cdots a_3 a_2)$.

11. Prove that the cycle $(a_1 a_2 \cdots a_k)$ is even if and only if k is odd.

12. Let $\sigma = (a_1 a_2 \cdots a_k)$ and $\tau = (b_1 b_2 \cdots b_r)$ be disjoint cycles in S_n. Prove that $\sigma\tau = \tau\sigma$. [*Hint:* You must show that $\sigma\tau$ and $\tau\sigma$ agree as functions on each i in $\{1, 2, \ldots, n\}$. Consider three cases: i is one of the a's; i is one of the b's; i is neither.]

13. If $\tau \in S_n$, prove that the order of τ is the least common multiple of the lengths of the disjoint cycles whose product is τ. [*Hint:* Exercises 9 and 12 may be helpful.]

14. Find the order of σ^{1000}, where σ is the permutation $\begin{pmatrix} 1\,2\,3\,4\,5\,6\,7\,8\,9 \\ 3\,7\,8\,9\,4\,5\,2\,1\,6 \end{pmatrix}$.

15. Show that the subgroup G of S_4 generated by the elements $\sigma = (1234)$ and $\tau = (24)$ has order 8.

16. Prove that the center of $S_n (n > 2)$ is the identity subgroup.

17. If σ is a k-cycle with k odd, prove that there is a cycle τ such that $\tau^2 = \sigma$.

18. Let σ be a k-cycle in S_n.

(a) Prove that σ^2 is a cycle if and only if k is odd.

(b) If $k = 2t$, prove that there are t-cycles τ and β such that $\sigma^2 = \tau\beta$.

19. Let σ be a product of disjoint cycles of the same length. Prove that σ is a power of a cycle.

20. Prove that the decomposition of a permutation as a product of disjoint cycles is unique except for the order in which the cycles are listed.

21. Use induction on n to give an alternate proof of Corollary 7.48: Every element of S_n is a product of transpositions. [*Hint:* If the statement is true for $n = k - 1$ and if $\tau \in S_k$, consider the transposition (kr), where $r = \tau(k)$. Note that $(kr)\tau$ fixes k and hence may be considered as a permutation of $\{1, 2, \ldots, k - 1\}$.]

22. If $n \geq 3$; prove that every element of S_n can be written as a product of at most $n - 1$ transpositions.

23. Let τ be a transposition and let $\sigma \in S_n$. Prove that $\sigma\tau\sigma^{-1}$ is a transposition.

24. If τ is the k-cycle $(a_1 a_2 \cdots a_k)$ and if $\sigma \in S_n$, prove that $\sigma \tau \sigma^{-1} = (\sigma(a_1) \sigma(a_2) \cdots \sigma(a_k))$.

25. Let G be a subgroup of S_n that contains at least one odd permutation. Prove that the number of even permutations in G is the same as the number of odd permutations in G (that is, $G \cap A_n$ is a subgroup of index 2 in G).

26. Without using Theorem 7.51, prove that A_n has index 2 in S_n. [*Hint:* Show that A_n and $A_n(12)$ are distinct cosets whose union is S_n.]

27. Prove that $\{(1), (12)(34), (13)(24), (14)(23)\}$ is a normal subgroup of A_4.

28. Show that the converse of Lagrange's theorem is false by proving that A_4 is a group of order 12 that has no subgroup of order 6. [*Hint:* If N is a subgroup of order 6, then N is normal in A_4 and A_4/N has order 2 (see Lagrange's Theorem and Exercise 20 in Section 7.6). Hence, $\sigma^2 \in N$ for every $\sigma \in A_4$ (Why?). Show that this is impossible.]

29. Let σ and τ be transpositions in S_n with $n \geq 3$. Prove that $\sigma \tau$ is a product of (not necessarily disjoint) 3-cycles.

30. Prove that every element of A_n is a product of 3-cycles.

31. Show that D_4 is isomorphic to G. [*Hint:* Exercise 15 may be helpful. Note that every element of D_4 produces a permutation of the vertices of the square (see pages 165–166). If the vertices are numbered 1, 2, 3, 4, then this permutation can be considered as an element of S_4. Define a function $f: D_4 \to S_4$ by mapping each element of D_4 to its permutation of the vertices. Verify that f is an injective homomorphism with image G.]

C. 32. Prove that every element of A_n is a product of n-cycles.

33. Prove that the transpositions $(12), (13), (14), \ldots, (1n)$ generate S_n.

34. Prove that (12) and $(123 \cdots n)$ generate S_n.

35. If f is an automorphism of S_3, prove that there exists $\sigma \in S_3$ such that $f(\tau) = \sigma \tau \sigma^{-1}$ for every $\tau \in S_3$.

36. Use the following steps to prove Theorem 7.47: Every permutation τ in S_n is a product of disjoint cycles.

    (a) Let a_1 be any element of $\{1, 2, \ldots, n\}$ such that $\tau(a_1) \neq a_1$. Let $a_2 = \tau(a_1), a_3 = \tau(a_2), a_4 = \tau(a_3)$, and so on. Let k be the first index such that $\tau(a_k)$ is one of $a_1, \ldots, a_{k-1}$. Prove that $\tau(a_k) = a_1$. Conclude that τ has the same effect on $a_1, \ldots, a_k$ as the cycle $(a_1 a_2 \cdots a_k)$.

    (b) Let b_1 be any element of $\{1, 2, \ldots, n\}$ other than $a_1, \ldots, a_k$ that is not mapped to itself by τ. Let $b_2 = \tau(b_1), b_3 = \tau(b_2)$, and so on. Show that

$\tau(b_i)$ is never one of $a_1, \ldots, a_k$. Repeat the argument in part (a) to find a b_r such that $\tau(b_r) = b_1$ and τ agrees with the cycle $(b_1 b_2 \cdots b_r)$ on the b's.

(c) Let c_1 be any element of $\{1, 2, \ldots, n\}$ other than the a's or b's above such that $\tau(c_1) \neq c_1$. Let $c_2 = \tau(c_1)$, and so on. As above, find c_s such that τ agrees with the cycle $(c_1 c_2 \cdots c_s)$ on the c's.

(d) Continue in this fashion until the only elements unaccounted for are those that are mapped to themselves by τ. Verify that τ is the product of the cycles

$$(a_1 \cdots a_k)(b_1 \cdots b_r)(c_1 \cdots c_s) \cdots$$

and that these cycles are disjoint.

7.10 THE SIMPLICITY OF A_n*

As we saw at the end of Section 7.8, simple groups appear to be the key to solving the classification problem for finite groups. This fact and the following theorem are one reason that the alternating groups A_n are important.

THEOREM 7.52 *For each $n \neq 4$, the alternating group A_n is a simple group.*

The group A_4 is not simple (Exercise 27 in Section 7.9). Although the entire proof of Theorem 7.52 is rather long, it requires only basic facts about the symmetric groups and normal subgroups. There will be many instances in the proof where we will deal with permutations such as $(abcd)$ or $(a2b)$ or $(ab)(cd)$. In all such cases,

distinct letters represent distinct elements of $\{1, 2, \ldots, n\}$.

The proof of the theorem requires two lemmas.

LEMMA 7.53 *Every element of A_n (with $n \geq 3$) is a product of 3-cycles.*

Proof Every element of A_n is by definition the product of *pairs* of transpositions. But every such pair must be of one of these forms: $(ab)(cd)$ or $(ab)(ac)$ or $(ab)(ab)$. In the first case verify that $(ab)(cd) = (adb)(adc)$, in the second that $(ab)(ac) = (acb)$, and in the last that $(ab)(ab) = (1) = (abc)(acb)$. Thus every pair of transpositions is either a 3-cycle or a product of two 3-cycles. Hence, every product of pairs of transpositions is a product of 3-cycles. ◆

---

* This section is not used in the sequel and may be omitted if desired.

LEMMA 7.54 *If N is a normal subgroup of A_n (with $n \geq 3$) and N contains a 3-cycle, then $N = A_n$.*

Proof   For notational convenience, assume that $(123) \in N$ [the argument when $(rst) \in N$ is the same; just replace 1, 2, 3 by r, s, t, respectively]. Since $(123) \in N$, we see that $(123)(123) = (132)$ is also in N. For $k \geq 4$, let $x = (12)(3k)$ and verify that $x^{-1} = (3k)(12)$. The normality of N implies that $x(132)x^{-1} \in N$ by Theorem 7.34. But

$$x(132)x^{-1} = (12)(3k)(132)(3k)(12) = (12k).$$

Therefore,

(∗) N contains all 3-cycles of the form $(12k)$ with $k \geq 3$.

Verify that every other 3-cycle can be written in one of these forms:

$$(1a2), \quad (1ab), \quad (2ab), \quad (abc)$$

where a, b, $c \geq 3$. By (∗) and closure in N,

$$(1a2) = (12a)(12a) \in N;$$
$$(1ab) = (12b)(12a)(12a) \in N;$$
$$(2ab) = (12b)(12b)(12a) \in N;$$
$$(abc) = (12a)(12a)(12c)(12b)(12b)(12a) \in N.$$

Thus N contains all 3-cycles, and, hence, N contains all products of 3-cycles by closure. Therefore, $N = A_n$ by Lemma 7.53.   ◆

We are now ready to prove Theorem 7.52. The following fact will be used frequently:

The inverse of the cycle $(a_1 a_2 a_3 \cdots a_k)$ is the cycle $(a_1 a_k a_{k-1} \cdots a_3 a_2)$.

For example, $(12345)^{-1} = (15432)$ and $(678)^{-1} = (687)$, as you can easily verify.

Proof of Theorem 7.52 A_2 and A_3 are simple abelian groups (Exercise 2). So assume $n \geq 5$. We must prove that A_n has no *proper* normal subgroups. Let N be any normal subgroup of A_n, with $N \neq (1)$. We need only show that $N = A_n$. When all the nonidentity elements of N are written as products of disjoint cycles, then there are three possibilities for the lengths of these cycles:

1. Some cycle has length ≥ 4.
2. Every cycle has length ≤ 3, and some have length 3.
3. Every cycle has length ≤ 2.

We shall show that in each of these cases, $N = A_n$.

Case 1 N contains an element σ that is the product of disjoint cycles, at least one of which has length $r \geq 4$. For notational convenience we assume that $\sigma = (1234 \cdots r)\tau$, where τ is a product of disjoint cycles, none of which involve the symbols 1,2,3,4, . . . , r.* Let $\delta = (123) \, \epsilon \, A_n$. Since N is a normal subgroup and $\sigma \, \epsilon \, N$, we have $\sigma^{-1}(\delta \sigma \delta^{-1}) \, \epsilon \, N$ by Theorem 7.34. Using Corollary 7.6 and Theorem 7.46, we have

$$\sigma^{-1}(\delta \sigma \delta^{-1}) = [(1234 \cdots r)\tau]^{-1}(123)[(1234 \cdots r)\tau](123)^{-1}$$
$$= \tau^{-1}(1r \cdots 432)(123)(1234 \cdots r)\tau(132)$$
$$= \tau^{-1}\tau(1r \cdots 432)(123)(1234 \cdots r)(132) = (13r).$$

Therefore, $(13r) \, \epsilon \, N$, and hence, $N = A_n$ by Lemma 7.54.

Case 2A N contains an element σ that is the product of disjoint cycles, at least two of which have length 3. For convenience we assume that $\sigma = (123)(456)\tau$, where τ is a product of disjoint cycles, none of which involve the symbols 1, 2, . . . , 6.* Let $\delta = (124) \, \epsilon \, A_n$. Then, as in Case 1, N contains $\sigma^{-1}(\delta \sigma \delta^{-1})$. But

$$\sigma^{-1}(\delta \sigma \delta^{-1}) = [(123)(456)\tau]^{-1}(124)(123)(456)\tau(124)^{-1}$$
$$= \tau^{-1}(465)(132)(124)(123)(456)\tau(142)$$
$$= \tau^{-1}\tau(465)(132)(124)(123)(456)(142) = (14263).$$

Therefore, $(14263) \, \epsilon \, N$, and $N = A_n$ by Case 1.

Case 2B N contains an element σ that is the product of one 3-cycle and some 2-cycles. We assume that $\sigma = (123)\tau$, where τ is a product of disjoint transpositions, none of which involve the symbols 1, 2, 3.* Since a product of disjoint transpositions is its own inverse (Exercise 5),

$$\sigma^2 = (123)\tau(123)\tau = (123)(123)\tau\tau = (123)(123) = (132).$$

But $\sigma^2 \, \epsilon \, N$ since $\sigma \, \epsilon \, N$. Therefore, $(132) \, \epsilon \, N$, and $N = A_n$ by Lemma 7.54.

Case 2C N contains a 3-cycle. Then $N = A_n$ by Lemma 7.54.

Case 3   Every element of N is the product of an even number of disjoint 2-cycles. Then a typical element σ of N has the form $(12)(34)\tau$, where τ is a product of disjoint transpositions, none of which involve the symbols 1, 2, 3, 4.* Let $\delta = (123) \, \epsilon \, A_n$. Then, as above, $\sigma^{-1}(\delta \sigma \delta^{-1}) \, \epsilon \, N$ and

---

* The same argument works with an arbitrary r-cycle $(abcd \cdots t)$ in place of $(1234 \cdots r)$; just replace 1 by a, 2 by b, etc. *Analogous remarks apply in the other cases,* where specific cycles will also be used to make the argument easier to follow.

$$\sigma^{-1}(\delta\sigma\delta^{-1}) = \tau^{-1}(34)(12)(123)(12)(34)\tau(132) = (13)(24).$$

Since $n \geq 5$, there is an element k in $\{1, 2, \ldots, n\}$ distinct from $1, 2, 3, 4$. Let $\alpha = (13k) \in A_n$. Let $\beta = (13)(24)$, which was just shown to be in N. Then by the normality of N and closure, $\beta(\alpha\beta\alpha^{-1}) \in N$. But

$$\beta(\alpha\beta\alpha^{-1}) = (13)(24)(13k)(13)(24)(1k3) = (13k).$$

Therefore, $(13k) \in N$, and $N = A_n$ by Lemma 7.54.   ◆

Theorem 7.52 leads to an interesting fact about the normal subgroups of S_n:

COROLLARY 7.55   *If $n \geq 5$, then (1), A_n, and S_n are the only normal subgroups of S_n.*

Proof   See Exercise 9.   ◆

◆ **EXERCISES**

A.   **1. (a)** List all the 3-cycles in S_4.

(b) List all the elements of A_4 and express each as a product of 3-cycles.

2. (a) Verify that $A_2 = (1)$.

(b) Show that A_3 is a cyclic group of order 3 and hence simple by Theorem 7.45.

3. Find the center of the group A_4.

4. If $n \geq 5$, what is the center of A_n?

B.   **5.** If $\sigma \in S_n$ is a product of disjoint transpositions, prove that $\sigma^2 = (1)$.

6. Show that for $n \geq 3$, the center of S_n is (1).

7. Let N be a subgroup of S_n such that $\sigma\tau = (1)$ for all nonidentity elements σ, $\tau \in N$. Prove that $N = (1)$ or N is cyclic of order 2.

8. Prove that no subgroup of order 2 in $S_n (n \geq 3)$ is normal.

9. Prove Corollary 7.55. [*Hint:* If N is a normal subgroup of S_n, then $N \cap A_n$ is a normal subgroup of A_n by Exercise 17 of Section 7.6. By Theorem 7.52, $N \cap A_n$ is A_n or (1). In the first case, $A_n \subseteq N$; use Lagrange's Theorem and Theorem 7.51 to show that $N = A_n$ or S_n. In the second case, all the nonidentity elements of N are odd, which implies that the product of any two of them is (1). Use Exercises 7 and 8.]

10. Prove that A_n is the only subgroup of index 2 in S_n. [*Hint:* Exercise 20 of Section 7.6 and Corollary 7.55.]

11. If $f: S_n \to S_n$ is a homomorphism, prove that $f(A_n) \subseteq A_n$.

2

OTHER TOPICS

CHAPTER **8**

Topics in Group Theory

◆

T his chapter takes a deeper look at various aspects of the classification problem for finite groups, which was introduced in Section 7.5. After the necessary preliminaries are developed in Section 8.1, all finite abelian groups are classified up to isomorphism in Section 8.2 (the Fundamental Theorem of Finite Abelian Groups). Nonabelian groups are considerably more complicated. The basic tools for analyzing them (the Sylow Theorems and conjugacy) are presented in Sections 8.3 and 8.4. Applications of these results and several other facts about the structure of finite groups are considered in Seciton 8.5, where groups of small order are classified.

Sections 8.3 and 8.4 are independent of Sections 8.1 and 8.2 and may be read first if desired. Sections 8.1–8.4 are prerequisites for Section 8.5.

8.1 DIRECT PRODUCTS

If G and H are groups, then their Cartesian product $G \times H$ is also a group, with the operation defined coordinatewise (Theorem 7.4). In this section we extend this notion to more than two groups. Then we examine the conditions under which a group is (isomorphic to) a direct product of certain of its subgroups. When these subgroups are of a particularly simple kind, then the structure of the group can be completely determined, as will be demonstrated in Section 8.2. Throughout the general discussion, all groups are written multiplicatively, but specific examples of familiar additive groups are written additively as usual.

If $G_1, G_2, \ldots, G_n$ are groups, we define a coordinatewise operation on the Cartesian product $G_1 \times G_2 \times \cdots \times G_n$ as follows:

$$(a_1, a_2, \ldots, a_n)(b_1, b_2, \ldots, b_n) = (a_1 b_1, a_2 b_2, \ldots, a_n b_n).$$

It is easy to verify that $G_1 \times G_2 \times \cdots \times G_n$ is a group under this operation: If e_i is the identity element of G_i, then $(e_1, e_2, \ldots, e_n)$ is the identity element of $G_1 \times G_2 \times \cdots \times G_n$ and $(a_1^{-1}, a_2^{-1}, \ldots, a_n^{-1})$ is the inverse of $(a_1, a_2, \ldots, a_n)$. This group is called the **direct product** of $G_1, G_2, \ldots, G_n$.*

> **EXAMPLE**   Recall that U_n is the multiplicative group of units in $\mathbb{Z}_n$ and that $U_4 = \{1, 3\}$ and $U_6 = \{1, 5\}$. The direct product $U_4 \times U_6 \times \mathbb{Z}_3$ consists of the 12 triples
>
> | | | | | | |
> |---|---|---|---|---|---|
> | $(1, 1, 0)$, | $(1, 1, 1)$, | $(1, 1, 2)$, | $(1, 5, 0)$, | $(1, 5, 1)$, | $(1, 5, 2)$, |
> | $(3, 1, 0)$, | $(3, 1, 1)$, | $(3, 1, 2)$, | $(3, 5, 0)$, | $(3, 5, 1)$, | $(3, 5, 2)$. |
>
> Note that U_4 has order 2, U_6 has order 2, $\mathbb{Z}_3$ has order 3, and the direct product $U_4 \times U_6 \times \mathbb{Z}_3$ has order $2 \cdot 2 \cdot 3 = 12$. Similarly, in the general case,

if $G_1, G_2, \ldots, G_n$ are finite groups, then

$G_1 \times G_2 \times \cdots \times G_n$ has order $|G_1| \cdot |G_2| \cdots |G_n|$.

In the preceding example it is important to note that the groups U_4, U_6, and $\mathbb{Z}_3$ are *not* contained in the direct product $U_4 \times U_6 \times \mathbb{Z}_3$. For instance, 5 is an element of U_6, but 5 is *not* in $U_4 \times U_6 \times \mathbb{Z}_3$ because the elements of $U_4 \times U_6 \times \mathbb{Z}_3$ are *triples*. In general, for $1 \le i \le n$

G_i is *not* a subgroup of the direct product $G_1 \times G_2 \times \cdots \times G_n$.**

This situation is not entirely satisfactory, but by changing our viewpoint slightly we can develop a notion of direct product in which the component groups may be considered as subgroups.

> **EXAMPLE**   It is easy to verify that $M = \{0, 3\}$ and $N = \{0, 2, 4\}$ are normal subgroups of $\mathbb{Z}_6$ (do it!). Observe that every element of $\mathbb{Z}_6$ can be written as a sum of an element in M and an element in N in *one and only one* way:
>
> $$0 = 0 + 0 \qquad 1 = 3 + 4 \qquad 2 = 0 + 2$$
> $$3 = 3 + 0 \qquad 4 = 0 + 4 \qquad 5 = 3 + 2.$$

---

* When each G_i is an additive abelian group, the direct product of $G_1, \ldots, G_n$ is sometimes called the **direct sum** and denoted $G_1 \oplus G_2 \oplus \cdots \oplus G_n$.

** It is true, however, that an isomorphic copy of G_i is a subgroup of $G_1 \times G_2 \times \cdots \times G_n$ (see Exercise 12).

Verify that, when the elements of $\mathbb{Z}_6$ are written as sums in this way, then the addition table for $\mathbb{Z}_6$ looks like this:

	$0 + 0$	$3 + 4$	$0 + 2$	$3 + 0$	$0 + 4$	$3 + 2$
$0 + 0$	$0 + 0$	$3 + 4$	$0 + 2$	$3 + 0$	$0 + 4$	$3 + 2$
$3 + 4$	$3 + 4$	$0 + 2$	$3 + 0$	$0 + 4$	$3 + 2$	$0 + 0$
$0 + 2$	$0 + 2$	$3 + 0$	$0 + 4$	$3 + 2$	$0 + 0$	$3 + 4$
$3 + 0$	$3 + 0$	$0 + 4$	$3 + 2$	$0 + 0$	$3 + 4$	$0 + 2$
$0 + 4$	$0 + 4$	$3 + 2$	$0 + 0$	$3 + 4$	$0 + 2$	$3 + 0$
$3 + 2$	$3 + 2$	$0 + 0$	$3 + 4$	$0 + 2$	$3 + 0$	$0 + 4$

Compare the $\mathbb{Z}_6$ table with the operation table for the direct product $M \times N$:

	$(0, 0)$	$(3, 4)$	$(0, 2)$	$(3, 0)$	$(0, 4)$	$(3, 2)$
$(0, 0)$	$(0, 0)$	$(3, 4)$	$(0, 2)$	$(3, 0)$	$(0, 4)$	$(3, 2)$
$(3, 4)$	$(3, 4)$	$(0, 2)$	$(3, 0)$	$(0, 4)$	$(3, 2)$	$(0, 0)$
$(0, 2)$	$(0, 2)$	$(3, 0)$	$(0, 4)$	$(3, 2)$	$(0, 0)$	$(3, 4)$
$(3, 0)$	$(3, 0)$	$(0, 4)$	$(3, 2)$	$(0, 0)$	$(3, 4)$	$(0, 2)$
$(0, 4)$	$(0, 4)$	$(3, 2)$	$(0, 0)$	$(3, 4)$	$(0, 2)$	$(3, 0)$
$(3, 2)$	$(3, 2)$	$(0, 0)$	$(3, 4)$	$(0, 2)$	$(3, 0)$	$(0, 4)$

The only difference in these two tables is that elements are written $a + b$ in the first and (a, b) in the second. Among other things, the tables show that the direct product $M \times N$ is isomorphic to $\mathbb{Z}_6$ under the isomorphism that assigns each pair $(a, b) \in M \times N$ to the sum of its coordinates $a + b \in \mathbb{Z}_6$.

Consequently, we can express $\mathbb{Z}_6$ as a direct product in a purely *internal* fashion, without looking at the set $M \times N$, which is external to $\mathbb{Z}_6$: Write each element uniquely as a sum $a + b$, with $a \in M$ and $b \in N$. We now develop this same idea in the general case, with multiplicative notation in place of addition in $\mathbb{Z}_6$.

THEOREM 8.1   *Let $N_1, N_2 \ldots , N_k$ be normal subgroups of a group G such that every element in G can be written uniquely in the form $a_1 a_2 \cdots a_k$, with $a_i \in N_i$.* Then G is isomorphic to the direct product $N_1 \times N_2 \times \cdots \times N_k$.*

The proof depends on this useful fact:

* Uniqueness means that if $a_1 a_2 \cdots a_k = b_1 b_2 \cdots b_k$ with each $a_i, b_i \in N_i$, then $a_i = b_i$ for every i.

LEMMA 8.2 *Let M and N be normal subgroups of a group G such that $M \cap N = \langle e \rangle$. If $a \in M$ and $b \in N$, then $ab = ba$.*

Proof Consider $a^{-1}b^{-1}ab$. Since M is normal, $b^{-1}ab \in M$ by Theorem 7.34. Closure in M shows that $a^{-1}b^{-1}ab = a^{-1}(b^{-1}ab) \in M$. Similarly, the normality of N implies that $a^{-1}b^{-1}a \in N$ and, hence, $a^{-1}b^{-1}ab = (a^{-1}b^{-1}a)b \in N$. Thus $a^{-1}b^{-1}ab \in M \cap N = \langle e \rangle$. Multiplying both sides of $a^{-1}b^{-1}ab = e$ on the left by ba shows that $ab = ba$. $\blacklozenge$

Proof of Theorem 8.1 Guided by the example preceding the theorem (but using multiplicative notation), we define a map

$$f : N_1 \times N_2 \times \cdots \times N_k \to G \qquad \text{by} \qquad f(a_1, a_2, \ldots, a_k) = a_1 a_2 \cdots a_k.$$

Since every element of G can be written in the form $a_1 a_2 \cdots a_k$ (with $a_i \in N_i$) by hypothesis, f is surjective. If $f(a_1, a_2, \ldots, a_k) = f(b_1, b_2, \ldots, b_k)$, then $a_1 a_2 \cdots a_k = b_1 b_2 \cdots b_k$. By the uniqueness hypothesis, $a_i = b_i$ for each i $(1 \le i \le k)$. Therefore,

$$(a_1, a_2, \ldots, a_k) = (b_1, b_2, \ldots, b_k) \text{ in } N_1 \times N_2 \times \cdots \times N_k,$$

and f is injective.

In order to prove that f is a homomorphism we must first show that the N's are mutually disjoint subgroups, that is, $N_i \cap N_j = \langle e \rangle$ when $i \ne j$. If $a \in N_i \cap N_j$, then a can be written as a product of elements of the N's in two different ways:

$$\underset{N_1}{\underset{\uparrow}{ee}} \cdots \underset{N_i}{\underset{\uparrow}{eae}} \cdots \underset{N_j}{\underset{\uparrow}{e}} \cdots \underset{N_k}{\underset{\uparrow}{e}} = a = \underset{N_1}{\underset{\uparrow}{ee}} \cdots \underset{N_i}{\underset{\uparrow}{e}} \cdots \underset{N_j}{\underset{\uparrow}{eae}} \cdots \underset{N_k}{\underset{\uparrow}{e}}.$$

The uniqueness hypothesis implies that the components in N_i must be equal: $a = e$. Therefore, $N_i \cap N_j = \langle e \rangle$ for $i \ne j$. In showing that f is a homomorphism, we shall make repeated use of this fact, which together with Lemma 8.2, implies that $a_i b_j = b_j a_i$ for $a_i \in N_i$ and $b_j \in N_j$:

$$
\begin{aligned}
f[(a_1, \ldots, a_k)(b_1, \ldots, b_k)] &= f(a_1 b_1, \ldots, a_k b_k) \\
&= a_1 b_1\, a_2 b_2\, a_3 b_3 \cdots a_k b_k \\
&= a_1 a_2\, b_1 b_2\, a_3 b_3 \cdots a_k b_k \\
&= a_1 a_2\, b_1 a_3\, b_2 b_3 \cdots a_k b_k \\
&= a_1 a_2\, a_3 b_1\, b_2 b_3 \cdots a_k b_k.
\end{aligned}
$$

Continuing in this way we successively move $a_4, a_5, \ldots, a_k$ to the left until we obtain

$$f[(a_1, \ldots, a_k)(b_1, \ldots, b_k)] = (a_1 a_2 \cdots a_k)(b_1 b_2 \cdots b_k)$$
$$= f(a_1, \ldots, a_k)f(b_1, \ldots, b_k).$$

Therefore, f is homomorphism and, hence, an isomorphism.   ◆

Whenever G is a group and $N_1, \ldots, N_k$ are subgroups satisfying the hypotheses of Theorem 8.1 we shall say that **G is the direct product of $N_1, \ldots, N_k$** and write $G = N_1 \times \cdots \times N_k$. Each N_i is said to be a **direct factor** of G. Depending on the context, we can think of G as the *external* direct product of the N_i (each element a k-tuple $(a_1, \ldots, a_k) \in N_1 \times \cdots \times N_k$) or as an *internal* direct product (each element written uniquely in the form $a_1 a_2 \cdots a_k \in G$).

The next theorem is often easier to use than Theorem 8.1 to prove that a group is the direct product of certain of its subgroups. The statement of the theorem uses the following notation. If M and N are subgroups of a group G, then MN denotes the set of all products mn, with $m \in M$ and $n \in N$.

THEOREM 8.3   *If M and N are normal subgroups of a group G such that $G = MN$ and $M \cap N = \langle e \rangle$, then $G = M \times N$.*

For the case of more than two subgroups, see Exercise 25.

Proof of Theorem 8.3   By hypothesis every element of G is of the form mn, with $m \in M$, $n \in N$. Suppose that an element had two such representations, say $mn = m_1 n_1$, with $m, m_1 \in M$ and $n, n_1 \in N$. Then multiplying on the left by m_1^{-1} and on the right by n^{-1} shows that $m_1^{-1}m = n_1 n^{-1}$. But $m_1^{-1}m \in M$ and $n_1 n^{-1} \in N$ and $M \cap N = \langle e \rangle$. Thus $m_1^{-1}m = e$ and $m = m_1$; similarly, $n = n_1$. Therefore, every element of G can be written *uniquely* in the form mn ($m \in M$, $n \in N$), and, hence, $G = M \times N$ by Theorem 8.1.   ◆

> **EXAMPLE**   The multiplicative group of units in $\mathbb{Z}_{15}$ is $U_{15} = \{1, 2, 4, 7, 8, 11, 13, 14\}$. The groups $M = \{1, 11\}$ and $N = \{1, 2, 4, 8\}$ are normal subgroups whose intersection is $\langle 1 \rangle$. Every element of N is in MN (for instance, $2 = 1 \cdot 2$), and similarly for M. Since $11 \cdot 2 = 7$, $11 \cdot 8 = 13$, and $11 \cdot 4 = 14$, we see that $U_{15} = MN$. Therefore, $U_{15} = M \times N$ by Theorem 8.3. Since N is cyclic of order 2 and M cyclic of order 4 (2 is a generator), we conclude that U_{15} is isomorphic to $\mathbb{Z}_2 \times \mathbb{Z}_4$ (see Exercise 10 and Theorem 7.18).

◆ **EXERCISES**

NOTE: *Unless stated otherwise, $G_1, \ldots, G_n$ are groups.*

A.   1. Find the order of each element in the given group:

(a) $\mathbb{Z}_2 \times \mathbb{Z}_4$ (b) $\mathbb{Z}_3 \times \mathbb{Z}_3 \times \mathbb{Z}_2$ (c) $D_4 \times \mathbb{Z}_2$

2. What is the order of the group $U_5 \times U_6 \times U_7 \times U_8$?

3. (a) List all subgroups of $\mathbb{Z}_2 \times \mathbb{Z}_2$. (There are more than two.)

   (b) Do the same for $\mathbb{Z}_2 \times \mathbb{Z}_2 \times \mathbb{Z}_2$.

4. If G and H are groups, prove that $G \times H \cong H \times G$.

5. Give an example to show that the direct product of cyclic groups need not be cyclic.

6. (a) Write $\mathbb{Z}_{12}$ as a direct sum of two of its subgroups.

   (b) Do the same for $\mathbb{Z}_{15}$.

   (c) Write $\mathbb{Z}_{30}$ in three different ways as a direct sum of two or more of its subgroups [*Hint:* Theorem 8.3.]

7. Let $G_1, \ldots, G_n$ be groups. Prove that $G_1 \times \cdots \times G_n$ is abelian if and only if every G_i is abelian.

8. Let i be an integer with $1 \le i \le n$. Prove that the function

$$\pi_i : G_1 \times G_2 \times \cdots \times G_n \to G_i$$

given by $\pi_i(a_1, a_2, a_3, \ldots, a_n) = a_i$ is a surjective homomorphism of groups.

9. Is $\mathbb{Z}_8$ isomorphic to $\mathbb{Z}_4 \times \mathbb{Z}_2$?

B. 10. (a) If $f : G_1 \to H_1$ and $g : G_2 \to H_2$ are isomorphisms of groups, prove that the map $\theta : G_1 \times G_2 \to H_1 \times H_2$ given by $\theta(a, b) = (f(a), g(b))$ is an isomorphism.

   (b) If $G_i \cong H_i$ for $i = 1, 2, \ldots, n$, prove that

$$G_1 \times \cdots \times G_n \cong H_1 \times \cdots \times H_n.$$

11. Let H, K, M, N be groups such that $K \cong M \times N$. Prove that $H \times K \cong H \times M \times N$.

12. Let i be an integer with $1 \le i \le n$. Let G_i^* be the subset of $G_1 \times \cdots \times G_n$ consisting of those elements whose ith coordinate is any element of G_i and whose other coordinates are each the identity element, that is,

$$G_i^* = \{(e_1, \ldots, e_{i-1}, a_i, e_{i+1}, \ldots, e_n) \mid a_i \in G_i\}.$$

Prove that

(a) G_i^* is a normal subgroup of $G_1 \times \cdots \times G_n$.

(b) $G_i^* \cong G_i$.

(c) $G_1 \times \cdots \times G_n$ is the (internal) direct product of its subgroups $G_1^*, \ldots, G_n^*$. [*Hint:* Show that every element of $G_1 \times \cdots \times G_n$ can be written uniquely in the form $a_1 a_2 \cdots a_n$, with $a_i \in G_i^*$; apply Theorem 8.1.]

13. Let G be a group and let $D = \{(a, a, a) \mid a \in G\}$.

    (a) Prove that D is a subgroup of $G \times G \times G$.

    (b) Prove that D is normal in $G \times G \times G$ if and only if G is abelian.

14. If $G_1, \ldots, G_n$ are finite groups, prove that the order of $(a_1, a_2, \ldots, a_n)$ in $G_1 \times \cdots \times G_n$ is the least common multiple of the orders $|a_1|, |a_2|, \ldots, |a_n|$.

15. Let $i_1, i_2, \ldots, i_n$ be a permutation of the integers $1, 2, \ldots, n$. Prove that
$$G_{i_1} \times G_{i_2} \times \cdots \times G_{i_n}$$
is isomorphic to
$$G_1 \times G_2 \times \cdots \times G_n.$$
[Exercise 4 is the case $n = 2$.]

16. If N, K are subgroups of a group G such that $G = N \times K$ and M is a normal subgroup of N, prove that M is a normal subgroup of G. [Compare this with Exercise 14 in Section 7.6.]

17. Let $\mathbb{Q}^*$ be the multiplicative group of nonzero rational numbers, $\mathbb{Q}^{**}$ the subgroup of positive rationals, and H the subgroup $\{1, -1\}$. Prove that $\mathbb{Q}^* = \mathbb{Q}^{**} \times H$.

18. Let $\mathbb{C}^*$ be the multiplicative group of nonzero complex numbers and $\mathbb{R}^{**}$ be the multiplicative group of positive real numbers. Prove that $\mathbb{C}^* \cong \mathbb{R}^{**} \times \mathbb{R}/\mathbb{Z}$, where $\mathbb{R}$ is the additive group of real numbers.

19. Let G be a group and $f_1 : G \to G_1, f_2 : G \to G_2, \ldots, f_n : G \to G_n$ homomorphisms. For $i = 1, 2, \ldots, n$, let π_i be the homomorphism of Exercise 8. Let $f^* : G \to G_1 \times \cdots \times G_n$ be the map defined by $f^*(a) = (f_1(a_1), f_2(a_2), \ldots, f_n(a_n))$.

    (a) Prove that f^* is a homomorphism such that $\pi_i \circ f^* = f_i$ for each i.

    (b) Prove that f^* is the unique homomorphism from G to $G_1 \times \cdots \times G_n$ such that $\pi_i \circ f^* = f_i$ for every i.

20. Let $N_1, \ldots, N_k$ be subgroups of an abelian group G. Assume that every element of G can be written in the form $a_1 \cdots a_n$ (with $a_i \in N_i$) and that whenever $a_1 a_2 \cdots a_n = e$, then $a_i = e$ for every i. Prove that $G = N_1 \times N_2 \times \cdots \times N_k$.

21. Let G be an additive abelian group with subgroups H and K. Prove that $G = H \times K$ if and only if there are homomorphisms
$$H \underset{\delta_1}{\overset{\pi_1}{\rightleftarrows}} G \underset{\delta_2}{\overset{\pi_2}{\rightleftarrows}} K$$
such that $\delta_1(\pi_1(x)) + \delta_2(\pi_2(x)) = x$ for every $x \in G$ and $\pi_1 \circ \delta_1 = \iota_H$, $\pi_2 \circ \delta_2 = \iota_K$, $\pi_1 \circ \delta_2 = 0$, and $\pi_2 \circ \delta_1 = 0$, where ι_X is the identity map on X, and 0 is the

map that sends every element onto the zero (identity) element. [*Hint:* Let π_i be as in Exercise 8.]

22. Let G and H be finite cyclic groups. Prove that $G \times H$ is cyclic if and only if $(|G|, |H|) = 1$.

23. (a) Show by example that Lemma 8.2 may be false if N is not normal.

(b) Do the same for Theorem 8.3.

24. Let N, K be subgroups of a group G, with N normal in G. If N and K are abelian groups and $G = NK$, is G the direct product of N and K?

25. Let $N_1, \ldots, N_k$ be normal subgroups of a group G. Let $N_1 N_2 \cdots N_k$ denote the set of all elements of the form $a_1 a_2 \cdots a_k$ with $a_j \in N_j$. Assume that $G = N_1 N_2 \cdots N_k$ and that

$$N_i \cap N_1 \cdots N_{i-1} N_{i+1} \cdots N_k = \langle e \rangle$$

for each i $(1 \le i \le n)$. Prove that $G = N_1 \times N_2 \times \cdots \times N_k$.

26. Let $N_1, \ldots, N_k$ be normal subgroups of a finite group G. If $G = N_1 N_2 \cdots N_k$ (notation as in Exercise 25) and $|G| = |N_1| \cdot |N_2| \cdots |N_k|$, prove that $G = N_1 \times N_2 \times \cdots \times N_k$.

27. Let N, H be subgroups of a group G. G is called the **semidirect product** of N and H if N is normal in G, $G = NH$, and $N \cap H = \langle e \rangle$. Show that each of the following groups is the semidirect product of two of its subgroups:

(a) S_3 **(b)** D_4 **(c)** S_4

28. A group G is said to be **indecomposable** if it is *not* the direct product of two of its proper normal subgroups. Prove that each of these groups is indecomposable:

(a) S_3 **(b)** D_4 **(c)** $\mathbb{Z}$

29. If p is prime and n is a positive integer, prove that $\mathbb{Z}_{p^n}$ is indecomposable.

30. Prove that $\mathbb{Q}$ is an indecomposable group.

31. Show by example that a homomorphic image of an indecomposable group need not be indecomposable.

32. Prove that a group G is indecomposable if and only if whenever H and K are normal subgroups such that $G = H \times K$, then $H = \langle e \rangle$ or $K = \langle e \rangle$.

33. Let I be the set of positive integers and assume that for each $i \in I$, G_i is a group.* The **infinite direct product** of the G_i is denoted $\prod_{i \in I} G_i$ and consists

---

* Any infinite index set I may be used here, but the restriction to the positive integers simplifies the notation.

of all sequences $(a_1, a_2, \ldots)$ with $a_i \in G_i$. Prove that $\prod_{i \in I} G_i$ is a group under the coordinatewise operation

$$(a_1, a_2, \ldots)(b_1, b_2, \ldots) = (a_1 b_1, a_2 b_2, \ldots).$$

C. 34. With the notation as in Exercise 33, let $\sum_{i \in I} G_i$ denote the subset of $\prod_{i \in I} G_i$ consisting of all sequences $(c_1, c_2, \ldots)$ such that there are at most a finite number of coordinates with $c_j \neq e_j$, where e_j is the identity element of G_j. Prove that $\sum_{i \in I} G_i$ is a normal subgroup of $\prod_{i \in I} G_i$. $\sum_{i \in I} G_i$ is called the **infinite direct sum** of the G_i.

35. Let G be a group and assume that for each positive integer i, N_i is a normal subgroup of G. If every element of G can be written uniquely in the form $n_{i_1} \cdot n_{i_2} \cdots n_{i_k}$, with $i_1 < i_2 < \cdots < i_k$ and $n_{i_j} \in N_{i_j}$, prove that $G \cong \sum_{i \in I} N_i$ (see Exercise 34).* [*Hint:* Adapt the proof of Theorem 8.1 by defining $f(a_1, a_2, \ldots)$ to be the product of those a_i that are not the identity element.]

36. If $(m, n) = 1$, prove that $U_{mn} \cong U_m \times U_n$.

37. Let H be a group and $\tau_1 : H \to G_1$, $\tau_2 : H \to G_2$, $\ldots$, $\tau_n : H \to G_n$ homomorphism with this property: Whenever G is a group and $g_1 : G \to G_1$, $g_2 : G \to G_2$, $\ldots$, $g_n : G \to G_n$ are homomorphisms, then there exists a unique homomorphism $g^* : G \to H$ such that $\tau_i \circ g^* = g_i$ for every i. Prove that $H \cong G_1 \times G_2 \times \cdots \times G_n$. [See Exercise 19.]

8.2 FINITE ABELIAN GROUPS

All finite abelian groups will now be classified. We shall prove that every finite abelian group G is a direct sum of cyclic subgroups and that the orders of these cyclic subgroups are uniquely determined by G. The only prerequisites for the proof other than Section 8.1 are basic number theory (Section 1.2) and elementary group theory (Chapter 7, through Section 7.8).

Following the usual custom with abelian groups, **all groups are written in additive notation** in this section. The following dictionary may be helpful for translating from multiplicative to additive notation:

---

* Uniqueness means that if $a_{i_1} \cdots a_{i_k} = b_{j_1} \cdots b_{j_t}$, with $i_1 < i_2 < \cdots < i_k$ and $j_1 < j_2 < \cdots < j_t$, then $k = t$ and for $r = 1, 2, \ldots, k : i_r = j_r$ and $a_{i_r} = b_{i_r}$.

MULTIPLICATIVE NOTATION	ADDITIVE NOTATION
ab	$a + b$
e	0
a^k	ka
$a^k = e$	$ka = 0$
$MN = \{mn \mid m \in M, n \in N\}$	$M + N = \{m + n \mid m \in M, n \in N\}$
direct product $M \times N$	direct sum $M \oplus N$
direct factor M	direct summand M

Here is a restatement in additive notation of several earlier results that will be used frequently here:

THEOREM 7.8 *Let G be an additive group and let $a \in G$.*

(3) *If a has order n, then $ka = 0$ if and only if $n \mid k$.*

(4) *If a has order td, with $d > 0$, then ta has order d.*

THEOREM 8.1 *If $N_1, \ldots, N_k$ are normal subgroups of an additive group G such that every element of G can be written uniquely in the form $a_1 + a_2 + \cdots + a_k$ with $a_i \in N_i$, then $G \cong N_1 \oplus N_2 \oplus \cdots \oplus N_k$.*

THEOREM 8.3 *If M and N are normal subgroups of an additive group G such that $G = M + N$ and $M \cap N = \langle 0 \rangle$, then $G = M \oplus N$.*

Finally we note that Exercise 11 of Section 8.1 will be used without explicit mention at several points.

If G is an abelian group and p is a prime, then **$G(p)$** denotes the set of elements in G whose order is some power of p; that is,

$$G(p) = \{a \in G \mid |a| = p^n \text{ for some } n \geq 0\}.$$

It is easy to verify that $G(p)$ is closed under addition and that the inverse of any element in $G(p)$ is also in $G(p)$ (Exercise 1). Therefore, $G(p)$ is a subgroup of G.

> **EXAMPLE** If $G = \mathbb{Z}_{12}$, then $G(2)$ is the set of elements having orders $2^0, 2^1, 2^2$, etc. Verify that $G(2)$ is the subgroup $\{0, 3, 6, 9\}$; similarly, $G(3) = \{0, 4, 8\}$. If $G = \mathbb{Z}_3 \oplus \mathbb{Z}_3$, then $G(3) = G$ since every nonzero element in G has order 3.

The first step in proving that a finite abelian group G is the direct sum of cyclic subgroups is to show that G is the direct sum of its subgroups $G(p)$, one for each of the distinct primes dividing the order of G. In order to do this, we need

LEMMA 8.4   *Let G be an abelian group and $a \in G$ an element of finite order. Then $a = a_1 + a_2 + \cdots + a_k$, with $a_i \in G(p_i)$, where $p_1, \ldots, p_k$ are the distinct positive primes that divide the order of a.*

Proof   The proof is by induction on the number of distinct primes that divide the order of a. If $|a|$ is divisible only by the single prime p_1, then the order of a is a power of p_1 and, hence, $a \in G(p_1)$. So the lemma is true in this case. Assume inductively that the lemma is true for all elements whose order is divisible by at most $k - 1$ distinct primes and that $|a|$ is divisible by the distinct primes $p_1, \ldots, p_k$. Then $|a| = p_1^{r_1} \cdots p_k^{r_k}$, with each $r_i > 0$. Let $m = p_2^{r_2} \cdots p_k^{r_k}$ and $n = p_1^{r_1}$, so that $|a| = mn$. Then $(m, n) = 1$ and by Theorem 1.3 there are integers u, v such that $1 = mu + nv$. Consequently,

$$a = 1a = (mu + nv)a = mua + nva.$$

But $mua \in G(p_1)$ because a has order mn, and, hence, $p_1^{r_1} (mua) = (nm)ua = u(mna) = u0 = 0$. Similarly, $m(nva) = 0$ so that by Theorem 7.8 the order of nva divides m, an integer with only $k - 1$ distinct prime divisors. Therefore, by the induction assumption $nva = a_2 + a_3 + \cdots + a_k$, with $a_i \in G(p_i)$. Let $a_1 = mua$; then $a = mua + nva = a_1 + a_2 + \cdots + a_k$, with $a_i \in G(p_i)$.   ◆

THEOREM 8.5   *If G is a finite abelian group, then*

$$G = G(p_1) \oplus G(p_2) \oplus \cdots \oplus G(p_t),$$

where $p_1, \ldots, p_t$ are the distinct positive primes that divide the order of G.

Proof   If $a \in G$, then its order divides $|G|$ by Corollary 7.27. Hence, $a = a_1 + \cdots + a_t$, with $a_i \in G(p_i)$ by Lemma 8.4 (where $a_j = 0$ if the prime p_j does not divide $|a|$). To prove that this expression is unique, suppose $a_1 + a_2 + \cdots + a_t = b_1 + b_2 + \cdots + b_t$, with $a_i, b_i \in G(p_i)$. Since G is abelian

$$a_1 - b_1 = (b_2 - a_2) + (b_3 - a_3) + \cdots + (b_t - a_t).$$

For each i, $b_i - a_i \in G(p_i)$ and, hence, has order a power of p_i, say $p_i^{r_i}$. If $m = p_2^{r_2} \cdots p_t^{r_t}$, then $m(b_i - a_i) = 0$ for $i \geq 2$, so that

$$m(a_1 - b_1) = m(b_2 - a_2) + \cdots + m(b_t - a_t) = 0 + \cdots + 0 = 0.$$

Consequently, the order of $a_1 - b_1$ must divide m by Theorem 7.8. But $a_1 - b_1 \in G(p_1)$, so its order is a power of p_1. The only power of p_1 that divides $m = p_2^{r_2} \cdots p_t^{r_t}$ is $p_1^0 = 1$. Therefore, $a_1 - b_1 = 0$ and $a_1 = b_1$. Similar arguments for $i = 2, \ldots, t$ show that $a_i = b_i$ for every i. Therefore, every element of G can be written uniquely in the form $a_1 + \cdots + a_t$, with $a_i \in G(p_i)$ and, hence, $G = G(p_1) \oplus \cdots \oplus G(p_t)$ by Theorem 8.1.   ◆

If p is a prime, then a group in which every element has order a power of p is called a **p-group.** Each of the $G(p_i)$ in Theorem 8.5 is a p-group by its very definition. An element a of a p-group B is called an **element of maximal order** if $|b| \leq |a|$ for every $b \in B$. If $|a| = p^n$ and $b \in B$, then b has order p^j with $j \leq n$. Since $p^n = p^j p^{n-j}$ we see that $p^n b = p^{n-j}(p^j b) = 0$ for every $b \in B$. Note that elements of maximal order always exist in a *finite* p-group.

The next step in classifying finite abelian groups is to prove that every finite abelian p-group has a cyclic direct summand, after which we will be able to prove that every finite abelian p-group is a direct sum of cyclic groups.

LEMMA 8.6 *Let G be a finite abelian p-group and a an element of maximal order in G. Then there is a subgroup K of G such that $G = \langle a \rangle \oplus K$.*

The following proof is more intricate than most of the proofs earlier in the book. Nevertheless, it uses only elementary group theory, so if you read it carefully, you shouldn't have trouble following the argument.

Proof of Lemma 8.6 Consider those subgroups H of G such that $\langle a \rangle \cap H = \langle 0 \rangle$. There is at least one ($H = \langle 0 \rangle$), and since G is finite, there must be a largest subgroup K with this property. Then $\langle a \rangle \cap K = \langle 0 \rangle$, and by Theorem 8.3 we need only show that $G = \langle a \rangle + K$. If this is *not* the case, then there is a nonzero b such that $b \notin \langle a \rangle + K$. Let k be the smallest positive integer such that $p^k b \in \langle a \rangle + K$ (there must be one since G is a p-group and, hence, $p^j b = 0 = 0 + 0 \in \langle a \rangle + K$ for some positive j). Then

(1) $c = p^{k-1}b$ is *not* in $\langle a \rangle + K$

and $pc = p^k b$ *is* in $\langle a \rangle + K$, say

(2) $pc = ta + k$ $(t \in \mathbb{Z}, k \in K)$.

If a has order p^n, then $p^n x = 0$ for all $x \in G$ because a has maximal order. Consequently, by **(2)**

$$p^{n-1}ta + p^{n-1}k = p^{n-1}(ta + k) = p^{n-1}(pc) = p^n c = 0.$$

Therefore, $p^{n-1}ta = -p^{n-1}k \in \langle a \rangle \cap K = \langle 0 \rangle$ and $p^{n-1}ta = 0$. Theorem 7.8 shows that p^n (the order of a) divides $p^{n-1}t$, and it follows that $p \mid t$. Therefore, $pc = ta + k = pma + k$ for some m, and, consequently, $k = pc - pma = p(c - ma)$. Let

(3) $d = c - ma$.

Then $pd = p(c - ma) = k \in K$, but $d \notin K$ (since $c - ma = k' \in K$ would imply that $c = ma + k' \in \langle a \rangle + K$, contradicting **(1)**). Use Theorem 7.11 to verify that $H = \{x + zd \mid x \in K, z \in \mathbb{Z}\}$ is a subgroup of G with $K \subseteq H$. Since $d = 0 + 1d \in H$ and $d \notin K$, H is larger than K. But K is the largest group such that $\langle a \rangle \cap K = \langle 0 \rangle$, so we must have $\langle a \rangle \cap H \neq \langle 0 \rangle$. If w is a nonzero element of $\langle a \rangle \cap H$, then

(4) $w = sa = k_1 + rd$ $(k_1 \epsilon K; r, s \epsilon \mathbb{Z})$.

We claim that $p\!\!\not|\,r$; for if $r = py$, then since $pd \, \epsilon \, K$, $0 \ne w = sa = k_1 + ypd \, \epsilon$ $\langle a \rangle \cap K$, a contradiction. Consequently, $(p, r) = 1$, and by Theorem 1.3 there are integers u, v with $pu + rv = 1$. Then

$$c = 1c = (pu + rv)c = u(pc) + v(rc)$$
$$= u(ta + k) + v(r(d + ma)) \qquad [\text{by } (2) \text{ and } (3)]$$
$$= u(ta + k) + v(rd + rma)$$
$$= u(ta + k) + v(sa - k_1 + rma) \qquad [\text{by } (4)]$$
$$= (ut + vs + rm)a + (uk - vk_1) \, \epsilon \, \langle a \rangle + K.$$

This contradicts **(1)**. Therefore, $G = \langle a \rangle + K$, and, hence, $G = \langle a \rangle \oplus K$ by Theorem 8.3. ◆

THEOREM 8.7 (THE FUNDAMENTAL THEOREM OF FINITE ABELIAN GROUPS) *Every finite abelian group G is the direct sum of cyclic groups, each of prime power order.*

Proof By Theorem 8.5, G is the direct sum of its subgroups $G(p)$, one for each prime p that divides $|G|$. Each $G(p)$ is a p-group. So to complete the proof, we need only show that every finite abelian p-group H is a direct sum of cyclic groups, each of order a power of p. We prove this by induction on the order of H. The assertion is true when H has order 2 by Theorem 7.28. Assume inductively that it is true for all groups whose order is less than $|H|$ and let a be an element of maximal order p^n in H. Then $H = \langle a \rangle \oplus K$ by Lemma 8.6. By induction, K is a direct sum of cyclic groups, each with order a power of p. Therefore, the same is true of $H = \langle a \rangle \oplus K$. ◆

> **EXAMPLE** The number 36 can be written as a product of prime powers in just four ways: $36 = 2 \cdot 2 \cdot 3 \cdot 3 = 2 \cdot 2 \cdot 3^2 = 2^2 \cdot 3 \cdot 3 = 2^2 \cdot 3^2$. Consequently, by Theorem 8.7 every abelian group of order 36 must be isomorphic to one of the following groups:
>
> $\mathbb{Z}_2 \oplus \mathbb{Z}_2 \oplus \mathbb{Z}_3 \oplus \mathbb{Z}_3,$ $\mathbb{Z}_2 \oplus \mathbb{Z}_2 \oplus \mathbb{Z}_9,$ $\mathbb{Z}_4 \oplus \mathbb{Z}_3 \oplus \mathbb{Z}_3,$ $\mathbb{Z}_4 \oplus \mathbb{Z}_9.$
>
> You can easily verify that no two of these groups are isomorphic (the number of elements of order 2 or 3 is different for each group). Thus we have a complete classification of all abelian groups of order 36 up to isomorphism.

You probably noticed that a familiar group of order 36, namely $\mathbb{Z}_{36}$, doesn't appear explicitly on the list in the preceding example. However, it is isomorphic to $\mathbb{Z}_4 \oplus \mathbb{Z}_9$, as we now prove.

LEMMA 8.8 *If $(m, k) = 1$, then $\mathbb{Z}_m \oplus \mathbb{Z}_k \cong \mathbb{Z}_{mk}$.*

Proof The order of $(1, 1)$ in $\mathbb{Z}_m \oplus \mathbb{Z}_k$ is the smallest positive integer t such that $(0, 0) = t(1, 1) = (t, t)$. Thus $t \equiv 0 \pmod{m}$ and $t \equiv 0 \pmod{k}$, so that $m \mid t$ and $k \mid t$. But $(m, k) = 1$ implies that $mk \mid t$ by Exercise 17 in Section 1.2. Hence, $mk \leq t$. Since $mk(1, 1) = (mk, mk) = (0, 0)$, we must have $mk = t = |(1, 1)|$. Therefore, $\mathbb{Z}_m \oplus \mathbb{Z}_k$ (a group of order mk) is the cyclic group generated by $(1, 1)$ and, hence, is isomorphic to $\mathbb{Z}_{mk}$ by Theorem 7.18. ◆

THEOREM 8.9 *If $n = p_1^{n_1} p_2^{n_2} \cdots p_t^{n_t}$, with $p_1, \ldots, p_t$ distinct primes, then $\mathbb{Z}_n \cong \mathbb{Z}_{p_1^{n_1}} \oplus \cdots \oplus \mathbb{Z}_{p_t^{n_t}}$.*

Proof The theorem is true for groups of order 2. Assume inductively that it is true for groups of order less than n. Apply Lemma 8.8 with $m = p_1^{n_1}$ and $k = p_2^{n_2} \cdots p_t^{n_t}$. Then $\mathbb{Z}_n \cong \mathbb{Z}_{p_1^{n_1}} \oplus \mathbb{Z}_k$, and the induction hypothesis shows that $\mathbb{Z}_k \cong \mathbb{Z}_{p_2^{n_2}} \oplus \cdots \oplus \mathbb{Z}_{p_t^{n_t}}$. ◆

Combining Theorems 8.7 and 8.9 yields a second way of expressing a finite abelian group as a direct sum of cyclic groups.

> **EXAMPLE** Consider the group
> $$G = \mathbb{Z}_2 \oplus \mathbb{Z}_2 \oplus \mathbb{Z}_4 \oplus \mathbb{Z}_8 \oplus \mathbb{Z}_3 \oplus \mathbb{Z}_3 \oplus \mathbb{Z}_3 \oplus \mathbb{Z}_5 \oplus \mathbb{Z}_{25}.$$
> Arrange the prime power orders of the cyclic factors by size, with one row for each prime:
>
> $$
> \begin{array}{cccc}
> 2 & 2 & 2^2 & 2^3 \\
>   & 3 & 3 & 3 \\
>   &   & 5 & 5^2
> \end{array}
> $$
>
> Now rearrange the cyclic factors of G using the *columns* of this array as a guide (see Exercise 15 of Section 8.1) and apply Theorem 8.9:
>
> $$G \cong (\mathbb{Z}_2) \oplus (\mathbb{Z}_2 \oplus \mathbb{Z}_3) \oplus (\mathbb{Z}_4 \oplus \mathbb{Z}_3 \oplus \mathbb{Z}_5) \oplus (\mathbb{Z}_8 \oplus \mathbb{Z}_3 \oplus \mathbb{Z}_{25})$$
> $$G \cong \mathbb{Z}_2 \oplus \quad \mathbb{Z}_6 \quad \oplus \quad \mathbb{Z}_{60} \quad \oplus \quad \mathbb{Z}_{600}.$$
>
> This last decomposition of G as a sum of cyclic groups is sometimes more convenient than the original prime power decomposition: There are fewer cyclic factors, and the order of each cyclic factor *divides* the order of the next one. Although the notation is a bit more involved, the same process works in the general case and proves

THEOREM 8.10 *Every finite abelian group is the direct sum of cyclic groups of orders $m_1, m_2, \ldots, m_t$, where $m_1 \mid m_2, m_2 \mid m_3, m_3 \mid m_4, \ldots, m_{t-1} \mid m_t$.*

We pause briefly here to present an interesting corollary that will be used in Chapter 10. It was proved earlier as Theorem 7.15.

COROLLARY 8.11   *If G is a finite subgroup of the multiplicative group of nonzero elements of a field F, then G is cyclic.*

Proof   Since G is a finite abelian group, Theorem 8.10 implies that $G \cong \mathbb{Z}_{m_1} \oplus \cdots \oplus \mathbb{Z}_{m_t}$, where each m_i divides m_t. Every element b in $\mathbb{Z}_{m_1} \oplus \cdots \oplus \mathbb{Z}_{m_t}$ satisfies $m_t b = 0$ (Why?). Consequently, every element g of the multiplicative group G must satisfy $g^{m_t} = 1_F$ (that is, must be a root of the polynomial $x^{m_t} - 1_F$). Since G has order $m_1 m_2 \cdots m_t$ and $x^{m_t} - 1_F$ has at most m_t distinct roots in F by Corollary 4.16, we must have $t = 1$ and $G \cong \mathbb{Z}_{m_t}$.   ◆

If G is a finite abelian group, then the integers $m_1, \ldots, m_t$ in Theorem 8.10 are called the **invariant factors** of G. When G is written as a direct sum of cyclic groups of prime power orders, as in Theorem 8.7, the prime powers are called the **elementary divisors** of G. Theorems 8.7 and 8.10 show that the order of G is the product of its elementary divisors and also the product of its invariant factors.

> **EXAMPLE**   All abelian groups of order 36 can be classified up to isomorphism in terms of their elementary divisors (as in the example preceding Lemma 8.8) or in terms of their invariant factors:
>
> | GROUP | ELEMENTARY DIVISORS | INVARIANT FACTORS | ISOMORPHIC GROUP |
> |---|---|---|---|
> | $\mathbb{Z}_2 \oplus \mathbb{Z}_2 \oplus \mathbb{Z}_3 \oplus \mathbb{Z}_3$ | 2, 2, 3, 3 | 6, 6 | $\mathbb{Z}_6 \oplus \mathbb{Z}_6$ |
> | $\mathbb{Z}_2 \oplus \mathbb{Z}_2 \oplus \mathbb{Z}_9$ | $2, 2, 3^2$ | 2, 18 | $\mathbb{Z}_2 \oplus \mathbb{Z}_{18}$ |
> | $\mathbb{Z}_4 \oplus \mathbb{Z}_3 \oplus \mathbb{Z}_3$ | $2^2, 3, 3$ | 3, 12 | $\mathbb{Z}_3 \oplus \mathbb{Z}_{12}$ |
> | $\mathbb{Z}_4 \oplus \mathbb{Z}_9$ | $2^2, 3^2$ | 36 | $\mathbb{Z}_{36}$ |

The Fundamental Theorem 8.7 can be used to obtain a list of all possible abelian groups of a given order. To complete the classification of such groups, we must show that no two groups on the list are isomorphic, that is, that the elementary divisors of a group are uniquely determined.*

THEOREM 8.12   *Let G and H be finite abelian groups. Then G is isomorphic to H if and only if G and H have the same elementary divisors.*

---

* The remainder of this section is optional. Theorem 8.12 is often considered to be part of the Fundamental Theorem of Finite Abelian Groups.

It is also true that $G \cong H$ if and only if G and H have the same invariant factors (Exercise 24).

Proof of Theorem 8.12   If G and H have the same elementary divisors, then both G and H are isomorphic to the same direct sum of cyclic groups and, hence, are isomorphic to each other. Conversely, if $f : G \rightarrow H$ is an isomorphism, then a and $f(a)$ have the same order for each $a \in G$. It follows that for each prime p, $f(G(p)) = H(p)$ and, hence, $G(p) \cong H(p)$. The elementary divisors of G that are powers of the prime p are precisely the elementary divisors of $G(p)$, and similarly for H. So we need only prove that isomorphic p-groups have the same elementary divisors. In other words, we need to prove this half of the theorem only when G and H are p-groups.

Assume G and H are isomorphic p-groups. We use induction on the order of G to prove that G and H have the same elementary divisors. All groups of order 2 obviously have the same elementary divisor, 2, by Theorem 7.28. So assume that the statement is true for all groups of order less than $|G|$. Suppose that the elementary divisors of G are

$$p^{n_1}, p^{n_2}, \ldots, p^{n_t}, \underbrace{p, p, \ldots, p}_{r \text{ copies}} \qquad \text{with } n_1 \geq n_2 \geq \cdots \geq n_t > 1$$

and that the elementary divisors of H are

$$p^{m_1}, p^{m_2}, \ldots, p^{m_k}, \underbrace{p, p, \ldots, p}_{s \text{ copies}} \qquad \text{with } m_1 \geq m_2 \geq \cdots \geq m_k > 1.$$

Verify that $pG = \{px \mid x \in G\}$ is a subgroup of G (Exercise 2). If G is the direct sum of groups C_i, verify that pG is the direct sum of the groups pC_i (Exercise 4). If C_i is cyclic with generator a of order p^n, then pC_i is the cyclic group generated by pa. Since pa has order p^{n-1} by part (4) of Theorem 7.8, pC_i is cyclic of order p^{n-1}. Note that when $n = 1$ (that is, when C_i is cyclic of order p), then $pC_i = \langle 0 \rangle$. Consequently, the elementary divisors of pG are

$$p^{n_1-1}, p^{n_2-1}, \ldots, p^{n_t-1}.$$

A similar argument shows that the elementary divisors of pH are

$$p^{m_1-1}, p^{m_2-1}, \ldots, p^{m_k-1}.$$

If $f : G \rightarrow H$ is an isomorphism, verify that $f(pG) = pH$ so that $pG \cong pH$. Furthermore, $pG \neq G$ (Exercise 9), so that $|pG| < |G|$. Hence pG and pH have the same elementary divisors by the induction hypothesis; that is, $t = k$ and

$$p^{n_i-1} = p^{m_i-1}, \qquad \text{so that } n_i - 1 = m_i - 1 \text{ for } i = 1, 2, \ldots, t.$$

Therefore, $n_i = m_i$ for each i. So the only possible difference in elementary divisors of G and H is the number of copies of p that appear on each list. Since $|G|$ is the product of its elementary divisors, and similarly for $|H|$, and since $G \cong H$, we have

$$p^{n_1}p^{n_2} \cdots p^{n_r}p^r = |G| = |H| = p^{m_1}p^{m_2} \cdots p^{m_k}p^s.$$

Since $m_i = n_i$ for each i, we must have $p^r = p^s$ and, hence, $r = s$. Thus G and H have the same elementary divisors. ◆

◆ **EXERCISES**

NOTE: *All groups are written additively, and p always denotes a positive prime, unless noted otherwise.*

A. **1.** If G is an abelian group, prove that $G(p)$ is a subgroup.

2. If G is an abelian group, prove that $pG = \{px \mid x \in G\}$ is a subgroup of G.

3. List all abelian groups (up to isomorphism) of the given order:

(a) 12   (b) 15   (c) 30   (d) 72

(e) 90   (f) 144   (g) 600   (h) 1160

4. If G and G_i $(1 \le i \le n)$ are abelian groups such that $G = G_1 \oplus \cdots \oplus G_n$, show that $pG = pG_1 \oplus \cdots \oplus pG_n$.

5. Find the elementary divisors of the given group:

(a) $\mathbb{Z}_{250}$ (b) $\mathbb{Z}_6 \oplus \mathbb{Z}_{12} \oplus \mathbb{Z}_{18}$

(c) $\mathbb{Z}_{10} \oplus \mathbb{Z}_{20} \oplus \mathbb{Z}_{30} \oplus \mathbb{Z}_{40}$ (d) $\mathbb{Z}_{12} \oplus \mathbb{Z}_{30} \oplus \mathbb{Z}_{100} \oplus \mathbb{Z}_{240}$

6. Find the invariant factors of each of the groups in Exercise 5.

B. **7.** Find the elementary divisors and the invariant factors of the given group. Note that the group operation is multiplication in the first three and addition in the last.

(a) U_8 (b) U_{17} (c) U_{15} (d) $M(\mathbb{Z}_2)$

8. If G is the additive group $\mathbb{Q}/\mathbb{Z}$, what are the elements of the subgroup $G(2)$? Of $G(p)$ for any positive prime p?

9. (a) If G is a finite abelian p-group, prove that $pG \ne G$.

(b) Show that part (a) may be false if G is infinite. [*Hint:* Consider the group $G(2)$ in Exercise 8.]

10. If G is an abelian p-group and $(n, p) = 1$ prove that the map $f: G \to G$ given by $f(a) = na$ is an isomorphism.

11. If G is a finite abelian p-group such that $pG = \langle 0 \rangle$, prove that $G \cong \mathbb{Z}_p \oplus \cdots \oplus \mathbb{Z}_p$ for some finite number of copies of $\mathbb{Z}_p$.

12. **(Cauchy's Theorem for Abelian Groups)**   If G is a finite abelian group and p is a prime that divides $|G|$, prove that G contains an element of order p. [*Hint:* Use the Fundamental Theorem to show that G has a cyclic subgroup of order p^k; use Theorem 7.8 to find an element of order p.]

13. Prove that a finite abelian p-group has order a power of p.

14. If G is an abelian group of order $p^t m$, with $(p, m) = 1$, prove that $G(p)$ has order p^t.

15. If G is a finite abelian group and p is a prime such that p^n divides $|G|$, then prove that G has a subgroup of order p^n.

16. For which positive integers n is there exactly one abelian group of order n (up to isomorphism)?

17. Let G, H, K be finite abelian groups.

(a) If $G \oplus G \cong H \oplus H$, prove that $G \cong H$.

(b) If $G \oplus H \cong G \oplus K$, prove that $H \cong K$.

18. If G is an abelian group of order n and $k \mid n$, prove that there exist a group H of order k and a surjective homomorphism $G \to H$.

19. Let G be an abelian group and T the set of elements of finite order in G. Prove that

(a) T is a subgrup of G (called the **torsion subgroup**).

(b) Every nonzero element of the quotient group G/T has infinite order.

20. If G is an abelian group, do the elements of infinite order in G (together with 0) form a subgroup? [*Hint:* Consider $\mathbb{Z} \oplus \mathbb{Z}_3$.]

C. 21. If G is an abelian group and $f: G \to \mathbb{Z}$ a surjective homomorphism with kernel K, prove that G has a subgroup H such that $H \cong \mathbb{Z}$ and $G = K \oplus H$.

22. Let G and H be finite abelian groups with this property: For each positive integer m the number of elements of order m in G is the same as the number of elements of order m in H. Prove that $G \cong H$.

23. Let G be finite abelian group with this property: For each positive integer m such that $m \mid |G|$, there are exactly m elements in G with order dividing m. Prove that G is cyclic.

24. Let G and H be finite abelian groups. Prove that $G \cong H$ if and only if G and H have the same invariant factors.

25. If G is an infinite abelian **torsion group** (meaning that every element in G

has finite order), prove that G is the infinite direct sum $\Sigma\, G(p)$, where the sum is taken over all positive primes p. [*Hint:* See Exercises 34 and 35 in Section 8.1 and adapt the proof of Theorem 8.5.]

8.3 THE SYLOW THEOREMS

Nonabelian finite groups are vastly more complicated than finite abelian groups, which were classified in the last section. The Sylow Theorems are the first basic step in understanding the structure of nonabelian finite groups. Since the proofs of these theorems are largely unrelated to the way the theorems are actually used to analyze groups, the proofs will be postponed to the next section.* In this section we shall try to give you a sound understanding of the meaning of the Sylow Theorems and some examples of their applications.

Throughout the general discussion in this section *all groups are written multiplicatively* and all integers are assumed to be nonnegative.

Once again the major theme is the close connection between the structure of a group G and the arithmetical properties of the integer $|G|$. One of the most important results of this sort is Lagrange's Theorem, which states that if G has a subgroup H, then the integer $|H|$ divides $|G|$. The First Sylow Theorem provides a partial converse:

THEOREM 8.13   (FIRST SYLOW THEOREM)   *Let G be a finite group. If p is a prime and p^k divides $|G|$, then G has a subgroup of order p^k.*

> **EXAMPLE**   The symmetric group S_6 has order $6! = 720 = 2^4 \cdot 3^2 \cdot 5$. The First Sylow Theorem (with $p = 2$) guarantees that S_6 has subgroups of orders 2, 4, 8, and 16. There may well be more than one subgroup of each of these orders. For instance, there are at least 60 subgroups of order 4 (Exercise 1). Applying the theorem with $p = 3$ shows that S_6 has subgroups of orders 3 and 9. Similarly, S_6 has at least one subgroup of order 5.

If p is a prime that divides the order of a group G, then G contains a subgroup K of order p by the First Sylow Theorem. Since K is cyclic by Theorem 7.28, its generator is an element of order p in G. This proves

COROLLARY 8.14   (CAUCHY'S THEOREM)   *If G is a finite group whose order is divisible by a prime p, then G contains an element of order p.*

---

* Puritans who believe that the work *must* come before the fun should read Section 8.4 before proceeding further.

Let G be a finite group and p a prime. If p^n is the largest power of p that divides $|G|$, then a subgroup of G of order p^n is called a **Sylow p-subgroup.** The existence of Sylow p-subgroups is an immediate consequence of the First Sylow Theorem.

EXAMPLE Since S_4 has order $4! = 24 = 2^3 \cdot 3$, every subgroup of order 8 is a Sylow 2-subgroup. You can readily verify that

$$\{(1), (1234), (13)(24), (1432), (24), (12)(34), (13), (14)(32)\}$$

is a subgroup of order 8 and, hence, a Sylow 2-subgroup. There are two other Sylow 2-subgroups (Exercise 2). Any subgroup of S_4 of order 3 is a Sylow 3-subgroup. Two of the four Sylow 3-subgroups are $\{(123), (132), (1)\}$ and $\{(134), (143), (1)\}$.

EXAMPLE* Let p be a prime and G a finite *abelian* group of order $p^n m$, where $p \nmid m$. Then

$$G(p) = \{a \in G \mid |a| = p^k \text{ for some } k \geq 0\}$$

is a Sylow p-subgroup of G since $G(p)$ has order p^n by Exercise 14 of Section 8.2. As we shall see, $G(p)$ is the unique Sylow p-subgroup of G. Theorem 8.5 shows that G is the direct sum of all its Sylow subgroups (one for each of the distinct primes that divide $|G|$).

Let G be a group and $x \in G$. The second example on page 193 shows that the map $f : G \to G$ given by $f(a) = x^{-1}ax$ is an isomorphism. If K is a subgroup of G, then the image of K under f is $x^{-1}Kx = \{x^{-1}kx \mid k \in K\}$. Hence, *$x^{-1}Kx$ is a subgroup of G that is isomorphic to K.* In particular, $x^{-1}Kx$ has the same order as K. Consequently,

if K is a Sylow p-subgroup of G, then so is $x^{-1}Kx$.

The next theorem shows that every Sylow p-subgroup of G can be obtained from K in this fashion.

THEOREM 8.15 (SECOND SYLOW THEOREM) *If P and K are Sylow p-subgroups of a group G, then there exists $x \in G$ such that $P = x^{-1}Kx$.*

Theorem 8.15, together with the italicized statement in the preceding paragraph, shows that

any two Sylow p-subgroups of G are isomorphic.

* Skip this example if you haven't read Section 8.2.

COROLLARY 8.16   *Let G be a finite group and K a Sylow p-subgroup for some prime p. Then K is normal in G if and only if K is the only Sylow p-subgroup in G.*

Proof   We know that $x^{-1}Kx$ is a Sylow p-subgroup for every $x \in G$. If K is the only Sylow p-subgroup of G, then we must have $x^{-1}Kx = K$ for every $x \in G$. Therefore, K is normal by Theorem 7.34. Conversely, suppose K is normal and let P be any Sylow p-subgroup. By the Second Sylow Theorem there exists $x \in G$ such that $P = x^{-1}Kx$. Since K is normal, $P = x^{-1}Kx = K$. Therefore, K is the unique Sylow p-subgroup.   ◆

The preceding theorems establish the existence of Sylow p-subgroups and the relationship between any two such subgroups. The next theorem tells us how many Sylow p-subgroups a given group may have.

THEOREM 8.17   (THIRD SYLOW THEOREM)   *The number of Sylow p-subgroups of a finite group G divides |G| and is of the form $1 + pk$ for some nonnegative integer k.*

Applications of the Sylow Theorems

Simple groups (those with no proper normal subgroups) are the basic building blocks for all groups. So it is useful to be able to tell if there are any simple groups of a particular order. The Third Sylow Theorem, together with appropriate counting arguments and Corollary 8.16, can often be used to establish the existence of a proper normal subgroup of a group G, thus showing that G is not simple.

> **EXAMPLE**   If G is a group of order $63 = 3^2 \cdot 7$, then each Sylow 7-subgroup has order 7 and the number of such subgroups is a divisor of 63 of the form $1 + 7k$ by the Third Sylow Theorem. The divisors of 63 are 1, 3, 7, 9, 21, 63 and the numbers of the form $1 + 7k$ (with $k \geq 0$) are 1, 8, 15, 22, 29, 36, 43, 50, 57, 64, etc. Since 1 is the only number on both lists, G has exactly one Sylow 7-subgroup. This subgroup is normal by Corollary 8.16. Consequently, no group of order 63 is simple.

> **EXAMPLE**   We shall show that there is no simple group of order $56 = 2^3 \cdot 7$. The only divisors of 56 of the form $1 + 7k$ are 1 and 8. So G has either one or eight Sylow 7-subgroups, each of order 7. If there is just one Sylow 7-group, it has to be normal by Corollary 8.16. So G is not simple in that case. If G has eight Sylow 7-groups, then each of them has six non-identity elements, and each nonidentity element has order 7 by Corollary 7.27. Furthermore, the intersection of any two of these subgroups is $\langle e \rangle$ by

Exercise 9 of Section 7.5. Consequently, there are $8 \cdot 6 = 48$ elements of order 7 in G. Every Sylow 2-subgroup of G has order 8. Each element of a Sylow 2-subgroup must have order dividing 8 by Corollary 7.27 and, therefore, cannot be in the set of 48 elements of order 7. Thus there is room in G for only one group of order 8. In this case, therefore, the single Sylow 2-subgroup of order 8 is normal by Corollary 8.16, and G is not simple.

In the preceding examples, the Sylow Theorems were used to reach a negative conclusion (the group is not simple). But the same techniques can also lead to positive results. In particular, they allow us to classify certain finite groups.

COROLLARY 8.18 *Let G be a group of order pq, where p and q are primes such that $p > q$. If $q \nmid (p - 1)$, then $G \cong \mathbb{Z}_{pq}$.*

Proof Since the only divisors of $|G|$ are $1, p, q$, and pq, there are either one or q Sylow p-subgroups by the Third Sylow Theorem. But q is not of the form $1 + pk$ since $p > q$. So there is a unique Sylow p-subgroup H of order p, which is normal by Corollary 8.16. Similarly, there are either one or p Sylow q-subgroups. We cannot have $p = 1 + qk$ since $q \nmid (p - 1)$. So there is a unique Sylow q-subgroup K of order q, which is normal by Corollary 8.16. Since $H \cap K$ is a subgroup of both H and K, its order must divide both $|H| = p$ and $|K| = q$ by Lagrange's Theorem. Hence, $H \cap K = \langle e \rangle$. Exercise 15 shows that $G = HK$. Therefore, $G = H \times K$ by Theorem 8.3. But $H \cong \mathbb{Z}_p$ and $K \cong \mathbb{Z}_q$ by Theorem 7.28. Consequently, by Lemma 8.8, $G = H \times K \cong \mathbb{Z}_p \times \mathbb{Z}_q \cong \mathbb{Z}_{pq}$.* ◆

EXAMPLE It is now easy to classify all groups of order $15 = 5 \cdot 3$. Apply Corollary 8.18 with $p = 5, q = 3$ to conclude that every group of order 15 is isomorphic to $\mathbb{Z}_{15}$. Similarly, there is a single group (up to isomorphism) for each of these orders: $33 = 11 \cdot 3$, $35 = 7 \cdot 5$, $65 = 13 \cdot 5$, $77 = 11 \cdot 7$, and $91 = 13 \cdot 7$.

Other applications of the Sylow Theorems are given in Section 8.5.

◆ **EXERCISES**

NOTE: *Unless stated otherwise, G is a finite group and p is a prime.*

A. **1.** Show that S_6 has at least 60 subgroups of order 4. [*Hint:* Consider cyclic subgroups generated by a 4-cycle (such as $\langle (1234) \rangle$) or by the product of a 4-cycle

---

* The proof of Lemma 8.8 is independent of the rest of Section 8.2 and may be read now if you skipped that section.

and a disjoint transposition (such as $\langle(1234)(56)\rangle$); also look at noncyclic subgroups, such as $\{(1), (12), (34), (12)(34)\}$.]

2. (a) List three Sylow 2-subgroups of S_4.

(b) List four Sylow 3-subgroups of S_4.

3. List the Sylow 2-subgroups and Sylow 3-subgroups of A_4.

4. List the Sylow 2-subgroups, Sylow 3-subgroups, and Sylow 5-subgroups of $\mathbb{Z}_{12} \times \mathbb{Z}_{12} \times \mathbb{Z}_{10}$. [Section 8.2 is a prerequisite for this exercise.]

5. How many Sylow p-subgroups can G possibly have when

(a) $p = 3$ and $|G| = 72$ **(b)** $p = 5$ and $|G| = 60$

6. Classify all groups of the given order:

(a) 115   **(b)** 143   **(c)** 391

7. Prove that there are no simple groups of the given order:

(a) 42   **(b)** 200   **(c)** 231   **(d)** 255

B.   8. Use Cauchy's Theorem to prove that a finite p-group has order p^n for some $n \geq 0$.

9. If N is a normal subgroup of a (not necessarily finite) group G and both N and G/N are p-groups, then prove that G is a p-group.

10. If H is a normal subgroup of G and $|H| = p^k$, show that H is contained in every Sylow p-subgroup of G. [You may assume Exercise 24 on page 275.]

11. If f is an automorphism of G and K is a Sylow p-subgroup of G, is it true that $f(K) = K$?

12. Let K be a Sylow p-subgroup of G and H any subgroup of G. Is $K \cap H$ a Sylow p-subgroup of H? [*Hint:* Consider S_4.]

13. If every Sylow subgroup of G is normal, prove that G is the direct product of its Sylow subgroups (one for each prime that divides $|G|$). A group with this property is said to be **nilpotent.**

14. Let K be a Sylow p-subgroup of G and N a normal subgroup of G. Prove that $K \cap N$ is a Sylow p-subgroup of N.

15. (a) If H and K are subgroups of G, then HK denotes the set $\{hk \in G \mid h \in H, k \in K\}$. If $H \cap K = \langle e \rangle$, prove that $|HK| = |H| \cdot |K|$. [*Hint:* If $hk = h_1 k_1$, then $h_1^{-1}h = k_1 k^{-1}$.]

(b) If H and K are any subgroups of G, prove that

$$|HK| = \frac{|H| \cdot |K|}{|H \cap K|}.$$

16. If G is a group of order 60 that has a normal Sylow 3-subgroup, prove that G also has a normal Sylow 5-subgroup.

17. If G is a noncyclic group of order 21, how many Sylow 3-subgroups does G have?

18. If G is a simple group of order 168, how many Sylow 7-subgroups does G have?

19. If p and q are distinct primes, prove that there are no simple groups of order pq.

20. If G has order $p^k m$ with $m < p$, prove that G is not simple.

21. Prove that there are no simple groups of order 30.

22. If p is prime, prove that there are no simple groups of order $2p$.

23. If p and q are distinct primes, prove that there is no simple group of order $p^2 q$.

24. **(a)** If $|G| = 105$, prove that G has a subgroup of order 35.

    (b) If $|G| = 375$, prove that G has a subgroup of order 15.

C. 25. If p, q, r are primes with $p < q < r$, prove that a group of order pqr has a normal Sylow r-subgroup and, hence, is not simple.

8.4 CONJUGACY AND THE PROOF OF THE SYLOW THEOREMS

Appendix D (Equivalence Relations) is a prerequisite for this section. The proofs of the Sylow Theorems depend heavily on the concept of conjugacy, which we now develop.

Let G be a group and $a, b \in G$. We say that a is **conjugate** to b if there exists $x \in G$ such that $b = x^{-1}ax$. For example, (12) is conjugate to (13) in S_3 because

$$(123)^{-1}(12)(123) = (132)(12)(123) = (13).$$

The key fact about conjugation is

THEOREM 8.19 *Conjugacy is an equivalence relation on G.*

Proof We write $a \sim b$ if a is conjugate to b. *Reflexive:* $a \sim a$ since $a = eae = e^{-1}ae$. *Symmetric:* If $a \sim b$, then $b = x^{-1}ax$ for some x in G. Multiplying on the left by x and on the right by x^{-1} shows that $a = xbx^{-1} = (x^{-1})^{-1}bx^{-1}$. Hence, $b \sim a$. *Transitive:* If $a \sim b$ and $b \sim c$, then $b = x^{-1}ax$ and $c = y^{-1}by$ for some $x, y \in G$. Hence, $c = y^{-1}(x^{-1}ax)y = (y^{-1}x^{-1})a(xy) = (xy)^{-1}a(xy)$. Thus $a \sim c$; therefore, $\sim$ is an equivalence relation. $\blacklozenge$

The equivalence classes in G under the relation of conjugacy are called **conjugacy classes.** The discussion of equivalence relations in Appendix D shows that

The conjugacy class of an element a consists of all the elements in G that are conjugate to a.

Two conjugacy classes are either disjoint or identical.

The group G is the union of its distinct conjugacy classes.

EXAMPLE The conjugacy class of (12) in S_3 consists of all elements $x^{-1}(12)x$, with $x \in S_3$. A straightforward computation shows that for any $x \in S_3$, $x^{-1}(12)x$ is one of (12), (13), or (23); for instance,

$$(23)^{-1}(12)(23) = (23)(12)(23) = (13)$$
$$(132)^{-1}(12)(132) = (123)(12)(132) = (23).$$

Thus the conjugacy class of (12) is $\{(12), (13), (23)\}$. Similar computations show that there are three distinct conjugacy classes in S_3:

$$\{(1)\} \qquad \{(123), (132)\} \qquad \{(12), (13), (23)\}.$$

Although these conjugacy classes are of different sizes, note that the number of elements in any conjugacy class (1, 2, or 3) is a divisor of 6, the order of S_3. We shall see that this phenomenon occurs in the general case as well.

Let G be a group and $a \in G$. The **centralizer** of a is denoted $C(a)$ and consists of all elements in G that commute with a, that is,

$$C(a) = \{g \in G \mid ga = ag\}.$$

If $G = S_3$ and $a = (123)$, for example, you can readily verify that $C(a) = \{(1), (123), (132)\}$ and that $C(a)$ is a subgroup of S_3. If a is a nonzero rational number in the multiplicative group $\mathbb{Q}^*$, every element of $\mathbb{Q}^*$ commutes with a, so $C(a)$ is the entire group $\mathbb{Q}^*$. These examples are illustrations of

THEOREM 8.20 *If G is a group and $a \in G$, then $C(a)$ is a subgroup of G.*

Proof Since $ea = ae$, we have $e \in C(a)$, so that $C(a)$ is nonempty. If $g, h \in C(a)$, then

$$(gh)a = g(ha) = g(ah) = (ga)h = (ag)h = a(gh).$$

So $gh \in C(a)$, and $C(a)$ is closed. Multiplying $ga = ag$ on both the left and right by g^{-1} shows that $ag^{-1} = g^{-1}a$. Hence, $g \in C(a)$ implies that $g^{-1} \in C(a)$. Therefore, $C(a)$ is a subgroup by Theorem 7.10. ◆

The centralizer leads to a very useful fact about the size of conjugacy classes:

THEOREM 8.21 *Let G be a group and $a \in G$. The number of distinct conjugates of a (that is, the number of elements in the conjugacy class of a) is $[G:C(a)]$, the index of $C(a)$ in G and, therefore, divides $|G|$.*

Proof Denote $C(a)$ by C and let $x, y \in G$. Then

$$x^{-1}ax = y^{-1}ay \Leftrightarrow a = xy^{-1}ayx^{-1} \qquad \text{[left multiply by } x \text{; right multiply by } x^{-1} \text{]}$$

$$\Leftrightarrow a = (yx^{-1})^{-1}a(yx^{-1}) \qquad \text{[Corollary 7.6]}$$

$$\Leftrightarrow (yx^{-1})a = a(yx^{-1}) \qquad \text{[left multiply by } yx^{-1} \text{]}$$

$$\Leftrightarrow yx^{-1} \in C \qquad \text{[definition of } C \text{]}$$

$$\Leftrightarrow Cy = Cx \qquad \text{[Theorem 7.23]}$$

Thus x and y produce the same conjugate of a if and only if x and y are in the same coset of C. Equivalently, if x and y are in different cosets of C, then $x^{-1}ax \neq y^{-1}ay$. Now G is the union of the distinct cosets of C, say $G = Cx_1 \cup Cx_2 \cup \cdots \cup Cx_t$, and the conjugates of a are all the elements $z^{-1}az$, as z runs over G. So the distinct conjugates of a will be the t elements

$$x_1^{-1}ax_1, x_2^{-1}ax_2, x_3^{-1}ax_3, \ldots , x_t^{-1}ax_t.$$

But t is the number of distinct cosets of C in G, that is, $t = [G:C]$. This index divides $|G|$ by Lagrange's Theorem 7.26. ◆

Let G be a finite group and let $C_1, C_2, \ldots , C_t$ be the distinct conjugacy classes of G. Then $G = C_1 \cup C_2 \cup \cdots \cup C_t$. Since distinct conjugacy classes are mutually disjoint,

(1) $$|G| = |C_1 \cup C_2 \cup \cdots \cup C_t| = |C_1| + |C_2| + \cdots + |C_t|,$$

where $|C_i|$ denotes the number of elements in the class C_i. Now choose one element, say a_i, in each class C_i. Then C_i consists of all the conjugates of a_i. By Theorem 8.21, $|C_i|$ is precisely $[G:C(a_i)]$, a divisor of $|G|$. So equation (1) becomes

(2) $$|G| = [G:C(a_1)] + [G:C(a_2)] + \cdots + [G:C(a_t)].$$

This equation (in either version (1) or (2)) is called the **class equation** of the group G. It will be the basic tool for proving the Sylow Theorems. Other applications of the class equation are discussed in Section 8.5.

EXAMPLE In the example on page 268 we saw that S_3 has three distinct conjugacy classes of sizes 1, 2, and 3. Since $|S_3| = 6$, the class equation of S_3 is $6 = 1 + 2 + 3$.

If c and x are elements of a group G, then $cx = xc$ if and only if $x^{-1}cx = c$. Thus c is in the center of G [$cx = xc$ for every $x \in G$] if and only if c has exactly one conjugate, itself [$x^{-1}cx = c$ for every $x \in G$]. Therefore, the center $Z(G)$ of G is the union of all the one-element conjugacy classes of G, so that the class equation can be written in a third form:

(3) $$|G| = |Z(G)| + |C_1| + |C_2| + \cdots + |C_r|,$$

where $C_1, \ldots, C_r$ are the distinct conjugacy classes of G that contain more than one element each and each $|C_i|$ divides $|G|$.

In addition to the class equation, one more result is needed for the proof of the Sylow Theorems.

LEMMA 8.22 (CAUCHY'S THEOREM FOR ABELIAN GROUPS) *If G is a finite abelian group and p is a prime that divides the order of G, then G contains an element of order p.*

The lemma is an immediate consequence of the Fundamental Theorem of Abelian Groups (Exercise 12 in Section 8.2). The following proof, however, depends only on Chapter 7.

Proof of Lemma 8.22 The proof is by induction on the order of G, using the Principle of Complete Induction.* To do this, we must first show that the theorem is true when $|G| = 2$. In this case, if p divides $|G|$, then $p = 2$. The nonidentity element of G must have order 2 by part (1) of Corollary 7.27, and so the theorem is true.

Now assume that the theorem is true for all abelian groups of order less than n and suppose $|G| = n$. Let a be any nonidentity element of G. Then the order of a is a positive integer and is therefore divisible by some prime q (Theorem 1.11), say $|a| = qt$. The element $b = a^t$ has order q by Theorem 7.8. If $q = p$, the theorem is proved. If $q \neq p$, let N be the cyclic subgroup $\langle b \rangle$. N is normal since G is abelian and N has order q by Theorem 7.14. By Theorem 7.36 the quotient group G/N has order $|G|/|N| = n/q < n$. Consequently, by the induction hypothesis, *the theorem is true for G/N*. The prime p divides $|G|$, and $|G| = |N| |G/N| = q|G/N|$. Since q is a prime other than p, p must divide $|G/N|$ by Theorem 1.8. Therefore, G/N contains an element of order p, say Nc. Since Nc has order p in G/N, we have $Nc^p = (Nc)^p = Ne$ and, hence, $c^p \in N$. Since N has order q, $c^{pq} = (c^p)^q = e$ by part (2) of Corollary 7.27.

Therefore, c must have order dividing pq by Theorem 7.8. However, c cannot have order 1 because then Nc would have order 1 instead of p in G/N. Nor can c have order q because then $(Nc)^q = Nc^q = Ne$ in G/N, so that p (the order of Nc) would divide q by Theorem 7.8. The only possibility is that c has

---

* See Appendix C.

order p or pq; in the latter case, c^q has order p by Theorem 7.8. In either case, G contains an element of order p. Therefore, the theorem is true for abelian groups of order n and, hence, by induction for all finite abelian groups. ◆

Proofs of the Sylow Theorems

We now have all the tools needed to prove the Sylow Theorems.

Proof of the First Sylow Theorem 8.13   The proof is by induction on the order of G. If $|G| = 1$, then p^0 is the only prime power that divides $|G|$, and G itself is a subgroup of order p^0. Suppose $|G| > 1$ and assume inductively that the theorem is true for all groups of order less than $|G|$. Combining the second and third forms of the class equation of G shows that

$$|G| = |Z(G)| + [G:C(a_1)] + [G:C(a_2)] + \cdots + [G:C(a_r)],$$

where for each i, $[G:C(a_i)] > 1$. Furthermore, $|Z(G)| \geq 1$ (since $e \in Z(G)$), and $|C(a_i)| < |G|$ (otherwise, $[G:C(a_i)] = 1$).

Suppose there is an index j such that p does *not* divide $[G:C(a_j)]$. Then by Theorem 1.8 p^k must divide $|C(a_j)|$ because p^k divides $|G|$ by hypothesis and $|G| = |C(a_j)| \cdot [G:C(a_j)]$ by Lagrange's Theorem. Since the subgroup $C(a_j)$ has order less than $|G|$, the induction hypothesis implies that $C(a_j)$, and, hence, G has a subgroup of order p^k.

On the other hand, if p divides $[G:C(a_i)]$ for every i, then since p divides $|G|$, p must also divide $|G| - [G:C(a_1)] - \cdots - [G:C(a_r)] = |Z(G)|$. Since $Z(G)$ is abelian, $Z(G)$ contains an element c of order p by Lemma 8.22. Let N be the cyclic subgroup generated by c. Then N has order p and is normal in G (Exercise 8). Consequently, the order of the quotient group G/N, namely $|G|/p$, is less than $|G|$ and divisible by p^{k-1}. By the induction hypothesis G/N has a subgroup T of order p^{k-1}. There is a subgroup H of G such that $N \subseteq H$ and $T = H/N$ by Theorem 7.44. Lagrange's Theorem shows that

$$|H| = |N| \cdot |H/N| = |N| \cdot |T| = pp^{k-1} = p^k.$$

So G has a subgroup of order p^k in this case, too. ◆

The basic tools needed to prove the last two Sylow Theorems are very similar to those used above, except that we will now deal with conjugate subgroups rather than conjugate elements. More precisely, let H be a fixed subgroup of a group G and let A and B be any subgroups of G. We say that A is **H-conjugate** to B if there exists an $x \in H$ such that

$$B = x^{-1}Ax = \{x^{-1}ax \mid a \in A\}.$$

In the special case when H is the group G itself, we simply say that A is **conjugate** to B.

THEOREM 8.23 *Let H be a subgroup of a group G. Then H-conjugacy is an equivalence relation on the set of all subgroups of G.*

Proof Copy the proof of Theorem 8.19, using subgroups A, B, C in place of elements a, b, c. ◆

Let A be a subgroup of a group G. The **normalizer** of A is the set $N(A)$ defined by

$$N(A) = \{g \in G \mid g^{-1}Ag = A\}.$$

THEOREM 8.24 *If A is a subgroup of a group G, then N(A) is a subgroup of G and A is a normal subgroup of N(A).*

Proof Exercise 7 shows that $A \subseteq N(A)$ and that $g \in N(A)$ if and only if $Ag = gA$. Using this fact, the proof of Theorem 8.20 can be readily adapted to prove that $N(A)$ is a subgroup. The definition of $N(A)$ shows that A is normal in $N(A)$. ◆

THEOREM 8.25 *Let H and A be subgroups of a finite group G. The number of distinct H-conjugates of A (that is, the number of elements in the equivalence class of A under H-conjugacy) is $[H:H \cap N(A)]$ and, therefore, divides $|H|$.*

Proof The proof of Theorem 8.21 carries over to the present situation if you replace G by H, a by A, and C by $H \cap N(A)$. ◆

LEMMA 8.26 *Let Q be a Sylow p-subgroup of a finite group G. If $x \in G$ has order a power of p and $x^{-1}Qx = Q$, then $x \in Q$.*

Proof Since Q is normal in $N(Q)$ by Theorem 8.24, the quotient group $N(Q)/Q$ is defined. By hypothesis, $x \in N(Q)$. Since $|x|$ is some power of p, the coset Qx in $N(Q)/Q$ also has order a power of p. Now Qx generates a cyclic subgroup T of $N(Q)/Q$ whose order is a power of p. By Theorem 7.44, $T = H/Q$, where H is a subgroup of G that contains Q. Since the orders of the groups Q and T are each powers of p and $|H| = |Q| \cdot |T|$ by Lagrange's Theorem, $|H|$ must be a power of p. But $Q \subseteq H$, and $|Q|$ is the largest power of p that divides $|G|$ by the definition of a Sylow p-subgroup. Therefore, $Q = H$, and, hence, $T = H/Q$ is the identity subgroup. So the generator Qx of T must be the identity coset Qe. The equality $Qx = Qe$ implies that $x \in Q$. ◆

Proof of the Second Sylow Theorem 8.15 Since K is a Sylow p-subgroup, K has order p^n, where $|G| = p^nm$ and $p \nmid m$. Let $K = K_1, K_2, \ldots , K_t$ be the distinct conjugates of K in G. By Theorem 8.25 (with $H = G$ and $K = A$), $t = [G:N(K)]$. Note that p *does not divide* t [reason: $p^nm = |G| = |N(K)| \cdot [G:N(K)] = $

$|N(K)| \cdot t$ and p^n divides $|N(K)|$ because K is a subgroup of $N(K)$]. We must prove that the Sylow p-subgroup P is conjugate to K, that is, that P is one of the K_i. To do so we use the relation of P-conjugacy.

Since each K_i is a conjugate of K_1 and conjugacy is transitive, every conjugate of K_i in G is also a conjugate of K_1. In other words, every conjugate of K_i is some K_j. Consequently, the equivalence class of K_i under P-conjugacy contains only various K_j. So the set $S = \{K_1, K_2, \ldots, K_t\}$ of all conjugates of K is a union of distinct equivalence classes under P-conjugacy. The number of subgroups in each of these equivalence classes is a power of p because by Theorem 8.25 the number of subgroups that are P-conjugate to K_i is $[P : P \cap N(K_i)]$, which is a divisor of $|P| = p^n$ by Lagrange's Theorem. Therefore, t (the number of subgroups in the set S) is the sum of various powers of p (each being the number of subgroups in one of the distinct equivalence classes whose union is S). Since p doesn't divide t, at least one of these powers of p must be $p^0 = 1$. Thus some K_i is in an equivalence class by itself, meaning that $x^{-1}K_ix = K_i$ for every $x \in P$. Lemma 8.26 (with $Q = K_i$) implies that $x \in K_i$ for every such x, so that $P \subseteq K_i$. Since both P and K_i are Sylow p-subgroups, they have the same order. Hence, $P = K_i$.   ◆

Proof of the Third Sylow Theorem 8.17   Let $S = \{K_1, \ldots, K_t\}$ be the set of all Sylow p-subgroups of G. By the Second Sylow Theorem, they are all conjugates of K_1. Let P be one of the K_i and consider the relation of P-conjugacy. The only P-conjugate of P is P itself by closure. The proof of the Second Sylow Theorem shows that the only equivalence class consisting of a single subgroup is the class consisting of P itself. The proof also shows that S is the union of distinct equivalence classes and that the number of subgroups in each class is a power of p. Just one of these classes contains P, so the number of subgroups in each of the others is a *positive* power of p. Hence, the number t of Sylow p-subgroups is the sum of 1 and various positive powers of p and, therefore, can be written in the form $1 + kp$ for some integer k.   ◆

◆ EXERCISES

NOTE: *Unless stated otherwise, G is a finite group and p is a prime.*

A.   1. List the distinct conjugacy classes of the given group.

(a) D_4 (b) S_4 (c) A_4

2. If $a \in G$, then show by example that $C(a)$ may not be abelian. [*Hint:* If $a = (12)$ in S_5, then (34) and (345) are in $C(a)$.]

3. If H is a subgroup of G and $a \in H$, show by example that the conjugacy class of a in H may not be the same as the conjugacy class of a in G.

4. What is the center of S_3?

5. List all conjugates of the Sylow 3-subgroup $\langle(123)\rangle$ in S_4.

6. If H and K are subgroups of G and H is normal in K, prove that K is a subgroup of $N(H)$. In other words, $N(H)$ is the largest subgroup of G in which H is a normal subgroup.

7. If A is a subgroup of G, prove that

   (a) $A \subseteq N(A)$;

   (b) $g \in N(A)$ if and only if $Ag = gA$.

8. If N is a subgroup of $Z(G)$, prove that N is a normal subgroup of G.

B. 9. If C is a conjugacy class in G and f is an automorphism of G, prove that $f(C)$ is also a conjugacy class of G.

10. Let G be an infinite group and H the subset of all elements of G that have only a finite number of distinct conjugates in G. Prove that H is a subgroup of G.

11. If G is a nilpotent group (see Exercise 13 of Section 8.3), prove that G has this property: If m divides $|G|$, then G has a subgroup of order m. [You may assume Exercise 22.]

12. Let K be a Sylow p-subgroup of G and N a normal subgroup of G. If K is a normal subgroup of N, prove that K is normal in G.

13. Prove Theorem 8.23.

14. Let N be a normal subgroup of G, $a \in G$, and C the conjugacy class of a in G.

    (a) Prove that $a \in N$ if and only if $C \subseteq N$.

    (b) If C_i is any conjugacy class in G, prove that $C_i \subseteq N$ or $C_i \cap N = \varnothing$.

    (c) Use the class equation to show that $|N| = |C_1| + \cdots + |C_k|$, where $C_1, \ldots, C_k$ are all the conjugacy classes of G that are contained in N.

15. If $N \neq \langle e \rangle$ is a normal subgroup of G and $|G| = p^n$, prove that $N \cap Z(G) \neq \langle e \rangle$. [*Hint:* Exercise 14(c) may be helpful.]

16. Complete the proof of Theorem 8.24.

17. Prove Theorem 8.25.

18. If K is a Sylow p-subgroup of G and H is a subgroup that contains $N(K)$, prove that $[G:H] \equiv 1 \pmod{p}$.

19. If K is a Sylow p-subgroup of G, prove that $N(N(K)) = N(K)$.

20. If H is a proper subgroup of G, prove that G is *not* the union of all the conjugates of H. [*Hint:* Remember that H is a normal subgroup of $N(H)$; theorem 8.25 may be helpful.]

21. If H is a normal subgroup of G and $|H| = p^k$, prove that H is contained in every Sylow p-subgroup of G. [You may assume Exercise 24.]

C. 22. If $|G| = p^n$, prove that G has a normal subgroup of order p^{n-1}. [*Hint:* You may assume Theorem 8.27 below. Use induction on n. Let $N = \langle a \rangle$, where $a \in Z(G)$ has order p (Why is there such an a?); then G/N has a subgroup of order p^{n-2}; use Theorem 7.44.]

23. If $|G| = p^n$, prove that every subgroup of G of order p^{n-1} is normal.

24. If H is a subgroup of G and H has order some power of p, prove that H is contained in a Sylow p-subgroup of G. [*Hint:* Proceed as in the proofs of the Second and Third Sylow Theorems but use the relation of H-conjugacy instead of P-conjugacy on the set $\{K_1, \ldots, K_t\}$ of all Sylow p-subgroups.]

8.5 THE STRUCTURE OF FINITE GROUPS

The tools developed in Sections 8.1–8.4 are applied here to various aspects of the classification problem. In particular, all groups of orders ≤ 15 are classified. We begin with some useful facts about p-groups.

THEOREM 8.27   *If G is a group of order p^n, with p prime and $n \geq 1$, then the center $Z(G)$ contains more than one element. In particular, $|Z(G)| = p^k$ with $1 \leq k \leq n$.*

Proof   By Lagrange's Theorem, $|Z(G)| = p^k$ with $0 \leq k \leq n$. We now show that $k \geq 1$, that is, that $|Z(G)| \geq p$. Form (3) of the class equation (page 270) shows that

$$|Z(G)| = |G| - |C_1| - |C_2| - \cdots - |C_r|$$

where each $|C_i|$ is a number larger than 1 that divides $|G|$. Since $|G| = p^n$, the divisors of $|G|$ larger than 1 are positive powers of p. Therefore, each $|C_i|$ is divisible by p. Since $|G|$ is also divisible by p, it follows that p divides $|Z(G)|$ and, hence, $|Z(G)| \geq p$.   ◆

COROLLARY 8.28   *If p is a prime and $n > 1$, then there is no simple group of order p^n.*

Proof   If G is a group of order p^n, then $Z(G)$ is a normal subgroup. If $Z(G) \neq G$, then G is not simple. If $Z(G) = G$, then G is abelian and not simple by Theorem 7.45.   ◆

COROLLARY 8.29   *If G is a group of order p^2, with p prime, then G is abelian. Hence, G is isomorphic to $\mathbb{Z}_{p^2}$ or $\mathbb{Z}_p \times \mathbb{Z}_p$.*

Proof $Z(G)$ has order p or p^2 by Lagrange's Theorem and Theorem 8.27. If $Z(G)$ has order p^2, then $G = Z(G)$, which means that G is abelian. If $Z(G)$ has order p, then the quotient group $G/Z(G)$ has order $|G|/|Z(G)| = p^2/p = p$ by Theorem 7.36. Hence, $G/Z(G)$ is cyclic by Theorem 7.28. Therefore, G is abelian by Theorem 7.38. The last statement of the theorem now follows immediately from the Fundamental Theorem of Finite Abelian Groups. ◆

In Corollary 8.18 certain groups of order pq (with p, q prime) were characterized. We can now extend that argument to some groups of order p^2q.

THEOREM 8.30 *Let p and q be distinct primes such that $q \not\equiv 1$ (mod p) and $p^2 \not\equiv 1$ (mod q). If G is a group of order p^2q, then G is isomorphic to $\mathbb{Z}_{p^2q}$ or $\mathbb{Z}_p \times \mathbb{Z}_p \times \mathbb{Z}_q$.*

Applying Theorem 8.30 with $p = 3$ and $q = 5$ shows that every group of order $45 = 3^2 \cdot 5$ is isomorphic to $\mathbb{Z}_{45}$ or $\mathbb{Z}_3 \times \mathbb{Z}_3 \times \mathbb{Z}_5$. Similarly, the theorem may be used to classify all groups of orders 99, 153, 175, 207, 325, 425, etc.

Proof of Theorem 8.30 By the Third Sylow Theorem, the number of Sylow p-subgroups of G is congruent to 1 modulo p and divides $|G|$. Since the divisors of $|G|$ are $1, p, p^2, q, pq,$ and p^2q, the only possibilities are 1 and q. There cannot be q of them because $q \not\equiv 1$ (mod p). Hence, there is a unique Sylow p-subgroup H, which is normal by Corollary 8.16. Similarly, G has 1, p, or p^2 Sylow q-subgroups, and neither p nor p^2 is possible since $p^2 \not\equiv 1$ (mod q). Hence, there is a unique normal Sylow q-subgroup K. The order of the subgroup $H \cap K$ must divide both $|H| = p^2$ and $|K| = q$ by Lagrange's Theorem. Hence, $H \cap K = \langle e \rangle$. Furthermore, $HK = G$ by Exercise 15 in Section 8.3. Therefore, $G = H \times K$ by Theorem 8.3. Now H is isomorphic to $\mathbb{Z}_{p^2}$ or $\mathbb{Z}_p \times \mathbb{Z}_p$ by Corollary 8.29 and $K \cong \mathbb{Z}_q$ by Theorem 7.28. Consequently, by Lemma 8.8, $G = H \times K \cong \mathbb{Z}_{p^2} \times \mathbb{Z}_q \cong \mathbb{Z}_{p^2q}$ or $G = H \times K \cong \mathbb{Z}_p \times \mathbb{Z}_p \times \mathbb{Z}_q$. ◆

COROLLARY 8.31 *If p and q are distinct primes, then there is no simple group of order p^2q.*

Proof Suppose G is a group of order p^2q. If either $p^2 \not\equiv 1$ (mod q) or $q \not\equiv 1$ (mod p), then the proof of Theorem 8.30 shows that G has a normal Sylow subgroup and, hence, is not simple. If both $p^2 \equiv 1$ (mod q) and $q \equiv 1$ (mod p), then $q \mid (p^2 - 1)$ and $p \mid (q - 1)$, which implies that $p \leq q - 1$ or, equivalently, $q \geq p + 1$. Since $p^2 - 1 = (p - 1)(p + 1)$, we know that $q \mid (p - 1)$ or $q \mid (p + 1)$ by Theorem 1.8. The former is impossible because $q \geq p + 1$, and the latter implies that $q \leq p + 1$, so that $q = p + 1$. Since p and q are primes, the only possibility is $p = 2$ and $q = 3$. Exercise 2 shows that no group of order $2^2 \cdot 3 = 12$ is simple. ◆

Dihedral Groups

We now introduce a family of groups that play a crucial role in the classification of groups of order $2p$. Recall that the group D_4 consists of various rotations and reflections of the square (see Section 7.1). This idea can be generalized as follows. Let P be a regular polygon of n sides ($n \geq 3$).* For convenient reference, assume that P has its center at the origin and a vertex on the negative x-axis, with the other vertices numbered counterclockwise from this one, as illustrated here in the cases $n = 5$ and $n = 6$.

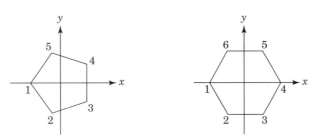

Think of the plane as a thin sheet of hard plastic. Cut out P, pick it up, and replace it, not necessarily in the same position, but so that it fits exactly in the cut-out space. Such a motion is called a **symmetry** of P.** By considering a symmetry as a function from P to itself and using composition of functions as the operation (gf means motion f followed by motion g), the set D_n of all symmetries of P forms a group, called the **dihedral group of degree n.**

THEOREM 8.32   *The dihedral group D_n is a group of order $2n$ generated by elements r and d such that*

$$|r| = n, \qquad |d| = 2, \qquad and \qquad dr = r^{-1}d.$$

Proof   The proof that D_n is a group is left to the reader. Let r be the counterclockwise rotation of $360/n$ degrees about the center of P; r sends vertex 1 to vertex 2, vertex 2 to vertex 3, and so on. Note that r has order n because r^n is a $360°$ rotation that returns P to its initial position (the identity symmetry). Let d be the reflection in the x-axis. As shown in the figure on the next page, d "reverses the orientation" of P: vertices that were formerly numbered counterclockwise from vertex 1 are now numbered clockwise:

---

* "Regular" means that all sides of P have the same length and all its vertex angles (each formed by two adjacent sides) are the same size. It can be shown that the perpendicular bisectors of the n sides all intersect at a single point, which is called the **center** of P.

** All motions that result in the same final position for P are considered to be the same.

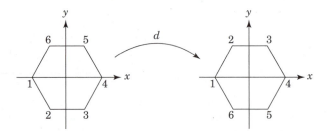

The element d has order 2 because reflecting twice in the x-axis also returns P to its initial position.

Since adjacent vertices of P remain adjacent under any symmetry, the final position of P is completely determined by two factors: the new orientation of P (whether the vertices are numbered clockwise or counterclockwise from vertex 1) and the new location of vertex 1. Consequently, every symmetry is the same as either

r^i $(0 \le i < n)$ [counterclockwise rotation of $i(360/n)$ degrees that preserves orientation and moves vertex 1 to the position originally occupied by vertex $i + 1$]

or

$r^i d$ $(0 \le i < n)$ [reflection in the x-axis that reverses orientation followed by a counterclockwise rotation that moves vertex 1 to the position originally occupied by vertex $i + 1$].

Therefore

$$D_n = \{e = r^0, r, r^2, \ldots, r^{n-1}; d = r^0 d, rd, r^2 d, \ldots, r^{n-1} d\}.$$

Furthermore, the $2n$ elements listed here are all distinct (r^i and r^j move vertex 1 to different positions and $r^i = r^j d$ is impossible since r^i preserves the vertex orientation, but $r^j d$ reverses it). Hence, D_n is a group of order $2n$.

Finally, verify that drd moves vertex 1 to the position originally occupied by vertex n and leaves the vertices in counterclockwise order. In other words, drd is the rotation that moves vertex 1 to vertex n, that is, $drd = r^{n-1}$. Since r has order n, $r^{-1} = r^{n-1}$ and, hence, $drd = r^{-1}$. Multiplying on the right by d shows that $dr = r^{-1}d$. ◆

We can now classify another family of groups.

THEOREM 8.33 *If G is a group of order $2p$, where p is an odd prime, then G is isomorphic to the cyclic group $\mathbb{Z}_{2p}$ or the dihedral group D_p.*

The theorem provides another proof that there are exactly two nonisomorphic groups of order 6 (see Theorem 7.30), as well as providing a complete list of groups of orders 10, 14, 22, 26, 34, etc.

Proof of Theorem 8.33 G contains an element a of order p and an element b of order 2 by Cauchy's Theorem (Corollary 8.14). Note that $b^2 = e$ implies $b^{-1} = b$. Let H be the cyclic group $\langle a \rangle$. Since $|G| = 2p$, the subgroup H has index 2 and is, therefore, normal by Exercise 20 of Section 7.6. Consequently, $bab = bab^{-1} \in H$. Since H is cyclic, $bab = a^t$ for some t. Using this and the fact that $b^2 = e$, we see that

$$a^{t^2} = (a^t)^t = (bab)^t = (bab)(bab)(bab) \cdots (bab) = ba^t b = b(bab)b = a$$

Hence, $t^2 \equiv 1 \pmod{p}$ by part (3) of Theorem 7.8. Consequently, p divides $t^2 - 1 = (t - 1)(t + 1)$, which implies that $p \mid (t - 1)$ or $p \mid (t + 1)$ by Theorem 1.8. Thus $t \equiv 1 \pmod{p}$ or $t \equiv -1 \pmod{p}$.

If $t \equiv 1 \pmod{p}$, then $bab = a^t = a$ by Theorem 7.8. Multiplying both sides by b shows that $ba = ab$. It follows that ab has order $2p = |G|$ (Exercise 31 of Section 7.2). Therefore, G is cyclic and isomorphic to $\mathbb{Z}_{2p}$ by Theorem 7.18.

If $t \equiv -1 \pmod{p}$, then $bab = a^{-1}$. Exercise 9 shows that the map $f : D_p \rightarrow G$ given by $f(r^i d^j) = a^i b^j$ is a homomorphism. Let K be the subgroup $\langle b \rangle$. Since $|H| = p$ (with p odd) and $|K| = 2$, $H \cap K = \langle e \rangle$ by Lagrange's Theorem and $G = HK$ by Exercise 15 in Section 8.3. Thus every element of G can be written in the form $a^i b^j$, which implies that f is surjective. Since D_p and G have the same order, f must be injective and, hence, an isomorphism. ◆

Groups of Small Order

We are now in a position to complete the classification of groups of small order that was begun in Section 7.5, where groups of orders ≤ 7 were classified. We already know three abelian groups of order 8 ($\mathbb{Z}_2 \times \mathbb{Z}_2 \times \mathbb{Z}_2$, $\mathbb{Z}_4 \times \mathbb{Z}_2$, and $\mathbb{Z}_8$) and one nonabelian one (D_4). Another nonabelian group of order 8, the quaternion group Q, was introduced in Exercise 14 of Section 7.1. It is not isomorphic to D_4 by Exercise 31 of Section 7.4. These five groups are the only ones:

THEOREM 8.34 *If G is a group of order 8, then G is isomorphic to one of the following groups: $\mathbb{Z}_8$, $\mathbb{Z}_4 \times \mathbb{Z}_2$, $\mathbb{Z}_2 \times \mathbb{Z}_2 \times \mathbb{Z}_2$, the dihedral group D_4, or the quaternion group Q.*

Proof If G is abelian, then G is isomorphic to $\mathbb{Z}_8$, $\mathbb{Z}_4 \times \mathbb{Z}_2$, or $\mathbb{Z}_2 \times \mathbb{Z}_2 \times \mathbb{Z}_2$ by the Fundamental Theorem of Finite Abelian Groups. So suppose G is a nonabelian group of order 8. The nonidentity elements of G must have order 2, 4, or 8 by Lagrange's Theorem. However, G cannot contain an element of order 8 (because then G would be cyclic and abelian), nor can all the nonidentity elements of G have order 2 (see Exercise 25 of Section 7.2). Hence, G contains an element a of order 4. Let b be any element of G such that $b \notin \langle a \rangle = \{e, a, a^2, a^3\}$. Then the eight elements $e, a, a^2, a^3, b, ab, a^2b, a^3b$ are all distinct because $|a| = 4$ and $a^i = a^jb$ implies $b = a^{i-j} \in \langle a \rangle$, contrary to the choice of b. Thus $G = \{e, a, a^2, a^3, b, ab, a^2b, a^3b\}$.

The subgroup $\langle a \rangle$ has order 4 and index 2 in G. Hence, $\langle a \rangle$ is normal by Exercise 20 of Section 7.6. Now the element bab^{-1} has order 4 by Exercise 17 of Section 7.2 and $bab^{-1} \in \langle a \rangle$ by normality. Therefore, bab^{-1} is either a or a^3 (because e has order 1 and a^2 has order 2). If $bab^{-1} = a$, however, then $ba = ab$, which implies that G is abelian. Therefore, $bab^{-1} = a^3 = a^{-1}$ so that $ba = a^{-1}b$. This fact can be used to construct most of the multiplication table of G. For instance, $(ab)a^2 = a(ba)a = a(a^{-1}b)a = ba = a^{-1}b = a^3b$. You can use similar arguments to verify that the table must look like this:

	e	a	a^2	a^3	b	ab	a^2b	a^3b
e	e	a	a^2	a^3	b	ab	a^2b	a^3b
a	a	a^2	a^3	e	ab	a^2b	a^3b	b
a^2	a^2	a^3	e	a	a^2b	a^3b	b	ab
a^3	a^3	e	a	a^2	a^3b	b	ab	a^2b
b	b	a^3b	a^2b	ab				
ab	ab	b	a^3b	a^2b				
a^2b	a^2b	ab	b	a^3b				
a^3b	a^3b	a^2b	ab	b				

In order to complete the table, we must find b^2. Since $b^2 = a^ib$ implies $b = a^i \in \langle a \rangle$, which is a contradiction, b^2 must be one of e, a, a^2, or a^3. If $b^2 = a$, however, then $ab = b^2b = bb^2 = ba$, which implies that G is abelian. Similarly, $b^2 = a^3$ implies that G is abelian (Exercise 13). Therefore, $b^2 = e$ or $b^2 = a^2$. Each of these possibilities leads to a different table for G. Completing the table when $b^2 = e$ and comparing it to the table for D_4 on page 167 shows that $G \cong D_4$ under the correspondence

$$a^i \longrightarrow r_i, \quad b \longrightarrow d, \quad ab \longrightarrow h, \quad a^2b \longrightarrow t, \quad a^3b \longrightarrow v$$

(Exercise 4). Similarly, completing the table when $b^2 = a^2$ and comparing it to the table for the quaternion group Q shows that $G \cong Q$ (Exercise 5). ◆

According to the Fundamental Theorem of Finite Abelian Groups there are two abelian groups of order 12: $\mathbb{Z}_4 \times \mathbb{Z}_3 \cong \mathbb{Z}_{12}$ and $\mathbb{Z}_2 \times \mathbb{Z}_2 \times \mathbb{Z}_3$. We have also seen two nonabelian groups of order 12: the alternating group A_4 and the dihedral group D_6. It can be shown that there is a third nonabelian group T of order 12, which is generated by elements a and b such that $|a| = 6$, $b^2 = a^3$, and $ba = a^{-1}b$ and that no two of these three nonabelian groups are isomorphic (Exercise 14).

THEOREM 8.35   *If G is a group of order 12, then G is isomorphic to one of the following groups: $\mathbb{Z}_{12}$, $\mathbb{Z}_2 \times \mathbb{Z}_2 \times \mathbb{Z}_3$, the alternating group A_4, the dihedral group D_6, or the group T described in the preceding paragraph.*

Proof   An argument similar to the proof of Theorem 8.34 can be used to prove the theorem. See Theorem II.6.4 in Hungerford [7].   ◆

The preceding results provide a complete classification of all groups of orders ≤ 15, that is, a list of groups such that every group of order ≤ 15 is isomorphic to exactly one group on the list.

ORDER	GROUPS	REFERENCE
2	$\mathbb{Z}_2$	Theorem 7.28
3	$\mathbb{Z}_3$	Theorem 7.28
4	$\mathbb{Z}_4, \mathbb{Z}_2 \times \mathbb{Z}_2$	Theorem 7.29
5	$\mathbb{Z}_5$	Theorem 7.28
6	$\mathbb{Z}_6, S_3$	Theorem 7.30
7	$\mathbb{Z}_7$	Theorem 7.28
8	$\mathbb{Z}_8, \mathbb{Z}_4 \times \mathbb{Z}_2, \mathbb{Z}_2 \times \mathbb{Z}_2 \times \mathbb{Z}_2, D_4, Q$	Theorem 8.34
9	$\mathbb{Z}_9, \mathbb{Z}_3 \times \mathbb{Z}_3$	Corollary 8.29
10	$\mathbb{Z}_{10}, D_5$	Theorem 8.33
11	$\mathbb{Z}_{11}$	Theorem 7.28
12	$\mathbb{Z}_{12}, \mathbb{Z}_2 \times \mathbb{Z}_2 \times \mathbb{Z}_3, A_4, D_6, T$	Theorem 8.35
13	$\mathbb{Z}_{13}$	Theorem 7.28
14	$\mathbb{Z}_{14}, D_7$	Theorem 8.33
15	$\mathbb{Z}_{15}$	Corollary 8.18

This list could be continued to order 100 and beyond. For more than half of the orders between 2 and 100, the techniques presented above provide a complete classification of groups of that order. For other orders, however, a great deal of additional work would be necessary. For instance, there are 14 different groups of order 16 and 267 of order 64. There is no known formula giving the number of distinct groups of order n.

◆ EXERCISES

A.   **1.**   If p and q are primes with $p < q$ and $q \not\equiv 1 \pmod{p}$ and G is a group of order p^2q, prove that G is abelian.

2. Prove that there is no simple group of order 12. [*Hint:* Show that one of the Sylow subgroups must be normal.]

3. Prove that the dihedral group D_6 is isomorphic to $S_3 \times \mathbb{Z}_2$.

4. **(a)** In the proof of Theorem 8.34, complete the operation table for the group G in the case when $b^2 = e$.

   (b) Show that $G \cong D_4$ under the correspondence

   $$a^i \longrightarrow r_i, \qquad b \longrightarrow d, \qquad ab \longrightarrow h, \qquad a^2b \longrightarrow t, \qquad a^3b \longrightarrow v$$

   by comparing the table in part (a) with the table for D_4 on page 167.

5. **(a)** In the proof of Theorem 8.34, complete the operation table for the group G in the case when $b^2 = a^2$.

   (b) Show that $G \cong Q$ under the correspondence

   $$a^r b^s \longrightarrow i^r j^s \qquad (0 \leq r \leq 3, 0 \leq s \leq 1)$$

   by comparing the table in part (a) with the table for Q (see Exercise 14 in Section 7.1).

6. The theorems of Chapters 7 and 8 (without any additional argument) are sufficient to classify groups of many orders. For instance, groups of orders 25 and 49 are classified by Corollary 8.29 and groups of order 99 by Theorem 8.30. List all such orders from 2 to 100.

B. 7. If G is a group such that every one of its Sylow subgroups (for every prime p) is cyclic and normal, prove that G is a cyclic group.

8. Let $n \geq 3$ be a positive integer and let G be the set of all matrices of the forms

   $$\begin{pmatrix} 1 & a \\ 0 & 1 \end{pmatrix} \quad \text{or} \quad \begin{pmatrix} -1 & a \\ 0 & 1 \end{pmatrix} \quad \text{with } a \in \mathbb{Z}_n.$$

   (a) Prove that G is a group of order $2n$ under matrix multiplication.

   (b) Prove that G is isomorphic to D_n.

9. Complete the proof of Theorem 8.33 by showing that when $bab = a^{-1}$, the map $f: D_p \to G$ given by $f(r^i d^j) = a^i b^j$ is a homomorphism. [*Hint:* $bab = a^{-1}$ is equivalent to $ba = a^{-1}b$. Use this fact and Theorem 8.32 to compute products in G and D_p.]

10. **(a)** If $n = 2k$, show that r^k is in the center of D_n.

    (b) If n is even, show that $Z(D_n) = \{e, r^k\}$.

    (c) If n is odd, show that $Z(D_n) = \{e\}$.

11. What is the center of the quaternion group Q?

12. Show that every subgroup of the quaternion group Q is normal.

13. If G is a group of order 8 generated by elements a and b such that $|a| = 4$, $b \notin \langle a \rangle$, and $b^2 = a^3$, then G is abelian. [This fact is used in the proof of Theorem 8.34, so don't use Theorem 8.34 to prove it.]

14. Let G be the group $S_3 \times \mathbb{Z}_4$ and let $a = ((123), 2)$ and $b = ((12), 1)$.

   (a) Show that $|a| = 6$, $b^2 = a^3$, and $ba = a^{-1}b$.

   (b) Verify that the set $T = \{e = a^0, a^1, a^2, a^3, a^4, a^5, b, ab, a^2b, a^3b, a^4b, a^5b\}$ consists of 12 distinct elements.

   (c) Show that T is a nonabelian subgroup of G. [*Hint:* Use part (a) and Theorem 7.11.]

   (d) Show that T is not isomorphic to D_6 or to A_4.

15. Let n be a composite positive integer and p a prime such that p divides n and 1 is the only divisor of n that is congruent to 1 modulo p. If G is a group of order n, prove that G is not simple.

16. If G is a simple group that has a subgroup K of index n, prove that $|G|$ divides $n!$. [*Hint:* Let T be the set of distinct right cosets of K and consider the homomorphism $\varphi : G \to A(T)$ of Exercise 25 in Section 7.8. Show that φ is injective and note that $A(T) \cong S_n$ (Why?).]

C. 17. Classify all groups of order 21 up to isomorphism.

18. Classify all groups of order 66 up to isomorphism.

19. Prove that there is no simple nonabelian group of order less than 60. [*Hint:* Exercise 16 may be helpful.]

CHAPTER **9**

Arithmetic in Integral Domains

$\blacklozenge$

In Chapters 1 and 4 we saw that the ring $\mathbb{Z}$ of integers and the ring $F[x]$ of polynomials over a field F have very similar structures: both have division algorithms, greatest common divisors, and unique factorization into primes (irreducibles). In this chapter we find conditions under which these properties carry over to arbitrary integral domains, with particular emphasis on unique factorization.

Unique factorization turns out to be closely related to the ideals of a domain. On the one hand, unique factorization is not possible unless the principal ideals of the domain satisfy certain conditions (Section 9.2). On the other hand, ideals can be used to restore a kind of unique factorization to some domains that lack it. Indeed, ideals were originally invented just for this purpose, as we shall see in Section 9.3.

Section 9.4 (The Field of Quotients of an Integral Domain) is independent of the rest of the chapter and may be read at any point after Chapter 3. Sections 9.2 and 9.3 depend on Chapter 6, but the rest of the chapter may be read after Chapter 4.

The interdependence of the sections of this chapter is shown below. The dashed arrows indicate that Sections 9.2, 9.3, and 9.5 depend only on the first part of Section 9.1 (pages 285–287) and that Section 9.5 uses only three results in Section 9.2, all of which can be read independently of the rest of that section.

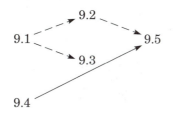

A shortened version of Sections 9.1 and 9.2 that contains all the basic information may be obtained by omitting the last parts of each of these sections (see the notes on pages 288 and 300).

9.1 EUCLIDEAN DOMAINS

In early chapters we analyzed the structure of $\mathbb{Z}$ and the polynomial ring $F[x]$ by using divisibility, units, associates, and primes (irreducibles). We begin by defining these concepts in the more general setting of an integral domain.*

> *Throughout this chapter, R is an integral domain.*

Let $a, b \in R$, with a nonzero. We say that **a divides b** (or **a is a factor of b**) and write $a \mid b$ if $b = ac$ for some $c \in R$. Recall that an element u in R is a **unit** provided that $uv = 1_R$ for some $v \in R$. Thus the units in R are precisely the divisors of 1_R.

> **EXAMPLE**   The only units in $\mathbb{Z}$ are 1 and -1. If F is a field, then the units in the polynomial ring $F[x]$ are the nonzero constant polynomials (Corollary 4.9).

> **EXAMPLE**   The set $\mathbb{Z}[\sqrt{2}] = \{r + s\sqrt{2} \mid r, s \in \mathbb{Z}\}$ is a subring of the real numbers (Exercise 1). The element $1 + \sqrt{2}$ is a unit in $\mathbb{Z}[\sqrt{2}]$ because
>
> $$(1 + \sqrt{2})(-1 + \sqrt{2}) = 1.$$

The ring in the preceding example is one of many similar rings that will frequently be used as examples later. If d is a fixed integer, then it is easy to verify that the set $\mathbb{Z}[\sqrt{d}] = \{r + s\sqrt{d} \mid r, s \in \mathbb{Z}\}$ is an integral domain that is contained in the complex numbers. If $d \geq 0$, then $\mathbb{Z}[\sqrt{d}]$ is a subring of the real numbers (Exercise 1). When $d = -1$, then the ring $\mathbb{Z}[\sqrt{-1}]$ is usually denoted $\mathbb{Z}[i]$ and is called the **ring of Gaussian integers.**

---

* The basic definitions apply in any commutative ring with identity. We restrict our attention to integral domains because most of the theorems fail in nondomains.

Remark   Let $u \in R$ be a unit with inverse v, so that $uv = 1_R$. For any $b \in R$ we have $u(vb) = (uv)b = 1_R b = b$. Therefore,

a unit divides every element of R

An element $a \in R$ is an **associate** of $b \in R$ provided $a = bu$ for some unit u. Now, u has an inverse, say $uv = 1_R$, and v is also a unit. Multiplying both sides of $a = bu$ by v shows that $av = buv = b1_R = b$. Use these facts to verify that

a is an associate of b if and only if b is an associate of a

and

a nonzero element of R is divisible by each of its associates.

> **EXAMPLE**   Every nonzero integer n has exactly two associates in $\mathbb{Z}$, n and $-n$. If F is a field, the associates of $f(x) \in F[x]$ are the nonzero constant multiples of $f(x)$. In the ring $\mathbb{Z}[\sqrt{2}]$, the elements $\sqrt{2}$ and $2 - \sqrt{2}$ are associates because $\sqrt{2} = (2 - \sqrt{2})(1 + \sqrt{2})$ and $1 + \sqrt{2}$ is a unit.

A nonzero element $p \in R$ is said to be **irreducible** provided that p is not a unit and the only divisors of p are its associates and the units of R.

> **EXAMPLE**   The irreducible elements in $\mathbb{Z}$ are just the prime integers because the only divisors of a prime p are $\pm p$ (its associates) and ± 1 (the units in $\mathbb{Z}$). The definition of irreducible given above is identical to the definition of an irreducible polynomial in the integral domain $F[x]$, when F is a field (page 95). In Section 9.3 we shall see that $1 + i$ is irreducible in the ring $\mathbb{Z}[i]$.

The next theorem is usually the easiest way to prove that an element is irreducible and is sometimes used as a definition. Theorem 4.11 is the special case when $R = F[x]$.

THEOREM 9.1   *Let p be a nonzero, nonunit element in an integral domain R. Then p is irreducible if and only if*

$$\text{whenever } p = rs, \text{ then } r \text{ or } s \text{ is a unit.}$$

Proof   If p is irreducible and $p = rs$, then r is a divisor of p. So r must be either a unit or an associate of p. If r is a unit, there is nothing to prove. If r is an associate of p, say $r = pv$, then $p = rs = pvs$. Cancelling p on the two ends (Theorem 3.10) shows that $1_R = vs$. Therefore, s is a unit.

To prove the converse, suppose p has the stated property. Let c be any divisor of p, say $p = cd$. Then by hypothesis either c or d is a unit. If d is a unit, then so is d^{-1}. Multiplying both sides of $p = cd$ by d^{-1} shows that $c = d^{-1}p$.

Thus in every case c is either a unit or an associate of p. Therefore, p is irreducible. ◆

Euclidean Domains

The Division Algorithm was a key tool in analyzing the arithmetic of both $\mathbb{Z}$ and $F[x]$. So we now look at domains that have some kind of analogue of the Division Algorithm. To see how to describe such an analogue, note that the degree of a polynomial in $F[x]$ can be thought of as defining a function from the nonzero polynomials in $F[x]$ to the nonnegative integers. By identifying the key properties of this function we obtain this

> **DEFINITION** • *An integral domain R is a **Euclidean domain** if there is a function δ from the nonzero elements of R to the nonnegative integers with these properties:*
>
> *(i) If a and b are nonzero elements of R, then $\delta(a) \leq \delta(ab)$.*
>
> *(ii) If $a, b \in R$ and $b \neq 0_R$, then there exist $q, r \in R$ such that $a = bq + r$ and either $r = 0_R$ or $\delta(r) < \delta(b)$.*

EXAMPLE If F is a field, then the polynomial domain $F[x]$ is a Euclidean domain with the function δ given by $\delta(f(x)) = $ degree of $f(x)$. Property (i) follows from Theorem 4.2 because

$$\delta(f(x)g(x)) = \deg f(x)g(x) = \deg f(x) + \deg g(x)$$
$$\geq \deg f(x) = \delta(f(x)),$$

and property (ii) is just the Division Algorithm (Theorem 4.4).

EXAMPLE $\mathbb{Z}$ is a Euclidean domain with the function δ given by $\delta(a) = |a|$. Property (i) holds because $|ab| = |a||b| \geq |b|$ for all $a \neq 0$. The Division Algorithm, as stated in Corollary 1.2, shows that property (ii) holds.

EXAMPLE We shall prove that the ring of Gaussian integers $\mathbb{Z}[i] = \{s + ti \mid s, t \in \mathbb{Z}\}$ is a Euclidean domain with the function δ given by $\delta(s + ti) = s^2 + t^2$. Since $s + ti = 0$ if and only if both s and t are 0, we see that $\delta(s + ti) \geq 1$ when $s + ti \neq 0$. Verify that for any $a = s + ti$ and $b = u + vi$ in $\mathbb{Z}[i]$, $\delta(ab) = \delta(a)\delta(b)$ (Exercise 15). Then when $b \neq 0$ we have

$$\delta(a) = \delta(a) \cdot 1 \leq \delta(a)\delta(b) = \delta(ab),$$

so that property (i) holds. If $b \neq 0$, verify that a/b is a complex number that can be written in the form $c + di$, where $c, d \in \mathbb{Q}$ (Exercise 9). Since $c \in \mathbb{Q}$, it lies between two consecutive integers; and similarly for d. Hence, there are integers m and n such that $|m - c| \leq 1/2$ and $|n - d| \leq 1/2$. Since $a/b = c + di$.

$$
\begin{aligned}
a &= b[c + di] = b[(c - m + m) + (d - n + n)i] \\
&= b[(m + ni) + ((c - m) + (d - n)i)] \\
&= b[m + ni] + b[(c - m) + (d - n)i] \\
&= bq + r,
\end{aligned}
$$

where $q = m + ni \in \mathbb{Z}[i]$ and $r = b[(c - m) + (d - n)i]$. Since $r = a - bq$ and $a, b, q \in \mathbb{Z}[i]$, we see that $r \in \mathbb{Z}[i]$. Property (ii) holds because

$$
\begin{aligned}
\delta(r) &= \delta(b)\delta[(c - m) + (d - n)i] = \delta(b)[(c - m)^2 + (d - n)^2] \\
&\leq \delta(b)[(1/2)^2 + (1/2)^2] = (1/2) \cdot \delta(b) < \delta(b).
\end{aligned}
$$

NOTE: *The remainder of this section is optional. The development here is elementary and assumes only the basic facts about rings in Section 3.1. A more sophisticated approach is presented in Section 9.2, where ideals are used to develop the key facts about a wider class of domains that includes Euclidean domains as a special case. Thus this section develops some remarkably strong results with a minimum of mathematical tools, whereas Section 9.2 obtains the same results more efficiently in a wider setting.*

It is possible that a given integral domain may be made into a Euclidean domain in more than one way by defining the function δ differently (see Exercises 10–11). Whenever the Euclidean domains in the preceding examples are mentioned, however, you may assume that the function δ is the one defined above.

In $F[x]$, the units are the polynomials of degree 0 (Corollary 4.9), that is, the polynomials that have the same degree as the identity polynomial 1_F. Furthermore, if k is a constant (unit in $F[x]$), then $f(x)$ and $kf(x)$ have the same degree. Analogous facts hold in any Euclidean domain.

THEOREM 9.2 *Let R be a Euclidean domain and u a nonzero element of R. Then the following conditions are equivalent:*

(1) u *is a unit.*

(2) $\delta(u) = \delta(1_R)$.

(3) $\delta(c) = \delta(uc)$ *for some nonzero $c \in R$.*

Proof (1) $\Rightarrow$ (2) Exercise 13.

(2) $\Rightarrow$ (3) Statement (3) holds with $c = 1_R$ because $\delta(1_R) = \delta(u) = \delta(u \cdot 1_R)$.

(3) $\Rightarrow$ (1) According to (ii) in the definition of a Euclidean domain (with c and uc in place of a and b), there exist $q, r \in R$ such that

$$c = (uc)q + r \qquad \text{and either} \qquad r = 0_R \qquad \text{or} \qquad \delta(r) < \delta(uc).$$

If $\delta(r) < \delta(uc)$, then by part (i) of the definition and statement (2)

$$\delta(c) \leq \delta(c(1_R - uq)) = \delta(c - ucq) = \delta(r) < \delta(uc) = \delta(c),$$

so that $\delta(c) < \delta(c)$, a contradiction. Hence, we must have $r = 0_R$. Thus $c = (uc)q$, which implies that $1_R = uq$. Therefore, u is a unit. ◆

In the remainder of this section we shall develop the basic facts about greatest common divisors, irreducibles, and unique factorization in Euclidean domains. The development here parallels the ones given in Chapter 1 for $\mathbb{Z}$ and in Chapter 4 for $F[x]$ and most of the arguments are the same ones used there, with appropriate modifications. Alternatively, the major results in Sections 1.2–1.3 and 4.2–4.3 may be considered as special cases of the theorems proved here.

Greatest Common Divisors

The integers are ordered by $\leq$ and polynomials in $F[x]$ are partially ordered by their degrees. This made it natural to define greatest common divisors in these domains in terms of size or degree. The same idea carries over to Euclidean domains, where "size" is measured by the function δ.

> **DEFINITION•** *Let R be a Euclidean domain and $a, b \in R$ (not both zero). A **greatest common divisor** of a and b is an element d such that*
>
> *(i) $d \mid a$ and $d \mid b$;*
>
> *(ii) if $c \mid a$ and $c \mid b$, then $\delta(c) \leq \delta(d)$.*

Any two elements of a Euclidean domain R have at least one common divisor, namely 1_R. If $c \mid a$, say $a = ct$, then $\delta(c) \leq \delta(ct) = \delta(a)$. Consequently, every common divisor c of a and b satisfies $\delta(c) \leq \max\{\delta(a), \delta(b)\}$, which implies that there is a common divisor of largest possible δ value. In other words, greatest common divisors always exist.

When gcd's were defined in $\mathbb{Z}$ and $F[x]$, an extra condition was included in each case: The gcd of two integers is the *positive* common divisor of largest absolute value and the gcd of two polynomials is the *monic* common divisor of highest degree. These extra conditions guarantee that greatest common divisors in $\mathbb{Z}$ and $F[x]$ are unique. In arbitrary Euclidean domains there are no such extra conditions and greatest common divisors are not unique. Thus the preceding definition is consistent with, but not identical to, what was done in $\mathbb{Z}$ and $F[x]$.

EXAMPLE $\mathbb{Z}$ is a Euclidean domain with $\delta(a) = |a|$. Under the preceding definition, 2 is the gcd of 10 and 18 just as before. However, -2 also satisfies this definition because -2 divides both 10 and 18 and any common divisor of 10 and 18 has absolute value $\leq |-2|$. Note that the greatest common divisors 2 and -2 are associates in $\mathbb{Z}$.

THEOREM 9.3   *Let R be a Euclidean domain and a, b ϵ R (not both zero).*

(1) *If d is a greatest common divisor of a and b, then every associate of d is also a greatest common divisor of a and b.*

(2) *Any two greatest common divisors of a and b are associates.*

(3) *If d is a greatest common divisor of a and b, then there exit u, v ϵ R such that d = au + bv.*

Proof   The proof of statement (1) is Exercise 14. We begin by finding a particular greatest common divisor of a and b that will then be used to prove statements (2) and (3). Let

$$S = \{\delta(w) \mid 0_R \neq w \ \epsilon \ R \text{ and } w = as + bt \text{ for some } s, t \ \epsilon \ R\}.$$

Since at least one of $a = a1_R + b0_R$ and $b = a0_R + b1_R$ is nonzero by hypothesis, S is a nonempty set of nonnegative integers. By the Well-Ordering Axiom, S contains a smallest element, that is, there are elements d^*, u^*, v^* of R such that $d^* = au^* + bv^*$ and

(∗)   for every nonzero w of the form $as + bt$ (with $s, t \ \epsilon \ R$), $\delta(d^*) \leq \delta(w)$.

We claim that d^* is a greatest common divisor of a and b. To prove this we first show that $d^* \mid a$. By the definition of Euclidean domain, there are elements q, r such that $a = d^*q + r$ and either $r = 0_R$ or $\delta(r) < \delta(d^*)$. Note that

$$r = a - d^* q = a - (au^* + bv^*)q$$
$$= a - aqu^* - bv^*q = a(1_R - qu^*) + b(-v^*q).$$

Thus r is a linear combination of a and b, and, hence, we cannot have $\delta(r) < \delta(d^*)$ by (∗). Therefore, $r = 0_R$, so that $a = d^*q$ and $d^* \mid a$. A similar argument shows that $d^* \mid b$ and, hence, d^* is a common divisor of a and b.

Let c be any other common divisor of a and b. Then $a = cs$ and $b = ct$ for some $s, t \ \epsilon \ R$ and hence

(∗∗) $d^* = au^* + bv^* = (cs)u^* + (ct)v^* = c(su^* + tv^*).$

Thus by part (i) of the definition of Euclidean domain $\delta(c) \leq \delta(c(su^* + tv^*)) = \delta(d^*)$. Therefore, d^* is a greatest common divisor of a and b. Note that (∗∗) also shows that

(∗∗∗)   every common divisor c of a and b divides d^*.

This completes the preliminaries. We now prove the rest of the theorem.

(2) Let d be any greatest common divisor of a and b. Since d divides both a and b and d^* is a greatest common divisor, we must have $\delta(d) \leq \delta(d^*)$ by part (ii) of the definition. The same definition with the roles of d and d^* reversed shows that $\delta(d^*) \leq \delta(d)$. Hence, $\delta(d) = \delta(d^*)$. By (***) we know that $d \mid d^*$, say $d^* = dk$. Therefore, $\delta(d) = \delta(d^*) = \delta(dk)$. Hence, k is a unit by Theorem 9.2 and d is an associate of d^*. Since every gcd is an associate of d^*, any two of them must be associates of each other by Exercise 6.

(3) If d is a greatest common divisor of a and b, then as we saw in the previous paragraph $d^* = dk$, with k a unit. Since $d^* = au^* + bv^*$, we have

$$d = d^*k^{-1} = (au^* + bv^*)k^{-1} = a(u^*k^{-1}) + b(v^*k^{-1}).$$

Hence, $d = au + bv$, with $u = u^*k^{-1}$ and $v = v^*k^{-1}$. ◆

COROLLARY 9.4 *Let R be a Euclidean domain and a, $b \in R$ (not both zero). Then d is a greatest common divisor of a and b if and only if d satisfies these conditions:*

    (i) $d \mid a$ and $d \mid b$;

    (ii) if $c \mid a$ and $c \mid b$, then $c \mid d$.

Proof If d is a greatest common divisor of a and b, then d satisfies (i) by definition. Suppose c is a common divisor of a and b. Let d^* be as in (***) in the proof of Theorem 9.3. Then $c \mid d^*$, say $d^* = ct$. Furthermore, d^* is an associate of d by Theorem 9.3 so that $d^* = dk$, with k a unit. Hence, $d = d^*k^{-1} = (ct)k^{-1} = c(tk^{-1})$, so that $c \mid d$. Therefore, condition (ii) holds. The proof of the converse is Exercise 16. ◆

The Euclidean Algorithm (Theorem 1.6) provides the most efficient way of calculating the greatest common divisor of two integers. With minor modification its proof carries over to Euclidean domains and provides a constructive method of finding both greatest common divisors and the coefficients needed to write the gcd of a and b as a linear combination of a and b. See Exercise 29.

Unique Factorization

Elements a and b of a Euclidean domain are said to be **relatively prime** if one of their greatest common divisors is 1_R. In any domain the units are the associates of 1_R. Thus by Theorem 9.3, a and b are relatively prime if and only if one of their greatest common divisors is a unit.

THEOREM 9.5 *Let R be a Euclidean domain and a, b, $c \in R$. If $a \mid bc$ and a and b are relatively prime, then $a \mid c$.*

Proof Copy the proof of Theorem 1.5, using Theorem 9.3 in place of Theorem 1.3. ◆

COROLLARY 9.6 *Let p be an irreducible element in a Euclidean domain R.*

(1) *If $p \mid bc$, then $p \mid b$ or $p \mid c$.*

(2) *If $p \mid a_1 a_2 \cdots a_n$, then p divides at least one of the a_r.*

Proof (1) Let d be a greatest common divisor of p and b. Since d divides p, we know that d is either an associate of p or a unit. If d is an associate of p, then p is also a greatest common divisor of p and b by theorem 9.3; in particular, $p \mid b$. If d is a unit, then p and b are relatively prime and, hence, $p \mid c$ by Theorem 9.5.

(2) Copy the proof of Corollary 1.9, using (1) in place of Theorem 1.8. ◆

THEOREM 9.7 *Let R be a Euclidean domain. Every nonzero, nonunit element of R is the product of irreducible elements,* and this factorization is unique up to associates; that is, if*

$$p_1 p_2 \cdots p_r = q_1 q_2 \cdots q_s$$

with each p_i and q_j irreducible, then $r = s$ and, after reordering and relabeling if necessary,

$$p_i \text{ is an associate of } q_i \text{ for } i = 1, 2, \ldots, r.$$

Proof Let S be the set of all nonzero nonunit elements of R that are *not* the product of irreducibles. We shall show that S is empty, which proves that every nonzero nonunit element has at least one factorization as a product of irreducibles. Suppose, on the contrary, that S is nonempty. Then the set $\{\delta(s) \mid s \epsilon S\}$ is a nonempty set of nonnegative integers, which contains a smallest element by the Well-Ordering Axiom. That is, there exists $a \epsilon S$ such that

(∗) $\delta(a) \leq \delta(s)$ for every $s \epsilon S.$

Since $a \epsilon S$, a is not itself irreducible. By the definition of irreducibility, $a = bc$ with both b and c nonunits. Now $\delta(b) \leq \delta(bc)$ by the definition of Euclidean domain. If $\delta(b) = \delta(bc)$, then b would be a unit by Theorem 9.2, which is a contradiction. Hence, $\delta(b) < \delta(bc) = \delta(a)$, so that $b \notin S$ by (∗). A similar argument shows that $c \notin S$. By the definition of S, both b and c are the product of irreducibles and, hence, so is $a = bc$. This contradicts the fact that $a \epsilon S$. Therefore, S is empty, and every nonzero nonunit element of R is the product of irreducibles. To show that this factorization is unique up to associates, copy the proof of Theorem 4.13, replacing *constant* by *unit* and Corollary 4.12 by Corollary 9.6. ◆

---

* We allow the possibility of a product with just one factor in case the original element is itself irreducible.

◆ **EXERCISES**

NOTE: *Unless stated otherwise, R is an integral domain.*

A. **1.** Show that $\mathbb{Z}[\sqrt{d}\,]$ is a subring of $\mathbb{C}$. If $d \geq 0$, show that $\mathbb{Z}[\sqrt{d}\,]$ is a subring of $\mathbb{R}$.

2. Let $d \neq \pm 1$ be a square-free integer (that is, d has no integer divisors of the form c^2 except $(\pm 1)^2$). Prove that in $\mathbb{Z}[\sqrt{d}\,]$, $r + s\sqrt{d} = r_1 + s_1\sqrt{d}$ if and only if $r = r_1$ and $s = s_1$. Give an example to show that this result may be false if d is not square-free.

3. If the statement is true, prove it; if it is false, give a counterexample:

    (a) If $a \mid b$ and $c \mid d$ in R, then $ac \mid bd$.

    (b) If $a \mid b$ and $c \mid d$ in R, then $(a + c) \mid (b + d)$.

4. Prove that c and d are associates in R if and only if $c \mid d$ and $d \mid c$.

5. If $a = bc$ with $a \neq 0$ and b and c nonunits, show that a is not an associate of b.

6. Denote the statement "a is an associate of b" by $a \sim b$. Prove that $\sim$ is an equivalence relation; that is, for all $r, s, t \in R$: (i) $r \sim r$. (ii) If $r \sim s$, then $s \sim r$. (iii) If $r \sim s$ and $s \sim t$, then $r \sim t$.

7. **(a)** Prove that every associate of an irreducible element is irreducible.

    (b) If u and v are units, prove that u and v are associates.

8. Is $2x + 2$ irreducible in $\mathbb{Z}[x]$? Why not?

9. If $a = s + ti$ and $b = u + vi$ are in $\mathbb{Z}[i]$ and $b \neq 0$, show that $a/b = c + di$, where $c = \dfrac{su + tv}{u^2 + v^2}$ and $d = \dfrac{tu - sv}{u^2 + v^2}$.

10. **(a)** Show that $\mathbb{Z}$ is a Euclidean domain with the function δ given by $\delta(n) = n^2$.

    (b) Is $\mathbb{Q}$ a Euclidean domain when δ is defined by $\delta(r) = r^2$?

11. Let R be a Euclidean domain with function δ and let k be a positive integer.

    (a) Show that R is also a Euclidean domain under the function θ given by $\theta(r) = \delta(r) + k$.

    (b) Show that R is also a Euclidean domain under the function β given by $\beta(r) = k\delta(r)$.

12. Let F be a field. Prove that F is a Euclidean domain with the function δ given by $\delta(a) = 0$ for each nonzero $a \in F$.

13. Let R be a Euclidean domain and $u \in R$. Prove that u is a unit if and only if $\delta(u) = \delta(1_R)$.

14. If d is the greatest common divisor of a and b in a Euclidean domain, prove that every associate of d is also a greatest common divisor of a and b.

15. **(a)** If $a = s + ti$ and $b = u + vi$ are nonzero elements of $\mathbb{Z}[i]$, show that $\delta(ab) = \delta(a)\delta(b)$, where $\delta(r + si) = r^2 + s^2$.

    (b) If R is a Euclidean domain, is it true that $\delta(ab) = \delta(a)\delta(b)$ for all nonzero $a, b \in R$?

16. Complete the proof of Corollary 9.4 by showing that an element d satisfying conditions (i) and (ii) is a greatest common divisor of a and b.

17. Show that the elements q and r in the definition of a Euclidean domain are not necessarily unique. [*Hint:* In $\mathbb{Z}[i]$, let $a = -4 + i$ and $b = 5 + 3i$; consider $q = -1$ and $q = -1 + i$.]

B. 18. If any two nonzero elements of R are associates, prove that R is a field.

19. If every nonzero element of R is either irreducible or a unit, prove that R is a field.

20. **(a)** Show that $1 + i$ is not a unit in $\mathbb{Z}[i]$. [*Hint:* What is the inverse of $1 + i$ in $\mathbb{C}$?]

    (b) Show that 2 is not irreducible in $\mathbb{Z}[i]$.

21. Let p be a nonzero, nonunit element of R such that whenever $p \mid cd$, then $p \mid c$ or $p \mid d$. Prove that p is irreducible.

22. If $f : R \to S$ is a surjective homomorphism of integral domains, p is irreducible in R, and $f(p) \neq 0_S$, is $f(p)$ irreducible in S?

23. Let R be a Euclidean domain. Prove that

    (a) $\delta(1_R) \leq \delta(a)$ for all nonzero $a \in R$.

    (b) If a and b are associates, then $\delta(a) = \delta(b)$.

    (c) If $a \mid b$ and $\delta(a) = \delta(b)$, then a and b are associates.

24. Show that $\mathbb{Z}[\sqrt{-2}]$ is a Euclidean domain with $\delta(r + s\sqrt{-2}) = r^2 + 2s^2$.

25. Let $\omega = (-1 + \sqrt{-3})/2$ and $\mathbb{Z}[\omega] = \{r + s\omega \mid r, s \in \mathbb{Z}\}$. Prove that $\mathbb{Z}[\omega]$ is a Euclidean domain with $\delta(r + s\omega) = (r + s\omega)(r + s\omega^2) = r^2 - rs + s^2$. [*Hint:* Note that $\omega^3 = 1$ and $\omega^2 + \omega + 1 = 0$ (Why?).]

26. Prove or disprove: Let R be a Euclidean domain; then $I = \{a \in R \mid \delta(a) > \delta(1_R)\}$ is an ideal in R.

27. Let R be a Euclidean domain. If the function δ is a constant function, prove that R is a field.

28. **(a)** Prove that $1 - i$ is irreducible in $\mathbb{Z}[i]$. [*Hint:* If $a \mid (1 - i)$, then $1 - i = ab$; see Exercises 15(a) and 23.]

    (b) Write 2 as a product of irreducibles in $\mathbb{Z}[i]$. [*Hint:* Try $1 - i$ as a factor.]

C. 29. State and prove the Euclidean Algorithm for finding the gcd of two elements of a Euclidean domain.

30. Let R be a Euclidean domain such that $\delta(a + b) \le \max\{\delta(a), \delta(b)\}$ for all non-zero $a, b \in R$. Prove that q and r in the definition of Euclidean domain are unique.

9.2 PRINCIPAL IDEAL DOMAINS AND UNIQUE FACTORIZATION DOMAINS

A Euclidean domain is, in effect, a domain that has an analogue of the division algorithm. Consequently, all the proofs used for the integers and polynomial rings, most of which ultimately depended on the division algorithm, can be readily carried over to Euclidean domains. We now consider domains that may not have an analogue of the division algorithm but do have the other important arithmetic properties of $\mathbb{Z}$, such as unique factorization and greatest common divisors.

> **DEFINITION•** *A **principal ideal domain (PID)** is an integral domain in which every ideal is principal.*

The next theorem shows, for example, that $\mathbb{Z}$, $\mathbb{Q}[x]$, and $\mathbb{Z}[i]$ are all principal ideal domains because all of them are Euclidean domains (see pages 287–288). The example on page 137 shows that the polynomial ring $\mathbb{Z}[x]$ is not a PID.

THEOREM 9.8 *Every Euclidean domain is a principal ideal domain.*

Proof Suppose I is a nonzero ideal in a Euclidean domain R. Then the set $\{\delta(i) \mid i \in I\}$ is a nonempty set of nonnegative integers, which contains a smallest element by the Well-Ordering Axiom. That is, there exists $b \in I$ such that

$$(*) \qquad\qquad \delta(b) \le \delta(i) \qquad \text{for every} \qquad i \in I.$$

We claim that I is the principal ideal $(b) = \{rb \mid r \in R\}$. Since $b \in I$ and I is an ideal, $rb \in I$ for every $r \in R$; hence, $(b) \subseteq I$. Conversely, suppose $c \in I$. Then there exist $q, r \in R$ such that

$$c = bq + r \qquad \text{and} \qquad r = 0_R \qquad \text{or} \qquad \delta(r) < \delta(b).$$

Since $r = i - bq$ and both i and b are in I, we must have $r \in I$. Hence, it is impossible to have $\delta(r) < \delta(b)$ by $(*)$. Consequently, $r = 0_R$ and $c = bq + r = bq \in (b)$. Thus $I \subseteq (b)$ and, hence, $I = (b)$. Therefore, R is a PID. ◆

The converse of Theorem 9.8 is false: There are principal ideal domains that are not Euclidean domains (see Wilson and Williams [23]). Thus the class of Euclidean domains is strictly contained in the class of principal ideal domains.

In our development of the integers, polynomial rings, and Euclidean domains we first considered greatest common divisors and used them to prove unique factorization. Although this approach could also be used with principal ideal domains, it is just as easy to proceed directly to unique factorization.* We begin by developing the connection between divisibility and principal ideals in any integral domain.

LEMMA 9.9 *Let a and b be elements of an integral domain R. Then*

    (1) $a \subseteq (b)$ if and only if $b \mid a$.

    (2) $(a) = (b)$ if and only if $b \mid a$ and $a \mid b$.

    (3) $(a) \subsetneqq (b)$ if and only if $b \mid a$ and b is not an associate of a.

Proof (1) Note first that the principal ideal (b) consists of all multiples of b, that is, all elements divisible by b. Hence,

$$a \in (b) \qquad \text{if and only if} \qquad b \mid a.$$

Now if $(a) \subseteq (b)$, then a is in the ideal (b), so that $b \mid a$. Conversely, if $b \mid a$, then $a \in (b)$, which implies that every multiple of a is also in the ideal (b). Hence, $(a) \subseteq (b)$.

(2) $(a) = (b)$ if and only if $(a) \subseteq (b)$ and $(b) \subseteq (a)$. By (1), $(a) \subseteq (b)$ and $(b) \subseteq (a)$ if and only if $b \mid a$ and $a \mid b$.

(3) Exercise; use (1), (2), and Exercise 4 in Section 9.1, which shows that $a \mid b$ and $b \mid a$ if and only if b is an associate of a. ◆

To understand the origin of the next definition, it may help to recall the typical process for factoring an integer a_1 as a product of primes. Find a prime divisor p_1 of a_1 and factor: $a_1 = p_1 a_2$. Next find a prime divisor p_2 of a_2 and factor: $a_2 = p_2 a_3$, so that $a_1 = p_1 p_2 a_3$. Now find a prime divisor p_3 of a_3 and factor again: $a_3 = p_3 a_4$ and $a_1 = p_1 p_2 p_3 a_4$. Continue in this manner. Since a_1 has only a finite number of prime divisors, we must eventually have some a_k prime so that $a_k = p_k \cdot 1$ and $a_1 = p_1 p_2 p_3 \cdots p_k \cdot 1$. The only way to continue factoring (with positive factors and without changing the p's) is to use the fact that $1 = 1 \cdot 1$ repeatedly to write a_1 as

$$a_1 = p_1 p_2 p_3 \cdots p_k \cdot 1 \cdot 1 \cdot 1 \cdots 1.$$

Now look at the same procedure from the point of view of ideals. We have $a_2 \mid a_1$,

---

* Greatest common divisors are discussed at the end of this section; also see Exercises 20–22.

$a_3 \mid a_2, a_4 \mid a_3, \ldots, 1 \mid a_k, 1 \mid 1, 1 \mid 1$, and so on. Consequently, by Lemma 9.9 this factorization process leads to a chain of ideals

$$(a_1) \subseteq (a_2) \subseteq (a_3) \subseteq \cdots \subseteq (a_k) \subseteq (1) \subseteq (1) \subseteq (1) \subseteq \cdots$$

in which all the ideals are equal after some point. This suggests that factorization as a product of irreducibles is somehow related to chains of principal ideals in which all the ideals are equal after some point and motivates the following definition.

> **DEFINITION•** *An integral domain R satisfies the **ascending chain condition (ACC) on principal ideals** provided that whenever $(a_1) \subseteq (a_2) \subseteq (a_3) \subseteq \cdots$, then there exists a positive integer n such that $(a_i) = (a_n)$ for all $i \geq n$.*

Note that in this definition the identical ideals beginning with (a_n) may not be the ideal (1_R), unlike the preceding case. Nevertheless, the preceding discussion suggests the possibility that $\mathbb{Z}$ has the ACC on principal ideals. This is indeed the case as we now prove.

LEMMA 9.10 *Every principal ideal domain R satisfies the ascending chain condition on principal ideals.*

Proof If $(a_1) \subseteq (a_2) \subseteq \cdots$ is an ascending chain of ideals in R, let A be the set-theoretic union $\bigcup_{t \geq 1} (a_t)$. We claim that A is an ideal. Suppose $a, b \in A$; then $a \in (a_j)$ and $b \in (a_k)$ for some $j, k \geq 1$. Either $j \leq k$ or $k \leq j$, say $j \leq k$. Then $(a_j) \subseteq (a_k)$, so that $a, b \in (a_k)$. Since (a_k) is an ideal, we know that $a - b \in (a_k) \subseteq A$ and $ra \in (a_k) \subseteq A$ for any $r \in R$. Therefore, A is an ideal by Theorem 6.1. Since R is a PID, $A = (c)$ for some $c \in R$. Since $A = \bigcup_{t \geq 1} (a_t)$, we know that $c \in (a_n)$ for some n. Consequently, $(c) \subseteq (a_n)$ and for each $i \geq n$

$$(a_n) \subseteq (a_i) \subseteq \bigcup_{t \geq 1} (a_t) = A = (c) \subseteq (a_n).$$

Therefore, $(a_i) = (a_n)$ for each $i \geq n$. ◆

As we shall see, Lemma 9.10 is the key to showing that every nonzero nonunit element in a PID can be factored as a product of irreducibles. The fact that this factorization is essentially unique is a consequence of the next lemma.

LEMMA 9.11 *Let R be a principal ideal domain. If p is irreducible in R and $p \mid bc$, then $p \mid b$ or $p \mid c$.*

Proof*   If $p \mid bc$, then bc is in the ideal (p). If (p) were known to be a prime ideal, we could conclude that $b \in (p)$ or $c \in (p)$, that is, that $p \mid b$ or $p \mid c$. Since every maximal ideal is prime by Corollary 6.16, we need only show that (p) is a maximal ideal. Suppose I is any ideal with $(p) \subseteq I \subseteq R$. Since R is a PID, $I = (d)$ for some $d \in R$. Then $(p) \subseteq (d) = I$ implies that $d \mid p$. Since p is irreducible, d must be either a unit or an associate of p. If d is a unit, then $I = (d) = R$ by Exercise 13 of Section 6.1. If d is an associate of p, say $d = pu$, then $p \mid d$ and, hence, $(d) \subseteq (p)$. In this case, $(p) \subseteq (d) \subseteq (p)$, so that $(p) = (d) = I$. Therefore, (p) is maximal, and the proof is complete.   ◆

THEOREM 9.12   *Let R be a principal ideal domain. Every nonzero, nonunit element of R is the product of irreducible elements,*** and this factorization is unique up to associates; that is, if*

$$p_1 p_2 \cdots p_r = q_1 q_2 \cdots q_s$$

with each p_i and q_j irreducible, then $r = s$ and, after reordering and relabeling if necessary,

$$p_i \text{ is an associate of } q_i \text{ for } i = 1, 2, \ldots, r.$$

Proof   Let a be a nonzero, nonunit element in R. We must show that a has at least one factorization. Suppose, on the contrary, that a is *not* a product of irreducibles. Then a is not itself irreducible. So $a = a_1 b_1$ for some nonunits a_1 and b_1 (otherwise every factorization of a would include a unit and a would be irreducible by Theorem 9.1). If both a_1 and b_1 are products of irreducibles, then so is a. Thus at least one of them, say a_1, is not a product of irreducibles. Since b_1 is not a unit, a_1 is not an associate of a (Exercise 5 in Section 9.1). Consequently, $(a) \subsetneq (a_1)$ by part (3) of Lemma 9.9.

Now repeat the preceding argument with a_1 in place of a. This leads to a nonzero nonunit a_2 such that $(a_1) \subsetneq (a_2)$ and a_2 is not a product of irreducibles. Continuing this process indefinitely would lead to a strictly ascending chain of principal ideals $(a_1) \subsetneq (a_2) \subsetneq (a_3) \subsetneq \cdots$, contradicting Lemma 9.10. Therefore, a must have at least one factorization as a product of irreducibles.

Now we must show that this factorization is unique up to associates. To do this, adapt the proof of Theorem 4.13 (the case when $R = F[x]$) to the general situation by replacing the word *constant* by *unit* and using Lemma 9.11 and Exercise 2 in place of Corollary 4.12.   ◆

---

* For an alternate proof using greatest common divisors in place of Corollary 6.16, see Exercise 23.
** We allow the possibility of a product with just one factor in case the original element is itself irreducible.

To appreciate the importance of Theorem 9.12, it may be beneficial to examine a domain in which unique factorization fails.

> **EXAMPLE**   Let $\mathbb{Q}_Z[x]$ denote the set of polynomials with rational coefficients and integer constant terms. For instance, x, $\frac{1}{2}x$, and 2 are in $\mathbb{Q}_Z[x]$, but $x^2 + \frac{1}{2}$ and $\frac{1}{4}$ are not. Verify that $\mathbb{Q}_Z[x]$ is an integral domain and that the constant polynomial 2 is irreducible in $\mathbb{Q}_Z[x]$ (Exercise 16). The irreducible element 2 is a factor of $x \in \mathbb{Q}_Z[x]$ because $x = 2 \cdot (\frac{1}{2}x)$. Similarly, 2 is an irreducible factor of $\frac{1}{2}x$ because $\frac{1}{2}x = 2 \cdot (\frac{1}{4}x)$. Hence, $x = 2 \cdot 2 \cdot (\frac{1}{4}x)$. In fact, the process of factoring out irreducible 2's *never ends* because
>
> $$(*) \quad x = 2 \cdot \left(\frac{1}{2}x\right) = 2 \cdot 2 \cdot \left(\frac{1}{4}x\right) = 2 \cdot 2 \cdot 2 \cdot \left(\frac{1}{8}x\right) = \cdots$$
>
> $$= 2 \cdot 2 \cdots 2 \cdot \left(\frac{1}{2^n}x\right) = \cdots .$$
>
> In view of this, it should not be surprising that x cannot be factored as a product of irreducibles of $\mathbb{Q}_Z[x]$ (Exercise 17).
>
> Compare this situation with the prime factorization of a_1 in $\mathbb{Z}$ as described on pages 296–297. In $\mathbb{Z}$ the factorization becomes trivial after a finite number of steps (the only remaining factors are 1's), and all the ideals in the corresponding chain are equal after that point. In the factorization $(*)$ in $\mathbb{Q}_Z[x]$, however, things are different. The remaining factors each time a 2 is factored from x are the elements
>
> $$x, \frac{1}{2}x, \frac{1}{4}x, \frac{1}{8}x \ \ldots , \frac{1}{2^n}x \ \ldots .$$
>
> No two of these elements are associates (Exercise 3) and each element is 2 times the following one, that is, each element is divisible by the following one. Therefore, by part (3) of Lemma 9.9
>
> $$(x) \subsetneq \left(\frac{1}{2}x\right) \subsetneq \left(\frac{1}{4}x\right) \subsetneq \left(\frac{1}{8}x\right) \subsetneq \cdots .$$
>
> Hence, the ACC for principal ideals does not hold in $\mathbb{Q}_Z[x]$.

Unique Factorization Domains

In our study of Euclidean domains and principal ideal domains, the main result was that unique factorization held. Now we reverse the process and consider domains in which unique factorization always holds to see what other properties from ordinary arithmetic they may have.

D E F I N I T I O N • *An integral domain R is a **unique factorization domain (UFD)** provided that every nonzero, nonunit element of R is the product of irreducible elements,* and this factorization is unique up to associates; that is, if*

$$p_1 p_2 \cdots p_r = q_1 q_2 \cdots q_s$$

with each p_i and q_j irreducible, then $r = s$ and, after reordering and relabeling if necessary,

$$p_i \text{ is an associate of } q_i \text{ for } i = 1, 2, \ldots, r.$$

EXAMPLE   Theorem 9.12 shows that every PID is a unique factorization domain. In particular, the ring $\mathbb{Z}[i]$ of Gaussian integers is a UFD.

EXAMPLE   As noted in the example after Theorem 9.12, $\mathbb{Q}_{\mathbb{Z}}[x]$ is not a unique factorization domain because the element x has no factorization as a product of a finite number of irreducibles. In Section 9.3 we shall see that $\mathbb{Z}[\sqrt{-5}]$ fails to be a UFD for a different reason: Every element is a product of irreducibles, but this factorization is not unique.

EXAMPLE   A proof that the polynomial ring $\mathbb{Z}[x]$ is a UFD is given in Section 9.5. Since $\mathbb{Z}[x]$ is not a principal ideal domain (see page 137), we see that the class of all unique factorization domains is strictly larger than the class of all principal ideal domains.

NOTE: *The remainder of this section is optional and is not needed for the sequel.*

When working with two integers, you can always arrange things so that the same primes appear in the factorizations of both elements. For instance, consider the prime factorizations $-18 = 2 \cdot 3 \cdot (-3)$ and $40 = 2 \cdot (-2) \cdot (-2) \cdot 5$. The list of all primes that appear in both factorizations is 2, 3, -3, 2, -2, -2, 5, but several of these primes are associates of each other. By eliminating any prime on the list that is an associate of an earlier number on the list we obtain the list 2, 3, 5 in which no two numbers are associates. We can write both 18 and 40 as products of these three primes and the units ± 1:

$$-18 = 2 \cdot 3 \cdot (-3) = (-1) \cdot 2 \cdot 3 \cdot 3 = (-1) \cdot 2^0 \cdot 3^2 \cdot 5^0$$
$$40 = 2 \cdot (-2) \cdot (-2) \cdot 5 = (-1)(-1) \cdot 2 \cdot 2 \cdot 2 \cdot 5 = (1) \cdot 2^3 \cdot 3^0 \cdot 5^1$$

Essentially the same procedure works in any UFD.

---

* We allow the possibility of a product with just one factor in case the original element is itself irreducible.

THEOREM 9.13 *If c and d are nonzero elements in a unique factorization domain R, then there exist units u and v and irreducibles $p_1, p_2, \ldots, p_k$, no two of which are associates, such that*

$$c = up_1^{m_1}p_2^{m_2}\cdots p_k^{m_k} \quad and \quad d = vp_1^{n_1}p_2^{n_2}\cdots p_k^{n_k},$$

where each m_i and n_i is a nonnegative integer. Furthermore,

$$c \mid d \quad if\ and\ only\ if \quad m_i \leq n_i \quad for\ each \quad i = 1, 2, \cdots, k.$$

In the example preceding the theorem, with $c = -18$ and $d = 40$, we had $u = -1, v = 1, p_1 = 2, p_2 = 3$, and $p_3 = 5$.

Proof of Theorem 9.13 Since R is a UFD, both c and d have can be factored, say $c = q_1q_2\cdots q_s$ and $d = r_1r_2\cdots r_t$ with each q_i and r_j irreducible. In the list q_1, $q_2, \ldots, q_s, r_1, r_2, \ldots, r_t$ delete any element that has an associate appearing earlier on the list and denote the remaining elements by $p_1, p_2, \ldots, p_k$. Then each p_i is irreducible, no two of them are associates of each other, and each one of the q's and r's is an associate of some p_i. Consequently, in the factorization $c = q_1q_2\cdots q_s$ each q_j is of the form wp_i with w a unit. By rearranging terms, c can be written (product of units) (product of p's). The product of these units is itself a unit, call it u. By rearranging the p's in this product and inserting other p's with zero exponents if necessary, we can write $c = up_1^{m_1}p_2^{m_2}\cdots p_k^{m_k}$, with each $m_i \geq 0$. A similar procedure works for d and proves the first part of the theorem.

To prove first half of the last statement, suppose $c \mid d$. Then $d = cb$ for some $b \in R$. Since the irreducible p_i appears exactly n_i times in the factorization of d, it must also appear exactly n_i times in the factorization of cb. But p_i already appears m_i in the factorization of c and may possibly appear in the factorization of b, so we must have $m_i \leq n_i$. Conversely, suppose that $m_i \leq n_i$ for every i. Verify that $d = ca$, where

$$a = (u^{-1}v)(p_1^{n_1-m_1}p_2^{n_2-m_2}\cdots p_k^{n_k-m_k}).$$

Therefore, $c \mid d$. ◆

COROLLARY 9.14 *Every unique factorization domain satisfies the ascending chain condition on principal ideals.*

Proof First, suppose (c) and (d) are principal ideals in a UFD R such that $(d) \subsetneq (c)$. Then $c \mid d$ and c is not an associate of d by Lemma 9.9. If c and d are written in the form given by Theorem 9.13, then each $m_i \leq n_i$. If $m_i = n_i$ for every i, then $c = uv^{-1}d$, which means that c is an associate of d, a contradiction. Hence, there must be some index j for which $m_j < n_j$.

Suppose $(a_1) \subseteq (a_2) \subseteq (a_3) \subseteq \cdots$ is a chain of principal ideals in R. Lemma 9.9 shows that each a_i divides a_1. By Theorem 9.13 we may assume that $a_1 = vp_1^{n_1}p_2^{n_2} \cdots p_k^{n_k}$ and that each a_i is of the form $a_i = up_1^{m_1}p_2^{m_2} \cdots p_k^{m_k}$, where the p_j are nonassociate irreducibles. If there are just a finite number of strict inclusions ($\subsetneq$) in the chain of ideals, then there are only equalities after a certain point and the ACC holds. There cannot be an infinite number of strict inclusions because the first paragraph shows that each time a strict inclusion occurs, one of the exponents on one of the p's must decrease. Consequently, after a finite number of strict inclusions, there would be an a_n of the form $a_n = up_1^0 \cdots p_k^0 = u$. Thus a_n is a unit, which implies that $(a_n) = R$ by Exercise 13 of Section 6.1. For each $i \geq n$ we have $(a_n) \subseteq (a_i) \subseteq R = (a_n)$, so that $(a_n) = (a_i)$. Therefore, R satisfies the ACC on principal ideals. ◆

Irreducibles in a unique factorization domain have a property that we have used frequently in the special cases of Euclidean domains and principal ideal domains.

THEOREM 9.15 *Let p be an irreducible element in a unique factorization domain R. If $p \mid bc$, then $p \mid b$ or $p \mid c$.*

Proof If b or c is 0_R, then there is nothing to prove because $p \mid 0_R$. If c is a unit and $p \mid bc$, then $pt = bc$ for some $t \in R$ and $ptc^{-1} = b$. Hence, $p \mid b$; similarly, if b is a unit, then $p \mid c$. If both b and c are nonzero nonunits, then $b = q_1 \cdots q_k$ and $c = q_{k+1} \cdots q_s$ with the q_i (not necessarily distinct) irreducibles. Since $p \mid bc$, we have $pr = bc = q_1 \cdots q_s$ for some $r \in R$. The irreducible p must be an associate of some q_i by unique factorization. Therefore, p divides q_i and, hence, divides b or c. ◆

We are now in a position to characterize unique factorization domains.

THEOREM 9.16 *An integral domain R is a unique factorization domain if and only if*

(1) R has the ascending chain condition on principal ideals; and

(2) whenever p is irreducible in R and $p \mid cd$, then $p \mid c$ or $p \mid d$.

As the proof of the theorem shows, condition (1) corresponds to the existence of an irreducible factorization for each nonzero nonunit element and condition (2), to the uniqueness of this factorization. The two conditions are independent: (1) fails and (2) holds in $\mathbb{Q}_\mathbb{Z}[x]$ (see page 299 and Exercise 33), whereas (1) holds and (2) fails in $\mathbb{Z}[\sqrt{-5}]$ (see page 312 and Exercise 21 in Section 9.3).

Proof of Theorem 9.16   If R is a UFD, then R satisfies (1) and (2) by Corollary 9.14 and Theorem 9.15. Conversely, assume R satisfies (1) and (2) and let a be a nonzero nonunit element of R. The argument used in the proof of Theorem 9.12, which depends only on the ACC, is valid here and shows that a can be factored as a product of irreducibles. To show that this factorization is unique, adapt the proof of Theorem 4.13 (the case when $R = F[x]$) to the general situation by replacing the word *constant* by *unit* and using (2) and Exercise 2 in place of Corollary 4.12.   ◆

Greatest Common Divisors

Greatest common divisors were a useful tool in our study of $\mathbb{Z}$, $F[x]$, and other Euclidean domains. In each case the gcd of two elements was defined to be a common divisor of "largest size," where size was measured by absolute value in $\mathbb{Z}$, by polynomial degrees in $F[x]$, and by the function δ in an arbitrary Euclidean domain. Unfortunately, there may be no similar way to measure "size" in an arbitrary integral domain, so greatest common divisors must be defined in terms of divisibility properties alone:

> **DEFINITION•**   *Let $a_1, a_2, \ldots , a_n$ be elements (not all zero) of an integral domain R. A **greatest common divisor** of a_1, $a_2 \ldots , a_n$ is an element d of R such that*
>
> *(i) d divides each of the a_i;*
> *(ii) if $c \in R$ and c divides each of the a_i, then $c \mid d$.*

Corollaries 1.4, 4.6, and 9.4 show that this definition is equivalent to the definitions used previously in $\mathbb{Z}$, $F[x]$, and other Euclidean domains. The only difference is that greatest common divisors in $\mathbb{Z}$ and $F[x]$, are no longer unique (see the discussion on page 289).

THEOREM 9.17   *Let d be a greatest common divisor of $a_1, a_2, \ldots , a_n$ in an integral domain R. Then*

(1) Every associate of d is also a gcd of $a_1, \ldots , a_n$.
(2) Any two greatest common divisors of $a_1, \ldots , a_n$ are associates.

Proof   (1) Exercise 7.
   (2) Suppose both d and t are gcd's of $a_1, \ldots , a_n$. Then t divides each a_i, and, therefore, $t \mid d$ by (ii) in the definition of the greatest common divisor d. But d also divides each a_i, and, hence, $d \mid t$ by (ii) in the definition of the gcd t. Since $t \mid d$ and $d \mid t$, we know that d and t are associates by Exercise 4 of Section 9.1.   ◆

Warning   In some integral domains a finite set of elements may not have a greatest common divisor (see Exercise 13 in Section 9.3).

THEOREM 9.18 *Let $a_1, a_2, \ldots, a_n$ (not all zero) be elements in a unique factorization domain R. Then $a_1, \ldots, a_n$ have a greatest common divisor in R.*

Proof The gcd of any set of elements is the gcd of the nonzero members of the set, so we may assume that each a_i is nonzero. By Theorem 9.13 there are irreducibles $p_1, \ldots, p_t$ (no two of which are associates), units $u_1, \ldots, u_n$, and nonnegative integers m_{ij} such that

$$a_1 = u_1 p_1^{m_{11}} p_2^{m_{12}} p_3^{m_{13}} \cdots p_t^{m_{1t}}$$
$$a_2 = u_2 p_1^{m_{21}} p_2^{m_{22}} p_3^{m_{23}} \cdots p_t^{m_{2t}}$$
$$\vdots$$
$$a_n = u_n p_1^{m_{n1}} p_2^{m_{n2}} p_3^{m_{n3}} \cdots p_t^{m_{nt}}.$$

Let k_1 be the smallest exponent that appears on p_1; that is, k_1 is the minimum of $m_{11}, m_{21}, m_{31}, \ldots, m_{n1}$. Similarly, let k_2 be the smallest exponent that appears on p_2, and so on. Use Theorem 9.13 to verify that $d = p_1^{k_1} p_2^{k_2} \cdots p_t^{k_t}$ is a gcd of $a_1, \ldots, a_n$. ◆

In an arbitrary unique factorization domain, it may not be possible to write the gcd of elements a and b as a linear combination of a and b as it was in $\mathbb{Z}$ and $F[x]$. In Section 9.5, for example, we shall see that 1 is a gcd of the polynomials x and 2 in the UFD $\mathbb{Z}[x]$, but 1 is not a linear combination of x and 2 in $\mathbb{Z}[x]$ (Exercise 6). In a principal ideal domain, however, the gcd of a and b can always be written as a linear combination of a and b (Exercise 20).

◆ EXERCISES

A. **1.** If a, b are nonzero elements of an integral domain and a is a nonunit, prove that $(ab) \subsetneq (b)$.

2. Suppose p is an irreducible element in an integral domain R such that whenever $p \mid bc$, then $p \mid b$ or $p \mid c$. If $p \mid a_1 a_2 \cdots a_n$, prove that p divides at least one a_i.

3. (a) Prove that the only units in $\mathbb{Q}_{\mathbb{Z}}[x]$ are 1 and -1. [*Hint:* Theorem 4.2.]

(b) If $f(x) \in \mathbb{Q}_{\mathbb{Z}}[x]$, show that its only associates are $f(x)$ and $-f(x)$.

4. Is a field a UFD?

5. Give an example to show that a subdomain of a unique factorization domain need not be a UFD.

6. Prove that 1 is not a linear combination of the polynomials 2 and x in $\mathbb{Z}[x]$, that is, prove it is impossible to find $f(x), g(x) \in \mathbb{Z}[x]$ such that $2f(x) + xg(x) = 1$.

7. Let d be a gcd of $a_1, \ldots, a_k$ in an integral domain. Prove that every associate of d is also a gcd of $a_1, \ldots, a_k$.

8. Let p be an irreducible element in an integral domain. Prove that 1_R is a gcd of p and a if and only if $p \nmid a$.

B. 9. Let R be a PID. If (c) is a nonzero ideal in R, then show that there are only finitely many ideals in R that contain (c). [*Hint:* Consider the divisors of c.]

10. Prove that an ideal (p) in a PID is maximal if and only if p is irreducible.

11. Prove that every ideal in a principal ideal domain R (except R itself) is contained in a maximal ideal. [*Hint:* Exercise 10.]

12. Prove that an ideal in a PID is prime if and only if it is maximal. [*Hint:* Exercise 10.]

13. Let $f: R \to S$ be a surjective homomorphism of rings with identity.

    (a) If R is a PID, prove that every ideal in S is principal.

    (b) Show by example that S need not be an integral domain.

14. Let p be a fixed prime integer and let R be the set of all rational numbers that can be written in the form a/b with b not divisible by p. Prove that

    (a) R is an integral domain containing $\mathbb{Z}$. [Note $n = n/1$].

    (b) If $a/b \in R$ and $p \nmid a$, then a/b is a unit in R.

    (c) If I is a nonzero ideal in R and $I \neq R$, then I contains p^t for some $t > 0$.

    (d) R is a PID. (If I is an ideal, show that $I = (p^k)$, where p^k is the smallest power of p in I.)

15. Let I be a nonzero ideal in $\mathbb{Z}[i]$. Show that the quotient ring $\mathbb{Z}[i]/I$ is finite.

16. **(a)** If p is prime in $\mathbb{Z}$, prove that the constant polynomial p is irreducible in $\mathbb{Q}_{\mathbb{Z}}[x]$. [*Hint:* Theorem 4.2 and Exercise 3.]

    (b) If p and q are positive primes in $\mathbb{Z}$ with $p \neq q$, prove that p and q are not associates in $\mathbb{Q}_{\mathbb{Z}}[x]$.

17. **(a)** Show that the only divisors of x in $\mathbb{Q}_{\mathbb{Z}}[x]$ are the integers (constant polynomials) and first-degree polynomials of the form $\dfrac{1}{n}x$ with $0 \neq n \in \mathbb{Z}$.

    (b) For each nonzero $n \in \mathbb{Z}$, show that the polynomial $\dfrac{1}{n}x$ is not irreducible in $\mathbb{Q}_{\mathbb{Z}}[x]$. [*Hint:* Theorem 9.1.]

(c) Show that x cannot be written as a finite product of irreducible elements in $\mathbb{Q}_{\mathbb{Z}}[x]$.

18. A ring R is said to satisfy the **ascending chain condition (ACC) on ideals** if whenever $I_1 \subseteq I_2 \subseteq I_3 \subseteq \cdots$ is a chain of ideals in R (not necessarily principal ideals), then there is an integer n such that $I_j = I_n$ for all $j \geq n$. Prove that if every ideal in a commutative ring R is finitely generated, then R satisfies the ACC. [*Hint:* See Theorem 6.3 and adapt the proof of Lemma 9.10.]

19. A ring R is said to satisfy the **descending chain condition (DCC) on ideals** if whenever $I_1 \supseteq I_2 \supseteq I_3 \supseteq \cdots$ is a chain of ideals in R, then there is an integer n such that $I_j = I_n$ for all $j \geq n$.

    (a) Show that $\mathbb{Z}$ does not satisfy the DCC.

    (b) Show that an integral domain R is a field if and only if R satisfies the DCC. [*Hint:* If $0 \neq a \in R$ is not a unit, what can be said about the chain of ideals $(a) \supseteq (a^2) \supseteq (a^3) \supseteq \cdots$?]

20. Let R be a PID and $a, b \in R$, not both zero. Prove that a, b have a greatest common divisor that can be written as a linear combination of a and b. [*Hint:* Let I be the ideal generated by a and b (see Theorem 6.3); then $I = (d)$ for some $d \in R$. Show that d is a gcd of a and b.]

21. Let R be a PID and S an integral domain that contains R. Let $a, b, d \in R$. If d is a gcd of a and b in R, prove that d is a gcd of a and b in S. [*Hint:* See Exercise 20.]

22. Extend Exercise 20 to any finite number of elements.

23. Give an alternative proof of Lemma 9.11 as follows. If $p \mid b$, there is nothing to prove. If $p \nmid b$, then 1_R is a gcd of p and b by Exercise 8. Now show that $p \mid c$ by copying the proof of Theorem 1.5 with p in place of a and Exercise 20 in place of Theorem 1.3.

24. Let R be an integral domain. Prove that R is a PID if and only if (i) every ideal of R is finitely generated (Theorem 6.3) and (ii) whenever $a, b \in R$, the sum ideal $(a) + (b)$ is principal. [Sum is defined in Exercise 18 of Section 6.1.]

25. Let (r,s) denote any gcd of r and s. Use $\sim$ to denote associates as in Exercise 6 of Section 9.1. Prove that for all $r, s, t \in R$:

    (a) If $s \sim t$, then $rs \sim rt$.

    (b) If $s \sim t$, then $(r,s) \sim (r,t)$.

    (c) $r(s,t) \sim (rs,rt)$.

    (d) $(r,(s,t)) \sim ((r,s),t)$. [*Hint:* Show that both are gcd's of r, s, t.]

26. With the notation as in Exercise 25, prove that if $(b,c) \sim 1_R$ and $(b,d) \sim 1_R$,

then $(b,d) \sim (b,(bd,cd))$. Apply (d), (c), and (a) of Exercise 25 to show $(b,(bd,cd)) \sim (b,cd)$.]

27. Let R be an integral domain in which any two elements (not both zero) have a gcd. Let p be an irreducible element of R. Prove that whenever $p \mid cd$, then $p \mid c$ or $p \mid d$. [*Hint:* Exercises 8 and 26.]

28. If R is a UFD, if a, b, and c are elements such that $a \mid c$ and $b \mid c$, and if 1_R is a gcd of a and b, prove that $ab \mid c$.

29. Let R be a UFD. If $a \mid bc$ and if 1_R is a gcd of a and b, prove that $a \mid c$.

30. A **least common multiple (lcm)** of the nonzero elements $a_1, \ldots, a_k$ is an element b such that (i) each a_i divides b and (ii) if each a_i divides an element c, then $b \mid c$. Prove that any finite set of nonzero elements in a UFD has a least common multiple.

31. Prove that nonzero elements a and b in R have a least common multiple if and only if the intersection of the principal ideals (a) and (b) is also a principal ideal.

C. 32. Prove that every ideal I in $\mathbb{Z}[\sqrt{d}\,]$ is finitely generated (Theorem 6.3) as follows. Let $I_0 = I \cap \mathbb{Z}$ and let $I_1 = \{b \in \mathbb{Z} \mid a + b\sqrt{d} \in I$ for some $a \in \mathbb{Z}\}$.

   (a) Prove that I_0 and I_1 are ideals in $\mathbb{Z}$. Therefore, $I_0 = (r_0)$ and $I_1 = (r_1)$ for some $r_i \in \mathbb{Z}$.

   (b) Prove that $I_0 \subseteq I_1$.

   (c) By the definition of I_1 there exists $a_1 \in \mathbb{Z}$ such that $a_1 + r_1\sqrt{d}$ is in I. Prove that I is the ideal generated by r_0 and $a_1 + r_1\sqrt{d}$. [*Hint:* If $r + s\sqrt{d} \in I$, then $s \in I_1$ so that $s = r_1 s_1$. Show that $(r + s\sqrt{d}) - s_1(a_1 + r_1\sqrt{d}) \in I_0$; use this to write $r + s\sqrt{d}$ as a linear combination of r_0 and $a_1 + r_1\sqrt{d}$.]

33. Prove that $p(x)$ is irreducible in $\mathbb{Q}_{\mathbb{Z}}[x]$ if and only if $p(x)$ is either a prime integer or an irreducible polynomial in $\mathbb{Q}[x]$ with constant term ± 1. Conclude that every irreducible $p(x)$ in $\mathbb{Q}_{\mathbb{Z}}[x]$ has the property that whenever $p(x) \mid c(x)d(x)$, then $p(x) \mid c(x)$ or $p(x) \mid d(x)$.

34. Show that every nonzero $f(x)$ in $\mathbb{Q}_{\mathbb{Z}}[x]$ can be written in the form $cx^n p_1(x) \cdots p_k(x)$, with $c \in \mathbb{Q}$, $n \geq 0$, and each $p_i(x)$ nonconstant irreducible in $\mathbb{Q}_{\mathbb{Z}}[x]$ and that this factorization is unique in the following sense: If $f(x) = dx^m q_1(x) \cdots q_t(x)$ with $d \in \mathbb{Q}$, $m \geq 0$, and each $q_i(x)$ nonconstant irreducible in $\mathbb{Q}_{\mathbb{Z}}[x]$, then $c = \pm d$, $m = n$, $k = t$, and, after relabeling if necessary, each $p_i(x) = \pm q_i(x)$.

35. Prove that any two nonzero polynomials in $\mathbb{Q}_{\mathbb{Z}}[x]$ have a gcd.

36. (a) Prove that $f(x)$ is irreducible in $\mathbb{Z}[x]$ if and only if $f(x)$ is either a prime integer or an irreducible polynomial in $\mathbb{Q}[x]$ such that the gcd in $\mathbb{Z}$ of the coefficients of $f(x)$ is 1.

   (b) Prove that $\mathbb{Z}[x]$ is a UFD. [*Hint:* See Theorems 4.13 and 4.22.]

9.3 FACTORIZATION OF QUADRATIC INTEGERS*

In this section we take a closer look at the domains $\mathbb{Z}[\sqrt{d}]$. Because unique factorization frequently fails in these domains, they provide a simplified model of the kinds of difficulties that played a crucial role in the historical origin of the concept of an ideal. These domains also illustrate how ideals can be used to "restore" unique factorization in some domains that lack it. We begin with a brief sketch of the relevant history.

Early in the last century, Gauss proved the "Law of Biquadratic Reciprocity," which provides a fast way of determining whether or not a congruence of the form $x^4 \equiv c \pmod{n}$ has a solution. Although the statement of this theorem involves only integers, Gauss's proof was set in the larger domain $\mathbb{Z}[i]$. He proved and used the fact that $\mathbb{Z}[i]$ is a unique factorization domain.

Since Gauss's proof involved $\mathbb{Z}[i]$ and i is a complex fourth root of 1, the German mathematician E. Kummer thought that analogous theorems for congruences of degree p might involve unique factorization in the domain.

$$\mathbb{Z}[\omega] = \{a_0 + a_1\omega + a_2\omega^2 + \cdots + a_{p-1}\omega^{p-1} \mid a_i \in \mathbb{Z}\},$$

where $\omega = \cos(2\pi/p) + i \sin(2\pi/p)$ is a complex pth root of 1. He was unable to develop higher-order reciprocity theorems because he discovered that $\mathbb{Z}[\omega]$ may not be a UFD.**

Later in the century questions about unique factorization arose in connection with the following problem. It is easy to find many nonzero integer solutions of the equation $x^2 + y^2 = z^2$, such as 3, 4, 5, or 5, 12, 13. But no one has ever found nonzero integer solutions for $x^3 + y^3 = z^3$ or $x^4 + y^4 = z^4$, which suggests that

$$x^n + y^n = z^n \text{ has no nonzero integer solutions when } n > 2.$$

This statement is known as **Fermat's Last Theorem** because in the late 1630s Fermat wrote it in the margin of his copy of Diophantus' *Arithmetica* and added "I have discovered a truly remarkable proof, but the margin is too small to contain it." Fermat's "proof" has never been found. Most mathematicians today doubt that he actually had a valid one.

In 1847 the French mathematician G. Lame thought he had found a proof of Fermat's Last Theorem in the case when n is prime.† His proof used the fact that for any odd positive prime p, $x^p + y^p$ can be factored in the domain $\mathbb{Z}[\omega]$ described above:

---

* The prerequisites for this section are pages 285–287 of Section 9.1 and the definition of unique factorization domain (page 300).

** The domain $\mathbb{Z}[\omega]$ is a UFD for every prime p less than 23 and fails to be a UFD for every larger prime.

† If the theorem is true for prime exponents, then it is true for all exponents; see Exercise 1.

$$x^p + y^p = (x + y)(x + \omega y)(x + \omega^2 y) \cdots (x + \omega^{p-1} y).$$

Lame's purported proof depended on the assumption that $\mathbb{Z}[\omega]$ is a unique factorization domain. When he became aware of Kummer's work, he realized that his proof could not be carried through.

Kummer had already found a way to avoid the difficulty. He invented what he called "ideal numbers" and proved that unique factorization *does* hold for these ideal numbers. This work eventually led to a proof that Fermat's Theorem is true for a large class of primes, including almost all the primes less than 100. This was a remarkable breakthrough and deeply influenced later work on the problem.* But it had even greater significance in the development of modern algebra. For Kummer's "ideal numbers" were what we now call ideals.

We shall return to ideals at the end of the section. Now we consider factorization in the domains $\mathbb{Z}[\sqrt{d}]$. These domains are similar to the ones that Kummer used and illustrate in simplified form the problems he faced and his method of solution. We shall assume that the integer d is **square-free,** meaning that $d \neq 1$ and d has no integer factors of the form c^2 except $(\pm 1)^2$. The following function is the key to factorization in $\mathbb{Z}[\sqrt{d}]$.

> **DEFINITION •** *The function $N: \mathbb{Z}[\sqrt{d}] \to \mathbb{Z}$ given by*
>
> $$N(s + t\sqrt{d}) = (s + t\sqrt{d})(s - t\sqrt{d}) = s^2 - dt^2$$
>
> *is called the **norm**.*

For example, in $\mathbb{Z}[\sqrt{3}]$,

$$N(5 + 2\sqrt{3}) = 5^2 - 3 \cdot 2^2 = 13 \qquad \text{and} \qquad N(2 - 4\sqrt{3}) = 2^2 - 3(-4)^2 = -44.$$

Note that

when $d < 0$, the norm of every element is nonnegative.

For instance, in $\mathbb{Z}[\sqrt{-5}]$,

$$N(s + t\sqrt{-5}) = s^2 - (-5)t^2 = s^2 + 5t^2 \geq 0.$$

In the last section we saw that the norm makes $\mathbb{Z}[i] = \mathbb{Z}[\sqrt{-1}]$ into a Euclidean domain. This is not true in general, but we do have

THEOREM 9.19   *If d is a square-free integer, then for all $a, b \in \mathbb{Z}[\sqrt{d}]$*

    (1) $N(a) = 0$ if and only if $a = 0$.

    (2) $N(ab) = N(a)N(b)$.

---

* Fermat's Last Theorem was finally proved in 1994 by Andrew Wiles. His proof uses results and techniques not available until relatively recently.

Proof (1) If $a = s + t\sqrt{d}$, then $N(a) = s^2 - dt^2$ so that $N(a) = 0$ if and only if $s^2 = dt^2$. If $d = -1$, then $s^2 = -t^2$ can occur in $\mathbb{Z}$ if and only if $s = 0 = t$, that is, if and only if $a = 0$. So suppose $d \neq -1$. Every prime in the factorization of s^2 and t^2 must occur an even number of times. But the prime factors of d do not repeat because d is square-free. So if p is a prime factor of d, it must occur an odd number of times in the factorization of dt^2. By unique factorization in $\mathbb{Z}$, the equation $s^2 = dt^2$ is impossible unless $s = 0 = t$, that is, unless $a = 0$.

(2) Let $a = r + s\sqrt{d}$ and $b = m + n\sqrt{d}$. The proof is a straightforward computation (Exercise 3). ◆

THEOREM 9.20 *Let d be a square-free integer. Then $u \in \mathbb{Z}[\sqrt{d}]$ is a unit if and only if $N(u) = \pm 1$.*

Proof If u is a unit, then $uv = 1$ for some $v \in \mathbb{Z}[\sqrt{d}]$. By Theorem 9.19, $N(u)N(v) = N(uv) = N(1) = 1^2 - d \cdot 0^2 = 1$. Since $N(u)$ and $N(v)$ are integers, the only possibilities are $N(u) = \pm 1$ and $N(v) = \pm 1$. Conversely, if $u = s + t\sqrt{d}$ and $N(u) = \pm 1$, let $\overline{u} = s - t\sqrt{d} \in \mathbb{Z}[\sqrt{d}]$. Then by the definition of the norm, $u\overline{u} = N(u) = \pm 1$. Hence, $u(\pm \overline{u}) = 1$ and u is a unit. ◆

> **EXAMPLE** In $\mathbb{Z}[\sqrt{2}]$ the element $3 + 2\sqrt{2}$ is a unit because $N(3 + 2\sqrt{2}) = 3^2 - 2 \cdot 2^2 = 1$. As noted in the preceding proof, the inverse of $3 + 2\sqrt{2}$ is $3 - 2\sqrt{2}$. Every power of a unit is also a unit, so $\mathbb{Z}[\sqrt{2}]$ has infinitely many units, including $(3 + 2\sqrt{2})$, $(3 + 2\sqrt{2})^2$, $(3 + 2\sqrt{2})^3$, . . .

According to Theorem 9.20 we can determine every unit $s + t\sqrt{d}$ in $\mathbb{Z}[\sqrt{d}]$ by finding all the integer solutions (for s and t) of the equations $s^2 - dt^2 = \pm 1$. When $d > 1$, these equations have infinitely many solutions (see the preceding example and Burton [14, Section 14.4]). When $d = -1$, the equations reduce to $s^2 + t^2 = 1$.* The only integer solutions are $s = \pm 1$, $t = 0$, and $s = 0$, $t = \pm 1$. So the only units in $\mathbb{Z}[i] = \mathbb{Z}[\sqrt{-1}]$ are ± 1 and $\pm i$. If $d < -1$, say $d = -k$ with $k > 1$, then the equations reduce to $s^2 + kt^2 = 1$.* Since $k > 1$, the only integer solutions are $s = \pm 1$, $t = 0$. Thus we have

COROLLARY 9.21 *Let d be a square-free integer. If $d > 1$, then $\mathbb{Z}[\sqrt{d}]$ has infinitely many units. The units in $\mathbb{Z}[\sqrt{-1}]$ are ± 1 and $\pm i$. If $d < -1$, then the units in $\mathbb{Z}[\sqrt{d}]$ are ± 1.*

COROLLARY 9.22 *Let d be a square-free integer. If $p \in \mathbb{Z}[\sqrt{d}]$ and $N(p)$ is a prime integer in $\mathbb{Z}$, then p is irreducible in $\mathbb{Z}[\sqrt{d}]$.*

Proof Since $N(p)$ is prime, $N(p) \neq \pm 1$, so p is not a unit in $\mathbb{Z}[\sqrt{d}]$ by Theorem

---

* The solution -1 is impossible since the left side of the equation is always nonnegative.

9.20. If $p = ab$ in $\mathbb{Z}[\sqrt{d}\,]$, then by Theorem 9.19, $N(p) = N(a)N(b)$ in $\mathbb{Z}$. Since $N(a)$, $N(b)$, $N(p)$ are integers and $N(p)$ is prime, we must have $N(a) = \pm 1$ or $N(b) = \pm 1$. So a or b is a unit by Theorem 9.20. Therefore, p is irreducible by Theorem 9.1. ◆

> **EXAMPLE** The element $1 - i$ is irreducible in $\mathbb{Z}[i]$ because $N(1 - \sqrt{-1}) = 2$. Similarly, $1 + i$ is also irreducible. Therefore, a factorization of 2 as a product of irreducibles in $\mathbb{Z}[i]$ is given by $2 = (1 + i)(1 - i)$.

The converse of Corollary 9.22 is false. For instance, in $\mathbb{Z}[\sqrt{-5}\,]$ the norm of $1 + \sqrt{-5}$ is 6, which is not prime. But the next example shows that $1 + \sqrt{-5}$ *is* irreducible.

> **EXAMPLE** To show that $1 + \sqrt{-5}$ is irreducible in $\mathbb{Z}[\sqrt{-5}\,]$, suppose $1 + \sqrt{-5} = ab$. By Theorem 9.1 we need only show that a or b is a unit. By Theorem 9.19, $N(a)N(b) = N(ab) = N(1 + \sqrt{-5}) = 6$. Since $N(a)$ and $N(b)$ are nonnegative integers, the only possibilities are $N(a) = 1, 2, 3,$ or 6. If $a = s + t\sqrt{-5}$ and $N(a) = 2$, then $s^2 + 5t^2 = 2$. It is easy to see that this equation has no integer solutions for s and t; so $N(a) = 2$ is impossible. A similar argument shows that $N(a) = 3$ is impossible. If $N(a) = 1$, then a is a unit by Theorem 9.20. If $N(a) = 6$, then $N(b) = 1$ and b is a unit. Therefore, $1 + \sqrt{-5}$ is irreducible.

We have seen an example of an integral domain in which a nonzero, non-unit element could not be factored as a product of irreducibles (Exercise 17 in Section 9.2). We shall now see that $\mathbb{Z}[\sqrt{d}\,]$ may fail to be a UFD for a different reason: Although factorization as a product of irreducibles is always possible in $\mathbb{Z}[\sqrt{d}\,]$, it may not be unique.

THEOREM 9.23 *Let d be a square-free integer. Then every nonzero, nonunit element in $\mathbb{Z}[\sqrt{d}\,]$ is a product of irreducible elements.**

Proof Let S be the set of all nonzero, nonunits in $\mathbb{Z}[\sqrt{d}\,]$ that are *not* the product of irreducibles. We must show that S is empty. So suppose, on the contrary, that S is nonempty. Then the set $W = \{\,|N(t)| \mid t \in S\}$ is a nonempty set of positive integers. By the Well-Ordering Axiom, W contains a smallest integer. Thus there is an element $a \in S$ such that $|N(a)| \leq |N(t)|$ for every $t \in S$. Since $a \in S$ we know that a is not itself irreducible. So there exist nonunits $b, c \in \mathbb{Z}[\sqrt{d}\,]$ such that $a = bc$. At least one of b, c must be in S (otherwise a would be a product of irreducibles and, hence, not in S), say $b \in S$. Since b and c are nonunits,

---

* As usual, we allow a "product" with just one factor.

$|N(b)| > 1$ and $|N(c)| > 1$ by Theorem 9.20. But $|N(a)| = |N(b)||N(c)|$ by Theorem 9.19, so we must have $1 < |N(b)| < |N(a)|$. But $b \in S$, so $|N(a)| \leq |N(b)|$ by the choice of a. This is a contraction. Therefore, S is empty, and the theorem is proved. ◆

> **EXAMPLE** The domain $\mathbb{Z}[\sqrt{-5}]$ is not a unique factorization domain. The element 6 in $\mathbb{Z}[\sqrt{-5}]$ has two factorizations:
>
> $$6 = 2 \cdot 3 \quad \text{and} \quad 6 = (1 + \sqrt{-5})(1 - \sqrt{-5}).$$
>
> The proof that $1 + \sqrt{-5}$ is irreducible was given in the example preceding Theorem 9.23. The proofs that 2, 3, and $1 - \sqrt{-5}$ are irreducible are similar. For instance, if $2 = ab$, then $N(a)N(b) = N(ab) = N(2) = 4$ so that $N(a) = 1$, 2, or 4. But $N(a) = 2$ is impossible because the equation $s^2 + 5t^2 = 2$ has no integer solutions. So either $N(a) = 1$ and a is a unit, or $N(a) = 4$. In the latter case $N(b) = 1$ and b is a unit. Therefore, 2 is irreducible by Theorem 9.1. Since the only units in $\mathbb{Z}[\sqrt{-5}]$ are ± 1, it is clear that neither 2 nor 3 is an associate of $1 + \sqrt{-5}$ or $1 - \sqrt{-5}$. Thus the factorization of 6 as a product of irreducibles is not unique up to associates and $\mathbb{Z}[\sqrt{-5}]$ is not a UFD.

The preceding example demonstrates that the irreducible 2 divides $(1 + \sqrt{-5})(1 - \sqrt{-5}]$ in $\mathbb{Z}[\sqrt{-5}]$ but does not divide either $1 + \sqrt{-5}$ or $1 - \sqrt{-5}$. So when unique factorization fails, an irreducible element p may not have the property that when $p \mid cd$, then $p \mid c$ or $p \mid d$.* Another consequence of the failure of unique factorization is the possible absence of greatest common divisors (Exercise 13).

Unique Factorization of Ideals

We are now in the position that Kummer was in a century and a half ago and the question is: How can some kind of unique factorization be restored in domains such as $\mathbb{Z}[\sqrt{-5}]$? Kummer's answer was to change the focus from elements to ideals.** The product IJ of ideals I and J is defined to be the set of all sums of elements of the form ab, with $a \in I$ and $b \in J$; that is,

$$IJ = \{a_1b_1 + a_2b_2 + \cdots + a_nb_n \mid n \geq 1, a_k \in I, b_k \in J\}.$$

Exercise 34 in Section 6.1 shows that IJ is an ideal. Instead of factoring an element a as a product of irreducibles, Kummer factored the principal ideal (a) as a product of prime ideals.

---

* This is not particularly surprising in view of Theorem 9.16.

** Kummer used different terminology, but the ideas here are essentially his. We use the modern terminology of ideals that was introduced by R. Dedekind, who generalized Kummer's theory.

EXAMPLE We shall express the principal ideal (6) in $\mathbb{Z}[\sqrt{-5}]$ as a product of prime ideals. The irreducible factorization of elements $6 = 2 \cdot 3$ seems a natural place to start, and it is easy to prove that the ideal (6) is the product ideal (2)(3) (Exercise 16). But (2) is not a prime ideal (for instance, the product $(1 + \sqrt{-5})(1 - \sqrt{-5}) = 6$ is in (2) but neither of the factors is in (2)). So we must look elsewhere. Let P be the ideal in $\mathbb{Z}[\sqrt{-5}]$ generated by 2 and $1 + \sqrt{-5}$, that is,

$$P = \{2a + (1 + \sqrt{-5})b \mid a, b \in \mathbb{Z}[\sqrt{-5}]\}.$$

Then P is an ideal by Theorem 6.3. Exercise 17 shows that $r + s\sqrt{-5} \in P$ if and only if r and s are both even or both odd. This implies that the only distinct cosets in $\mathbb{Z}[\sqrt{-5}]/P$ are $0 + P$ and $1 + P$; for if $m + n\sqrt{-5}$ has m odd and n even, then $(m + n\sqrt{-5}) - 1 = (m - 1) + n\sqrt{-5} \in P$ because $m - 1$ and n are even. Hence, $(m + n\sqrt{-5}) + P = 1 + P$. Similarly, if m is even and n is odd, then $(m - 1) + n\sqrt{-5} \in P$ because $m - 1$ and n are odd. It follows that the quotient ring $\mathbb{Z}[\sqrt{-5}]/P$ is isomorphic to $\mathbb{Z}_2$. Therefore, P is a prime ideal in $\mathbb{Z}[\sqrt{-5}]$ by Theorem 6.14. A similar argument (Exercise 19) shows that Q_1 and Q_2 are prime ideals, where

$$Q_1 = \{3a + (1 + \sqrt{-5})b \mid a, b \in \mathbb{Z}[\sqrt{-5}]\},$$
$$Q_2 = \{3a + (1 - \sqrt{-5})b \mid a, b \in \mathbb{Z}[\sqrt{-5}]\}.$$

Exercises 18 and 19 show that the product ideal $P^2 = PP$ is precisely the ideal (2) and that $Q_1 Q_2 = (3)$. Therefore, the ideal (6) is a product of four prime ideals: $(6) = (2)(3) = P^2 Q_1 Q_2$.

Kummer went on to show that in the domains he was considering, the factorization of an ideal as a product of prime ideals is unique except for the order of the factors. This result was later generalized by R. Dedekind. In order to state this generalization precisely, we need to fill in some background.

An **algebraic number** is a complex number that is the root of some monic polynomial with rational coefficients. If t is an algebraic number and t is the root of a polynomial degree n in $\mathbb{Q}[x]$, then

$$\mathbb{Q}(t) = \{a_0 + a_1 t + a_2 t^2 + \cdots + a_{n-1} t^{n-1} \mid a_i \in \mathbb{Q}\}$$

is a subfield of $\mathbb{C}$ and every element in $\mathbb{Q}(t)$ is an algebraic number.* An **algebraic integer** is a complex number that is the root of some monic polynomial with *integer* coefficients. It can be shown that the set of all algebraic integers in $\mathbb{Q}(t)$ is an integral domain (see Birkhoff and MacLane [2; page 445]). If ω is a complex root of $x^p - 1$, then the domain $\mathbb{Z}[\omega]$ that Kummer used is in fact the domain of all algebraic integers in $\mathbb{Q}(\omega)$ (see Ireland and Rosen [15; page 199]). So Kummer's results are a special case of

_____
* For a proof see Theorems 10.7 and 10.9.

THEOREM 9.24 *Let t be an algebraic number and R the domain of all algebraic integers in* $\mathbb{Q}(t)$. *Then every ideal in R (except 0 and R) is the product of prime ideals and this factorization is unique up to the order of the factors.*

For a proof see Ireland and Rosen [15; page 174].

Most of the rings $\mathbb{Z}[\sqrt{d}]$ are also special cases of Theorem 9.24. For if d is a square-free integer, then $t = \sqrt{d}$ is an algebraic number (because it is a root of $x^2 - d$) and $\mathbb{Q}(\sqrt{d}) = \{a_0 + a_1\sqrt{d} \mid a_i \in \mathbb{Q}\}$. The algebraic integers in the field $\mathbb{Q}(\sqrt{d})$ are called **quadratic integers.** Every element $r + s\sqrt{d}$ of $\mathbb{Z}[\sqrt{d}]$ is a quadratic integer in $\mathbb{Q}(\sqrt{d})$ because it is a root of this monic polynomial in $\mathbb{Z}[x]$:

$$x^2 - 2rx + (r^2 - ds^2) = (x - (r + s\sqrt{d}))(x - (r - s\sqrt{d})).$$

When $d \equiv 2$ or $3 \pmod 4$, then $\mathbb{Z}[\sqrt{d}]$ is the domain R of *all* quadratic integers in $\mathbb{Q}(\sqrt{d})$, but when $d \equiv 1 \pmod 4$, there are quadratic integers in R that are not in $\mathbb{Z}[\sqrt{d}]$ (see Exercise 22).* An elementary proof of Theorem 9.24 for the case of quadratic integers is given in Robinson [21].

Theorem 9.24 has proved very useful in algebraic number theory. But it does not answer many questions about unique factorization of elements, such as: If R is the domain of all quadratic integers in $\mathbb{Q}(\sqrt{d})$, for what values of d is R a UFD? When $d < 0$, R is a UFD if and only if $d = -1, -2, -3, -7, -11, -19, -43, -67$, or -163 (see Stark [22]). When $d > 0$, R is known to be a UFD for $d = 2, 3, 5, 6, 7, 11, 13, 17, 19, 21, 22, 23, 29$, and many other values. But there is no complete list as there is when d is negative. It is conjectured that R is a UFD for infinitely many values of d.

◆ **EXERCISES**

A. **1.** If $x^k + y^k = z^k$ has no nonzero integer solutions and $k \mid n$, then show that $x^n + y^n = z^n$ has no nonzero integer solutions.

2. Let ω be a complex number such that $\omega^p = 1$. Show that

$$\mathbb{Z}[\omega] = \{a_0 + a_1\omega + a_2\omega^2 + \cdots + a_{p-1}\omega^{p-1} \mid a_i \in \mathbb{Z}\}$$

is an integral domain. [*Hint:* $\omega^p = 1$ implies $\omega^{p+1} = \omega$, $\omega^{p+2} = \omega^2$, etc.]

3. If $a = r + s\sqrt{d}$ and $b = m + n\sqrt{d}$ in $\mathbb{Z}[\sqrt{d}]$, show that $N(ab) = N(a)N(b)$.

4. Explain why $\mathbb{Z}[\sqrt{-5}]$ is not a Euclidean domain for any function δ.

5. If $a \in \mathbb{Q}$ is an algebraic integer, as defined on page 313, show that $a \in \mathbb{Z}$. [*Hint:* Theorem 4.20.]

---

* Since d is square-free, $d \not\equiv 0 \pmod 4$.

B. **6.** In which of these domains is 5 an irreducible element?

(a) $\mathbb{Z}$ (b) $\mathbb{Z}[i]$ (c) $\mathbb{Z}[\sqrt{-2}]$

7. In $\mathbb{Z}[\sqrt{-7}]$, factor 8 as a product of two irreducible elements and as a product of three irreducible elements. [*Hint:* Consider $(1 + \sqrt{-7})(1 - \sqrt{-7})$.]

8. Factor each of the elements below as a product of irreducibles in $\mathbb{Z}[i]$, [*Hint:* Any factor of a must have norm dividing $N(a)$.]

(a) 3   (b) 7   (c) $4 + 3i$ (d) $11 + 7i$

9. (a) Verify that each of $5 + \sqrt{2}$, $2 - \sqrt{2}$, $11 - 7\sqrt{2}$, and $2 + \sqrt{2}$ is irreducible in $\mathbb{Z}[\sqrt{2}]$.

(b) Explain why the fact that

$$(5 + \sqrt{2})(2 - \sqrt{2}) = (11 - 7\sqrt{2})(2 + \sqrt{2})$$

does *not* contradict unique factorization in $\mathbb{Z}[\sqrt{2}]$.

10. Find two different factorizations of 9 as a product of irreducibles in $\mathbb{Z}[\sqrt{-5}]$.

11. Show that $\mathbb{Z}[\sqrt{-6}]$ is not a UFD. [*Hint:* Factor 10 in two ways.]

12. Show that $\mathbb{Z}[\sqrt{10}]$ is not a UFD. [*Hint:* Factor 6 in two ways.]

13. Show that 6 and $2 + 2\sqrt{-5}$ have no greatest common divisor in $\mathbb{Z}[\sqrt{-5}]$. [*Hint:* A common divisor a of 6 and $2 + 2\sqrt{-5}$ must have norm dividing both $N(6) = 36$ and $N(2 + 2\sqrt{-5}) = 24$; hence, $a = r + s\sqrt{-5}$ with $r^2 + 5s^2 = N(a) = 1, 2, 3, 4, 6,$ or 12. Use this to find the common divisors. Verify that none of them is divisible by all the others, as required of a gcd. Also see the example on page 312.]

14. Show that 1 is a gcd of 2 and $1 + \sqrt{-5}$ in $\mathbb{Z}[\sqrt{-5}]$, but 1 cannot be written in the form $2a + (1 + \sqrt{-5})b$ with $a, b \in \mathbb{Z}[\sqrt{-5}]$.

15. Prove that every principal ideal in a UFD is a product of prime ideals uniquely except for the order of the factors.

16. Show that $(6) = (2)(3)$ in $\mathbb{Z}[\sqrt{-5}]$. (The product of ideals is defined on page 312.)

17. Let P be the ideal $\{2a + (1 + \sqrt{-5})b \mid a, b \in \mathbb{Z}[\sqrt{-5}]\}$ in $\mathbb{Z}[\sqrt{-5}]$. Prove that $r + s\sqrt{-5} \in P$ if and only if $r \equiv s \pmod{2}$ (that is, r and s are both even or both odd).

18. Let P be as in Exercise 17. Prove that P^2 is the principal ideal (2).

19. Let Q_1 be the ideal $\{3a + (1 + \sqrt{-5})b \mid a, b \in \mathbb{Z}[\sqrt{-5}]\}$ and Q_2 the ideal $\{3a + (1 - \sqrt{-5})b \mid a, b \in \mathbb{Z}[\sqrt{-5}]\}$ in $\mathbb{Z}[\sqrt{-5}]$.

(a) Prove that $r + s\sqrt{-5} \in Q_1$ if and only if $r \equiv s \pmod{3}$.

(b) Show that $\mathbb{Z}[\sqrt{-5}]/Q_1$ has exactly three distinct cosets.

(c) Prove that $\mathbb{Z}[\sqrt{-5}]/Q_1$ is isomorphic to $\mathbb{Z}_3$; conclude that Q_1 is a prime ideal.

(d) Prove that Q_2 is a prime ideal. [*Hint*: Adapt (a)–(c).]

(e) Prove that $Q_1Q_2 = (3)$.

20. If $r + s\sqrt{-5} \in \mathbb{Z}[\sqrt{-5}]$ with $s \neq 0$, then prove that 2 is not in the principal ideal $(r + s\sqrt{-5})$.

21. If d is a square-free integer, prove that $\mathbb{Z}[\sqrt{d}]$ satisfies the ascending chain condition on principal ideals.

C. 22. Let d be a square-free integer and let $\mathbb{Q}(\sqrt{d})$ be as defined on page 313. We know that $\mathbb{Z}[\sqrt{d}] \subseteq \mathbb{Q}(\sqrt{d})$ and every element of $\mathbb{Z}[\sqrt{d}]$ is a quadratic integer. Determine all the quadratic integers in $\mathbb{Q}(\sqrt{d})$ as follows.

(a) Show that every element of $\mathbb{Q}(\sqrt{d})$ is of the form $(r + s\sqrt{d})/t$, where $r, s, t \in \mathbb{Z}$ and the gcd (r, s, t) of r, s, t is 1. Hereafter, let $a = (r + s\sqrt{d})/t$ denote such an arbitrary element of $\mathbb{Q}(\sqrt{d})$.

(b) Show that a is a root of

$$p(x) = x^2 - \left(\frac{2r}{t}\right) x + \left(\frac{r^2 - ds^2}{t^2}\right) \in \mathbb{Q}[x].$$

[*Hint*: Show that $p(x) = (x - a)(x - \bar{a})$, where $\bar{a} = (r - s\sqrt{d})/t$.]

(c) If $s \neq 0$, show that $p(x)$ is irreducible in $\mathbb{Q}[x]$.

(d) Prove that a is a quadratic integer if and only if $p(x)$ has integer coefficients. [*Hint*: If $s = 0$, use Exercise 5; if $s \neq 0$ and a is a root of a monic polynomial $f(x) \in \mathbb{Z}[x]$, use Theorem 4.22 to show that a is a root of some monic $g(x) \in \mathbb{Z}[x]$, with $g(x)$ irreducible in $\mathbb{Q}[x]$. Apply (c) and Theorem 4.13 to show $g(x) = p(x)$.]

(e) If a is a quadratic integer, show that $t \mid 2r$ and $t^2 \mid 4ds^2$. Use this fact to prove that t must be 1 or 2. [*Hint*: d is square-free, $(r, s, t) = 1$; use (b) and (d).]

(f) If $d \equiv 2$ or $3 \pmod 4$, show that a is a quadratic integer if and only if $t = 1$. [*Hint*: If $t = 2$, then $r^2 \equiv ds^2 \pmod 4$ by (b) and (d). If s is even, reach a contradiction to the fact that $(r, s, t) = 1$; if s is odd, use Exercise 3 of Section 2.1 to get a contradiction.]

(g) If $d \equiv 1 \pmod 4$ and $a \in \mathbb{Q}(\sqrt{d})$, show that a is a quadratic integer if and only if $t = 1$, or $t = 2$ and both r and s are odd. [*Hint*: Use (d).]

(h) Use (f) and (g) to show that the set of all quadratic integers in $\mathbb{Q}(\sqrt{d})$ is $\mathbb{Z}[\sqrt{d}]$ if $d \equiv 2$ or $3 \pmod 4$ and $\left\{ \dfrac{m + n\sqrt{d}}{2} \mid m, n \in \mathbb{Z} \text{ and } m \equiv n \pmod 2 \right\}$ if $d \equiv 1 \pmod 4$.

9.4 THE FIELD OF QUOTIENTS OF AN INTEGRAL DOMAIN*

For any integral domain R we shall construct a field F that contains R and consists of "quotients" of elements of R. When the domain R is $\mathbb{Z}$, then F will be the field $\mathbb{Q}$ of rational numbers. So you may view these proceedings either as a rigorous formalization of the construction of $\mathbb{Q}$ from $\mathbb{Z}$ or as a generalization of this construction to arbitrary integral domains. The field F will be the essential tool for studying factorization in $R[x]$ in Section 9.5.

Our past experience with rational numbers will serve as a guide for the formal development. But all the proofs will be independent of any prior knowledge of the rationals.

A rational number a/b is determined by the pair of integers a, b (with $b \neq 0$). But different pairs may determine the same rational number; for instance, $\frac{1}{2} = \frac{3}{6} = \frac{4}{8}$, and in general

$$\frac{a}{b} = \frac{c}{d} \quad \text{if and only if} \quad ad = bc.$$

This suggests that the rationals come from some kind of equivalence relation on pairs of integers (equivalent pairs determine the same rational number). We now formalize this idea.

Let R be an integral domain and let S be this set of pairs:

$$S = \{(a,b) \mid a, b \in R \text{ and } b \neq 0_R\}.$$

Define a relation $\sim$ on the set S by

$$(a,b) \sim (c,d) \quad \text{means} \quad ad = bc \text{ in } R.$$

THEOREM 9.25 *The relation $\sim$ is an equivalence relation on S.*

Proof *Reflexive:* Since r is commutative $ab = ba$, so that $(a,b) \sim (a,b)$ for every pair (a,b) in S. *Symmetric:* If $(a,b) \sim (c,d)$, then $ad = bc$. By commutativity $cb = da$, so that $(c,d) \sim (a,b)$. *Transitive:* Suppose that $(a,b) \sim (c,d)$ and $(c,d) \sim (r,s)$. Then $ad = bc$ and $cs = dr$. Multiplying $ad = bc$ by s and using $cs = dr$ we have $ads = (bc)s = b(cs) = bdr$. Since $d \neq 0_R$ by the definition of S and R is an integral domain we can cancel d from $ads = bdr$ and conclude that $as = br$. Therefore, $(a,b) \sim (r,s)$. ◆

The equivalence relation $\sim$ partitions S into disjoint equivalence classes by Corollary D.2 in Appendix D. For convenience we shall denote the equiva-

---

* This section is independent of the rest of Chapter 9. Its prerequisites are Chapter 3 and Appendix D.

lence class of (a,b) by $[a,b]$ rather than the more cumbersome $[(a,b)]$. Let F denote the set of all equivalence classes under $\sim$. Note that by Theorem D.1,

$$[a, b] = [c, d] \text{ in } F \qquad \text{if and only if} \qquad (a, b) \sim (c, d) \text{ in } S.$$

Therefore, by the definition of $\sim$,

$$\mathbf{[a, b] = [c, d] \text{ in } F \qquad \text{if and only if} \qquad ad = bc \text{ in } R.}$$

We want to make the set F into a field. Addition and multiplication of equivalence classes are defined by*

$$[a, b] + [c, d] = [ad + bc, bd]$$
$$[a, b][c, d] = [ac, bd].$$

In order for this definition to make sense, we must first show that the quantities on the right side of the equal sign are actually elements of the set F. Now $[a, b]$ is the equivalence class of the pair (a,b) in S. By the definition of S we have $b \neq 0_R$; similarly, $d \neq 0_R$. Since R is an integral domain, $bd \neq 0_R$. Thus $(ad + bc, bd)$ and (ac, bd) are in the set S, so that the equivalence classes $[ad + bc, bd]$ and $[ac, bd]$ *are* elements of F. But more is required in order to guarantee that addition and multiplication in F are well defined.

In ordinary arithmetic, $\frac{1}{2} \cdot \frac{3}{5} = \frac{3}{10}$ and replacing $\frac{1}{2}$ by $\frac{4}{8}$ produces the same answer because $\frac{4}{8} \cdot \frac{3}{5} = \frac{12}{40} = \frac{3}{10}$. The answer doesn't depend on how the fractions are represented. Similarly, in F we must show that arithmetic does not depend on the way the equivalence classes are written:

LEMMA 9.26 *Addition and multiplication in F are independent of the choice of equivalence class representatives. In other words, if $[a, b] = [a', b']$ and $[c, d] = [c', d']$, then*

$$[ad + bc, bd] = [a'd' + b'c', b'd']$$

and

$$[ac, bd] = [a'c', b'd'].$$

Proof As noted above $[ad + bc, bd] = [a'd' + b'c', b'd']$ in F if and only if $(ad + bc)b'd' = bd(a'd' + b'c')$ in R. So we shall prove this last statement. Since $[a, b] = [a', b']$ and $[c, d] = [c', d']$ we know that

$$(*) \qquad\qquad ab' = ba' \qquad \text{and} \qquad cd' = dc'.$$

---

* These definitions are motivated by the arithmetical rules for rational numbers (just replace the fraction r/s by the equivalence class $[r, s]$):

$$\frac{a}{b} + \frac{c}{d} = \frac{ad + bc}{bd} \qquad \frac{a}{b} \cdot \frac{c}{d} = \frac{ac}{bd}$$

Multiplying the first equation by dd' and the second by bb' and adding the results show that

$$ab'dd' = ba'dd'$$
$$\underline{cd'bb' = dc'bb'}$$
$$ab'dd' + cd'bb' = ba'dd' + dc'bb'$$
$$(ad + bc)b'd' = bd(a'd' + b'c').$$

Therefore, $[ad + bc, bd] = [a'd' + b'c', b'd']$.

For the second part of the proof multiply the first equation in (∗) by cd' and the second by ba' so that

$$ab'cd' = ba'cd' \quad \text{and} \quad cd'ba' = dc'ba'.$$

By commutativity the right side of the first equation is the same as the left side of the second equation so that the other sides of the two equations are equal: $ab'cd' = dc'ba'$. Consequently,

$$(ac)(b'd') = ab'cd' = dc'ba' = (bd)(a'c').$$

The two ends of this equation show that $[ac, bd] = [a'c', b'd']$. ◆

LEMMA 9.27 *If R is an integral domain and F is as above, then for all nonzero $a, b, c, d, k \in R$:*

　　(1) $[0_R, b] = [0_R, d]$;

　　(2) $[a, b] = [ak, bk]$;

　　(3) $[a, a] = [c, c]$.

Proof Exercise 1. ◆

LEMMA 9.28 *With the addition and multiplication defined above, F is a field.*

Proof Closure of addition and multiplication follows from Lemma 9.26 and the remarks preceding it. Addition is commutative in F because addition and multiplication in R are commutative:

$$[a, b] + [c, d] = [ad + bc, bd] = [cb + da, db] = [c, d] + [a, b].$$

Let 0_F be the equivalence class $[0_R, b]$ for *any* nonzero $b \in R$ (by (1) in Lemma 9.27 *all* pairs of the form $(0_R, b)$ with $b \neq 0_R$ are in the same equivalence class). If $[a, b] \in F$, then by (2) in Lemma 9.27 (with $k = b$):

$$[a, b] + 0_F = [a, b] + [0_R, b] = [ab + b0_R, bb] = [ab, bb] = [a, b].$$

Therefore, 0_F is the zero element of F. The negative of $[a, b]$ in F is $[-a, b]$ because

$$[a, b] + [-a, b] = (ab - ba, b^2] = [0_R, b^2] = 0_F.$$

The proofs that addition is associative and that multiplication is associative and commutative are left to the reader (Exercise 2), as is the verification that $[1_R, 1_R]$ is the multiplicative identity element in F. If $[a, b]$ is a nonzero element of F, then $a \neq 0_R$. Hence, $[b, a]$ is a well-defined element of F and by (3) in Lemma 9.27

$$[a, b][b, a] = [ab, ba] = [1_R ab, 1_R ab] = [1_R, 1_R].$$

Therefore, $[b, a]$ is the multiplicative inverse of $[a, b]$. To see that the distributive law holds in F, note that

$$[a, b]([c, d] + [r, s]) = [a, b][cs + dr, ds]$$
$$= [a(cs + dr), b(ds)]$$
$$= [acs + adr, bds].$$

On the other hand, by (2) in Lemma 9.27 (with $k = b$)

$$[a, b][c, d] + [a, b][r, s] = [ac, bd] + [ar, bs]$$
$$= [(ac)(bs) + (bd)(ar), (bd)(bs)]$$
$$= [(acs + adr)b, (bds)b]$$
$$= [acs + adr, bds].$$

Therefore, $[a, b]([c, d] + [r, s]) = [a, b][c, d] + [a, b][r, s]$. ◆

We usually identify the integers with rational numbers of the form $a/1$. The same idea works in the general case:

LEMMA 9.29 *Let R be an integral domain and F the field of Lemma 9.28. Then the subset $R^* = \{[a, 1_R] \mid a \in R\}$ of F is an integral domain that is isomorphic to R.*

Proof Verify that R^* is a subring of F (Exercise 3). Clearly $[1_R, 1_R]$, the identity element of F, is in R^*, so R^* is an integral domain. Define a map $f: R \rightarrow R^*$ by $f(a) = [a, 1_R]$. Then f is a homomorphism:

$$f(a) + f(c) = [a, 1_R] + [c, 1_R] = [a1_R + 1_R c, 1_R 1_R]$$
$$= [a + c, 1_R] = f(a + c)$$
$$f(a)f(c) = [a, 1_R][c, 1_R] = [ac, 1_R] = f(ac).$$

If $f(a) = f(c)$, then $[a, 1_R] = [c, 1_R]$, which implies that $a1_R = 1_R c$. Thus $a = c$ and f is injective. Since f is obviously surjective, f is an isomorphism. ◆

The equivalence class notation for elements of F is awkward and doesn't convey the promised idea of "quotients." This is easily remedied by a change of notation. Instead of denoting the equivalence class of (a,b) by $[a, b]$,

denote the equivalence class of (*a,b*) by *a/b*.

If we translate various statements above from the brackets notation to the new quotient notation, things begin to look quite familiar:

THEOREM 9.30   *Let R be an integral domain. Then there exists a field F whose elements are of the form a/b with a, b ϵ R and b $\neq$ 0_R, subject to the equality condition*

$$\frac{a}{b} = \frac{c}{d} \text{ in } F \quad \text{if and only if} \quad ad = bc \text{ in } R.$$

Addition and multiplication in F are given by

$$\frac{a}{b} + \frac{c}{d} = \frac{ad + bc}{bd}, \quad \frac{a}{b} \cdot \frac{c}{d} = \frac{ac}{bd}.$$

The set of elements in F of the form a/1_R (a ϵ R) is an integral domain isomorphic to R.

Proof   Lemmas 9.28 and 9.29 and the notation change preceding the theorem.*   ◆

It is now clear that if $R = \mathbb{Z}$, then the field F is precisely $\mathbb{Q}$. So Theorem 9.30 may be taken as a formal construction of $\mathbb{Q}$ from $\mathbb{Z}$. In the general case, we shall follow the same custom we use with $\mathbb{Q}$: The ring R will be *identified* with its isomorphic copy in F. Then we can say that R is the subset of F consisting of elements of the form $a/1_R$. The field F is called the **field of quotients** of R.

> **EXAMPLE**   Let F be a field. The field of quotients of the polynomial domain $F[x]$ is denoted by $F(x)$ and consists of all $f(x)/g(x)$, where $f(x), g(x) \in F[x]$ and $g(x) \neq 0_K$. The field $F(x)$ is called the **field of rational functions** over F.

The field of quotients of an integral domain R is the smallest field that contains R in the following sense.**

THEOREM 9.31   *Let R be an integral domain and F its field of quotients. If K is a field containing R, then K contains a subfield E such that R $\subseteq$ E $\subseteq$ K and E is isomorphic to F.*

---

* At this point you may well ask, "Why didn't we adopt the quotient notation sooner?" The reason is psychological rather than mathematical. The quotient notation makes things look so much like the familiar rationals that there is a tendency to *assume* everything works like it always did, instead of actually carrying out the formal (and tiresome) details of the rigorous development.
** Theorem 9.31 is not used in the sequel.

Proof If $a/b \in F$, then $a, b \in R$ and b is nonzero. Since $R \subseteq K$, b^{-1} exists. Define a map $f: F \to K$ by $f(a/b) = ab^{-1}$. Exercise 9 shows that f is well defined, that is, $a/b = c/d$ in F implies $f(a/b) = f(c/d)$ in K. Exercise 10 shows that f is an injective homomorphism. If E is the image of F under f, then $F \cong E$. For each $a \in R$, $a = a1_R^{-1} = f(a/1_R) \in E$, so $R \subseteq E \subseteq K$. ◆

◆ **EXERCISES**

NOTE: *Unless noted otherwise, R is an integral domain and F its field of quotients.*

A. **1.** Prove Lemma 9.27.

2. Complete the proof of Lemma 9.28 by showing that

(a) Addition of equivalence classes is associative.

(b) Multiplication of equivalence classes is associative.

(c) Multiplication of equivalence classes is commutative.

3. Show that $R^* = \{[a, 1_R] \mid a \in R\}$ is a subring of F.

B. **4.** If R is itself a field, show that $R = F$.

5. If $R = \mathbb{Z}[i]$, then show that $F \cong \{r + si \mid r, s \in \mathbb{Q}\}$.

6. If $R = \mathbb{Z}[\sqrt{d}]$, then show that $F \cong \{r + s\sqrt{d} \mid r, s \in \mathbb{Q}\}$.

7. Show that there are infinitely many integral domains R such that $\mathbb{Z} \subseteq R \subseteq \mathbb{Q}$, each of which has $\mathbb{Q}$ as its field of quotients. [*Hint:* Exercise 23 in Section 3.1.]

8. Let $f: R \to R_1$ be an isomorphism of integral domains. Let F be the field of quotients of R and F_1 the field of quotients of R_1. Prove that the map $f^*: F \to F_1$ given by $f^*(a/b) = f(a)/f(b)$ is an isomorphism.

9. If R is contained in a field K and $a/b = c/d$ in F, show that $ab^{-1} = cd^{-1}$ in K. [*Hint: a/b = c/d* implies $ad = bc$ in K.]

10. (a) Prove that the map f in the proof of Theorem 9.31 is injective. [*Hint: f(a/b) = f(c/d)* implies $ab^{-1} = cd^{-1}$; show that $ad = bc$.]

(b) Use a straightforward calculation to show that f is a homomorphism.

11. Let $a, b \in R$. Assume there are positive integers m, n such that $a^m = b^m$, $a^n = b^n$, and $(m, n) = 1$. Prove that $a = b$. [Remember that negative powers of a and b are not necessarily defined in R, but they do make sense in the field F; for instance, $a^{-2} = 1_R/a^2$.]

12. Let R be an integral domain of characteristic 0 (see Exercises 31–33 in Section 3.2).

(a) Prove that R has a subring isomorphic to $\mathbb{Z}$ [*Hint:* Consider $\{n1_R \mid n \in \mathbb{Z}\}$.]

(b) Prove that a field of characteristic 0 contains a subfield isomorphic to $\mathbb{Q}$. [*Hint:* Theorem 9.31.]

13. Prove that Theorem 9.30 is valid when R is a commutative ring with no zero divisors (not necessarily an integral domain). [*Hint:* Show that for any non-zero $a \in R$, the class $[a, a]$ acts as a multiplicative identity for F and the set $\{[ra, a] \mid r \in R\}$ is a subring of F that is isomorphic to R. The even integers are a good model of this situation.]

9.5 UNIQUE FACTORIZATION IN POLYNOMIAL DOMAINS*

Throughout this section R is a unique factorization domain. We shall prove that the polynomial ring $R[x]$ is also a UFD. The basic idea of the proof is quite simple: Given a polynomial $f(x)$, factor it repeatedly as a product of polynomials of lower degree until $f(x)$ is written as a product of irreducibles. To prove uniqueness, consider $f(x)$ as a polynomial in $F[x]$, where F is the field of quotients of R. Use the fact that $F[x]$ is a UFD (Theorem 4.13) to show that factorization in $R[x]$ is unique. There are some difficulties, however, in carrying out this program.

> **EXAMPLE**   The polynomial $3x^2 + 6$ cannot be factored as a product of two polynomials of lower degree in $\mathbb{Z}[x]$ and is irreducible in $\mathbb{Q}[x]$. But $3x^2 + 6$ is *reducible* in $\mathbb{Z}[x]$ because $3x^2 + 6 = 3(x^2 + 2)$ and neither 3 nor $x^2 + 2$ is a unit in $\mathbb{Z}[x]$.

So the first step is to examine the role of constant polynomials in $R[x]$. By Theorem 4.8 and Exercise 1

the units in $R[x]$ are the units in R

and

**the irreducible constant polynomials in $R[x]$ are
the irreducible elements of R.**

For example, the units of $\mathbb{Z}[x]$ are ± 1. The constant polynomial 3 is irreducible in $\mathbb{Z}[x]$ even though it is a unit in $\mathbb{Q}[x]$.

The constant irreducible factors of a polynomial in $R[x]$ may be found by

---

* The prerequisites for this section are pages 285–287 of Section 9.1, the definition of unique factorization domain (together with Theorems 9.13, 9.15, and 9.18), and Section 9.4. Theorems 9.13, 9.15, and 9.18 depend only on the definition of UFD and may be read independently of the rest of Section 9.2.

factoring out any constants and expressing them as products of irreducible elements in R.

> **EXAMPLE**   In $\mathbb{Z}[x]$,
>
> $$6x^2 + 18x + 12 = 6(x^2 + 3x + 2) = 2 \cdot 3(x^2 + 3x + 2).$$
>
> Note that $x^2 + 3x + 2$ is a polynomial whose only *constant* divisors in $\mathbb{Z}[x]$ are the units ± 1. This example suggests a strategy for the general case.

Let R be a unique factorization domain. A nonzero polynomial in $R[x]$ is said to be **primitive** if the only constants that divide it are the units in R. For instance, $x^2 + 3x + 2$ and $3x^4 - 5x^3 + 2x$ are primitive in $\mathbb{Z}[x]$. Primitive polynomials of degree 0 are units. Every primitive polynomial of degree 1 must be irreducible by Theorem 9.1 (because every factorization includes a constant (Theorem 4.2) and every such constant must be a unit). However, primitive polynomials of higher degree need not be irreducible (such as $x^2 + 3x + 2 = (x + 1)(x + 2)$ in $\mathbb{Z}[x]$). On the other hand, an irreducible polynomial of positive degree has no constant divisors except units by Theorems 4.2 and 9.1. So

an irreducible polynomial of positive degree is primitive.

Furthermore, as the example illustrates,

every nonzero polynomial $f(x) \in R[x]$ factors as $f(x) = cg(x)$ with $g(x)$ primitive.

To prove this claim, let c be a greatest common divisor of the coefficients of $f(x)$.* Then $f(x) = cg(x)$ for some $g(x)$. If $d \in R$ divides $g(x)$, then $g(x) = dh(x)$ so that $f(x) = cdh(x)$. Since cd is a constant divisor of $f(x)$, it must divide the coefficients of $f(x)$ and, hence, must divide the gcd c. Thus $cdu = c$ for some $u \in R$. Since $c \neq 0_R$ we see that $du = 1_R$ and d is a unit. Therefore, $g(x)$ is primitive.

Using these facts about primitive polynomials, we can now modify the argument given at the beginning of the section and prove the first of the two conditions necessary for $R[x]$ to be a UFD.

THEOREM 9.32   *Let R be a unique factorization domain. Then every nonzero, nonunit $f(x)$ in $R[x]$ is a product of irreducible polynomials.***

Proof   Let $f(x) = cg(x)$ with $g(x)$ primitive. Since R is a UFD c is either a unit or a product of irreducible elements in R (and, hence, in $R[x]$). So we need to prove only that $g(x)$ is either a unit or a product of irreducibles in $R[x]$. If $g(x)$ is a unit or is itself irreducible, there is nothing to prove. If not, then by Theorem 9.1

---

* The gcd c exists by Theorem 9.18.

** As usual we allow a "product" with just one factor.

$g(x) = h(x)k(x)$ with neither $h(x)$ or $k(x)$ a unit. Since $g(x)$ is primitive, its only divisors of degree 0 are units, so we must have $0 < \deg h(x) < \deg g(x)$ and $0 < \deg k(x) < \deg g(x)$. Furthermore, $h(x)$ and $k(x)$ are primitive (any constant that divides one of them must divide $g(x)$ and hence be a unit). If they are irreducible, we're done. If not, we can repeat the preceding argument and factor them as products of primitive polynomials of lower degree, and so on. This process must stop after a finite number of steps because the degrees of the factors get smaller at each stage and every primitive polynomial of degree 1 is irreducible. So $g(x)$ is a product of irreducibles in $R[x]$. ◆

The proof that factorization in $R[x]$ is unique depends on several technical facts that will be developed next. But to get an idea of how all the pieces fit together, you may want to read the proof of Theorem 9.38 now, referring to the intermediate results as needed and accepting them without proof. Then you can return to this point and read the proofs, knowing where the argument is headed.

LEMMA 9.33 *Let R be a unique factorization domain and $g(x)$, $h(x) \in R[x]$. If p is an irreducible element of R that divides $g(x)h(x)$, then p divides $g(x)$ or p divides $h(x)$.*

Proof Copy the proof of Lemma 4.21, which is the special case $R = \mathbb{Z}$. Just replace $\mathbb{Z}$ by R and *prime* by *irreducible* and use Theorem 9.15 in place of Theorem 1.8. ◆

COROLLARY 9.34 (GAUSS'S LEMMA) *Let R be a unique factorization domain. Then the product of primitive polynomials in $R[x]$ is primitive.*

Proof If $g(x)$ and $h(x)$ are primitive and $g(x)h(x)$ is not, then $g(x)h(x)$ is divisible by some nonunit $c \in R$. Consequently, each irreducible factor p of c divides $g(x)h(x)$. By Lemma 9.33, p divides $g(x)$ or $h(x)$, contradicting the fact that they are primitive. Therefore, $g(x)h(x)$ is primitive. ◆

THEOREM 9.35 *Let R be a unique factorization domain and r, s nonzero elements of R. Let $f(x)$ and $g(x)$ be primitive polynomials in $R[x]$ such that $rf(x) = sg(x)$. Then r and s are associates in R and $f(x)$ and $g(x)$ are associates in $R[x]$.*

Proof If r is a unit, then $f(x) = r^{-1}sg(x)$. Since $r^{-1}s$ divides the primitive polynomial $f(x)$, it must be a unit, say $(r^{-1}s)u = 1_R$. Hence, $f(x)$ and $g(x)$ are associates in $R[x]$. Furthermore, u is a unit in R and $su = r$ so that r and s are associates in R. If r is a nonunit, then $r = p_1 p_2 \cdots p_k$ with each p_i irreducible. Then $p_1 p_2 \cdots p_k f(x) = sg(x)$, so p_1 divides $sg(x)$. By Lemma 9.33 p_1 divides s or $g(x)$. Since p_1 is a nonunit and $g(x)$ is primitive, p_1 must divide s, say $s = p_1 t$.

Then $p_1 p_2 \cdots p_k f(x) = sg(x) = p_1 tg(x)$.   Cancelling p_1 shows that $p_2 \cdots p_k f(x) = tg(x)$.   Repeating the argument with p_2 shows that $p_3 \cdots p_k f(x) = zg(x)$, where $p_2 z = t$ and, hence, $p_1 p_2 z = p_1 t = s$. After k such steps we have $f(x) = wg(x)$ and $s = p_1 p_2 \cdots p_k w$ for some $w \in R$. Since w divides the primitive polynomial $f(x)$, w is a unit. Therefore, $f(x)$ and $g(x)$ are associates in $R[x]$. Since $s = p_1 \cdots p_k w = rw$, r and s are associates in R.   ◆

COROLLARY 9.36   *Let R be a unique factorization domain and F its field of quotients. Let $f(x)$, $g(x)$ be primitive polynomials in $R[x]$. If $f(x)$ and $g(x)$ are associates in $F[x]$, then they are associates in $R[x]$.*

Proof   If $f(x)$ and $g(x)$ are associates in $F[x]$, then $g(x) = \dfrac{r}{s} f(x)$ for some nonzero $\dfrac{r}{s} \in F$ by Corollary 4.9. Consequently, $sg(x) = rf(x)$ in $R[x]$. Therefore, $f(x)$ and $g(x)$ are associates in $R[x]$ by Theorem 9.35.   ◆

COROLLARY 9.37   *Let R be a unique factorization domain and F its field of quotients. If $f(x) \in R[x]$ has positive degree and is irreducible in $R[x]$, then $f(x)$ is irreducible in $F[x]$.*

Proof   If $f(x)$ is not irreducible in $F[x]$, then $f(x) = g(x)h(x)$ for some $g(x)$, $h(x) \in F[x]$ with positive degree. Let b be a least common denominator of the coefficients of $g(x)$. Then $bg(x)$ has coefficients in R. So $bg(x) = ag_1(x)$ with $a \in R$ and $g_1(x)$ primitive of positive degree in $R[x]$. Hence, $g(x) = \dfrac{a}{b} g_1(x)$. Similarly $h(x) = \dfrac{c}{d} h_1(x)$ with c, $d \in R$ and $h_1(x)$ primitive of positive degree in $R[x]$. Therefore, $f(x) = g(x)h(x) = \dfrac{a}{b} g_1(x) \dfrac{c}{d} h_1(x) = \dfrac{ac}{bd} g_1(x)h_1(x)$, so that $bdf(x) = acg_1(x)h_1(x)$ in $R[x]$. Now $f(x)$ is primitive because it is irreducible and $g_1(x)h_1(x)$ is primitive by Corollary 9.34. So bd is an associate of ac by Theorem 9.35, say $bdu = ac$ for some unit $u \in R$. Therefore, $f(x) = \dfrac{ac}{bd} g_1(x)h_1(x) = ug_1(x)h_1(x)$. Since $ug_1(x)$ and $h_1(x)$ are polynomials of positive degree in $R[x]$, this contradicts the irreducibility of $f(x)$. Therefore, $f(x)$ must be irreducible in $F[x]$.   ◆

THEOREM 9.38   *If R is a unique factorization domain, then so is $R[x]$.*

Proof   Every nonzero nonunit $f(x)$ in $R[x]$ is a product of irreducibles by Theorem 9.32. Any such factorization consists of irreducible constants (that is, irreducibles in R) and irreducible polynomials of positive degree. Suppose

$$c_1 \cdots c_m p_1(x) \cdots p_k(x) = d_1 \cdots d_n q_1(x) \cdots q_t(x)$$

with each c_i, d_j irreducible in R and each $p_i(x)$, $q_j(x)$ irreducible of positive degree in $R[x]$ (and, hence, primitive).* Then $p_1(x) \cdots p_k(x)$ and $q_1(x) \cdots q_t(x)$ are primitive by Corollary 9.34. So Theorem 9.35 shows that $c_1 \cdots c_m$ is an associate of $d_1 \cdots d_n$ in R and $p_1(x) \cdots p_k(x)$ is an associate of $q_1(x) \cdots q_t(x)$ in $R[x]$. Hence, $c_1 \cdots c_m = u d_1 d_2 \cdots d_n$ for some unit $u \in R$. Associates of irreducibles are irreducible (Exercise 7 of Section 9.1), so $u d_1$ is irreducible. Since R is a UFD, we must have $m = n$ and (after relabeling if necessary) c_1 is an associate of $u d_1$ (and hence of d_1), and c_i is an associate of d_i for $i \geq 2$. Let F be the field of quotients of R. Each of the $p_i(x)$, $q_j(x)$ is irreducible in $F[x]$ by Corollary 9.37. Unique factorization in $F[x]$ (Theorem 4.13) and an argument similar to the one just given for R show that $k = t$ and (after relabeling if necessary) each $p_i(x)$ is an associate of $q_i(x)$ in $F[x]$. Consequently, $p_i(x)$ and $q_i(x)$ are associates in $R[x]$ by Corollary 9.36. Therefore, $R[x]$ is a UFD. ◆

An immediate consequence of Theorems 1.11 and 9.38 and the example on page 137 is

COROLLARY 9.39 $\mathbb{Z}[x]$ *is a unique factorization domain that is not a principal ideal domain.*

As illustrated in the preceding discussion, theorems about $\mathbb{Z}[x]$ and $\mathbb{Q}[x]$ are quite likely to carry over to an arbitrary UFD and its field of quotients. among such results are the Rational Root Test and Eisenstein's Criterion (Exercises 9–11).

◆ **EXERCISES**

NOTE: *Unless stated otherwise R is a UFD and F its field of quotients.*

A.   **1.** Let R be any integral domain and $p \in R$. Prove that p is irreducible in R if and only if the constant polynomial p is irreducible in $R[x]$. [*Hint:* Theorem 4.8 may be helpful.]

2. Give an example of polynomials $f(x), g(x) \in R[x]$ such that $f(x)$ and $g(x)$ are associates in $F[x]$ but not in $R[x]$. Does this contradict Corollary 9.36?

3. If $c_1 \cdots c_m f(x) = g(x)$ with $c_i \in R$ and $g(x)$ primitive in $R[x]$, prove that each c_i is a unit.

4. If $g(x)$ is primitive in $R[x]$, prove that every nonconstant polynomial in $R[x]$ that divides $g(x)$ is also primitive.

---

* It may be that neither factorization contains constants, but this doesn't affect the argument. It is not possible to have irreducible constants in one factorization but not in the other (Exercise 5).

B. **5.** Prove that a polynomial is primitive if and only if 1_R is a greatest common divisor of its coefficients. This property is often taken as the definition of primitive.

6. If $f(x)$ is primitive in $R[x]$ and irreducible in $F[x]$, prove that $f(x)$ is irreducible in $R[x]$.

7. If R is a ring such that $R[x]$ is a UFD, prove that R is a UFD.

8. If R is a ring such that $R[x]$ is a principal ideal domain, prove that R is a field.

9. Verify that the Rational Root Test (Theorem 4.20) is valid with $\mathbb{Z}$ and $\mathbb{Q}$ replaced by R and F.

10. Verify that Theorem 4.22 is valid with $\mathbb{Z}$ and $\mathbb{Q}$ replaced by R and F.

11. Verify that Eisenstein's Criterion (Theorem 4.23) is valid with $\mathbb{Z}$ and $\mathbb{Q}$ replaced by R and F and *prime* replaced by *irreducible*.

12. Show that $x^3 - 6x^2 + 4ix + 1 + 3i$ is irreducible in $(\mathbb{Z}[i])[x]$. [*Hint:* Exercise 11.]

CHAPTER **10**

Field Extensions

◆

High-school algebra deals primarily with the three fields $\mathbb{Q}$, $\mathbb{R}$, and $\mathbb{C}$ and plane geometry, with the set $\mathbb{R} \times \mathbb{R}$. Calculus is concerned with functions from $\mathbb{R}$ to $\mathbb{R}$. Indeed, most classical mathematics is set in the field $\mathbb{C}$ and its subfields. Other fields play an equally important role in more recent mathematics. They are used in analysis, algebraic geometry, and parts of number theory, for example, and have numerous applications, including coding theory and algebraic cryptography.

In this chapter we develop the basic facts about fields that are needed to prove some famous results in the theory of equations (Chapter 11) and to study some of the topics listed above. The principal theme is the relationship of a field with its various subfields.

10.1 VECTOR SPACES

An essential tool for the study of fields is the concept of a vector space, which is introduced in this section. Vector spaces are treated in detail in books and courses on Linear Algebra. Here we present only those topics that are needed for our study of fields. If you have had a course in linear algebra, you can probably skip most of this section. Nevertheless, it would be a good idea to review the main results, particularly Theorems 10.4 and 10.5.

Consider the additive abelian group* $M(\mathbb{R})$ of all 2×2 matrices over the

---

* Except for the last two results in the chapter, group theory is not a prerequisite for this chapter. In this section you need only know that an additive abelian group is a set with an addition operation that satisfies axioms 1–5 in the definition of a ring (page 42).

field $\mathbb{R}$ of real numbers. If r is a real number and $A = \begin{pmatrix} a & b \\ c & d \end{pmatrix}$ is an element of $M(\mathbb{R})$, then the product of the number r and the matrix A is defined to be the matrix $rA = \begin{pmatrix} ra & rb \\ rc & rd \end{pmatrix}$. This operation, which is called *scalar multiplication*, takes a real number (field element) and a matrix (group element) and produces another matrix (group element). This is an example of a more general concept. Let F be a field and G an additive abelian group.* Then a **scalar multiplication** is an operation such that for each $a \, \epsilon \, F$ and each $v \, \epsilon \, G$ there is a unique element $av \, \epsilon \, G$.

DEFINITION• *Let F be a field. A **vector space over F** is an additive abelian group* V equipped with a scalar multiplication such that for all $a, a_1, a_2 \, \epsilon \, F$ and $v, v_1, v_2 \, \epsilon \, V$:*

(i) $a(v_1 + v_2) = av_1 + av_2$;

(ii) $(a_1 + a_2)v = a_1v + a_2v$;

(iii) $a_1(a_2v) = (a_1a_2)v$;

(iv) $1_F v = v$.

EXAMPLE Scalar multiplication in $M(\mathbb{R})$, as defined above, makes $M(\mathbb{R})$ into a vector space over $\mathbb{R}$ (Exercise 1).

EXAMPLE Consider the set $\mathbb{Q}^2 = \mathbb{Q} \times \mathbb{Q}$, where $\mathbb{Q}$ is the field of rational numbers. Then $\mathbb{Q}^2$ is a group under addition (Theorem 3.1 or 7.4); its zero element is $(0, 0)$ and the negative of (s, t) is $(-s, -t)$. For $a \, \epsilon \, \mathbb{Q}$ and $(s, t) \, \epsilon \, \mathbb{Q}^2$, scalar multiplication is defined by $a(s, t) = (as, at)$. Under these operations $\mathbb{Q}^2$ is a vector space over $\mathbb{Q}$ (Exercise 2).

EXAMPLE The preceding example can be generalized as follows. If F is any field and $n \geq 1$ an integer, let $F^n = F \times F \times \cdots \times F$ (n summands). Then F^n is a vector space over F, with addition defined coordinatewise:

$$(s_1, s_2, \ldots, s_n) + (t_1, t_2, \ldots, t_n) = (s_1 + t_1, s_2 + t_2, \ldots, s_n + t_n)$$

and scalar multiplication defined by:

$$a(s_1, s_2, \ldots, s_n) = (as_1, as_2, \ldots, as_n) \qquad a \, \epsilon \, F$$

(see Exercise 5).

---

* Except for the last two results in the chapter, group theory is not a prerequisite for this chapter. In this section you need only know that an additive abelian group is a set with an addition operation that satisfies axioms 1–5 in the definition of a ring (page 42).

> **EXAMPLE** The complex numbers $\mathbb{C}$ form a vector space over the real numbers $\mathbb{R}$, with addition of complex numbers (vectors) defined as usual and with scalar multiplication being ordinary multiplication (the product of a real number and a complex number is a complex number).

Special terminology is used in situations like the preceding example. If F and K are fields with $F \subseteq K$, we say that K is an **extension field** of F. For instance, the complex numbers $\mathbb{C}$ are an extension field of the field $\mathbb{R}$ of real numbers. As the preceding example shows, the extension field $\mathbb{C}$ can be considered as a vector space over $\mathbb{R}$. The same thing is true in the general case.

If K is an extension field of F, then K is a vector space over F, with addition of vectors being ordinary addition in K and scalar multiplication being ordinary multiplication in K

(the product of an element the subfield F and an element of K is an element of K). For the purposes of this chapter, extension fields are the most important examples of vector spaces.

If V is a vector space over a field F, then the following properties hold for any $v \in V$ and $a \in F$ (Exercise 21):

$$0_F v = 0_V, \qquad a0_V = 0_V, \qquad -(av) = (-a)v = a(-v).$$

Spanning Sets

Suppose V is a vector space over a field F and that w and $v_1, v_2, \ldots, v_n$ are elements of V. We say that w is a **linear combination** of $v_1, v_2, \ldots, v_n$ if w can be written in the form

$$w = a_1 v_1 + a_2 v_2 + \cdots + a_n v_n$$

for some $a_i \in F$.

> **DEFINITION•** *If every element of a vector space V over a field F is a linear combination of $v_1, v_2, \ldots, v_n$, we say that the set $(v_1, v_2, \ldots, v_n\}$ **spans** V over F.*

> **EXAMPLE** The set $\{(1, 0, 0), (0, 1, 0), (0, 0, 1)\}$ spans the vector space $\mathbb{Q}^3$ over $\mathbb{Q}$ because every element (a, b, c) of $\mathbb{Q}^3$ is a linear combination of these three vectors:
>
> $$(a, b, c) = a(1, 0, 0) + b(0, 1, 0) + c(0, 0, 1).$$

> **EXAMPLE** Every element of $\mathbb{C}$ (considered as a vector space over $\mathbb{R}$) is a linear combination of 1 and i because every element can be written in the form $a1 + bi$, with $a, b \in \mathbb{R}$. Thus the set $\{1, i\}$ spans $\mathbb{C}$ over $\mathbb{R}$. The set

$\{1 + i, 5i, 2 + 3i\}$ also spans $\mathbb{C}$ because any $a + bi \in \mathbb{C}$ is a linear combination of these three elements with coefficients in $\mathbb{R}$:

$$a + bi = 3a(1 + i) + \frac{b}{5}(5i) + (-a)(2 + 3i).$$

Linear Independence and Bases

The set $\{1, i\}$ not only spans the extension field $\mathbb{C}$ of $\mathbb{R}$, but it also has this property: If $a1 + bi = 0$, then $a = 0$ and $b = 0$. In other words, when a linear combination of 1 and i is 0, then all the coefficients are 0. On the other hand, the set $\{1 + i, 5i, 2 + 3i\}$ does not have this property because some linear combinations of these elements are 0 even though the coefficients are not; for instance,

$$2(1 + i) + \frac{1}{5}(5i) - 1(2 + 3i) = 0.$$

The distinction between these two situations will be crucial in our study of field extensions.

> **DEFINITION•** *A subset $\{v_1, v_2, \ldots, v_n\}$ of a vector space V over a field F is said to be **linearly independent** over F provided that whenever*
>
> $$c_1 v_1 + c_2 v_2 + \cdots + c_n v_n = 0_V$$
>
> *with each $c_i \in F$, then $c_i = 0_F$ for every i. A set that is not linearly independent is said to be **linearly dependent**.*

Thus a set $\{u_1, u_2, \ldots, u_m\}$ is linearly dependent over F if there exist elements $b_1, b_2, \ldots, b_m$ of F, at least one of which is nonzero, such that $b_1 u_1 + b_2 u_2 + \cdots + b_m u_m = 0_V$.

EXAMPLE The remarks preceding the definition show that the subset $\{1, i\}$ of $\mathbb{C}$ is linearly independent over $\mathbb{R}$ and that the set $\{1 + i, 5i, 2 + 3i\}$ is linearly dependent. Note, however, that both of these sets span $\mathbb{C}$.

EXAMPLE Consider the subset $\{(3, 0, 0), (0, 0, 4)\}$ of the vector space $\mathbb{Q}^3$ over $\mathbb{Q}$ and suppose $c_1, c_2 \in \mathbb{Q}$ are such that $c_1(3, 0, 0) + c_2(0, 0, 4) = (0, 0, 0)$. Then

$$(0, 0, 0) = c_1(3, 0, 0) + c_2(0, 0, 4) = (3c_1, 0, 4c_2),$$

which implies that $c_1 = 0 = c_2$. Hence, $\{(3, 0, 0), (0, 0, 4)\}$ is linearly independent over $\mathbb{Q}$. However, the set $\{(3, 0, 0), (0, 0, 4)\}$ does not span $\mathbb{Q}^3$

because there is no way to write the vector $(0, 5, 0)$, for example, in the form $a_1(3, 0, 0) + a_2(0, 0, 4) = (3a_1, 0, 4a_2)$ with $a_i \in \mathbb{Q}$.

Let V be a vector space over a field F. The preceding examples show that linear independence and spanning do not imply each other; a subset of V may have one, both, or neither of these properties. A subset that has both properties is given a special name.

> **DEFINITION •** *A subset $\{v_1, v_2, \ldots, v_n\}$ of a vector space V over a field F is said to be a **basis** of V if it spans V and is linearly independent over F.*

EXAMPLE The second example on page 331 shows that the subset $\{(1, 0, 0), (0, 1, 0), (0, 0, 1)\}$ spans the vector space $\mathbb{Q}^3$ over $\mathbb{Q}$. This set is also linearly independent over $\mathbb{Q}$ (Exercise 8) and, hence, is a basis.

EXAMPLE The examples on pages 331 and 332 show that the set $\{1, i\}$ is a basis of $\mathbb{C}$ over $\mathbb{R}$. We claim that the set $\{1 + i, 2i\}$ is also a basis of $\mathbb{C}$ over $\mathbb{R}$. If $c_1(1 + i) + c_2(2i) = 0$, with $c_1, c_2 \in \mathbb{R}$, then $c_1 1 + (c_1 + 2c_2)i = 0$. This can happen only if $c_1 = 0$ and $c_1 + 2c_2 = 0$. But this implies that $2c_2 = 0$ and, hence, $c_2 = 0$. Therefore, $\{1 + i, 2i\}$ is linearly independent. In order to see that $\{1 + i, 2i\}$ spans $\mathbb{C}$, note that the element $a + bi \in \mathbb{C}$ can be written as $a(1 + i) + \left(\dfrac{b - a}{2}\right) 2i$.

One situation always leads to linear dependence. Let V be a vector space over a field F and S a subset of V. Suppose that $v, u_1, u_2, \ldots, u_t$ are some of the elements of S and that v is a linear combination of $u_1, u_2, \ldots, u_t$, say $v = a_1 u_1 + \cdots + a_t u_t$, with each $a_i \in F$. If $w_1, \ldots, w_r$ are the rest of the elements of S, then

$$v = a_1 u_1 + \cdots + a_t u_t + 0_F w_1 + \cdots + 0_F w_r$$

and, hence,

$$-1_F v + a_1 u_1 + \cdots + a_t u_t + 0_F w_1 + \cdots + 0_F w_r = 0_V.$$

Since at least one of these coefficients is nonzero (namely -1_F), S is linearly dependent. We have proved this useful fact:

If $v \in V$ is a linear combination of $u_1, u_2, \ldots, u_t \in V$, then any set containing v and all the u_i is linearly dependent.

In fact, somewhat more is true.

LEMMA 10.1   *Let V be a vector space over a field F. The subset $\{u_1, u_2, \ldots, u_n\}$ of V is linearly dependent over F if and only if some u_k is a linear combination of the preceding ones, $u_1, u_2, \ldots, u_{k-1}$.*

Proof   If some u_k is a linear combination of the preceding ones, then the set is linearly dependent by the remarks preceding the lemma. Conversely, suppose $\{u_1, \ldots, u_n\}$ is linearly dependent. Then there must exist elements $c_1, \ldots, c_n \in F$, *not all zero*, such that $c_1u_1 + c_2u_2 + \cdots + c_nu_n = 0_V$. Let k be the largest index such that c_k is nonzero. Then $c_i = 0_F$ for $i > k$ and

$$c_1u_1 + c_2u_2 + \cdots + c_ku_k = 0_V$$
$$c_ku_k = -c_1u_1 - c_2u_2 - \cdots - c_{k-1}u_{k-1}.$$

Since F is a field and $c_k \neq 0$, c_k^{-1} exists; multiplying the preceding equation by c_k^{-1} shows that u_k is a linear combination of the preceding u's:

$$u_k = (-c_1c_k^{-1})u_1 + (-c_2c_k^{-1})u_2 + \cdots + (-c_{k-1}c_k^{-1})u_{k-1}. \quad \blacklozenge$$

The next lemma gives an upper limit on the size of a linearly independent set. It says, in effect, that if V can be spanned by n elements over F, then every linearly independent subset of V contains at most n elements.

LEMMA 10.2   *Let V be a vector space over the field F that is spanned by the set $\{v_1, v_2, \ldots, v_n\}$. If $\{u_1, u_2, \ldots, u_m\}$ is any linearly independent subset of V, then $m \leq n$.*

Proof   By the definition of spanning, every element of V (in particular u_1) is a linear combination of $v_1, \ldots, v_n$. So the set $\{u_1, v_1, v_2, \ldots, v_n\}$ is linearly dependent. Therefore, one of its elements is a linear combination of the preceding ones by Lemma 10.1, say $v_i = a_1u_1 + b_1v_1 + \cdots + b_{i-1}v_{i-1}$. If v_i is deleted, then the remaining set

$$(*) \qquad \{u_1, v_1, \ldots, v_{i-1}, v_{i+1}, \ldots, v_n\}$$

still spans V since every element of V is a linear combination of the v's and any appearance of v_i can be replaced by $a_1u_1 + b_1v_1 + \cdots + b_{i-1}v_{i-1}$. In particular, u_2 is a linear combination of the elements of the set (*). Consequently, the set

$$\{u_1, u_2, v_1, \ldots, v_{i-1}, v_{i+1}, \ldots, v_n\}$$

is linearly dependent. By Lemma 10.1 one of its elements is a linear combination of the preceding ones. This element can't be one of the u's because this would imply that the u's were linearly dependent. So some v_j is a linear combination of u_1, u_2, and the v's that precede it. Deleting v_j produces the set

$$\{u_1, u_2, v_1, \ldots, v_{i-1}, v_{i+1}, \ldots, v_{j-1}, v_{j+1}, \ldots, v_n\}.$$

This set still spans V since every element of V is a linear combination of the v's and v_i, v_j can be replaced by linear combinations of u_1, u_2, and the other v's. In particular, u_3 is a linear combination of the elements in this new set. We can continue this process, at each stage adding a u, deleting a v, and producing a set that spans V. If $m > n$, we will run out of v's before all the u's are inserted, resulting in a set of the form $\{u_1, u_2, \ldots, u_n\}$ that spans V. But this would mean that u_m would be a linear combination of $u_1, \ldots, u_n$, contradicting the linear independence of $\{u_1, \ldots, u_m\}$. Therefore, $m \leq n$. ◆

THEOREM 10.3   *Let V be a vector space over a field F. Then any two finite bases of V over F have the same number of elements.*

Proof   Suppose $\{u_1, \ldots, u_m\}$ and $\{v_1, \ldots, v_n\}$ are bases of V over F. Then the v's span V and the u's are linearly independent, so $m \leq n$ by Lemma 10.2. Now reverse the roles: The u's span V and the v's are linearly independent, so $n \leq m$ by Lemma 10.2 again. Therefore, $m = n$. ◆

According to Theorem 10.3, the *number* of elements in a basis of V over F does not depend on which basis is chosen. So this number is a property of V.

> **DEFINITION•** *If a vector space V over a field F has a finite basis, then V is said to be **finite dimensional** over F. The **dimension of V over F** is the number of elements in any basis of V and is denoted $[V:F]$. If V does not have a finite basis, then V is said to be **infinite dimensional** over F.*

EXAMPLE   The dimension of $\mathbb{Q}^3$ over $\mathbb{Q}$ is 3 because $\{(1, 0, 0), (0, 1, 0), (0, 0, 1)\}$ is a basis. More generally, if F is a field, then F^n is an n-dimensional vector space over F (Exercise 27).

EXAMPLE $[\mathbb{C}:\mathbb{R}] = 2$ since $\{1, i\}$ is a basis of $\mathbb{C}$ over $\mathbb{R}$. On the other hand, the extension field $\mathbb{R}$ of $\mathbb{Q}$ is an infinite-dimensional vector space over $\mathbb{Q}$. The proof of this fact is omitted here because it requires some nontrivial facts about the cardinality of infinite sets.

Applications to Extension Fields

In the remainder of this section, K is an extension field of a field F. We say that K is a **finite-dimensional extension** of F if K, considered as a vector space over F, is finite dimensional over F.

Remark   If $[K:F] = 1$ and $\{u\}$ is a basis, then every element of K is of the form cu for some $c \in F$. In particular, $1_F = cu$, and, hence, $u = c^{-1}$ is in F. Thus,

$K = F$. On the other hand, if $K = F$, it is easy to see that $\{1_F\}$ is a basis and, hence, $[K:F] = 1$. Therefore,

$$[K:F] = 1 \qquad \text{if and only if} \qquad K = F.$$

If F, K, and L are fields with $F \subseteq K \subseteq L$, then both K and L can be considered as vector spaces over F, and L can be considered as a vector space over K. It is reasonable to ask how the dimensions $[K:F]$, $[L:K]$, and $[L:F]$ are related. Here is the answer.

THEOREM 10.4 *Let F, K, and L be fields with $F \subseteq K \subseteq L$. If $[K:F]$ and $[L:K]$ are finite, then L is a finite dimensional extension of F and $[L:F] = [L:K][K:F]$.*

Proof Suppose $[K:F] = m$ and $[L:K] = n$. Then there is a basis $\{u_1, \ldots, u_m\}$ of K over F and a basis $\{v_1, \ldots, v_n\}$ of L over K. Each u_i and v_j is nonzero by Exercise 19; hence, all the products $u_i v_j$ are nonzero. The set of all products $\{u_i v_j \mid 1 \le i \le m,\ 1 \le j \le n\}$ has exactly mn elements (no two of them can be equal because $u_i v_j = u_k v_t$ implies that $u_i v_j - u_k v_t = 0_K$ with u_i, $u_k \in K$, contradicting the linear independence of the v's over K). We need to show only that this set of mn elements is a basis of L over F because in that case $[L:K][K:F] = nm = [L:F]$.

If w is any element of L, then w is a linear combination of the basis elements $v_1, \ldots, v_n$, say

(∗) $\qquad w = b_1 v_1 + b_2 v_2 + \cdots + b_n v_n, \qquad$ with each $b_j \in K$.

Each $b_j \in K$ is a linear combination of the basis elements $u_1, \ldots, u_m$; so there are $a_{ij} \in F$ such that

$$b_1 = a_{11}u_1 + a_{21}u_2 + \cdots + a_{m1}u_m$$
$$b_2 = a_{12}u_1 + a_{22}u_2 + \cdots + a_{m2}u_m$$
$$\cdot$$
$$\cdot$$
$$\cdot$$
$$b_n = a_{1n}u_1 + a_{2n}u_2 + \cdots + a_{mn}u_m.$$

Substituting the right side of each of these expressions in (∗) shows that w is a sum of terms of the form $a_{ij}u_i v_j$ with $a_{ij} \in F$. Therefore, the set of all products $u_i v_j$ spans L over F.

To show linear independence, suppose $c_{ij} \in F$ and

(∗∗) $\qquad \sum_{i,j} c_{ij}u_i v_j = c_{11}u_1 v_1 + c_{12}u_1 v_2 + \cdots + c_{mn}u_m v_n = 0_F.$

By collecting all the terms involving v_1, then all those involving v_2, and so on, we can rewrite (∗∗) as

$$(c_{11}u_1 + c_{21}u_2 + \cdots + c_{m1}u_m)v_1$$
$$+ (c_{12}u_1 + c_{22}u_2 + \cdots + c_{m2}u_m)v_2$$
$$+ \cdots + (c_{1n}u_1 + c_{2n}u_2 + \cdots + c_{mn}u_m)v_n = 0_F.$$

The coefficients of the v's are elements of K, so the linear independence of the v's implies that for each $j = 1, 2, \ldots, n$

$$c_{1j}u_1 + c_{2j}u_2 + \cdots + c_{mj}u_m = 0_F.$$

Since each $c_{ij} \in F$ and the u's are linearly independent over F, we must have $c_{ij} = 0_F$ for all i, j. This completes the proof of linear independence, and the theorem is proved. ◆

The following result will be needed for the proof of Theorem 10.15 in Section 10.4.

THEOREM 10.5 *Let K and L be finite dimensional extension fields of F and let $f : K \to L$ be an isomorphism such that $f(c) = c$ for every $c \in F$. Then $[K:F] = [L:F]$.*

Proof Suppose $[K:F] = n$ and $\{u_1, \ldots, u_n\}$ is a basis of K over F. In order to prove that $[L:F] = n$ also, we need only show that $\{f(u_1), \ldots, f(u_n)\}$ is a basis of L over F. Let $v \in L$; then since f is an isomorphism, $v = f(u)$ for some $u \in K$. By the definition of basis, $u = c_1u_1 + \cdots + c_nu_n$ with each $c_i \in F$. Hence, $v = f(u) = f(c_1u_1 + \cdots + c_nu_n) = f(c_1)f(u_1) + \cdots + f(c_n)f(u_n)$. But $f(c_i) = c_i$ for every i, so that $v = c_1f(u_1) + \cdots + c_nf(u_n)$. Therefore, $\{f(u_1), \ldots, f(u_n)\}$ spans L. To show linear independence, suppose that

$$d_1f(u_1) + \cdots + d_nf(u_n) = 0_F$$

with each $d_i \in F$. Then since $f(d_i) = d_i$ we have

$$f(d_1u_1 + \cdots + d_nu_n) = f(d_1)f(u_1) + \cdots + f(d_n)f(u_n)$$
$$= d_1f(u_1) + \cdots + d_nf(u_n) = 0_F.$$

Since the isomorphism f is injective, $d_1u_1 + \cdots + d_nu_n = 0_F$ by Theorem 6.11. But the u's are linearly independent in K, and, hence, every $d_i = 0_F$. Thus $\{f(u_1), \ldots, f(u_n)\}$ is linearly independent and, therefore, a basis. ◆

◆ **EXERCISES**

NOTE: *V denotes a vector space over a field F, and K denotes an extension field of F.*

A. **1.** Show that $M(\mathbb{R})$ is a vector space over $\mathbb{R}$.

2. Show that $\mathbb{Q}^2$ is a vector space over $\mathbb{Q}$.

3. Show that the polynomial ring $\mathbb{R}[x]$ (with the usual addition of polynomials and product of a constant and a polynomial) is a vector space over $\mathbb{R}$.

4. If $n \geq 1$ is an integer, let $\mathbb{R}_n[x]$ denote the set consisting of the constant polynomial 0 and all polynomials in $\mathbb{R}[x]$ of degree $\leq n$. Show that $\mathbb{R}_n[x]$ (with the usual addition of polynomials and product of a constant and a polynomial) is a vector space over $\mathbb{R}$.

5. If $n \geq 1$ is an integer, show that F^n is a vector space over F.

6. If $\{v_1, v_2, \ldots, v_n\}$ spans K over F and w is any element of K, show that $\{w, v_1, v_2, \ldots, v_n\}$ also spans K.

7. Show that $\{i, 1 + 2i, 1 + 3i\}$ spans $\mathbb{C}$ over $\mathbb{R}$.

8. Show that the subset $\{(1, 0, 0), (0, 1, 0), (0, 0, 1)\}$ of $\mathbb{Q}^3$ is linearly independent over $\mathbb{Q}$.

9. Show that $\{\sqrt{2}, \sqrt{2} + i, \sqrt{3} - i\}$ is linearly dependent over $\mathbb{R}$.

10. If v is a nonzero element of V, prove that $\{v\}$ is linearly independent over F.

11. Prove that any subset of V that contains 0_V is linearly dependent over F.

12. If the subset $\{u, v, w\}$ of V is linearly independent over F, prove that $\{u, u + v, u + v + w\}$ is linearly independent.

13. If $S = \{v_1, \ldots, v_k\}$ is a linearly dependent subset of V, then prove that any subset of V that contains S is also linearly dependent over F.

14. If the subset $T = \{u_1, \ldots, u_t\}$ of V is linearly independent over F, then prove that any nonempty subset of T is also linearly independent.

15. Let b and d be distinct nonzero real numbers and c any real number. Prove that $\{b, c + di\}$ is a basis of $\mathbb{C}$ over $\mathbb{R}$.

16. If K is an n-dimensional extension field of $\mathbb{Z}_p$, what is the maximum possible number of elements in K?

17. Let $\{v_1, \ldots, v_n\}$ be a basis of V over F and let $c_1, \ldots, c_n$ be nonzero elements of F. Prove that $\{c_1v_1, c_2v_2, \ldots, c_nv_n\}$ is also a basis of V over F.

18. Show that $\{1, [x]\}$ is a basis of $\mathbb{Z}_2[x]/(x^2 + x + 1)$ over $\mathbb{Z}_2$.

19. If $\{v_1, v_2 \cdots, v_n\}$ is a basis of v, prove that $v_i \neq 0_V$ for every i.

20. Let F, K, and L be fields such that $F \subseteq K \subseteq L$. If $S = \{v_1, v_2, \ldots, v_n\}$ spans L over F, explain why S also spans L over K.

B. 21. For any vector $v \in V$ and any element $a \in F$, prove that

(a) $0_F v = 0_V$. [*Hint:* Adapt the proof of Theorem 3.5.]

(b) $a0_V = 0_V$.

(c) $-(av) = (-a)v = a(-v)$.

22. (a) Prove that the subset $\{1, \sqrt{2}\}$ of $\mathbb{R}$ is linearly independent over $\mathbb{Q}$.

(b) Prove that $\sqrt{3}$ is not a linear combination of 1 and $\sqrt{2}$ with coefficients in $\mathbb{Q}$. Conclude that $\{1, \sqrt{2}\}$ does not span $\mathbb{R}$ over $\mathbb{Q}$.

23. (a) Show that $\{1, \sqrt{2}, \sqrt{3}\}$ is linearly independent over $\mathbb{Q}$.

(b) Show that $\{1, \sqrt{2}, \sqrt{3}, \sqrt{6}\}$ is linearly independent over $\mathbb{Q}$.

24. Let v be a nonzero real number. Prove that $\{1, v\}$ is linearly independent over $\mathbb{Q}$ if and only if v is irrational.

25. (a) Let $k \geq 1$ be an integer. Show that the subset $\{1, x, x^2, x^3, \ldots, x^k\}$ of $\mathbb{R}[x]$ is linearly independent over $\mathbb{R}$ (see Exercise 3).

(b) Show that $\mathbb{R}[x]$ is infinite dimensional over $\mathbb{R}$.

26. Show that the vector space $\mathbb{R}_n[x]$ of Exercise 4 has dimension $n + 1$ over $\mathbb{R}$.

27. If F is a field, show that the vector space F^n has dimension n over F.

28. Prove that K has exactly one basis over F if and only if $K = F \cong \mathbb{Z}_2$.

29. Assume $1_F + 1_F \neq 0_F$. If $\{u, v, w\}$ is a basis of V over F, prove that the set $\{u + v, v + w, u + w\}$ is also a basis.

30. Prove that $\{v_1, \ldots, v_n\}$ is a basis of V over F if and only if every element of V can be written in a unique way as a linear combination of $v_1, \ldots, v_n$ ("unique" means that if $w = c_1 v_1 + \cdots + c_n v_n$ and $w = d_1 v_1 + \cdots + d_n v_n$, then $c_i = d_i$ for every i).

31. Let $p(x) = a_0 + a_1 x + \cdots + a_n x^n$ be irreducible in $F[x]$ and let L be the extension field $F[x]/(p(x))$ of F. Prove that L has dimension n over F. [*Hint:* Corollary 5.5, Theorems 5.8 and 5.10, and Exercise 30 may be helpful.]

32. If $S = \{v_1, \ldots, v_t\}$ spans V over F, prove that some subset of S is a basis of K over F. [*Hint:* Use Lemma 10.1 repeatedly to eliminate v's until you reduce to a set that still spans V and is linearly independent.]

33. If the subset $\{u_1, \ldots, u_t\}$ of V is linearly independent over F and $w \in V$ is *not* a linear combination of the u's, prove that $\{u_1, \ldots, u_t, w\}$ is linearly independent.

34. If V is infinite dimensional over F, then prove that for any positive integer k, V contains a set of k vectors that is linearly independent over F. [*Hint:* Use induction; Exercise 10 is the case $k = 1$, and Exercise 33 can be used to prove the inductive step.]

35. Assume that the subset $\{v_1, \ldots, v_n\}$ of V is linearly independent over F and that $w = c_1 v_1 + \cdots + c_n v_n$, with $c_i \in F$. Prove that the set $\{w - v_1, w - v_2, \ldots, w - v_n\}$ is linearly independent over F if and only if $c_1 + \cdots + c_n \neq 1_F$.

36. Assume that V is finite-dimensional over F and S is a linearly independent subset of V. Prove that S is contained in a basis of V. [*Hint:* Let $[V:F] = n$ and $S = \{u_1, \ldots, u_m\}$; then $m \le n$ by Lemma 10.2. If S does not span V, then there must be some w that is not a linear combination of the u's. Apply Exercise 33 to obtain a larger independent set; if it doesn't span, repeat the argument. Use Lemma 10.2 to show that the process must end with a basis that contains S.]

37. Assume that $[V:F] = n$ and prove that the following conditions are equivalent:

   (i) $\{v_1, \ldots, v_n\}$ spans V over F.

   (ii) $\{v_1, \ldots, v_n\}$ is linearly independent over F.

   (iii) $\{v_1, \ldots, v_n\}$ is a basis of V over F.

38. Let F, K, and L be fields such that $F \subseteq K \subseteq L$. If $[L:F]$ is finite, then prove that $[L:K]$ and $[K:F]$ are also finite and both are $\le [L:F]$. [*Hint:* Use Exercises 20 and 32 to show that $[L:K]$ is finite. To show that $[K:F]$ is finite, suppose $[L:F] = n$. The set $\{1_K\}$ is linearly independent by Exercise 10; if it doesn't span K, proceed as in the hint to Exercise 36 to build larger and larger linearly independent subsets of K. Use Lemma 10.2 and the fact that $[L:F] = n$ to show that the process must end with a basis of K containing at most n elements.]

39. If $[K:F] = p$, with p prime, prove that there is no field E such that $F \subsetneq E \subsetneq K$. [*Hint:* Exercise 38 and Theorem 10.4.]

10.2 SIMPLE EXTENSIONS

Field extensions can be considered from two points of view. You can look upward from a field to its extensions or downward to its subfields. Chapter 5 provided an example of the upward point of view. We took a field F and an irreducible polynomial $p(x)$ in $F[x]$ and formed the field of congruence classes (that is, the quotient field) $F[x]/(p(x))$. Theorem 5.11 shows that $F[x]/(p(x))$ is an extension field of F that contains a root of $p(x)$.

In this section we take the downward view, starting with a field K and a subfield F. If $u \in K$, what can be said about the subfields of K that contain both u and F? Is there a smallest such subfield? If u is the root of some irreducible $p(x)$ in $F[x]$, how is this smallest subfield related to the extension field $F[x]/(p(x))$, which also contains a root of $p(x)$?

The theoretical answer to the first two questions is quite easy. Let K be an extension field of F and $u \in K$. Let $F(u)$ denote the intersection of all subfields of K that contain both F and u (this family of subfields is nonempty since K at least is in it). Since the intersection of any family of subfields of K is itself a field

(Exercise 1), $F(u)$ is a field. By its definition, $F(u)$ is contained in every subfield of K that contains F and u, and, hence, $F(u)$ is the smallest subfield of K containing F and u. $F(u)$ is said to be a **simple extension** of F.

As a practical matter, this answer is not entirely satisfactory. A more explicit description of the simple extension field $F(u)$ is needed. It turns out that the structure of $F(u)$ depends on whether or not u is the root of some polynomial in $F[x]$. So we pause to introduce some terminology.

> **DEFINITION•** *An element u of an extension field K of F is said to be **algebraic** over F if u is the root of some nonzero polynomial in $F[x]$. An element of K that is not the root of any nonzero polynomial in $F[x]$ is said to be **transcendental** over F.*

EXAMPLE In the extension field $\mathbb{C}$ of $\mathbb{R}$, i is algebraic over $\mathbb{R}$ because i is the root of $x^2 + 1 \in \mathbb{R}[x]$. You can easily verify that element $2 + i$ of $\mathbb{C}$ is a root of $x^3 - x^2 - 7x + 15 \in \mathbb{Q}[x]$. Thus $2 + i$ is algebraic over $\mathbb{Q}$. Similarly, $\sqrt[5]{3}$ is algebraic over $\mathbb{Q}$ since it is a root of $x^5 - 3$.

EXAMPLE Every element c in a field F is algebraic over F because c is the root of $x - c \in F[x]$.

EXAMPLE The real numbers π and e are transcendental over $\mathbb{Q}$; see Niven [31] for a proof. Hereafter we shall concentrate on algebraic elements. For more information on transcendental elements, see Exercises 10 and 24–26.

If u is an algebraic element of an extension field K of F, then there may be many polynomials in $F[x]$ that have u as a root. The next theorem shows that all of them are multiples of a single polynomial; this polynomial will enable us to give a precise description of the simple extension field $F(u)$.

THEOREM 10.6 *Let K be an extension field of F and $u \in K$ an algebraic element over F. Then there exists a unique monic irreducible polynomial $p(x)$ in $F[x]$ that has u as a root. Furthermore, if u is a root of $g(x) \in F[x]$, then $p(x)$ divides $g(x)$.*

Proof Let S be the set of all nonzero polynomials in $F[x]$ that have u as a root. Then S is nonempty because u is algebraic over F. The degrees of polynomials in S form a nonempty set of nonnegative integers, which must contain a smallest element by the Well-Ordering Axiom. Let $p(x)$ be a polynomial of smallest degree in S. Every nonzero constant multiple of $p(x)$ is a polynomial of the same degree with u as a root. So we can choose $p(x)$ to be monic (if it isn't, multiply by the inverse of its leading coefficient).

If $p(x)$ were not irreducible in $F[x]$, there would be polynomials $k(x)$ and $t(x)$ such that $p(x) = k(x)t(x)$, with $\deg k(x) < \deg p(x)$ and $\deg t(x) < \deg p(x)$. Consequently, $k(u)t(u) = p(u) = 0_F$ in K. Since K is a field either $k(u) = 0_F$ or $t(u) = 0_F$, that is, either $k(x)$ or $t(x)$ is in S. This is impossible since $p(x)$ is a polynomial of smallest degree in S. Hence, $p(x)$ is irreducible.

Next we show that $p(x)$ divides every $g(x)$ in S. By the Division Algorithm, $g(x) = p(x)q(x) + r(x)$, where $r(x) = 0_F$ or $\deg r(x) < \deg p(x)$. Since u is a root of both $g(x)$ and $p(x)$,

$$r(u) = g(u) - p(u)q(u) = 0_F + 0_F q(u) = 0_F.$$

So u is a root of $r(x)$. If $r(x)$ were nonzero, then $r(x)$ would be in S, contradicting the fact that $p(x)$ is a polynomial of smallest degree in S. Therefore, $r(x) = 0_F$, so that $g(x) = p(x)q(x)$. Hence, $p(x)$ divides every polynomial in S.

To show that $p(x)$ is unique, suppose $t(x)$ is a monic irreducible polynomial in S. Then $p(x) \mid t(x)$. Since $p(x)$ is irreducible (and, hence, nonconstant) and $t(x)$ is irreducible, we must have $t(x) = cp(x)$ for some $c \in F$. But $p(x)$ is monic, so c is the leading coefficient of $cp(x)$ and, hence, of $t(x)$. Since $t(x)$ is monic, we must have $c = 1_F$. Therefore, $p(x) = t(x)$ and $p(x)$ is unique.   ◆

If K is an extension field of F and $u \in K$ is algebraic over F, then the monic, irreducible polynomial $p(x)$ in Theorem 10.6 is called the **minimal polynomial** of u over F. The uniqueness statement in Theorem 10.6 means that once we have found any monic, irreducible polynomial in $F[x]$ that has u as a root, it must be the minimal polynomial of u over F.

EXAMPLE $x^2 - 3$ is a monic, irreducible polynomial in $\mathbb{Q}[x]$ that has $\sqrt{3} \in \mathbb{R}$ as a root. Therefore, $x^2 - 3$ is the minimal polynomial of $\sqrt{3}$ over $\mathbb{Q}$. Note that $x^2 - 3$ is reducible over $\mathbb{R}$ since it factors as $(x - \sqrt{3})(x + \sqrt{3})$ in $\mathbb{R}[x]$. So the minimal polynomial of $\sqrt{3}$ over $\mathbb{R}$ is $x - \sqrt{3}$, which is monic and irreducible in $\mathbb{R}[x]$.

EXAMPLE   Let $u = \sqrt{3} + \sqrt{5} \in \mathbb{R}$. Then $u^2 = 3 + 2\sqrt{3}\sqrt{5} + 5 = 8 + 2\sqrt{15}$. Hence, $u^2 - 8 = 2\sqrt{15}$ so that $(u^2 - 8)^2 = 60$, or, equivalently, $(u^2 - 8)^2 - 60 = 0$. Therefore, $u = \sqrt{3} + \sqrt{5}$ is a root of $(x^2 - 8)^2 - 60 = x^4 - 16x^2 + 4 \in \mathbb{Q}[x]$. Verify that this polynomial is irreducible in $\mathbb{Q}[x]$ (Exercise 14). Hence, it must be the minimal polynomial of $\sqrt{3} + \sqrt{5}$ over $\mathbb{Q}$.

The minimal polynomial of u provides the connection between the upward and downward views of simple field extensions and allows us to give a useful description of $F(u)$.

THEOREM 10.7   *Let K be an extension field of F and $u \in K$ an algebraic element over F with minimal polynomial $p(x)$ of degree n. Then*

(1) $F(u) \cong F[x]/(p(x))$.

(2) $\{1_F, u, u^2, \ldots, u^{n-1}\}$ is a basis of the vector space $F(u)$ over F.

(3) $[F(u):F] = n$.

Theorem 10.7 shows that when u is algebraic over F, then $F(u)$ does not depend on K but is completely determined by $F[x]$ and the minimal polynomial $p(x)$. Consequently, we sometimes say that $F(u)$ is the **field obtained by adjoining u to F**.

Proof of Theorem 10.7 (1) Since $F(u)$ is a field containing u, it must contain every positive power of u. Since $F(u)$ also contains F, $F(u)$ must contain every element of the form $b_0 + b_1u + b_2u^2 + \cdots + b_tu^t$ with $b_i \in F$, that is, $F(u)$ contains the element $f(u)$ for every $f(x) \in F[x]$. Verify that the map $\varphi:F[x] \to F(u)$ given by $\varphi(f(x)) = f(u)$ is a homomorphism of rings. A polynomial in $F[x]$ is in the kernel of φ precisely when it has u as a root. By Theorem 10.6 the kernel of φ is the principal ideal $(p(x))$. The First Isomorphism Theorem 6.13 shows that $F[x]/(p(x))$ is isomorphic to Im φ under the map that sends congruence class (coset) $[f(x)]$ to $f(u)$. Furthermore, since $p(x)$ is irreducible, the quotient ring $F[x]/(p(x))$, and, hence, Im φ, are fields by Theorem 5.10. Every constant polynomial is mapped to itself by φ and $\varphi(x) = u$. So Im φ is a subfield of $F(u)$ that contains both F and u. Since $F(u)$ is the smallest subfield of K containing F and u, we must have $F(u) = $ Im $\varphi \cong F[x]/(p(x))$.

   (2) and (3) Since $F(u) = $ Im φ, every nonzero element of $F(u)$ is of the form $f(u)$ for some $f(x) \in F[x]$. If deg $p(x) = n$, then by the Division Algorithm $f(x) = p(x)q(x) + r(x)$, where $r(x) = b_0 + b_1x + \cdots + b_{n-1}x^{n-1} \in F[x]$. Consequently, $f(u) = p(u)q(u) + r(u) = 0_Fq(u) + r(u) = r(u) = b_01_F + b_1u + \cdots + b_{n-1}u^{n-1}$. Therefore, the set $\{1_F, u, u^2, \ldots, u^{n-1}\}$ spans $F(u)$. To show that this set is linearly independent, suppose $c_0 + c_1u + \cdots + c_{n-1}u^{n-1} = 0_F$ with each $c_i \in F$. Then u is a root of $c_0 + c_1x + \cdots + c_{n-1}x^{n-1}$, so this polynomial (which has degree $\leq n - 1$) must be divisible by $p(x)$ (which has degree n). This can happen only when $c_0 + c_1x + \cdots + c_{n-1}x^{n-1}$ is the zero polynomial; that is, each $c_i = 0_F$. Thus $\{1_F, u, u^2, \ldots, u^{n-1}\}$ is linearly independent over F and, therefore, a basis of $F(u)$. Hence, $[F(u):F] = n$. ◆

> **EXAMPLE** The minimal polynomial of $\sqrt{3}$ over $\mathbb{Q}$ is $x^2 - 3$. Applying Theorem 10.7 with $n = 2$ we see that $\{1, \sqrt{3}\}$ is a basis of $\mathbb{Q}(\sqrt{3})$ over $\mathbb{Q}$, whence $[\mathbb{Q}(\sqrt{3}):\mathbb{Q}] = 2$. Similarly, the previous example shows that $\sqrt{3} + \sqrt{5}$ has minimal polynomial $x^4 - 16x^2 + 4$ over $\mathbb{Q}$ so that $[\mathbb{Q}(\sqrt{3} + \sqrt{5}):\mathbb{Q}] = 4$ and $\{1, \sqrt{3} + \sqrt{5}, (\sqrt{3} + \sqrt{5})^2, (\sqrt{3} + \sqrt{5})^3\}$ is a basis.

An immediate consequence of Theorem 10.7 is that

> **if u and v have the same minimal polynomial $p(x)$
> in $F[x]$, then $F(u)$ is isomorphic to $F(v)$.**

The reason is that both $F(u)$ and $F(v)$ are isomorphic to $F[x]/(p(x))$ and, hence, to each other. Note that this result holds even when u and v are not in the same extension field of F. The remainder of this section, which is not needed until Section 10.4, deals with generalizations of this idea. We shall consider not only simple extensions of the same field but also simple extensions of two different, but isomorphic, fields.

Suppose F and E are fields and that $\sigma: F \to E$ is an isomorphism. Verify that the map from $F[x]$ to $E[x]$ that maps $f(x) = a_0 + a_1 x + a_2 x^2 + \cdots + a_n x^n$ to the polynomial $\sigma f(x) = \sigma(a_0) + \sigma(a_1)x + \sigma(a_2)x^2 + \cdots + \sigma(a_n)x^n$ is an isomorphism of rings (Exercise 19 in Section 4.1). Note that if $f(x) = c$ is a constant polynomial in $F[x]$ (that is, an element of F), then this isomorphism maps it onto $\sigma(c) \epsilon E$. Consequently, we say that the isomorphism $F[x] \to E[x]$ **extends** the isomorphism $\sigma: F \to E$, and we denote the extended isomorphism by σ as well.

COROLLARY 10.8 *Let $\sigma: F \to E$ be an isomorphism of fields. Let u be an algebraic element in some extension field of F with minimal polynomial $p(x) \epsilon F[x]$. Let v be an algebraic element in some extension field of E, with minimal polynomial $\sigma p(x) \epsilon E[x]$. Then σ extends to an isomorphism of fields $\overline{\sigma}: F(u) \to E(v)$ such that $\overline{\sigma}(u) = v$ and $\overline{\sigma}(c) = \sigma(c)$ for every $c \epsilon F$.*

The special case when σ is the identity map $F \to F$ states whenever u and v have the same minimal polynomial, then $F(u) \cong F(v)$ under a function that maps u to v and every element of F to itself.

Proof of Corollary 10.8 The isomorphism σ extends to an isomorphism (also denoted σ) $F[x] \to E[x]$ by the remarks preceding the corollary. The proof of Theorem 10.7 shows that there is an isomorphism $\overline{\tau}: E[x]/(\sigma p(x)) \to E(v)$ given by $\overline{\tau}[g(x)] = g(v)$. Let π be the surjective homomorphism

$$E[x] \to E[x]/(\sigma p(x))$$

that maps $g(x)$ to $[g(x)]$ and consider the composition

$$F[x] \xrightarrow{\ \sigma\ } E[x] \xrightarrow{\ \pi\ } E[x]/(\sigma p(x)) \xrightarrow{\ \overline{\tau}\ } E(v)$$
$$f(x) \dashrightarrow \sigma f(x) \dashrightarrow [\sigma f(x)] \dashrightarrow \sigma f(v).$$

Since all three maps are surjective, so is the composite function. The kernel of the composite function consists of all $h(x) \epsilon F[x]$ such that $\sigma h(v) = 0_E$. Since $\overline{\tau}$ is an isomorphism, $\sigma h(v) = 0_E$ if and only if $[\sigma h(x)]$ is the zero class in $E[x]/(\sigma p(x))$, that is, if and only if $\sigma h(x)$ is a multiple of $\sigma p(x)$. But if $\sigma h(x) = k(x) \cdot \sigma p(x)$, then applying the inverse of the isomorphism σ shows that $h(x) = (\sigma^{-1}k(x))p(x)$. Thus the kernel of the composite function is the principal ideal $(p(x))$ in $F[x]$. Therefore, $F[x]/(p(x)) \cong E(v)$ by the First Isomorphism

Theorem 6.13; the proof of that theorem shows that this isomorphism (call it θ) is given by $\theta([\,f(x)]) = \sigma f(v)$. Note that $\theta([x]) = v$ and that for each $c \in F$, $\theta([c]) = \sigma(c)$. So we have the following situation, where $\overline{\varphi}$ is the isomorphism of Theorem 10.7:

$$F(u) \xleftarrow{\ \overline{\varphi}\ } F[x]/(p(x)) \xrightarrow{\ \theta\ } E(v)$$
$$f(u) \longleftarrow\text{---}\ [\,f(x)]\ \text{---}\longrightarrow \sigma f(v)$$
$$c \longleftarrow\text{----}\ [c]\ \text{----}\longrightarrow \sigma(c) \qquad c \in F.$$

The composite function $\theta \circ \overline{\varphi}^{-1} : F(u) \to E(v)$ is an isomorphism that extends σ and maps u to v. ◆

> **EXAMPLE** The polynomial $x^3 - 2$ is irreducible in $\mathbb{Q}[x]$ by Eisenstein's Criterion. It has a root in $\mathbb{R}$, namely $\sqrt[3]{2}$. Verify that $\sqrt[3]{2}\omega$ is also a root of $x^3 - 2$ in $\mathbb{C}$, where $\omega = \dfrac{-1 + \sqrt{3}i}{2}$ is a complex cube root of 1. Applying Corollary 10.8 to the identity map $\mathbb{Q} \to \mathbb{Q}$ we see that the real subfield $\mathbb{Q}(\sqrt[3]{2})$ is isomorphic to the complex subfield $\mathbb{Q}(\sqrt[3]{2}\omega)$ under a map that sends $\sqrt[3]{2}$ to $\sqrt[3]{2}\omega$ and each element of $\mathbb{Q}$ to itself.

◆ EXERCISES

NOTE: *Unless stated otherwise, K is an extension field of the field F.*

A. 1. Let $\{E_i | i \in I\}$ be a family of subfields of K. Prove that $\bigcap_{i \in I} E_i$ is a subfield of K.

2. If $u \in K$, prove that $F(u^2) \subseteq F(u)$.

3. If $u \in K$ and $c \in F$, prove that $F(u + c) = F(u) = F(cu)$.

4. Prove that $\mathbb{Q}(3 + i) = \mathbb{Q}(1 - i)$.

5. Prove that the given element is algebraic over $\mathbb{Q}$:

(a) $3 + 5i$ (b) $\sqrt{i - \sqrt{2}}$ (c) $1 + \sqrt[3]{2}$

6. If $u \in K$ and u^2 is algebraic over F, prove that u is algebraic over F.

7. If L is a field such that $F \subseteq K \subseteq L$ and $u \in L$ is algebraic over F, show that u is algebraic over K.

8. If $u, v \in K$ and $u + v$ is algebraic over F, prove that u is algebraic over $F(v)$.

9. Prove that $\sqrt{\pi}$ is algebraic over $\mathbb{Q}(\pi)$.

10. If $u \in K$ is transcendental over F and $0_F \neq c \in F$, prove that each of $u + 1_F$, cu, and u^2 is transcendental over F.

11. Find $[\mathbb{Q}(\sqrt[6]{2}):\mathbb{Q}]$.

12. If $a + bi \in \mathbb{C}$ and $b \neq 0$, prove that $\mathbb{C} = \mathbb{R}(a + bi)$.

13. If $[K:F]$ is prime and $u \in K$ is algebraic over F, show that either $F(u) = K$ or $F(u) = F$.

14. Prove that $x^4 - 16x^2 + 4$ is irreducible in $\mathbb{Q}[x]$.

B. **15.** Show that every element of $\mathbb{C}$ is algebraic over $\mathbb{R}$ [*Hint:* See Lemma 4.28.]

16. If $u \in K$ is algebraic over F and $c \in F$, prove that $u + 1_F$ and cu are algebraic over F.

17. Find the minimal polynomial of the given element over $\mathbb{Q}$:

   (a) $\sqrt{1 + \sqrt{5}}$ **(b)** $\sqrt{3}i + \sqrt{2}$

18. Find the minimal polynomial of $\sqrt{2} + i$ over $\mathbb{Q}$ and over $\mathbb{R}$.

19. Let u be an algebraic element of K whose minimal polynomial in $F[x]$ has prime degree. If E is a field such that $F \subseteq E \subseteq F(u)$, show that $E = F$ or $E = F(u)$.

20. Let u be an algebraic element of K whose minimal polynomial in $F[x]$ has odd degree. Prove that $F(u) = F(u^2)$.

21. Let $F = \mathbb{Q}(\pi^4)$ and $K = \mathbb{Q}(\pi)$. Show that π is algebraic over F and find a basis of K over F.

22. If r and s are nonzero, prove that $\mathbb{Q}(\sqrt{r}) = \mathbb{Q}(\sqrt{s})$ if and only if $r = t^2s$ for some $t \in \mathbb{Q}$.

23. If K is an extension field of $\mathbb{Q}$ such that $[K:\mathbb{Q}] = 2$, prove that $K = \mathbb{Q}(\sqrt{d})$ for some square-free integer d. [Square-free means d is not divisible by p^2 for any prime p.]

24. If $u \in K$ is transcendental over F, prove that $F(u) \cong F(x)$, where $F(x)$ is the field of quotients of $F[x]$, as constructed in Section 9.4 (see the example on page 321). [*Hint:* Consider the map from $F(x)$ to $F(u)$ that sends $f(x)/g(x)$ to $f(u)g(u)^{-1}$.]

25. If $u \in K$ is transcendental over F, prove that all elements of $F(u)$, except those in F, are transcendental over F.

26. Let $F(x)$ be as in Exercise 24. Show that $\dfrac{x^3}{x + 1} \in F(x)$ is transcendental over F.

10.3 ALGEBRAIC EXTENSIONS

The emphasis in the last section was on a single algebraic element. Now we consider extensions that consist entirely of algebraic elements.

DEFINITION • *An extension field K of a field F is said to be an **algebraic extension** of F if every element of K is algebraic over F.*

EXAMPLE If $a + bi \in \mathbb{C}$, then $a + bi$ is a root of

$$(x - (a + bi))(x - (a - bi)) = x^2 - 2ax + (a^2 + b^2) \in \mathbb{R}[x].$$

Therefore, $a + bi$ is algebraic over $\mathbb{R}$, and, hence, $\mathbb{C}$ is an algebraic extension of $\mathbb{R}$. On the other hand, neither $\mathbb{C}$ nor $\mathbb{R}$ is an algebraic extension of $\mathbb{Q}$ since there are real numbers (such as π and e) that are not algebraic over $\mathbb{Q}$.

Every algebraic element u over F lies in some finite dimensional extension field of F, namely $F(u)$, by Theorem 10.7. On the other hand, if we begin with a finite-dimensional extension of F we have

THEOREM 10.9 *If K is a finite-dimensional extension field of F, then K is an algebraic extension of F.*

Proof By hypothesis, K has a finite basis over F, say $\{v_1, v_2, \ldots, v_n\}$. Since these n elements span K, Lemma 10.2 implies that every linearly independent set in K must have n or fewer elements. If $u \in K$ and $u^i = u^j$ with $0 \le i < j$, then u is a root of $x^i - x^j \in F[x]$. For any other $u \in K$, $\{1_F, u, u^2, \ldots, u^n\}$ is a set of $n + 1$ elements in K and must, therefore, be linearly dependent over F. Consequently, there are elements c_i in F, not all zero, such that $c_0 1_F + c_1 u + c_2 u^2 + \cdots + c_n u^n = 0_F$. Therefore, u is the root of the nonzero polynomial $c_0 + c_1 x + c_2 x^2 + \cdots + c_n x^n$ in $F[x]$ and, hence, algebraic over F. ◆

If an extension field K of F contains a transcendental element u, then K must be infinite dimensional over F (otherwise u would be algebraic by Theorem 10.9). Nevertheless, the converse of Theorem 10.9 is false since there do exist infinite-dimensional algebraic extensions (Exercise 16).

Simple extensions have a nice property. You need only verify that the single element u is algebraic over F to conclude that the entire field $F(u)$ is an algebraic extension (because $F(u)$ is finite dimensional by Theorem 10.7 and, hence, algebraic by Theorem 10.9). This suggests that generalizing the notion of simple extension might lead to fields whose algebraicity could be determined by checking just a finite number of elements.

If $u_1, \ldots, u_n$ are elements of an extension field K of F, let

$$F(u_1, u_2, \cdots, u_n)$$

denote the intersection of all the subfields of K that contain F and every u_i. As in the case of simple extensions, $F(u_1, \ldots, u_n)$ is the smallest subfield of K

that contains F and all the u_i. $F(u_1, \ldots, u_n)$ is said to be a **finitely gener- ated extension** of F, generated by $u_1, \ldots, u_n$.

EXAMPLE The field $\mathbb{Q}(\sqrt{3}, i)$ is the smallest subfield of $\mathbb{C}$ that contains both the field $\mathbb{Q}$ and the elements $\sqrt{3}$ and i.

EXAMPLE A finitely generated extension may actually be a simple ex- tension. For instance, the field $\mathbb{Q}(i)$ contains both i and $-i$, so $\mathbb{Q}(i, -i) = \mathbb{Q}(i)$.

EXAMPLE Every finite-dimensional extension is also finitely generated. If $\{u_1, \ldots, u_n\}$ is a basis of K over F, then all linear combinations of the u_i (coefficients in F) are in $F(u_1, \ldots, u_n)$. Therefore, $K = F(u_1, \ldots, u_n)$.

The key to dealing with finitely generated extensions is to note that they can be obtained by taking successive simple extensions. For instance, if K is an extension field of F and $u, v \in K$, then $F(u, v)$ is a subfield of K that contains both F and u and, hence, must contain $F(u)$. Since v is in $F(u, v)$, this latter field must contain $F(u)(v)$, the smallest subfield containing both $F(u)$ and v. But $F(u)(v)$ is a field containing F, u, and v and, hence, must contain $F(u, v)$. Therefore, $F(u, v) = F(u)(v)$. Thus the finitely generated extension $F(u, v)$ can be obtained from a chain of simple extensions:

$$F \subseteq F(u) \subseteq F(u)(v) = F(u, v).$$

EXAMPLE The extension field $\mathbb{Q}(\sqrt{3}, i)$ can be obtained by this sequence of simple extensions:

$$\mathbb{Q} \subseteq \mathbb{Q}(\sqrt{3}) \subseteq \mathbb{Q}(\sqrt{3})(i) = \mathbb{Q}(\sqrt{3}, i).$$

On page 343 we saw that $[\mathbb{Q}(\sqrt{3}):\mathbb{Q}] = 2$. Furthermore, i is a root of $x^2 + 1$, whose coefficients are in $\mathbb{Q}(\sqrt{3})$. Therefore, i is algebraic over $\mathbb{Q}(\sqrt{3})$, and, hence, $[\mathbb{Q}(\sqrt{3})(i):\mathbb{Q}(\sqrt{3})]$ is finite by Theorem 10.7. Consequently, by Theorem 10.4

$$[\mathbb{Q}(\sqrt{3}, i):\mathbb{Q}] = [\mathbb{Q}(\sqrt{3})(i):\mathbb{Q}(\sqrt{3})][\mathbb{Q}(\sqrt{3}):\mathbb{Q}]$$

is finite. Thus the finitely generated extension $\mathbb{Q}(\sqrt{3}, i)$ is finite dimen- sional and, hence, algebraic over $\mathbb{Q}$ by Theorem 10.9.

Essentially the same argument works in the general case and provides a useful way to determine that an extension is algebraic:

THEOREM 10.10 *If $K = F(u_1, \ldots, u_n)$ is a finitely generated extension field of*

F and each u_i is algebraic over F, then K is a finite-dimensional algebraic extension of F.

Proof The field K can be obtained from this chain of extensions:

$$F \subseteq F(u_1) \subseteq F(u_1, u_2) \subseteq F(u_1, u_2, u_3) \subseteq \cdots$$
$$\subseteq F(u_1, \ldots, u_{n-1}) \subseteq F(u_1, \ldots, u_n) = K.$$

Furthermore, $F(u_1, u_2) = F(u_1)(u_2)$, $F(u_1, u_2, u_3) = F(u_1, u_2)(u_3)$, and in general $F(u_1, \ldots, u_i)$ is the simple extension $F(u_1, \ldots, u_{i-1})(u_i)$. Each u_i is algebraic over F and, hence, algebraic over $F(u_1, \ldots, u_{i-1})$ by Exercise 7 of Section 10.2. But every simple extension by an algebraic element is finite dimensional by Theorem 10.7. Therefore,

$$[F(u_1, \ldots, u_i):F(u_1, \ldots, u_{i-1})]$$

is finite for each $i = 2, \ldots, n$. Consequently, by repeated application of Theorem 10.4, we see that $[K:F]$ is the product

$$[K:F(u_1, \ldots, u_{n-1})] \cdots [F(u_1, u_2, u_3):F(u_1, u_2)][F(u_1, u_2):F(u_1)][F(u_1):F].$$

Thus $[K:F]$ is finite, and, hence, K is algebraic over F by Theorem 10.9. ◆

> **EXAMPLE** Both $\sqrt{3}$ and $\sqrt{5}$ are algebraic over $\mathbb{Q}$, so $\mathbb{Q}(\sqrt{3}, \sqrt{5})$ is a finite-dimensional algebraic extension field of $\mathbb{Q}$ by Theorem 10.10. We can calculate the dimension of $\mathbb{Q}(\sqrt{3}, \sqrt{5})$ over $\mathbb{Q}$ by considering this chain of simple extensions:
>
> $$\mathbb{Q} \subseteq \mathbb{Q}(\sqrt{3}) \subseteq \mathbb{Q}(\sqrt{3})(\sqrt{5}) = \mathbb{Q}(\sqrt{3}, \sqrt{5}).$$
>
> We know that $[\mathbb{Q}(\sqrt{3}):\mathbb{Q}] = 2$. To determine $[\mathbb{Q}(\sqrt{3})(\sqrt{5}):\mathbb{Q}(\sqrt{3})]$ we shall find the minimal polynomial of $\sqrt{5}$ over $\mathbb{Q}(\sqrt{3})$. The obvious candidate is $x^2 - 5$; it is irreducible in $\mathbb{Q}[x]$, but we must show that it is irreducible over $\mathbb{Q}(\sqrt{3})$, in order to conclude that it is the minimal polynomial. If $\sqrt{5}$ or $-\sqrt{5}$ is in $\mathbb{Q}(\sqrt{3})$, then $\pm\sqrt{5} = a + b\sqrt{3}$, with $a, b \in \mathbb{Q}$. Squaring both sides shows that $5 = a^2 + 2ab\sqrt{3} + 3b^2$, whence $\sqrt{3} = \dfrac{5 - a^2 - 3b^2}{2ab}$, contradicting the fact that $\sqrt{3}$ is irrational; a similar contradiction results if $a = 0$ or $b = 0$. Therefore, $\pm\sqrt{5}$ are not in $\mathbb{Q}(\sqrt{3})$, and, hence, $x^2 - 5$ is irreducible over $\mathbb{Q}(\sqrt{3})$ by Corollary 4.18. So $x^2 - 5$ is the minimal polynomial of $\sqrt{5}$ over $\mathbb{Q}(\sqrt{3})$, and $[\mathbb{Q}(\sqrt{3})(\sqrt{5}):\mathbb{Q}(\sqrt{3})] = 2$ by Theorem 10.7. Consequently, by Theorem 10.4.
>
> $$[\mathbb{Q}(\sqrt{3}, \sqrt{5}):\mathbb{Q}] = [\mathbb{Q}(\sqrt{3})(\sqrt{5}):\mathbb{Q}(\sqrt{3})][\mathbb{Q}(\sqrt{3}):\mathbb{Q}] = 2 \cdot 2 = 4.$$

The remainder of this section is not used in the sequel. Theorem 10.4 tells us that the top field in a chain of finite dimensional extensions is finite dimensional over the ground field. Here is an analogous result for algebraic extensions that may not be finite dimensional.

COROLLARY 10.11   *If L is an algebraic extension field of K and K is an algebraic extension field of F, then L is an algebraic extension of F.*

Proof   Let $u \in L$. Since u is algebraic over K, there exist $a_i \in K$ such that $a_0 + a_1 u + a_2 u^2 + \cdots + a_m u^m = 0_K$. Since each of the a_i is in the field $F(a_1, \ldots, a_m)$, u is actually algebraic over $F(a_1, \ldots a_m)$. Consequently, in the extension chain

$$F \subseteq F(a_1, \ldots, a_m) \subseteq F(a_1, \ldots, a_m)(u) = F(a_1, \ldots, a_m, u)$$

$F(a_1, \ldots, a_m)(u)$ is finite dimensional over $F(a_1, \ldots, a_m)$ by Theorem 10.10 (with $F(a_1, \ldots, a_m)$ in place of F, $n = 1$, and $u = u_1$). Furthermore, $[F(a_1, \ldots, a_m):F]$ is finite by Theorem 10.10 since each a_i is algebraic over F. Therefore, $F(a_1, \ldots, a_m, u)$ is finite dimensional over F by Theorem 10.4 and, hence, is algebraic over F by Theorem 10.9. Thus u is algebraic over F. Since u was an arbitrary element of L, L is an algebraic extension of F.   ◆

COROLLARY 10.12   *Let K be an extension field of F and let E be the set of all elements of K that are algebraic over F. Then E is a subfield of K and an algebraic extension field of F.*

Proof   Every element of F is algebraic over F, so $F \subseteq E$. If $u, v \in E$, then u and v are algebraic over F by definition. The subfield $F(u, v)$ is an algebraic extension of F by Theorem 10.10, and, hence, $F(u, v) \subseteq E$. Since $F(u, v)$ is a field, $u + v$, uv, $-u$, $-v \in F(u, v) \subseteq E$. Similarly, if u is nonzero, then $u^{-1} \in F(u, v) \subseteq E$. Therefore, E is closed under addition and multiplication; negatives and inverses of elements of E are also in E. Hence, E is a field.   ◆

> **EXAMPLE**   If $K = \mathbb{C}$ and $F = \mathbb{Q}$ in Corollary 10.12, then the field E is called the field of **algebraic numbers.** The field E is an infinite-dimensional algebraic extension of $\mathbb{Q}$ (Exercise 16). Algebraic numbers were discussed in a somewhat different context on page 313.

◆ **EXERCISES**

NOTE: *Unless stated otherwise, K is an extension field of the field F.*

A.   **1.**   If $u, v \in K$, verify that $F(u)(v) = F(v)(u)$.

2. If K is a finite field, show that K is an algebraic extension of F.

3. Find a basis of the given extension field of $\mathbb{Q}$.

   (a) $\mathbb{Q}(\sqrt{5}, i)$ (b) $\mathbb{Q}(\sqrt{5}, \sqrt{7})$

   (c) $\mathbb{Q}(\sqrt{2}, \sqrt{3}, \sqrt{5})$ (d) $\mathbb{Q}(\sqrt[3]{2}, \sqrt{3})$

4. Find a basis of $\mathbb{Q}(\sqrt{2} + \sqrt{3})$ over $\mathbb{Q}(\sqrt{3})$.

5. Show that $[\mathbb{Q}(\sqrt{3}, i):\mathbb{Q}] = 4$.

6. Verify that $[\mathbb{Q}(\sqrt{2}, \sqrt{5}, \sqrt{10}):\mathbb{Q}] = 4$.

7. If $[K:F]$ is finite and u is algebraic over K, prove that $[K(u):K] \le [F(u):F]$.

8. If $[K:F]$ is finite and u is algebraic over K, prove that $[K(u):F(u)] \le [K:F]$. [*Hint:* Show that any basis of K over F spans $K(u)$ over $F(u)$.]

9. If $[K:F]$ is finite and u is algebraic over K, prove that $[F(u):F]$ divides $[K(u):K]$.

B. 10. Prove that $[K:F]$ is finite if and only if $K = F(u_1, \ldots, u_n)$, with each u_i algebraic over F. [This is a stronger version of Theorem 10.10.]

11. Assume that $u, v \in K$ are algebraic over F, with minimal polynomials $p(x)$ and $q(x)$, respectively.

   (a) If $\deg p(x) = m$ and $\deg q(x) = n$ and $(m,n) = 1$, prove that $[F(u,v):F] = mn$.

   (b) Show by example that the conclusion of part (a) may be false if m and n are not relatively prime.

   (c) What is $[\mathbb{Q}(\sqrt{2}, \sqrt[3]{2}):\mathbb{Q}]$?

12. Let D be a ring such that $F \subseteq D \subseteq K$. If K is algebraic over F, prove that D is a field. [*Hint:* To find the inverse of a nonzero $u \in D$, use Theorem 10.7 to show that $F(u) \subseteq D$.]

13. Let $p(x)$ and $q(x)$ be irreducible in $F[x]$ and assume that $\deg p(x)$ is relatively prime to $\deg q(x)$. Let u be a root of $p(x)$ and v a root of $q(x)$ in some extension field of F. Prove that $q(x)$ is irreducible over $F(u)$.

14. (a) Let $F_1 \subseteq F_2 \subseteq F_3 \subseteq \cdots$ be a chain of fields. Prove that the union of all the F_i is also a field.

   (b) If each F_i is algebraic over F_1, show that the union of the F_i is an algebraic extension of F_1.

15. Let E be the field of all elements of K that are algebraic over F, as in Corollary 10.12. Prove that every element of the set $K - E$ is transcendental over E.

16. Let E be the field of algebraic numbers (see the example after Corollary

10.12). Prove that E is an infinite dimensional algebraic extension of $\mathbb{Q}$.
[*Hint:* It suffices to show that $[E:\mathbb{Q}] \geq n$ for every positive integer n. Consider roots of the polynomial $x^n - 2$ and Eisenstein's Criterion.]

17. Assume that $1_F + 1_F \neq 0_F$. If $u \in F$, let $\sqrt{u}$ denote a root of $x^2 - u$ in K. Prove that $F(\sqrt{u} + \sqrt{v}) = F(\sqrt{u}, \sqrt{v})$. [*Hint:* 1, $(\sqrt{u} + \sqrt{v})$, $(\sqrt{u} + \sqrt{v})^2$, $(\sqrt{u} + \sqrt{v})^3$, etc., must span $F(\sqrt{u} + \sqrt{v})$ by Theorem 10.7. Use this to show that $\sqrt{u}$ and $\sqrt{v}$ are in $F(\sqrt{u} + \sqrt{v})$.]

18. If $n_1, \ldots, n_t$ are distinct positive integers, show that
$[\mathbb{Q}(\sqrt{n_1}, \ldots, \sqrt{n_t}):\mathbb{Q}] \leq 2^t$.

C. 19. If each n_i is prime in Exercise 18, show that $\leq$ may be replaced by $=$.

10.4 SPLITTING FIELDS

Let F be a field and $f(x)$ a polynomial in $F[x]$. Previously we considered extension fields of F that contained a root of $f(x)$. Now we investigate extension fields that contain *all* the roots of $f(x)$.

The word "all" in this context needs some clarification. Suppose $f(x)$ has degree n. Then by Corollary 4.16, $f(x)$ has at most n roots in any field. So if an extension field K of F contains n distinct roots of $f(x)$, one can reasonably say that K contains "all" the roots of $f(x)$, even though there may be another extension of F that also contains n roots of $f(x)$. On the other hand, suppose that K contains fewer than n roots of $f(x)$. It might be possible to find an extension field of K that contains additional roots of $f(x)$. But if no such extension of K exists, it is reasonable to say that K contains "all" the roots. We can express this condition in a usable form as follows.

Let K be an extension field of F and $f(x)$ a nonconstant polynomial of degree n in $F[x]$. If $f(x)$ factors in $K[x]$ as

$$f(x) = c(x - u_1)(x - u_2) \cdots (x - u_n)$$

then we say that **$f(x)$ splits over the field K.** In this case, the (not necessarily distinct) elements $u_1, \ldots, u_n$ are the only roots of $f(x)$ in K or in any extension field of K. For if v is in some extension of K and $f(v) = 0_F$, then $c(v - u_1)(v - u_2) \cdots (v - u_n) = 0_F$. Now c is nonzero since $f(x)$ is nonconstant. Hence, one of the $v - u_i$ must be zero, that is, $v = u_i$. So if $f(x)$ splits over K, we can reasonably say that K contains all the roots of $f(x)$. The next step is to consider the *smallest* extension field that contains all the roots of $f(x)$.

DEFINITION• *If F is a field and f(x) ∈ F[x], then an extension field K of F is said to be a **splitting field** (or **root field**) of f(x) over F provided that*

(i)   f(x) splits over K, say $f(x) = c(x - u_1)(x - u_2) \cdots (x - u_n)$;

(ii) $K = F(u_1, u_2, \ldots , u_n)$.

EXAMPLE If $x^2 + 1$ is considered as a polynomial in $\mathbb{R}[x]$, then $\mathbb{C}$ is a splitting field since $x^2 + 1 = (x + i)(x - i)$ in $\mathbb{C}[x]$ and $\mathbb{C} = \mathbb{R}(i) = \mathbb{R}(i, -i)$. Similarly, $\mathbb{Q}(\sqrt{2})$ is a splitting field of the polynomial $x^2 - 2$ in $\mathbb{Q}[x]$ since $x^2 - 2 = (x + \sqrt{2})(x - \sqrt{2})$ and $\mathbb{Q}(\sqrt{2}) = \mathbb{Q}(\sqrt{2}, -\sqrt{2})$.

EXAMPLE The polynomial $f(x) = x^4 - x^2 - 2$ in $\mathbb{Q}[x]$ factors as $(x^2 - 2)(x^2 + 1)$, so its roots in $\mathbb{C}$ are $\pm\sqrt{2}$ and $\pm i$. Therefore, $\mathbb{Q}(\sqrt{2}, i)$ is a splitting field of $f(x)$ over $\mathbb{Q}$.

EXAMPLE Every first-degree polynomial $cx + d$ in $F[x]$ splits over F since $cx + d = c(x - (-c^{-1}d))$ with $-c^{-1}d \in F$. Obviously, F is the smallest field containing both F and $c^{-1}d$, that is, $F = F(c^{-1}d)$. So F itself is the splitting field of $cx + d$ over F.

EXAMPLE The concept of splitting field depends on the polynomial *and* the base field. For instance, $\mathbb{C}$ is a splitting field of $x^2 + 1$ over $\mathbb{R}$ but not over $\mathbb{Q}$ because $\mathbb{C}$ is not the extension $\mathbb{Q}(i, -i) = \mathbb{Q}(i)$. See Exercise 1 for a proof.

At this point we need to answer two major questions about splitting fields: Does every polynomial in $F[x]$ have a splitting field over F? If it has more than one splitting field over F, how are they related?

The informal answer to the first question is easy. Given $f(x) \in F[x]$, we can find an extension $F(u)$ that contains a root u of $f(x)$ by Corollary 5.12. By the Factor Theorem in $F(u)[x]$, we know that $f(x) = (x - u)g(x)$. By Corollary 5.12 again there is an extension $F(u)(v)$ of $F(u)$ that contains a root v of $g(x)$. Continuing this, we eventually get a splitting field of $f(x)$. We can formalize this argument via induction and prove slightly more:

THEOREM 10.13 *Let F be a field and f(x) a nonconstant polynomial of degree n in F[x]. Then there exists a splitting field K of f(x) over F such that $[K:F] \leq n!$.*

Proof The proof is by induction on the degree of $f(x)$. If $f(x)$ has degree 1, then F itself is a splitting field of $f(x)$ and $[F:F] = 1 \leq 1!$. Suppose the theorem is true for all polynomials of degree $n - 1$ and that $f(x)$ has degree n. By Theorem 4.13 $f(x)$ has an irreducible factor in $F[x]$. Multiplying this polynomial by the inverse of its leading coefficient produces a monic irreducible factor $p(x)$ of $f(x)$. By Theorem 5.11 there is an extension field that contains a root u of $p(x)$ (and, hence, of $f(x)$). Furthermore, $p(x)$ is necessarily the minimal polynomial of u. Consequently, by Theorem 10.7 $[F(u):F] = \deg p(x) \le \deg f(x) = n$. The Factor Theorem 4.15 shows that $f(x) = (x - u)g(x)$ for some $g(x) \in F(u)[x]$. Since $g(x)$ has degree $n - 1$, the induction hypothesis guarantees the existence of a splitting field K of $g(x)$ over $F(u)$ such that $[K:F(u)] \le (n - 1)!$. In $K[x]$,

$$g(x) = c(x - u_1)(x - u_2) \cdots (x - u_{n-1})$$

and, hence, $f(x) = c(x - u)(x - u_1) \cdots (x - u_{n-1})$. Since

$$K = F(u)(u_1, \ldots, u_{n-1}) = F(u, u_1, \ldots, u_{n-1})$$

we see that K is a splitting field of $f(x)$ over F such that $[K:F] = [K:F(u)][F(u):F] \le ((n - 1)!)n = n!$. This completes the inductive step and the proof of the theorem.   ◆

The relationship between two splitting fields of the same polynomial is quite easy to state:

Any two splitting fields of a polynomial in $F[x]$ are isomorphic.

Surprisingly, the easiest way to prove this fact is to prove a stronger result of which this is a special case.

THEOREM 10.14   *Let $\sigma:F \to E$ be an isomorphism of fields, $f(x)$ a nonconstant polynomial in $F[x]$, and $\sigma f(x)$ the corresponding polynomial in $E[x]$. If K is a splitting field of $f(x)$ over F and L is a splitting field of $\sigma f(x)$ over E, then σ extends to an isomorphism $K \cong L$.*

If $F = E$ and σ is the identity map $F \to F$, then the theorem states that any two splitting fields of $f(x)$ are isomorphic.

Proof of Theorem 10.14   The proof is by induction on the degree of $f(x)$. If $\deg f(x) = 1$, then by the definition of splitting field $f(x) = c(x - u)$ in $K[x]$ and $K = F(u)$. But $f(x) = cx - cu$ is in $F[x]$, so we must have c and cu in F. Hence, $u = c^{-1}cu$ is also in F. Therefore, $K = F(u) = F$. On page 344 we saw that σ extends to an isomorphism $F[x] \cong E[x]$; hence, $\sigma f(x)$ also has degree 1, and a similar argument shows that $E = L$. In this case, σ itself is an isomorphism with the required properties.

Suppose the theorem is true for polynomials of degree $n - 1$ and that $f(x)$ has degree n. As in the proof of Theorem 10.13, $f(x)$ has a monic irreducible factor $p(x)$ in $F[x]$ by Theorem 4.13. Since σ extends to an isomorphism $F[x] \cong E[x]$, (page 344), $\sigma p(x)$ is a monic irreducible factor of $\sigma f(x)$ in $E[x]$. Every root of $p(x)$ is also a root of $f(x)$, so K contains all the roots of $p(x)$, and similarly L contains all the roots of $\sigma p(x)$. Let u be a root of $p(x)$ in K and v a root of $\sigma p(x)$ in L. Then σ extends to an isomorphism $F(u) \rightarrow E(v)$ that maps u to v by Corollary 10.8, and the situation looks like this:

$$
\begin{array}{ccc}
K & & L \\
\cup| & & \cup| \\
F(u) & \xrightarrow{\;\cong\;} & E(v) \\
\cup| & & \cup| \\
F & \xrightarrow{\;\sigma\;} & E.
\end{array}
$$

The Factor Theorem 4.15 shows that $f(x) = (x - u)g(x)$ in $F(u)[x]$ and, hence, in $E(v)[x]$

$$\sigma f(x) = \sigma(x - u)\sigma g(x) = (x - \sigma u)\sigma g(x) = (x - v)\sigma g(x).$$

Now $f(x)$ splits over K, say $f(x) = c(x - u)(x - u_2) \cdots (x - u_n)$. Since $f(x) = (x - u)g(x)$, we have $g(x) = c(x - u_2) \cdots (x - u_n)$. The smallest subfield containing all the roots of $g(x)$ and the field $F(u)$ is $F(u, u_2, \ldots, u_n) = K$, so K is a splitting field of $g(x)$ over $F(u)$. Similarly, L is a splitting field of $\sigma g(x)$ over $E(v)$. Since $g(x)$ has degree $n - 1$, the induction hypothesis implies that the isomorphism $F(u) \cong E(v)$ can be extended to an isomorphism $K \cong L$. This completes the inductive step and the proof of the theorem. ◆

A splitting field of some polynomial over F contains all the roots of that polynomial by definition. Surprisingly, however, splitting fields have a much stronger property, which we now define.

> **DEFINITION•** *An algebraic extension field K of F is* ***normal*** *provided that whenever an irreducible polynomial in F[x] has one root in K, then it splits over K (that is, has all its roots in K).*

THEOREM 10.15 *The field K is a splitting field over the field F of some polynomial in F[x] if and only if K is a finite-dimensional, normal extension of F.*

Proof If K is a splitting field of $f(x) \in F[x]$, then $K = F(u_1, \ldots, u_n)$, where the u_i are all the roots of $f(x)$. Consequently, $[K:F]$ is finite by Theorem 10.10. Let $p(x)$ be an irreducible polynomial in $F[x]$ that has a root v in K. Consider $p(x)$ as a polynomial in $K[x]$ and let L be a splitting field of $p(x)$ over K, so that $F \subseteq K \subseteq L$. To prove that $p(x)$ splits over K, we need only show that every root of $p(x)$ in L is actually in K.

Let $w \in L$ be any root of $p(x)$ other than v. By Corollary 10.8 (with $E = F$

and σ the identity map), there is an isomorphism $F(v) \cong F(w)$ that maps v to w and maps every element of F to itself. Consider the subfield $K(w)$ of L; the situation looks like this:

$$
\begin{array}{ccc}
K & & K(w) \\
\cup| & & \cup| \\
F(v) & \cong & F(w) \\
\cup| & & \cup| \\
F & = & F.
\end{array}
$$

Since

$$K(w) = F(u_1, \ldots, u_n)(w) = F(u_1, \ldots, u_n, w) = F(w)(u_1, \ldots, u_n)$$

we see that $K(w)$ is a splitting field of $f(x)$ over $F(w)$. Furthermore, since $v \epsilon K$ and K is a splitting field of $f(x)$ over F, K is also a splitting field of $f(x)$ over the subfield $F(v)$. Consequently, by Theorem 10.14 the isomorphism $F(v) \cong F(w)$ extends to an isomorphism $K \rightarrow K(w)$ that maps v to w and every element of F to itself. Therefore, $[K:F] = [K(w):F]$ by Theorem 10.5. In the extension chain $F \subseteq K \subseteq K(w)$, $[K(w):K]$ is finite by Theorem 10.7 and $[K:F]$ is finite by the remarks in the first paragraph of the proof. So Theorem 10.4 implies that

$$[K:F] = [K(w):F] = [K(w):K][K:F].$$

Cancelling $[K:F]$ on each end shows that $[K(w):K] = 1$, and, therefore, $K(w) = K$. But this means that w is in K. Thus every root of $p(x)$ in L is in K, and $p(x)$ splits over K. Therefore, K is normal over F.

Conversely, assume K is a finite-dimensional, normal extension of F with basis $\{u_1, \ldots, u_n\}$. Then $K = F(u_1, \ldots, u_n)$. Each u_i is algebraic over F by Theorem 10.9 with minimal polynomial $p_i(x)$. Since each $p_i(x)$ splits over K by normality, $f(x) = p_1(x) \cdots p_n(x)$ also splits over K. Therefore, K is the splitting field of $f(x)$. ◆

> **EXAMPLE** The field $\mathbb{Q}(\sqrt[3]{2})$ contains the real root $\sqrt[3]{2}$ of the irreducible polynomial $x^3 - 2 \ \epsilon \ \mathbb{Q}[x]$ but does not contain the complex root $\sqrt[3]{2}\omega$ (as described in the example on page 345). Therefore, $\mathbb{Q}(\sqrt[3]{2})$ is not a normal extension of $\mathbb{Q}$ and, hence, cannot be the splitting field of *any* polynomial in $\mathbb{Q}[x]$.

At this point it is natural to ask if a field F has an extension field over which *every* polynomial in $F[x]$ splits. In other words, is there an extension field that contains all the roots of all the polynomials in $F[x]$? The answer is "yes," but the proof is beyond the scope of this book. A field over which every nonconstant polynomial splits is said to be **algebraically closed.** For example, the

Fundamental Theorem of Algebra and Corollary 4.27 show that the field $\mathbb{C}$ of complex numbers is algebraically closed.

If K is an algebraic extension of F and K is algebraically closed, then K is called the **algebraic closure** of F. The word "the" is justified by a theorem analogous to Theorem 10.14 that says any two algebraic closures of F are isomorphic. For example, $\mathbb{C}$ is the algebraic closure of $\mathbb{R}$ since $\mathbb{C} = \mathbb{R}(i)$ is an algebraic extension of $\mathbb{R}$ that is algebraically closed. The field $\mathbb{C}$ is not the algebraic closure of $\mathbb{Q}$, however, since $\mathbb{C}$ is not algebraic over $\mathbb{Q}$. The subfield E of algebraic numbers (see the example on page 350) is the algebraic closure of $\mathbb{Q}$ (Exercise 20).

◆ EXERCISES

A. 1. Show that $\sqrt{2}$ is not in $\mathbb{Q}(i)$ and, hence, $\mathbb{C} \neq \mathbb{Q}(i)$. [*Hint:* Show that $\sqrt{2} = a + bi$, with $a, b \in \mathbb{Q}$, leads to a contradiction.]

2. Show that $x^2 - 3$ and $x^2 - 2x - 2$ are irreducible in $\mathbb{Q}[x]$ and have the same splitting field, namely $\mathbb{Q}(\sqrt{3})$.

3. Find a splitting field of $x^4 - 4x^2 - 5$ over $\mathbb{Q}$ and show that it has dimension 4 over $\mathbb{Q}$.

4. If $f(x) \in \mathbb{R}[x]$, prove that $\mathbb{R}$ or $\mathbb{C}$ is a splitting field of $f(x)$ over $\mathbb{R}$.

5. Let K be a splitting field of $f(x)$ over F. If E is a field such that $F \subseteq E \subseteq K$, show that K is a splitting field of $f(x)$ over E.

6. Let K be a splitting field of $f(x)$ over F. If $[K:F]$ is prime, $u \in K$ is a root of $f(x)$, and $u \notin F$, show that $K = F(u)$.

7. If u is algebraic over F and $K = F(u)$ is a normal extension of F, prove that K is a splitting field over F of the minimal polynomial of u.

8. Which of the following are normal extensions of $\mathbb{Q}$?

   (a) $\mathbb{Q}(\sqrt{3})$ **(b)** $\mathbb{Q}(\sqrt[3]{3})$ **(c)** $\mathbb{Q}(\sqrt{5}, i)$

9. Prove that no finite field is algebraically closed. [*Hint:* If the elements of the field F are $a_1, \ldots, a_n$, with a_1 nonzero, consider $a_1 + (x - a_1)(x - a_2) \cdots (x - a_n) \in F[x]$.]

B. 10. By finding quadratic factors, show that $\mathbb{Q}(\sqrt{2}, \sqrt{3})$ is a splitting field of $x^4 + 2x^3 - 8x^2 - 6x - 1$ over $\mathbb{Q}$.

11. Find and describe a splitting field of $x^4 + 1$ over $\mathbb{Q}$.

12. Find a splitting field of $x^4 - 2$

   (a) over $\mathbb{Q}$.   **(b)** over $\mathbb{R}$.

13. Find a splitting field of $x^6 + x^3 + 1$ over $\mathbb{Q}$.

14. Show that $\mathbb{Q}(\sqrt{2}, i)$ is a splitting field of $x^2 - 2\sqrt{2}x + 3$ over $\mathbb{Q}(\sqrt{2})$.

15. Find a splitting field of $x^2 + 1$ over $\mathbb{Z}_3$.

16. Find a splitting field of $x^3 + x + 1$ over $\mathbb{Z}_2$.

17. If K is an extension field of F such that $[K:F] = 2$, prove that K is normal.

18. Let F, E, K be fields such that $F \subseteq E \subseteq K$ and $E = F(u_1, \ldots, u_t)$, where the u_i are some of the roots of $f(x) \in F[x]$. Prove that K is a splitting field of $f(x)$ over F if and only if K is a splitting field of $f(x)$ over E.

19. Prove that the following conditions on a field K are equivalent:

   (i) Every nonconstant polynomial in $K[x]$ has a root in K.

   (ii) Every nonconstant polynomial in $K[x]$ splits over K (that is, K is algebraically closed).

   (iii) Every irreducible polynomial in $K[x]$ has degree 1.

   (iv) There is no algebraic extension field of K except K itself.

20. Let K be an extension field of F and E the subfield of all elements of K that are algebraic over F, as in Corollary 10.12. If K is algebraically closed, prove that E is an algebraic closure of F. [The special case when $F = \mathbb{Q}$ and $K = \mathbb{C}$ shows that the field E of algebraic numbers is an algebraic closure of $\mathbb{Q}$.]

21. Let K be an algebraic extension field of F such that every polynomial in $F(x)$ splits over K. Prove that K is an algebraic closure of F.

C. 22. If K is a finite-dimensional extension field of F and $\sigma: F \to K$ is a homomorphism of fields, prove that there exists an extension field L of K and a homomorphism $\tau: K \to L$ such that $\tau(a) = \sigma(a)$ for every $a \in F$.

23. Prove that a finite-dimensional extension field K of F is normal if and only if it has this property: Whenever L is an extension field of K and $\sigma: K \to L$ an injective homomorphism such that $\sigma(c) = c$ for every $c \in F$, then $\sigma(K) \subseteq K$.

10.5 SEPARABILITY

Every polynomial has a splitting field that contains all its roots. These roots may all be distinct, or there may be repeated roots.* In this section we consider the case when the roots are distinct and use the information obtained to prove a very useful fact about finite-dimensional extensions.

---

* A repeated root occurs when $f(x) = (x - u_1) \cdots (x - u_n)$ in the splitting field and some $u_i = u_j$ with $i \neq j$.

Let F be a field. A polynomial $f(x) \in F[x]$ of degree n is said to be **separable** if it has n distinct roots in some splitting field.* Equivalently, $f(x)$ is separable if it has no repeated roots in any splitting field. If K is an extension field of F, then an element $u \in K$ is said to be **separable** over F if u is algebraic over F and its minimal polynomial $p(x) \in F[x]$ is separable. The extension field K is said to be a **separable extension** (or to be **separable over F**) if every element of K is separable over F. Thus a separable extension is necessarily algebraic.

> **EXAMPLE** The polynomial $x^2 + 1 \in \mathbb{Q}[x]$ is separable since it has distinct roots i and $-i$ in $\mathbb{C}$. But $f(x) = x^4 - x^3 - x + 1$ is not separable because it factors as $(x - 1)^2(x^2 + x + 1)$. Hence, $f(x)$ has one repeated root and a total of three distinct roots in $\mathbb{C}$.

There are several tests for separability that make use of the following concept. The **derivative** of

$$f(x) = c_0 + c_1 x + c_2 x^2 + \cdots + c_n x^n \in F[x]$$

is defined to be the polynomial

$$f'(x) = c_1 + 2c_2 x + 3c_3 x^2 + \cdots + nc_n x^{n-1} \in F[x]**.$$

You should use Exercises 4 and 5 to verify that derivatives defined in this algebraic fashion have these familiar properties.

$$(f + g)'(x) = f'(x) + g'(x)$$
$$(fg)'(x) = f(x)g'(x) + f'(x)g(x).$$

LEMMA 10.16 *Let F be a field and $f(x) \in F[x]$. If $f(x)$ and $f'(x)$ are relatively prime in $F[x]$, then $f(x)$ is separable.*

Note that the lemma operates entirely in $F[x]$ and does not require any knowledge of the splitting field to determine separability. For other separability criteria, see Exercises 8–10.

Proof of Lemma 10.16 Let K be a splitting field of $f(x)$ and suppose, on the contrary, that $f(x)$ is not separable. Then $f(x)$ must have a repeated root u in K. Hence, $f(x) = (x - u)^2 g(x)$ for some $g(x) \in K[x]$ and

---

* Since any two splitting fields are isomorphic, this means that $f(x)$ has n distinct roots in every splitting field.

** When $F = \mathbb{R}$, this is the usual derivative of elementary calculus. But our definition is purely algebraic and applies to polynomials over any field, whereas the limits used in calculus may not be defined in some fields.

$$f'(x) = (x - u)^2 g'(x) + 2(x - u)g(x).$$

Therefore, $f'(u) = 0_F g'(u) + 0_F g(u) = 0_F$ and u is also a root of $f'(x)$. If $p(x) \in F[x]$ is the minimal polynomial of u, then $p(x)$ is nonconstant and divides both $f(x)$ and $f'(x)$. Therefore, $f(x)$ and $f'(x)$ are not relatively prime, a contradiction. Hence, $f(x)$ is separable. ◆

Recall that for a positive integer n and $c \in F$,

$$nc \text{ is the element } c + c + \cdots + c \ (n \text{ summands}).$$

A field F is said to have **characteristic 0** if $n1_F \neq 0_F$ for every positive n. For example, $\mathbb{Q}$, $\mathbb{R}$, and $\mathbb{C}$ all have characteristic 0, but $\mathbb{Z}_3$ does not (since $3 \cdot 1 = 0$ in $\mathbb{Z}_3$). Every field of characteristic 0 is infinite (Exercise 3). If F has characteristic 0, then for every positive n and $c \in F$,

$$nc = c + \cdots + c = (1_F + \cdots + 1_F)c = (n1_F)c \qquad \text{with } n1_F \neq 0_F.$$

So $nc = 0_F$ if and only if $c = 0_F$. This fact is the key to separability in fields of characteristic 0:

THEOREM 10.17   *Let F be a field of characteristic 0. Then every irreducible polynomial in $F[x]$ is separable, and every algebraic extension field K of F is a separable extension.*

The theorem may be false if F does not have characteristic 0 (Exercise 15).

Proof of Theorem 10.17   An irreducible $p(x) \in F[x]$ is nonconstant and, hence,

$$p(x) = cx^n + (\text{lower-degree terms}), \qquad \text{with } c \neq 0_F \text{ and } n \geq 1.$$

Then

$$p'(x) = (nc)x^{n-1} + (\text{lower-degree terms}), \qquad \text{with } nc \neq 0_F.$$

Therefore, $p'(x)$ is a nonzero polynomial of lower degree than the irreducible $p(x)$. So $p(x)$ and $p'(x)$ must be relatively prime. Hence, $p(x)$ is separable by Lemma 10.16. In particular, the minimal polynomial of each $u \in K$ is separable. So K is a separable extension. ◆

Separable extensions are particularly nice because every finitely generated (in particular, every finite-dimensional) separable extension is actually simple:

THEOREM 10.18*   *If K is a finitely generated separable extension field of F, then $K = F(u)$ for some $u \in K$.*

---

* This theorem will be used only in Section 11.2.

Proof By hypothesis $K = F(u_1, \ldots, u_n)$. The proof is by induction on n. There is nothing to prove when $n = 1$ and $K = F(u_1)$. In the next paragraph we shall show that the theorem is true for $n = 2$. Assume inductively that it is true for $n = k - 1$ and suppose $n = k$. By induction and the case $n = 2$, there exist t, $u \in K$ such that

$$K = F(u_1, \ldots, u_k) = F(u_1, \ldots, u_{k-1})(u_k) = F(t)(u_k) = F(t, u_k) = F(u).$$

To complete the proof, we assume $K = F(v, w)$ and show that K is a simple extension of F. Assume first that F is infinite (which is always the case in characteristic 0 by Exercise 3). Let $p(x) \in F[x]$ be the minimal polynomial of v and $q(x) \in F[x]$ the minimal polynomial of w. Let L be a splitting field of $p(x)q(x)$ over F. Let $w = w_1, w_2, \ldots, w_n$ be the roots of $q(x)$ in L. By the definition of separability, all the w_i are distinct. Let $v = v_1, v_2, \ldots, v_m$ be the roots of $p(x)$ in L. Since F is infinite, there exists $c \in F$ such that

$$(*) \qquad\qquad c \neq \frac{v_i - v}{w - w_j} \qquad \text{for all } 1 \leq i \leq m, 1 < j \leq n.$$

Let $u = v + cw$. We claim that $K = F(u)$. To show that $w \in F(u)$, let $h(x) = p(u - cx) \in F(u)[x]$ and note that w is a root of $h(x)$:

$$h(w) = p(u - cw) = p(v) = 0_F.$$

Suppose some w_j (with $j \neq 1$) is also a root of $h(x)$. Then $p(u - cw_j) = 0_F$, so that $u - cw_j$ is one of the roots of $p(x)$, say $u - cw_j = v_i$. Since $u = v + cw$, we would have

$$v + cw - cw_j = v_i \qquad \text{or, equivalently,} \qquad c = \frac{v_i - v}{w - w_j}.$$

This contradicts $(*)$. Therefore, w is the only common root of $q(x)$ and $h(x)$.

Let $r(x)$ be the minimal polynomial of w over $F(u)$. Then $r(x)$ divides $q(x)$, so that every root of $r(x)$ is a root of $q(x)$. But $r(x)$ also divides $h(x)$, so all its roots are roots of $h(x)$. By the preceding paragraph, $r(x)$ has a single root w in L. Therefore, $r(x) \in F(u)[x]$ must have degree 1, and, hence, its root w is in $F(u)$. Since $v = u - cw$, with $u, w \in F(u)$, we see that $v \in F(u)$ and, hence, $K = F(v, w) \subseteq F(u)$. But $u = v + cw \in K$, so $F(u) \subseteq K$, whence $K = F(u)$. This completes the proof when F is infinite. For the case of finite F, see Theorem 10.28 in the next section. ◆

EXAMPLE Applying the proof of the theorem to $\mathbb{Q}(\sqrt{3}, \sqrt{5})$, we have $v = \sqrt{3}$, $v_2 = -\sqrt{3}$, $w = \sqrt{5}$, $w_2 = -\sqrt{5}$, so we can choose $c = 1$. Then $u = \sqrt{3} + \sqrt{5}$ and $\mathbb{Q}(\sqrt{3}, \sqrt{5})$ is the simple extension $\mathbb{Q}(\sqrt{3} + \sqrt{5})$.

◆ **EXERCISES**

NOTE: *K is an extension field of the field F.*

A. **1.** If K is separable over F and E is a field with $F \subseteq E \subseteq K$, show that K is separable over E.

2. If F has characteristic 0, show that K has characteristic 0.

3. Prove that every field of characteristic 0 is infinite. [*Hint:* Consider the elements $n1_F$ with $n \in \mathbb{Z}, n > 0$.]

B. **4.** If $f(x), g(x) \in F[x]$, prove:

  (a) $(f + g)'(x) = f'(x) + g'(x)$.

  (b) If $c \in F$, then $(cf)'(x) = cf'(x)$.

5. (a) If $f(x) = cx^n \in F[x]$ and $g(x) = b_0 + b_1x + \cdots + b_kx^k \in F[x]$, prove that $(fg)'(x) = f(x)g'(x) + f'(x)g(x)$.

  (b) If $f(x), g(x)$ are any polynomials in $F[x]$, prove that $(fg)'(x) = f(x)g'(x) + f'(x)g(x)$. [*Hint:* If $f(x) = a_0 + a_1x + \cdots + a_nx^n$, then $(fg)(x) = a_0g(x) + a_1xg(x) + \cdots + a_nx^ng(x)$; use part (a) and Exercise 4.]

6. If $f(x) \in F[x]$ and n is a positive integer, prove that the derivative of $f(x)^n$ is $nf(x)^{n-1}f'(x)$. [*Hint:* Use induction on n and Exercise 5.]

7. (a) If F has characteristic 0, $f(x) \in F[x]$, and $f'(x) = 0_F$, prove that $f(x) = c$ for some $c \in F$.

  (b) Give an example in $\mathbb{Z}_2[x]$ to show that part (a) may be false if F does not have characteristic 0.

8. Prove that $u \in K$ is a repeated root of $f(x) \in F[x]$ if and only if u is a root of both $f(x)$ and $f'(x)$. [*Hint:* $f(x) = (x - u)^m g(x)$ with $m \geq 1, g(x) \in K[x]$, and $g(u) \neq 0_F$, u is a repeated root of $f(x)$ if and only if $m > 1$. Use Exercises 5 and 6 to compute $f'(x)$.]

9. Prove that $f(x) \in F[x]$ is separable if and only if $f(x)$ and $f'(x)$ are relatively prime. [*Hint:* See Lemma 10.16 and Exercise 8.]

10. Let $p(x)$ be irreducible in $F[x]$. Prove that $p(x)$ is separable if and only if $p'(x) \neq 0_F$.

11. Assume F has characteristic 0 and K is a splitting field of $f(x) \in F[x]$. If $d(x)$ is the greatest common divisor of $f(x)$ and $f'(x)$ and $h(x) = f(x)/d(x) \in F[x]$, prove

  (a) $f(x)$ and $h(x)$ have the same roots in K.

  (b) $h(x)$ is separable.

12. Use the proof of Theorem 10.18 to express each of these as simple extensions of $\mathbb{Q}$:

(a) $\mathbb{Q}(\sqrt{2}, \sqrt{3})$ **(b)** $\mathbb{Q}(\sqrt{3}, i)$ **(c)** $\mathbb{Q}(\sqrt{2}, \sqrt{3}, \sqrt{5})$

13. If p and q are distinct primes, prove that $\mathbb{Q}(\sqrt{p}, \sqrt{q}) = \mathbb{Q}(\sqrt{p} + \sqrt{q})$.

14. Assume that F is infinite, that $v, w \in K$ are algebraic over F, and that w is the root of a separable polynomial in $F[x]$. Prove that $F(v, w)$ is a simple extension of F. [*Hint:* Adapt the proof of Theorem 10.18.]

15. Here is an example of an irreducible polynomial that is not separable. Let $F = \mathbb{Z}_2(t)$ be the quotient field of $\mathbb{Z}_2[t]$ (the ring of polynomials in the indeterminate t with coefficients in $\mathbb{Z}_2$), as defined in Section 9.4 (see page 321).

(a) Prove that $x^2 - t$ is an irreducible polynomial in $F[x]$. [*Hint:* If $x^2 - t$ has a root in F, then there are polynomials $g(t), h(t)$ in $\mathbb{Z}_2[t]$ such that $[g(t)/h(t)]^2 = t$; this leads to a contradiction; apply Corollary 4.18.]

(b) Prove that $x^2 - t \in F[x]$ is *not* separable. [*Hint:* Show that its derivative is zero and use Exercise 10.]

10.6 FINITE FIELDS

Finite fields have applications in many areas, including projective geometry, combinatories, experimental design, and cryptography. In this section, finite fields are characterized in terms of field extensions and splitting fields, and their structure is completely determined up to isomorphism.

We begin with some definitions and results that apply to rings that need *not* be fields or even finite. But our primary interest will be in their implications for finite fields.

Let R be a ring with identity. Recall that for a positive integer m and $c \in R$, mc is the element $c + c + \cdots + c$ (m summands). The ring R is said to have **characteristic 0** if $m1_R \neq 0_R$ for every positive m. On the other hand, if $m1_R = 0_R$ for some positive m, then there is a smallest such m by the Well-Ordering Axiom. Then R is said to have **characteristic n** if n is the smallest positive integer such that $n1_R = 0_R$.* For example, $\mathbb{Q}$ has characteristic 0 and $\mathbb{Z}_3$ has characteristic 3.

LEMMA 10.19 *If R is an integral domain, then the characteristic of R is either 0 or a positive prime.*

Proof If R has characteristic 0, there is nothing to prove. So assume R has characteristic $n > 0$. If n were not prime, then there would exist positive in-

---

* If you have read Chapter 7, you will recognize that when the characteristic of R is positive, it is simply the order of the element 1_R in the additive group of R.

tegers k, t such that $n = kt$, with $k < n$ and $t < n$. The distributive laws show that

$$(k1_R)(t1_R) = \underbrace{(1_R + \cdots + 1_R)}_{k \text{ summands}} \underbrace{(1_R + \cdots + 1_R)}_{t \text{ summands}}$$

$$= 1_R1_R + \cdots + 1_R1_R = 1_R + \cdots + 1_R \qquad [kt \text{ summands}]$$

$$= (kt)1_R = n1_R = 0_R.$$

Since R is an integral domain either $k1_R = 0_R$ or $t1_R = 0_R$, contradicting the fact that n is the smallest positive integer such that $n1_R = 0_R$. Therefore, n is prime. ◆

LEMMA 10.20 *Let R be a ring with identity of characteristic $n > 0$. Then $k1_R = 0_R$ if and only if $n \mid k$.* *

Proof If $n \mid k$, say $k = nd$, then $k1_R = nd1_R = (n1_R)(d1_R) = 0_R(d1_R) = 0_R$. Conversely, suppose $k1_R = 0_R$. By the Division Algorithm, $k = nq + r$ with $0 \le r < n$. Now $n1_R = 0_R$, so that

$$r1_R = r1_R + 0_R = r1_R + nq1_R = (r + nq)1_R = k1_R = 0_R.$$

Since $r < n$ and n is the smallest positive integer such that $n1_R = 0_R$ by the definition of characteristic, we must have $r = 0$. Therefore, $k = nq$ and $n \mid k$. ◆

THEOREM 10.21 *Let R be a ring with identity Then*

    (1) The set $P = \{k1_R \mid k \in \mathbb{Z}\}$ is a subring of R.

    (2) If R has characteristic 0, then $P \cong \mathbb{Z}$.

    (3) If R has characteristic $n > 0$, then $P \cong \mathbb{Z}_n$.

Proof Define $f : \mathbb{Z} \to R$ by $f(k) = k1_R$. Then

$$f(k + t) = (k + t)1_R = k1_R + t1_R = f(k) + f(t).$$

The distributive laws (as in the proof of Lemma 10.19) show that

$$f(kt) = (kt)1_R = (k1_R)(t1_R) = f(k)f(t).$$

Therefore, f is a homomorphism. The image of f is precisely the set P, and, therefore, P is a ring by Corollary 3.13. Consequently, f can be considered as a surjective homomorphism from $\mathbb{Z}$ onto P. Then $P \cong \mathbb{Z}/(\text{Ker } f)$ by the First Isomorphism Theorem 6.13. If R has characteristic 0, then the only integer k such

---

* This lemma is just a special case (in additive notation) of part (3) of Theorem 7.8, with $a = 1_R$ and $e = 0_R$.

that $k1_R = 0_R$ is $k = 0$. So the kernel of f is the ideal (0) in $\mathbb{Z}$, and $P \cong \mathbb{Z}/(0) \cong \mathbb{Z}$. If R has characteristic $n > 0$, then Lemma 10.20 shows that the kernel of f is the principal ideal (n) consisting of all multiples of n. Hence, $P \cong \mathbb{Z}/(n) = \mathbb{Z}_n$. ◆

According to Theorem 10.21 a field of characteristic 0 contains a copy of $\mathbb{Z}$ and, hence, must be infinite. Therefore, by Lemma 10.19 we have

COROLLARY 10.22 *Every finite field has characteristic p for some prime p.*

The converse of Corollary 10.22 is false, however, since there are infinite fields of characteristic p (Exercise 8).

If K is a field of prime characteristic p (in particular, if K is finite), then Theorem 10.21 shows that K contains a subfield P isomorphic to $\mathbb{Z}_p$. This field P is called the **prime subfield** of K and is contained in every subfield of K (because every subfield contains 1_K and, hence, contains $t1_K$ for every integer t).* See Exercise 4 for another description of P. We shall identify the prime subfield P with its isomorphic copy $\mathbb{Z}_p$; then

every field of characteristic p contains $\mathbb{Z}_p$.

The number of elements in a finite field K is called the **order** of K. To determine the order of a finite field K of characteristic p, we consider K as an extension field of its prime subfield $\mathbb{Z}_p$:

THEOREM 10.23 *A finite field K has order p^n, where p is the characteristic of K and $n = [K:\mathbb{Z}_p]$.*

Proof There is certainly a finite set of elements that spans K over $\mathbb{Z}_p$ (the set K itself, for example). Consequently, by Exercise 32 of Section 10.1, K has a finite basis $\{u_1, u_2, \ldots, u_n\}$ over $\mathbb{Z}_p$. Every element of K can be written uniquely in the form

(∗) $$c_1 u_1 + c_2 u_2 + \cdots + c_n u_n$$

with each $c_i \in \mathbb{Z}_p$ by Exercise 30 of Section 10.1. Since there are exactly p possibilities for each c_i, there are precisely p^n distinct linear combinations of the form (∗). So K has order p^n, with n = number of elements in the basis = $[K:\mathbb{Z}_p]$. ◆

Theorem 10.23 limits the possible size of a finite field. For instance, there cannot be a field of order 6 since 6 is not a power of any prime. It also suggests

---

* If K has characteristic 0, then K contains an isomorphic copy P of $\mathbb{Z}$. Since K contains the multiplicative inverse of every nonzero element of P, it follows that K contains a copy of the field $\mathbb{Q}$. As in the case of characteristic p, this field (called the **prime subfield**) is contained in every subfield of K. See Theorem 9.31 (with $R = P \cong \mathbb{Z}$ and $F \cong \mathbb{Q}$) for a more precise statement and proof.

several questions: Is there a field of order p^n for every prime p and every positive integer n? How are two fields of order p^n related? The answers to these questions are given in Theorem 10.25 and its corollaries. In order to prove that theorem, we need a technical lemma.

LEMMA 10.24 (THE FRESHMAN'S DREAM)* *Let p be a prime and R a commutative ring with identity of characteristic p. Then for every $a, b \in R$ and every positive integer n,*

$$(a + b)^{p^n} = a^{p^n} + b^{p^n}.$$

Proof The proof is by induction on n. If $n = 1$, then the Binomial Theorem in Appendix E shows that

$$(a + b)^p = a^p + \binom{p}{1} a^{p^1} b + \cdots + \binom{p}{r} a^{p-r} b^r$$

$$+ \cdots + \binom{p}{p-1} ab^{p-1} + b^p.$$

Each of the middle coefficients $\binom{p}{r} = \dfrac{p!}{r!\,(p-r)!}$ is an integer by Exercise 6 in Appendix E. Since every term in the denominator is strictly less than the prime p, the factor of p in the numerator does not cancel, and, therefore, $\binom{p}{r}$ is divisible by p, say $\binom{p}{r} = tp$. Since R has characteristic p,

$$\binom{p}{r} a^{p-r} b^r = tp1_R a^{p-r} b^r = t(p1_R) a^{p-r} b^r = t0_R a^{p-r} b^r = 0_R.$$

Thus all the middle terms are zero and $(a + b)^p = a^p + b^p$. So the theorem is true when $n = 1$. Assume the theorem is true when $n = k$. Using this assumption and the case when $n = 1$ shows that

$$(a + b)^{p^{k+1}} = ((a + b)^{p^k})^p$$
$$= (a^{p^k} + b^{p^k})^p = (a^{p^k})^p + (a^{p^k})^p = a^{p^{k+1}} + b^{p^{k+1}}.$$

Therefore, the theorem is true when $n = k + 1$ and, hence, for all n by induction. ◆

THEOREM 10.25 *Let K be an extension field of $\mathbb{Z}_p$ and n a positive integer. Then K has order p^n if and only if K is a splitting field of $x^{p^n} - x$ over $\mathbb{Z}_p$.*

---

* Terminology due to Vincent O. McBrien.

Proof Assume K is a splitting field of $f(x) = x^{p^n} - x \in \mathbb{Z}_p[x]$. Since $f'(x) = p^n x^{p^n - 1} - 1 = 0x^{p^n - 1} - 1 = -1$, $f(x)$ is separable by Lemma 10.16. Let E be the subset of K consisting of the p^n distinct roots of $x^{p^n} - x$. Note that $c \in E$ if and only if $c^{p^n} = c$. We shall show that the set E is actually a subfield of K. If $a, b \in E$, then by Lemma 10.24

$$(a + b)^{p^n} = a^{p^n} + b^{p^n} = a + b.$$

Therefore, $a + b \in E$, and E is closed under addition. The set E is closed under multiplication since $(ab)^{p^n} = a^{p^n} b^{p^n} = ab$. Obviously, 0_K and 1_K are in E. If a is a nonzero element of E, then $-a$ and a^{-1} are in E because, for example,

$$(a^{-1})^{p^n} = a^{-p^n} = (a^{p^n})^{-1} = a^{-1}.$$

The argument for $-a$ is similar (Exercise 7). Therefore, E is a subfield of K. Since the splitting field K is the smallest subfield containing the set E of roots, we must have $K = E$. Therefore, K has order p^n.

Conversely, suppose K has order p^n. We need only show that every element of K is a root of $x^{p^n} - x$, for in that case, the p^n distinct elements of K are all the possible roots and K is a splitting field of $x^{p^n} - x$.* Clearly 0_K is a root, so let c be any nonzero element of K. Let $c_1, c_2, \ldots, c_k$ be all the nonzero elements of K (where $k = p^n - 1$ and c is one of the c_i) and let u be the product $u = c_1 c_2 c_3 \cdots c_k$. The k elements $cc_1, cc_2, \ldots, cc_k$ are all distinct (since $cc_i = cc_j$ implies $c_i = c_j$), so they are just the nonzero elements of K in some other order, and their product is the element u. Therefore,

$$u = (cc_1)(cc_2) \cdots (cc_k) = c^k(c_1 c_2 c_3 \cdots c_k) = c^k u.$$

Cancelling u shows that $c^k = 1_K$ and, hence, $c^{k+1} = c$, or equivalently $c^{k+1} - c = 0_K$. Since $k + 1 = p^n$, c is a root of $x^{p^n} - x$. ◆

Theorem 10.25 has several important consequences; together with the theorem they provide a complete characterization of all finite fields.

COROLLARY 10.26 *For each positive prime p and positive integer n, there exists a field of order p^n.*

Proof A splitting field of $x^{p^n} - x$ over $\mathbb{Z}_p$ exists by Theorem 10.13; it has order p^n by Theorem 10.25. ◆

COROLLARY 10.27 *Two finite fields of the same order are isomorphic.*

Proof If K and L are fields of order p^n, then both are splitting fields of $x^{p^n} - x$

---

* A short proof, using group theory, is given in Exercise 22.

over $\mathbb{Z}_p$ by Theorem 10.25 and, hence, are isomorphic by Theorem 10.14 (with σ the identity map on $\mathbb{Z}_p$). ◆

According to Corollary 10.27, there is (up to isomorphism) a unique field of order p^n. This field is called the **Galois field of order p^n**. We complete our study of finite fields with two results whose proofs depend on group theory.

THEOREM 10.28   *Let K be a finite field and F a subfield. Then K is a simple extension of F.*

Proof   By Theorem 7.15, the multiplicative group of nonzero elements of K is cyclic. If u is a generator of this group, then the subfield $F(u)$ contains 0_F and all powers of u and, hence, contains every element of K. Therefore, $K = F(u)$. ◆

COROLLARY 10.29   *Let p be a positive prime. For each positive integer n, there exists an irreducible polynomial of degree n in $\mathbb{Z}_p[x]$.*

Proof   There is an extension field K of $\mathbb{Z}_p$ of order p^n by Corollary 10.26. By Theorem 10.28, $K = \mathbb{Z}_p(u)$ for some $u \in K$. The minimal polynomial of u in $\mathbb{Z}_p[x]$ is irreducible of degree $[K:\mathbb{Z}_p]$ by Theorem 10.7. Theorem 10.23 shows that $[K:\mathbb{Z}_p] = n$. ◆

◆ **EXERCISES**

A.   **1.** If R is a ring with identity and $m, n \in \mathbb{Z}$, prove that $(m1_R)(n1_R) = (mn)1_R$. [The case of positive m, n was done in the proof of Lemma 10.19.]

   2. What is the characteristic of

   (a) $\mathbb{Q}$ **(b)** $\mathbb{Z}_2 \times \mathbb{Z}_6$ **(c)** $\mathbb{Z}_3[x]$

   (d) $M(\mathbb{R})$ **(e)** $M(\mathbb{Z}_3)$

   3. Let R be a ring with identity of characteristic $n \geq 0$. Prove that $na = 0_R$ for every $a \in R$.

   4. If K is a field of prime characteristic p, prove that its prime subfield is the intersection of all the subfields of K.

   5. Let F be a subfield of a finite field K. If F has order q, show that K has order q^n, where $n = [K:F]$.

   6. Show that a field K of order p^n contains all kth roots of 1_K, where $k = p^n - 1$.

   7. Let E be the set of roots of $x^{p^n} - x \in \mathbb{Z}_p[x]$ in some splitting field. If $a \in E$, prove that $-a \in E$.

B.   **8.** Let p be prime and let $\mathbb{Z}_p(x)$ be the field of quotients of the polynomial ring $\mathbb{Z}_p[x]$ (as in the example on page 321). Show that $\mathbb{Z}_p(x)$ is an infinite field of characteristic p.

9. Let R be a commutative ring with identity of prime characteristic p. If $a, b \in R$ and $n \geq 1$, prove that $(a - b)^{p^n} = a^{p^n} - b^{p^n}$.

10. Let K be a finite field of characteristic p. Prove that the map $f : K \to K$ given by $f(a) = a^p$ is an isomorphism. Conclude that every element of K has a pth root in K.

11. Show that the Freshman's Dream (Lemma 10.24) may be false if the characteristic p is not prime or if R is noncommutative. [*Hint:* Consider $\mathbb{Z}_6$ and $M(\mathbb{Z}_2)$.]

12. If c is a root of $f(x) \in \mathbb{Z}_p[x]$, prove that c^p is also a root.

13. Prove **Fermat's Little Theorem:** If p is a prime and $a \in \mathbb{Z}$, then $a^p \equiv a \pmod{p}$. If a is relatively prime to p, then $a^{p-1} \equiv 1 \pmod{p}$. [*Hint:* Translate congruence statements in $\mathbb{Z}$ into equality statements in $\mathbb{Z}_p$ and use Theorem 10.25.]

14. Let F be a field and $f(x)$ a monic polynomial in $F[x]$, whose roots are all distinct in any splitting field K. Let E be the set of roots of $f(x)$ in K. If the set E is actually a subfield of K, prove that F has characteristic p for some prime p and that $f(x) = x^{p^n} - x$ for some $n \geq 1$.

15. (a) Show that $x^3 + x + 1$ is irreducible in $\mathbb{Z}_2[x]$ and construct a field of order 8.

(b) Show that $x^3 - x + 1$ is irreducible in $\mathbb{Z}_3[x]$ and construct a field of order 27.

(c) Show that $x^4 + x + 1$ is irreducible in $\mathbb{Z}_2[x]$ and construct a field of order 16.

16. Let K be a finite field of characteristic p, F a subfield of K, and m a positive integer. If $L = \{a \in K \mid a^{p^m} \in F\}$, prove that

(a) L is a subfield of K that contains F.

(b) $L = F$. [*Hint:* If $\{u, \ldots, u_n\}$ is a basis of L over F, use Exercise 10 and Lemma 10.24 to show that $\{u_1^{p^m}, \ldots, u_n^{p^m}\}$ is linearly independent over F, which implies $n = 1$.]

17. If E and F are subfields of a finite field K and E is isomorphic to F, prove that $E = F$.

18. Let K be a field and k, n positive integers.

(a) Prove that $x^k - 1_K$ divides $x^n - 1_K$ in $K[x]$ if and only if $k \mid n$ in $\mathbb{Z}$. [*Hint:* $n = kq + r$ by the Division Algorithm; show that $x^n - 1_K = (x^k - 1_K)h(x) + (x^r - 1_K)$, where $h(x) = x^{n-k} + x^{n-2k} + \cdots + x^{n-qk}$.]

(b) If $p \geq 2$ is an integer, prove that $(p^k - 1) \mid (p^n - 1)$ if and only if $k \mid n$. [*Hint:* Copy the proof of part (a) with p in place of x.]

19. Let K be a finite field of order p^n.

   (a) If F is a subfield of K, prove that F has order p^d for some d such that $d \mid n$. [*Hint:* Exercise 18 may be helpful.]

   (b) If $d \mid n$, prove that K has a unique subfield of order p^d. [*Hint:* See Exercise 17 and Corollary 10.27 for the uniqueness part.]

20. Let p be prime and $f(x)$ an irreducible polynomial of degree 2 in $\mathbb{Z}_p[x]$. If K is an extension field of $\mathbb{Z}_p$ of order p^3, prove that $f(x)$ is irreducible in $K[x]$.

21. Prove that every element in a finite field can be written as the sum of two squares.

22. Use part (2) of Corollary 7.27 to prove that every nonzero element c of a finite field K of order p^n satisfies $c^{p^n - 1} = 1_K$. Conclude that c is a root of $x^{p^n} - x$ and use this fact to prove Theorem 10.25.

Application   BCH codes (Section 16.3) may be covered at this point if desired.

CHAPTER **11**

Galois Theory

◆

$\mathbf{A}$ major question in classical algebra was whether or not there were formulas for the solution of higher-degree polynomial equations (analogous to the quadratic formula for second-degree equations). Although formulas for third- and fourth-degree equations were found in the sixteenth century, no further progress was made for almost 300 years. Then Ruffini and Abel provided the surprising answer: There is no formula for the solution of *all* polynomial equations of degree n when $n \geq 5$. This result did not rule out the possibility that the solutions of special types of equations might be obtainable from a formula. Nor did it give any clue as to which equations might be solvable by formula.

It was the amazingly original work of Galois that provided the full explanation, including a criterion for determining which polynomial equations *can* be solved by a formula. Galois' ideas had a profound influence on the development of later mathematics, far beyond the scope of the original solvability problem.

The solutions of the equation $f(x) = 0$ lie in some extension of the coefficient field of $f(x)$. Galois' remarkable discovery was the close connection between such field extensions and groups (Section 11.1). A detailed description of the connection is given by the Fundamental Theorem of Galois Theory in Section 11.2. This theorem is the principal tool for proving Galois' Criterion for the solvability of equations by formula (Section 11.3).

11.1 THE GALOIS GROUP

The key to studying field extensions is to associate with each extension a certain group, called its *Galois group*. The properties of the Galois group and theorems of group theory can then be used to establish important facts about the field extension. In this section we define the Galois group and develop its basic properties.

> **DEFINITION•** *Let K be an extension of field of F. An **F-automorphism of K** is an isomorphism $\sigma:K \to K$ that fixes F elementwise (that is, $\sigma(c) = c$ for every $c \in F$). The set of all F-automorphisms of K is denoted $\mathbf{Gal_F K}$ and is called the **Galois group** of K over F.*

The use of the word "group" in the definition is justified by:

THEOREM 11.1   *If K is an extension field of F, then $Gal_F K$ is a group under the operation of composition of functions.*

Proof $Gal_F K$ is nonempty since the identity map $\iota:K \to K$ is an automorphism.* If $\sigma, \tau \in Gal_F K$ then $\sigma \circ \tau$ is an isomorphism from K to K by Exercise 25 of Section 3.3. For each $c \in F$, $(\sigma \circ \tau)(c) = \sigma(\tau(c)) = \sigma(c) = c$. Hence, $\sigma \circ \tau \in Gal_F K$, and $Gal_F K$ is closed. Composition of functions is associative, and the identity map ι is the identity element of $Gal_F K$. Every bijective function has an inverse function by Theorem B.1 in Appendix B. If $\sigma \in Gal_F K$, then σ^{-1} is an isomorphism from K to K by Exercise 27 of Section 3.3. Verify that $\sigma^{-1}(c) = c$ for every $c \in F$ (Exercise 1). Therefore, $\sigma^{-1} \in Gal_F K$, and $Gal_F K$ is a group.   ◆

> **EXAMPLE 1.A****   The complex conjugation map $\sigma:\mathbb{C} \to \mathbb{C}$ given by $\sigma(a + bi) = a - bi$ is an automorphism of $\mathbb{C}$, as shown in the example on page 70. For every real number a,
>
> $$\sigma(a) = \sigma(a + 0i) = a - 0i = a.$$
>
> So σ is in $Gal_{\mathbb{R}}\mathbb{C}$. Note that i and $-i$ are the roots of $x^2 + 1 \in \mathbb{R}$ and that σ maps these roots onto each other: $\sigma(i) = -i$ and $\sigma(-i) = i$. This is an example of

---

* Throughout this chapter, ι denotes the identity map on the field under discussion.

** Throughout this section and the next, three basic examples appear repeatedly. The first appearance of Example 1 is labeled 1.A, its second appearance 1.B, etc.; the first appearance of Example 2 is labeled 2.A, and so on.

THEOREM 11.2   *Let K be an extension field of F and $f(x) \in F[x]$. If $u \in K$ is a root of $f(x)$ and $\sigma \in \text{Gal}_F K$, then $\sigma(u)$ is also a root of $f(x)$.*

Proof   If $f(x) = c_0 + c_1 x + c_2 x^2 + \cdots + c_n x^n$, then

$$c_0 + c_1 u + c_2 u^2 + \cdots + c_n u^n = 0_F.$$

Since σ is a homomorphism and $\sigma(c_i) = c_i$ for each $c_i \in F$,

$$
\begin{aligned}
0_F = \sigma(0_F) &= \sigma(c_0 + c_1 u + c_2 u^2 + \cdots + c_n u^n) \\
&= \sigma(c_0) + \sigma(c_1)\sigma(u) + \sigma(c_2)\sigma(u)^2 + \cdots + \sigma(c_n)\sigma(u)^n \\
&= c_0 + c_1\sigma(u) + c_2\sigma(u)^2 + \cdots + c_n\sigma(u)^n = f(\sigma(u)).
\end{aligned}
$$

Therefore, $\sigma(u)$ is a root of $f(x)$.   ◆

Let $u \in K$ be algebraic over F with minimal polynomial $p(x) \in F[x]$. Theorem 11.2 states that every image of u under an automorphism of the Galois group must also be a root of $p(x)$. Conversely, is *every* root of $p(x)$ in K the image of u under some automorphism of $\text{Gal}_F K$? Here is one case where the answer is yes.

THEOREM 11.3   *Let K be the splitting field of some polynomial over F and let u, $v \in K$. Then there exists $\sigma \in \text{Gal}_F K$ such that $\sigma(u) = v$ if and only if u and v have the same minimal polynomial in $F[x]$.*

Proof   If u and v have the same minimal polynomial, then by Corollary 10.8 there is an isomorphism $\sigma : F(u) \cong F(v)$ such that $\sigma(u) = v$, and σ fixes F elementwise. Since K is a splitting field of some polynomial over F, it is a splitting field of the same polynomial over both $F(u)$ and $F(v)$. Therefore, σ extends to an F-automorphism of K (also denoted σ) by Theorem 10.14. In other words, $\sigma \in \text{Gal}_F K$ and $\sigma(u) = v$. The converse is an immediate consequence of Theorem 11.2.   ◆

> **EXAMPLE 1.B**   Example 1.A shows that $\text{Gal}_\mathbb{R} \mathbb{C}$ has at least two elements, the identity map ι and the complex conjugation map σ. We now prove that these are the only elements in $\text{Gal}_\mathbb{R} \mathbb{C}$. Let τ be any automorphism in $\text{Gal}_\mathbb{R} \mathbb{C}$. Since i is a root of $x^2 + 1$, $\tau(i) = \pm i$ by Theorem 11.2. If $\tau(i) = i$, then since τ fixes every element of $\mathbb{R}$,
>
> $$\tau(a + bi) = \tau(a) + \tau(b)\tau(i) = a + bi,$$
>
> and, hence, $\tau = \iota$. Similarly, if $\tau(i) = -i$, then
>
> $$\tau(a + bi) = \tau(a) + \tau(b)\tau(i) = a + b(-i) = a - bi,$$
>
> and, therefore, $\tau = \sigma$. Thus $\text{Gal}_\mathbb{R} \mathbb{C} = \{\iota, \sigma\}$ is a group of order 2 and, hence, isomorphic to $\mathbb{Z}_2$ by Theorem 7.28.

The preceding example shows that an $\mathbb{R}$-automorphism of $\mathbb{C} = \mathbb{R}(i)$ is completely determined by its action on i. The same thing is true in the general case:

THEOREM 11.4 *Let $K = F(u_1, \ldots, u_n)$ be an algebraic extension field of F. If $\sigma, \tau \in Gal_F K$ and $\sigma(u_i) = \tau(u_i)$ for each $i = 1, 2, \ldots, n$, then $\sigma = \tau$. In other words, an automorphism in $Gal_F K$ is completely determined by its action on $u_1, \ldots, u_n$.*

Proof Let $\beta = \tau^{-1} \circ \sigma \in Gal_F K$. We shall show that β is the identity map ι. Since $\sigma(u_i) = \tau(u_i)$ for every i,

$$\beta(u_i) = (\tau^{-1} \circ \sigma)(u_i) = \tau^{-1}(\sigma(u_i)) = \tau^{-1}(\tau(u_i)) = (\tau^{-1} \circ \tau)(u_i) = \iota(u_i) = u_i.$$

Let $v \in F(u_1)$. By Theorem 10.7 there exist $c_i \in F$ such that $v = c_0 + c_1 u_1 + c_2 u_1{}^2 + \cdots + c_{m-1} u_1{}^{m-1}$, where m is the degree of the minimal polynomial of u_1. Since β is a homomorphism that fixes u_1 and every element of F,

$$\begin{aligned} \beta(v) &= \beta(c_0 + c_1 u_1 + c_2 u_1{}^2 + \cdots + c_{m-1} u_1{}^{m-1}) \\ &= \beta(c_0) + \beta(c_1)\beta(u_1) + \beta(c_2)\beta(u_1{}^2) + \cdots + \beta(c_{m-1})\beta(u_1{}^{m-1}) \\ &= c_0 + c_1 u_1 + c_2 u_1{}^2 + \cdots + c_{m-1} u_1{}^{m-1} \end{aligned}$$

Therefore, $\beta(v) = v$ for every $v \in F(u_1)$. Repeating this argument with $F(u_1)$ in place of F and u_2 in place of u_1 shows that $\beta(v) = v$ for every $v \in F(u_1)(u_2) = F(u_1, u_2)$. Another repetition, with $F(u_1, u_2)$ in place of F and u_3 in place of u_1, shows that $\beta(v) = v$ for every $v \in F(u_1, u_2, u_3)$. After a finite number of repetitions we have $\beta(v) = v$ for every $v \in F(u_1, u_2, \ldots, u_n) = K$, that is, $\iota = \beta = \tau^{-1} \circ \sigma$. Therefore,

$$\tau = \tau \circ \iota = \tau \circ (\tau^{-1} \circ \sigma) = (\tau \circ \tau^{-1}) \circ \sigma = \iota \circ \sigma = \sigma. \quad \blacklozenge$$

EXAMPLE 2.A By Theorem 11.2 any automorphism in the Galois group of $\mathbb{Q}(\sqrt{3}, \sqrt{5})$ over $\mathbb{Q}$ takes $\sqrt{3}$ to $\sqrt{3}$ or $-\sqrt{3}$, the roots of $x^2 - 3$. Similarly, it must take $\sqrt{5}$ to $\pm\sqrt{5}$, the roots of $x^2 - 5$. Since an automorphism is completely determined by its action on $\sqrt{3}$ and $\sqrt{5}$ by Theorem 11.4, there are *at most* four automorphisms in $Gal_{\mathbb{Q}}\mathbb{Q}(\sqrt{3}, \sqrt{5})$, corresponding to the four possible actions on $\sqrt{3}$ and $\sqrt{5}$:

$$\begin{array}{cccc} \sqrt{3} \xrightarrow{\ \iota\ } \sqrt{3} & \sqrt{3} \xrightarrow{\ \tau\ } -\sqrt{3} & \sqrt{3} \xrightarrow{\ \alpha\ } \sqrt{3} & \sqrt{3} \xrightarrow{\ \beta\ } -\sqrt{3} \\ \sqrt{5} \longrightarrow \sqrt{5} & \sqrt{5} \longrightarrow \sqrt{5} & \sqrt{5} \longrightarrow -\sqrt{5} & \sqrt{5} \longrightarrow -\sqrt{5}. \end{array}$$

We now show that $Gal_{\mathbb{Q}}\mathbb{Q}(\sqrt{3}, \sqrt{5})$ is a group of order 4 by constructing nonidentity automorphisms τ, α, β with these actions. To construct τ, note that $x^2 - 3$ is the minimal polynomial of both $\sqrt{3}$ and $-\sqrt{3}$ over $\mathbb{Q}$. By Corollary 10.8, there is an isomorphism $\sigma : \mathbb{Q}(\sqrt{3}) \cong \mathbb{Q}(-\sqrt{3})$ such that

$\sigma(\sqrt{3}) = -\sqrt{3}$, and σ fixes $\mathbb{Q}$ elementwise. The example on page 349 shows that $x^2 - 5$ is the minimal polynomial of $\sqrt{5}$ over $\mathbb{Q}(\sqrt{3})$. By Corollary 10.8 again, σ extends to a $\mathbb{Q}$-automorphism τ of $\mathbb{Q}(\sqrt{3})(\sqrt{5}) = \mathbb{Q}(\sqrt{3},\sqrt{5})$ such that $\tau(\sqrt{5}) = \sqrt{5}$. Therefore, $\tau \in \text{Gal}_{\mathbb{Q}}\mathbb{Q}(\sqrt{3},\sqrt{5})$ and $\tau(\sqrt{3}) = \sigma(\sqrt{3}) = -\sqrt{3}$ and $\tau(\sqrt{5}) = \sqrt{5}$. A similar two-step argument produces automorphisms α and β with the actions listed above. Furthermore, each of τ, α, β has order 2 in $\text{Gal}_{\mathbb{Q}}\mathbb{Q}(\sqrt{3},\sqrt{5})$; for instance,

$$(\tau \circ \tau)(\sqrt{3}) = \tau(\tau(\sqrt{3})) = \tau(-\sqrt{3}) = -\tau(\sqrt{3}) = -(-\sqrt{3}) = \sqrt{3} = \iota(\sqrt{3})$$

and $(\tau \circ \tau)(\sqrt{5}) = \sqrt{5} = \iota(\sqrt{5})$. Therefore, $\tau \circ \tau = \iota$ by Theorem 11.4. Use Theorem 7.29 to conclude that $\text{Gal}_{\mathbb{Q}}\mathbb{Q}(\sqrt{3},\sqrt{5}) \cong \mathbb{Z}_2 \times \mathbb{Z}_2$ or compute the operation table directly (Exercise 4). For instance, you can readily verify that $(\tau \circ \alpha)(\sqrt{3}) = \beta(\sqrt{3})$ and $(\tau \circ \alpha)(\sqrt{5}) = \beta(\sqrt{5})$ and, hence, $\tau \circ \alpha = \beta$ by Theorem 11.4.

In the preceding example, $\mathbb{Q}(\sqrt{3},\sqrt{5})$ is the splitting field of $f(x) = (x^2 - 3)(x^2 - 5)$, and every automorphism in the Galois group permutes the four roots $\sqrt{3}, -\sqrt{3}, \sqrt{5}, -\sqrt{5}$ of $f(x)$. This is an illustration of

COROLLARY 11.5 *If K is the splitting field of a separable polynomial $f(x)$ of degree n in $F[x]$, then $\text{Gal}_F K$ is isomorphic to a subgroup of S_n.*

Proof By separability $f(x)$ has n distinct roots in K, say $u_1, \dots, u_n$. Consider S_n to be the group of permutations of the set $R = \{u_1, \dots, u_n\}$. If $\sigma \in \text{Gal}_F K$, then $\sigma(u_1), \sigma(u_2), \dots, \sigma(u_n)$ are roots of $f(x)$ by Theorem 11.2. Furthermore, since σ is injective, they are all distinct and, hence, must be $u_1, u_2, \dots, u_n$ in some order. In other words, the restriction of σ to the set R (denoted $\sigma \mid R$) is a permutation of R. Define a map $\theta : \text{Gal}_F K \to S_n$ by $\theta(\sigma) = \sigma \mid R$. Since the operation in both groups is composition of functions, it is easy to verify that θ is a homomorphism of groups. $K = F(u_1, \dots, u_n)$ by the definition of splitting field. If $\sigma \mid R = \tau \mid R$, then $\sigma(u_i) = \tau(u_i)$ for every i, and, hence, $\sigma = \tau$ by Theorem 11.4. Therefore, θ is an injective homomorphism, and thus $\text{Gal}_F K$ is isomorphic to Im θ, a subgroup of S_n, by Theorem 7.19 ◆

If K is the splitting field of $f(x)$, we shall usually

identify $\text{Gal}_F K$ with its isomorphic subgroup in S_n

by identifying each automorphism with the permutation it induces on the roots of $f(x)$.

EXAMPLE 3.A Let K be the splitting field of $x^3 - 2$ over $\mathbb{Q}$. Verify that the roots of $x^3 - 2$ are $\sqrt[3]{2}, \sqrt[3]{2}\omega, \sqrt[3]{2}\omega^2$, where $\omega = (-1 + \sqrt{3}i)/2$ is a complex cube root of 1. Then $\text{Gal}_{\mathbb{Q}} K$ is a subgroup of S_3. By Theorem 11.3, there is at least one automorphism σ that maps the first root $\sqrt[3]{2}$ to the second $\sqrt[3]{2}\omega$; it must take the third root $\sqrt[3]{2}\omega^2$ to itself *or* to the first root $\sqrt[3]{2}$ by Theorem 11.2. So σ is either the permutation (12) or (123) in S_3.

Warning When K is the splitting field of a polynomial $f(x) \in F[x]$, then by Corollary 11.5 every element of $\text{Gal}_F K$ produces a permutation of the roots of $f(x)$, but not vice versa: a permutation of the roots need not come from an F-automorphism of K. For example, $\mathbb{Q}(\sqrt{3},\sqrt{5})$ is a splitting field of $f(x) = (x^2 - 3)(x^2 - 5)$, but by Example 2.A there is no $\mathbb{Q}$-automorphism of $\mathbb{Q}(\sqrt{3},\sqrt{5})$ that gives this permutation of the roots

$$\begin{array}{cccc} \sqrt{3} & -\sqrt{3} & \sqrt{5} & -\sqrt{5} \\ \downarrow & \downarrow & \downarrow & \downarrow \\ \sqrt{5} & -\sqrt{5} & \sqrt{3} & -\sqrt{3} \end{array}$$

Let K be an extension field of F. A field E such that $F \subseteq E \subseteq K$ is called an **intermediate field** of the extension. In this case, we can consider K as an extension of E. The Galois group $\text{Gal}_E K$ consists of all automorphisms of K that fix E elementwise. Every such automorphism automatically fixes each element of F since $F \subseteq E$. Hence, every automorphism in $\text{Gal}_E K$ is in $\text{Gal}_F K$, that is,

if E is an intermediate field, $\text{Gal}_E K$ is a subgroup of $\text{Gal}_F K$.

EXAMPLE 2.B $\mathbb{Q}(\sqrt{3})$ is an intermediate field of the extension $\mathbb{Q}(\sqrt{3},\sqrt{5})$ of $\mathbb{Q}$. Example 2.A shows that $\text{Gal}_\mathbb{Q}\mathbb{Q}(\sqrt{3},\sqrt{5}) = \{\iota, \tau, \alpha, \beta\}$. The automorphisms that fix every element of $\mathbb{Q}(\sqrt{3})$ are exactly the ones that map $\sqrt{3}$ to itself by Theorem 11.4. Therefore,

$$\text{Gal}_{\mathbb{Q}(\sqrt{3})}\mathbb{Q}(\sqrt{3},\sqrt{5})$$

is the subgroup $\{\iota, \alpha\}$ of $\{\iota, \tau, \alpha, \beta\}$.

We now have a natural way of associating a subgroup of the Galois group with each intermediate field of the extension. Conversely, if H is a subgroup of the Galois group, we can associate an intermediate field with H by using

THEOREM 11.6 *Let K be an extension field of F. If H is a subgroup of $\text{Gal}_F K$, let*

$$E_H = \{k \in K \mid \sigma(k) = k \text{ for every } \sigma \in H\}.$$

Then E_H is an intermediate field of the extension.

The field E_H is called the **fixed field** of the subgroup H.

Proof of Theorem 11.6   If $c, d \in E_H$ and $\sigma \in H$, then

$$\sigma(c + d) = \sigma(c) + \sigma(d) = c + d \qquad \text{and} \qquad \sigma(cd) = \sigma(c)\sigma(d) = cd.$$

Therefore, E_H is closed under addition and multiplication. Since $\sigma(0_F) = 0_F$ and $\sigma(1_F) = 1_F$ for every automorphism, 0_F and 1_F are in E_H. Theorem 3.12 shows that for any nonzero c in E_H and any σ in H,

$$\sigma(-c) = -\sigma(c) = -c \qquad \text{and} \qquad \sigma(c^{-1}) = \sigma(c)^{-1} = c^{-1}.$$

Therefore, $-c \in E_H$ and $c^{-1} \in E_H$. Hence, E_H is a subfield of K. Since H is a subgroup of $\mathrm{Gal}_F K$, $\sigma(c) = c$ for every $c \in F$ and every $\sigma \in H$. Therefore, $F \subseteq E_H$.   ◆

EXAMPLE 2.C   Consider the subgroup $H = \{\iota, \alpha\}$ of the Galois group $\{\iota, \tau, \alpha, \beta\}$ of $\mathbb{Q}(\sqrt{3}, \sqrt{5})$ over $\mathbb{Q}$. Since $\alpha(\sqrt{3}) = \sqrt{3}$, the subfield $\mathbb{Q}(\sqrt{3})$ is contained in the fixed field E_H of H. To prove that $E_H = \mathbb{Q}(\sqrt{3})$, you must show that the elements of $\mathbb{Q}(\sqrt{3})$ are the *only* ones that are fixed by ι and α; see Exercise 14.

EXAMPLE 1.C   As we saw in Example 1.B, $\mathrm{Gal}_{\mathbb{R}}\mathbb{C} = \{\iota, \sigma\}$, where σ is the complex conjugation map. Obviously, the fixed field of the identity subgroup is the entire field $\mathbb{C}$. Since σ fixes every real number and moves every nonreal one, the fixed field of $\mathrm{Gal}_{\mathbb{R}}\mathbb{C}$ is the field $\mathbb{R}$.

Unlike the situation in the preceding example, the ground field F need not always be the fixed field of the group $\mathrm{Gal}_F K$.

EXAMPLE 3.B   Every automorphism in the Galois group of $\mathbb{Q}(\sqrt[3]{2})$ over $\mathbb{Q}$ must map $\sqrt[3]{2}$ to a root of $x^3 - 2$ by Theorem 11.2. Example 3.A shows that $\sqrt[3]{2}$ is the only real root of this polynomial. Since $\mathbb{Q}(\sqrt[3]{2})$ consists entirely of real numbers by Theorem 10.7, every automorphism in $\mathrm{Gal}_{\mathbb{Q}}\mathbb{Q}(\sqrt[3]{2})$ must map $\sqrt[3]{2}$ to itself. Therefore, $\mathrm{Gal}_{\mathbb{Q}}\mathbb{Q}(\sqrt[3]{2})$ consists of the identity automorphism alone by Theorem 11.4. So the fixed field of $\mathrm{Gal}_{\mathbb{Q}}\mathbb{Q}(\sqrt[3]{2})$ is the entire field $\mathbb{Q}(\sqrt[3]{2})$.

◆ **EXERCISES**

NOTE: *Unless stated otherwise, K is an extension field of the field F.*

A.   **1.** If σ is an F-automorphism of K, show that σ^{-1} is also an F-automorphism of K.

   2. Assume $[K\!:\!F]$ is finite. Is it true that every F-automorphism of K is completely determined by its action on a basis of K over F?

3. If $[K:F]$ is finite, $\sigma \epsilon \operatorname{Gal}_F K$, and $u \epsilon K$ is such that $\sigma(u) = u$, show that $\sigma \epsilon \operatorname{Gal}_{F(u)} K$.

4. Write out the operation table for the group

$$\operatorname{Gal}_{\mathbb{Q}} \mathbb{Q}(\sqrt{3},\sqrt{5}) = \{\iota, \tau, \alpha, \beta\}.$$

[See Example 2.A.]

5. Let $f(x) \epsilon F[x]$ be separable of degree n and K a splitting field of $f(x)$. Show that the order of $\operatorname{Gal}_F K$ divides $n!$.

6. If K is an extension field of $\mathbb{Q}$ and σ is an automorphism of K, prove that σ is a $\mathbb{Q}$-automorphism. [*Hint:* $\sigma(1) = 1$ implies that $\sigma(n) = n$ for all $n \epsilon \mathbb{Z}$.]

B. 7. (a) Show that $\operatorname{Gal}_{\mathbb{Q}} \mathbb{Q}(\sqrt{2})$ has order 2 and, hence, is isomorphic to $\mathbb{Z}_2$. [*Hint:* The minimal polynomial is $x^2 - 2$; see Theorem 10.7.]

   (b) If $d \epsilon \mathbb{Q}$ and $\sqrt{d} \notin \mathbb{Q}$, show that $\operatorname{Gal}_{\mathbb{Q}} \mathbb{Q}(\sqrt{d})$ is isomorphic to $\mathbb{Z}_2$.

8. Show that $\operatorname{Gal}_{\mathbb{Q}} \mathbb{Q}(\sqrt[4]{2}) \neq \langle \iota \rangle$.

9. (a) Let $\omega = (-1 + \sqrt{3}i)/2$ be a complex cube root of 1. Find the minimal polynomial $p(x)$ of ω over $\mathbb{Q}$ and show that ω^2 is also a root of $p(x)$. [*Hint:* ω is a root of $x^3 - 1$.]

   (b) What is $\operatorname{Gal}_{\mathbb{Q}} \mathbb{Q}(\omega)$?

10. (a) Find $\operatorname{Gal}_{\mathbb{Q}} \mathbb{Q}(\sqrt{2},\sqrt{3})$. [*Hint:* See Example 2.A.]

    (b) If p, q are distinct positive primes, find $\operatorname{Gal}_{\mathbb{Q}} \mathbb{Q}(\sqrt{p},\sqrt{q})$.

11. Find $\operatorname{Gal}_{\mathbb{Q}} \mathbb{Q}(\sqrt{2},i)$. [*Hint:* Consider $\mathbb{Q} \subseteq \mathbb{Q}(\sqrt{2}) \subseteq \mathbb{Q}(\sqrt{2},i)$ and proceed as in Example 2.A.]

12. Show that $\operatorname{Gal}_{\mathbb{Q}} \mathbb{Q}(\sqrt{2},\sqrt{3},\sqrt{5}) \cong \mathbb{Z}_2 \times \mathbb{Z}_2 \times \mathbb{Z}_2$.

13. If F has characteristic 0 and K is the splitting field of $f(x) \epsilon F[x]$, prove that the order of $\operatorname{Gal}_F K$ is $[K:F]$. [*Hint:* $K = F(u)$ by Theorems 10.17 and 10.18.]

14. Let H be the subgroup $\{\iota, \alpha\}$ of $\operatorname{Gal}_{\mathbb{Q}} \mathbb{Q}(\sqrt{3},\sqrt{5}) = \{\iota, \tau, \alpha, \beta\}$. Show that the fixed field of H is $\mathbb{Q}(\sqrt{3})$. [*Hint:* Verify that $\mathbb{Q}(\sqrt{3}) \subseteq E_H \subseteq \mathbb{Q}(\sqrt{3},\sqrt{5})$; what is $[\mathbb{Q}(\sqrt{3},\sqrt{5}):\mathbb{Q}(\sqrt{3})]$?]

15. (a) Show that every automorphism of $\mathbb{R}$ maps positive elements to positive elements. [*Hint:* Every positive element of $\mathbb{R}$ is a square.]

    (b) If $a, b \epsilon \mathbb{R}$, $a < b$, and $\sigma \epsilon \operatorname{Gal}_{\mathbb{Q}} \mathbb{R}$, prove that $\sigma(a) < \sigma(b)$. [*Hint:* $a < b$ if and only if $b - a > 0$.]

    (c) Prove that $\operatorname{Gal}_{\mathbb{Q}} \mathbb{R} = \langle \iota \rangle$. [*Hint:* If $c < r < d$, with $c, d \epsilon \mathbb{Q}$, then $c < \sigma(r) < d$; show that this implies $\sigma(r) = r$.]

C. 16. Suppose $\zeta, \zeta^2, \ldots, \zeta^n = 1$ are n distinct roots of $x^n - 1$ in some extension field of $\mathbb{Q}$. Prove that $\operatorname{Gal}_{\mathbb{Q}} \mathbb{Q}(\zeta)$ is abelian.

17. Let E be an intermediate field that is normal over F and $\sigma \in \mathrm{Gal}_F K$. Prove that $\sigma(E) = E$.

11.2 THE FUNDAMENTAL THEOREM OF GALOIS THEORY

Throughout this section *K is a finite-dimensional extension field of F.* The essential idea of Galois theory is to relate properties of the extension to properties of the Galois group $\mathrm{Gal}_F K$. The key to doing this is the Fundamental Theorem of Galois Theory, which will be proved in this section. It states (in part) that under suitable hypotheses, *there is a one-to-one correspondence between the intermediate fields of the extension and the subgroups of the Galois group.**

The first step in establishing this correspondence is easy: Given an intermediate field E, there is a subgroup of $\mathrm{Gal}_F K$ that is naturally associated with E, namely $\mathrm{Gal}_E K$, the subgroup of automorphisms of K that fix E elementwise. So we can define a function from the set S of all intermediate fields to the set T of all subgroups of $\mathrm{Gal}_F K$ by assigning the intermediate field E to the subgroup $\mathrm{Gal}_E K$. This function is called the **Galois correspondence.**

Under the Galois correspondence the trivial intermediate fields, F and K, correspond to the trivial subgroups of the Galois group: K to the identity subgroup $\mathrm{Gal}_K K$ and F to the full Galois group $\mathrm{Gal}_F K$.

> **EXAMPLE 2.D****   Consider the Galois correspondence for the extension $\mathbb{Q}(\sqrt 3,\sqrt 5)$ of $\mathbb{Q}$ and the intermediate field $\mathbb{Q}(\sqrt 3)$. By the preceding remarks and Example 2.B on page 376, we have
>
> $$\mathbb{Q}(\sqrt 3,\sqrt 5) \longrightarrow \mathrm{Gal}_{\mathbb{Q}(\sqrt 3,\sqrt 5)}\mathbb{Q}(\sqrt 3,\sqrt 5) = \{\iota\}.$$
> $$\mathbb{Q}(\sqrt 3) \longrightarrow \mathrm{Gal}_{\mathbb{Q}(\sqrt 3)}\mathbb{Q}(\sqrt 3,\sqrt 5) = \{\iota, \alpha\}.$$
> $$\mathbb{Q} \longrightarrow \mathrm{Gal}_{\mathbb{Q}}\mathbb{Q}(\sqrt 3,\sqrt 5) = \{\iota, \tau, \alpha, \beta\}.$$
>
> Example 2.C shows that $E = \mathbb{Q}(\sqrt 3)$ is the fixed field of the subgroup $H = \{\iota, \alpha\} = \mathrm{Gal}_{\mathbb{Q}(\sqrt 3)}\mathbb{Q}(\sqrt 3,\sqrt 5)$. Furthermore, $K = \mathbb{Q}(\sqrt 3,\sqrt 5) = \mathbb{Q}(\sqrt 3)(\sqrt 5)$ is a normal, separable extension of the fixed field $E = \mathbb{Q}(\sqrt 3)$ because it's the splitting field of $x^2 - 5$ (Theorem 10.15) and has characteristic 0 (Theorem 10.17). This is an example of

LEMMA 11.7 *Let K be a finite-dimensional extension field of F. If H is a subgroup of the Galois group $\mathrm{Gal}_F K$ and E is the fixed field of H, then K is a simple, normal, separable extension of E.*

---

* That is, a bijective map from the set of all intermediate fields to the set of all subgroups of the Galois group.

** The numbering scheme for examples in Sections 11.1 and 11.2 is explained on page 372.

Proof   Each $u \in K$ is algebraic over F by Theorem 10.9 and, hence, algebraic over E by Exercise 7 in Section 10.2. Every automorphism in H must map u to some root of its minimal polynomial $p(x) \in E[x]$ by Theorem 11.2. Therefore, u *has a finite number of distinct images under automorphisms in H,* say $u = u_1$, $u_2, \ldots, u_t \in K$.

If $\sigma \in H$ and $u_i = \tau(u)$ (with $\tau \in H$), then $\sigma(u_i) = \sigma(\tau(u))$. Since $\sigma \circ \tau \in H$, we see that $\sigma(u_i)$ is also an image of u and, hence, must be in the set $\{u_1, u_2, \ldots, u_t\}$. Since σ is injective, the elements $\sigma(u_i), \ldots, \sigma(u_t)$ are t distinct images of u and, hence, must be the elements $u_1, u_2, \ldots, u_t$ in some order. In other words, *every automorphism in H permutes $u_1, u_2, \ldots, u_t$.* Let

$$f(x) = (x - u_1)(x - u_2) \cdots (x - u_t).$$

Since the u_i are distinct, $f(x)$ is separable. We claim that $f(x)$ is actually in $E[x]$. To prove this, let $\sigma \in H$ and recall that σ induces an isomorphism $K[x] \cong K[x]$ (also denoted σ), as described on page 344. Then

$$\sigma f(x) = (x - \sigma(u_1))(x - \sigma(u_2)) \cdots (x - \sigma(u_t)).$$

Since σ permutes the u_i, it simply rearranges the factors of $f(x)$, and, hence, $\sigma f(x) = f(x)$. Therefore, every automorphism of H maps the coefficients of the separable polynomial $f(x)$ to themselves, and, hence, these coefficients are in E, the fixed field of H. Since $u = u_1$ is a root of $f(x) \in E[x]$, u is separable over E. Hence, K is a separable extension of E.

The field K is finitely generated over F (since $[K:F]$ is finite; see the example on page 348). Consequently, K is finitely generated over E, and, hence, $K = E(u)$ for some $u \in K$ by Theorem 10.18. Let $f(x)$ be as in the preceding paragraph. Then $f(x)$ splits in $K[x]$, and, hence, $K = E(u)$ is the splitting field of $f(x)$ over E. Therefore, K is normal over E by Theorem 10.15.   ◆

THEOREM 11.8   *Let K be a finite-dimensional extension field of F. If H is a subgroup of the Galois group $\mathrm{Gal}_F K$ and E is the fixed field of H, then $H = \mathrm{Gal}_E K$ and $|H| = [K:E]$. Therefore, the Galois correspondence is surjective.*

Proof   Lemma 11.7 shows that $K = E(u)$ for some $u \in K$. If $p(x)$, the minimal polynomial of u over E, has degree n, then $[K:E] = n$ by Theorem 10.7. Distinct automorphisms of $\mathrm{Gal}_E K$ map u onto distinct roots of $p(x)$ by Theorems 11.2 and 11.4. So the number of distinct automorphisms in $\mathrm{Gal}_E K$ is at most n, the number of roots of $p(x)$. Now $H \subseteq \mathrm{Gal}_E K$ by the definition of the fixed field E. Consequently,

$$|H| \leq |\mathrm{Gal}_E K| \leq n = [K:E].$$

Let $f(x)$ be as in the proof of Lemma 11.7. Then H contains at least t automorphisms (the number of distinct images of u under H). Since $u = u_1$ is a root of $f(x)$, $p(x)$ divides $f(x)$. Hence,

$$|H| \geq t = \deg f(x) \geq \deg p(x) = n = [K{:}E].$$

Combining these inequalities, we have

$$|H| \leq |\mathrm{Gal}_E K| \leq [K{:}E] \leq |H|.$$

Therefore, $|H| = |\mathrm{Gal}_E K| = [K{:}E]$, and, hence, $H = \mathrm{Gal}_E K.$ ◆

EXAMPLE 3.C The Galois group $\mathrm{Gal}_{\mathbb{Q}}\mathbb{Q}(\sqrt[3]{2}) = \langle \iota \rangle$ by Examples 3.B, so both of the intermediate fields $\mathbb{Q}(\sqrt[3]{2})$ and $\mathbb{Q}$ are associated with $\langle \iota \rangle$ under the Galois correspondence. Note that $\mathbb{Q}(\sqrt[3]{2})$ is *not* a normal extension of $\mathbb{Q}$ [it doesn't contain the complex roots of $x^3 - 2$, so this polynomial has a root but doesn't split in $\mathbb{Q}(\sqrt[3]{2})$].

Galois Extensions

Although the Galois correspondence is surjective by Theorem 11.8, the preceding example shows that it may not be injective. In order to guarantee injectivity, additional hypotheses on the extension are necessary. The preceding proofs and example suggest that normality and separability are likely candidates.

> **DEFINITION•** *If K is a finite-dimensional, normal, separable extension field of the field F, we say that K is a **Galois extension** of F or that K is **Galois over F**.*

A Galois extension of characteristic 0 is simply a splitting field by Theorems 10.15 and 10.17.

THEOREM 11.9 *Let K be a Galois extension of F and E an intermediate field. Then E is the fixed field of the subgroup $\mathrm{Gal}_E K$.*

If E and L are intermediate fields with $\mathrm{Gal}_E K = \mathrm{Gal}_L K$, then Theorem 11.9 shows that both E and L are the fixed field of the *same* group, and, hence, $E = L$. Therefore, *the Galois correspondence is injective for Galois extensions.*

Proof of Theorem 11.9 The fixed field E_0 of $\mathrm{Gal}_E K$ contains E by definition. To show that $E_0 \subseteq E$, we prove the contrapositive: If $u \notin E$, then u is moved by some automorphism in $\mathrm{Gal}_E K$, and, hence, $u \notin E_0$. Since K is a Galois extension of the intermediate field E (normal by Theorem 10.15 and Exercise 5 of Section 10.4; separable by Exercise 1 of Section 10.5), it is an algebraic extension of E. Consequently, u is algebraic over E with minimal polynomial $p(x) \in E[x]$ of degree ≥ 2 (if $\deg p(x) = 1$, then u would be in E). The roots of $p(x)$ are distinct by separability, and all of them are in K by normality. Let v be a root of $p(x)$

other than u. Then there exists $\sigma \epsilon \operatorname{Gal}_E K$ such that $\sigma(u) = v$ by Theorem 11.3. Therefore, $u \notin E_0$, and, hence, $E_0 = E$. ◆

COROLLARY 11.10   *Let K be a finite-dimensional extension field of F. Then K is Galois over F if and only if F is the fixed field of the Galois group $\operatorname{Gal}_F K$.*

Proof   If K is Galois over F, then Theorem 11.9 (with $E = F$) shows that F is the fixed field of $\operatorname{Gal}_F K$. Conversely, if F is the fixed field of $\operatorname{Gal}_F K$, then Lemma 11.7 (with $E = F$) shows that K is Galois over F. ◆

In view of Corollary 11.10, a Galois extension is often *defined* to be a finite-dimensional one in which F is the fixed field of $\operatorname{Gal}_F K$. When reading other books on Galois theory, it's a good idea to check which definition is being used so that you don't make unwarranted assumptions.

EXAMPLE 2.E   The field $\mathbb{Q}(\sqrt{3},\sqrt{5})$ is a Galois extension of $\mathbb{Q}$ because it is the splitting field of $f(x) = (x^2 - 3)(x^2 - 5)$. So the Galois correspondence is bijective by Theorem 11.8 and the remarks after Theorem 11.9. The Galois group $\operatorname{Gal}_{\mathbb{Q}} \mathbb{Q}(\sqrt{3},\sqrt{5}) = \{\iota, \tau, \alpha, \beta\}$ by Example 2.A. Verify the accuracy of the chart below, in which subfields and subgroups in the same relative position correspond to each other under the Galois correspondence. For instance, $\mathbb{Q}(\sqrt{3})$ corresponds to $\{\iota, \alpha\}$ by Example 2.B.

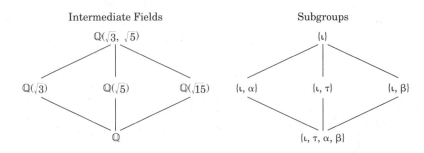

Note that *all* the intermediate fields are themselves Galois extensions of $\mathbb{Q}$ (for instance, $\mathbb{Q}(\sqrt{5})$ is the splitting field of $x^2 - 5$). Furthermore, the corresponding subgroups of the Galois group are normal. A similar situation holds in the general case:

THEOREM 11.11   (THE FUNDAMENTAL THEOREM OF GALOIS THEORY) *If K is a Galois extension field of F, then*

1. *There is bijection between the set S of all intermediate fields of the extension and the set T of all subgroups of the Galois group $\operatorname{Gal}_F K$, given by assigning each intermediate field E to the subgroup $\operatorname{Gal}_E K$. Furthermore,*

$$[K{:}E] = |\operatorname{Gal}_E K| \quad and \quad [E{:}F] = [\operatorname{Gal}_F K {:} \operatorname{Gal}_E K].$$

2. *An intermediate field E is a normal extension of F if and only if the corresponding group $Gal_E K$ is a normal subgroup of $Gal_F K$, and in this case $Gal_F E \cong Gal_F K / Gal_E K$.*

Proof   Theorem 11.8 and the remarks after Theorem 11.9 prove the first statement in part 1. Each intermediate field E is the fixed field of $Gal_E K$ by Theorem 11.9. Consequently, $[K:E] = |Gal_E K|$ by Theorem 11.8. In particular, if $F = E$, then $[K:F] = |Gal_F K|$. Therefore, by Lagrange's Theorem 7.26 and Theorem 10.4,

$$[K:E][E:F] = [K:F] = |Gal_F K| = |Gal_E K|[Gal_F K : Gal_E K].$$

Dividing this equation by $[K:E] = |Gal_E K|$ shows that

$$[E:F] = [Gal_F K : Gal_E K].$$

To prove part 2, assume first that $Gal_E K$ is a normal subgroup of $Gal_F K$. If $p(x)$ is an irreducible polynomial in $F[x]$ with a root u in E, we must show that $p(x)$ splits in $E[x]$. Since K is normal over F, we know that $p(x)$ splits in $K[x]$. So we need to show only that each root v of $p(x)$ in K is actually in E. There is an automorphism σ in $Gal_F K$ such that $\sigma(u) = v$ by Theorem 11.3. If τ is any element of $Gal_E K$, then normality implies $\tau \circ \sigma = \sigma \circ \tau_1$ for some $\tau_1 \in Gal_E K$. Since $u \in E$, we have $\tau(v) = \tau(\sigma(u)) = \sigma(\tau_1(u)) = \sigma(u) = v$. Hence, v is fixed by every element τ in $Gal_E K$ and, therefore, must be in the fixed field of $Gal_E K$, namely E (see Theorem 11.9).

Conversely, assume that E is a normal extension of F. Then E is finite-dimensional over F by part 1. By Lemma 11.12, which is proved below, there is a surjective homomorphism of groups $\theta : Gal_F K \rightarrow Gal_F E$ whose kernel is $Gal_E K$. Then $Gal_E K$ is a normal subgroup of $Gal_F K$ by Theorem 7.39, and $Gal_F K / Gal_E K \cong Gal_F E$ by the First Isomorphism Theorem 7.42.   ◆

> **EXAMPLE 3.D**   The splitting field K of $x^3 - 2$ is a Galois extension of $\mathbb{Q}$ whose Galois group is a subgroup of S_3 by Example 3.A.* Note that $\mathbb{Q} \subseteq \mathbb{Q}(\sqrt[3]{2}) \subseteq K$. Since $x^3 - 2$ is the minimal polynomial of $\sqrt[3]{2}$, $[\mathbb{Q}(\sqrt[3]{2}) : \mathbb{Q}] = 3$ by Theorem 10.7. Neither of the other roots ($\sqrt[3]{2}\,\omega$ and $\sqrt[3]{2}\,\omega^2$) is a real number, and, hence, neither is in $\mathbb{Q}(\sqrt[3]{2})$. So $[K:\mathbb{Q}] > 3$. Since $[K:\mathbb{Q}] \leq 6$ (Theorems 10.13, 10.14) and $[K:\mathbb{Q}]$ is divisible by 3 (Theorem 10.4), we must have $[K:\mathbb{Q}] = 6$. Thus $Gal_\mathbb{Q} K$ has order 6 by Theorem 11.11 and is S_3.
>
> The only proper subgroups of S_3 are the cyclic group $\langle(123)\rangle$ of order 3 and three cyclic groups of order 2: $\langle(12)\rangle$, $\langle(13)\rangle$, $\langle(23)\rangle$. Verify that the Galois correspondence is as follows, where subgroups and subfields in the

---

* We consider S_3 as the group of permutations of the roots $\sqrt[3]{2}$, $\sqrt[3]{2}\,\omega$, $\sqrt[3]{2}\,\omega^2$ in this order. For instance, (12) interchanges $\sqrt[3]{2}$ and $\sqrt[3]{2}\,\omega$ and fixes $\sqrt[3]{2}\,\omega^2$.

same relative position correspond to each other. The integer by the line connecting two subfields is the dimension of the larger over the smaller. The integer by the line connecting two subgroups is the index of the smaller in the larger.

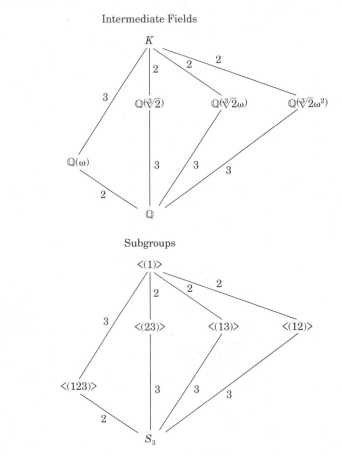

Intermediate Fields

Subgroups

The field $\mathbb{Q}(\omega)$ is an intermediate field because $\omega = (\frac{1}{2})(\sqrt[3]{2})^2(\sqrt[3]{2}\,\omega)$ $\epsilon\,K$. $\mathbb{Q}(\omega)$ is the splitting field of $x^2 + x + 1$ (Exercise 3) and, hence, Galois over $\mathbb{Q}$. The corresponding subgroup is the normal subgroup $\langle(123)\rangle$. On the other hand, Example 3.C shows that $\mathbb{Q}(\sqrt[3]{2})$ is *not* Galois over $\mathbb{Q}$; the corresponding subgroup $\langle(23)\rangle$ is not normal in S_3.

The preceding example illustrates an important fact:

The Galois correspondence is inclusion-reversing.

For instance, $\mathbb{Q} \subseteq \mathbb{Q}(\omega)$, but the corresponding subgroups satisfy the reverse inclusion: $S_3 \supseteq \langle(123)\rangle$.

Finally, we complete the proof of the Fundamental Theorem by proving

LEMMA 11.12 *Let K be a finite-dimensional normal extension field of F and E an intermediate field, which is normal over F. Then there is a surjective homomorphism of groups $\theta : Gal_F K \to Gal_F E$ whose kernel is $Gal_E K$.*

Proof Let $\sigma \in Gal_F K$ and $u \in E$. Then u is algebraic over F with minimal polynomial $p(x)$. Since E is a normal extension of F, $p(x)$ splits in $E[x]$, that is, all the roots of $p(x)$ are in E. Since $\sigma(u)$ must be some root of $p(x)$ by Theorem 11.2, we see that $\sigma(u) \in E$. Therefore, $\sigma(E) \subseteq E$ for every $\sigma \in Gal_F K$. Thus the restriction of σ to E (denoted $\sigma \mid E$) is an F-isomorphism $E \cong \sigma(E)$. Hence, $[E:F] = [\sigma(E):F]$ by Theorem 10.5. Since $F \subseteq \sigma(E) \subseteq E$, we have $[E:F] = [E:\sigma(E)][\sigma(E):F]$ by Theorem 10.4, which forces $[E:\sigma(E)] = 1$. Therefore, $E = \sigma(E)$, and $\sigma \mid E$ is actually an automorphism in $Gal_F E$.

Define a function $\theta : Gal_F K \to Gal_F E$ by $\theta(\sigma) = \sigma \mid E$. It is easy to verify that θ is a homomorphism of groups. Its kernel consists of the automorphisms of K whose restriction to E is the identity map, that is, the subgroup $Gal_E K$.

To show that θ is surjective, note that K is a splitting field over F by Theorem 10.15, and, hence, K is a splitting field of the same polynomial over E. Consequently, every $\tau \in Gal_F E$ can be extended to an F-automorphism σ in $Gal_F K$ by Theorem 10.14. This means that $\sigma \mid E = \tau$, that is, $\theta(\sigma) = \tau$. Therefore, θ is surjective. ◆

In the preceding proof, the normality of K was not used until the last paragraph. So the first paragraph proves this useful fact:

COROLLARY 11.13 *Let K be an extension field of F and E an intermediate field that is normal over F. If $\sigma \in Gal_F K$, then $\sigma \mid E \in Gal_F E$.*

◆ **EXERCISES**

NOTE: *K is an extension field of the field F.*

A. 1. If K is Galois over F, show that there are only finitely many intermediate fields.

2. If K is a normal extension of $\mathbb{Q}$ and $[K:\mathbb{Q}] = p$, with p prime, show that $Gal_\mathbb{Q} K \cong \mathbb{Z}_p$.

3. **(a)** Show that $\omega = (-1 + \sqrt{3}i)/2$ is a root of $x^3 - 1$.

   (b) Show that ω and ω^2 are roots of $x^2 + x + 1$. Hence, $\mathbb{Q}(\omega)$ is the splitting field of $x^2 + x + 1$.

4. Exhibit the Galois correspondence of intermediate fields and subgroups for the given extension of $\mathbb{Q}$:

(a) $\mathbb{Q}(\sqrt{d})$, where $d \in \mathbb{Q}$, but $\sqrt{d} \notin \mathbb{Q}$.

(b) $\mathbb{Q}(\omega)$, where ω is as in Exercise 3.

5. If K is Galois over F and $\mathrm{Gal}_F K$ is an abelian group of order 10, how many intermediate fields does the extension have and what are their dimensions over F?

6. Give an example of extension fields K and L of F such that both K and L are Galois over F, $K \neq L$, and $\mathrm{Gal}_F K \cong \mathrm{Gal}_F L$.

B. 7. Exhibit the Galois correspondence for the given extension of $\mathbb{Q}$:

(a) $\mathbb{Q}(\sqrt{2},\sqrt{3})$ (b) $\mathbb{Q}(i,\sqrt{2})$

8. If K is Galois over F, $\mathrm{Gal}_F K$ is abelian, and E is an intermediate field that is normal over F, prove that $\mathrm{Gal}_E K$ and $\mathrm{Gal}_F E$ are abelian.

9. Let K be Galois over F and assume $\mathrm{Gal}_F K \cong \mathbb{Z}_n$.

(a) If E is an intermediate field that is normal over F, prove that $\mathrm{Gal}_E K$ and $\mathrm{Gal}_F E$ are cyclic.

(b) Show that there is exactly one intermediate field for each positive divisor of n and that these are the only intermediate fields.

10. Two intermediate fields E and L are said to be **conjugate** if there exists $\sigma \in \mathrm{Gal}_F K$ such that $\sigma(E) = L$. Prove that E and L are conjugate if and only if $\mathrm{Gal}_E K$ and $\mathrm{Gal}_L K$ are conjugate subgroups of $\mathrm{Gal}_F K$ (as defined on page 271).

11. (a) Show that $K = \mathbb{Q}(\sqrt[4]{2},i)$ is a splitting field of $x^4 - 2$ over $\mathbb{Q}$.

(b) Prove that $[K:\mathbb{Q}] = 8$ and conclude from Theorem 11.11 that $\mathrm{Gal}_{\mathbb{Q}} K$ has order 8. [*Hint:* $\mathbb{Q} \subseteq \mathbb{Q}(\sqrt[4]{2}) \subseteq \mathbb{Q}(\sqrt[4]{2},i)$.]

(c) Prove that there exists $\sigma \in \mathrm{Gal}_{\mathbb{Q}} K$ such that $\sigma(\sqrt[4]{2}) = (\sqrt[4]{2})i$ and $\sigma(i) = i$ and that σ has order 4.

(d) By Corollary 11.13 restriction of the complex conjugation map to K is an element τ of $\mathrm{Gal}_{\mathbb{Q}} K$. Show that

$$\mathrm{Gal}_{\mathbb{Q}} K = \{\sigma, \sigma^2, \sigma^3, \sigma^4 = \iota, \tau, \sigma\tau, \sigma^2\tau, \sigma^3\tau\}.$$

[*Hint:* Use Theorem 11.4 to show these elements are distinct.]

(e) Prove that $\mathrm{Gal}_{\mathbb{Q}} K \cong D_4$. [*Hint:* Map σ to r_1 to τ to v.]

12. Let K be as in Exercise 11. Prove that $\mathrm{Gal}_{\mathbb{Q}(i)} K \cong \mathbb{Z}_4$.

C. 13. Let K be as in Exercise 11. Exhibit the Galois correspondence for this extension. [Among the intermediate fields are $\mathbb{Q}((1 + i)\sqrt[4]{2})$ and $\mathbb{Q}((1 - i)\sqrt[4]{2})$.]

14. Exhibit the Galois correspondence for the extension $\mathbb{Q}(\sqrt{2}, \sqrt{3}, \sqrt{5})$ of $\mathbb{Q}$. [The Galois group has seven subgroups of order 2 and seven of order 4.]

11.3 SOLVABILITY BY RADICALS

The solutions of the quadratic equation $ax^2 + bx + c = 0$ are given by the well-known formula

$$x = \frac{-b \pm \sqrt{b^2 - 4ac}}{2a}.$$

This fact was known in ancient times. In the sixteenth century, formulas for the solution of cubic and quartic equations were discovered. For instance, the solutions of $x^3 + bx + c = 0$ are given by

$$x = \sqrt[3]{(-c/2) + \sqrt{d}} + \sqrt[3]{(-c/2) - \sqrt{d}}$$
$$x = \omega(\sqrt[3]{(-c/2) + \sqrt{d}}) + \omega^2(\sqrt[3]{(-c/2) - \sqrt{d}})$$
$$x = \omega^2(\sqrt[3]{(-c/2) + \sqrt{d}} + \omega(\sqrt[3]{(-c/2) - \sqrt{d}}),$$

where $d = (b^3/27) + (c^2/4)$, $\omega = (-1 + \sqrt{3}i)/2$ is a complex cube root of 1, and the other cube roots are chosen so that

$$(\sqrt[3]{(-c/2) + \sqrt{d}})(\sqrt[3]{(-c/2) - \sqrt{d}}) = -b/3.*$$

In the early 1800s Ruffini and Abel independently proved that, for $n \geq 5$, there is no formula for solving *all* equations of degree n. But the complete analysis of the problem is due to Galois, who provided a criterion for determining which polynomial equations *are* solvable by formula. This criterion, which is presented here, will enable us to exhibit a fifth-degree polynomial equation that cannot be solved by a formula. To simplify the discussion, we shall assume that *all fields have characteristic 0*.

As illustrated above, a "formula" is a specific procedure that starts with the coefficients of the polynomial $f(x) \in F[x]$ and arrives at the solutions of the equation $f(x) = 0_F$ by using only the field operations (addition, subtraction, multiplication, division) *and* the extraction of roots (square roots, cube roots, fourth roots, etc.). In this context, an **nth root** of an element c in F is any root of the polynomial $x^n - c$ in some extension field of F.

If $f(x) \in F[x]$, then performing field operations does not get you out of the coefficient field F (closure!). But taking an nth root may land you in an extension field. Taking an mth root after that may move you up to still another extension field. Thus the existence of a formula for the solutions of $f(x) = 0_F$ implies that these solutions lie in a special kind of extension field of F.

> **EXAMPLE** Applying the cubic formula above with $b = 3$, $c = 2$ shows that the solutions of $x^3 + 3x + 2 = 0$ are

---

* The formulas for the general cubic and the quartic are similar but more complicated.

$$\sqrt[3]{-1 + \sqrt{2}} + \sqrt[3]{-1 - \sqrt{2}},$$

$$\omega\sqrt[3]{-1 + \sqrt{2}} + (\omega^2)\sqrt[4]{-1 - \sqrt{2}},$$

$$(\omega^2)\sqrt[3]{-1 + \sqrt{2}} + \omega\sqrt[4]{-1 - \sqrt{2}}.$$

All of these solutions lie in the extension chain:

$$\mathbb{Q} \subseteq \mathbb{Q}(\omega) \subseteq \mathbb{Q}(\omega,\sqrt{2}) \subseteq \mathbb{Q}(\omega,\sqrt{2},\sqrt[3]{-1 + \sqrt{2}}) \subseteq \mathbb{Q}(\omega,\sqrt{2},\sqrt[3]{-1 + \sqrt{2}}, \sqrt[3]{-1 - \sqrt{2}})$$

$$F_0 \subseteq F_1 \subseteq F_2 \subseteq F_3 \subseteq F_4.$$

Each field in this chain is a simple extension of the preceding one and is of the form $F_j(u)$, where $u^n \in F_j$ for some n (that is, u is an nth root of some element of F_j):

$F_1 = F_0(\omega),$ where $\omega^3 = 1 \in F_0$.

$F_2 = F_1(\sqrt{2}),$ where $(\sqrt{2})^2 = 2 \in F_0 \subseteq F_1$.

$F_3 = F_2(\sqrt[3]{-1 + \sqrt{2}}),$ where $(\sqrt[3]{-1 + \sqrt{2}})^3 = -1 + \sqrt{2} \in F_2$.

$F_4 = F_3(\sqrt[3]{-1 - \sqrt{2}}),$ where $(\sqrt[3]{-1 - \sqrt{2}})^3 = -1 - \sqrt{2} \in F_2 \subseteq F_3$.

Since F_4 contains all the solutions of $x^3 + 3x + 2 = 0$, it also contains a splitting field of $x^3 + 3x + 2$.

The preceding example is an illustration of the next definition (with $K = F_4$).

> **DEFINITION** • *A field K is said to be a **radical extension** of a field F if there is a chain of fields*
>
> $$F = F_0 \subseteq F_1 \subseteq F_2 \subseteq \cdots \subseteq F_t = K$$
>
> *such that for each $i = 1, 2, \ldots, t$,*
>
> $$F_i = F_{i-1}(u_i) \text{ and some power of } u_i \text{ is in } F_{i-1}.$$

Let $f(x) \in F[x]$. The equation $f(x) = 0_F$ is said to be **solvable by radicals** if there is a radical extension of F that contains a splitting field of $f(x)$. The example above shows that $x^3 + 3x + 2 = 0$ is solvable by radicals.

The preceding discussion shows that if there is a formula for its solutions, then the equation $f(x) = 0_F$ is solvable by radicals. Contrapositively, if $f(x) = 0_F$ is *not* solvable by radicals, then there cannot be a formula (in the sense discussed above) for finding its solutions.

Solvable Groups

Before stating Galois' Criterion for an equation to be solvable by radicals, we need to introduce a new class of groups. A group G is said to be **solvable** if it has a chain of subgroups

$$G = G_0 \supseteq G_1 \supseteq G_2 \supseteq \cdots \supseteq G_{n-1} \supseteq G_n = \langle e \rangle$$

such that each G_i is a normal subgroup of the preceding group G_{i-1} and the quotient group G_{i-1}/G_i is abelian.

> **EXAMPLE** Every abelian group G is solvable because every quotient group of G is abelian, so the sequence $G \supseteq \langle e \rangle$ fulfills the conditions in the definition.

> **EXAMPLE** Let $\langle (123) \rangle$ be the cyclic subgroup of order 3 in S_3. The chain $S_3 \supseteq \langle (123) \rangle \supseteq \langle (1) \rangle$ shows that S_3 is solvable. But for other symmetric groups we have

THEOREM 11.14 *For $n \geq 5$ the group S_n is not solvable.*

Proof Suppose, on the contrary, that S_n is solvable and that

$$S_n = G_0 \supseteq G_1 \supseteq G_2 \supseteq \cdots \supseteq G_t = \langle (1) \rangle$$

is the chain of subgroups required by the definition. Let (rst) be any 3-cycle in S_n and let u, v be any elements of $\{1, 2, \ldots, n\}$ other than r, s, t (u and v exist because $n \geq 5$). Since S_n/G_1 is abelian, Theorem 7.37 (with $a = (tus)$, $b = (srv)$) shows that G_1 must contain

$$(tus)(srv)(tus)^{-1}(srv)^{-1} = (tus)(srv)(tsu)(svr) = (rst).$$

Therefore, G_1 contains all the 3-cycles. Since G_1/G_2 is abelian, we can repeat the argument with G_1 in place of S_n and G_2 in place of G_1 and conclude that G_2 contains all the 3-cycles. The fact that each G_{i-1}/G_i is abelian and continued repetition lead to the conclusion that the identity subgroup G_t contains all the 3-cycles, which is a contradiction. Therefore, S_n is not solvable. ◆

This key property of solvable groups will be needed below:

THEOREM 11.15 *Every homomorphic image of a solvable group G is solvable.*

Proof Suppose that $f : G \to H$ is a surjective homomorphism and that $G = G_0 \supseteq G_1 \supseteq G_2 \supseteq \cdots \supseteq G_t = \langle e_G \rangle$ is the chain of subgroups in the definition of solvability. For each i, let $H_i = f(G_i)$ and consider this chain of subgroups:

$$H = H_0 \supseteq H_1 \supseteq H_2 \supseteq \cdots \supseteq H_t = f(\langle e_G \rangle) = \langle e_H \rangle.$$

Exercise 21 of Section 7.6 shows that H_i is a normal subgroup of H_{i-1} for each $i = 1, 2, \ldots, t$. Let $a, b \in H_{i-1}$. Then there exist $c, d \in G_{i-1}$ such that $f(c) = a$ and $f(d) = b$. Since G_{i-1}/G_i is abelian by solvability, $cdc^{-1}d^{-1} \in G_i$ by Theorem 7.37. Consequently,

$$aba^{-1}b^{-1} = f(c)f(d)f(c^{-1})f(d^{-1}) = f(cdc^{-1}d^{-1}) \in f(G_i) = H_i.$$

Therefore, H_{i-1}/H_i is abelian by Theorem 7.37, and H is solvable. ◆

Galois' Criterion

If $f(x) \in F[x]$, then the **Galois group of the polynomial $f(x)$** is $\text{Gal}_F K$, where K is a splitting field of $f(x)$ over F.* **Galois' Criterion** states that

$f(x) = 0_F$ is solvable by radicals if and only if the Galois group of $f(x)$ is a solvable group.

In order to prove Galois' solvability criterion, we need more information about radical extensions and nth roots. If F is a field and ζ is a root of $x^n - 1_F$ in some extension field of F (so that $\zeta^n = 1_F$), then ζ is called an **nth root of unity.** The derivative nx^{n-1} of $x^n - 1_F$ is nonzero (since F has characteristic 0) and relatively prime to $x^n - 1_F$. Therefore, $x^n - 1_F$ is separable by Lemma 10.16. So there are exactly n distinct nth roots of unity in any splitting field K of $x^n - 1_F$. If ζ and τ are nth roots of unity in K, then

$$(\zeta\tau)^n = \zeta^n \tau^n = 1_F 1_F = 1_F,$$

so that $\zeta\tau$ is also an nth root of unity. Since the set of nth roots of unity is closed under multiplication, it is a subgroup of order n of the multiplicative group of the field K (Theorem 7.11) and is, therefore, cyclic by Theorem 7.15. A generator of this cyclic group of nth roots of unity in K is called a **primitive nth root of unity.** Thus ζ is a primitive nth root of unity if and only if $\zeta, \zeta^2, \zeta^3, \ldots, \zeta^n = 1_F$ are the n distinct nth roots of unity.

> **EXAMPLE** The fourth roots of unity in $\mathbb{C}$ are $1, -1, i, -i$. Since $i^2 = -1$, $i^3 = -i$, and $i^4 = 1$, i is a primitive fourth root of unity. Similarly, $-i$ is also a primitive fourth root of unity. DeMoivre's Theorem shows that for any positive n,
>
> $\cos(2\pi/n) + i\sin(2\pi/n)$ is a primitive nth root of unity in $\mathbb{C}$.
>
> When $n = 3$, this states that
>
> $$\omega = \cos(2\pi/3) + i\sin(2\pi/3) = (-1/2) + (\sqrt{3}/2)i$$
>
> is a primitive cube root of unity.

LEMMA 11.16 *Let F be a field and ζ a primitive nth root of unity in F. Then F contains a primitive dth root of unity for every positive divisor d of n.*

Proof By hypothesis ζ has order n in the multiplicative group of F. If $n = dt$, then ζ^t has order d by Theorem 7.8. So ζ^t generates a subgroup of order d, each of whose elements must have order dividing d by Corollary 7.27. In other words,

---

* Since any two splitting fields of $f(x)$ are isomorphic by Theorem 10.14, it follows that the corresponding Galois groups are isomorphic. So the Galois group of $f(x)$ is independent of the choice of K.

$((\zeta^t)^k)^d = 1_F$ for every k. Thus the d distinct powers ζ^t, $(\zeta^t)^2$, ... , $(\zeta^t)^{d-1}$, $(\zeta^t)^d = 1_F$ are roots of $x^d - 1_F$. Since $x^d - 1_F$ has at most d roots and every dth root of unity is a root of $x^d - 1_F$, ζ^t is a primitive dth root of unity. ◆

We can now tie together the preceding themes and prove two theorems that are special cases of Galois' Criterion as well as essential tools for proving the general case.

THEOREM 11.17 *Let F be a field of characteristic 0 and ζ a primitive nth root of unity in some extension field of F. Then $K = F(\zeta)$ is a normal extension of F, and $Gal_F K$ is abelian.*

Proof The field $K = F(\zeta)$ contains all the powers of ζ and is, therefore, a splitting field of $x^n - 1_F$.* Hence, K is normal over F by Theorem 10.15. Every automorphism in the Galois group must map ζ onto a root of $x^n - 1_F$ by Theorem 11.2. So if σ, $\tau \in Gal_F K$, then $\sigma(\zeta) = \zeta^k$ and $\tau(\zeta) = \zeta^t$ for some positive integers k, t. Consequently,

$$(\sigma \circ \tau)(\zeta) = \sigma(\tau(\zeta)) = \sigma(\zeta^t) = \sigma(\zeta)^t = (\zeta^k)^t = \zeta^{kt}.$$
$$(\tau \circ \sigma)(\zeta) = \tau(\sigma(\zeta)) = \tau(\zeta^k) = \tau(\zeta)^k = (\zeta^t)^k = \zeta^{kt}.$$

Therefore, $\sigma \circ \tau = \tau \circ \sigma$ by Theorem 11.4, and $Gal_F K$ is abelian. ◆

THEOREM 11.18 *Let F be a field of characteristic 0 that contains a primitive nth root of unity. If u is a root of $x^n - c \in F[x]$ in some extension field of F, then $K = F(u)$ is a normal extension of F, and $Gal_F K$ is abelian.*

Proof* By hypothesis, $u^n = c$. If ζ is a primitive nth root of unity in F, then for any k,

$$(\zeta^k u)^n = (\zeta^k)^n u^n = (\zeta^n)^k u^n = 1_F c = c.$$

Consequently, since ζ, ζ^2, ... , $\zeta^n = 1_F$ are distinct elements of F, the elements ζu, $\zeta^2 u$, $\zeta^3 u$, ... , $\zeta^n u = u$ are the n distinct roots of $x^n - c$. Hence, $K = F(u)$ is a splitting field of $x^n - c$ over F and is, therefore, normal over F by Theorem 10.15.† If σ, τ, $\in Gal_F K$, then $\sigma(u) = \zeta^k u$ and $\tau(u) = \zeta^t u$ for some k, t by Theorem 11.2. Consequently,

---

* The field $K = F(\zeta)$ is a radical extension of F since $\zeta^n = 1_F$. Thus $x^n - 1_F = 0_F$ is solvable by radicals. So the theorem, which says that $Gal_F K$ (the Galois group of $x^n - 1_F$) is abelian (and hence, solvable) is a special case of Galois' Criterion.

** For an alternate proof showing that $Gal_F K$ is actually cyclic, see Exercise 22.

† The field $K = F(u)$ is also a radical extension of F since $u^n = c \in F$, so $x^n - c = 0_F$ is solvable by radicals. Hence, the theorem is another special case of Galois' Criterion.

$$(\sigma \circ \tau)(u) = \sigma(\tau(u)) = \sigma(\zeta^t u) = \sigma(\zeta^t)\sigma(u) = \zeta^t(\zeta^k u) = \zeta^{t+k}u.$$
$$(\tau \circ \sigma)(u) = \tau(\sigma(u)) = \tau(\zeta^k u) = \tau(\zeta^k)\tau(u) = \zeta^k(\zeta^t u) = \zeta^{t+k}u.$$

Therefore, $\sigma \circ \tau = \tau \circ \sigma$ by Theorem 11.4, and $\text{Gal}_F K$ is abelian. ◆

THEOREM 11.19   (GALOIS' CRITERION)   *Let F be a field of characteristic 0 and f(x) ∈ F[x]. Then f(x) = 0_F is solvable by radicals if and only if the Galois group of f(x) is solvable.*

We shall prove only the half of the theorem that is needed below; see Hungerford [7] or Jacobson [8] for the other half.

Proof of Theorem 11.19   Assume that $f(x) = 0_F$ is solvable by radicals. The proof, whose details are on pages 393–395, is in three steps:

1. Theorem 11.21: There is a *normal* radical extension K of F that contains a splitting field E of $f(x)$.*
2. The field E is normal over F by Theorem 10.15.
3. Theorem 11.22: Any intermediate field of K that is normal over F has a solvable Galois group; in particular, $\text{Gal}_F E$ (the Galois group of $f(x)$) is solvable. ◆

Before completing the proof of Theorem 11.19, we use it to demonstrate the insolvability of the quintic.

EXAMPLE   We claim that the Galois group of the polynomial $f(x) = 2x^5 - 10x + 5 \in \mathbb{Q}[x]$ is S_5, which is not solvable by Theorem 11.14. Consequently, the equation $2x^5 - 10x + 5 = 0$ is *not* solvable by radicals by Theorem 11.19. So, as explained on page 388,

there is no formula (involving only field operations and extraction of roots) for the solution of all fifth-degree polynomial equations.

To prove our claim, note that the derivative of $f(x)$ is $10x^4 - 10$, whose only real roots are ± 1 (the others being $\pm i$). Then $f''(x) = 40x^3$, and the second-derivative test of elementary calculus shows that $f(x)$ has exactly one relative maximum at $x = -1$, one relative minimum at $x = 1$, and one point of inflection at $x = 0$. So its graph must have the general shape shown below. In particular, $f(x)$ has exactly three real roots.

---

* This is a crucial technical detail. The definition of solvability by radicals guarantees only a radical extension of F containing E. But a radical extension need not be normal over F (Exercise 19), and if it is not, the Fundamental Theorem 11.11 can't be used.

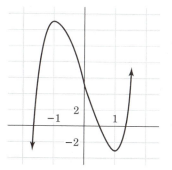

Note that $f(x)$ is irreducible in $\mathbb{Q}[x]$ by Eisenstein's Criterion (with $p = 5$). If K is a splitting field of $f(x)$ in $\mathbb{C}$, then $\text{Gal}_F K$ has order $[K:\mathbb{Q}]$ by the Fundamental Theorem. If r is any root of $f(x)$, then $[K:\mathbb{Q}] = [K:\mathbb{Q}(r)][\mathbb{Q}(r):\mathbb{Q}]$ by Theorem 10.4 and $[\mathbb{Q}(r):\mathbb{Q}] = 5$ by Theorem 10.7. So the order of $\text{Gal}_F K$ is divisible by 5. It follows that $\text{Gal}_F K$ contains an element of order 5.*

The group $\text{Gal}_F K$, considered as a group of permutations of the roots of $f(x)$, is a subgroup of S_5 (Corollary 11.5). But the only elements of order 5 in S_5 are the 5-cycles (see Exercise 13 in Section 7.9). So $\text{Gal}_F K$ contains a 5-cycle. Complex conjugation induces an automorphism on K (Corollary 11.13). This automorphism interchanges the two nonreal roots of $f(x)$ and fixes the three real ones. Thus $\text{Gal}_F K$ contains a transposition. Exercise 8 shows that the only subgroup of S_5 that contains both a 5-cycle and a transposition is S_5 itself. Therefore, $\text{Gal}_F K = S_5$ as claimed.

We now complete the proof of Galois' Criterion, beginning with a technical lemma whose import will become clear in the next theorem.

LEMMA 11.20 *Let F, E, L be fields of characteristic 0 with*

$$F \subseteq E \subseteq L = E(v) \qquad and \qquad v^k \in E.$$

If L is finite-dimensional over F and E is normal over F, then there exists an extension field M of L, which is a radical extension of E and a normal extension of F.

Proof By Theorem 10.15, E is the splitting field over F of some $g(x) \in F[x]$. Let $p(x) \in F[x]$ be the minimal polynomial of v over F and let M be a splitting field of $g(x)p(x)$ over F. Then M is normal over F by Theorem 10.15. Furthermore, $F \subseteq E \subseteq L \subseteq M$ (since $L = E(v)$ and E is generated over F by the roots of $g(x)$). Let $v = v_1, v_2, \ldots, v_r$ be all the roots of $p(x)$ in M. For each i there exists

* If you have read Chapter 8, use Corollary 8.14; otherwise, use Exercise 9 in this section.

$\sigma_i \in \text{Gal}_F M$ such that $\sigma_i(v) = v_i$ by Theorem 11.3. Corollary 11.13 shows that $\sigma_i(E) \subseteq E$. By hypothesis, $v^k = b \in E$; so for each i,

$$(v_i)^k = \sigma_i(v)^k = \sigma_i(v^k) = \sigma_i(b) \in E \subseteq E(v_1, \ldots, v_{i-1}).$$

Consequently,

$$E \subseteq L = E(v_1) \subseteq E(v_1, v_2) \subseteq E(v_1, v_2, v_3) \subseteq \cdots \subseteq E(v_1, v_2, \ldots, v_r) = M$$

is a radical extension of E. ◆

THEOREM 11.21   *Let F be a field of characteristic 0 and $f(x) \in F[x]$. If $f(x) = 0_F$ is solvable by radicals, then there is a normal radical extension field of F that contains a splitting field of $f(x)$.*

Proof   By definition some splitting field K of $f(x)$ is contained in a radical extension

$$F = F_0 \subseteq F_1 \subseteq F_2 \subseteq F_3 \subseteq \cdots \subseteq F_t,$$

where $F_i = F_{i-1}(u_i)$ and $(u_i)^{n_i}$ is in F_{i-1} for each $i = 1, 2, \ldots, t$. Applying Lemma 11.20 with $E = F$, $L = F_1$, and $v = u_1$ produces a normal radical extension field M_1 of F that contains F_1. By hypothesis $(u_2)^{n_2} \in F_1 \subseteq M_1$. Applying Lemma 11.20 with $E = M_1$, $v = u_2$, and $L = M_1(u_2)$ produces a normal extension field M_2 of F that is a radical extension of M_1 and, hence, a radical extension of F. Furthermore, M_2 contains $F_2 = F_1(u_2)$. Continued repetition of this argument leads to a normal radical extension field M_t of F that contains F_t and, hence, contains K. ◆

THEOREM 11.22   *Let K be a normal radical extension field of F and E an intermediate field, all of characteristic 0. If E is normal over F, then $\text{Gal}_F E$ is a solvable group.*

Proof   By hypothesis there is a chain of subfields

$$F = F_0 \subseteq F_1 \subseteq F_2 \subseteq F_3 \subseteq \cdots \subseteq F_t = K,$$

where $F_i = F_{i-1}(u_i)$ and $(u_i)^{n_i}$ is in F_{i-1} for each $i = 1, 2, \ldots, t$. Let n be the least common multiple of $n_1, n_2, \ldots, n_t$ and let ζ be a primitive nth root of unity. For each $i \geq 0$, let $E_i = F_i(\zeta)$. Then for each $i \geq 1$

$$E_i = F_i(\zeta) = F_{i-1}(u_i)(\zeta) = F_{i-1}(u_i, \zeta) = F_{i-1}(\zeta)(u_i) = E_{i-1}(u_i).$$

Since $(u_i)^{n_i} \in F_{i-1} \subseteq E_{i-1}$ for $i \geq 1$ and $\zeta^n \in F$,

$$F \subseteq E_0 \subseteq E_1 \subseteq E_2 \subseteq E_3 \subseteq \cdots \subseteq E_t = L$$

is a radical extension of F that contains K (and, hence, E).* The normal extension $K = F_t$ is the splitting field of some polynomial $p(x) \in F[x]$ by Theorem 10.15, and, hence, $L = E_t = F_t(\zeta)$ is the splitting field of $p(x)(x^n - 1_F)$ over F. Therefore, L is Galois over F by Theorems 10.15 and 10.17.

Consider the following chain of subgroups of $\mathrm{Gal}_F L$:

$$\mathrm{Gal}_F L \supseteq \mathrm{Gal}_{E_0} L \supseteq \mathrm{Gal}_{E_1} L \supseteq \mathrm{Gal}_{E_2} L \supseteq \cdots \supseteq \mathrm{Gal}_{E_{t-1}} L \supseteq \mathrm{Gal}_L L = \langle \iota \rangle.$$

We shall show that each subgroup is normal in the preceding one and that each quotient is abelian. Since each n_i divides n, E_0 contains a primitive n_ith root of unity by Lemma 11.16. Consequently, by Theorem 11.18 each E_i (with $i \geq 1$) is a normal extension of E_{i-1}, and the Galois group $\mathrm{Gal}_{E_{i-1}} E_i$ is abelian. Since L is Galois over F, it is Galois over every E_j. Applying the Fundamental Theorem 11.11 to the extension L of E_{i-1}, we see that $\mathrm{Gal}_{E_i} L$ is a normal subgroup of $\mathrm{Gal}_{E_{i-1}} L$ and that the quotient group $\mathrm{Gal}_{E_{i-1}} L / \mathrm{Gal}_{E_i} L$ is isomorphic to the abelian group $\mathrm{Gal}_{E_{i-1}} E_i$. Similarly by Theorems 11.11 and 11.17, E_0 is normal over F, $\mathrm{Gal}_{E_0} L$ is normal in $\mathrm{Gal}_F L$, and $\mathrm{Gal}_F L / \mathrm{Gal}_{E_0} L$ is isomorphic to the abelian group $\mathrm{Gal}_F E_0$. Therefore, $\mathrm{Gal}_F L$ is a solvable group.

Since E is normal over F, the Fundamental Theorem shows that $\mathrm{Gal}_E L$ is normal in $\mathrm{Gal}_F L$ and $\mathrm{Gal}_F L / \mathrm{Gal}_E L$ is isomorphic to $\mathrm{Gal}_F E$. So $\mathrm{Gal}_F E$ is the homomorphic image of the solvable group $\mathrm{Gal}_F L$ (see Theorem 7.41) and is, therefore, solvable by Theorem 11.15. ◆

◆ EXERCISES

NOTE: *F denotes a field, and all fields have characteristic 0.*

A. **1.** Find a radical extension of $\mathbb{Q}$ containing the given number:

   (a) $\sqrt[4]{1 + \sqrt{7}} - \sqrt[5]{2 + \sqrt{5}}$

   (b) $(\sqrt[5]{\sqrt{2} + i})/(\sqrt[3]{5})$

   (c) $(\sqrt[3]{3 - \sqrt{2}})/(4 + \sqrt{2})$

2. Show that $x^2 - 3$ and $x^2 - 2x - 2 \in \mathbb{Q}[x]$ have the same Galois group. [*Hint:* What is the splitting field of each?]

3. If K is a radical extension of F, prove that $[K:F]$ is finite. [*Hint:* Theorems 10.7 and 10.4.]

---

* The construction of L does not use the hypothesis that K is normal over F, and, as we shall see below, every field in the chain is a normal extension of the immediately preceding one. But this is *not* enough to guarantee that L is normal (hence Galois) over F (Exercise 19). We need the hypothesis that K is normal over F to guarantee this, so that we can use the Fundamental Theorem on L.

4. Prove that for $n \geq 5$, A_n is not solvable. [*Hint:* Adapt the proof of Theorem 11.14.]

5. (a) Show that S_4 is a solvable group. [*Hint:* Consider the subgroup $H = \{(12)(34), (13)(24), (14)(23), (1)\}$ of A_4.]

   (b) Show that D_4 is a solvable group.

6. If G is a simple nonabelian group, prove that G is not solvable. [This fact and Theorem 7.52 provide another proof that A_n is not solvable for $n \geq 5$.]

7. List all the nth roots of unity in $\mathbb{C}$ when $n =$

   (a) 2   **(b)** 3   **(c)** 4   **(d)** 5   **(e)** 6

B. **8.** Let G be a subgroup of S_5 that contains a transposition $\sigma = (rs)$ and a 5-cycle α. Prove that $G = S_5$ as follows.

   (a) Show that for some k, α^k is of the form $(rsxyz)$. Let $\tau = \alpha^k \in G$; by relabeling we may assume that $\sigma = (12)$ and $\tau = (12345)$.

   (b) Show that $(12), (23), (34), (45) \in G$. [*Hint:* Consider $\tau^k \sigma \tau^{-k}$ for $k \geq 1$].

   (c) Show that $(13), (14), (15) \in G$. [*Hint:* $(12)(23)(12) = ?$]

   (d) Show that every transposition is in G. Therefore, $G = S_5$ by Corollary 7.48.

9. Let G be a group of order n. If $5|n$, prove that G contains an element of order 5 as follows. Let S be the set of all ordered 5-tuples (r,s,t,u,v) with r, s, t, u, $v \in G$ and $rstuv = e$.

   (a) Show that S contains exactly n^4 5-tuples. [*Hint:* If r, s, t, u, $\in G$ and $v = (rstu)^{-1}$, then $(r,s,t,u,v) \in S$.]

   (b) Two 5-tuples in S are said to be *equivalent* if one is a cyclic permutation of the other.* Prove that this relation is an equivalence relation on S.

   (c) Prove that an equivalence class in S either has exactly five 5-tuples in it or consists of a single 5-tuple of the form (r,r,r,r,r).

   (d) Prove that there are at least two equivalence classes in S that contain a single 5-tuple. [*Hint:* One is $\{(e,e,e,e,e)\}$. If this is the only one, show that $n^4 \equiv 1 \pmod 5$. But $5 \mid n$, so $n^4 \equiv 0 \pmod 5$, which is a contradiction.]

   (e) If $\{(c,c,c,c,c)\}$, with $c \neq e$, is a single-element equivalence class, prove that c has order 5.

10. If N is a normal subgroup of G, N is solvable, and G/N is solvable, prove that G is solvable.

---

* For instance, (r,s,t,u,v) is equivalent to each of (s,t,u,v,r), (t,u,v,r,s), (u,v,r,s,t), (v,r,s,t,u), (r,s,t,u,v) and to no other 5-tuples in S.

11. Prove that a subgroup H of a solvable group G is solvable. [*Hint:* If $G = G_0 \supseteq G_1 \supseteq \cdots \supseteq G_n = \langle e \rangle$ is the solvable series for G, consider the groups $H_i = H \cap G_i$. To show that H_{i-1}/H_i is abelian, verify that the map $H_{i-1}/H_i \to G_{i-1}/G_i$ given by $H_i x \to G_i x$ is a well-defined injective homomorphism.]

12. Prove that the Galois group of an irreducible quadratic polynomial is isomorphic to $\mathbb{Z}_2$.

13. Prove that the Galois group of an irreducible cubic polynomial is isomorphic to $\mathbb{Z}_3$ or S_3.

14. Prove that the Galois group of an irreducible quartic polynomial is solvable. [*Hint:* Corollary 11.5 and Exercises 5 and 11.]

15. Let $p(x), q(x)$ be irreducible quadratics. Prove that the Galois group of $f(x) = p(x)q(x)$ is isomorphic to $\mathbb{Z}_2 \times \mathbb{Z}_2$ or $\mathbb{Z}_2$. [*Hint:* If u is a root of $p(x)$ and v a root of $q(x)$, then there are two cases: $v \notin F(u)$ and $v \in F(u)$.]

16. Use Galois' Criterion to prove that every polynomial of degree ≤ 4 is solvable by radicals. [*Hint:* Exercises 12–15.]

17. Find the Galois group G of the given polynomial in $\mathbb{Q}[x]$:

    (a) $x^6 - 4x^3 + 4$ [*Hint:* Factor.]

    (b) $x^4 - 5x^2 + 6$

    (c) $x^5 + 6x^3 + 9x$

    (d) $x^4 + 3x^3 - 2x - 6$

    (e) $x^5 - 10x - 5$ [*Hint:* See the example after Theorem 11.19.]

18. Determine whether the given equation over $\mathbb{Q}$ is solvable by radicals:

    (a) $x^6 + 2x^3 + 1 = 0$ **(b)** $3x^5 - 15x + 5 = 0$

    (c) $2x^5 - 5x^4 + 5 = 0$ **(d)** $x^5 - x^4 - 16x + 16 = 0$

19. **(a)** Prove that $\mathbb{Q}(\sqrt{2}i)$ is normal over $\mathbb{Q}$ by showing it is the splitting field of $x^2 + 2$.

    (b) Prove that $\mathbb{Q}(\sqrt[4]{2}(1 - i))$ is normal over $\mathbb{Q}(\sqrt{2}i)$ by showing that it is the splitting field of $x^2 + 2\sqrt{2}i$.

    (c) Show that $\mathbb{Q} \subseteq \mathbb{Q}(\sqrt{2}i) \subseteq \mathbb{Q}(\sqrt[4]{2}(1 - i))$ is a radical extension of $\mathbb{Q}$ with $[\mathbb{Q}(\sqrt[4]{2}(1 - i)):\mathbb{Q}] = 4$ and note that $\mathbb{Q}$ contains all second roots of unity (namely ± 1).

    (d) Let $L = \mathbb{Q}(\sqrt[4]{2}(1 - i))$. Show that $v = \sqrt[4]{2}(1 + i)$ is *not* in L. [*Hint:* If $v \in L$ and $u = \sqrt[4]{2}(1 - i) \in L$, show that $v/u = i$ and $(v - u)/2i = \sqrt[4]{2} \in L$, which implies that $[L:\mathbb{Q}] \geq [\mathbb{Q}(\sqrt[4]{2},i):\mathbb{Q}]$, contradicting (c) and Exercise 11(b) in Section 11.2.]

(e) Prove that $L = \mathbb{Q}(\sqrt[4]{2}(1 - i))$ is *not* normal over $\mathbb{Q}$ [*Hint: u* and *v* (as in (d)) are roots of the irreducible polynomial $x^4 + 8$.]

20. Let ζ be a primitive fifth root of unity. Assume Exercise 20 in Section 4.5 and prove that $\text{Gal}_\mathbb{Q}\mathbb{Q}(\zeta)$, the Galois group of $x^5 - 1$, is cyclic of order 4.

21. What is the Galois group of $x^5 + 32$ over $\mathbb{Q}$? [*Hint:* Show that $\mathbb{Q}(\zeta)$ is a splitting field, where ζ is a primitive fifth root of unity; see Exercise 20.]

22. Prove that the group $\text{Gal}_F K$ in Theorem 11.18 is cyclic. [*Hint:* Define a map f from $\text{Gal}_F K$ to the additive group $\mathbb{Z}_n$ by $f(\sigma) = k$, where $\sigma(u) = \zeta^k u$. Show that f is a well-defined injective homomorphism and use Theorem 7.16.]

C. 23. If p is prime and G is a subgroup of S_p that contains a transposition and a p-cycle, prove that $G = S_p$. [Exercise 8 is the case $p = 5$.]

24. If $f(x) \in \mathbb{Q}[x]$ is irreducible of prime degree p and $f(x)$ has exactly two nonreal roots, prove that the Galois group of $f(x)$ is S_p. [The example after Theorem 11.19 is essentially the case $p = 5$.]

25. Construct a polynomial in $\mathbb{Q}[x]$ of degree 7 whose Galois group is S_7.

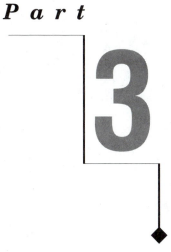
EXCURSIONS AND APPLICATIONS

CHAPTER **12**

Public-Key Cryptography

Prerequisite: Section 2.3.

Codes have been used for centuries by merchants, spies, armies, and diplomats to transmit secret messages. In recent times, the large volume of sensitive material in government and corporate computerized data banks (much of which is transmitted by satellite or over telephone lines) has increased the need for efficient, high-security codes.

It is easy to construct unbreakable codes for one-time use. Consider this "code pad":

Actual Word:	morning	evening	Monday	Tuesday	attack
Code Word:	bat	glxt	king	button	figle

If I send you the message FIGLE BUTTON BAT, there is no way an enemy can know for certain that it means "attack on Tuesday morning" unless he or she has a copy of the pad. Of course, if the same code is used again, the enemy might well be able to break it by analyzing the events that occur after each message.

Although one-time code pads are unbreakable, they are cumbersome and inefficient when many long messages must be routinely sent. Even if the encoding and decoding are done by a computer, it is still necessary to design and supply a new pad (at least as long as the message) to each participant for every message and to make all copies of these pads secure from unauthorized persons. This is expensive and impractical when hundreds of thousands of words must be encoded and decoded every day.

For frequent computer-based communication among several parties, the ideal code system would be one in which

1. Each person has efficient, reusable, computer algorithms for encoding and decoding messages.

2. Each person's decoding algorithm is *not* obtainable from his or her encoding algorithm in any reasonable amount of time.

A code system with these properties is called a **public-key system.** Although it may not be clear how condition 2 could be satisfied, it is easy to see the advantages of a public-key system.

The *encoding* algorithm of each participant could be publicly announced— perhaps published in a book (like a telephone directory)—thus eliminating the need for couriers and the security problems associated with the distribution of code pads. This would not compromise secrecy because of condition 2: Knowing a person's *encoding* algorithm would not enable you to determine his or her *decoding* algorithm. So you would have no way of decoding messages sent to another person in his or her code, even though you could send coded messages to that person.

Since the encoding algorithms for a public-key system are available to everyone, forgery appears to be a possibility. Suppose, for example, that a bank receives a coded message claiming to be from Anne and requesting the bank to transfer money from Anne's account into Tom's account. How can the bank be sure the message was actually sent by Anne?

The answer is as simple as it is foolproof. Coding and decoding algorithms are inverses of each other: Applying one after the other (in either order) produces the word you started with. So Anne first uses her secret *decoding* algorithm to write her name; say it becomes Gybx. She then applies the bank's public encoding algorithm to Gybx and sends the result (her "signature") along with her message. The bank uses its secret decoding algorithm on this "signature" and obtains Gybx. It then applies Anne's public *encoding* algorithm to Gybx, which turns it into Anne. The bank can then be sure the message is from Anne, because no one else could use her *decoding* algorithm to produce the word Gybx that is encoded as Anne.

One public-key system was developed by R. Rivest, A. Shamir, and L. Adleman in 1977. Their system, now called the RSA system, is based on elementary number theory. Its security depends on the difficulty of factoring large integers (as described in Section 1.4). Here are the mathematical preliminaries needed to understand the RSA system.

LEMMA 12.1 Let $p, r, s, c \in \mathbb{Z}$ with p prime. If $p \nmid c$ and $rc \equiv sc \pmod{p}$, then $r \equiv s \pmod{p}$.

Proof Since $rc \equiv sc \pmod{p}$, p divides $rc - sc = (r - s)c$. By Theorem 1.8, $p \mid (r - s)$ or $p \mid c$. Since $p \nmid c$, we have $p \mid (r - s)$, and, hence, $r \equiv s \pmod{p}$. ◆

LEMMA 12.2 (FERMAT'S LITTLE THEOREM) *If p is prime, $a \in \mathbb{Z}$, and $p \nmid a$, then $a^{p-1} \equiv 1 \pmod{p}$.*

Proof* None of the numbers $a, 2a, 3a, \ldots, (p-1)a$ is congruent to 0 modulo p by Exercise 1. Consequently, each of them must be congruent to one of 1, 2, 3, $\ldots$, $p-1$ by Corollary 2.5 and Theorem 2.3. If two of them were congruent to the same one, say $ra \equiv i \equiv sa \pmod{p}$ with

$$1 \le i, r, s \le p - 1,$$

then we would have $r \equiv s \pmod{p}$ by Lemma 12.1 (with $c = a$). This is impossible because no two of the numbers 1, 2, 3, $\ldots$, $p-1$ are congruent modulo p (the difference of any two is less than p and, hence, not divisible by p). Therefore, in some order $a, 2a, 3a, \ldots, (p-1)a$ are congruent to 1, 2, 3, $\ldots$, $p-1$. By repeated use of Theorem 2.2,

$$a \cdot 2a \cdot 3a \cdots (p-1)a \equiv 1 \cdot 2 \cdot 3 \cdots (p-1) \pmod{p}.$$

Rearranging the left side shows that

$$a \cdot a \cdot a \cdots a \cdot 1 \cdot 2 \cdot 3 \cdots (p-1) \equiv 1 \cdot 2 \cdot 3 \cdots (p-1) \pmod{p}$$
$$a^{p-1}(1 \cdot 2 \cdot 3 \cdots (p-1)) \equiv 1(1 \cdot 2 \cdot 3 \cdots (p-1)) \pmod{p}.$$

Now $p \nmid (1 \cdot 2 \cdot 3 \cdots (p-1))$ (if it did, p would divide one of the factors by Corollary 1.9). Therefore, $a^{p-1} \equiv 1 \pmod{p}$ by Lemma 12.1 (with $c = 1 \cdot 2 \cdot 3 \cdots (p-1)$). ◆

Throughout the rest of this discussion p and q are distinct positive primes. Let $n = pq$ and $k = (p-1)(q-1)$. Choose d such that $(d,k) = 1$. Then the equation $dx = 1$ has a solution in $\mathbb{Z}_k$ by Corollary 2.10. Therefore, the congruence $dx \equiv 1 \pmod{k}$ has a solution in $\mathbb{Z}$; call it e.

THEOREM 12.3 *Let p, q, n, k, e, d be as in the preceding paragraph. Then $b^{ed} \equiv b \pmod{n}$ for every $b \in \mathbb{Z}$.*

Proof Since e is a solution of $dx \equiv 1 \pmod{k}$, $de - 1 = kt$ for some t. Hence, $ed = kt + 1$, so that

$$b^{ed} = b^{kt+1} = b^{kt}b^1 = b^{(p-1)(q-1)t}b = (b^{p-1})^{(q-1)t}b.$$

If $p \nmid b$, then by Lemma 12.2,

$$b^{ed} = (b^{p-1})^{(q-1)t}b \equiv (1)^{(q-1)t}b \equiv b \pmod{p}.$$

If $p \mid b$, then b and every one of its powers are congruent to 0 modulo p. There-

---

* A proof based on group theory is outlined in Exercise 22 of Section 7.3, and one based on field theory is in Exercise 13 of Section 10.6.

fore, in every case, $b^{ed} \equiv b$. (mod p). A similar argument shows that $b^{ed} \equiv b$ (mod q). By the definition of congruence,

$$p \mid (b^{ed} - b) \qquad \text{and} \qquad q \mid (b^{ed} - b).$$

Therefore, $pq \mid (b^{ed} - b)$ by Exercise 2. Since $pq = n$, this means that n divides $(b^{ed} - b)$, and, hence, $b^{ed} \equiv b$ (mod n). ◆

The **least residue modulo n** of an integer c is the remainder r when c is divided by n. By the Division Algorithm, $c = nq + r$, so that $c - r = nq$, and, hence, $c \equiv r$ (mod n). Since two numbers strictly between 0 and n cannot be congruent modulo n, the least residue of c is the only integer between 0 and n that is congruent to c modulo n.

We can now describe the mechanics of the RSA system, after which we shall show how it satisfies the conditions for a public-key system. The message to be sent is first converted to numerical form by replacing each letter or space by a two-digit number:*

$$\text{space} = 00, \text{A} = 01, \text{B} = 02, \ \ldots \ , \text{Y} = 25, \text{Z} = 26.$$

For instance, the word GO is written as the number 0715 and WEST is written 23051920, so that the message "GO WEST" becomes the number 07150023051920, which we shall denote by B.

Let p, q, n, k, d, e, be as in Theorem 12.3, with p and q chosen so that $B < pq = n$. To encode message B, compute the least residue of B^e modulo n; denote it by C. Then C is the coded form of B. Send C in any convenient way.

The person who receives C decodes it by computing the least residue of C^d modulo n. This produces the original message for the following reasons. Since B^e, is congruent modulo n to its least residue C, Theorem 12.3 shows that

$$C^d \equiv (B^e)^d = B^{ed} \equiv B \ (\text{mod } n).$$

The least residue of C^d is the only number between 0 and n that is congruent to C^d modulo n and $0 < B < n$. So the original message B is the least residue of C^d.

Before presenting a numerical example, we show that the RSA system satisfies the conditions for a public-key system:

1. When the RSA system is used in practice, p and q are large primes (200 or more digits each). As noted in Section 1.4 such primes can be quickly identified by a computer. Even though B, e, C, d are large numbers, there are fast algorithms for finding the least residues of B^e and C^d modulo n. They are based on binary representation of the ex-

---

* More numbers could be used for punctuation marks, numerals, special symbols, etc. But this will be sufficient for illustrating the basic concepts.

ponent and do *not* require direct computation of B^e or C^d (which would be gigantic numbers). See Knuth [33] for details. So the encoding and decoding algorithms of the RSA system *are* computationally efficient.

2. To use the RSA system, each person in the network uses a computer to choose appropriate p, q, d and then determines n, k, e. The numbers e and n for the encoding algorithm are publicly announced, but the prime factors p, q of n and the numbers d and k are kept secret. Anyone with a computer can encode messages by using e and n. But there is no practical way for outsiders to determine d (and, hence, the decoding algorithm) without first finding p and q by factoring n.* With present technology this would take thousands of years, as explained in Section 1.4. So the RSA system appears secure, as long as new and very fast methods of factoring are not developed.

Even when n is chosen as above, there may be some messages that in numerical form are larger than n. In such cases the original message is broken into several blocks, each of which is less than n. Here is an example, due to Rivest-Shamir-Adleman.

EXAMPLE   Let $p = 47$ and $q = 59$. Then $n = pq = 47 \cdot 59 = 2773$ and $k = (p - 1)(q - 1) = 46 \cdot 58 = 2668$.** *Let* $d = 157$, which is easily shown to be relatively prime to 2668. Solving the congruence $157x \equiv 1$ (mod 2668) as in the proof of Corollary 2.10 shows that $e = 17$. We shall encode the message "IT'S ALL GREEK TO ME." We can encode only numbers less than $n = 2773$. So we write the message in two-letter blocks (and denote spaces by #):

I T	S #	A L	L #	G R
0920	1900	0112	1200	0718

E E	K #	T O	# M	E #
0505	1100	2015	0013	0500.

Then each block is a number less than 2773. The first block, 0920, is encoded by using $e = 17$ and a computer to calculate the least residue of 920^{17} modulo 2773:

$$920^{17} \equiv 948 \pmod{2773}.$$

---

* Alternatively, one might try to find k and then solve the congruence $ex \equiv 1 \pmod{k}$ to get d. But this can be shown to be computationally equivalent to factoring n, so no time is saved.

** These numbers will illustrate the concepts. But they are too small to provide a secure code since 2773 can be factored by hand.

The other blocks are encoded similarly, so the coded form of the message is

$$0948 \quad 2342 \quad 1084 \quad 1444 \quad 2663$$
$$2390 \quad 0778 \quad 0774 \quad 0219 \quad 1655.$$

A person receiving this message would use $d = 157$ to decode each block. For instance, to decode 0948, the computer calculates

$$948^{157} \equiv 920 \ (\text{mod } 2773).$$

This is the original first block $0920 = \text{IT}$.

For more information on cryptography and the RSA system, see DeMillo-Davida [36], Diffie-Hellman [37], Rivest-Shamir-Adleman [38], and Simmons [39].

◆ EXERCISES

A. 1. Let p be a prime and $k, a \in \mathbb{Z}$ such that $p \nmid a$ and $0 < k < p$. Prove that $ka \not\equiv 0$ (mod p). [*Hint:* Theorem 1.8.]

2. If p and q are distinct primes such that $p \mid c$ and $q \mid c$, prove that $pq \mid c$. [*Hint:* If $c = pk$, then $q \mid pk$; use Theorem 1.8.]

3. Use a calculator and the RSA encoding algorithm with $e = 3$, $n = 2773$ to encode these messages:

(a) GO HOME   **(b)** COME BACK   **(c)** DROP DEAD

[*Hint:* Use 2-letter blocks and don't omit spaces.]

4. Prove this version of Fermat's Little Theorem: If p is a prime and $a \in \mathbb{Z}$, then $a^p \equiv a$ (mod p). [*Hint:* Consider two cases, $p \mid a$ and $p \nmid a$; use Lemma 12.2 in the second case.]

B. 5. Find the decoding algorithm for the code in Exercise 3.

6. Let C be the coded form of a message that was encoded by using the RSA algorithm. Suppose that you discover that C and the encoding modulus n are *not* relatively prime. Explain how you could factor n and thus find the decoding algorithm. [The probability of such a C occurring is less than 10^{-99} when the prime factors p, q, of n have more than 100 digits.]

CHAPTER **13**

◆

The Chinese Remainder Theorem

Prerequisites: Section 2.1 and Appendix C for Section 13.1; Section 3.1 for Section 13.2; Section 6.2 for Section 13.3.

T he Chinese Remainder Theorem (Section 13.1) is a famous result in number theory that was known to Chinese mathematicians in the first century. It also has practical applications in computer arithmetic (Section 13.2). An extension of the theorem to rings other than $\mathbb{Z}$ has interesting consequences in ring theory (Section 13.3). Although obviously motivated by Section 13.1, Section 13.3 is independent of the rest of the chapter and may be read at any time after you have read Section 6.2.

13.1 PROOF OF THE CHINESE REMAINDER THEOREM

A **congruence** is an equation with integer coefficients in which "=" is replaced by "≡ (mod n)." The same equation can lead to different congruences, such as

$$6x + 5 \equiv 7 \ (\mathrm{mod}\ 3) \qquad \text{or} \qquad 6x + 5 \equiv 7 \ (\mathrm{mod}\ 5).$$

Only integers make sense as solutions of congruences, so the techniques of solving equations are not always applicable to congruences. For instance, the equation $6x + 5 = 7$ has $x = 1/3$ as a solution, but the congruence $6x + 5 \equiv 7$ (mod 3) has no solutions (Exercise 3), and $6x + 5 \equiv 7$ (mod 5) has infinitely many solutions (Exercise 4).

A number of theoretical problems and practical applications require the solving of a **system of linear congruences,** such as

$$x \equiv 2 \pmod 4$$
$$x \equiv 5 \pmod 7$$
$$x \equiv 0 \pmod{11}$$
$$x \equiv 8 \pmod{15}$$

A **solution of the system** is an integer that is a solution of *every* congruence in the system. We shall examine some cases in which a system of linear congruences must have a solution.

LEMMA 13.1   *If m and n are relatively prime positive integers and a, b ∈ ℤ, then the system*

$$x \equiv a \pmod m$$
$$x \equiv b \pmod n$$

has a solution.

Proof   Since $(m,n) = 1$, there exist (by Theorem 1.3) integers u and v such that $mu + nv = 1$. Multiplying this equation by $b - a$, we have

$$mu(b - a) + nv(b - a) = b - a$$
(∗) $$a + mu(b - a) = b - nv(b - a).$$

Let $t = a + mu(b - a)$. Then t is a solution of the first congruence in the system because

$$t - a = (a + mu(b - a)) - a = mu(b - a) = m(ub - ua),$$

and, hence, by the definition of congruence, $t \equiv a \pmod m$. Using equation (∗), we see that

$$t - b = (a + mu(b - a)) - b = (b - nv(b - a)) - b$$
$$= n(va - vb).$$

Therefore, $t \equiv b \pmod n$, and t is a solution of the system.   ◆

The proof of Lemma 13.1 provides an explicit **solution algorithm** for systems of two congruences:

1. Find u and v such that $mu + nv = 1$.*
2. Then $x = a + mu(b - a)$ is a solution of the system.

---

* This can always be done by using Euclidean Algorithm (Theorem 1.6); see the example on pages 11–12.

EXAMPLE   To solve the system

$$x \equiv 2 \ (\text{mod } 4)$$
$$x \equiv 5 \ (\text{mod } 7),$$

apply the algorithm with $m = 4$, $n = 7$, $a = 2$, $b = 5$:

1. It is easy to see that $u = 2$, $v = -1$ satisfy $4u + 7v = 1$.
2. Therefore, a solution of the system is:

$$x = a + mu(b - a) = 2 + 4 \cdot 2(5 - 2) = 2 + 8(3) = 26.$$

THEOREM 13.2   (THE CHINESE REMAINDER THEOREM)   *Let* $m_1, m_2, \ldots$, m_r *be pairwise relatively prime positive integers (meaning that* $(m_i, m_j) = 1$ *whenever* $i \neq j$). *Then for any* $a_i \in \mathbb{Z}$, *the system*

$$x \equiv a_1 \ (\text{mod } m_1)$$
$$x \equiv a_2 \ (\text{mod } m_2)$$
$$x \equiv a_3 \ (\text{mod } m_3)$$
$$\cdot$$
$$\cdot$$
$$\cdot$$
$$x \equiv a_r \ (\text{mod } m_r)$$

has a solution. Any two solutions of the system are congruent modulo $m_1 m_2 m_3 \cdots m_r$. *If* t *is one solution of the system, then an integer* z *is a solution if and only if* $z \equiv t \ (\text{mod } m_1 m_2 m_3 \cdots m_r)$.

Proof   Existence of a Solution   The proof is by induction on the number r of congruences in the system. If $r = 2$, then there is a solution by Lemma 13.1 (with $m = m_1$, $n = m_2$, $a = a_1$, $b = a_2$). So suppose inductively that there is a solution when $r = k$ and consider the system

$$x \equiv a_1 \ (\text{mod } m_1)$$
$$x \equiv a_2 \ (\text{mod } m_2)$$
$$x \equiv a_3 \ (\text{mod } m_3)$$
$$\cdot$$

(**)

$$\cdot$$
$$\cdot$$
$$x \equiv a_k \ (\text{mod } m_k)$$
$$x \equiv a_{k+1} \ (\text{mod } m_{k+1})$$

By the induction hypothesis, the system consisting of the first k congruences in (**) has a solution s. Furthermore, $m_1 m_2 m_3 \cdots m_k$ and m_{k+1} are relatively prime (Exercise 5). Consequently, by Lemma 13.1, the system

$$x \equiv s \quad (\mathrm{mod}\ m_1 m_2 m_3 \cdots m_k)$$
(∗∗∗)
$$x \equiv a_{k+1} (\mathrm{mod}\ m_{k+1})$$

has a solution t. The number t necessarily satisfies

$$t \equiv s\ (\mathrm{mod}\ m_1 m_2 m_3 \cdots m_k).$$

Consequently, for each $i = 1, 2, 3, \ldots, k$,

$$t \equiv s\ (\mathrm{mod}\ m_i).$$

(Reason: If $t - s$ is divisible by $m_1 m_2 m_3 \cdots m_k$, then it is divisible by each m_i).
Now s is a solution of the first k congruences in (∗∗), so for each $i \leq k$

$$t \equiv s\ (\mathrm{mod}\ m_i) \quad \text{and} \quad s \equiv a_i\ (\mathrm{mod}\ m_i).$$

By transitivity (Theorem 2.1),

$$t \equiv a_i\ (\mathrm{mod}\ m_i) \quad \text{for } i = 1, 2, \ldots, k.$$

Since t is a solution of (∗∗∗), it must also satisfy $t \equiv a_{k+1} (\mathrm{mod}\ m_{k+1})$. Hence, t is a solution of the system (∗∗), so that there is a solution when $r = k + 1$. Therefore, by induction, every such system has a solution.

Complete Set of Solutions   If z is any other solution of the system, then for each $i = 1, 2, \ldots, r$,

$$z \equiv a_i\ (\mathrm{mod}\ m_i) \quad \text{and} \quad t \equiv a_i\ (\mathrm{mod}\ m_i).$$

By transitivity (Theorem 2.1), $z \equiv t\ (\mathrm{mod}\ m_i)$. Thus

$$m_1 \mid (z - t), m_2 \mid (z - t), m_3 \mid (z - t), \ldots, m_r \mid (z - t).$$

Therefore, $m_1 m_2 m_3 \cdots m_r \mid (z - t)$ by Exercise 7. Hence,

$$z \equiv t\ (\mathrm{mod}\ m_1 m_2 m_3 \cdots m_r).$$

Conversely, if $z \equiv t\ (\mathrm{mod}\ m_1 m_2 m_3 \cdots m_r)$, then, as above, $z \equiv t\ (\mathrm{mod}\ m_i)$ for each $i = 1, 2, \ldots, r$. Since $t \equiv a_i\ (\mathrm{mod}\ m_i)$, transitivity shows that $z \equiv a_i$ $(\mathrm{mod}\ m_i)$ for each i. Therefore, z is a solution of the system. ◆

The proof of Theorem 13.2 actually provides an effective computational algorithm for solving large systems: Solve the first two by Lemma 13.1, then repeat the inductive step as often as needed to determine a solution of the entire system.

EXAMPLE To solve the system

$$x \equiv 2\ (\mathrm{mod}\ 4)$$
$$x \equiv 5\ (\mathrm{mod}\ 7)$$
$$x \equiv 0\ (\mathrm{mod}\ 11)$$
$$x \equiv 8\ (\mathrm{mod}\ 15),$$

begin with the first two congruences. The preceding example shows that $x = 26$ is a solution of both of them. By the last part of Theorem 13.2, any number that is congruent to 26 modulo $4 \cdot 7 = 28$ is also a solution. Since $-2 \equiv 26 \pmod{28}$, -2 is a solution of the first two congruences.*

The proof of Theorem 13.2 shows that the solutions of the first three congruences are the same as the solutions of this system:

$$x \equiv -2 \pmod{4 \cdot 7} \qquad \text{or, equivalently,} \qquad x \equiv -2 \pmod{28}$$
$$x \equiv 0 \pmod{11} \qquad\qquad\qquad\qquad x \equiv 0 \pmod{11}.$$

Applying Lemma 13.1 with $m = 28$ and $n = 11$, we find that $u = 2$, $v = -5$ satisfy $28u + 11v = 1$, and, hence,

$$s = a + mu(b - a) = -2 + 28 \cdot 2(0 - (-2)) = 110$$

is a solution of the first three congruences.

The solutions of all four congruences are the same as the solutions of this system:

$$x \equiv 110 \pmod{4 \cdot 7 \cdot 11} \qquad \text{or, equivalently,} \qquad x \equiv 110 \pmod{308}$$
$$x \equiv 8 \pmod{15} \qquad\qquad\qquad\qquad x \equiv 8 \pmod{15}.$$

A bit of work shows that $308(2) + 15(-41) = 1$, so $u = 2$ and

$$t = a + mu(b - a) = 110 + 308 \cdot 2(8 - 110) = -62,722$$

is a solution of all four congruences. Since $4 \cdot 7 \cdot 11 \cdot 15 = 4620$ and $-62,722 \equiv 1958 \pmod{4620}$, 1958 is also a solution of all four congruences.* Therefore, the solutions of the system are all numbers that are congruent to 1958 modulo 4620.

◆ **EXERCISES**

A. **1.** If $u \equiv v \pmod{n}$ and u is a solution of $6x + 5 \equiv 7 \pmod{n}$, then show that v is also a solution. [*Hint*: Theorem 2.2.]

2. If $6x + 5 \equiv 7 \pmod{n}$ has a solution, show that one of the numbers 1, 2, 3, . . . , $n - 1$ is also a solution. [*Hint*: Exercise 1 and Corollary 2.5.]

3. Show that $6x + 5 \equiv 7 \pmod{3}$ has no solutions. [*Hint:* Exercise 2.]

4. Show that $6x + 5 \equiv 7 \pmod{5}$ has infinitely many solutions. [*Hint:* Exercises 1 and 2.]

5. If $m_1, m_2, \ldots, m_k, m_{k+1}$ are pairwise relatively prime positive integers

---

* To simplify computations, it is customary to use solutions that are as small as possible in absolute value.

(that is, $(m_i, m_j) = 1$ when $i \neq j$), prove that $m_1 m_2 \cdots m_k$ and m_{k+1} are relatively prime. [*Hint*: If they aren't, then some prime p divides both of them (Why?). Use Corollary 1.9 to reach a contradiction.]

6. If $(m,n) = 1$ and $m \mid d$ and $n \mid d$, prove that $mn \mid d$. [*Hint*: If $d = mk$, then $n \mid mk$; use Theorem 1.5.]

7. Let $m_1, m_2, \ldots, m_r$ be pairwise relatively prime positive integers (that is, $(m_i, m_j) = 1$ when $i \neq j$). Assume that $m_i \mid d$ for each i. Prove that $m_1 m_2 m_3 \cdots m_r \mid d$. [*Hint*: Use Exercises 5 and 6 repeatedly.]

In Exercises 8–13, solve the system of congruences.

8. $x \equiv 5 \pmod 6$ **9.** $x \equiv 3 \pmod{11}$
$\ x \equiv 7 \pmod{11}$ $\ x \equiv 4 \pmod{17}$

10. $x \equiv 1 \pmod 2$ **11.** $x \equiv 2 \pmod 5$
$\ x \equiv 2 \pmod 3$ $\ x \equiv 0 \pmod 6$
$\ x \equiv 3 \pmod 5$ $\ x \equiv 3 \pmod 7$

12. $x \equiv 1 \pmod 5$ **13.** $x \equiv 1 \pmod 7$
$\ x \equiv 3 \pmod 6$ $\ x \equiv 6 \pmod{11}$
$\ x \equiv 5 \pmod{11}$ $\ x \equiv 0 \pmod{12}$
$\ x \equiv 10 \pmod{13}$ $\ x \equiv 9 \pmod{13}$
$\phantom{\ x \equiv 10 \pmod{13}}$ $\ x \equiv 0 \pmod{17}$

B. 14. (Ancient Chinese Problem) A gang of 17 bandits stole a chest of gold coins. When they tried to divide the coins equally among themselves, there were three left over. This caused a fight in which one bandit was killed. When the remaining bandits tried to divide the coins again, there were ten left over. Another fight started, and five of the bandits were killed. When the survivors divided the coins, there were four left over. Another fight ensued in which four bandits were killed. The survivors then divided the coins equally among themselves, with none left over. What is the smallest possible number of coins in the chest?

15. If $(a,n) = d$ and $d \mid b$, show that $ax \equiv b \pmod n$ has a solution. [*Hint*: $b = dc$ for some c, and $au + nv = d$ for some u, v (Why?). Multiply the last equation by c; what is auc congruent to modulo n?]

16. If $(a,n) = d$ and $d \nmid b$, show that $ax \equiv b \pmod n$ has no solutions.

17. If $(a,n) = 1$ and s, t are solutions of $ax \equiv b \pmod n$, prove that $s \equiv t \pmod n$. [*Hint*: Show that $n \mid (as - at)$ and use Theorem 1.5.]

18. If $(a,n) = d$ and s, t are solutions of $ax \equiv b \pmod n$, prove that $s \equiv t \pmod{n/d}$.

19. If $(m,n) = d$, prove that the system

$$x \equiv a \pmod m$$
$$x \equiv b \pmod n$$

has a solution if and only if $a \equiv b \pmod d$.

20. If s, t are solutions of the system in Exercise 19, prove that $s \equiv t \pmod{r}$, where r is the least common multiple of m and n.

13.2 APPLICATIONS OF THE CHINESE REMAINDER THEOREM

Every computer has a limit on the size of integers that can be used in machine arithmetic, called the **word size.** In a large computer this might be 2^{35}. Computer arithmetic with integers larger than the word size requires time-consuming multiprecision techniques. In such cases an alternate method of addition and multiplication, based on the Chinese Remainder Theorem, is often faster.

For any numbers r, s, t, n less than the word size, a large computer can quickly calculate

$r + s$ and $r \cdot s$ (even when the answer is larger than the word size);

the least residue of t modulo $n*$ (including the case when t exceeds the word size—see Exercise 2);

sums and products in $\mathbb{Z}_n$.

Finally, a computer can use a slight variation of the Chinese Remainder Theorem solution algorithm (Theorem 13.2) to solve systems of congruences. But this may involve numbers larger than the word size and, hence, require slower multiprecision techniques.

To get an idea of how the alternate method works, imagine that the word size of our computer is 100, so that multiprecision techniques must be used for larger numbers. The following example shows how to multiply two four-digit numbers on such a computer, with minimal use of multiprecision techniques.

> **EXAMPLE**   We shall multiply 3456 by 7982 by considering various systems of congruences and using the Chinese Remainder Theorem. We begin by choosing several numbers as moduli and finding the least residues of 3456 and 7982 for each modulus:**

---

* The **least-residue modulo** n of a number t is the remainder r when t is divided by n. By the Division Algorithm, $t = nq + r$ so that $t - r = nq$ and $t \equiv r \pmod{n}$.
** The reason why 89, 95, 97, 98, and 99 were chosen will be explained below.

(∗)

$$3456 \equiv 74 \pmod{89} \qquad 7982 \equiv 61 \pmod{89}$$
$$3456 \equiv 36 \pmod{95} \qquad 7982 \equiv 2 \pmod{95}$$
$$3456 \equiv 61 \pmod{97} \qquad 7982 \equiv 28 \pmod{97}$$
$$3456 \equiv 26 \pmod{98} \qquad 7982 \equiv 44 \pmod{98}$$
$$3456 \equiv 90 \pmod{99} \qquad 7982 \equiv 62 \pmod{99}.$$

Then by Theorem 2.2 we know that $3456 \cdot 7982 \equiv 74 \cdot 61 \pmod{89}$. Taking the least residue of $74 \cdot 61$ modulo 89 and proceeding in similar fashion for the other congruences, we have

(∗∗)

$$3456 \cdot 7982 \equiv 74 \cdot 61 \equiv 64 \pmod{89}$$
$$3456 \cdot 7982 \equiv 36 \cdot 2 \equiv 72 \pmod{95}$$
$$3456 \cdot 7982 \equiv 61 \cdot 28 \equiv 59 \pmod{97}$$
$$3456 \cdot 7982 \equiv 26 \cdot 44 \equiv 66 \pmod{98}$$
$$3456 \cdot 7982 \equiv 90 \cdot 62 \equiv 36 \pmod{99}.$$

Therefore, $3456 \cdot 7982$ is a solution of this system:

(∗∗∗)

$$x \equiv 64 \pmod{89}$$
$$x \equiv 72 \pmod{95}$$
$$x \equiv 59 \pmod{97}$$
$$x \equiv 66 \pmod{98}$$
$$x \equiv 36 \pmod{99}.$$

The Chinese Remainder Theorem* shows that one solution of (∗∗∗) is 27,585,792 and that every solution (including $3456 \cdot 7982$) is congruent to this one modulo $89 \cdot 95 \cdot 97 \cdot 98 \cdot 99 = 7{,}956{,}949{,}770$ (which we denote hereafter by M). Since no two numbers between 0 and M can be congruent modulo M, 27,585,792 is the *only* solution between 0 and M. We know that $0 < 3456 \cdot 7982 < 10^4 \cdot 10^4 = 10^8 < M$. Since $3456 \cdot 7982$ is a solution, we must have $3456 \cdot 7982 = 27{,}585{,}792$.

Now look at this example from a different perspective. If you think of the least residue of a number modulo n as an element of $\mathbb{Z}_n$, then the congruences in (∗) say that the integer 3456 may be represented by the element $(74,36,61,26,90)$ in $\mathbb{Z}_{89} \times \mathbb{Z}_{95} \times \mathbb{Z}_{97} \times \mathbb{Z}_{98} \times \mathbb{Z}_{99}$. Similarly, 7982 is represented by $(61,2,28,44,62)$. Saying that $74 \cdot 61 \equiv 64 \pmod{89}$ in (∗∗) is the same as saying $74 \cdot 61 = 64$ in $\mathbb{Z}_{89}$. So the congruences in (∗∗) are equivalent to multiplication in $\mathbb{Z}_{89} \times \mathbb{Z}_{95} \times \mathbb{Z}_{97} \times \mathbb{Z}_{98} \times \mathbb{Z}_{99}$:

---

* Up to this point, all computations have been quickly performed by our imaginary computer. This is the first place where slower multiprecision calculations may be needed because of numbers that exceed the word size.

$$(74, 36, 61, 26, 90) \cdot (61, 2, 28, 44, 62)$$
$$= (74 \cdot 61, 36 \cdot 2, 61 \cdot 28, 26 \cdot 44, 90 \cdot 62)$$
$$= (64, 72, 59, 66, 36).$$

The solution of (∗∗∗) shows that the element $(64,72,59,66,36)$ of $\mathbb{Z}_{89} \times \mathbb{Z}_{95} \times \mathbb{Z}_{97} \times \mathbb{Z}_{98} \times \mathbb{Z}_{99}$ represents the integer 27,585,792.

The procedure in the case of a realistic word size is now clear. Let $m_1, \ldots, m_r$ be pairwise relatively prime positive integers:

1. Represent each integer t as an element of $\mathbb{Z}_{m_1} \times \cdots \times \mathbb{Z}_{m_r}$ by taking the congruence class of t modulo each m_i.
2. Do the arithmetic in $\mathbb{Z}_{m_1} \times \cdots \times \mathbb{Z}_{m_r}$.
3. Use the Chinese Remainder Theorem to convert the answer into integer form.

The m_i must be chosen so that their product M is larger than any number that will result from the computations. Otherwise, the conversion process in Step 3 may fail (Exercises 3–5). This is sometimes done, as in the example, by taking the m_i to be as large as possible without exceeding the word size of the computer. If smaller moduli are chosen, more of them may be necessary to ensure that M is large enough.

The conversion process from integer to modular representation and back (Steps 1 and 3) requires time that is not needed in conventional integer multiplication (especially Step 3, which may involve multiprecision techniques). But this need be done only once for each number, at input and output. The modular representation may be used for all intermediate calculations. It is much faster than direct computation with large integers, especially in a computer with parallel processing capability, which can work simultaneously in each $\mathbb{Z}_{m_i}$. Under appropriate conditions the speed advantage in Step 2 outweighs the disadvantage of the extra time required for Steps 1 and 3. For more details, see Knuth [33].

It is sometimes necessary to find an exact solution (not a decimal approximation) of a system of linear equations. When there are hundreds of equations or unknowns in the system and the coefficients are large integers, the usual computer methods will produce only approximate solutions because they round off very large numbers during the intermediate calculations. The Chinese Remainder Theorem is the basis of a method of finding exact solutions of such systems.

Very roughly, the idea is this. Let $m_1, \ldots, m_r$ be distinct primes (and, hence, pairwise relatively prime).* For each m_i, translate the given system of

---

* Considerations of size similar to those discussed above play a role in the selection of the m_i.

equations into a system over $\mathbb{Z}_{m_i}$ by replacing the integer coefficients by their congruence classes modulo m_i. Then solve each of these new systems by the usual methods (Gauss-Jordan elimination works equally well over the field $\mathbb{Z}_{m_i}$ as over $\mathbb{R}$, and round-off is not a problem with the smaller numbers in $\mathbb{Z}_{m_i}$). Finally, use the Chinese Remainder Theorem and matrix algebra to convert these solutions modulo m_i into a solution of the original system.* A readable explanation of this method (accessible to anyone who has had a first course in linear algebra) is given in Mackiw [35].

◆ **EXERCISES**

A. 1. Assume that your computer has word size 100. Use the method outlined in the text to find the sum $123{,}684 + 413{,}456$, using $m_1 = 95$, $m_2 = 97$, $m_3 = 98$, $m_4 = 99$.

2. (a) Find the least residue of $64{,}397$ modulo 12, using only arithmetic in $\mathbb{Z}_{12}$. [*Hint*: Use Theorems 2.2 and 2.3 and the fact that $64{,}397 = (((6 \cdot 10 + 4)10 + 3)10 + 9)10 + 7$.]

(b) Let n be a positive integer less than the word size of your computer and t any integer (possibly larger than the word size). Explain how you might find the least residue of t modulo n, using only arithmetic in $\mathbb{Z}_n$ (and thus avoiding the need for multiprecision methods).

3. Use the method outlined in the text to represent 7 and 8 as elements of $\mathbb{Z}_3 \times \mathbb{Z}_5$. Show that the product of these representatives in $\mathbb{Z}_3 \times \mathbb{Z}_5$ is $(2,1)$. If you use the Chinese Remainder Theorem as in the text to convert $(2,1)$ to integer form, do you get 56? Why not? This example shows why the method won't work when the product of the m_i is less than the answer to the arithmetic problem in question. Also see Exercise 5.

B. 4. Let $f: \mathbb{Z} \to \mathbb{Z}_3 \times \mathbb{Z}_4 \times \mathbb{Z}_5$ be given by $f(t) = ([t]_3, [t]_4, [t]_5)$, where $[t]_n$ is the congruence class of t in $\mathbb{Z}_n$. The function f may be thought of as representing t as an element of $\mathbb{Z}_3 \times \mathbb{Z}_4 \times \mathbb{Z}_5$ by taking its least residues.

(a) If $0 \le r, s < 60$, prove that $f(r) = f(s)$ if and only if $r = s$. [*Hint:* Theorem 13.2.]

(b) Give an example to show that if r or s is greater than 60, then part (a) may be false.

---

* This conversion is a bit trickier than may first appear. For instance, the system

$$8x + 5y = 12 \qquad\qquad x + 5y = 5$$
$$\text{becomes}$$
$$4x + 5y = 10 \qquad\qquad 4x + 5y = 3 \qquad \text{over } \mathbb{Z}_7.$$

You can verify that $x = 4$, $y = 3$ is a solution of the $\mathbb{Z}_7$ system. It is not immediately clear how to get from this to the solution of the original system, which is $x = 1/2$, $y = 8/5$.

5. Let $m_1, m_2, \ldots, m_r$ be pairwise relatively prime positive integers and $f \colon \mathbb{Z} \to \mathbb{Z}_{m_1} \times \mathbb{Z}_{m_2} \times \cdots \times \mathbb{Z}_{m_r}$ the function given by

$$f(t) = ([t]_{m_1}, [t]_{m_2}, \ldots, [t]_{m_r}),$$

where $[t]_{m_i}$ is the congruence class of t in $\mathbb{Z}_{m_i}$. Let $M = m_1 m_2 \cdots m_r$. If $0 \le r, s < M$, prove that $f(r) = f(s)$ if and only if $r = s$. [Exercise 4 is a special case.]

6. Assume Exercise 7(c). If your computer has word size 2^{35}, what m_i might you choose in order to do arithmetic with integers as large as 2^{184} (approximately 2.45×10^{55})?

C. 7. (a) If a and b are positive integers, prove that the least residue of $2^a - 1$ modulo $2^b - 1$ is $2^r - 1$, where r is the least residue of a modulo b.

(b) If a and b are positive integers, prove that the greatest common divisor of $2^a - 1$ and $2^b - 1$ is $2^t - 1$, where t is the gcd of a and b. [*Hint:* Use the Euclidean algorithm and part (a).]

(c) Let a and b be positive integers. Prove that $2^a - 1$ and $2^b - 1$ are relatively prime if and only if a and b are relatively prime.

13.3 THE CHINESE REMAINDER THEOREM FOR RINGS

The Chinese Remainder Theorem for two congruences can be extended from $\mathbb{Z}$ to other rings by expressing it in terms of ideals. The key to doing this is the definition of congruence modulo an ideal (Section 6.1) and the following fact: When A and B are ideals in a ring R, the set of sums $\{a + b \mid a \in A, b \in B\}$ is denoted $A + B$ and is itself an ideal (Exercise 18 of Section 6.1).

Let m and n be integers. Let I be the ideal of all multiples of m in $\mathbb{Z}$ and J the ideal of all multiples of n. Then *congruence modulo m is the same as congruence modulo the ideal I.* If $(m,n) = 1$, then $mu + nv = 1$ for some $u, v \in \mathbb{Z}$. Multiplying this equation by any integer r shows that $m(ur) + n(vr) = r$. Thus every integer is the sum of a multiple of m and a multiple of n, that is, the sum of an element of the ideal I and an element of the ideal J. Therefore, $I + J$ is the entire ring $\mathbb{Z}$. So *the condition $(m,n) = 1$ amounts to saying $I + J = \mathbb{Z}$.*

When $(m,n) = 1$, the intersection of the ideals I and J is the ideal consisting of all multiples of mn (Exercise 6 of Section 13.1). So *two integers are congruent modulo mn precisely when they are congruent modulo the ideal $I \cap J$.*

The italicized statements in the preceding paragraphs tell us how to translate the Chinese Remainder Theorem for two congruences into the language of ideals. By replacing the ideals in that discussion by ideals in *any* ring R, we obtain

THEOREM 13.3 (CHINESE REMAINDER THEOREM FOR RINGS) *Let I and J be ideals in a ring R such that $I + J = R$. Then for any a, b ϵ R, the system*

$$x \equiv a \pmod{I}$$
$$x \equiv b \pmod{J}$$

has a solution. Any two solutions of the system are congruent modulo $I \cap J$.

When R has an identity, the theorem can be extended to the case of r ideals $I_1, I_2, \ldots, I_r$ and congruences $x \equiv a_k \pmod{I_k}$, under the hypotheses that $I_i + I_j = R$ whenever $i \neq j$ (see Exercise 6 and Hungerford [7; p. 131]).

Proof of Theorem 13.3 Since $I + J = R$ and $b - a \epsilon R$, there exist $i \epsilon I, j \epsilon J$ such that $i + j = b - a$. Hence, $a + i = b - j$. Let $t = a + i$; then

$$t - a = (a + i) - a = i \epsilon I,$$

so that $t \equiv a \pmod{I}$. Similarly, since $a + i = b - j$

$$t - b = (a + i) - b = (b - j) - b = -j \epsilon J.$$

Hence, $t \equiv b \pmod{J}$, and t is a solution of the system. If z is also a solution, then

$$z \equiv a \pmod{I} \quad \text{and} \quad t \equiv a \pmod{I} \quad \text{imply that} \quad z \equiv t \pmod{I}$$

by Theorem 6.4. Similarly, $z \equiv t \pmod{J}$. This means that $z - t \epsilon I$ and $z - t \epsilon J$. Therefore, $z - t \epsilon I \cap J$ and $z \equiv t \pmod{I \cap J}$. ◆

One consequence of the Chinese Remainder Theorem is a useful isomorphism of rings.

THEOREM 13.4 *If I and J are ideals in a ring R and $I + J = R$, then there is an isomorphism of rings*

$$R/(I \cap J) \cong R/I \times R/J.$$

Proof Define a map $f: R \to R/I \times R/J$ by $f(r) = (r + I, r + J)$. Then f is a homomorphism because

$$f(r) + f(s) = (r + I, r + J) + (s + I, s + J)$$
$$= ((r + s) + I, (r + s) + J) = f(r + s)$$

and

$$f(r)f(s) = (r + I, r + J)(s + I, s + J)$$
$$= (rs + I, rs + J) = f(rs).$$

To show that f is surjective, let $(a + I, b + J) \epsilon R/I \times R/J$. We must find an element of R whose image under f is $(a + I, b + J)$. By Theorem 13.3 there is a solution $t \, \epsilon \, R$ for this system:

$$x \equiv a \pmod{I}$$
$$x \equiv b \pmod{J}.$$

But $t \equiv a \pmod{I}$ implies that $t + I = a + I$ by Theorem 6.6. Similarly, $t \equiv b$ $\pmod{J}$ implies $t + J = b + J$, so that

$$f(t) = (t + I, t + J) = (a + I, b + J).$$

Therefore, f is surjective.

Let K be the kernel of f. By the First Isomorphism Theorem 6.13, R/K is isomorphic to $R/I \times R/J$. Now K consists of all elements $r \, \epsilon \, R$ such that $f(r)$ is the zero element in $R/I \times R/J$, that is, all r such that

$$(r + I, r + J) = (0_R + I, 0_R + J),$$

or equivalently,

$$r + I = 0_R + I \quad \text{and} \quad r + J = 0_R + J.$$

But $r + I = 0_R + I$ means that $r \equiv 0_R \pmod{I}$, and, hence, $r \, \epsilon \, I$. Similarly, $r + J = 0_R + J$ implies $r \, \epsilon \, J$. Therefore, $r \, \epsilon \, I \cap J$. So $I \cap J$ is the kernel of f, and $R/(I \cap J) = R/\text{Ker } f \cong R/I \times R/J$. ◆

COROLLARY 13.5   *If $(m,n) = 1$, then there is an isomorphism of rings $\mathbb{Z}_{mn} \cong \mathbb{Z}_m \times \mathbb{Z}_n$.*

Proof   In the ring $\mathbb{Z}$, the ideal (m) consists of all multiples of m and the ideal (n) of all multiples of n. The first three paragraphs of this section show that $(m) + (n) = \mathbb{Z}$ and that $(m) \cap (n)$ is the ideal (mn) of all multiples of mn. Furthermore, the quotient rings $\mathbb{Z}/(mn)$, $\mathbb{Z}/(m)$, and $\mathbb{Z}/(n)$ are, respectively, $\mathbb{Z}_{mn}$, $\mathbb{Z}_m$, and $\mathbb{Z}_n$. Therefore, by Theorem 13.4 (with $R = \mathbb{Z}, I = (m), J = (n)$) there is an isomorphism

$$\mathbb{Z}_{mn} = \mathbb{Z}/(mn) = \mathbb{Z}((m) \cap (n)) \cong \mathbb{Z}/(m) \times \mathbb{Z}/(n) = \mathbb{Z}_m \times \mathbb{Z}_n. \quad ◆$$

COROLLARY 13.6   *If $n = p_1^{n_1} p_2^{n_2} p_3^{n_3} \cdots p_t^{n_t}$, where the p_i are distinct positive primes and each $n_i > 0$, then there is an isomorphism of rings*

$$\mathbb{Z}_n \cong \mathbb{Z}_{p_1^{n_1}} \times \mathbb{Z}_{p_2^{n_2}} \times \mathbb{Z}_{p_3^{n_3}} \times \cdots \times \mathbb{Z}_{p_t^{n_t}}.$$

Proof   Since the p_j are distinct primes, $p_i^{n_i}$ and the product $p_{i+1}^{n+1} \cdots p_t^{n_t}$ are relatively prime for each i. So repeated use of Corollary 13.5 shows that

$$\mathbb{Z}_n \cong \mathbb{Z}_{p_1^{n_1}} \times \mathbb{Z}_{p_2^{n_2} p_3^{n_3} \cdots p_t^{n_t}} \cong \mathbb{Z}_{p_1^{n_1}} \times \mathbb{Z}_{p_2^{n_2}} \times \mathbb{Z}_{p_3^{n_3} \cdots p_t^{n_t}} \cong \cdots$$
$$\cong \mathbb{Z}_{p_1^{n_1}} \times \mathbb{Z}_{p_2^{n_2}} \times \mathbb{Z}_{p_3^{n_3}} \times \cdots \times \mathbb{Z}_{p_t^{n_t}}. \quad ◆$$

◆ **EXERCISES**

A. 1. (a) Show that $\mathbb{Z}_5 \times \mathbb{Z}_{12}$ is isomorphic to $\mathbb{Z}_3 \times \mathbb{Z}_{20}$.

  (b) Is $\mathbb{Z}_4 \times \mathbb{Z}_{35}$ isomorphic to $\mathbb{Z}_5 \times \mathbb{Z}_{28}$?

2. If I and J are ideals in a ring R and $a \in I$, $b \in J$, show that $ab \in I \cap J$.

B. 3. If $(m,n) \neq 1$, show that $\mathbb{Z}_{mn}$ is *not* isomorphic to $\mathbb{Z}_m \times \mathbb{Z}_n$. [*Hint:* If $(m,n) = d$, then $\dfrac{mn}{d}$ is an integer (Why?). If there were an isomorphism, then $1 \in \mathbb{Z}_{mn}$ would be mapped to $(1,1) \in \mathbb{Z}_m \times \mathbb{Z}_n$. Reach a contradiction by showing that $\dfrac{mn}{d} \cdot 1 \neq 0$ in $\mathbb{Z}_{mn}$, but $\dfrac{mn}{d} \cdot (1,1) = (0,0)$ in $\mathbb{Z}_m \times \mathbb{Z}_n$.]

4. Which of the following rings are isomorphic: $\mathbb{Z}_2 \times \mathbb{Z}_6 \times \mathbb{Z}_7$, $\mathbb{Z}_3 \times \mathbb{Z}_4 \times \mathbb{Z}_7$, $\mathbb{Z}_{84}$, $\mathbb{Z}_7 \times \mathbb{Z}_{12}$, $\mathbb{Z}_2 \times \mathbb{Z}_3 \times \mathbb{Z}_{14}$, $\mathbb{Z}_4 \times \mathbb{Z}_{21}$?

5. If I_1, I_2, I_3 are ideals in a ring R with identity such that $I_1 + I_3 = R$ and $I_2 + I_3 = R$, prove that $(I_1 \cap I_2) + I_3 = R$. [*Hint:* If $r \in R$, then $r = i_1 + i_3$ and $1_R = t_2 + t_3$ for some $i_1 \in I_1$, $t_2 \in I_2$, $i_3, t_3 \in I_3$. Then $r = (i_1 + i_3)(t_2 + t_3)$; multiply this out to show that r is in $(I_1 \cap I_2) + I_3$. Exercise 2 may be helpful.]

6. Let I_1, I_2, I_3 be ideals in a ring R with identity such that $I_i + I_j = R$ whenever $i \neq j$. If $a_i \in R$, prove that the system

$$x \equiv a_1 \pmod{I_1}$$
$$x \equiv a_2 \pmod{I_2}$$
$$x \equiv a_3 \pmod{I_3}$$

has a solution and that any two solutions are congruent modulo $I_1 \cap I_2 \cap I_3$. [*Hint:* If s is a solution of the first two congruences, use Exercise 5 and Theorem 13.3 to show that the system

$$x \equiv s \pmod{I_1 \cap I_2}$$
$$x \equiv a_3 \pmod{I_3}$$

has a solution, and it is a solution of the original system.]

CHAPTER **14**

Lattices and Boolean Algebras

Prerequisites: Chapter 3; plus Appendix B for Sections 14.1 and 14.2, and Appendix A for Section 14.3. There are also optional examples and exercises with additional prerequisites. Those labeled "rings" require Section 6.1; those labeled "groups," Section 7.3; and those labeled "fields," Section 10.2.

Lattices (Section 14.1) are a generalization of the properties of the order relation $\leq$ in familiar number systems and of the order relation $\subseteq$ among sets. Boolean algebras (Section 14.2) are special types of lattices. They have applications in logic, circuit theory, probability, classical mechanics, and quantum mechanics, some of which are discussed in Section 14.3.

14.1 LATTICES

If S is a set, then any subset of $S \times S$ is called a **relation** on S. A relation T on S is called a **partial order** provided that the subset T is

(i) **Reflexive:** $(a,a) \in T$ for every $a \in S$.
(ii) **Antisymmetric:** If $(a,b) \in T$ and $(b,a) \in T$, then $a = b$.
(iii) **Transitive:** If $(a,b) \in T$ and $(b,c) \in T$, then $(a,c) \in T$.

A set equipped with a partial-order relation is called a **partially ordered set (or poset)**.

The symbol $\leq$ is usually used to denote an arbitrary partial order T:

$$a \leq b \quad \text{means} \quad (a,b) \, \epsilon \, T.$$

In this notation, the conditions defining a partial order become

(i) **Reflexive:** $a \leq a$ for every $a \, \epsilon \, S$.
(ii) **Antisymmetric:** If $a \leq b$ and $b \leq a$, then $a = b$.
(iii) **Transitive:** If $a \leq b$ and $b \leq c$, then $a \leq c$.

When this notation is used, a partial order on S is usually defined without explicit reference to a subset of $S \times S$. We shall also adopt the usual notation.

$$b \geq a \quad \text{means} \quad a \leq b.$$

EXAMPLE 1* The set $\mathbb{Z}$ of integers is a partially ordered set in which $a \leq b$ has its usual meaning. The same is true of the set $\mathbb{Q}$ of rational numbers and the set $\mathbb{R}$ of real numbers.

EXAMPLE 2 Let S be the set of all subsets of $\{x,y,z\}$ and define $A \leq B$ to mean A is a subset of B. The relation $\subseteq$ is reflexive, antisymmetric, and transitive (see Appendix B). So S is a partially ordered set. The ordering can be schematically displayed by this diagram, in which a line connecting two sets means that the lower of the two is a subset of the higher:

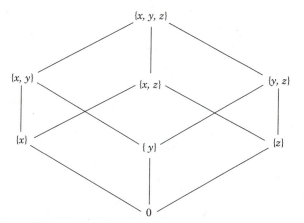

We don't draw a line from $\{x\}$ to $\{x,y,z\}$, for example, because we can use transitivity to move upward from $\{x\}$ to $\{x,y,z\}$ by means of the lines from $\{x\}$ to $\{x,y\}$ and from $\{x,y\}$ to $\{x,y,z\}$.

---

* The first five examples of this section are numbered because they will be referred to frequently.

EXAMPLE 3 If X is any set (possibly infinite), then the set $P(X)$ of all subsets of X is a partially ordered set in which $A \leq B$ means $A \subseteq B$.

EXAMPLE 4 Define a relation on the set P of positive integers by: $a \leq b$ means $a \mid b$.* Then $\mid$ is a partial order on P:

> *Reflexive:* $a \mid a$ for every $a \in P$ since $a = a \cdot 1$.
>
> *Antisymmetric:* If $a \mid b$ and $b \mid a$, then $b = au$ and $a = bv$. Hence, $b = (bv)u$, which implies $uv = 1$. Since u and v are positive integers, $v = 1$, and, hence, $a = bv = b$.
>
> *Transitive:* If $a \mid b$ and $b \mid c$, then $b = ar$ and $c = bs$. Hence, $c = (ar)s = a(rs)$ and $a \mid c$.

EXAMPLE 5 The set $S = \{r,s,t,u,v,w,x\}$ is a partially ordered set whose partial order is given by this diagram, in which $a \leq b$ means that either $a = b$ or a lies below b and there is a path of line segments from a to b that never moves downward:

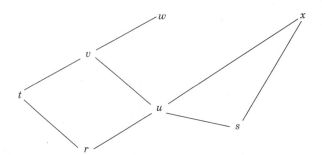

Thus $r \leq u$ and $r \leq w$, but it is *not* true that $r \leq s$. Similarly, $a \leq w$ for every $a \in S$ except x.

> **DEFINITION**• *Let B be a subset of a partially ordered set S. An element u of S is said to be an **upper bound** of B if $b \leq u$ for every $b \in B$.*

The set of B may have many upper bounds, some of which are not in B itself, or B may have no upper bounds.

---

* Although $\leq$ is used in general discussions of partially ordered sets, we use $\mid$ for the partial ordering of divisibility to avoid any confusion with the usual meaning of $\leq$ for integers.

> **EXAMPLE** In Example 5, the only upper bounds of the subset $B = \{t,u\}$ are v and w. The subset $\{r,u,s\}$ has four upper bounds (u,v,w,x). In the set $\mathbb{Z}$ of integers with the usual ordering (Example 1), the subset of even integers has no upper bound.

DEFINITION • *If u is an upper bound of B such that $u \leq v$ for every other upper bound v of B, then u is the **least upper bound** (or **l.u.b.** or **supremum**) of B.*

> **EXAMPLE** In Example 5, v is the least upper bound of $\{t,u\}$, and u is the least upper bound of $\{r,u,s\}$.

> **EXAMPLE** The subset $\{2,3,6\}$ of the set P of positive integers ordered by divisibility (Example 4) has every multiple of 6 as an upper bound (because 2, 3, and 6 divide every multiple of 6) and 6 itself is the least upper bound.

> **EXAMPLE** In the set $\mathbb{Q}$ with the usual ordering (Example 1), the subset $B = \{b \in \mathbb{Q} \mid b^2 < 2\}$ has many upper bounds (any rational larger than $\sqrt{2}$) but no least upper bound (Exercise 9).

DEFINITION • *Let B be a subset of a partially ordered set A. An element w of A is said to be a **lower bound** of B if $w \leq b$ for every $b \in B$. If w is a lower bound of B such that $v \leq w$ for every other lower bound v of B, then w is the **greatest lower bound** (or **g.l.b.** or **infimum**) of B.*

> **EXAMPLE** In the set $\mathbb{Z}$ with the usual ordering (Example 1), the set P of positive integers has 1, 0, and every negative integer as a lower bound and 1 as a greatest lower bound. The set B of negative integers has no lower bound and hence no g.l.b.

Observe that the definition of "lower bound" is identical to that of "upper bound," provided $\leq$ is replaced by $\geq$. Consequently, we have the

PRINCIPLE OF DUALITY *A true statement about partially ordered set remains true when $\leq$ is replaced by $\geq$ and "(least) upper bound" is replaced by "(greatest) lower bound."*

The preceding examples show that a set B need not have a least upper bound or a greatest lower bound. But if B does have an l.u.b., it has only one; and similarly for g.l.b.'s:

THEOREM 14.1   *Let B be a nonempty subset of a partially ordered set S. If B has a least upper bound, then this l.u.b. is unique. If B has a greatest lower bound, then this g.l.b. is unique.*

Proof   Suppose both u_1 and u_2 are least upper bounds of B. By the definition of least upper bound, $u_1 \leq v$ for every other upper bound v of B. In particular, since u_2 is an upper bound of B, $u_1 \leq u_2$. Conversely, since u_2 is a least upper bound of B, $u_2 \leq v$ for every other upper bound v of B. In particular, $u_2 \leq u_1$. Therefore, $u_1 = u_2$ by the antisymmetric property. So B has a unique least upper bound.

The corresponding statement for greatest lower bounds is now an immediate consequence of the Principle of Duality. (In other words, it can be proved by repeating the argument above with $\geq$ in place of $\leq$, "lower" in place of "upper," and "greatest" in place of "least.")   ◆

> **DEFINITION•** *A **lattice** is a partially ordered set L in which every pair of elements has both a least upper bound and a greatest lower bound. If a, b ∈ L, then their least upper bound is denoted $a \vee b$ and called the **join** of a and b. The greatest lower bound of a and b is denoted $a \wedge b$ and called their **meet**.*

EXAMPLE   The set of positive integers ordered by divisibility (Example 4) is a lattice. The l.u.b. of integers b and c is their least common multiple, and the g.l.b. of b, c is their greatest common divisor (Exercise 10). The partially ordered sets in Examples 1 and 2 are also lattices (Exercise 4). The set in Example 5 is *not* a lattice because w and x have no l.u.b.

EXAMPLE   The set $P(X)$ of all subsets of X (Example 3) is a lattice. If A and B are subsets of X, their l.u.b. is $A \cup B$ and their g.l.b. is $A \cap B$ (Exercise 11).

EXAMPLE (RINGS)* If R is a ring, then the set S of all ideals of R, partially ordered by set-theoretic inclusion ($\subseteq$), is a lattice (Exercise 19). The g.l.b. of ideals I and J is the ideal $I \cap J$. The union of two ideals may not be an ideal, so $I \cup J$ is *not* the least upper bound of I and J in this lattice. The l.u.b. of I and J is the ideal $I + J$ (defined in Exercise 18 of Section 6.1).

---

* Items labeled "rings," "groups," or "fields" are optional and should be omitted by those who have not covered the appropriate sections of Chapters 6, 7, and 10, respectively.

EXAMPLE (GROUPS) If G is a group, then the set S of all subgroups of G, partially ordered by set-theoretic inclusion, is a lattice (Exercise 20). The g.l.b. of subgroups H and K is the subgroup $H \cap K$. The set $H \cup K$ may not be a subgroup; the l.u.b. of H and K is the subgroup *generated by* the set $H \cup K$ (see Theorem 7.17).

Let L be a lattice. If $a, b \in L$, then $a \vee b$ is also an element of L. Hence, $\vee$ is a *binary operation* on L (as defined in Appendix B). Similarly, $\wedge$ is also a binary operation on L.

THEOREM 14.2 *If L is a lattice, then the binary operations $\vee$ and $\wedge$ satisfy these conditions for all $a, b, c \in L$:*

1. **Commutative Laws:**

$$a \vee b = b \vee a \quad \text{and} \quad a \wedge b = b \wedge a.$$

2. **Associative Laws:**

$$a \vee (b \vee c) = (a \vee b) \vee c \quad \text{and} \quad a \wedge (b \wedge c) = (a \wedge b) \wedge c.$$

3. **Absorption Laws:**

$$a \vee (a \wedge b) = a \quad \text{and} \quad a \wedge (a \vee b) = a.$$

4. **Idempotent Laws:**

$$a \vee a = a \quad \text{and} \quad a \wedge a = a.$$

Proof In view of the Principle of Duality, we need only prove the first statement in each part.

1. By the definition, $a \vee b$ is the least upper bound of $\{a, b\}$ and $b \vee a$ is the l.u.b. of $\{b, a\}$. Since $\{a, b\}$ is the *same* set as $\{b, a\}$, we must have $a \vee b = b \vee a$ by Theorem 14.1.

2. We first show that $(a \vee b) \vee c$ is an upper bound of $\{a, b, c\}$. Let $d = a \vee b$. By the definition of upper bound, $c \leq d \vee c = (a \vee b) \vee c$. Similarly,

$$a \leq a \vee b \quad \text{and} \quad a \vee b = d \leq d \vee c = (a \vee b) \vee c.$$

Hence, $a \leq (a \vee b) \vee c$ by transitivity. A similar argument shows that $b \leq (a \vee b) \vee c$. Therefore, $(a \vee b) \vee c$ is an upper bound of $\{a, b, c\}$.

Next we show that $(a \vee b) \vee c$ is a least upper bound of $\{a, b, c\}$. If v is any other upper bound of $\{a, b, c\}$, then $a \leq v$ and $b \leq v$, so that $d = a \vee b \leq v$ by the definition of the *least* upper bound $a \vee b$. Since $d \leq v$ and $c \leq v$, we have $d \vee c \leq v$ by the definition of l.u.b. once again. Hence, $(a \vee b) \vee c = d \vee c \leq v$. Therefore, $(a \vee b) \vee c$ is the least upper bound of $\{a, b, c\}$. A similar argument shows that $a \vee (b \vee c)$ is also a least upper bound of $\{a, b, c\}$ (Exercise 13). Hence, $a \vee (b \vee c) = (a \vee b) \vee c$ by Theorem 14.1 (with $B = \{a, b, c\}$).

3. Let $t = a \wedge b$. Then $a \vee t$ is an upper bound of $\{a,t\}$, so $a \leq a \vee t$. Since $a \wedge b$ is lower bound of $\{a,b\}$, we have $t = a \wedge b \leq a$. Since $a \leq a$, we must have $a \vee t \leq a$ by the definition of the *least* upper bound $a \vee t$. By the antisymmetric property, $a = a \vee t = a \vee (a \wedge b)$.

4. The join $a \vee a$ is the least upper bound of the set $\{a,a\} = \{a\}$. But a itself is obviously the least upper bound of $\{a\}$. So $a \vee a = a$ by Theorem 14.1. ◆

Theorem 14.2 says that a lattice is an algebraic system with two binary operations that satisfy certain conditions. Conversely, any such algebraic system is a lattice:

THEOREM 14.3 *Let L be a nonempty set equipped with two binary operations, $\vee$ and $\wedge$, that obey the commutative, associative, absorption, and idempotent laws. Define a relation $\leq$ on L by $a \leq b$ if and only if $a \vee b = b$. Then L is a lattice with respect to $\leq$ such that for all a, b ϵ L:*

$$l.u.b. \ \{a,b\} = a \vee b \qquad and \qquad g.l.b. \ \{a,b\} = a \wedge b.$$

Although we have denoted the operations on L by the same symbols used earlier for l.u.b. and g.l.b., we are not assuming that L comes equipped with a partial order. We could just as well have used other symbols (for instance $+$ and $\cdot$) for the two operations. The point of the theorem is that two operations with the stated *algebraic* properties can be used to construct a partial order under which L is a lattice.

Proof of Theorem 14.3 We first show that $\leq$ is a partial order on L.

Reflexive: $a \vee a = a$ by the idempotent law; hence, $a \leq a$.

Antisymmetric: If $a \leq b$ and $b \leq a$, then $a \vee b = b$ and $b \vee a = a$.
   Hence, by commutativity, $a = b \vee a = a \vee b = b$.

Transitive: If $a \leq b$ and $b \leq c$, then $a \vee b = b$ and $b \vee c = c$. Therefore, by associativity,

$$a \vee c = a \vee (b \vee c) = (a \vee b) \vee c = b \vee c = c.$$

Hence, $a \leq c$.

To complete the proof, we must show that for any a, b ϵ L, $a \vee b$ is the l.u.b. and $a \wedge b$ is the g.l.b. of $\{a,b\}$. By the commutative, associative, and idempotent laws,

$$b \vee (a \vee b) = (a \vee b) \vee b = a \vee (b \vee b) = a \vee b.$$

Hence, $b \leq a \vee b$. Similarly,

$$a \vee (a \vee b) = (a \vee a) \vee b = a \vee b,$$

so that $a \leq a \vee b$. Therefore, $a \vee b$ is an upper bound of $\{a,b\}$. If w is any other upper bound of $\{a,b\}$, then $a \leq w$ and $b \leq w$, that is, $a \vee w = w$ and $b \vee w = w$. Therefore, by associativity

$$(a \vee b) \vee w = a \vee (b \vee w) = a \vee w = w.$$

Hence, $a \vee b \leq w$, so that $a \vee b$ is the least upper bound of $\{a,b\}$. Finally, Exercise 18 shows that

$$a \vee b = b \quad \text{if and only if} \quad a \wedge b = a.$$

Use this fact to show that $a \wedge b$ is the greatest lower bound of $\{a,b\}$ (Exercise 18). ◆

Theorem 14.3 shows that lattices can be defined in purely algebraic terms as systems whose operations satisfy the commutative, associative, absorption, and idempotent laws. This is sometimes more useful than viewing lattices as special types of partially ordered sets.

◆ EXERCISES

A. **1.** Let $S = \{1, 2, 3, 4, 6, 12\}$ be the set of positive integer divisors of 12.

(a) Show that S is a partially ordered set under divisibility (that is, $a \leq b$ means $a \mid b$).

(b) Draw a diagram of the partial order on S, as in Examples 2 and 5.

(c) Verify that S is a lattice in which $a \vee b$ is the least common multiple of a, b and $a \wedge b$ is their greatest common divisor.

2. Do Exercise 1 when S is the set of all positive integers that divide 60.

3. Explain why divisibility ($a \leq b$ means $a \mid b$) is *not* a partial order on the set of all nonzero integers.

4. Show that $\mathbb{Z}$ with the usual meaning of $\leq$ is a lattice. [*Hint:* If $a \leq b$, show that a is the g.l.b. and b the l.u.b. of $\{a,b\}$.]

5. Do Exercise 4 with $\mathbb{Q}$ in place of $\mathbb{Z}$; do it with $\mathbb{R}$ in place of $\mathbb{Z}$.

6. (Groups) Sketch the partial order of the lattice of all subgroups of $\mathbb{Z}_6$ (as in Example 2).

7. (Groups) Do Exercise 6 with S_3 in place of $\mathbb{Z}_6$.

8. If S and T are partially ordered sets, define an order relation on $S \times T$ by

$$(s,t) \leq (u,v) \quad \text{if and only if} \quad s \leq u \text{ in } S \text{ and } t \leq v \text{ in } T.$$

(a) Show that this relation is a partial order on $S \times T$.

(b) If S and T are lattices, show that $S \times T$ is a lattice.

9. Assume that there exists a rational number between any two real numbers and that $\sqrt{2}$ is irrational. In the set $\mathbb{Q}$ with the usual ordering, prove that $B = \{b \in \mathbb{Q} \mid b^2 < 2\}$ has no least upper bound.

10. Prove that the partially ordered set P in Example 4 is a lattice by showing that for $a, b \in P$,

   (a) The greatest common divisor of a, b is the g.l.b. of $\{a,b\}$. [*Hint:* Corollary 1.4.]

   (b) The least common multiple of a, b is the l.u.b. of $\{a,b\}$. [*Hint:* Exercise 31 of Section 1.2.]

11. Let A be a set. Prove that the set $P(X)$ of all subsets of X, partially ordered by $\subseteq$, is a lattice in which $A \cup B$ is the l.u.b. of A and B and $A \cap B$ is their g.l.b.

12. Let L be a lattice and $a, b \in L$. Prove that the following statements are equivalent:

   (i) $a \le b$;　　(ii) $a \vee b = b$;　　(iii) $a \wedge b = a$.

13. Complete the proof of Theorem 14.2 by showing that $a \vee (b \vee c)$ is an l.u.b. of $\{a,b,c\}$.

14. Prove that every finite subset of a lattice L has both an l.u.b. and a g.l.b. [*Hint:* The proof of part 2 of Theorem 14.2 shows that $(a \vee b) \vee c$ is the l.u.b. of $\{a,b,c\}$. Adapt this proof and use induction to show that $(a_1 \vee \cdots \vee a_n) \vee a_{n+1}$ is the l.u.b. of $\{a_1, \ldots, a_{n+1}\}$. Use the Principle of Duality for the g.l.b.]

15. Show by example that Exercise 14 may be false if "finite" is replaced by "infinite."

16. If L is a lattice and $a, b, c \in L$, with $a \le b$, prove that

   (a) $a \vee c \le b \vee c$.

   (b) $a \wedge c \le b \wedge c$.

17. If L is a lattice and $a, b, c \in L$, prove that

   (a) $(a \wedge b) \vee (a \wedge c) \le a \wedge (b \vee c)$.

   (b) $a \vee (b \wedge c) \le (a \vee b) \wedge (a \vee c)$.

18. Let L be a nonempty set equipped with two binary operations, $\vee$ and $\wedge$, that obey the commutative, associative, absorption, and idempotent laws. Complete the proof of Theorem 14.3 by showing that

(a) $a \vee b = b$ if and only if $a \wedge b = a$.

(b) If $a \leq b$ means $a \vee b = b$, then $a \wedge b$ is the g.l.b. of $\{a,b\}$.

19. (Rings) Let S be the set of all ideals in a ring R, partially ordered by $\subseteq$. Prove that S is a lattice in which $I + J$ is the l.u.b. of I and J, and $I \cap J$ is their g.l.b. [See Exercises 15 and 18 in Section 6.1.]

20. (Groups) Let S be the set of all subgroups of a group G, partially ordered by $\subseteq$. Prove that S is a lattice in which $H \cap K$ is the g.l.b. of H and K, and the subgroup generated by the set $H \cup K$ is their l.u.b. [See Theorem 7.17.]

21. (Rings) Describe the lattice of ideals of the ring $\mathbb{Z}$.

22. (Fields) Let K be an extension field of F. Prove that the set of all intermediate fields is a lattice under $\subseteq$. [*Hint:* Show that the g.l.b. of E_1 and E_2 is $E_1 \cap E_2$, and that their l.u.b. is the intersection of all intermediate fields that contain both E_1 and E_2; see Exercise 1 in Section 10.2.]

23. Let S be a partially ordered set with this property:

$$\text{if } a, b \; \epsilon \; S, \text{ then either } a \leq b \text{ or } b \leq a.$$

Then S is said to be a **totally ordered set** or a **chain.**

(a) Verify that the sets $\mathbb{Z}$, $\mathbb{Q}$, $\mathbb{R}$ in Example 1 are totally ordered sets.

(b) Show that the set P in Example 4 is *not* a totally ordered set by exhibiting specific integers a, b such that $a \nmid b$ and $b \nmid a$.

(c) Show that the set S in Example 2 is *not* a totally ordered set.

24. Prove that a totally ordered set [Exercise 23] is a lattice.

25. An element c of a partially ordered set S is said to be **maximal** if there is no element b in S such that $b \neq c$ and $c \leq b$. An element d of S is said to be **minimal** if there is no $b \; \epsilon \; S$ such that $b \neq d$ and $b \leq d$.

(a) Show that the set P in Example 4 has one minimal element and no maximal elements.

(b) Show that the set S in Example 5 has exactly two maximal and two minimal elements.

(c) Give an example of a partially ordered set with exactly one maximal and exactly one minimal element.

(d) Give an example of a partially ordered set with no maximal or minimal elements.

26. If L and M are partially ordered sets, then a function $f : L \rightarrow M$ is **order-preserving** if $a \leq b$ in L implies that $f(a) \leq f(b)$ in M. If L and M are lat-

tices, then a **lattice homomorphism** is a function $f : L \to M$ such that $f(a \vee b) = f(a) \vee f(b)$ and $f(a \wedge b) = f(a) \wedge f(b)$ for all $a, b \in L$.

(a) Prove that every lattice homomorphism is order-preserving.

(b) Show by example that an order-preserving map of lattices need not be a lattice homomorphism.

14.2 BOOLEAN ALGEBRAS

A Boolean algebra is a special kind of lattice. The basic model that motivates the definitions is the lattice $P(X)$ of all subsets of the set X. Recall that

$$A \in P(X) \qquad \text{means} \qquad A \subseteq X,$$

and for $A, B \in P(X)$,

$$A \leq B \qquad \text{means} \qquad A \subseteq B.$$

In this lattice, $A \vee B = A \cup B$ and $A \wedge B = A \cap B$ (Exercise 11 of Section 14.1).

By the definition of $P(X)$, $A \subseteq X$ for every $A \in P(X)$. Thus X itself is the *largest* element in $P(X)$. Since $\emptyset \in P(X)$ and $\emptyset \subseteq A$ for every $A \in P(X)$, the empty set $\emptyset$ is the *smallest* element of $P(X)$. Furthermore, for any $A \in P(X)$,

$$A \cup \emptyset = A \qquad \text{and} \qquad A \cap X = A.$$

For each $A \in P(X)$, let A' denote the relative *complement* of A in X, that is

$$A' = X - A = \{x \mid x \in X \text{ and } x \notin A\}.$$

It follows immediately that

$$A' \vee A = A' \cup A = X \qquad \text{(the largest element in } P(X)),$$

and

$$A' \wedge A = A' \cap A = \emptyset \qquad \text{(the smallest element in } P(X)).$$

Finally, it is easy to verify that for any $A, B, C \in P(X)$,

$$A \cap (B \cup C) = (A \cap B) \cup (A \cap C).$$

If you write $\cap$ as multiplication and $\cup$ as addition, this identity is just the *distributive law* $A(B + C) = AB + AC$. Verify that reversing the roles of union and intersection produces another distributive law in $P(X)$:

$$A \cup (B \cap C) = (A \cup B) \cap (A \cup C).$$

Now we extend these ideas to arbitrary lattices. An element I of a partially ordered set S is a **largest element** if $a \leq I$ for every $a \in S$. An element O of a partially ordered set S is a **smallest element** if $O \leq a$ for every $a \in S$. A partially ordered set may not have a largest element, but if there is one, it is unique (Exercise 1). Similar remarks apply to smallest elements.

EXAMPLE   The largest element in $P(X)$ is X, and the smallest element is $\emptyset$, as explained previously.

EXAMPLE   In the lattice P of all positive integers ordered by divisibility, there is no largest element (that is, no integer that is divisible by every integer). But 1 is a smallest element since $1 \mid a$ for every $a \in P$.

Let L be a lattice with a largest element I and a smallest element O. If $a \in L$, then $O \leq a \leq I$. Hence, by Exercise 12 of Section 14.1,

$$a \vee O = a \quad \text{and} \quad a \wedge I = a.$$

DEFINITION• *A lattice L with a largest element I and a smallest element O is said to be **complemented** if for each $a \in L$, there exists $a' \in L$ such that*

$$a \vee a' = I \quad \text{and} \quad a \wedge a' = O.$$

*The element a' is called the **complement** of a.*

EXAMPLE   The lattice $P(X)$ is a complemented lattice, as explained previously.

EXAMPLE   The set $L = \{1, 2, 3, 4, 6, 12\}$ of all positive integer divisors of 12, ordered by divisibility, is a lattice in which $a \vee b$ = the least common multiple (l.c.m.) of a and b (Exercise 31 of Section 1.2) and $a \wedge b$ = gcd of a, b (Exercise 1 of Section 14.1). Clearly, 12 is the largest and 1 the smallest element of L. But L is *not* a complemented lattice because 6 has no complement (Exercise 4).

DEFINITION• *A lattice L is said to be **distributive** if the following **distributive law** holds:*

$$a \wedge (b \vee c) = (a \wedge b) \vee (a \wedge c) \quad \text{for all } a, b, c \in L.$$

We saw above that the lattice $P(X)$ is distributive. The fact that a second distributive law holds in $P(X)$ (with the roles of $\wedge$ and $\vee$ reversed) is a special case of

THEOREM 14.4   *A lattice L is distributive if and only if*

$$a \lor (b \land c) = (a \lor b) \land (a \lor c) \text{ for all } a, b, c \in L.$$

Proof   Assume L is distributive. Then

$$
\begin{aligned}
a \lor (b \land c) &= [a \lor (a \land c)] \lor (b \land c) && \text{[absorption law]} \\
&= a \lor [(a \land c) \lor (b \land c)] && \text{[associative law]} \\
&= a \lor [(c \land a) \lor (c \land b)] && \text{[commutative law]} \\
&= a \lor [c \land (a \lor b)] && \text{[distributive law]} \\
&= a \lor [(a \lor b) \land c] && \text{[commutative law]} \\
&= [a \land (a \lor b)] \lor [(a \lor b) \land c] && \text{[absorption law]} \\
&= [(a \lor b) \land a] \lor [(a \lor b) \land c] && \text{[commutative law]} \\
&= (a \lor b) \land (a \lor c). && \text{[distributive law applied to } a \lor b, a, \text{ and } c]
\end{aligned}
$$

The Principle of Duality proves the converse.   ◆

> **DEFINITION•** *A **Boolean algebra*** is a lattice with greatest element I and smallest element O that is both complemented and distributive.*

> **EXAMPLE**   The lattice $P(X)$ of subsets of X is a Boolean algebra. The set $\{1, 3, 5, 7, 9, 15, 21, 35, 105\}$ of positive integer divisors of 105, ordered by divisibility, is a Boolean algebra (Exercise 5). But the set $\{1, 2, 3, 4, 6, 12\}$ of positive divisors of 12 ordered by divisibility is *not* a Boolean algebra because it is not complemented (Exercise 4).

In Section 14.1 we saw that lattices may be described algebraically in terms of the binary operations $\lor$ and $\land$, without any reference to an order relation. The same is true of Boolean algebras:

THEOREM 14.5   *A nonempty set B is a Boolean algebra if and only if there are binary operations $\lor$ and $\land$ on B such that*

---

* Named for George Boole (1815–1864). Boolean algebras were the first lattices to be studied. Boole used them to formalize basic logic and the calculus of propositions. See Section 14.3 for more details.

1. For all a, b, c ∈ B, ∨ and ∧ obey the

Commutative Laws:

$$a \vee b = b \vee a \quad and \quad a \wedge b = b \wedge a.$$

Associative Laws:

$$a \vee (b \vee c) = (a \vee b) \vee c;$$
$$a \wedge (b \wedge c) = (a \wedge b) \wedge c.$$

Distributive Laws:

$$a \wedge (b \vee c) = (a \wedge b) \vee (a \wedge c);$$
$$a \vee (b \wedge c) = (a \vee b) \wedge (a \vee c).$$

2. There exists I, O ∈ B such that

$$a \vee O = a \quad and \quad a \wedge I = a \quad for\ every\ a \in B.$$

3. For each a ∈ B, there exists a' ∈ B such that

$$a \vee a' = I \quad and \quad a \wedge a' = O.$$

Proof If B is a Boolean algebra with greatest element I and least element O, then the definitions of distributive and complemented, together with Theorems 14.2 and 14.4, show that 1–3 hold with $a \vee b$ defined as l.u.b. $\{a, b\}$, $a \wedge b$ defined as g.l.b. $\{a, b\}$, and a' as the complement of a.

Conversely, if B is a set with binary operations satisfying 1, then B will be a distributive lattice by Theorem 14.3 provided that B satisfies the idempotent and absorption laws. The first idempotent law holds because

$$
\begin{aligned}
a = a \vee O &= a \vee (a' \wedge a) && \text{[2, 3, and commutative law]} \\
&= (a \vee a') \wedge (a \vee a) && \text{[distributive law]} \\
&= I \wedge (a \vee a) && \text{[3]} \\
&= a \vee a. && \text{[2 applied to } a \vee a\text{]}
\end{aligned}
$$

The other idempotent law is proved similarly. The first absorption law also holds:

$$
\begin{aligned}
a \vee (a \wedge b) &= (a \wedge I) \wedge (a \wedge b) && \text{[2]} \\
&= a \wedge (I \vee b). && \text{[distributive law]}
\end{aligned}
$$

But Exercise 8 shows that $I \vee b = I$. Hence, by 2,

$$a \vee (a \wedge b) = a \wedge (I \vee b) = a \wedge I = a.$$

The other absorption law is proved similarly. Therefore, by Theorem 14.3, B is a distributive lattice in which $a \leq b$ means $a \vee b = b$.

If $a \in B$, then $O \vee a = a$ by 2 and commutativity. Hence, $O \leq a$ for every a. The condition $a \vee b = b$ is equivalent to $a \wedge b = a$ by Exercise 18(a) of Section 14.1. Hence the second half of 2 ($a \wedge I = a$) implies that $a \leq I$ for every $a \in B$. Therefore, I is the greatest and O the smallest element in B. B is a complemented lattice by 3 and, hence, is a Boolean algebra. ◆

Theorem 14.5 is often used as a definition of Boolean algebras. Here are the other important algebraic facts about Boolean algebras.

THEOREM 14.6 *If B is a Boolean algebra and a, b, c $\in$ B, then*

1. $a \vee I = I$ *and* $a \wedge O = O$.
2. *This* **cancellation law** *holds:*

$$\text{If } a \vee b = a \vee c \text{ and } a \wedge b = a \wedge c, \text{ then } b = c.$$

3. *a has a unique complement; that is,*

$$\text{If } a \vee c = I \text{ and } a \wedge c = O, \text{ then } c = a'.$$

4. $(a')' = a$.
5. $O' = I$ *and* $I' = O$.
6. **De Morgan's laws** *hold:*

$$(a \vee b)' = a' \wedge b' \text{ and } (a \wedge b)' = a' \vee b'.$$

Proof 1. Exercise 8.

2. If $a \vee b = a \vee c$ and $a \wedge b = a \wedge c$, then

$b = b \vee (b \wedge a)$	[absorption law]
$= b \vee (a \wedge b)$	[commutative law]
$= b \vee (a \wedge c)$	[$a \wedge b = a \wedge c$]
$= (b \vee a) \wedge (b \vee c)$	[distributive law]
$= (a \vee b) \wedge (b \vee c)$	[commutative law]
$= (a \vee c) \wedge (b \vee c)$	[$a \vee b = a \vee c$]
$= (c \vee a) \wedge (c \vee b)$	[commutative law]
$= c \vee (a \wedge b)$	[distributive law]
$= c \vee (a \wedge c)$	[$a \wedge b = a \wedge c$]
$= c \vee (c \wedge a)$	[commutative law]
$= c$	[absorption law]

3. If $a \vee c = I$ and $a \wedge c = O$, then $a \vee c = a \vee a'$ and $a \wedge c = a \wedge a'$. Hence, $c = a'$ by the cancellation law.

4. We know that $a' \vee a = I$ and $a' \wedge a = O$. Applying 3 (with a' in place of a and a in place of c), we have $a = (a')'$.

5. Exercise 9.

6. According to 3 (with $a \vee b$ in place of a and $a' \wedge b'$ in place of c), we need to show only that

$$(a \vee b) \vee (a' \wedge b') = I \quad \text{and} \quad (a \vee b) \wedge (a' \wedge b') = O$$

to conclude that $a' \wedge b' = (a \vee b)'$. But

$(a \vee b) \vee (a' \wedge b')$

$= a \vee (b \vee (a' \wedge b'))$	[associative law]
$= a \vee ((b \vee a') \wedge (b \vee b'))$	[distributive law]
$= a \vee ((b \vee a') \wedge I)$	[definition of b']
$= a \vee (b \vee a')$	[definition of I]
$= (a \vee b) \vee a'$	[associative law]
$= a' \vee (a \vee b)$	[commutative law]
$= (a' \vee a) \vee b$	[associative law]
$= I \vee b$	[definition of a']
$= I.$	[1]

$(a \vee b) \wedge (a' \wedge b')$

$= (a' \wedge b') \wedge (a \vee b)$	[commutative law]
$= ((a' \wedge b') \wedge a) \vee ((a' \wedge b') \wedge b)$	[distributive law]
$= ((a' \wedge a) \wedge b') \vee (a' \wedge (b' \wedge b))$	[commutative and associative laws]
$= (O \wedge b') \vee (a' \wedge O)$	[definition of a', b']
$= O \vee O = O.$	[1 and definition of O]

The Principle of Duality now proves the second of De Morgan's laws.   ◆

NOTE: *The remainder of this section is not needed for Section 14.3 and may be omitted if desired.*

Two Boolean algebras, B and C, are said to be **isomorphic** if there is a bijective function $f : B \to C$ such that

$$f(a \vee b) = f(a) \vee f(b) \quad \text{and} \quad f(a \wedge b) = f(a) \wedge f(b)$$

for all $a, b \in B$; f is called an **isomorphism.** This definition is consistent with the definition of isomorphism for rings—in each case there is a bijective map that preserves the binary operations of the systems.

Here is one reason why the Boolean algebra $P(X)$ of all subsets of a set X is the primary example of Boolean algebras:

THEOREM 14.7   *If B is a finite Boolean algebra, then there exists a finite set X such that B is isomorphic to P(X).*

The proof of the theorem is outlined in Exercise 31. The example on page 521 in Appendix C shows that if a set X has n elements, then $P(X)$ has 2^n elements. Hence,

COROLLARY 14.8   *A finite Boolean algebra has 2^n elements for some integer n.*

Theorem 14.7 is false for infinite Boolean algebras. But M. H. Stone [44] proved an analogous result in that case: If B is any Boolean algebra, then there exists a set X such that B is isomorphic to the Boolean algebra formed by some collection of subsets of X (possibly not all of them).

The description of Boolean algebras in Theorem 14.5 shows that a Boolean algebra is *almost* a commutative ring with identity. If $\vee$ is taken as addition and $\wedge$ as multiplication, then O is the zero element and I the multiplicative identity. The only ring axiom that fails is the existence of negatives (additive inverses); see Exercise 32. Nevertheless, every Boolean algebra *does* have a commutative ring associated with it:

THEOREM 14.9   *Let B be a Boolean algebra and define binary operations $+$ and $\cdot$ on B by*

$$a + b = (a \wedge b') \vee (a' \wedge b) \qquad and \qquad a \cdot b = a \wedge b.$$

Under these operations B is a commutative ring with identity such that $a \cdot a = a$ for every $a \in B$.

Exercises 11 and 36 in Section 3.1 are special cases of this theorem (because for sets, $A \cap B' = A - B$).

Proof of Theorem 14.9 B is closed under $+$ and $\cdot$ because it is closed under $\vee$ and $\wedge$. Since $\wedge$ and $\vee$ are commutative, so is $+$:

$$a + b = (a \wedge b') \vee (a' \wedge b) = (a' \wedge b) \vee (a \wedge b')$$
$$= (b \wedge a') \vee (b' \wedge a) = b + a.$$

By the definition of $+$ we have

$$(a + b) + c = [((a \wedge b') \vee (a' \wedge b)) \wedge c'] \vee [((a \wedge b') \vee (a' \wedge b))' \wedge c].$$

Exercise 17 shows that

$$((a \wedge b') \vee (a' \wedge b))' = (a \wedge b) \vee (a' \wedge b').$$

This fact and the distributive, associative, and commutative laws for $\vee$ and $\wedge$ show that

$$(a + b) + c = [((a \wedge b') \vee (a' \wedge b)) \wedge c'] \vee [((a \wedge b) \vee (a' \wedge b')) \wedge c]$$
$$= [((a \wedge b') \wedge c') \vee ((a' \wedge b)) \wedge c')] \vee$$
$$[((a \wedge b) \wedge c) \vee ((a' \wedge b') \wedge c)]$$
$$= (a \wedge b' \wedge c') \vee (a' \wedge b \wedge c') \vee (a \wedge b \wedge c) \vee (a' \wedge b' \wedge c).$$

The commutativity of $+$, the preceding fact (applied to b, c, a in place of a, b, c), and the commutativity of $\vee, \wedge$ show that

$$a + (b + c) = (b + c) + a$$
$$= (b \wedge c' \wedge a') \vee (b' \wedge c \wedge a') \vee (b \wedge c \wedge a) \vee (b' \wedge c' \wedge a)$$
$$= (a \wedge b' \wedge c') \vee (a' \wedge b \wedge c') \vee (a \wedge b \wedge c) \vee (a' \wedge b' \wedge c)$$
$$= (a + b) + c.$$

Therefore, addition is associative. The zero element is O because by Theorems 14.5 and 14.6

$$a + O = (a \wedge O') \vee (a' \wedge O)$$
$$= (a \wedge I) \vee (a' \wedge O) = a \vee O = a.$$

Each $a \, \epsilon \, B$ is its own negative because by Theorem 14.5

$$a + a = (a \wedge a') \vee (a' \wedge a) = O \vee O = O.$$

Multiplication $\cdot$ $(= \wedge)$ is commutative and associative with identity element I by Theorem 14.5. By absorption and Theorem 14.5,

$$a \cdot a = a \wedge a = a \wedge (a \vee O) = a.$$

To complete the proof, we need only verify the distributive law:

$$ab + ac = [(a \wedge b) \wedge (a \wedge c)'] \vee [(a \wedge b)' \wedge (a \wedge c)]$$
$$= [(a \wedge b) \wedge (a' \vee c')] \vee [(a' \vee b') \wedge (a \wedge c)]$$
$$= [((a \wedge b) \wedge a') \vee ((a \wedge b) \wedge c')] \vee [(a' \wedge (a \wedge c)) \vee (b' \wedge (a \wedge c))]$$
$$= ((a \wedge a') \wedge b) \vee (a \wedge (b \wedge c')) \vee ((a' \wedge a) \wedge c) \vee (b' \wedge (a \wedge c))$$
$$= (O \wedge b) \vee (a \wedge (b \wedge c')) \vee (O \wedge c) \vee (a \wedge (b' \wedge c))$$
$$= O \vee (a \wedge (b \wedge c')) \vee O \vee (a \wedge (b' \wedge c))$$
$$= (a \wedge (b \wedge c')) \vee (a \wedge (b' \wedge c))$$
$$= a \wedge [(b \wedge c') \vee (b' \wedge c)] = a \cdot (b + c). \quad \blacklozenge$$

A ring with identity in which $aa = a$ for every element a is called a **Boolean ring**. Exercise 25 in Section 3.2 shows that every Boolean ring R is commutative and that $a + a = 0_R$ for every $a \, \epsilon \, R$. Theorem 14.9 says that every Boolean algebra defines a Boolean ring. The converse is also true:

THEOREM 14.10 *Let R be a Boolean ring and define binary operations $\vee$ and $\wedge$ on R by*

$$a \vee b = a + b - ab \quad and \quad a \wedge b = ab.$$

Under these operations R is a Boolean algebra.

Proof   Exercise 33.   ◆

◆ EXERCISES

A.   **1.** Show that a partially ordered set contains *at most* one largest element and one smallest element.

   2. Show that the five-element lattice whose ordering is given by the following diagram is *not* complemented.

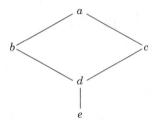

   3. In the five-element lattice whose ordering is given by the following diagram, find an element with two complements.

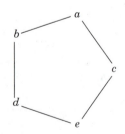

   4. Show that 6 does not have a complement in the lattice $\{1, 2, 3, 4, 6, 12\}$ ordered by divisibility. [*Hint:* A complement c of 6 must satisfy l.c.m. $\{6, c\} = 12$ and g.c.d. $\{6, c\} = 1$.]

   5. Show that lattice of all positive integer divisors of 105, ordered by divisibility, is a complemented lattice.

6. Show that the five-element lattice whose ordering is given by the following diagram is *not* distributive.

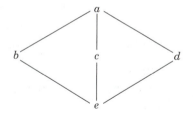

7. Show that the lattice of all positive integer divisors of 15, ordered by divisibility, is a Boolean algebra.

8. If B is a Boolean algebra, prove that $a \vee I = I$ and $a \wedge O = O$ for every $a \in B$.

9. If B is a Boolean algebra, prove that $O' = I$ and $I' = O$.

B. 10. Let p, q be distinct positive primes and let L be the lattice of all positive integer divisors of p^2q, ordered by divisibility. Show that pq has no complement. [Exercise 4 is a special case.]

11. Let p, q, r be distinct positive primes and let L be the lattice of all positive integer divisors of pqr, ordered by divisibility. Prove that L is a complemented lattice. [Exercise 5 is a special case.]

12. A lattice L is said to be **modular** if for all $a, b, c \in L$, $a \le c$ implies that $a \vee (b \wedge c) = (a \vee b) \wedge c$. Prove that every distributive lattice is modular. [The converse is false; see Exercise 35.]

13. Prove that the lattice P of positive integers ordered by divisibility is distributive.

14. Let S be a nonempty set, D a distributive lattice, and L the set of all functions from S to D. Define an order relation on L by $f \le g$ in L if and only if $f(x) \le g(x)$ in D for all $x \in S$. Prove that L is a distributive lattice.

15. If L is a distributive lattice, prove that the cancellation law holds in L (see Theorem 14.6).

16. Prove that in a Boolean algebra, $a \le b$ if and only if $b' \le a'$.

17. Let B be a Boolean algebra. Prove that for all $a, b \in B$,

$$((a \wedge b') \vee (a' \wedge b))' = (a \wedge b) \vee (a' \wedge b').$$

18. In a Boolean algebra, prove that $a = b$ if and only if

$$(a \wedge b') \vee (a' \wedge b) = O.$$

19. Let B be a Boolean algebra and $a, b, c \in B$. If $a \vee b = a \vee c$ and $a' \vee b = a' \vee c$, prove that $b = c$.

20. Let **B** be the set {0, 1}, with ordering given by $0 \leq 1$. For each positive integer n, let $\mathbf{B}^n$ be the Cartesian product $\mathbf{B} \times \cdots \times \mathbf{B}$ of n copies of **B**. Define an operation $\vee$ on $\mathbf{B}^n$ by

$$(a_1, \ldots, a_n) \vee (b_1, \ldots, b_n) = (c_1, \ldots, c_n),$$

where for each i, $c_i =$ maximum of a_i, b_i. Similarly, define

$$(a_1, \ldots, a_n) \wedge (b_1, \ldots, b_n) = (d_1, \ldots, d_n),$$

where for each i, $d_i =$ minimum of a_i, b_i. Finally, define

$$(a_1, \ldots, a_n)' = (a_1', \ldots, a_n'),$$

where $0' = 1$ and $1' = 0$. Prove that $\mathbf{B}^n$ is a Boolean algebra with $O = (0, 0, \ldots, 0)$ and $I = (1, 1, \ldots, 1)$.

NOTE: *The following definitions are used in Exercise 21–31. Let B be a Boolean algebra. Then $a < b$ means $a \leq b$ and $a \neq b$. An element $a \in B$ is an **atom** if $O < a$, and there is no element b such that $O < b < a$.*

21. Prove that a is an atom in a Boolean algebra B if and only if $a \neq O$ and for every $b \in B$, either $a \wedge b = a$ or $a \wedge b = O$.

22. Let B be a Boolean algebra and $a, b, c, d \in B$, with a an atom. Prove

    (a) $d \leq b \wedge c$ if and only if $d \leq b$ and $d \leq c$.

    (b) $a \leq b \vee c$ if and only if $a \leq b$ or $a \leq c$. [*Hint:* If $a \leq b \vee c$, then $a = a \wedge (b \vee c)$. Use distributivity and Exercise 21 to show that $a \wedge b = a$ or $a \wedge c = a$.]

23. In the Boolean algebra $P(X)$ of all subsets of X, show that A is an atom if and only if A is a one-element subset of X.

24. Let B be a finite Boolean algebra and $b \in B$, with $b \neq O$. Prove that there exists an atom $a \in B$ such that $a \leq b$. [*Hint:* If b is an atom, let $a = b$. If not, there exists $a_1 \in B$ such that $O < a_1 < b$ (Why?). If a_1 is not an atom, there exists $a_2 \in B$ such that $O < a_2 < a_1 < b$. Continue in this way; why must the process end with an atom?]

25. If a_1 and a_2 are atoms in a Boolean algebra B and $a_1 \wedge a_2 \neq O$, prove that $a_1 = a_2$. [*Hint:* Exercise 21.]

26. Let B be a Boolean algebra and $b, c \in B$. Prove that the following conditions are equivalent:

    (i) $b \leq c$;   (ii) $b \wedge c' = O$;   (iii) $b' \vee c = I$.

[*Hint:* (i) $\Rightarrow$ (ii) $b \leq c$ implies $b \vee c = c$; use substitution and De Morgan's laws. (ii) $\Rightarrow$ (iii) De Morgan. (iii) $\Rightarrow$ (i) $b = b \wedge I = b \wedge (b' \vee c)$; use distributivity.]

27. Let B be a finite Boolean algebra and $b, c \in B$, with $b \nleq c$. Prove that there exists an atom $a \in B$ such that $a \leq b$ and $a \nleq c$. [*Hint:* Exercises 24 and 26.]

28. Let B be a finite Boolean algebra and $b \in B$. If $a_1, a_2, \ldots, a_m$ are all the atoms in B that are $\leq b$, prove that $b = a_1 \vee a_2 \vee \cdots \vee a_m$. [*Hint:* Let $c = a_1 \vee \cdots \vee a_m$; then $c \leq b$ (Why?). To show that $b \leq c$, assume, on the contrary, that $b \nleq c$ and use Exercise 27 to reach a contradiction.]

29. Let B be a finite Boolean algebra and $b \in B$. If a and $a_1, a_2, \ldots, a_m$ are atoms in B such that $a \leq b$ and $b = a_1 \vee a_2 \vee \cdots \vee a_m$, prove that $a = a_i$ for some i. [*Hint:* $a = a \wedge b$ (Why?) $= a \wedge (a_1 \vee a_2 \vee \cdots \vee a_m)$. Use distributivity and Exercise 25.]

30. Let B be a finite Boolean algebra and $b \in B$. If $b = a_1 \vee a_2 \vee \cdots \vee a_m$ and $b = c_1 \vee c_2 \vee \cdots \vee c_k$, with the a_i distinct atoms and the c_j distinct atoms, prove that $m = k$ and (after reordering and reindexing, if necessary) $a_i = c_i$ for every i. [*Hint:* Exercise 29.]

31. Prove Theorem 14.7: If B is a finite Boolean algebra, then there exists a set X such that B is isomorphic to $P(X)$. [*Hint:* Let X be the set of atoms of B. By Exercises 28 and 30, every element of B can be written uniquely in the form $a_1 \vee a_2 \vee \cdots a_m$, with the a_i distinct elements of X. Define $\theta : B \rightarrow P(X)$ by $\theta(a_1 \vee a_2 \vee \cdots \vee a_m) = \{a_1, a_2, \ldots, a_m\}$. Why is θ surjective? Use Exercise 27 to show that θ is injective and Exercise 22 to show that $\theta(b \vee c) = \theta(b) \vee \theta(c)$ and $\theta(b \wedge c) = \theta(b) \wedge \theta(c)$.]

32. If X is a set with more than two elements, show that the Boolean algebra $P(X)$ is not a ring with the operations of $\cup$ as addition and $\cap$ as multiplication.

33. Prove Theorem 14.10: A Boolean ring R is a Boolean algebra with binary operations $\vee$ and $\wedge$ defined by

$$a \vee b = a + b - ab \qquad \text{and} \qquad a \wedge b = ab.$$

[*Hint:* Let $I = 1_R$, $O = 0_R$, and $a' = 1_R - a$; use Theorem 14.5.]

34. For each Boolean algebra B, Let $S(B)$ denote the ring defined by B, as in Theorem 14.9. For each Boolean ring R, let $T(R)$ denote the Boolean algebra defined by R, as in Theorem 14.10. Show that the concepts of Boolean algebra and Boolean ring are equivalent by proving

(a) For any Boolean algebra B, $T(S(B)) = B$.

(b) For any Boolean ring R, $S(T(R)) = R$.

C. 35. (Groups) Prove that the set of all *normal* subgroups of a group G is a modular lattice that may not be distributive. [See Exercise 12.]

14.3 APPLICATIONS OF BOOLEAN ALGEBRAS

This section consists of two independent subsections that may be read in either order. The first deals with symbolic logic, where Boolean algebras originated and still have useful applications. The second presents one of the most important practical applications of Boolean algebras: modeling, analyzing, and simplifying switching or relay circuits.

Propositional Logic

If p and q are statements (as defined in Appendix A), then the statement "p or q" will be denoted $p \lor q$ and the statement "p and q," by $p \land q$. The negation of the statement p will be denoted p'. The truth or falsity of such compound statements depends on the truth or falsity of the components p and q. The following charts (called **truth tables**) summarize the basic rules of logic that were developed in Appendix A:

p	q	$p \land q$
T	T	T
T	F	F
F	T	F
F	F	F

p	q	$p \lor q$
T	T	T
T	F	T
F	T	T
F	F	F

p	p'
T	F
F	T

For example, the first chart shows that "p and q" is true when both p and q are true, and false otherwise.

Let $p_1, p_2, \ldots, p_n$ be statements. Two compound statements, formed by using the p_i, $\lor$, $\land$, $'$, and parentheses, are said to be **equivalent** if they have the same truth value for every possible combination of truth values for p_1, $p_2, \ldots, p_n$.

EXAMPLE   The statement $p \land (q \lor r)$ is equivalent to $(p \land q) \lor (p \land r)$. To prove this we must determine the truth value of each statement for every possible choice of truth values for p, q, and r. For instance, when p, q are true and r false, then $q \lor r$ is true, and, hence, $p \land (q \lor r)$ is also true. Also when p, q are true and r false, then $p \land q$ is true, $p \land r$ is false, and, hence, $(p \land q) \lor (p \land r)$ is true. So the two statements have the same truth value in this case. This argument is summarized in row 2 of the following truth table, which includes all the possibilities for the truth values of p, q, and r:

p	q	r	$q \vee r$	$p \wedge (q \vee r)$	$p \wedge q$	$p \wedge r$	$(p \wedge q) \vee (p \wedge r)$
T	T	T	T	**T**	T	T	**T**
T	T	F	T	**T**	T	F	**T**
T	F	T	T	**T**	F	T	**T**
T	F	F	F	**F**	F	F	**F**
F	T	T	T	**F**	F	F	**F**
F	T	F	T	**F**	F	F	**F**
F	F	T	T	**F**	F	F	**F**
F	F	F	F	**F**	F	F	**F**

The two statements are equivalent because the two columns in boldface are identical.

If p and q are equivalent statements, we shall treat them as being the same and write $p = q$.* In this terminology, the preceding example states that $p \wedge (q \vee r) = (p \wedge q) \vee (p \wedge r)$. So the distributive law holds for statements. In fact, we have

THEOREM 14.11 *Let p_1, p_2, . . . p_n be statements and let B be the set of all statements that can be formed by using the p_i, $\vee$, $\wedge$, ', and parentheses.** Then B is a Boolean algebra under the operations of $\vee$ and $\wedge$.*

Proof If p and q are statements formed by using the p_i, $\vee$, $\wedge$, ', and parentheses, then the same is true of $p \vee q$ and $p \wedge q$. So $\vee$ and $\wedge$ are actually binary operations on B.

By Theorem 14.5 we must verify that $\vee$ and $\wedge$ satisfy the commutative, associative, and distributive laws. The preceding example shows that the first distributive law holds. The other laws are proved similarly, using appropriate truth tables (Exercise 5).

B contains statements that are always false (Exercise 1), and any two such statements are necessarily equivalent. Let O be a statement in B that is

---

* This is analogous to treating 1 and 4 as the same element of $\mathbb{Z}_3$ and writing $1 = 4$ instead of $[1] = [4]$. Readers who prefer a formal treatment can easily adapt the discussion as follows. The relation of equivalence on a set of statements is an equivalence relation, as defined in Appendix D. When p and q are equivalent statements, their equivalence *classes* are equal: $[p] = [q]$. Writing $p = q$ instead of $[p] = [q]$ is a convenient shorthand. Thus "statements" in the following discussion are actually "equivalence classes of statements."

** In a formal development, the elements of B are equivalence classes of statements formed in this way. The operations on B are defined by $[p] \vee [q] = [p \vee q]$ and $[p] \wedge [q] = [p \wedge q]$. These definitions do not depend on the choice of representatives in each class because by Exercise 11, if p is equivalent to p^* and q is equivalent to q^* (that is, $[p] = [p^*]$ and $[q] = [q^*]$), then $p \vee q$ is equivalent to $p^* \vee q^*$ (that is, $[p \vee q] = [p^* \vee q^*]$); and similarly for $\wedge$.

always false. If $p \in B$, then $p \lor O$ is true exactly when p is true and false exactly when p is false. So $p \lor O$ is equivalent to p; that is, $p \lor O = p$ for every $p \in B$. Furthermore, $p \land p' = O$ since $p \land p'$ is always false (because one component is always false). Similarly, any two statements that are always true are equivalent. Let I be a statement in B that is always true and verify that $p \land I = p$ and $p \lor p' = I$ for every $p \in B$. Therefore, B is a Boolean algebra by Theorem 14.5. ◆

The Boolean algebra B in Theorem 14.11 provides a useful mathematical model of the logic of compound statements discussed in Appendix A. This model also includes conditional statements. To see why, let $p, q \in B$ and compare the truth table for the conditional statement $p \Rightarrow q$ (as determined by the rules of logic in Appendix A) and the table for the statement $p' \lor q$:

p	q	$p \Rightarrow q$		p	q	p'	$p' \lor q$
T	T	T		T	T	F	T
T	F	F		T	F	F	F
F	T	T		F	T	T	T
F	F	T		F	F	T	T

Since the last columns are identical, we see that $p \Rightarrow q$ is equivalent to the statement $p' \lor q$ in B. The Boolean algebra B also contains biconditional statements ["$p \Leftrightarrow q$" means "$p \Rightarrow q$ and $q \Rightarrow p$" and thus is equivalent to $(p' \lor q) \land (q' \lor p)$].

The properties of Boolean algebras can be used to verify other logical equivalents. For example, the conditional statement $p \Rightarrow q$ is equivalent to its contrapositive $q' \Rightarrow p'$ because

$$(q' \Rightarrow p') = (q')' \lor p' = q \lor p' = p' \lor q = (p \Rightarrow q).$$

Similarly, the rules for negating statements involving the connectives "and" and "or" are just De Morgan's laws:

$$(p \land q)' = p' \lor q' \quad \text{and} \quad (p \lor q)' = p' \land q'.$$

Switching Circuits

We shall speak of switches as mechanical devices for controlling the flow of electricity through a circuit. If the switch is **closed,** the current can flow through the circuit, and if the switch is **open,** the current cannot flow, as schematically shown here:

open switch:
current does
not flow

closed switch:
current flows

But the discussion applies to a wide variety of two-state devices, such as magnetic dipoles, electronic transistors, fluid control valves, relay switches, traffic lights, and photocells. Depending on the context, the analogue of current flowing or not might be charged–uncharged, positively magnetized–negatively magnetized, high potential–low potential, traffic moves–traffic stops, etc.

In the schematic diagrams below, switches will usually be denoted by lowercase letters (a, b, c, . . .). Two different switches will be labeled with the same letter if they are **linked** in such a way that they are always both open or both closed. The prime notation will be used to indicate switches that are **oppositely linked,** meaning that, switch a' is open when switch a is closed, and a' is closed when a is open.

Two switches are said to be connected in **series** provided that the current flows when both switches are closed and does not flow when at least one switch is open:

series connection: ——— a ——— b ———

If switches a and b are connected in series, we write $a \cdot b$ (or simply ab). If 1 indicates that current flows (switch closed) and 0 that it doesn't (switch open), then the current flow for a pair of switches connected in series is given by this chart:

a	b	$a \cdot b$
0	0	0
1	0	0
0	1	0
1	1	1

Two switches are said to be connected in **parallel** provided that the current flows when at least one switch is closed and does not flow only when both switches are open:

parallel connection:

If switches a and b are connected in parallel, we write $a + b$. Here is the current-flow chart for a pair of switches connected in parallel:

a	b	$a + b$
0	0	0
1	0	1
0	1	1
1	1	1

EXAMPLE   The circuit given by this diagram is represented algebraically by $a + a'(b + c)$.

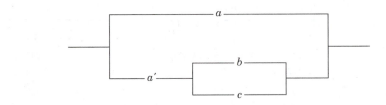

Let $a_1, a_2, \ldots, a_n$ denote switches and let B be the set of all circuits (connecting two terminals) that can be built from the a_i using series and parallel connections, and allowing linked and oppositely linked switches. The preceding example shows one of the many circuits that can be built from switches a, b, c. Two circuits C and D in B are said to be **equivalent** if, for every possible choice of open and closed positions for $a_1, a_2, \ldots, a_n$ current flows through C if and only if it flows through D. In this case we write $C = D$.*

EXAMPLE   These circuits are *not* equivalent because current flows through the left-hand one, but not the right-hand one, when a is closed and b, c are open:

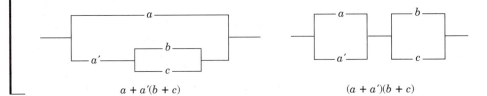

$a + a'(b + c)$ $\qquad\qquad\qquad\qquad$ $(a + a')(b + c)$

EXAMPLE   These two circuits *are* equivalent:

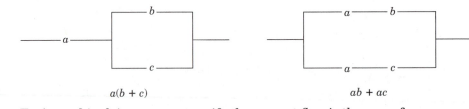

$a(b + c)$ $\qquad\qquad\qquad\qquad$ $ab + ac$

To prove this claim, we must verify the current flow is the same for every possible combination of open/closed for a, b, c:

---

* Strictly speaking, equivalence of circuits is an equivalence relation on B, and $C = D$ is shorthand for the statement that the equivalence class of C is equal to the equivalence class of D. See the footnote on page 443 (with "circuit" in place of "statement," $+$ in place of $\vee$, and $\cdot$ in place of $\wedge$).

a	b	c	$b + c$	$a(b + c)$	ab	ac	$ab + ac$
0	0	0	0	**0**	0	0	**0**
0	1	0	1	**0**	0	0	**0**
0	0	1	1	**0**	0	0	**0**
0	1	1	1	**0**	0	0	**0**
1	0	0	0	**0**	0	0	**0**
1	1	0	1	**1**	1	0	**1**
1	0	1	1	**1**	0	1	**1**
1	1	1	1	**1**	1	1	**1**

The two columns in boldface show that current flows through $a(b + c)$ if and only if it flows through $ab + ac$. Hence, $a(b + c) = ab + ac$, and the distributive law holds for circuits.

THEOREM 14.12   *Let $a_1, a_2, \ldots, a_n$ be switches and B the set of all circuits (connecting two terminals) that can be built from the a_i using series and parallel connections, and allowing linked and oppositely linked switches. Then B is a Boolean algebra under the operations $+$ and $\cdot$.*

Proof   It is easy to verify that $+$ and $\cdot$ are binary operations on B. By using current-flow charts, as in the preceding example, verify that the commutative, associative, and distributive laws hold. There exist circuits in B through which current *always* flows (for instance, $a + a'$ for any $a \in B$), and any two such circuits are equivalent. Let I be a circuit in B through which current always flows and verify that for every $a \in B$

$$a \cdot I = a \quad \text{and} \quad a + a' = I.$$

Similarly, let O be a circuit through which current never flows and verify that

$$a + O = a \quad \text{and} \quad a \cdot a' = O.$$

Therefore, B is a Boolean algebra by Theorem 14.5 (with $+$ as $\vee$ and $\cdot$ as $\wedge$).   ◆

The properties of Boolean algebras can now be used to simplify various circuits.

EXAMPLE   To simplify the circuit

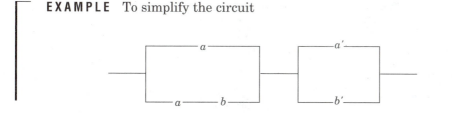

we express it algebraically and use Theorem 14.5 to compute:

$$(a + ab)(a' + b') = a(a' + b') \qquad \text{[absorption law]}$$
$$= aa' + ab'$$
$$= 0 + ab'$$
$$= ab'$$

So the original circuit may be replaced by this simpler equivalent one:

———— a ————b'————

◆ **EXERCISES**

A. **1.** Let p be a statement.

   (a) List a compound statement involving p that is always false. [See the proof of Theorem 14.11.]

   (b) List a compound statement involving p that is always true.

2. Write out the truth tables for each of these statements:

   (a) $(p \vee q) \Rightarrow q'$

   (b) $(p \wedge q') \Rightarrow (q \vee r)$

   (c) $(p \wedge q \wedge r) \Rightarrow ((p \vee q) \wedge r)$

3. A compound statement that is true in every case is called a **tautology.** Use truth tables or the properties of Boolean algebras to verify that each of these statements is a tautology.

   (a) $(p')' \Leftrightarrow p$

   (b) $(p \Rightarrow q)' \Leftrightarrow (p \wedge q')$

   (c) $((p \wedge (p \Rightarrow q)) \Rightarrow q$

   (d) $[p \Rightarrow (q \wedge r)] \Leftrightarrow [(p \Rightarrow q) \wedge (p \Rightarrow r)]$

   (e) $[(p \Rightarrow q) \wedge (q \Rightarrow r)] \Rightarrow (p \Rightarrow r)$

4. If a, b, c are switches, sketch the circuit:

   (a) $ab + c$

   (b) $(a + b)(a + c)$

   (c) $a(b + c') + a'c$

   (d) $ac + ab' + bc(a + b')$

B. **5.** Complete the proof of Theorem 14.11 by proving the commutative, associative, and distributive laws for the Boolean algebra of statements.

6. Sketch a diagram for a circuit (connecting two terminals) with switches a, b, and c that satisfies the given conditions:

(a) Current flows through the circuit if and only if at least one of a, b, c is closed.

(b) Current flows through the circuit if and only if at most one of a, b, c is closed.

(c) Current flows through the circuit if and only if exactly one of a, b, c is closed.

7. In each part, are the two circuits equivalent?

(a)

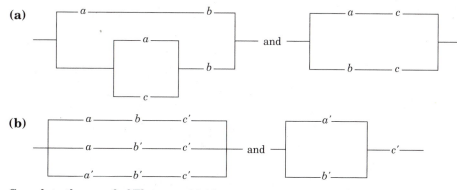

(b)

8. Complete the proof of Theorem 14.12.

9. Use the properties of Boolean algebras to simplify this circuit; then sketch the simplified circuit.

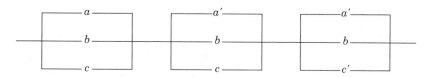

10. Let $p_1, p_2, \ldots, p_n$ be statements. Let p and p^* be compound statements formed by using the p_i, $\vee$, $\wedge$, $'$, and parentheses. Prove that p and p^* are equivalent if and only if the statement $p \Leftrightarrow p^*$ is always true.

11. Let $p_1, p_2, \ldots, p_n$ be statements and p, p^*, q, q^* compound statements formed by using the p_i, $\vee$, $\wedge$, $'$, and parentheses. If p is equivalent to p^* and q is equivalent to q^*, prove that $p \vee q$ is equivalent to $p^* \vee q^*$ and $p \wedge q$ is equivalent to $p^* \wedge q^*$. [*Hint:* Exercise 10.]

CHAPTER **15**

Geometric Constructions

Prerequisites: Sections 4.1, 4.4, and 4.5.

Since the sixth century B.C., mathematicians have studied geometric constructions with straightedge (unmarked ruler) and compass. Despite their prowess in geometry, the ancient Greeks were never able to perform certain constructions using only straightedge and compass, such as

> *Duplication of the Cube:* Construct the edge of a cube having twice the volume of a given cube.*
>
> *Trisection of the Angle:* Construct an angle one third the size of a given angle.
>
> *Squaring the Circle:* Construct a square whose area is equal to the area of a given circle.

Finally in the last century it was proved that each of these constructions is impossible. This chapter presents an elementary proof of the impossibility of the first two constructions listed above (the third is discussed in Exercise 21).

Many people remain fascinated by these problems, particularly angle trisection, and continue to publish what they say are "solutions," even though it has been proved that there are none (see, for example Dudley [45]). Consequently, it is important to understand just what we claim is impossible here and what constitutes a proof.

---

* This problem supposedly had its origin in an ancient legend: Athens was afflicted by a plague and its people were told by the oracle at Delos that the plague would end when they built a new altar to Apollo in the shape of a cube that had twice the volume of the old altar, which was also a cube.

The ancient Greeks knew that all the constructions listed above could readily be carried out provided that additional tools were permitted. For instance, any angle can be trisected using a compass and straightedge with just one mark on it. The Greeks also knew that some angles, such as 90°, *can* be trisected by straightedge and compass alone (Exercise 3). So the issue is not whether these constructions can ever be performed, but whether they can be performed in *every* possible case using *only* an (unmarked) straightedge and a compass. Furthermore, physical measurement alone is not sufficient to justify such constructions because no measuring device is absolutely accurate. Justification requires a valid mathematical proof based on accepted principles and the rules of logic.

The key to the impossibility proofs presented here (and to every other known proof of these facts) is to translate the geometric problem into a equivalent algebraic one. Under this translation process, as we shall see, constructions with a straightedge correspond to solving linear equations and constructions with a compass to solving quadratic equations. Before we can begin this translation process, we present a typical straightedge-and-compass construction to give you a feel for what we are dealing with.

EXAMPLE   Given points O and P, construct a line perpendicular to line OP through O as follows. Construct the circle with center O and radius OP; it intersects line OP at points R and P, as shown on the left side of Figure 1. Segments OR and OP are radii of the circle and thus have the same length. Now construct the circle with center R and radius RP and the circle with center P and radius RP. These circles intersect in points A and B as shown in the center of Figure 1. Segments RP, RA, and PA have the same length. (Why?)

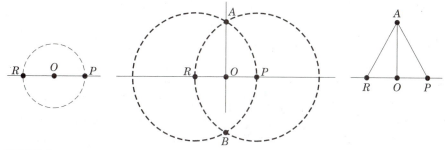

FIGURE 1

Draw the line *AO*. In triangle *RAP,* shown on the right of Figure 1, the sides *RA* and *PA* are congruent, as are the sides *OR* and *OP*. Side *OA* is congruent to itself. Therefore, triangles *ORA* and *OPA* are congruent by side-side-side. Since angles *ROA* and *POA* are congruent and supplementary, each of them must be a right angle. Therefore, line *AO* is perpendicular to line *OP* at *O*.

Outline of the Argument

Now we begin the translation from geometry to algebra. The following outline should help you to see where we're headed and to keep things straight as we go along. The capitalized headings here correspond to the headings on the subsections below.

CONSTRUCTIBLE POINTS   We begin with any two points and determine what additional points can be constructed from them by straightedge-and-compass constructions; these are the *constructible points.* Next we use the distance between the original two points as the unit length and coordinatize the plane.

CONSTRUCTIBLE NUMBERS   A number r is said to be *constructible* if the point $(r, 0)$ is a constructible point. We then examine the equations of lines and circles determined by constructible points and the coordinates of their intersection points. This leads to a characterization of constructible numbers in terms of certain subfields of $\mathbb{R}$ and square roots of positive elements of $\mathbb{R}$.

ROOTS OF POLYNOMIALS   The characterization of constructible numbers is then used to show that certain cubic polynomials have no constructible numbers as roots.

IMPOSSIBILITY PROOFS   Finally, we demonstrate the impossibility of the constructions in question by using proof by contradiction: If the construction were possible, then one of the cubic polynomials mentioned in the preceding paragraph would have a constructible number as a root, which is a contradiction.

Constructible Points

We first give a formal mathematical description of straightedge-and-compass constructions, such as those in the example above, that begin with two points *O* and *P*. Let *S* be the set $\{O, P\}$. Form the line determined by the two points of *S*. Form the two circles with centers *O* and *P* and radius *OP*. Let S_1 be the set of all points of intersection of this line and these circles, together with the points *O*, *P* in the original set *S*. Repeat this process with S_1. Form every line determined by pairs of points in S_1. Form every circle whose radius is the distance between some pair of points in S_1 and whose center is a point in S_1. Let S_2 be the set of all points of intersection of these lines and circles, together with the points in S_1. Repeat the process with S_2. Continuing in this way produces a sequence of sets

$$S \subseteq S_1 \subseteq S_2 \subseteq S_3 \subseteq \cdots$$

A **constructible point** is any point that lies in some S_i. A **constructible line** is a line that contains at least two constructible points. A **constructible circle** is one whose center is a constructible point and whose radius has length equal to the distance between some pair of constructible points. For example, all the labeled points and all the lines and circles in Figure 1 are constructible. Note that points of intersection of constructible lines and circles are constructible points.

Now we coordinatize the plane by taking O as the origin, the distance from O to P as the unit length, and the line OP as the x-axis, and P having coordinates $(1,0)$. Figure 1 shows that the y-axis (the line AO) is a constructible line. The point $(0,1)$ is constructible since it is the intersection of the y-axis and the constructible circle with center O and radius OP. A similar argument shows that

> **$(r,0)$ is constructible if and only if $(0,r)$ is constructible.**

Constructible Numbers

A real number r is said to be a **constructible number** if the point $(r,0)$ is a constructible point. Every integer is a constructible number (Exercise 4). If r is the distance between two constructible points A and B, then r is a constructible number because $(r,0)$ is the intersection of the constructible x-axis and the constructible circle with center O and radius r. Exercise 18 shows that

> **a point is constructible if and only if its coordinates are constructible numbers.**

THEOREM 15.1 *Let a, b, c, d be constructible numbers with $c \neq 0$ and $d > 0$. Then each of $a + b$, $a - b$, ab, a/c and $\sqrt{d}$ is a constructible number.*

Proof We first assume a and c are positive and show that a/c is a constructible number. Since a and c are constructible numbers, the points $(a,0)$ and $(0,c)$ are constructible and so is the line L they determine. The line through the constructible point $(0,1)$ parallel to L is constructible (Exercise 19). It intersects the x-axis at the constructible point $(x,0)$, as shown on the left side of Figure 2. Hence, x is a constructible number. Use similar triangles to show that $\dfrac{1}{c} = \dfrac{x}{a}$, which implies that $x = a/c$. When $a = 0$ or when a or c is negative, Exercise 13 shows that a/c is a constructible.

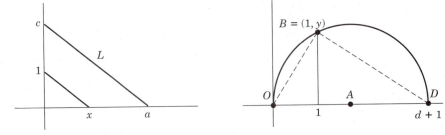

FIGURE 2

If $b = 0$, then $ab = 0$ is certainly constructible. If $b \neq 0$, then $1/b$ is constructible by the previous paragraph, and hence $a/(1/b) = ab$ is also constructible. Exercise 2 shows that $a + b$ and $a - b$ are constructible.

The number $d + 1$ is constructible by Exercise 2. So the midpoint A of the line segment joining the constructible points $(0,0)$ and $(d + 1,0)$ is constructible (Exercise 20). Hence, the circle with center A and radius $(d + 1)/2$ is constructible. The constructible line that is perpendicular to the x-axis at the point $(1,0)$ intersects this circle at the constructible point $B = (1,y)$, as shown on the right of Figure 2. A theorem in plane geometry states that an angle that is inscribed in a semi-circle (such as OBD) is a right angle. Use right triangles and the Pythagorean Theorem to show that $y^2 = d$ and, therefore, $y = \sqrt{d}$. It follows that $y = \sqrt{d}$ is a constructible number.   ◆

COROLLARY 15.2 *Every rational number is constructible.*

Proof   Every integer is constructible (Exercise 4). Therefore, every quotient of a pair of integers (rational number) is constructible by Theorem 15.1.   ◆

In order to determine exactly which real numbers are constructible, we must examine the equations of constructible lines and circles.

LEMMA 15.3 *Let F be a subfield of the field $\mathbb{R}$ of real numbers.*

(1) If a line contains two points whose coordinates are in F, then the line has an equation of the form

$$ax + by + c = 0, \qquad \textit{where } a, b, c \in F.$$

(2) If the center of a circle is a point whose coordinates are in F and the radius of the circle is a number whose square is in F, then the circle has an equation of the form

$$x^2 + y^2 + rx + sy + t = 0, \qquad \textit{where } r, s, t \in F.$$

Proof   (1) Suppose (x_1,y_1) and (x_2,y_2) are points on the line with $x_i, y_i \in F$. If

$x_1 \neq x_2$, the two-point formula for the equation of a line shows that the line has equation

$$y - y_1 = \frac{y_2 - y_1}{x_2 - x_1}(x - x_1)$$

$$\underbrace{\left(\frac{y_2 - y_1}{x_2 - x_1}\right)}_{}x - 1y + \underbrace{[-x_1\left(\frac{y_2 - y_1}{x_2 - x_1}\right) + y_1]}_{} = 0$$

$$ax + by + \qquad\qquad c \qquad\qquad = 0$$

Since F is a field and $x_i, y_i \in F$, each of a, b, c is in F. The case when $x_1 = x_2$ is left to the reader.

(2) If (x_1, y_1) is the center and k the radius, with $x_1, y_1, k^2 \in F$, then the equation of the circle is

$$(x - x_1)^2 + (y - y_1)^2 = k^2$$

$$x^2 + y^2 + (-2x_1)x + (-2y_1)y + [x_1{}^2 + y_1{}^2 + k^2] = 0.$$

The coefficients are in F.   ◆

LEMMA 15.4   *Let F be a subfield of $\mathbb{R}$ and k a positive element of F such that $\sqrt{k} \notin F$. Let $F(\sqrt{k})$ be the set $\{a + b\sqrt{k} \mid a, b \in F\}$. Then*

(1) *$F(\sqrt{k})$ is a subfield of $\mathbb{R}$ that contains F.*

(2) *Every element of $F(\sqrt{k})$ can be written uniquely in the form $a + b\sqrt{k}$, with $a, b \in F$.*

Proof   (1) Exercise 15. (2) If $a + b\sqrt{k} = a_1 + b_1\sqrt{k}$, with $a, b, a_1, b_1 \in F$, then $a - a_1 = (b_1 - b)\sqrt{k}$. If $b - b_1 \neq 0$, then $\sqrt{k} = (a - a_1)(b_1 - b)^{-1}$, which is an element of F. This contradicts the fact that $\sqrt{k} \notin F$. Hence, $b_1 - b_1 = 0$, and, therefore, $a - a_1 = (0)\sqrt{k} = 0$. Thus $a = a_1$ and $b = b_1$.   ◆

The field $F(\sqrt{k})$ is called a **quadratic extension field** of F. Quadratic extension fields play a crucial role in determining which numbers are constructible.

LEMMA 15.5   *Let F be a subfield of $\mathbb{R}$. Let L_1 and L_2 be lines whose equations have coefficients in F. Let C_1 and C_2 be circles whose equations have coefficients in F. Then*

(1) *If L_1 intersects L_2, then the point of intersection has coordinates in F.*

(2) *If C_1 intersects C_2, then the points of intersection have coordinates in F or in some quadratic extension field $F(\sqrt{k})$.*

(3) *If L_1 intersects C_1, then the points of intersection have coordinates in F or in some quadratic extension field $F(\sqrt{k})$.*

Proof (1) Suppose L_1 and L_2 have equations

$$L_1 : a_1 x + b_1 y = c_1$$
$$L_2 : a_2 x + b_2 y = c_2$$

with $a_i, b_i, c_i \in F$. Since L_1 intersects L_2, these equations have a simultaneous solution. By using elimination or determinants, we see that this solution is

$$x = \frac{b_2 c_1 - b_1 c_2}{a_1 b_2 - a_2 b_1} \quad \text{and} \quad y = \frac{a_1 c_2 - a_2 c_1}{a_1 b_2 - a_2 b_1}.$$

Since $a_i, b_i, c_i \in F$, the point of intersection (x,y) has coordinates in the field F.

   (2) Suppose C_1 and C_2 have equations

$$C_1 : x^2 + y^2 + r_1 x + s_1 y + t_1 = 0$$
$$C_2 : x^2 + y^2 + r_2 x + s_2 y + t_2 = 0$$

with $r_i, s_i, t_i \in F$. The coordinates of the intersection points satisfy both equations and, hence, must satisfy the equation obtained by subtracting the second equation from the first:

$$(r_1 - r_2)x + (s_1 - s_2)y + (t_1 - t_2) = 0.$$

This is the equation of a line, and its coefficients are in F. Since the intersection points of C_1 and C_2 lie on this line and on the circle C_1, we need only prove (3) to complete the proof of the theorem.

   (3) Let L_1 and C_1 have the equations given above. At least one of a_1, b_1 must be nonzero, say $b_1 \neq 0$. Solve the equation of L_1 for y and substitute this result in the equation for C_1. Verify that this leads to an equation of the form $ax^2 + bx + c = 0$, with $a, b, c \in F$. The solutions of this equation are

$$x = \frac{-b + \sqrt{b^2 - 4ac}}{2a} = A \pm B\sqrt{k},$$

where $A = -b/2a$, $B = 1/2a$, and $k = b^2 - 4ac$ are elements of F. Since L_1 and C_1 intersect, we know that $k \geq 0$. Using the equation for L_1, we see that the coordinates of the points of intersection of L_1 and C_1 are

$$x = A + B\sqrt{k} \quad \text{and} \quad y = \frac{c_1 - a_1 A}{b_1} - \frac{a_1 B}{b_1}\sqrt{k}$$

$$x = A - B\sqrt{k} \quad \text{and} \quad y = \frac{c_1 - a_1 A}{b_1} - \frac{a_1 B}{b_1}\sqrt{k}.$$

If $k = 0$, these reduce to a single point of intersection. Since $b_1 \neq 0$, all these coordinates lie either in F (if $\sqrt{k} \in F$) or in the quadratic extension $F(\sqrt{k})$ (if $\sqrt{k} \notin F$). ◆

THEOREM 15.6 *If a real number r is constructible, then there is a finite chain of fields $\mathbb{Q} = F_0 \subseteq F_1 \subseteq F_2 \subseteq \cdots \subseteq F_n \subseteq \mathbb{R}$ such that $r \in F_n$ and each F_i is a quadratic extension of the preceding field, that is,*

$$F_1 = \mathbb{Q}(\sqrt{c_0}), \qquad F_2 = F_1(\sqrt{c_1}) \qquad F_3 = F_2(\sqrt{c_2}), \ldots, F_n = F_{n-1}(\sqrt{c_{n-1}}),$$

where $c_i \in F_i$ but $\sqrt{c_i} \notin F_i$ for $i = 0, 1, 2, \ldots, n-1$.

A finite chain of fields as in the theorem is called a **quadratic extension chain.**

Proof of Theorem 15.6   Let r be a constructible number. Then the point $(r,0)$ can be constructed from the points $O = (0,0)$ and $P = (1,0)$ by a finite sequence of operations of the following types:

(i) Form the line determined by A and B, where A,B are previously constructed points or elements of $\{O,P\}$;

(ii) Form the circle with center A and radius the distance from B to C, where A, B, C are previously constructed points or elements of $\{O,P\}$;

(iii) Determine the points of intersection of lines and circles formed in (i) and (ii).

This process begins with the points O and P whose coordinates are in $\mathbb{Q}$. Lines or circles determined by them will have equations with rational coefficients by Lemma 15.3. The intersections of such lines and circles will be points whose coordinates are either in $\mathbb{Q}$ or in some quadratic extension $\mathbb{Q}(\sqrt{c_0})$ by Lemma 15.5. The lines and circles determined by these points will have equations with coefficients in the field $F_1 = \mathbb{Q}(\sqrt{c_0})$ by Lemma 15.3. The intersections of such lines and circles will have coefficients either in F_1 or in some quadratic extension $F_1(\sqrt{c_1})$ by Lemma 15.5. Continuing in this fashion, we see that at each stage of the construction of $(r,0)$ the points in question have coordinates in some field F_i and at the next stage the newly created points have coordinates in F_i or in a quadratic extension $F_i(\sqrt{c_i})$. After a finite number of such steps we reach the point $(r,0)$, which necessarily has coordinates in the last field of the quadratic extension chain $\mathbb{Q} = F_0 \subseteq F_1 \subseteq F_2 \subseteq \cdots \subseteq F_n$. ◆

Roots of Polynomials

There are two ways to show that some real numbers are *not* constructible. The method presented here is elementary and depends only on Chapter 4. But if you've covered Sections 10.1 and 10.2, read the footnote and skip to page 459.*

LEMMA 15.7 *Let F be a subfield of $\mathbb{R}$ and $f(x) \in F[x]$. Suppose that $k \in F$ but $\sqrt{k} \notin F$. If $a + b\sqrt{k}$ is a root of $f(x)$, then $a - b\sqrt{k}$ is also a root of $f(x)$.*

Proof If $u = r + s\sqrt{k} \in F(\sqrt{k})$, let $\overline{u}$ denote $r - s\sqrt{k}$. This operation is well defined because every element of $F(\sqrt{k})$ can be written uniquely in the form $r + s\sqrt{k}$ $(r, s \in F)$ by Lemma 15.4. Verify that for any $u, v \in F(\sqrt{k})$, $\overline{(u + v)} = \overline{u} + \overline{v}$ and $\overline{uv} = \overline{u} \cdot \overline{v}$. Also note that $u = \overline{u}$ if and only if $s = 0$, that is, if and only if $u \in F$. The rest of the proof is identical to the proof of Lemma 4.28, which is the special case when $F = \mathbb{R}$, $k = -1$, and $\sqrt{k} = i$. ◆

LEMMA 15.8 *Let F be a subfield of a field K. Let $f(x), g(x) \in F[x]$ and $h(x) \in K[x]$. If $f(x) = g(x)h(x)$, then $h(x)$ is actually in $F[x]$.*

Proof By the Division Algorithm in $F[x]$, there are polynomials $k(x)$ and $r(x)$ in $F[x]$ such that $f(x) = g(x)k(x) + r(x)$, with $r(x) = 0$ or $\deg r(x) < \deg g(x)$. Since $F \subseteq K$, all these polynomials are in $K[x]$. Now consider the Division Algorithm in $K[x]$, which says that there is a *unique* quotient and remainder. We have $f(x) = g(x)k(x) + r(x)$, and by hypothesis we also have $f(x) = g(x)h(x) + 0$. By uniqueness, we must have $r(x) = 0$ and $h(x) = k(x)$. Since $k(x) \in F[x]$, the lemma is proved. ◆

THEOREM 15.9 *Let $f(x)$ be a cubic polynomial in $\mathbb{Q}[x]$. If $f(x)$ has no roots in $\mathbb{Q}$, then $f(x)$ has no constructible numbers as roots.*

The theorem implies, for example, that $\sqrt[3]{2}$ is *not* a constructible number because it is a root of $x^3 - 2$, which has no rational roots by the Rational Root Test (Theorem 4.20).

Proof of Theorem 15.9 Suppose on the contrary that $f(x)$ has real roots that are constructible. Each such root lies in a quadratic extension chain of $\mathbb{Q}$ by Theorem 15.6. Among all the quadratic extension chains containing a root of $f(x)$, choose one of the smallest possible length, say $\mathbb{Q} = F_0 \subseteq F_1 \subseteq \cdots \subseteq F_n$. This means that $f(x)$ has a root r in F_n and that no quadratic extension chain of

---

* If $k \in F$ and $\sqrt{k} \notin F$, then $x^2 - k \in F[x]$ is the minimal polynomial of $\sqrt{k}$ over F, and, hence, $[F(\sqrt{k}):F] = 2$ by Theorem 10.7. If $\mathbb{Q} \subseteq \cdots \subseteq F_n$ is a quadratic extension chain, then $[F_n:\mathbb{Q}]$ must be a power of 2 by Theorem 10.4. Therefore, the minimal polynomial of a constructible number u has degree 2^k for some k (since this degree is the dimension $[\mathbb{Q}(u):\mathbb{Q}]$, which must divide $[F_n:\mathbb{Q}]$). Consequently, no constructible number can be the root of an irreducible cubic in $\mathbb{Q}[x]$. Since a cubic polynomial in $\mathbb{Q}[x]$ with no rational roots is irreducible by Corollary 4.18, no such polynomial can have a constructible number as a root.

length $n - 1$ or less contains any root of $f(x)$. Note that $F_n \neq \mathbb{Q}$ since $f(x)$ has no rational roots. By the Factor Theorem 4.15 $f(x) = (x - r)t(x)$ for some $t(x) \in F_n[x]$. Now $r \in F_n$, and by the definition of a quadratic extension chain $F_n = F_{n-1}(\sqrt{k})$ for some $b \in F_{n-1}$ with $\sqrt{k} \notin F_{n-1}$. Therefore $r = a + b\sqrt{k}$ with $a, b \in F_{n-1}$. We must have $b \neq 0$; otherwise, r would be in the chain $F_0 \subseteq F_1 \subseteq \cdots \subseteq F_{n-1}$, contradicting the fact that $f(x)$ has no roots in a chain of length $n - 1$. By Lemma 15.7 $\bar{r} = a - b\sqrt{k}$ is also a root of $f(x) = (x - r)t(x)$. Since $\bar{r} \neq r$ (because $b \neq 0$) $\bar{r}$ must be a root of $t(x)$. By the Factor Theorem

$$f(x) = (x - r)(x - \bar{r})h(x) \text{ for some } h(x) \in F_n[x].$$

Let $g(x) = (x - r)(x - \bar{r})$ and observe that the coefficients of $g(x)$ are in F_{n-1}:

$$g(x) = (x - (a + b\sqrt{k}))(x - (a - b\sqrt{k})) = x^2 - 2ax + (a^2 - kb^2).$$

Therefore, $f(x) = g(x)h(x)$ with $f(x), g(x) \in F_{n-1}[x]$. Consequently, $h(x) \in F_{n-1}[x]$ by Lemma 15.8. Now $f(x)$ has degree 3 and $g(x)$ has degree 2, so $h(x)$ must have degree 1 by Theorem 4.2. Since every first degree polynomial over a field has a root in that field, $h(x)$—and, hence, $f(x)$—has a root in F_{n-1}. This contradicts the choice of $F_0 \subseteq F_1 \subseteq \cdots \subseteq F_n$ as a quadratic extension chain of minimal length containing a root of $f(x)$. Therefore, $f(x)$ has no constructible numbers as roots. ◆

Impossibility Proofs

Finally, we are in a position to prove the impossibility of the constructions discussed at the beginning of the chapter. In what follows, it is assumed that whenever a point, line radius, etc., may be chosen arbitrarily, a *constructible* point, line, radius, etc., will be chosen. This guarantees that all points, lines, etc., produced by the construction process will be constructible ones.

DUPLICATION OF THE CUBE   Label the endpoints of one edge of the given cube as O and P and use this edge OP as the unit segment for coordinatizing the plane. Since the given cube has side length 1, its volume is also 1. If there were some way to construct with straightedge and compass the side of a cube of volume 2, then the length c of this side would be a constructible number such that $c^3 = 2$. Thus c would be a root of $x^3 - 2$. But this polynomial has no rational roots by the Rational Root Test and, hence, no constructible ones by Theorem 15.9. This contradiction shows that duplication of the cube by straightedge and compass is impossible.

TRISECTION OF THE ANGLE   It suffices to prove that an angle of $60°$ cannot be trisected by straightedge and compass. Choose two points O, P and coordinatize the plane with O as origin and $P = (1, 0)$. The point $Q = (1/2, \sqrt{3}/2)$ is constructible since its coordinates are constructible numbers by Theorem 15.1 and Corollary 15.2. Furthermore, Q lies on the unit circle $x^2 + y^2 = 1$. Therefore, angle POQ has cosine 1/2 (the first coordinate of Q) and, hence, has measure 60°. If it were possible to trisect this angle with straightedge and compass, there would be a finite sequence of constructions that would result in a constructible point R such that the angle ROP has measure 20°, as shown in Figure 3.

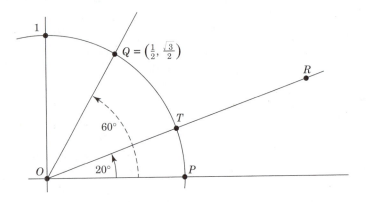

FIGURE 3

The point T where the constructible line OR meets the constructible unit circle is a constructible point. Hence, its first coordinate, which is cos 20°, is a constructible number. Therefore, 2 cos 20° is a constructible number by Theorem 15.1. But for any angle of t degrees, elementary trigonometry (Exercise 5) shows that

$$\cos 3t = 4 \cos^3 t - 3 \cos t.$$

If $t = 20°$, then this identity becomes

$$\cos 60° = 4 \cos^3 20° - 3 \cos 20°$$
$$\tfrac{1}{2} = 4 \cos^3 20° - 3 \cos 20°.$$

Multiplying by 2 and rearranging, we have

$$(2 \cos 20°)^3 - 3(2 \cos 20°) - 1 = 0.$$

Thus the supposedly constructible number 2 cos 20° is a root of $x^3 - 3x - 1$. The Rational Root Test shows that his polynomial has no rational roots and, hence, no constructible ones by Theorem 15.9. This is a contradiction. Therefore, an angle of 60° cannot be trisected by straightedge and compass.

◆ **EXERCISES**

A. **1.** Prove that r is a constructible number if and only if $-r$ is constructible.

   2. Let a, b be constructible numbers. Prove that $a + b$ and $a - b$ are constructible.

3. Use straightedge and compass to construct an angle of

    (a) 30°   **(b)** 45°

    (c) Show that angles of 90° and 45° can be trisected with straightedge and compass.

4. Prove that every integer is a constructible number. [*Hint:* 1 is constructible (Why?); construct a circle with center (1,0) and radius 1 to show 2 is constructible.]

5. Prove that $\cos 3t = 4 \cos^3 t - 3 \cos t$. [*Hint:* These identities may be helpful: (1) $\cos(t_1 + t_2) = \cos t_1 \cos t_2 - \sin t_1 \sin t_2$; (2) $\cos 2t = 2 \cos^2 t - 1$ and $\sin 2t = 2 \sin t \cos t$; (3) $\sin^2 t + \cos^2 t = 1$.]

6. Is it possible to trisect an angle of $3t$ degrees if $\cos 3t = 1/3$? What if $\cos 3t = 11/16$?

B. **7.** Consider a rectangular box with a square bottom of edge x and height y. Assume the volume of the box is 3 cubic units and its surface area is 7 square units. Can the edges of such a box be constructed with straightedge and compass?

8. Use straightedge and compass to construct a line segment of length $1 + \sqrt{3}$, beginning with the unit segment.

9. Is it possible to construct with straightedge and compass an isosceles triangle of perimeter 8 and area 1?

10. **(a)** Prove that the sum of two constructible angles is constructible. [A constructible angle is an angle whose sides are constructible lines.]

    (b) Prove that it is impossible to construct an angle of 1° with straightedge and compass, starting with the unit segment. [*Hint:* If it were possible, what could be said about an angle of 20°?]

11. Prove that an angle of t degrees is constructible if and only if $\cos t$ is a constructible number.

12. Prove that r is a constructible number if and only if a line segment of length $|r|$ can be constructed by straightedge and compass, beginning with a segment of length 1.

13. Let a, c be constructible numbers with $c \neq 0$. Prove that a/c is constructible. [*Hint:* The case when $a > 0, c > 0$ was done in the proof of Theorem 15.1.]

14. Prove that the set of all constructible numbers is a field.

15. Let F be a subfield of $\mathbb{R}$ and $k \in F$. Prove that $F(\sqrt{k}) = \{a + b\sqrt{k} \mid a, b \in F\}$ is a subfield of $\mathbb{C}$ that contains F. If $k > 0$, show that F is a subfield of $\mathbb{R}$. [*Hint:* Adapt the hint for Exercise 31 in Section 3.1.]

16. Prove the converse of Theorem 15.6: If r is in some quadratic extension chain, then r is a constructible number. [*Hint:* Theorem 15.1 and Corollary 15.2.]

17. Let C be a constructible point and L a constructible line. Prove that the line through C perpendicular to L is constructible. [*Hint:* The case when C is on L was done in the example on page 451. If C is not on L and D is a constructible point on L, the circle with center C and radius CD is constructible and meets L at the constructible points D and E. The circles with center D, radius CD and center E, radius CE intersect at constructible points C and Q. Show that line CQ is perpendicular to L.]

18. Prove that (r,s) is a constructible point if and only if r and s are constructible numbers. [*Hint:* The lines through (r,s) perpendicular to the axes are constructible by Exercise 17.]

19. Let A be a constructible point not on the constructible line L. Prove that the line through A parallel to L is constructible. [*Hint:* Use Exercise 17 to find a constructible line M through A, perpendicular to L. Then construct a line through A perpendicular to M.]

20. Prove that the midpoint of the line segment between two constructible points is a constructible point. [*Hint:* Adapt the hint to Exercise 17.]

C. 21. *Squaring the Circle* Given a circle of radius r, show that it is impossible to construct by straightedge and compass the side of a square whose area is the same as that of the given circle. You may assume the nontrivial fact that π is not the root of any polynomial in $\mathbb{Q}[x]$.

CHAPTER **16**

Algebraic Coding Theory

◆

Prerequisites: Section 7.4 and Appendix F for Section 16.1; Section 7.8 for Section 16.2; Section 10.6 for Section 16.3.

Coding theory deals with the fast and accurate transmission of messages over an electronic "channel" (telephone, telegraph, radio, TV, satellite, computer relay, etc.) that is subject to "noise" (atmospheric conditions, interference from nearby electronic devices, equipment failures, etc.). The noise may cause errors so that the message received is not the same as the one that was sent. The aim of coding theory is to enable the receiver to detect such errors and, if possible, to correct them.*

The use of abstract algebra to solve coding problems was pioneered by Richard W. Hamming, whose name appears several times in this chapter. In 1950 he developed a large class of error-correcting codes, some of which are presented here.

16.1 LINEAR CODES

Verbal messages are normally converted to numerical form for electronic transmission. When computers are involved, this is usually done by means of a binary code, in which messages are expressed as strings of 0's and 1's. Such

---

* Thus coding theory has virtually no connection with the secret codes discussed in Chapter 12. The purpose of the latter was to conceal the message, whereas the purpose here is to guarantee its clarity.

messages are easily handled because the internal processing units on most computers represent letters, numerals, and symbols in this way. The discussion here deals only with such binary codes.*

Throughout this chapter we assume that we have a **binary symmetric channel,** meaning that:

1. The probability of a 0 being incorrectly received as a 1 is the same as the probability of a 1 being incorrectly received as a 0;

2. The probability of a transmission error in a single digit is less than .5; and

3. Multiple transmission errors occur independently.**

Here is a simple example that gives a flavor of the subject.

> **EXAMPLE** Suppose that the message to be sent is a single digit, either 1 or 0. The message might be, for example, a signal to tell a satellite whether or not to orbit a distant planet. With a single-digit message, the receiver has no way to tell if an error has occurred. But suppose instead that a four-digit message is sent: 1111 for 1 or 0000 for 0. Then this code can correct single errors. For instance, if 1101 is received, then it seems likely that a single error has been made and that 1111 is the correct message. It's possible, of course, that three errors were made and the correct message is 0000. But this is much less likely than a single error.†† The code can *detect* double errors, but not correct them. For instance, if 1100 is received, then two errors probably have been made, but the intended message isn't clear.

This example illustrates in simplified form the basic components of coding theory. The numerical *message words* (0 and 1) are translated into *codewords* (0000 and 1111). Only codewords are transmitted, but in the example any four-digit string of 0's and 1's is a possible *received word.* By comparing received words with codewords and deciding the most likely error, a *decoder* detects errors and, when possible, corrects them.‡ Finally, the corrected codewords are translated back to message words, or an error is signaled for received words that can't be corrected.

---

* "Binary" refers to the fact that these codes are based on $\mathbb{Z}_2$. Although binary codes are the most common, other codes can be constructed by using any finite field in place of $\mathbb{Z}_2$.

** The accuracy rate of message transmission depends on these probabilities. Since elementary probability is not a prerequisite for this book, our discussion of such questions will be minimal; see Exercises 27–31.

† If the probability of receiving a wrong digit is .01, then three or four errors occur in a message word less than .0004% of the time (once in 250,000 transmissions); see Exercise 27.

‡ This is sometimes called maximum-likelihood decoding.

Any method of assigning each message word to a unique codeword can be used for translating between message and codewords. If we had decided, for example, that 1111 would be the codeword for 0 and 0000 the codeword for 1, the process for detecting and correcting transmission errors would be *exactly the same* as before because it depends only on the codewords, not on what they mean. In terms of transmission and decoding, this is the *same code* as before, even though the codewords have different meanings.

We can develop a usable definition of "code" in the general case by considering the preceding example from a different viewpoint. If we think of the message words 0 and 1 as elements of $\mathbb{Z}_2$, then the received words can be considered as elements of the additive group $\mathbb{Z}_2 \times \mathbb{Z}_2 \times \mathbb{Z}_2 \times \mathbb{Z}_2$ by writing $(1,0,1,1)$, for instance, as 1010 or $(0,0,0,0)$ as 0000. Addition in this group is performed coordinatewise; for example, $1010 + 1011 = 0001$ (remember $1 + 1 = 0$ in $\mathbb{Z}_2$). The set of codewords $C = \{0000, 1111\}$ is closed under addition, so it is a subgroup of $\mathbb{Z}_2 \times \mathbb{Z}_2 \times \mathbb{Z}_2 \times \mathbb{Z}_2$ by Theorem 7.11.

For each positive integer n, $B(n)$ denotes the Cartesian product $\mathbb{Z}_2 \times \mathbb{Z}_2 \times \mathbb{Z}_2 \times \cdots \times \mathbb{Z}_2$ of n copies of $\mathbb{Z}_2$. With coordinatewise addition, $B(n)$ is an additive group of order 2^n (Exercise 10). The elements of $B(n)$ will be written as strings of 0's and 1's of length n.

> **DEFINITION** • *If $0 < k < n$, then an **(n,k) binary linear code** consists of a subgroup C of $B(n)$ of order 2^k.*

For convenience, C is often called an (n,k) code, a linear code, or just a code.* The elements of C are called **codewords.** Only codewords are transmitted, but any element of $B(n)$ can be a **received word.**

In the preceding example, $C = \{0000, 1111\}$ is a $(4,1)$ code since C is a subgroup of order 2^1 of the group $B(4) = \mathbb{Z}_2 \times \mathbb{Z}_2 \times \mathbb{Z}_2 \times \mathbb{Z}_2$ of order 2^4. In this case, the set of message words is just $\mathbb{Z}_2$. Similarly, when dealing with any (n,k) code, we shall consider the group $B(k) = \mathbb{Z}_2 \times \mathbb{Z}_2 \times \cdots \times \mathbb{Z}_2$ (k copies of $\mathbb{Z}_2$), which has order 2^k, to be the set of **message words.**

> **EXAMPLE** In the $(6,5)$ **parity-check code,** the message words are the elements of $B(5)$, that is, all five-digit strings of 0's and 1's. Message words are converted to codewords (elements of $B(6)$) by adding an extra digit at the end of the string; the extra digit is the *sum* (in $\mathbb{Z}_2$) of the digits in the message word. For instance, $110110 \in B(6)$ is the codeword for 11011 because $1 + 1 + 0 + 1 + 1 = 0$ in $\mathbb{Z}_2$. Similarly, $10101 \in B(5)$ is converted to the codeword $101011 \in B(6)$.
>
> An element of $B(6)$ is a codeword if and only if the sum of its digits is 0. [Reason: If the sum of the message-word digits is 0, a 0 is added to make

* Linear codes are also called **block codes** or **group codes.**

the codeword; if the sum of the message-word digits is 1, a 1 is added for the codeword and $1 + 1 = 0$; see Exercise 12 for the converse.] Using this property, it is easy to show that the set C of codewords *is* a subgroup of $B(6)$ (Exercise 13).

This code can detect single transmission errors (1 is received as 0 or 0 as 1) because the sum of the digits in the received word is 1 instead of 0. The same is true for any *odd* number of errors. But it cannot detect an even number of errors, nor can it correct any errors. For each $n \geq 2$, an $(n, n-1)$ parity-check code can be constructed in the same way.

When retransmission of messages is easy, a parity-check code can be very useful. Such codes are frequently used in banking and in the internal arithmetic of computers. But when retransmission is expensive, difficult, or impossible, an error-correcting code is more desirable. We now develop the mathematical tools for determining the number of errors a code can detect or correct.

> **DEFINITION•** *The **Hamming weight** of an element u of B(n) is the number of nonzero coordinates in u; it is denoted Wt(u).*

EXAMPLE If $u = 11011$ in $B(5)$, then $\text{Wt}(u) = 4$. Similarly, $v = 1010010 \in B(7)$ has weight 3, and 0000000 has weight 0.

> **DEFINITION•** *Let u, v ∈ B(n). The **Hamming distance** between u and v, denoted d(u,v), is the number of coordinates in which u and v differ.**

EXAMPLE If $u = 00101$ and $v = 10111$ in $B(5)$, then $d(u,v) = 2$ because u and v differ in the first and fourth coordinates. In $B(4)$ the distance between 0000 and 1111 is 4.

LEMMA 16.1 *If u, v, w ∈ B(n), then*

(1) $d(u,v) = \text{Wt}(u - v)$;

(2) $d(u,v) \leq d(u,w) + d(w,v)$.

Proof (1) A coordinate of $u - v$ is nonzero if and only if u and v differ in that coordinate. So the number of nonzero coordinates in $u - v$, namely $\text{Wt}(u - v)$, is the same as the number of coordinates in which u and v differ, namely $d(u,v)$.

---

* In other words, if $u = u_1 u_2 \cdots u_n$ and $v = v_1 v_2 \cdots v_n$ (with each u_i, v_i either 1 or 0), then $d(u,v)$ is the *number* of indices i such that $u_i \neq v_i$.

(2) It suffices by (1) to prove that $\text{Wt}(u - v) \leq \text{Wt}(u - w) + \text{Wt}(w - v)$. The left side of this inequality is the number of nonzero coordinates of $u - v$, and the right side is the total number of nonzero coordinates in $u - w$ and $w - v$. So we need to verify only that whenever $u - v$ has nonzero ith coordinate, at least one of $u - w$ and $w - v$ also has nonzero ith coordinate. Using the subscript i to denote ith coordinates, suppose the ith coordinate $u_i - v_i$ of $u - v$ is nonzero. If the ith coordinate $u_i - w_i$ of $u - w$ is nonzero, then there is nothing to prove. If $u_i - w_i = 0$, then $u_i = w_i$, and, hence, $w_i - v_i = u_i - v_i \neq 0$. Therefore, the ith coordinate $w_i - v_i$ of $w - v$ is nonzero.   ◆

If a codeword u is transmitted and the word w is received, then the number of errors in the transmission is the number of coordinates in which u and w differ, that is, the Hamming distance from u to w. Since a large number of transmission errors is less likely than a small number (Exercise 27), the nearest codeword to a received word is most likely to be the codeword that was transmitted. Therefore, a *received word is decoded as the codeword that is nearest to it in Hamming distance.* If there is more than one codeword nearest to it, the decoder signals an error.* This process is called **nearest-neighbor decoding.****

> **DEFINITION•** *A linear code is said to **correct t errors** if every codeword that is transmitted with t or fewer errors is correctly decoded by nearest-neighbor decoding.*

THEOREM 16.2   *A linear code corrects t errors if and only if the Hamming distance between any two codewords is at least $2t + 1$.*

Proof   Assume that the distance between any two codewords is at least $2t + 1$. If the codeword u is transmitted with t or fewer errors and received as w, then $d(u,w) \leq t$. If v is any other codeword, then $d(u,v) \geq 2t + 1$ hypothesis. Hence, by Lemma 16.1,

$$2t + 1 \leq d(u,v) \leq d(u,w) + d(w,v) \leq t + d(w,v).$$

Subtracting t from both sides of $2t + 1 \leq t + d(w,v)$ shows that $d(w,v) \geq t + 1$. Since $d(u,w) \leq t$, u is the closest codeword to w, so nearest-neighbor decoding correctly decodes w as u. Hence, the code corrects t errors. The proof of the converse is Exercise 15.   ◆

---

* Alternatively, the decoder can be programmed to choose one of the nearest codewords arbitrarily. This is usually done when retransmission is difficult or impossible.

** Under our assumptions in this chapter, nearest-neighbor decoding coincides with maximum-likelihood decoding.

Since only codewords are transmitted, errors are detected whenever a received word is not a codeword.

> **DEFINITION•** *A linear code is said to **detect t errors** if the received word in any transmission with at least one, but no more than t errors, is not a codeword.*

THEOREM 16.3 *A linear code detects t errors if and only if the Hamming distance between any two codewords is at least t + 1.*

Proof Assume that the distance between any two codewords is at least $t + 1$. If the codeword u is transmitted with at least one, but not more than t errors, and received as w, then

$$0 < d(u,w) \leq t, \qquad \text{and hence} \qquad d(u,w) < t + 1.$$

So w cannot be a codeword. Therefore, the code detects t errors. The proof of the converse is Exercise 16. ◆

If u and v are distinct codewords, then $d(u,v)$ is the weight of the nonzero codeword $u - v$ by Lemma 16.1. Conversely, the weight of any nonzero codeword w is the distance between the distinct codewords w and $\mathbf{0} = 000 \cdots 0 \in B(n)$ because $\text{Wt}(w) = \text{Wt}(w - \mathbf{0}) = d(w,\mathbf{0})$. Therefore, *the minimum Hamming distance between any two codewords is the same as the smallest Hamming weight of all the nonzero codewords.* Combining this fact with Theorems 16.2 and 16.3 yields

COROLLARY 16.4 *A linear code detects 2t errors and corrects t errors if and only if the Hamming weight of every nonzero codeword is at least 2t + 1.*

> **EXAMPLE** Let the message words be 00, 10, 01, 11 $\in B(2)$ and construct a $(10,2)$ code by assigning to each message word the codeword (element of $B(10)$) obtained by repeating the message word five times:
>
> 0000000000,   1010101010,   0101010101,   1111111111.
>
> The set C of codewords is closed under addition and, hence, a subgroup of order 2^2 (Theorem 7.11). So C is a $(10,2)$ code. Every nonzero codeword has Hamming weight at least $5 = 2 \cdot 2 + 1$. By Corollary 16.4 (with $t = 2$), the code C corrects two errors and detects four errors.

By constructing codes that repeat the message words a large number of times (five in the last example), you can always guarantee a high degree of error detection and correction. The disadvantage to such repetition codes is their inefficiency when long messages must be sent. It is time-consuming and expensive to transmit a large number of digits for each message word. So the goal is to construct codes that achieve an acceptable accuracy rate without unnecessarily reducing the transmission rate.

One efficient technique for constructing linear codes is based on matrix multiplication. Codes constructed in this way are automatically equipped with an encoding algorithm that assigns each message word to a unique codeword.

EXAMPLE We shall construct a (7,4) code. The message words will be the elements of $B(4)$, and the codewords elements of $B(7)$. Message words are considered as row vectors and converted to codewords by right multiplying by the following matrix, whose entries are in $\mathbb{Z}_2$:

$$G = \begin{pmatrix} 1 & 0 & 0 & 0 & 0 & 1 & 1 \\ 0 & 1 & 0 & 0 & 1 & 0 & 1 \\ 0 & 0 & 1 & 0 & 1 & 1 & 0 \\ 0 & 0 & 0 & 1 & 1 & 1 & 1 \end{pmatrix}.$$

For instance, the message word 1101 is converted to the codeword 1101001 because

$$(1 \quad 1 \quad 0 \quad 1) \begin{pmatrix} 1 & 0 & 0 & 0 & 0 & 1 & 1 \\ 0 & 1 & 0 & 0 & 1 & 0 & 1 \\ 0 & 0 & 1 & 0 & 1 & 1 & 0 \\ 0 & 0 & 0 & 1 & 1 & 1 & 1 \end{pmatrix} = (1 \quad 1 \quad 0 \quad 1 \quad 0 \quad 0 \quad 1).$$

The complete set C of codewords may be found similarly:

Message Word	Codeword	Message Word	Codeword
0000	0000000	1000	1000011
0001	0001111	1001	1001100
0010	0010110	1010	1010101
0011	0011001	1011	1011010
0100	0100101	1100	1100110
0101	0101010	1101	1101001
0110	0110011	1110	1110000
0111	0111100	1111	1111111

Theorem 16.6 shows that C is actually a subgroup of $B(7)$. So C is a (7,4) code, called the **(7,4) Hamming code.** The preceding table shows that every nonzero codeword has Hamming weight at least $3 = 2 \cdot 1 + 1$. Hence, by Corollary 16.4 (with $t = 1$) this code corrects single errors and detects double errors.

The preceding table shows that codewords in the Hamming (7,4) code have a special form: The first four digits of each codeword form the corresponding message word. For instance, *1101*001 is the codeword for 1101.* An (n,k) code in which the first k digits of each codeword form the corresponding message word is called a **systematic code.** All the examples above are systematic codes. Systematic codes are convenient because codewords are easily translated back to message words: Just take the first k digits.

We can construct other systematic codes by following a procedure similar to that in the last example. A $k \times n$ **standard generator matrix** is a $k \times n$ matrix G with entries in $\mathbb{Z}_2$ of the form

$$\begin{pmatrix} 1 & 0 & 0 & \cdots & 0 & 0 & a_{11} & \cdots & a_{1\,n-k} \\ 0 & 1 & 0 & \cdots & 0 & 0 & a_{21} & \cdots & a_{2\,n-k} \\ \vdots & \vdots & \vdots & & \vdots & \vdots & \vdots & & \vdots \\ 0 & 0 & 0 & \cdots & 1 & 0 & a_{(k-1)1} & \cdots & a_{k-1\,n-k} \\ 0 & 0 & 0 & \cdots & 0 & 1 & a_{k1} & \cdots & a_{k\,n-k} \end{pmatrix} = (I_k \mid A),$$

where I_k is the $k \times k$ identity matrix and A is a $k \times (n-k)$ matrix. For instance, the matrix G in the preceding example is a 4×7 standard generator matrix. It has the form $(I_4 \mid A)$, where A is a 4×3 matrix.

A standard generator matrix can be used as an encoding algorithm to convert elements of $B(k)$ into codewords (elements of $B(n)$) by right multiplication. Each $u \in B(k)$ is considered as a row vector of length k. The matrix product uG is then a row vector of length n, that is, an element of $B(n)$. Because the first k columns of G form the identity matrix I_k, *the first k coordinates of the codeword uG form the corresponding message word $u \in B(k)$* (Exercise 23). In order to justify calling uG a "codeword," we must show that the set of all such elements is a subgroup of $B(n)$.

LEMMA 16.5 *If $f : B(k) \to B(n)$ is an injective homomorphism of groups, then the image of f is an (n,k) code.*

Proof Im f is a subgroup of $B(n)$ that is isomorphic to $B(k)$ by Theorem 7.19. Therefore, Im f has order 2^k and, hence, is an (n,k) code. ◆

THEOREM 16.6 *If G is a $k \times n$ standard generator matrix, then $\{uG \mid u \in B(k)\}$ is a systematic (n,k) code.*

Proof Define a function $f : B(k) \to B(n)$ by $f(u) = uG$. The image of f is $\{f(u) \mid u \in B(k)\} = \{uG \mid u \in B(k)\}$. By Lemma 16.5 and the italicized remarks

---

* The last three digits of each codeword are **check digits** that can be used to determine if a received word is a codeword; see Exercise 22.

preceding it, we need to show only that f is an injective homomorphism of groups. Since matrix multiplication is distributive,

$$f(u + v) = (u + v)G = uG + vG = f(u) + f(v).$$

Hence, f is a homomorphism of groups.

If $u = u_1 u_2 \cdots u_k \in B(k)$, then the first k coordinates of uG are $u_1 u_2 \cdots u_k$ because G is a standard generator matrix, and similarly for $v = v_1 v_2 \cdots v_k \in B(k)$. We use this fact to show that f is injective. If $f(u) = f(v)$, then in $B(n)$:

$$u_1 u_2 \cdots u_k \;*\!*\!*\!*\!* = uG = f(u) = f(v) = vG = v_1 v_2 \cdots v_k \;*\!*\!*\!*\!*,$$

where the $*$'s indicate the remaining coordinates of uG and vG. Since these elements of $B(n)$ are equal, they must be equal in every coordinate. In particular, $u_1 = v_1$, $u_2 = v_2$, . . . , $u_k = v_k$. Therefore, $u = v$ in $B(k)$, and f is injective. ◆

EXAMPLE By Theorem 16.6, the standard generator matrix

$$G = \begin{pmatrix} 1 & 0 & 0 & 0 & 1 & 1 \\ 0 & 1 & 0 & 1 & 0 & 1 \\ 0 & 0 & 1 & 1 & 1 & 0 \end{pmatrix}$$

generates the (6,3) code $\{uG \mid u \in B(3)\}$. Verify that the encoding algorithm $u \to uG$ produces these codewords:

Message Word	Codeword	Message Word	Codeword
000	000000	100	100011
001	001110	101	101101
010	010101	110	110110
011	011011	111	111000

Since the Hamming weight of every nonzero codeword is at least 3, this code corrects single errors and detects double errors by Corollary 16.4 (with $t = 1$).

Describing a large code by means by a standard generator matrix is much more efficient than listing all the codewords. For instance, in a (50,30) code there are only 1500 entries in the 30×50 generator matrix but more than a billion codewords.

Linear algebra can be used to show that every systematic linear code is given by a standard generator matrix. The standard generator matrices for the codes in the examples above are in Exercises 7–9.

◆ EXERCISES

A.   **1.** Show that $C = \{0000, 0101, 1010, 1111\}$ is a (4,2) code.

2. Find the Hamming weight of

   (a) $0110110 \in B(7)$ **(b)** $11110011 \in B(8)$

   (c) $000001 \in B(6)$ **(d)** $101101101101 \in B(12)$

3. Find the Hamming distance between

   (a) 0010101 and 1010101

   (b) 110010101 and 100110010

   (c) 111111 and 000011

   (d) 00001000 and 10001000

4. Use nearest-neighbor decoding in the Hamming (7,4) code to detect errors and, if possible, decode these received words:

   (a) 0111000   **(b)** 1101001

   (c) 1011100   **(d)** 0010010

5. List all codewords generated by the standard generator matrix:

   (a) $\begin{pmatrix} 1 & 0 & 0 & 0 \\ 0 & 1 & 1 & 1 \end{pmatrix}$ **(b)** $\begin{pmatrix} 1 & 0 & 1 & 1 & 1 \\ 0 & 1 & 0 & 1 & 0 \end{pmatrix}$

   (c) $\begin{pmatrix} 1 & 0 & 0 & 1 \\ 0 & 1 & 0 & 1 \\ 0 & 0 & 1 & 0 \end{pmatrix}$ **(d)** $\begin{pmatrix} 1 & 0 & 0 & 1 & 1 & 1 \\ 0 & 1 & 0 & 1 & 0 & 1 \\ 0 & 0 & 1 & 1 & 1 & 0 \end{pmatrix}$

6. Determine the number of errors that each of the codes in Exericse 5 will detect and the number of errors each will correct.

7. Show that the standard generator matrix

$$G = \begin{pmatrix} 1 & 0 & 0 & 0 & 0 & 1 \\ 0 & 1 & 0 & 0 & 0 & 1 \\ 0 & 0 & 1 & 0 & 0 & 1 \\ 0 & 0 & 0 & 1 & 0 & 1 \\ 0 & 0 & 0 & 0 & 1 & 1 \end{pmatrix}$$

generates the (6,5) parity-check code on page 465. [*Hint:* List all the codewords generated by G; then list all the codewords in the parity-check code; compare the two lists.]

8. Show that the standard generator matrix

$$G = \begin{pmatrix} 1 & 0 & 1 & 0 & 1 & 0 & 1 & 0 & 1 & 0 \\ 0 & 1 & 0 & 1 & 0 & 1 & 0 & 1 & 0 & 1 \end{pmatrix}$$

generates the (10,2) repetition code on page 468. [*Hint:* See the hint for Exercise 7.]

9. Show that 1×4 standard generator matrix $(1 \quad 1 \quad 1 \quad 1)$ generates the code in the example on page 464.

10. Prove that $B(n) = \mathbb{Z}_2 \times \mathbb{Z}_2 \times \mathbb{Z}_2 \times \cdots \times \mathbb{Z}_2$ (n factors) with coordinatewise addition is an abelian group of order 2^n.

B. 11. Prove that for any $u, v, w \in B(n)$,

    (a) $d(u,v) = d(v,u)$.

    (b) $d(u,v) = 0$ if and only if $u = v$.

    (c) $d(u,v) = d(u + w, v + w)$.

12. Prove that an element of $B(6)$ is a codeword in the (6,5) parity-check code if the sum of its digits is 0. [*Hint:* Compare the sum of the first five digits with the sixth digit.]

13. Prove that the set of all codewords in the (6,5) parity-check code is a subgroup of $B(6)$. [*Hint:* Use Exercise 12.]

14. If u and v are distinct codewords of a code that corrects t errors, explain why $d(u,v) \geq t$.

15. Complete the proof of Theorem 16.2 by showing that if a code corrects t errors, then the Hamming distance between any two codewords is at least $2t + 1$. [*Hint:* If u, v are codewords with $d(u,v) \leq 2t$, obtain a contradiction by constructing a word w that differs from u in exactly t coordinates and from v in t or fewer coordinates; see Exercise 14.]

16. Complete the proof of Theorem 16.3 by showing that if a code detects t errors, then the Hamming distance between any two codewords is at least $t + 1$.

17. Construct a (5,2) code that corrects single errors.

18. Show that no (6,3) code corrects double errors.

19. Construct a (7,3) code in which every nonzero codeword has Hamming weight at least 4.

20. Is there a (6,2) code in which every nonzero codeword has Hamming weight at least 4?

21. Suppose only three messages are needed (for instance, "go," "slow down," "stop"). Find the smallest possible n so that these messages may be transmitted in an (n,k) code that corrects single errors.

22. Let G be the standard generator matrix for the (7,4) Hamming code on page 469.

(a) If $u = (u_1, u_2, u_3, u_4)$ is a message word, show that the corresponding codeword uG is

$$(u_1, u_2, u_3, u_4, u_2 + u_3 + u_4, u_1 + u_3 + u_4, u_1 + u_2 + u_4).$$

(b) If $v = (v_1, v_2, v_3, v_4, v_5, v_6, v_7) \in B(7)$, show that v is a codeword if and only if its last three coordinates (the check digits) satisfy these equations:

$$v_5 = v_2 + v_3 + v_4$$
$$v_6 = v_1 + v_3 + v_4$$
$$v_7 = v_1 + v_2 + v_4$$

23. If G is a $k \times n$ standard generating matrix and $u = u_1 u_2 u_3 \cdots u_k$ is a message word, show that the first k digits of the codeword uG are $u_1, u_2, \ldots, u_k$.

24. If C is a linear code, prove that either every codeword has even Hamming weight or exactly half of the codewords have even Hamming weight.

25. Prove that the elements of even Hamming weight in $B(n)$ form an $(n, n - 1)$ code.

26. If $k < n$ and $f : B(k) \to B(n)$ is a homomorphism of groups, is Im f a linear code? Is Im f an (n, k) linear code?

NOTE: *A knowledge of elementary probability and a calculator are needed for Exercises 27–31.*

27. Assume that the probability of transmitting a single digit incorrectly is .01 and that a four-digit codeword is transmitted. Construct a suitable probability tree and compute the probability that the codeword is transmitted with

(a) no errors;   **(b)** one error;

(c) two errors;   **(d)** three errors;

(e) four errors;   **(f)** at least three errors.

28. Do Exercise 27 for a five-digit codeword.

29. Suppose the probability of transmitting a single digit incorrectly is greater than .5. Explain why "inverse decoding" (decoding 1 as 0 and 0 as 1) should be employed.

30. Assume that the probability of transmitting a single digit incorrectly is .01 and that M is a 500-digit message.

(a) What is the probability that M will be transmitted with no errors?

(b) Suppose each digit is transmitted three times (111 for each 1, 000 for each 0) and that each received digit is decoded by "majority rule" (111, 110, 101, 011 are decoded as 1 and 000, 001, 010, 100 as 0). What is the probability that the message received when M is transmitted will be correctly decoded? [*Hint:* Find the probability that a single digit will be correctly decoded after transmission.]

31. (a) Show that the number of ways that k errors can occur in an n-digit message is $\binom{n}{k}$, where $\binom{n}{k}$ is the binomial coefficient.

(b) If p is the probability that a single digit is transmitted incorrectly and q is the probability that it is transmitted correctly, show that the probability that k errors occur in an n-digit message is $\binom{n}{k} p^k q^{n-k}$.

16.2 DECODING TECHNIQUES

Nearest-neighbor decoding for an (n,k) code was implemented in Section 16.1 by comparing each received word with all 2^k codewords in order to decode it. But when k is very large, this brute-force technique may be impractical or impossible. So we now develop decoding techniques that are sometimes more efficient. One of them is based on groups and cosets.

> **EXAMPLE** Let C be the (5,2) code {00000, 10110, 01101, 11011}. From the elements of $B(5)$ *not* in C, choose one of smallest weight (which in this case is weight 1), say $e_1 = 10000$. Form its coset $e_1 + C$ by adding e_1 successively to the elements of C and list the coset elements, with $e_1 + c$ directly below c for each $c \in C$:
>
> | C: | 00000 | 10110 | 01101 | 11011 |
> |------|-------|-------|-------|-------|
> | $e_1 + C$: | 10000 | 00110 | 11101 | 01011 |
>
> Thus, for example, 11101 is directly below $01101 \in C$ because $e_1 + 01101 = 10000 + 01101 = 11101$. Among the elements not listed above, choose one of smallest weight, say $e_2 = 01000$, and list its coset in the same way (with $e_2 + c$ below $c \in C$):
>
> | C: | 00000 | 10110 | 01101 | 11011 |
> |------|-------|-------|-------|-------|
> | $e_1 + C$: | 10000 | 00110 | 11101 | 01011 |
> | $e_2 + C$: | 01000 | 11110 | 00101 | 10011 |

Among the elements not yet listed, choose one of smallest weight and list its coset, and continue in this way until every element of $B(5)$ is on the table. Verify that this is a complete table:

00000	10110	01101	11011	*Codewords*
10000	00110	11101	01011	
01000	11110	00101	10011	
00100	10010	01001	11111	*Received Words*
00010	10100	01111	11001	
00001	10111	01100	11010	
11000	01110	10101	00011	
10001	00111	11100	01010	

The decoding rule (which will be justified below) is: *Decode a received word w as the codeword at the top of the column in which w appears.* For instance, 01001 (fourth row) is decoded as 01101; and 01010 (last row) is decoded as 11011. Similarly, 11000 (seventh row) is decoded as 00000.

The decoding table in the example is called a **standard array,** and the decoding rule **standard-array decoding** or **coset decoding.** The same procedure can be used to construct a standard array for any code C. Its rows are the cosets of C, with C itself as the first row. Each is of the form $e + C$, where e is the **coset leader** (an element of smallest weight in the coset and listed first in the row). The element $e + c$ (with $c \in C$) is listed in the column below c and is decoded as c.

THEOREM 16.7 *Let C be an (n,k) code. Standard-array decoding for C is nearest-neighbor decoding.*

Proof If $w \in B(n)$, then $w = e + v \in e + C$, where e is a coset leader and v is the codeword at the top of the column containing w. Standard-array decoding decodes w as v. We must show that v is a nearest codeword to w. If $u \in C$ is any other codeword, then $w - u$ is an element of $w + C$. But $w + C$ is the coset of e (because $e = w - v \in w + C$). By construction, the coset leader e has smallest weight in its coset, so $\mathrm{Wt}(w - u) \geq \mathrm{Wt}(e)$. Therefore, by Lemma 16.1,

$$d(w,u) = \mathrm{Wt}(w - u) \geq \mathrm{Wt}(e) = \mathrm{Wt}(w - v) = d(w,v).$$

Thus v is a nearest codeword to w. ◆

When nearest-neighbor decoding is implemented by a standard array, a codeword is automatically chosen whenever there is more than one codeword that is nearest to a received word w (rather than an error being signaled). So incorrect decoding may occur in such cases. The code in the last example corrects single errors (every codeword has weight at least 3; see Corollary 16.4).

Since two or more errors are much less likely than a single one, standard-array decoding for this code has a high rate of accuracy (Exercise 18).

Once a standard array has been constructed, it's much more efficient for decoding than brute-force comparison with all codewords. Unfortunately, constructing a standard array for a large code may require as much computer time and memory as brute force. But when a code is given by a generator matrix, a much shorter decoding array is possible, as we now see.

Consider an (n,k) code with $k \times n$ standard generator matrix $G = (I_k \,|\, A)$. The **parity-check matrix** of the code is the $n \times (n - k)$ matrix $H = \left(\dfrac{A}{I_{n-k}} \right).$*

EXAMPLE   Verify that the standard generator matrix for the (5,2) code {00000, 10110, 01101, 11011} of the previous example is

$$G = \begin{pmatrix} 1 & 0 & 1 & 1 & 0 \\ 0 & 1 & 1 & 0 & 1 \end{pmatrix} = (I_2 \,|\, A).$$

Here $k = 2, n = 5, n - k = 3$, and A is 2×3. So the parity-check matrix is the 5×3 matrix

$$H = \begin{pmatrix} 1 & 1 & 0 \\ 1 & 0 & 1 \\ 1 & 0 & 0 \\ 0 & 1 & 0 \\ 0 & 0 & 1 \end{pmatrix} = \left(\dfrac{A}{I_3} \right)$$

Verify that the product matrix GH is the 2×3 zero matrix. The phenomenon occurs in the general case as well:

LEMMA 16.8   *If $G = (I_k \,|\, A)$ is the standard generator matrix for a linear code and $H = \left(\dfrac{A}{I_{n-k}} \right)$ is its parity-check matrix, then GH is the zero matrix.*

Proof   The entry in row i and column j of GH is the product of the ith row of G (see page 470) and the jth column of H:**

---

* Since the generator matrix can always be obtained from the parity-check matrix, many books on coding theory define a code in terms of its parity-check matrix rather than its generator matrix. In most books, the parity-check matrix is defined to be the transpose of our matrix H, that is, the $(k - n) \times n$ matrix whose ith row is the same as the ith column of H. The matrix H is more convenient here, and, in any case, all the results are easily translated from one notation to the other.

** The Kronecker delta symbol δ_{rs} is explained in Appendix F.

$$(\delta_{i1}\delta_{i2}\cdots\delta_{ij}\cdots\delta_{ik}a_{i1}a_{i2}\cdots a_{ij}\cdots a_{i(n-k)})\begin{pmatrix} a_{1j} \\ a_{2j} \\ \vdots \\ a_{ij} \\ \vdots \\ a_{kj} \\ \delta_{1j} \\ \delta_{2j} \\ \vdots \\ \delta_{ij} \\ \vdots \\ \delta_{(n-k)j} \end{pmatrix}$$

$$= \delta_{i1}a_{1j} + \delta_{i2}a_{2j} + \cdots + \delta_{ii}a_{ij} + \cdots + \delta_{ik}a_{kj}$$
$$+ a_{i1}\delta_{1j} + a_{i2}\delta_{2j} + \cdots + a_{ij}\delta_{jj} + \cdots + a_{i(n-k)}\delta_{(n-k)j}.$$

Since $\delta_{rs} = 0$ whenever $r \neq s$ and since addition is in $\mathbb{Z}_2$, this sum reduces to

$$\delta_{ii}a_{ij} + a_{ij}\delta_{jj} = 1a_{ij} + a_{ij}1 = a_{ij} + a_{ij} = 0. \quad \blacklozenge$$

In an (n,k) code with $k \times n$ standard generator matrix G, every received word $w \in B(n)$ is a row vector of length n. Since the parity-check matrix H is $n \times (n - k)$, the product wH is a row vector of length $n - k$, that is, an element of $B(n - k)$. Let **0** denote $000 \cdots 0 \in B(n - k)$.

EXAMPLE Let H be the 5×3 parity-check matrix for the $(5,2)$ code in the previous example. Then $11000H = 011$ and $10110H = \mathbf{0}$:

$$(1\ \ 1\ \ 0\ \ 0\ \ 0)\begin{pmatrix} 1 & 1 & 0 \\ 1 & 0 & 1 \\ 1 & 0 & 0 \\ 0 & 1 & 0 \\ 0 & 0 & 1 \end{pmatrix} = (0\ \ 1\ \ 1) \quad \text{and}$$

$$(1\ \ 0\ \ 1\ \ 1\ \ 0)\begin{pmatrix} 1 & 1 & 0 \\ 1 & 0 & 1 \\ 1 & 0 & 0 \\ 0 & 1 & 0 \\ 0 & 0 & 1 \end{pmatrix} = (0\ \ 0\ \ 0).$$

The fact that 10110 is a codeword in this code and $10110H = 0$ is an example of

THEOREM 16.9 *Let C be an (n,k) code with standard generator matrix G and parity-check matrix H. Then an element w in $B(n)$ is a codeword if and only if $wH = 0$.*

Proof   Define a function $f : B(n) \to B(n-k)$ by $f(w) = wH$. Then f is a homomorphism of groups (same argument as in the proof of Theorem 16.6). Now w is a codeword if and only if $w \in C$. Also, $w \in K$ (the kernel of f) if and only if $wH = \mathbf{0}$. So we must prove that $w \in C$ if and only if $w \in K$, that is, that $C = K$. By the definition of generator matrix, every element of C is of the form uG for some $u \in B(k)$. But $(uG)H = u(GH) = \mathbf{0}$ because GH is the zero matrix (Lemma 16.8). Therefore, $C \subseteq K$. Since C is a subgroup of order 2^k, we need to show only that K has order 2^k in order to conclude that $C = K$.

Exercise 14 shows that f is surjective. By the First Isomorphism Theorem 7.42, $B(n - k) \cong B(n)/K$, and, hence, by Lagrange's Theorem 7.26,

$$2^n = |B(n)| = |K|\,[B(n):K]$$
$$= |K| \cdot |B(n)/K| = |K| \cdot |B(n-k)| = |K| \cdot 2^{n-k}.$$

Dividing the first and last terms of this equation by 2^{n-k} shows that $|K| = 2^k$. ◆

COROLLARY 16.10   *Let C be a linear code with parity-check matrix H and let $u, v \in B(n)$. Then u and v are in the same coset of C if and only if $uH = vH$.*

Proof   To say that u and v are in the same coset means $u + C = v + C$. Theorem 7.23 in additive notation shows that

$$u + C = v + C \qquad \text{if and only if} \qquad u - v \in C.$$

By Theorem 16.9,

$$u - v \in C \qquad \text{if and only if} \qquad (u-v)H = \mathbf{0}.$$

Since matrix multiplication is distributive, $(u - v)H = uH - vH$. Also, $uH - vH = \mathbf{0}$ is equivalent to $uH = vH$. Hence,

$$(u - v)H = \mathbf{0} \qquad \text{if and only if} \qquad uH = vH.$$

Combining the three centered statements above proves the theorem. ◆

If $w \in B(n)$ and H is the parity-check matrix, then wH is called the **syndrome** of w. By Corollary 16.10, w and its coset leader e have the same syndrome. If $w = e + v$ with $v \in C$, the standard array decodes w, as $v = w - e$. Therefore, standard-array (nearest-neighbor) decoding can be implemented as follows:

1. If w is a received word, compute the syndrome of w (that is, wH).
2. Find the coset leader e with the same syndrome (that is, $eH = wH$).
3. Decode w as $w - e$.

Since this procedure (called **syndrome decoding**) requires only that you know the syndromes of the coset leaders, the standard array can be replaced by a much shorter table.

> **EXAMPLE** The coset leaders for the (5,2) code {00000, 10110, 01101, 11011}, as shown on pages 475−476, are
>
> 00000, 10000, 01000, 00100, 00010, 00001, 11000, 10001.
>
> Multiplying each of them by the parity-check matrix H given in the preceding example produces its syndrome:
>
> | Syndrome | 000 | 110 | 101 | 100 | 010 | 001 | 011 | 111 |
> |---|---|---|---|---|---|---|---|---|
> | Coset Leader | 00000 | 10000 | 01000 | 00100 | 00010 | 000001 | 11000 | 10001 |
>
> To decode $w = 01001$, for example, we compute $01001H = 100$. The table shows that the coset leader with this syndrome is $e = 00100$. So we decode w as $w - e = 01001 - 00100 = 01101$.

Depending on the size of the code and whether or not coset leaders can be determined without constructing the entire standard array, syndrome decoding may be more efficient than brute-force nearest-neighbor decoding. For example, a (56,48) code has 2^{48} (approximately 2.8×10^{14}) codewords but only $2^8 = 256$ cosets.

Standard-array and syndrome decoding are complete decoding schemes, meaning that they *always* find a nearest codeword for each received word. When retransmission of the message is impractical, complete decoding is a necessity. But when retransmission is feasible, it may be better to use an incomplete decoding scheme that corrects t errors and requests retransmission when more than t errors are detected. We now describe one such scheme.

Let $e_i \in H(n)$ denote the row vector with 1 in coordinate i and 0 in every other coordinate. In $B(3)$, for instance, $e_1 = 100$, $e_2 = 010$, and $e_3 = 001$. Each e_i has weight 1; in fact

$$e_1, e_2, \ldots, e_n \text{ are the only elements of weight 1 in } B(n).$$

Consider the product of $e_2 \in B(3)$ and this matrix H:

$$e_2 H = (0 \quad 1 \quad 0) \begin{pmatrix} 1 & 0 & 1 \\ 0 & 1 & 1 \\ 1 & 1 & 1 \end{pmatrix} = (0 \quad 1 \quad 1) = \text{row 2 of } H.$$

Exercise 10 shows that the same thing happens in the general case. If $e_i \in B(n)$ and H is a matrix with n rows, then

$$e_i H \text{ is the } i\text{th row of the matrix } H.$$

Now assume that C is a linear code with parity-check matrix H and that *the rows of H are nonzero and no two of them are the same.* Then $e_i H = i$th row of $H \neq \mathbf{0}$ by hypothesis; hence, by Theorem 16.9,

e_i is not a codeword.

Furthermore, if $i \neq j$, then e_i and e_j cannot be in the same coset of C (otherwise row i of $H = e_i H = e_j H =$ row j of H by Corollary 16.10). Thus

e_i is the only element of weight 1 in its coset.

So every other element in the coset of e_i has weight at least 2.* Consequently,

e_i is *always* the coset leader in its coset.

Finally, if the syndrome of a received word w is the ith row of H, then $wH = e_i H$, so w and e_i are in the same coset by Corollary 16.10.

The preceding paragraph suggests a convenient way to implement (possibly incomplete) syndrome decoding when the rows of H are nonzero and distinct:

1. If w is received, compute its syndrome wH.
2. If $wH = \mathbf{0}$, decode w as w (because w is a codeword by Theorem 16.9).
3. If $wH \neq \mathbf{0}$ and wH is the ith row of H, decode w by changing its ith coordinate (that is, decode w as $w - e_i$ because e_i is w's coset leader).
4. If $wH \neq \mathbf{0}$ and wH is not a row of H, do not decode and request a retransmission.

This scheme (called **parity-check matrix decoding**) can be easily implemented with large codes because there is no need to compute cosets or find coset leaders. Furthermore,

THEOREM 16.11   *Let C be a linear code with parity-check matrix H. If every row of H is nonzero and no two are the same, then parity-check matrix decoding corrects all single errors.*

Proof   When a codeword u is transmitted with exactly one error in coordinate i and received as w, then $w - u = e_i$. By Theorem 16.9, $wH = (e_i + u)H = e_i H + uH = e_i H + \mathbf{0} = e_i H$, which is the ith row of H. Therefore, w is correctly decoded as $w - e_i = u$.   ◆

$\lceil$ **EXAMPLE**   Let C be the (5,2) code whose parity-check matrix H is given on page 477. If 10011 is received, its syndrome is

---

* The only element of weight 0 is $000 \cdots 0$, whose coset is C. C is not the coset of e_i because e_i is not a codeword.

$$(1 \quad 0 \quad 0 \quad 1 \quad 1)H = (1 \quad 0 \quad 0 \quad 1 \quad 1)\begin{pmatrix} 1 & 1 & 0 \\ 1 & 0 & 1 \\ 1 & 0 & 0 \\ 0 & 1 & 0 \\ 0 & 0 & 1 \end{pmatrix}$$

$$= (1 \quad 0 \quad 1) = \text{row 2 of } H.$$

Therefore, 10011 is decoded as $10011 - e_2 = 10011 - 01000 = 11011$. If 11000 is received, verify that its syndrome is 011, which is not a row of H. Therefore, 11000 is not decoded, and a retransmission is requested.

In one important class of codes, parity-check matrix decoding is actually *complete* syndrome (nearest-neighbor) decoding.

EXAMPLE The standard generator matrix for the Hamming (7,4) code was given on page 469. Its parity-check matrix H has distinct, nonzero rows:

$$H = \begin{pmatrix} 0 & 1 & 1 \\ 1 & 0 & 1 \\ 1 & 1 & 0 \\ 1 & 1 & 1 \\ 1 & 0 & 0 \\ 0 & 1 & 0 \\ 0 & 0 & 1 \end{pmatrix}$$

The possible syndromes of a received word w in this code are 000 and the seven nonzero elements of $B(3)$. But all the nonzero elements of $B(3)$ appear as rows of H. So every syndrome either is 000 (decode w as itself) or is the ith row of H for some i (decode w by changing its ith coordinate). Therefore, every received word is decoded.

This example is one of an infinite class of codes that can be described by using the fact that a linear code is completely determined by its parity-check matrix (from which a standard generator matrix is easily found). Let $r \geq 2$ be an integer and let $n = 2^r - 1$ and $k = 2^r - 1 - r$. Then $n - k = r$. The preceding example is the case $r = 3$. Let H be the $n \times (n - k)$ matrix whose last r rows are the identity matrix I_r, and whose n rows consist of *all* the nonzero elements of $B(r)$. Since the number of nonzero elements in $B(r)$ is $2^r - 1 = n$, each nonzero element appears exactly once as a row of H. So the rows of H are distinct and nonzero. The code with this parity-check matrix is called a **Hamming code.**

In every Hamming code, all possible syndromes are rows of H. So parity-check matrix decoding is complete syndrome decoding that corrects all single errors.

◆ EXERCISES

A. **1.** Find the parity-check matrix of each standard generator matrix in Exercise 5 of Section 16.1.

2. Find the parity-check matrix for the code in the last example in Section 16.1.

3. Find the parity-check matrix for the parity-check code on page 465. [See Exercise 7 in Section 16.1.]

4. Find the parity-check matrix for the (10,2) repetition code on page 468. [See Exercise 8 in Section 16.1.]

5. Find a parity-check matrix for the (15,11) Hamming code.

6. Show that the linear code C with parity-check matrix $\begin{pmatrix} 1 & 0 \\ 0 & 1 \\ 1 & 0 \\ 0 & 1 \\ 1 & 0 \\ 0 & 1 \end{pmatrix}$ cannot correct every single error.

7. Let C be the (4,2) code with standard generator matrix $G = \begin{pmatrix} 1 & 0 & 1 & 1 \\ 0 & 1 & 0 & 1 \end{pmatrix}$. Construct a standard array for C and find the syndrome of each coset leader.

8. Construct a standard array for the (6,3) code in the last example in Section 16.1 and find the syndrome of each coset leader.

9. Choose new coset leaders (when possible) for the (5,2) code on pages 475–476 and use them to construct a standard array. How does this array compare with the one on page 476?

10. Let $e_i = 00 \cdots 010 \cdots 00 \in B(n)$ have 1 in coordinate i and 0 elsewhere. If H is a matrix with n rows, show that $e_i H$ is the ith row of H.

B. **11.** Suppose a codeword u is transmitted and w is received. Show that standard-array decoding will decode w as u if and only if $w - u$ is a coset leader.

12. If every element of weight $\leq t$ is a coset leader in a standard array for a code C, show that C corrects t errors.

13. If a codeword u is transmitted and w is received, then $e = w - u$ is called an **error pattern.** Prove that an error will be detected if and only if the corresponding error pattern is not a codeword.

14. Prove that the function $f: B(n) \to B(n - k)$ in the proof of Theorem 16.9 is surjective. [*Hint*: If $v = v_1 v_2 \cdots v_{n-k} \in B(n - k)$, show that $v = f(u)$, where $u = 000 \cdots 0 v_1 v_2 \cdots v_{n-k} \in B(n)$.]

15. Let C be a linear code with parity-check matrix H. Prove that C corrects single errors if and only if the rows of H are distinct and nonzero.

16. Show by example that parity-check matrix decoding with the Hamming (7,4) code cannot detect two or more errors.

17. Show that in any Hamming code, every nonzero codeword has weight at least 3.

18. [Probability required]   In the (5,2) code on pages 475−476, suppose that the probability of a transmission error in a single digit is .01.

(a) Show that the probability of a single codeword being transmitted without error is .95099.

(b) Show that the probability of a 100-word message being transmitted without error is less than .01.

(c) Show that the probability of a single codeword being transmitted with exactly one error is .04803.

(d) Show that the probability that a single codeword is correctly decoded by the standard array on page 476 is at least .99921.

(e) Show that the probability of a 100-word message being correctly decoded by the standard array is at least .92. [*Hint*: Compare with part (b).]

16.3 BCH CODES

The Hamming codes in the last section have efficient decoding algorithms that correct all single errors. The same is true of the BCH codes* presented here. But these codes are even more useful because they correct multiple errors.

The construction of a BCH code uses a finite *ring* whose additive group is (isomorphic to) some $B(n)$. Each ideal in such a ring is a linear code because its additive group is (isomorphic to) a subgroup of $B(n)$. The additional algebraic structure of the ring provides efficient error-correcting decoding algorithms for the code.

The finite rings in question are constructed as follows. Let n be a positive integer and $(x^n - 1)$ the principal ideal in $\mathbb{Z}_2[x]$ consisting of all multiples of $x^n - 1$. The elements of the quotient ring $\mathbb{Z}_2[x]/(x^n - 1)$ are the congruence classes (cosets) modulo $x^n - 1$. By Corollary 5.5, the distinct congruence classes in $\mathbb{Z}_2[x]/(x^n - 1)$ are in one-to-one correspondence with the polynomials of the form

---

* The initials BCH stand for Bose, Chaudhuri, and Hocquenghem, who invented these codes in 1959−1960.

(∗) $a_0 + a_1x + a_2x^2 + \cdots + a_{n-1}x^{n-1},$ with $a_i \in \mathbb{Z}_2.$

Each such polynomial has n coefficients, and there are two possibilities for each coefficient. Hence, $\mathbb{Z}_2[x]/(x^n - 1)$ is a ring with 2^n elements. Furthermore, the n coefficients $(a_0, a_1, a_2, \ldots, a_{n-1})$ of the polynomial (∗) may be considered as an element of the group $B(n) = \mathbb{Z}_2 \times \cdots \times \mathbb{Z}_2.$

THEOREM 16.12   *The function* $f : \mathbb{Z}_2[x]/(x^n - 1) \to B(n)$ *given by*

$$f([a_0 + a_1x + a_2x^2 + \cdots a_{n-1}x^{n-1}]) = (a_0, a_1, a_2, \ldots, a_{n-1})$$

is an isomorphism of additive groups.

Proof   Exercise 7.   ◆

Theorem 16.12 shows that every ideal of $\mathbb{Z}_2[x]/(x^n - 1)$ can be considered as a linear code since it is (up to isomorphism) a subgroup of $B(n)$. In particular, if $g(x) \in \mathbb{Z}_2[x]$, then the congruence class (coset) of $g(x)$ generates a principal ideal I in $\mathbb{Z}_2[x]/(x^n - 1)$. The ideal I consists of all congruence classes of the form $[h(x)g(x)]$ with $h(x) \in \mathbb{Z}_2[x]$. BCH codes are of this type.

In order to define a BCH code that corrects t errors, choose a positive integer r such that $t < 2^{r-1}$. Let $n = 2^r - 1$. Then $g(x)$ is determined by considering a finite field of order 2^r, as explained below.

> **EXAMPLE**   We let $t = 2$ and $r = 4$, so that $n = 2^4 - 1 = 15$. We shall construct a code in $\mathbb{Z}_2[x]/(x^{15} - 1)$ that corrects all double errors by finding an appropriate $g(x)$. To do this, we need a field of order $2^4 = 16$.
>
> The polynomial $1 + x + x^4$ is irreducible in $\mathbb{Z}_2[x]$ (Exercise 3). Hence, $K = \mathbb{Z}_2[x]/(1 + x + x^4)$ is a field of order 16 by Theorem 5.10 (and the remarks after it). By Theorem 5.11, K contains a root α of $1 + x + x^4$. Using the fact that
>
> $$1 + \alpha + \alpha^4 = 0 \quad \text{and, hence,} \quad a^4 = 1 + a^*$$
>
> we can compute the powers of α. For example, $\alpha^6 = \alpha^2\alpha^4 = \alpha^2(1 + \alpha) = \alpha^2 + \alpha^3$. Similarly, we obtain
>
> | | | |
> |---|---|---|
> | $\alpha^1 = \alpha$ | $\alpha^6 = \alpha^2 + \alpha^3$ | $\alpha^{11} = \alpha + \alpha^2 + \alpha^3$ |
> | $\alpha^2 = \alpha^2$ | $\alpha^7 = 1 + \alpha + \alpha^3$ | $\alpha^{12} = 1 + \alpha + \alpha^2 + \alpha^3$ |
> | $\alpha^3 = \alpha^3$ | $\alpha^8 = 1 + \alpha^2$ | $\alpha^{13} = 1 + \alpha^2 + \alpha^3$ |
> | $\alpha^4 = 1 + \alpha$ | $\alpha^9 = \alpha + \alpha^3$ | $\alpha^{14} = 1 + \alpha^3$ |
> | $\alpha^5 = \alpha + \alpha^2$ | $\alpha^{10} = 1 + \alpha + \alpha^2$ | $\alpha^{15} = 1$ |

* Remember, $1 = -1$ in $\mathbb{Z}_2$.

These elements are distinct and nonzero by Theorem 10.7. Therefore, they are all the nonzero elements of K, and α is a generator of the multiplicative group of K.

To construct the polynomial $g(x)$, we first find the minimum polynomials of α, α^2, α^3, α^4 over $\mathbb{Z}_2$. By the construction of K, the minimal polynomial of α is $m_1(x) = 1 + x + x^4$. This polynomial $m_1(x)$ is also the minimal polynomial of α^2 and α^4, for instance, by the Freshman's Dream (Lemma 10.24),

$$m_1(\alpha^2) = 1 + (\alpha^2) + (\alpha^2)^4$$
$$= 1^2 + (\alpha)^2 + (\alpha^4)^2 = (1 + \alpha + \alpha^4)^2 = 0^2 = 0.$$

Verify that the minimum polynomial of α^3 is $m_3(x) = 1 + x + x^2 + x^3 + x^4$ (Exercise 5). The polynomial $g(x)$ is defined as the product $m_1(x)m_3(x)$, so that

$$g(x) = (1 + x + x^4)(1 + x + x^2 + x^3 + x^4)$$
$$= 1 + x^4 + x^6 + x^7 + x^8 \in \mathbb{Z}_2[x].$$

Let C be the ideal generated by $[g(x)]$ in $\mathbb{Z}_2[x]/(x^{15} - 1)$. Then C is a code by Theorem 16.12. We shall see that C is a $(15,7)$ code that corrects all single and double errors.

Just what do the codewords of C look like? By Corollary 5.5, each congruence class in $\mathbb{Z}_2[x]/(x^{15} - 1)$ is the class of a unique polynomial of the form

$(**)$ $a_0 + a_1x + a_2x^2 + \cdots + a_{13}x^{13} + a_{14}x^{14}$,   with $a_i \in \mathbb{Z}_2$.

So we shall denote the class by this polynomial.* When convenient, this polynomial will be identified (as in Theorem 16.12) with the element $a_0a_1a_2 \cdots a_{14} = (a_0,a_1,a_2, \ldots ,a_{14})$ of $B(15)$. The codewords consist of the classes of polynomial multiples of $g(x)$. For example,

Codeword in Polynomial Form	In *B* (15) Form
$g(x) = 1 + x^4 + x^6 + x^7 + x^8$	100010111000000
$xg(x) = x(1 + x^4 + x^6 + x^7 + x^8)$	
$\quad = x + x^5 + x^7 + x^8 + x^9$	010001011100000
$(1 + x^6)g(x) = (1 + x^6)(1 + x^4 + x^6 + x^7 + x^8)$	
$\quad = 1 + x^4 + x^7 + x^8 + x^{10} + x^{12}$	
$\quad\quad + x^{13} + x^{14}$	100010011010111

If $g(x)$ is multiplied by a polynomial $h(x)$ of degree ≥ 7, then the

---

* This is analogous to what was done on page 35, when we began writing elements (classes) in $\mathbb{Z}_n$ in the form k rather than $[k]$.

codeword $h(x)g(x)$ has degree ≥ 15 and is not of the form (∗∗). For example, if $h(x) = x^8$, then

$$h(x)g(x) = x^8 g(x) = x^8(1 + x^4 + x^6 + x^7 + x^8)$$
$$= x^8 + x^{12} + x^{14} + x^{15} + x^{16}.$$

The polynomial of the form (∗∗) that is in the same class as $h(x)g(x)$ is the remainder when $h(x)g(x)$ is divided by $x^{15} - 1$ (see Corollary 5.5). Verify that

$$h(x)q(x) = (1 + x)(x^{15} - 1) + (1 + x + x^8 + x^{12} + x^{14}).$$

Hence, $[f(x)g(x)]$ is the codeword $1 + x + x^8 + x^{12} + x^{14}$ or, equivalently, 110000001000101.

The procedure in the example is readily generalized. If t is the number of errors the code should correct, let $n = 2^r - 1$, where r is chosen so that $t < 2^{r-1}$ (in the example, $t = 2$, $r = 4$). By Corollary 10.26, there is a finite field K of order 2^r. By Theorem 10.28, $K = \mathbb{Z}_2(\alpha)$, where α is a generator of the multiplicative group of nonzero elements of K (and so has multiplicative order $2^r - 1 = n$). Let

$$m_1(x), m_2(x), m_3(x), \ldots , m_{2t}(x) \in \mathbb{Z}_2[x]$$

be the minimal polynomials of the elements

$$\alpha, \alpha^2, \alpha^3, \ldots , \alpha^{2t} \in K.$$

Let $g(x)$ be the product in $\mathbb{Z}_2[x]$ of the distinct polynomials on the list $m_1(x)$, $m_2(x), \ldots , m_{2t}(x)$.

The ideal C generated by $[g(x)]$ in $\mathbb{Z}_2[x]/(x^n - 1)$ is called the (primitive narrow-sense) **BCH code of length n and designed distance $2t + 1$** with **generator polynomial** $g(x)$. So the code in the last example is a BCH code of length 15 and designed distance 5 (= $2 \cdot 2 + 1$). If $g(x)$ has degree m, then Exercise 14 shows that the code C is an (n,k) code, where $k = n - m$.

THEOREM 16.13 *A BCH code of length n and designed distance $2t + 1$ corrects t errors.*

Proof The proof requires a knowledge of determinants; see Lidl-Pilz [34; page 230] or Mackiw [35; page 60]. ◆

Theorem 16.13 shows that there are BCH codes that will correct any desired number of errors. More importantly, from a practical viewpoint, there are efficient algorithms for decoding large BCH codes.* A complete description of them would take us too far afield. But here, in simplified form, is the underlying idea of the error-correcting procedure.

Let C be a BCH code of designed distance $2t + 1$ and generator polynomial $g(x)$. By the definition of $g(x)$, each minimal polynomial $m_1(x)$ divides $g(x)$. Hence, $g(\alpha^i) = 0$ for each $i = 1, 2, \ldots, 2t$. If $[f(x)]$ is a codeword in C, then $f(x) = h(x)g(x)$ for some $h(x)$, and, therefore,

$$f(\alpha^i) = h(\alpha^i)g(\alpha^i) = h(\alpha^i) \cdot 0 = 0.$$

Conversely, if $f(x) \in \mathbb{Z}_2[x]$ has every α^i as a root, then every $m_i(x)$ divides $f(x)$ by Theorem 10.6. This implies that $g(x) \mid f(x)$ (Exercise 8). Therefore,

$[f(x)]$ is a codeword if and only if $f(\alpha^i) = 0$ for $1 \leq i \leq 2t$

The decoder receives the word $a_0 a_1 \cdots a_k$, which represents the (class of) the polynomial

$$r(x) = a_0 + a_1 x + a_2 x^2 + \cdots + a_k x^k.$$

The decoder computes these elements of the field $K = \mathbb{Z}_2(\alpha)$:

$$r(\alpha), r(\alpha^2), r(\alpha^3), \ldots, r(\alpha^{2t}).$$

If all of them are 0, then $r(x)$ is a codeword by the remarks above. If certain ones are nonzero, the decoder uses them (according to a specified procedure) to construct a polynomial $D(x) \in K[x]$, called the **error-locator polynomial,** Since K is finite, the nonzero roots of $D(x)$ in K can be found by substituting each $\alpha^i \in K$ in $D(x)$].

If no more than t errors have been made, the nonzero roots of $D(x)$ give the location of the transmission errors. For instance, if α^7 is a root, then a_7 is incorrect in the received word $r(x)$; similarly if $\alpha^0 = 1$ is a root, then an error occured in transmitting a_0.

If $D(x)$ has no roots in K or if certain of the $r(\alpha^i)$ are 0, so that $D(x)$ cannot be constructed, then more than t errors have been made. So the decoder follows set procedures (omitted here) to choose arbitrarily a nearest codeword to $r(x)$.

EXAMPLE   In the (15,7) BCH code of the previous example, suppose this word is received:

$$r(x) = x + x^7 + x^8 = 010000011000000.$$

Using the table on page 485 and the fact that $u + u = 0$ for every element u in K (Exercise 1), we have

---

* This is one reason BCH codes are widely used. For example, the European and trans-Atlantic communication system uses a BCH code with $t = 6$ and $r = 8$. It is a (255,231) code that corrects six errors with a failure probability of only 1 in 16 million.

$$r(\alpha) = \alpha + \alpha^7 + \alpha^8 = \alpha + (1 + \alpha + \alpha^3) + (1 + \alpha^2)$$
$$= \alpha^2 + \alpha^3 = \alpha^6.$$
$$r(\alpha^3) = \alpha^3 + (\alpha^3)^7 + (\alpha^3)^8$$
$$= \alpha^3 + \alpha^{21} + \alpha^{24} = \alpha^3 + \alpha^6 + \alpha^9$$
$$= \alpha^3 + (\alpha^2 + \alpha^3) + (\alpha + \alpha^3) = \alpha + \alpha^2 + \alpha^3 = \alpha^{11}.$$

Exercise 6 shows that

$$r(\alpha^2) = r(\alpha)^2 = (\alpha^6)^2 = \alpha^{12};$$
$$r(\alpha^4) = r(\alpha)^4 = (\alpha^6)^4 = \alpha^{24} = \alpha^9.$$

The error-locator polynomial is given by this formula (which is justified in Exercise 15):

$$D(x) = x^2 + r(\alpha)x + \left(r(\alpha^2) + \frac{r(\alpha^3)}{r(\alpha)} \right).$$

Using the table on page 485, we see that

$$D(x) = x^2 + \alpha^6 x + \left(\alpha^{12} + \frac{\alpha^{11}}{\alpha^6} \right) = x^2 + \alpha^6 x + (\alpha^{12} + \alpha^5)$$
$$= x^2 + \alpha^6 x + \alpha^{14}.$$

By substituting each of the nonzero elements of K in $D(x)$, we discover that

$$D(\alpha^5) = (\alpha^5)^2 + \alpha^6 \alpha^5 + \alpha^{14} = \alpha^{10} + \alpha^{11} + \alpha^{14}$$
$$= (1 + \alpha + \alpha^2) + (\alpha + \alpha^2 + \alpha^3) + (1 + \alpha^3) = 0;$$
$$D(\alpha^9) = (\alpha^9)^2 + \alpha^6 \alpha^9 + \alpha^{14} = \alpha^{18} + \alpha^{15} + \alpha^{14} = \alpha^3 + 1 + \alpha^{14}$$
$$= \alpha^3 + 1 + (1 + \alpha^3) = 0.$$

Therefore, α^5 and α^9 are the roots of $D(x)$, so errors occurred in the coefficients of x^5 and x^9. The received word

$$r(x) = x + x^7 + x^8 = 010000011000000$$

is corrected as

$$c(x) = x + x^5 + x^7 + x^8 + x^9 = 010001011100000,$$

which *is* a codeword (see page 486).

Similarly, if $r(x) = x^2 + x^6 + x^9 + x^{10} = 001000100110000$ is received, then

$$r(\alpha) = \alpha^8, \qquad r(\alpha^2) = \alpha, \qquad r(\alpha^3) = \alpha^9, \qquad \text{and}$$
$$D(x) = x^2 + r(\alpha)x + \left[r(\alpha^2) + \frac{r(\alpha^3)}{r(\alpha)} \right] = x^2 + \alpha^8 x + \left(\alpha + \frac{\alpha^9}{\alpha^8} \right)$$
$$= x^2 + \alpha^8 x + (\alpha + \alpha) = x^2 + \alpha^8 x = x(x + \alpha^8).$$

The only nonzero root of $D(x)$ is α^8, so a single error occurred in the coefficient of x^8, and the correct word is

$$c(x) = x^2 + x^6 + x^8 + x^9 + x^{10} = 001000101110000.$$

Finally, if $1 + x + x^4$ is received, then

$$r(\alpha) = 1 + \alpha + \alpha^4 = 0 \quad \text{and} \quad r(\alpha^3) = 1 + \alpha^3 + \alpha^{12} = \alpha^5.$$

So $D(x)$ cannot be constructed, and we conclude that more than two errors have occurred. Similarly, if $1 + x + x^3$ is received, then verify that $D(x) = x^2 + \alpha^7 x + \alpha^5$ and that $D(x)$ has no roots in K. Once again, more than two errors have occurred.

◆ EXERCISES

NOTE: *Unless stated otherwise, K is the field $\mathbb{Z}_2[x]/(1 + x + x^4)$ of order 16 and α is a root of $1 + x + x^4$, as in the example on pages 485–487.*

A. **1. (a)** Prove that $f(x) + f(x) = 0$ for every $f(x) \in \mathbb{Z}_2[x]$.

(b) Prove that $u + u = 0$ for every u in the field K.

2. Show that the only irreducible quadratic in $\mathbb{Z}_2[x]$ is $x^2 + x + 1$. [*Hint*: List all the quadratics and use Corollary 4.18.]

3. Prove that $1 + x + x^4$ is irreducible in $\mathbb{Z}_2[x]$. [*Hint*: Exercise 2 and Theorem 4.15.]

4. Prove that the minimal polynomial of α^5 over $\mathbb{Z}_2$ is $1 + x + x^2$. [*Hint*: Use the table on page 485.]

5. (a) Prove that the minimal polynomial of α^3 over $\mathbb{Z}_2$ is $1 + x + x^2 + x^3 + x^4$. [*Hint*: Exercise 2, Theorem 4.15, and the table on page 485.]

(b) Show that α^4 is also a root of $1 + x + x^4$.

B. **6.** If $f(x) \in \mathbb{Z}_2[x]$ and α is an element in some extension field of $\mathbb{Z}_2$, prove that for every $k \geq 1$, $f(\alpha^{2k}) = f(\alpha^k)^2$. [*Hint*: Lemma 10.24.]

7. (a) Show that the function $f : \mathbb{Z}_2[x]/(x^n - 1) \rightarrow B(n)$ given by

$$f([a_0 + a_1 x + a_2 x^2 + \cdots + a_{n-1} x^{n-1}]) = (a_0, a_1, a_2, \ldots, a_{n-1})$$

is surjective.

(b) Prove that f is a homomorphism of additive groups.

(c) Prove that f is injective. [*Hint*: Theorem 7.40 in additive notation.]

8. (a) Let F be a field and $f(x) \in F[x]$. If $p(x)$ and $q(x)$ are distinct monic irreducibles in $F[x]$ such that $p(x) \mid f(x)$ and $q(x) \mid f(x)$, prove that

$p(x)q(x) \mid f(x)$. [*Hint:* If $f(x) = q(x)h(x)$, then $p(x) \mid q(x)h(x)$; use part 2 of Theorem 4.11.]

(b) If $m_1(x)$, $m_2(x)$, . . . , $m_k(x)$ are distinct monic irreducibles in $F[x]$ such that each $m_i(x)$ divides $f(x)$, prove that $g(x) = m_1(x)m_2(x) \cdots m_k(x)$ divides $f(x)$.

9. Let C be the (15,7) BCH code of the examples in the text. Use the error-correction technique presented there to correct these received words or to determine that three or more errors have been made.

(a) $1 + x = 110000000000000$.

(b) $1 + x^3 + x^4 + x^5 = 100111000000000$.

(c) $1 + x^2 + x^4 + x^7 = 101010010000000$.

(d) $1 + x^6 + x^7 + x^8 + x^9 = 100000111100000$.

10. Show that the generator polynomial for the BCH code with $t = 3$, $r = 4$, $n = 15$ is $g(x) = 1 + x + x^2 + x^4 + x^5 + x^8 + x^{10}$. [*Hint:* Exercises 3–5 may be helpful.]

11. Let $K = \mathbb{Z}_2(\alpha)$ be a finite field of order 2^r, whose multiplicative group is generated by α. For each i, let $m_i(x)$ be the minimal polynomial of α^i over $\mathbb{Z}_2$. If $n = 2^r - 1$, prove that each $m_i(x)$ divides $x^n - 1$. [*Hint:* $\alpha^n = 1$ (Why?); use Theorem 10.6.]

12. If $g(x)$ is the generator polynomial of a BCH code in $\mathbb{Z}_2[x]/(x^n - 1)$, prove that $g(x)$ divides $x^n - 1$. [*Hint:* Exercises 11 and 8(b).]

13. Let $g(x) \in \mathbb{Z}_2[x]$ be a divisor of $x^n - 1$ and let C be the principal ideal generated by $[g(x)]$ in $\mathbb{Z}_2[x]/(x^n - 1)$. Then C is a code. Prove that C is **cyclic,** meaning that C (with codewords written as elements of $B(n)$) has this property: If $(c_0, c_1, \ldots, c_{n-1}) \in C$, then $(c_{n-1}, c_0, c_1, \ldots, c_{n-2}) \in C$. [*Hint:* $c_{n-1} + c_0 x + \cdots + c_{n-2} x^{n-1} = x(c_0 + c_1 x + \cdots + c_{n-1} x^{n-1}) - c_{n-1}(x^n - 1)$.]

C. 14. Let C be the code in Exercise 13. Assume $g(x)$ has degree m and let $k = n - m$. Let J be the set of all polynomials in $\mathbb{Z}_2[x]$ of the form $a_0 + a_1 x + a_2 x^2 + \cdots + a_{k-1} x^{k-1}$.

(a) Prove that every element in C is of the form $[s(x)g(x)]$ with $s(x) \in J$. [*Hint:* Let $[h(x)g(x)] \in C$. By the Division Algorithm, $h(x)g(x) = e(x)(x^n - 1) + r(x)$, with $\deg r(x) < n$ and $[h(x)g(x)] = [r(x)]$. Show that $r(x) = s(x)g(x)$, where $s(x) = h(x) - e(x)f(x)$ and $g(x)f(x) = x^n - 1$. Use Theorem 4.2 to show $s(x) \in J$.]

(b) Prove that C has order 2^k, and, hence, C is an (n, k) code. [*Hint:* Use Corollary 5.5 to show that if $s(x) \neq t(x)$ in J, then $[s(x)g(x)] \neq [t(x)g(x)]$ in C. How many elements are in J?]

15. Let C be the $(15,7)$ BCH code of the examples in the text, with codewords written as polynomials of degree ≤ 14. Suppose the codeword $c(x)$ is transmitted with errors in the coefficients of x^i and x^j and $r(x)$ is received. Then $D(x) = (x + \alpha^i)(x + \alpha^j) \in K[x]$, whose roots are α^i and α^j, is the error-locator polynomial. Express the coefficients of $D(x)$ in terms of $r(\alpha)$, $r(\alpha^2)$, $r(\alpha^3)$ as follows.

(a) Show that $r(x) - c(x) = x^i + x^j$.

(b) Show that $r(\alpha^k) = \alpha^{ki} + \alpha^{kj}$ for $k = 1, 2, 3$. [See the boldface statement on page 488.]

(c) Show that $D(x) = x^2 + (\alpha^i + \alpha^j)x + \alpha^{i+j} = x^2 + r(\alpha)x + \alpha^{i+j}$.

(d) Show that $\alpha^{i+j} = r(\alpha^2) + \dfrac{r(\alpha^3)}{r(\alpha)}$. [*Hint:* Show that $r(\alpha)^3 = (\alpha^i + \alpha^j)^3 = \alpha^{3i} + \alpha^{3j} + \alpha^{i+j}(\alpha^i + \alpha^j) = r(\alpha^3) + r(\alpha)\alpha^{i+j}$ and solve for α^{i+j}; note that $r(\alpha)^2 = r(\alpha^2)$.]

16. Show that a BCH code with $t = 1$ is actually a Hamming code (see page 482).

APPENDIX **A**

$\blacklozenge$

Logic and Proof

This Appendix summarizes the basic facts about logic and proof that are needed to read this book. For a complete discussion of these topics see Galovich [9], Lucas [11], Smith-Eggen-St. Andre [12], or Solow [13].

LOGIC

A **statement** is a declarative sentence that is either true or false. For instance, each of these sentences is a statement:

π is a real number.

Every triangle is isosceles.

103 bald eagles were born in the United States last year.

Note that the last sentence *is* a statement even though we may not be able to verify its truth or falsity. Neither of the following sentences is a statement:

What time is it?   Wow!

Compound Statements

We frequently deal with **compound statements** that are formed from other statements by using the connectives "and" and "or." The truth of the compound statement will depend on the truth of its components. If P and Q are statements, then

"P and Q" is a true statement when *both*
P and Q are true, and false otherwise.

For example,

$$\pi \text{ is a real number and } 9 < 10$$

is a true statement because both of its components are true. But

$$\pi \text{ is a real number and } 7 - 5 = 18$$

is a false statement since one of its components is false.

In ordinary English the word "or" is most often used in exclusive sense, meaning "one or the other but not both," as in

He is at least 21 years old or he is younger than 21.

But "or" can also be used in an inclusive sense, meaning "one or the other, or possibly both," as in the sentence

They will win the first game or they will win the second.

Thus the inclusive "or" has the same meaning as "and/or" in everyday language. In mathematics, *"or" is always used in the inclusive sense,* which allows the possibility that both components might be true but does not require it. Consequently, if P and Q are statements, then

**"P or Q" is a true statement when at least one of P or Q
is true and false when both P and Q are false.**

For example, both

$$7 > 5 \qquad \text{or} \qquad 3 + 8 = 11$$

and

$$7 > 5 \qquad \text{or} \qquad 3 + 8 = 23$$

are true statements because at least one component is true in each case, but

$$4 < 2 \qquad \text{or} \qquad 5 + 3 = 12$$

is false since both components are false.

Negation

The **negation** of a statement P is the statement "it is not the case that P," which we can conveniently abbreviate as "not-P." Thus the negation of

7 is a positive integer

is the statement "it is not the case that 7 is a positive integer," which we would normally write in the less awkward form "7 is not a positive integer." If P is a statement, then

**The negation of P is true exactly when P is false, and
the negation of P is false exactly when P is true.**

The negation of the statement "*P* and *Q*" is the statement "it is not the case that *P* and *Q*." Now "*P* and *Q*" is true exactly when both *P* and *Q* are true, so to say that this is not the case means that at least one of *P* or *Q* is false. But this occurs exactly when at least one of not-*P* or not-*Q* is true. Thus

The negation of the statement "*P* and *Q*" is the statement
"not-*P* *or* not-*Q*."

For example, the negation of

f is continuous and *f* is differentiable at $x = 5$

is the statement

f is not continuous or *f* is not differentiable at $x = 5$.

The negation of the statement "*P* or *Q*" is the statement "it is not the case that *P* or *Q*." Now "*P* or *Q*" is true exactly when at least one of *P* or *Q* is true. To say that this is not the case means that both *P* and *Q* are false. But *P* and *Q* are both false exactly when not-*P* and not-*Q* are both true. Hence,

The negation of the statement "*P* or *Q*" is the statement
"not-*P* *and* not-*Q*."

For instance, the negation of

119 is prime or $\sqrt{3}$ is a rational number

is the statement

119 is not prime and $\sqrt{3}$ is not a rational number.

Quantifiers

Many mathematical statements involve quantifiers. The **universal quantifier** states that a property is true *for all* the items under discussion. There are several grammatical variations of the universal quantifier, such as

For all real numbers c, $c^2 - 1$.

Every integer is a real number.

All integers are rational numbers.

For each real number a, the number $a^2 + 1$ is positive.

The **existential quantifier** asserts that *there exists* at least one object with certain properties. For example,

There exist positive rational numbers.

There exists a number x such that $x^2 - 5x + 6 = 0$.

There is an even prime number.

In mathematics, the word "some" means "at least one" and is, in effect, an existential quantifier. For instance,

Some integers are prime

is equivalent to saying "at least one integer is prime," that is,

There exists a prime integer.

Care must be used when forming the negation of statements involving quantifiers. For example, the negation of

All real numbers are rational

is "it is not the case that all real numbers are rational," which means that there is at least one real number that is irrational (= not rational). So the negation is

There exists an irrational real number.

In particular, the statements "all real numbers are not rational" and "all real numbers are irrational" are *not* negations of "all real numbers are rational." This example illustrates the general principle:

The negation of a statement with a universal quantifier is a statement with an existential quantifier.

The negation of the statement

There exists a positive integer

is "it is not the case that there is a positive integer," which means that "every integer is nonpositive" or, equivalently, "no integer is positive." Thus

The negation of a statement with an existential quantifier is a statement with a universal quantifier.

Conditional and Biconditional Statements

In mathematical proofs we deal primarily with **conditional statements** of the form

If P, then Q

which is written symbolically as $P \Rightarrow Q$. The statement P is called the

hypothesis or **premise,** and Q is called the **conclusion.** Here are some examples:

> If c and d are integers, then cd is an integer.
>
> If f is continuous at $x = 3$, then f is differentiable there.
>
> $a \neq 0 \Rightarrow a^2 > 0$.

There are several grammatical variations, all of which mean the same thing as "if P, then Q":

> P implies Q.
>
> P is sufficient for Q.
>
> Q provided that P.
>
> Q whenever P.

In ordinary usage the statement "if P, then Q" means that the truth of P guarantees the truth of Q. Consequently,

**"$P \Rightarrow Q$" is a true statement when both P and Q are
true and false when P is true and Q is false.**

Although the situation rarely occurs, we must sometimes deal with the statement "$P \Rightarrow Q$" when P is false. For example, consider this campaign promise: "If I am elected, then taxes will be reduced." If the candidate is elected (P is true), the truth or falsity of this statement depends on whether or not taxes are reduced. But what if the candidate is *not* elected (P is false)? Regardless of what happens to taxes, you can't fairly call the campaign promise a lie. Consequently, it is customary in symbolic logic to adopt this rule:

When P is false, the statement "$P \Rightarrow Q$" is true.

The **contrapositive** of the conditional statement "$P \Rightarrow Q$" is the statement "not-$Q \Rightarrow$ not-P." For instance, the contrapositive of this statement about integers

If c is a multiple of 6, then c is even

is the statement

If c is not even, then c is not a multiple of 6.

Notice that both the original statement and its contrapositive are true. Two statements are said to be **equivalent** if one is true exactly when the other is. We claim that

The conditional statement "$P \Rightarrow Q$" is equivalent to its contrapositive "not-$Q \Rightarrow$ not-P."

To prove this equivalence, suppose $P \Rightarrow Q$ is true and consider the statement not-$Q \Rightarrow$ not-P. Suppose not-Q is true. Then Q is false. Now if P were true, then Q would necessarily be true, which is not the case. So P must be false, and, hence, not-P is true. Thus not-$Q \Rightarrow$ not-P is true. A similar argument shows that when not-$Q \Rightarrow$ not-P is true, then $P \Rightarrow Q$ is also true.

The **converse** of the conditional statement "$P \Rightarrow Q$" is the statement "$Q \Rightarrow P$." For example, the converse of the statement

If b is a positive real number, then b^2 is positive

is the statement

If b^2 is positive, then b is a positive real number.

This last statement is false since, for example, $(-3)^2$ is the positive number 9, but -3 is not positive. Thus

The converse of a true statement may be false.

There are some situations in which a conditional statement and its converse are both true. For example,

If the integer k is odd, then the integer $k + 1$ is even

is true, as is its converse

If the integer $k + 1$ is even, then the integer k is odd.

We can state this fact in succinct form by saying that "k is odd *if and only if* $k + 1$ is even." More generally, the statement

P if and only if Q,

which is abbreviated as "P iff Q" or "$P \Leftrightarrow Q$," means

$$P \Rightarrow Q \quad and \quad Q \Rightarrow P.$$

"P if and only if Q" is called a **biconditional statement.** The rules for compound statements show that "P if and only if Q" is true exactly when both $P \Rightarrow Q$ and $Q \Rightarrow P$ are true. In this case, the truth of P implies the truth of Q and vice-versa, so that P is true exactly when Q is true. In other words, "P if and only if Q" means that P and Q are equivalent statements.

THEOREMS AND PROOF

The formal development of a mathematical topic begins with certain undefined terms and **axioms** (statements about the undefined terms that are assumed to be true). These undefined terms and axioms are used to define new terms and to construct **theorems** (true statements about these objects). The **proof** of a theorem is a complete justification of the truth of the statement.

Most theorems are conditional statements. A theorem that is not stated in conditional form is often equivalent to a conditional statement. For instance, the statement

> Every integer greater than 1 is a product of primes

is equivalent to

> If n is an integer and $n > 1$, then n is a product of primes.

The first step in proving a theorem that can be phrased in conditional form is to identify the hypothesis P and the conclusion Q. In order to prove the theorem "$P \Rightarrow Q$," one assumes that the hypothesis P is true and then uses it, together with axioms, definitions, and previously proved theorems, to argue that the conclusion Q is necessarily true.

Methods of Proof

Some common proof techniques are described below. While such summaries are helpful, there are no hard and fast rules that give a precise procedure for proving every possible mathematical statement. The methods of proof to be discussed here are in the nature of maps to guide you in analyzing and constructing proofs. A map may not reveal all the difficulties of the terrain, but it usually makes the route clearer and the journey easier.

DIRECT METHOD   This method of proof depends on the basic rule of logic called *modus ponens:* If R is a true statement and "$R \Rightarrow S$" is a true conditional statement, then S is a true statement. To prove the theorem "$P \Rightarrow Q$" by the direct method, you find a series of statements $P_1, P_2, \ldots, P_n$ and then verify that each of the implications $P \Rightarrow P_1, P_1 \Rightarrow P_2, P_2 \Rightarrow P_3, \ldots, P_{n-1} \Rightarrow P_n$, and $P_n \Rightarrow Q$ is true. Then the assumption that P is true and repeated use of *modus ponens* show that Q is true.

The direct method is the most widely used method of proof. In actual practice, it may be quite difficult to figure out the various intermediate statements that allow you to proceed from P to Q. In order to find them, most mathematicians use a thought process that is sometimes called the **forward-backward technique.** You begin by working forward and asking yourself,

What do I know about the hypothesis P? What facts does it imply? What statements follow from these facts? And so on. At this point you may have a list of statements implied by P whose connection with the conclusion Q, if any, is not yet clear.

Now work backward from Q by asking, What facts would guarantee that Q is true? What statements would imply these facts? And so on. You now have a list of statements that imply Q. Compare it with the first list. If you are fortunate some statement will be on both lists, or more likely, there will be a statement S on the first list and a statement T on the second, and you may be able to show that $S \Rightarrow T$. Then you have $P \Rightarrow S$ and $S \Rightarrow T$ and $T \Rightarrow Q$, so that $P \Rightarrow Q$.

When you have used the forward-backward technique successfully to find a proof that $P \Rightarrow Q$, you should write the proof in finished form. This finished form may look quite different from the thought processes that led you to the proof. Your thought process jumped forward and backward, but the finished proof normally should begin with P and proceed in step-by-step logical order from P to S to T to Q. The finished proof should contain only those facts that are needed in the proof. Many statements that arise in the forward-backward process turn out to be irrelevant to the final argument, and they should *not* be included in the finished proof. As illustrated in most of the proofs in this book, the finished proof is usually written as a narrative rather than a series of conditional statements.

CONTRAPOSITIVE METHOD   Since every conditional statement is equivalent to its contrapositive, you may prove "not-Q $\Rightarrow$ not-P" in order to conclude that "$P \Rightarrow Q$" is true. For example, instead of proving that for a certain function f,

$$\text{If } a \neq b, \text{ then } f(a) \neq f(b)$$

you can prove the contrapositive

$$\text{If } f(a) = f(b), \text{ then } a = b.$$

PROOF BY CONTRADICTION   Suppose that you assume the truth of a statement R and that you make a valid argument that $R \Rightarrow S$ (that is, $R \Rightarrow S$ is a true statement). If the statement S is in fact a *false* statement, there is only one possible conclusion: The original statement R must have been false, because a true premise R and a true statement $R \Rightarrow S$ lead to the truth of S by *modus ponens*.

In order to use this fact to prove the theorem "$P \Rightarrow Q$," assume as usual that P is a true statement. Then apply the argument in the preceding paragraph with $R = $ not-Q. In other words, assume that *not-Q* is true and find an argument (presumably using P and previously proved results) that shows not-$Q \Rightarrow S$, where S is a statement known to be false. Conclude that not-Q must be false. But not-Q is false exactly when Q is true. Therefore, Q is true, and we have proved that $P \Rightarrow Q$. Once again, the hard part will usually be finding the statement S and proving that not-Q implies S.

EXAMPLE   Recall that an integer is even if it is a multiple of 2 and that an integer that is not even is said to be odd. We shall use proof by contradiction to prove this statement

If m^2 is even, then m is even.

Here P is the statement "m^2 is even" and Q is the statement "m is even." We assume "m is not even" or equivalently "m is odd" (statement not-Q). But every odd integer is 1 more than some even integer. Since every even integer is a multiple of 2, we must have $m = 2k + 1$ for some integer k. Then the basic laws of arithmetic show that

$$m^2 = (2k + 1)^2 = 4k^2 + 4k + 1 = 2(2k^2 + 2k) + 1.$$

This last statement says that m^2 is 1 more than a multiple of 2, that is, m^2 is odd. But we are given that m^2 is even (statement P), and, hence, "m^2 is both odd and even" (statement S). This statement is false since no integer is both odd and even. Therefore, our original assumption (not-Q) has led to a contradiction (the false statement S). Consequently, not-Q must be false, and, hence, the statement "m is even" (statement Q) is true.

In the preceding example various statements were labeled by letters so that you could easily relate the example to the general discussion. This is not usually done in proofs by contradiction, and such proofs may not be given in as much detail as in this example.

The choice of a method of proof is partly a matter of taste and partly a question of efficiency. Although any of those listed above may be used, one method may lead to a much shorter or easier-to-follow proof than another, depending on the circumstances. In addition there are methods of proof that can be applied only to certain types of statements.

PROOF BY INDUCTION   This method is discussed in detail in Appendix C.

CONSTRUCTION METHOD   This method is appropriate for theorems that include a statement of the type "There exists a such-and-such with property so-and-so." For instance,

There is an integer d such that $d^2 - 4d - 5 = 0$.

If r and s are distinct rational numbers, then there is a rational number between r and s.

If r is a positive real number, then there is a positive integer m such that $\dfrac{1}{m} < r$.

To prove such a statement, you must construct (find, build, guess, etc.) an object with the desired property. When you are reading the proof of such a statement, you need only verify that the object presented in the proof does in fact have the stated property. An existence proof may amount to nothing more than presenting an example (for instance, the integer 2 provides a proof of "there exists a positive integer"). But more often a nontrivial argument will be needed to produce the required object.

Warning   Although an example is sufficient to prove an existence statement, examples can never *prove* a statement that directly or indirectly involves a universal quantifier. For instance, even if you have a million examples for which this statement is true:

If c is an integer, then $c^2 - c + 11$ is prime,

you will not have proved it. For the statement says, in effect, that for *every* integer c, a certain other integer is prime. This is *not* the case when $c = 12$ since $12^2 - 12 + 11 = 143 = 13 \cdot 11$. So the statement is false. This example demonstrates that

A counterexample is sufficient to *disprove* a statement.

The moral of the story is that when you are uncertain if a statement is true, try to find some examples where it holds or fails. If you find just one example where it fails, you have disproved the statement. If you can find only examples where the statement holds, you haven't proved it, but you do have encouraging evidence that it may be true.

Proofs of Multiconditional Statements

In order to prove the biconditional statement "P if and only if Q," you must prove *both* "$P \Rightarrow Q$" *and* "$Q \Rightarrow P$." Proving one of these statements and failing to prove the other is a common student mistake. For example, the proof of

A triangle with sides a, b, c is a right triangle with hypotenuse c if and only if $c^2 = a^2 + b^2$

consists of two separate parts. *First* you must assume that you have a right triangle with sides a, b and hypotenuse c and prove that $c^2 = a^2 + b^2$. Then you must give a *second* argument: Assume that the sides of a triangle satisfy $c^2 = a^2 + b^2$ and prove that this is a right triangle with hypotenuse c.

A statement of the form

The following conditions are equivalent: P, Q, R, S, T

is called a **multiconditional statement** and means that any one of the statements $P, Q, R, S,$ or T implies every other one. Thus a multiconditional statement is just shorthand for a list of biconditional statements; $P \Leftrightarrow Q$ and $P \Leftrightarrow R$ and $P \Leftrightarrow S$ and $P \Leftrightarrow T$ and $Q \Leftrightarrow R$ and $Q \Leftrightarrow S$, etc. To prove this multiconditional statement you need only prove

$$P \Rightarrow Q \text{ and } Q \Rightarrow R \text{ and } R \Rightarrow S \text{ and } S \Rightarrow T \text{ and } T \Rightarrow P.$$

All the other required implications then follow immediately; for instance, from $T \Rightarrow P$ and $P \Rightarrow Q$, we know that $T \Rightarrow Q$, and similarly in the other cases.

> **EXAMPLE** In order to prove this theorem about integers:
>
> *The following conditions on a positive integer p are equivalent:*
>
> *(1) p is prime.*
>
> *(2) If p is a factor of ab, then p is a factor of a or p is a factor of b.*
>
> *(3) If p = rs, then r = ± 1 or s = ± 1.*
>
> you must make *three* separate arguments. First, assume (1) and prove (2), so that $(1) \Rightarrow (2)$ is true. Second, you assume (2) and prove (3), so that $(2) \Rightarrow (3)$ is true. Finally, you must assume (3) and prove (1), so that $(3) \Rightarrow (1)$ is true. *Be careful:* At each stage you assume only one of the three statements and use it to prove another; the third statement does not play a role in that part of the argument.

APPENDIX B

Sets and Functions

◆

For our purposes, a **set** is any collection of objects; for example,

> The set $\mathbb{Z}$ of integers.
>
> The set of right triangles with area 24.
>
> The set of positive irrational numbers.

The objects in a set are called **elements** or **members** of the set. If B is a set, the statement "b is an element of B" is abbreviated as "$b \in B$." Similarly, "$b \notin B$" means "b is not an element of B." For example, if $\mathbb{Z}$ is the set of integers,* then

$$2 \in \mathbb{Z} \qquad \text{and} \qquad \pi \notin \mathbb{Z}.$$

There are several methods of describing sets. A set may be defined by **verbal description** as in the examples above. A small finite set can be described by **listing** all its elements. Such a list is customarily placed between curly brackets; for instance,

$$\{3, 7, -4, 9\} \qquad \text{or} \qquad \{a, b, c, r, s, t\}.$$

Listing notation is sometimes used for infinite sets as well. For example, $\{2, 4, 6, 8, \ldots\}$ indicates the set of positive even integers. Strictly speaking, this notation is ambiguous in the infinite case since it relies on everyone's seeing the same pattern and understanding that it is to continue forever. But when the context is clear, no confusion will result.

Finally, a set can be described in terms of properties that are satisfied by its elements, and by these elements only. This is usually done with **set-builder notation.** For example,

$$\{x \mid x \text{ is an integer and } x > 9\}$$

---

* Throughout this book boldface capital $\mathbb{Z}$ always denotes the set of integers.

denotes the set of all elements x *such that* x is an integer greater than 9. In general, the vertical line is shorthand for "such that" and "$\{y \mid P\}$" is read "the set of all elements y such that P." Thus each of the following is the set of even integers:

> $\{x \mid x$ is an even integer$\}$.
>
> $\{t \mid t \in \mathbb{Z}$ and t is even$\}$.
>
> $\{r \mid r \in \mathbb{Z}$ and r is a multiple of 2$\}$.
>
> $\{y \mid y \in \mathbb{Z}$ and $y = 2k$ for some integer $k\}$.

The Empty Set

Some special cases of set-builder notation lead to an unusual set. For instance, the set

$$\{x \mid x \text{ is an integer and } 0 < x < 1\}$$

has no elements since there is no integer between 0 and 1. The set with no elements is called the **empty set** or **null set** and is denoted $\varnothing$. For every element c,

$$c \in \varnothing \text{ is false} \quad \text{and} \quad c \notin \varnothing \text{ is true.}$$

The empty set is a very convenient concept to have around, but some care must be taken when dealing with theorems that are true only for **nonempty sets** (that is, sets that have at least one element).

Subsets

A set B is said to be a **subset** of a set C (written $B \subseteq C$) provided that every element of B is also an element of C. In other words, $B \subseteq C$ exactly when this statement is true:

$$x \in B \Rightarrow x \in C.$$

For example, the set of even integers is a subset of the set $\mathbb{Z}$ of all integers, and the set of rational numbers is a subset of the set of real numbers.

The definition of "$B \subseteq C$" allows the possibility that $B = C$ (since it is certainly true in this case that every element of B is also an element of C). In other words,

$$B \subseteq B \text{ for every set } B.$$

If B is a subset of C and $B \neq C$ we say that B is a **proper subset** of C and write $B \subsetneq C$.

The subset relation is easily seen to be *transitive,* that is,

$$\textbf{If } B \subseteq C \textbf{ and } C \subseteq D, \textbf{ then } B \subseteq D.$$

Two sets B and C are **equal** when they have exactly the same elements. In this case every element of B is an element of C *and* every element of C is an element of B. Thus,

$$B = C \quad \text{if and only if} \quad B \subseteq C \text{ and } C \subseteq B.$$

This fact is the most commonly used method of proving that two sets are equal: Prove that each is a subset of the other.

Basic logic leads to a surprising fact about the empty set. Since the statement $x \in \varnothing$ is always false, the implication

$$x \in \varnothing \Rightarrow x \in C$$

is always true (see Appendix A). But this is precisely the definition of "$\varnothing$ is a subset of C." So

the empty set $\varnothing$ is a subset of every set.

Operations on Sets

We now review the standard ways of constructing new sets from given ones. If B and C are sets, then the **relative complement** of C in B is denoted $B - C$ and consists of the elements of B that are not in C. Thus

$$B - C = \{x \mid x \in B \text{ and } x \notin C\}.$$

For example, if E is the set of even integers, then $\mathbb{Z} - E$ is the set of odd integers.

The **intersection** of sets B and C consists of all the elements that are in *both* B and C and is denoted $B \cap C$. Thus

$$B \cap C = \{x \mid x \in B \text{ and } x \in C\}.$$

For example, if $B = \{-2, 1, \sqrt{2}, 5, \pi\}$ and C is the set of positive rational numbers, then $B \cap C = \{1, 5\}$ since 1 and 5 are the only elements in both sets. If B is the set of positive integers and C the set of negative integers, then $B \cap C = \varnothing$ since there are no elements in both sets. When B and C are sets such that $B \cap C = \varnothing$, we say that B and C are **disjoint.**

The **union** of sets B and C consists of all elements that are in at least one of B or C and is denoted $B \cup C$. Thus,

$$B \cup C = \{x \mid x \in B \text{ or } x \in C\}.$$

For example, the union of $B = \{1, 3, 5, 7\}$ and $C = \{-1, 1, 4, 9\}$ is $B \cup C = \{-1, 1, 3, 4, 5, 7, 9\}$. If B is the set of rational numbers and C is the set of irrational numbers, then $B \cup C$ is the set of all real numbers.

You should verify that union and intersection have the following properties. For any sets B, C, and D:

$$B \cup B = B \qquad B \cap B = B$$
$$B \cup \varnothing = B \qquad B \cap \varnothing = \varnothing$$
$$B \cup C = C \cup B \qquad B \cap C = C \cap B$$
$$B \subseteq B \cup C \qquad B \cap C \subseteq B$$
$$B \subseteq C \quad \text{if and only if} \quad B \cup C = C$$
$$B \subseteq C \quad \text{if and only if} \quad B \cap C = B$$
$$B \cup (C \cup D) = (B \cup C) \cup D \qquad B \cap (C \cap D) = (B \cap C) \cap D$$
$$B \cap (C \cup D) = (B \cap C) \cup (B \cap D)$$
$$B \cup (C \cap D) = (B \cup C) \cap (B \cup D).$$

The concepts of union and intersection extend readily to large, possibly infinite, collections of sets. Suppose that I is some nonempty set (called an **index set**) and that for each $i \in I$, we are given a set A_i. Then the intersection of this family of sets (denoted $\bigcap_{i \in I} A_i$) is the set of elements that are in *all* the sets A_i, that is,

$$\bigcap_{i \in I} A_i = \{x \mid x \in A_i \text{ for every } i \in I\}.$$

Similarly, the union of this family of sets (denoted $\bigcup_{i \in I} A_i$) is the set of elements that are in at least one of the sets A_i, that is,

$$\bigcup_{i \in I} A_i = \{x \mid x \in A_j \text{ for some } j \in I\}.$$

The **Cartesian product** of sets B and C is denoted $B \times C$ and consists of all ordered pairs (x,y) with $x \in B$ and $y \in C$. Equality of ordered pairs is defined by this rule:

$$(x,y) = (u,v) \qquad \text{if and only if} \qquad x = u \text{ in } B \text{ and } y = v \text{ in } C.$$

For example, if $B = \{r, s, t\}$ and $C = \{5, 7\}$, then $B \times C$ is the set

$$\{(r,5),\ (r,7),\ (s,5),\ (s,7),\ (t,5),\ (t,7)\}.$$

The set $\mathbb{R}$ of real numbers is sometimes identified with the number line. When this is done, the Cartesian product $\mathbb{R} \times \mathbb{R}$ is just the ordinary coordinate plane, the set of all points with coordinates (x,y) where $x, y \in \mathbb{R}$.

The Cartesian product of any finite number of sets $B_1, B_2, \ldots, B_n$ is defined in a similar fashion. $B_1 \times B_2 \times \cdots \times B_n$ is the set of all ordered n-tuples $(x_1, x_2, \ldots, x_n)$ where $x_i \in B_i$ for each $i = 1, 2, \ldots, n$. For example, if $B = \{0, 1\}$, $\mathbb{Z}$ is the set of integers, and $\mathbb{R}$ the set of real numbers, then $B \times \mathbb{Z} \times \mathbb{R}$ is the set of all ordered triples of the form $(0,k,r)$ and $(1,k,r)$ with $k \in \mathbb{Z}$ and $r \in \mathbb{R}$. The product $B \times \mathbb{Z} \times \mathbb{R}$ is an infinite set; among its elements are $(0,-5,3)$, $(1,24,\pi)$, and $(1,1,-\sqrt{3})$.

FUNCTIONS

A **function** (or **map** or **mapping**) f from a set B to a set C (denoted $f: B \to C$) is a rule that assigns to each element b of B exactly one element c of C; c is called the **image** of b or the **value** of the function f at b and is usually denoted $f(b)$. The set B is called the **domain** and the set C the **range** of the function f.

Your previous mathematics courses dealt with a wide variety of functions. For instance, if $\mathbb{R}$ is the set of real numbers, then each of the following rules defines a function from $\mathbb{R}$ to $\mathbb{R}$:

$$f(x) = \cos x, \qquad g(x) = x^2 + 1, \qquad h(x) = x^3 - 5x + 2.$$

The rule of a function need not be given by an algebraic formula. For instance, consider the function $f: \mathbb{Z} \to \{0, 1\}$, whose rule is

$$f(x) = 0 \text{ if } x \text{ is even and } f(x) = 1 \text{ if } x \text{ is odd.}$$

If B is a set, then the function from B to B defined by the rule "map every element to itself" is called the **identity map** on B and is denoted ι_B. Thus $\iota_B: B \to B$ is defined by

$$\iota_B(x) = x \text{ for every } x \in B.$$

Composition of Functions

Let f and g be functions such that the range of f is the same as the domain of g, say $f: B \to C$ and $g: C \to D$. Then the **composite** of f and g is the function $h: B \to D$ whose rule is

$$h(x) = g(f(x)).$$

In other words, the composite function is obtained by first applying f and then applying g:

$$B \xrightarrow{\ \ f\ \ } C \xrightarrow{\ \ g\ \ } D$$
$$x \dashrightarrow f(x) \dashrightarrow g(f(x)).$$

Instead of h, the usual notation for the composite function of f and g is $g \circ f$ (note the order). Thus $(g \circ f)(x) = g(f(x))$.

> **EXAMPLE**   Let E be the set of even integers and $\mathbb{N}$ the set of nonnegative integers. Let $f: E \to \mathbb{Z}$ be defined by $f(x) = x/2$ (since x is even, $x/2$ *is* an integer). Let $g: \mathbb{Z} \to \mathbb{N}$ be given by $g(n) = n^2$. Then the composite function $g \circ f: E \to \mathbb{N}$ has this rule:
>
> $$(g \circ f)(x) = g(f(x)) = g(x/2) = (x/2)^2 = x^2/4.$$

The composite function in the opposite order, $f \circ g$ (first apply g, then f), is *not defined* since the range of g is not the same as the domain of f. For instance, $g(3) = 9$, but the domain of f is the set of even integers; even though the rule of f makes sense for odd integers, $f(g(3)) = f(9) = 9/2$, which is not in $\mathbb{Z}$.

EXAMPLE Let $f:\mathbb{Z} \to \mathbb{Z}$ and $g:\mathbb{Z} \to \mathbb{Z}$ be given by $f(x) = x - 1$ and $g(x) = x^2$. Then the composite function $f \circ g$ is given by the rule

$$(f \circ g)(x) = f(g(x)) = f(x^2) = x^2 - 1.$$

In this case the composite function in the opposite order $g \circ f$ is also defined; its rule is

$$(g \circ f)(x) = g(f(x)) = g(x - 1) = (x - 1)^2 = x^2 - 2x + 1.$$

Thus we have, for instance,

$$(f \circ g)(3) = 9 - 1 = 8 \qquad \text{but} \qquad (g \circ f)(3) = 9 - 6 + 1 = 4.$$

So even though both are defined, $f \circ g$ *is not the same function as* $g \circ f$.

Two functions $h:B \to C$ and $k:B \to C$ are said to be **equal** provided that $h(b) = k(b)$ for every $b \in B$.

EXAMPLE Let $f:B \to C$ be any function and $\iota_C:C \to C$ the identity map on C. Then $\iota_C \circ f:B \to C$, and for every $b \in B$

$$(\iota_C \circ f)(b) = \iota_C(f(b)) = f(b).$$

Therefore $\iota_C \circ f = f$. Similarly, if ι_B is the identity map on B, then $f \circ \iota_B:B \to C$, and for every $b \in B$

$$(f \circ \iota_B)(b) = f(\iota_B(b)) = f(b).$$

Consequently,

$$\text{If } f:B \longrightarrow C, \text{ then} \qquad \iota_C \circ f = f \qquad \text{and} \qquad f \circ \iota_B = f.$$

If $f:B \to C$, $g:C \to D$, and $h:D \to E$ are functions, then each of the composite functions $(f \circ g) \circ h$ and $f \circ (g \circ h)$ is a map from B to E. We claim that

$$(f \circ g) \circ h = f \circ (g \circ h).$$

The proof of this statement is simply an exercise in using the definition of composite function. For each $b \in B$

$$[(f \circ g) \circ h](b) = (f \circ g)(h(b)) = f[g(h(b))]$$

and

$$[f \circ (g \circ h)](b) = f[(g \circ h)(b)] = f[g(h(b))].$$

Since the right sides of the two equalities are identical, the composite functions $(f \circ g) \circ h$ and $f \circ (g \circ h)$ have the same effect on each $b \in B$, which proves the claim.

Binary Operations

Informally we can think of a binary operation on the integers, for example, as a rule for producing a new integer from two given ones. Ordinary addition and multiplication are operations in this sense: Given a and b we get $a + b$ and ab. Producing a new integer from a pair of given ones also suggests the idea of a function. Addition of integers may be thought of as the function f from $\mathbb{Z} \times \mathbb{Z}$ to $\mathbb{Z}$ whose rule is

$$f(a,b) = a + b.$$

Similarly, multiplication can be thought of as the function $g : \mathbb{Z} \times \mathbb{Z} \to \mathbb{Z}$ given by $g(a,b) = ab$.

With the preceding examples in mind we make this formal definition. A **binary operation** on a nonempty set B (usually called simply an **operation** on B) is a function $f : B \times B \to B$. The familiar examples suggest a new notation for the general case. We use some symbol, say $*$, to denote the operation and write $a * b$ instead of $f(a,b)$.

EXAMPLE As we saw above, ordinary addition and multiplication are operations on $\mathbb{Z}$. Another operation on $\mathbb{Z}$ is defined by the function $f : \mathbb{Z} \times \mathbb{Z} \to \mathbb{Z}$ whose rule is $f(a,b) = ab - 1$. If we denote this operation by $*$, then $3 * 5 = 15 - 1 = 14$, and, similarly,

$$12 * 4 = 47 \qquad -7 * 4 = -29 \qquad 0 * 8 = -1.$$

Note that $a * b = ab - 1 = ba - 1 = b * a$, so that the order of the elements doesn't matter when applying $*$, as is the case with ordinary addition and multiplication (the technical term for this property is *commutativity*). On the other hand,

$$(1 * 2) * 3 = 1 * 3 = 2 \qquad \text{but} \qquad 1 * (2 * 3) = 1 * 5 = 4,$$

so that $(a * b) * c \neq a * (b * c)$ in general. Thus $*$ is not *associative* as are addition and multiplication (meaning that $(a + b) + c = a + (b + c)$ and $(ab)c = a(bc)$ always).

EXAMPLE Let S be a nonempty set. If $f : S \to S$ and $g : S \to S$ are functions, then their composite $f \circ g$ is also a function from S to S. So if B is the set of all functions from S to S, then composition of functions is an operation on the set B. In other words, the map that sends (f,g) to $f \circ g$ is a function from $B \times B$ to B. The discussion of composite functions above shows that the operation $\circ$ on B is associative (that is, $(f \circ g) \circ h = f \circ (g \circ h)$ always) but not commutative ($f \circ g$ need not equal $g \circ f$).

Let $*$ be an operation on a set B and $C \subseteq B$. The subset C is said to be **closed** under the operation $*$ provided that

Whenever $a, b \in C$, then $a * b \in C$.

Consider, for example, the operation of ordinary multiplication on the set B of positive real numbers. Let C be the subset of positive integers. Then C is closed under the operation since ab is a positive integer whenever a and b are. But when the operation on B is ordinary division, then C is not closed: If a and b are integers, $a \div b$ need not be an integer (for instance, $3 \div 7 = 3/7 \notin C$).

If $*$ is an operation on a set B, then B (considered as a subset of itself) is closed under $*$ by the definition of an operation. Nevertheless many texts, including this one, routinely list the **closure** of B under $*$ as one of the properties of the operation. Although this isn't logically necessary, it calls your attention to the importance of closure and reminds you that closure cannot be taken for granted for subsets other than B.

Injective and Surjective Functions

A function $f : B \to C$ is said to be **injective** (or **one-to-one**) provided f maps distinct elements of B to distinct elements of C, or in functional notation: If $a \neq b$ in B, then $f(a) \neq f(b)$ in C. This rather awkward statement is equivalent to its contrapositive, so that we have this useful description:

$f : B \longrightarrow C$ **is injective provided that
whenever** $f(a) = f(b)$ **in** C**, then** $a = b$ **in** B**.**

EXAMPLE   Let $\mathbb{R}$ be the set of real numbers. In order to show that the function $f : \mathbb{R} \to \mathbb{R}$ given by $f(x) = 2x + 3$ is injective, we assume that $f(a) = f(b)$, that is,

$$2a + 3 = 2b + 3.$$

Subtracting 3 from each side shows that $2a = 2b$; dividing both sides by 2 we conclude that $a = b$. Therefore, f is injective.

EXAMPLE   The map $f : \mathbb{Z} \to \mathbb{Z}$ given by $f(x) = x^2$ is *not* injective because we have $f(-3) = 9 = f(3)$, but $-3 \neq 3$. Alternatively, the distinct elements 3 and -3 have the same image.

A function $f: B \to C$ is said to be **surjective** (or **onto**) provided that every element of C is the image under f of at least one element of B, that is,

If $c \in C$, then there exists $b \in B$ such that $f(b) = c$.

> **EXAMPLE**   Let $\mathbb{N}$ be the set of nonnegative integers and $f: \mathbb{Z} \to \mathbb{N}$ the function given by $f(x) = |x|$. Then f is surjective since every element of $\mathbb{N}$ is the image under f of at least one element of $\mathbb{Z}$ (namely itself). Note, however, that f is not injective since, for example, $f(1) = f(-1)$.

> **EXAMPLE**   Let E be the set of even integers and consider the map $g: \mathbb{Z} \to E$ given by $g(x) = 4x$. We claim that the element 2 in E is *not* the image under g of any element of $\mathbb{Z}$. If $2 = g(b)$ for some $b \in \mathbb{Z}$, then $2 = 4b$, so that $1 = 2b$. This is impossible since 1 is not an integer multiple of 2. Therefore, g is *not* surjective. Note, however, that g *is* injective since $4a = 4b$ (that is, $g(a) = g(b)$) implies that $a = b$.

> **EXAMPLE**   Let $\mathbb{R}$ be the set of real numbers and $f: \mathbb{R} \to \mathbb{R}$ the function given by $f(x) = 2x + 3$. To prove that f is surjective, let $c \in \mathbb{R}$; we must find $b \in \mathbb{R}$ such that $f(b) = c$. In other words, we must find a number b such that $2b + 3 = c$. To do so, we solve this last equation for b and find $b = \dfrac{c - 3}{2}$. Then $f(b) = 2\left(\dfrac{c - 3}{2}\right) + 3 = c - 3 + 3 = c$. Therefore, f is surjective. The map f is also injective (see the Example on page 511).

The preceding examples demonstrate that *injectivity and surjectivity are independent concepts.* One does not imply the other, and a particular map might have one, both, or neither of these properties.

If $f: B \to C$ is a function, then the **image of f** is this subset of C:

$$\text{Im } f = \{c \mid c = f(b) \text{ for some } b \in B\} = \{f(b) \mid b \in B\}.$$

For example, if $f: \mathbb{Z} \to \mathbb{Z}$ is given by $f(x) = 2x$, then Im f is the set of even integers since Im $f = \{f(x) \mid x \in \mathbb{Z}\} = \{2x \mid x \in \mathbb{Z}\}$. Similarly, if $g: \mathbb{Z} \to \mathbb{Z}$ is given by $g(x) = |x|$, then Im g is the set of nonnegative integers. A map $f: B \to C$ is surjective exactly when every element of C is the image of an element of B. Thus

$$f: B \to C \text{ is surjective if and only if Im } f = C.$$

If $f: B \to C$ is a function and S is a subset of B, then the **image of the subset S** is the set

$$f(S) = \{c \mid c = f(b) \text{ for some } b \in S\} = \{f(b) \mid b \in S\}.$$

If $f:\mathbb{Z} \to \mathbb{Z}$ is given by $f(x) = 2x$, for example, and S is the set of odd integers, then $f(S) = \{2x \mid x \text{ is odd}\}$ is the set of even integers that are not multiples of 4. If the subset S is the entire set B, then $f(B)$ is precisely Im f.

Bijective Functions

A function $f:B \to C$ is **bijective** (or a **bijection** or **one-to-one correspondence**) provided that f is both injective and surjective.

EXAMPLE The Examples on pages 511 and 512 show that the map $f:\mathbb{R} \to \mathbb{R}$ given by $f(x) = 2x + 3$ is bijective.

EXAMPLE The map f from the set $\{1, 2, 3, 4, 5\}$ to the set $\{v, w, x, y, z\}$ given by

$$f(1) = v \qquad f(2) = w \qquad f(3) = x \qquad f(4) = y \qquad f(5) = z$$

is easily seen to be bijective.

The last example illustrates the fact that for any *finite* sets B and C, there is a bijection from B to C if and only if B and C have the same number of elements. In particular, if B is finite and $C \subsetneq B$, then there cannot be a bijection from B to C. But the situation is quite different with infinite sets.

EXAMPLE Let E be the set of even integers and consider the map $f:\mathbb{Z} \to E$ given by $f(x) = 2x$. By definition every even integer is 2 times some integer, so f is surjective. Furthermore, $2a = 2b$ implies that $a = b$, so f is injective. Therefore, f is a bijection. In this case, a bit more is true. Define a map $g:E \to \mathbb{Z}$ by $g(u) = u/2$; this makes sense since $u/2$ is an integer when u is even. Consider the composite function $g \circ f:\mathbb{Z} \to \mathbb{Z}$:

$$(g \circ f) = g(f(x)) = g(2x) = 2x/2 = x.$$

Thus $(g \circ f)(x) = x = \iota_{\mathbb{Z}}(x)$ for every x, and the composite map $g \circ f$ is just the identity map $\iota_{\mathbb{Z}}$ on $\mathbb{Z}$. Now look at the other composite, $f \circ g:E \to E$:

$$(f \circ g)(u) = f(g(u)) = f(u/2) = 2(u/2) = u.$$

Therefore, the composite map $f \circ g$ is the identity map ι_E.

The preceding example illustrates a property that all bijective functions have, as we now prove.

THEOREM B.1 *A function $f:B \to C$ is bijective if and only if there exists a function $g:C \to B$ such that*

$$g \circ f = \iota_B \qquad and \qquad f \circ g = \iota_C.$$

Proof   Assume first that f is bijective. Define $g : C \to B$ as follows. If $c \in C$, then there exists $b \in B$ such that $f(b) = c$ because f is surjective. Furthermore, since f is also injective, there is only one element b such that $f(b) = c$ (for if $f(b') = c$, then $f(b) = f(b')$ implies $b = b'$). So we can define a function $g : C \to B$ by this rule:

$g(c) = b$, where b is the unique element of B such that $f(b) = c$.

Then $g(c) = b$ exactly when $f(b) = c$. Thus for any $c \in C$

$$(f \circ g)(c) = f(g(c)) = f(b) = c,$$

from which we conclude that $f \circ g = \iota_C$. Similarly, for each $u \in B$, $f(u)$ is an element of C, say $f(u) = v$, and, hence, by the definition of g, we have $g(v) = u$. Therefore,

$$(g \circ f)(u) = g(f(u)) = g(v) = u$$

and $g \circ f = \iota_B$. This proves the first half of our biconditional theorem.

To prove the other half, we assume that a map $g : C \to B$ with the stated properties is given. We must show that f is bijective. Suppose $f(a) = f(b)$. Then

$$g(f(a)) = g(f(b))$$
$$(g \circ f)(a) = (g \circ f)(b)$$
$$\iota_B(a) = \iota_B(b)$$
$$a = b.$$

Therefore, $f(a) = f(b)$ implies $a = b$, and f is injective. To show that f is surjective, let c be any element of C. Then $g(c) \in B$ and $f(g(c)) = (f \circ g)(c) = \iota_C(c) = c$. So we have found an element of B that f maps onto c (namely $g(c)$); hence, f is surjective. Therefore, f is bijective, and the theorem is proved.   ◆

If $f : B \to C$ is a bijection, then the map g in Theorem B.1 is called the **inverse** of f and is sometimes denoted by f^{-1}. Reversing the roles of f and g in Theorem B.1 shows that the inverse map g of a bijection f is itself a bijection.

◆ EXERCISES

NOTE: $\mathbb{Z}$ *is the set of integers,* $\mathbb{Q}$ *the set of rational numbers, and* $\mathbb{R}$ *the set of real numbers.*

A.   **1.** Describe each set by listing:

   (a) The integers strictly between -3 and 9.

   (b) The negative integers greater than -10.

   (c) The positive integers whose square roots are less than or equal to 4.

2. Describe each set in set-builder notation:

(a) All positive real numbers.

(b) All negative irrational numbers.

(c) All points in the coordinate plane with rational first coordinate.

(d) All negative even integers greater than -50.

3. Which of the following sets are nonempty?

(a) $\{r \in \mathbb{Q} \mid r^2 = 2\}$

(b) $\{r \in \mathbb{R} \mid r^2 + 5r - 7 = 0\}$

(c) $\{t \in \mathbb{Z} \mid 6t^2 - t - 1 = 0\}$

4. Is B a subset of C when

(a) $B = \mathbb{Z}$ and $C = \mathbb{Q}$?

(b) $B =$ all solutions of $x^2 + 2x - 5 = 0$ and $C = \mathbb{Z}$?

(c) $B = \{a, b, 7, 9, 11, -6\}$ and $C = \mathbb{Q}$?

5. If $A \subseteq B$ and $B \subseteq C$, prove that $A \subseteq C$.

6. In each part find $B - C$, $B \cap C$, and $B \cup C$:

(a) $B = \mathbb{Z}$, $C = \mathbb{Q}$.   **(b)** $B = \mathbb{R}$, $C = \mathbb{Q}$.

(c) $B = \{a, b, c, 1, 2, 3, 4, 5\}$, $C = \{a, c, e, 2, 4, 6, 8\}$.

7. List the elements of $B \times C$ when $B = \{a, b, c\}$ and $C = \{0, 1, c\}$.

8. List the elements of $A \times B \times C$ when $A = \{0, 1\}$ and B, C are as in Exercise 7.

9. Let $A = \{1, 2, 3, 4\}$. Exhibit functions f and g from A to A such that $f \circ g \neq g \circ f$.

10. Do Exercise 9 when $A = \mathbb{Z}$.

11. Is the subset B closed under the given operation?

(a) $B =$ even integers; operation: multiplication in $\mathbb{Z}$.

(b) $B =$ odd integers; operation: addition in $\mathbb{Z}$.

(c) $B =$ nonzero rational numbers; operation: division in the set of nonzero real numbers.

(d) $B =$ odd integers; operation $*$ on $\mathbb{Z}$, where $a * b$ is defined to be the number $ab - (a + b) + 2$.

12. Find the image of the function f when

(a) $f : \mathbb{R} \to \mathbb{R}$; $f(x) = x^2$.   **(b)** $f : \mathbb{Z} \to \mathbb{Q}$; $f(x) = x - 1$.

(c) $f : \mathbb{R} \to \mathbb{R}$; $f(x) = -x^2 + 1$.

13. Let $B = \{1, 2, 3, 4\}$ and $C = \{a, b, c\}$.

    (a) List four different surjective functions from B to C.

    (b) List four different injective functions from C to B.

    (c) List all bijective functions from C to C.

14. (a) Give an example of a function f that is injective but not surjective.

    (b) Give an example of a function g that is surjective but not injective.

15. Let B and C be nonempty sets. Prove that the function

$$f : B \times C \longrightarrow C \times B$$

given by $f(x,y) = (y,x)$ is a bijection.

B. 16. List all the subsets of $\{1, 2\}$. Do the same for $\{1, 2, 3\}$ and $\{1, 2, 3, 4\}$. Make a conjecture as to the number of subsets of an n-element set. [Don't forget the empty set.]

17. Verify each of the properties of sets listed on page 507.

18. If $a, b \in \mathbb{R}$ with $a < b$, then the set $\{r \in \mathbb{R} \mid a \leq r < b\}$ is denoted $[a,b)$. Let N denote the nonnegative integers and P the positive integers. Find these unions and intersections:

    (a) $\displaystyle\bigcup_{n \in N} [n, n+1)$
    (b) $\displaystyle\bigcap_{n \in P} \left[-\frac{1}{n}, 0 \right)$

    (c) $\displaystyle\bigcup_{n \in P} \left[\frac{1}{n}, 2 + \frac{1}{n} \right)$
    (d) $\displaystyle\bigcap_{n \in P} \left[\frac{1}{n}, 2 + \frac{1}{n} \right)$

19. Prove that for any sets A, B, C:

$$A \times (B \cup C) = (A \times B) \cup (A \times C)$$

20. Let A, B be subsets of U. Prove **De Morgan's laws:**

    (a) $U - (A \cap B) = (U - A) \cup (U - B)$

    (b) $U - (A \cup B) = (U - A) \cap (U - B)$

21. Prove that for any sets A, B, C:

$$(A - B) \cup (B - A) = (A \cup B) - (A \cap B)$$

22. If C is a finite set, then $|C|$ denotes the number of elements in C. If A and B are finite sets, is it true that $|A \cup B| = |A| + |B|$?

23. Let $\mathbb{R}^{**}$ denote the positive real numbers. Does the following rule define a function from $\mathbb{R}^{**}$ to $\mathbb{R}$: assign to each positive real number c the real number whose square is c?

24. Determine whether the given operation on $\mathbb{R}$ is commutative (that is, $a * b = b * a$ for all a, b) or associative (that is, $a * (b * c) = (a * b) * c$ for all a, b, c).

(a) $a * b = 2^{ab}$ **(b)** $a * b = ab^2$

(c) $a * b = 0$ **(d)** $a * b = (a + b)/2$

(e) $a * b = 1$ **(f)** $a * b = b$

(g) $a * b = a^2 + b^2$

25. Prove that the given function is injective.

(a) $f: \mathbb{Z} \to \mathbb{Z}; f(x) = 2x$ **(b)** $f: \mathbb{R} \to \mathbb{R}; f(x) = x^3$

(c) $f: \mathbb{Z} \to \mathbb{Q}; f(x) = x/7$ **(d)** $f: \mathbb{R} \to \mathbb{R}; f(x) = -3x + 5$

26. Prove that the given function is surjective.

(a) $f: \mathbb{R} \to \mathbb{R}; f(x) = x^3$

(b) $f: \mathbb{Z} \to \mathbb{Z}; f(x) = x - 4$

(c) $f: \mathbb{R} \to \mathbb{R}; f(x) = -3x + 5$

(d) $f: \mathbb{Z} \times \mathbb{Z} \to \mathbb{Q}; f(a,b) = a/b$ when $b \neq 0$ and 0 when $b = 0$.

27. Let $f: B \to C$ and $g: C \to D$ be functions. Prove:

(a) If f and g are injective, then $g \circ f: B \to D$ is injective.

(b) If f and g are surjective, then $g \circ f$ is surjective.

28. **(a)** Let $f: B \to C$ and $g: C \to D$ be functions such that $g \circ f$ is injective. Prove that f is injective.

(b) Give an example of the situation in part (a) in which g is not injective.

29. **(a)** Let $f: B \to C$ and $g: C \to D$ be functions such that $g \circ f$ is surjective. Prove that g is surjective.

(b) Give an example of the situation in part (a) in which f is not surjective.

30. Let $g: B \times C \to C$ (with $B \neq \varnothing$) be the function given by $g(x,y) = y$.

(a) Prove that g is surjective.

(b) Under what conditions, if any, is g injective?

31. If $f: B \to C$ is a function, then f can be considered as a map from B to $\text{Im } f$ since $f(b) \in \text{Im } f$ for every $b \in B$. Show that the map $f: B \to \text{Im } f$ is surjective.

32. Let B be a finite set and $f: B \to B$ is a function. Prove that f is injective if and only if f is surjective.

33. Let $f : B \to C$ be a function and let S, T be subsets of B.

    (a) Prove that $f(S \cup T) = f(S) \cup f(T)$.

    (b) Prove that $f(S \cap T) \subseteq f(S) \cap f(T)$.

    (c) Give an example where $f(S \cap T) \neq f(S) \cap f(T)$.

34. Prove that $f : B \to C$ is injective if and only if $f(S \cap T) = f(S) \cap f(T)$ for every pair of subsets S, T of B.

35. Let $f : B \to C$ and $g : C \to D$ be bijective functions. Then the composite function $g \circ f : B \to D$ is bijective by Exercise 27. Prove that $(g \circ f)^{-1} = f^{-1} \circ g^{-1}$.

APPENDIX **C**

Well Ordering and Induction

We assume that you are familiar with ordinary arithmetic in the set $\mathbb{Z}$ of integers and with the usual order relation $(<)$ on $\mathbb{Z}$. The subset of nonnegative integers will be denoted by $\mathbb{N}$. Thus

$$\mathbb{N} = \{0, 1, 2, 3, \ldots\}.$$

Finally, we assume this fundamental axiom:

WELL-ORDERING AXIOM *Every nonempty subset of $\mathbb{N}$ contains a smallest element.*

Most people find this axiom quite plausible, but it is important to note that it may not hold if $\mathbb{N}$ is replaced by some other set of numbers; see page 2 of the text for examples.

An important consequence of the Well-Ordering Axiom is the method of proof known as mathematical induction. It can be used to prove statements such as

A set of n elements has 2^n subsets.

Denote this statement by the symbol $P(n)$ and observe that there are really infinitely many statements, one for each possible value of n:

$P(0)$: A set of 0 elements has $2^0 = 1$ subset.

$P(1)$: A set of 1 element has $2^1 = 2$ subsets.

$P(2)$: A set of 2 elements has $2^2 = 4$ subsets.

$P(3)$: A set of 3 elements has $2^3 = 8$ subsets.

And so on. To prove the original proposition we must prove that

$P(n)$ is a true statement for every $n \in \mathbb{N}$.

Here's how it can be done.

THEOREM C.1   (THE PRINCIPLE OF MATHEMATICAL INDUCTION)   *Assume that for each nonnegative integer n, a statement P(n) is given. If*

(i) *P(0) is a true statement; and*

(ii) *Whenever P(k) is a true statement, then P(k + 1) is also true,*

then P(n) is a true statement for every $n \in \mathbb{N}$.

The example of the number of subsets of a set of n elements is continued after the proof of the theorem. You may want to read that example now to see how Theorem C.1 is *applied,* which is quite different from the manner in which it is proved.

Proof of Theorem C.1   Let S be the subset of $\mathbb{N}$ consisting of those integers j for which $P(j)$ is *false.* To prove the theorem we need only show that S is empty; we shall use proof by contradiction to do this. Suppose S is nonempty. Then by the Well-Ordering Axiom, S contains a smallest element, say d. Since $P(d)$ is false by the definition of S and $P(0)$ is true by property (i), we must have $d \neq 0$. Consequently, $d \geq 1$ (because d is a nonnegative integer), and, hence, $d - 1 \geq 0$, that is, $d - 1 \in \mathbb{N}$. Since $d - 1 < d$ and d is the smallest element in S, $d - 1$ cannot be in S. Therefore, $P(d - 1)$ must be true (otherwise $d - 1$ would be in S). Property (ii) (with $k = d - 1$) implies that $P((d - 1) + 1) = P(d)$ is also a true statement. This is a contradiction since $d \in S$. Therefore, S is the empty set, and the theorem is proved.   ◆

In order to apply the Principle of Mathematical Induction to a series of statements, you must verify that these statements satisfy *both* properties (i) and (ii). Note that property (ii) does *not* assert that any particular $P(k)$ is actually true, but only that a conditional relationship holds: *If $P(k)$ is true, then $P(k + 1)$ must also be true.* So to verify property (ii), you assume the truth of $P(k)$ and use this assumption to prove that $P(k + 1)$ is true. As we shall see in the examples below, it is often possible to prove this conditional statement even though you may not be able to prove directly that a particular $P(j)$ is true. The assumption that $P(k)$ is true is called the **induction assumption** or the **induction hypothesis.**

You may have seen induction used to prove statements such as "the sum of the first n nonnegative integers is $\dfrac{n(n + 1)}{2}$"; here $P(n)$ is the statement: "$0 + 1 + 2 + 3 + \cdots + n = \dfrac{n(n + 1)}{2}$." Although such examples make nice exercises for beginners, they are not typical of the way induction is used in advanced mathematics. The examples below will give you a more comprehensive picture of inductive proof. They are a bit more complicated than the usual elementary examples but are well within your reach.

EXAMPLE We shall use the Principle of Mathematical Induction to prove that for each $n \geq 0$,

> A set of n elements has 2^n subsets.

If $n = 0$, then the set must be the empty set (the only set with no elements). Its one and only subset is itself (since $\varnothing$ is a subset of every set). So the statement

$$P(0)\text{: A set of 0 elements has } 2^0 = 1 \text{ subset}$$

is true (property (i) holds).

In order to verify property (ii) of Theorem C.1, we assume the truth of

$$P(k)\text{: A set of } k \text{ elements has } 2^k \text{ subsets}$$

and use this induction hypothesis to prove

$$P(k+1)\text{: A set of } k+1 \text{ elements has } 2^{k+1} \text{ subsets.}$$

To do this, let T be any set of $k+1$ elements and choose some element c of T. Every subset of T either contains c or does not contain c. The subsets of T that do not contain c are precisely the subsets of $T - \{c\}$. Since the set $T - \{c\}$ has one fewer element than T, it is a set of k elements and, therefore, has exactly 2^k subsets (because the induction hypothesis $P(k)$ is assumed true). Now every subset of T that contains c must be of the form $\{c\} \cup D$, where D is a subset of $T - \{c\}$. There are 2^k possible choices for D and, hence, 2^k subsets of T that contain c. Consequently, the total number of subsets of T is

$$
\begin{pmatrix} \text{Number of subsets} \\ \text{that contain } c \end{pmatrix} + \begin{pmatrix} \text{Number of subsets that} \\ \text{do not contain } c \end{pmatrix} = 2^k + 2^k
$$

$$
= 2(2^k)
$$

$$
= 2^{k+1}.
$$

Thus any set T of $k+1$ elements has 2^{k+1} subsets, that is, $P(k+1)$ is a true statement. We have now verified property (ii) and can, therefore, apply Theorem C.1 to conclude that $P(n)$ is true for every $n \in \mathbb{N}$; that is, every set of n elements has 2^n subsets.

The Principle of Mathematical Induction cannot be conveniently used on certain propositions, even though they appear to be suitable for inductive proof. In such cases a variation on the procedure is needed:

THEOREM C.2 (THE PRINCIPLE OF COMPLETE INDUCTION) *Assume that for each nonnegative integer n, a statement P(n) is given. If*

(i) *P(0) is a true statement; and*

(ii) *Whenever P(j) is a true statement for all j such that $0 \leq j < t$, then P(t) is also true,*

then P(n) is a true statement for every $n \in \mathbb{N}$.

Although commonly used, the title "complete induction" is a bit of a misnomer since, as we shall see, this form of induction is equivalent to the previous one.

Proof of Theorem C.2 For each $n \in \mathbb{N}$, let $Q(n)$ be the statement

$$P(j) \text{ is true for all } j \text{ such that } 0 \leq j \leq n.$$

Note carefully that the last inequality sign in this statement is $\leq$ and not $<$. We shall use the Principle of Mathematical Induction (Theorem C.1) to show that $Q(n)$ is true for every $n \in \mathbb{N}$. This will mean, in particular, that $P(n)$ is true for every $n \in \mathbb{N}$. Now $Q(0)$ is the statement

$$P(j) \text{ is true for all } j \text{ such that } 0 \leq j \leq 0.$$

In other words, $Q(0)$ is just the statement "$P(0)$ is true." But we know that this is the case by hypothesis (i) in the theorem. Suppose that $Q(k)$ is true, that is,

$$P(j) \text{ is true for all } j \text{ such that } 0 \leq j \leq k.$$

By hypothesis (ii) (with $t = k + 1$), we conclude the $P(k + 1)$ is also true. Therefore, $P(j)$ is true for all j such that $0 \leq j \leq k + 1$, that is, $Q(k + 1)$ is a true statement. Thus we have shown that whenever $Q(k)$ is true, then $Q(k + 1)$ is also true. By the Principle of Mathematical Induction, $Q(n)$ is true for every $n \in \mathbb{N}$, and the proof is complete. ◆

In the formal description of induction (either principle), the notation $P(n)$ is quite convenient. But it is rarely used in actual proofs by induction. The next example is more typical of the way inductive proofs are usually phrased. But even here we include more detail than is customary in such proofs.

EXAMPLE We shall use the Principle of Complete Induction to prove:

If $n, b \in \mathbb{N}$ and $b > 0$, then there exist $q, r \in \mathbb{N}$ such that

(∗) $n = bq + r$ and $0 \leq r < b$.

This statement (called the **Division Algorithm** for nonnegative integers) is just a formalization of grade-school long division: When n is divided by

b, there is a quotient q and remainder r (smaller than the divisor b) such that $n = bq + r$; see the discussion on page 3 of the text.

　　Statement (∗) is true for $n = 0$ and any positive b (let $q = 0$ and $r = 0$). So property (i) of Theorem C.2 holds. Suppose that (∗) is true for all n such that $0 \le n < t$ (this is the induction hypothesis). We must show that (∗) is true for $n = t$. If $t < b$, then $t = b0 + t$, so (∗) is true with $q = 0$ and $r = t$. If $b \le t$, then $0 \le t - b < t$, and by the induction hypothesis, (∗) is true for $n = t - b$. Therefore, there exist integers q_1 and r_1 such that

$$t - b = q_1 b + r_1 \qquad \text{and} \qquad 0 \le r_1 < b.$$

Consequently,

$$t = b + q_1 b + r_1 = (1 + q_1)b + r_1 \qquad \text{and} \qquad 0 \le r_1 < b.$$

Therefore, (∗) is true for $n = t$ (with $q = 1 + q_1$ and $r = r_1$). Hence, property (ii) of Theorem C.2 is satisfied. By the Principle of Complete Induction, (∗) is true for every $n \in \mathbb{N}$.

Some mathematical statements are false (or undefined) for $n = 0$ or other small values of n but are true for $n = r$ and all subsequent integers. For instance, it can be shown that

$$3n > n + 1 \text{ for every integer } n \ge 1.$$
$$2^n > n^2 + 2 \text{ for every integer } n \ge 5.$$

Such statements can often be proved by using a variation of mathematical induction (either principle):

In order to prove that statement $P(n)$ is true for each integer $n \ge r$, follow the same basic procedure as before, starting with $P(r)$ instead of $P(0)$.

The validity of this procedure is a consequence of

THEOREM C.3　*Let r be a positive integer and assume that for each $n \ge r$ a statement $P(n)$ is given. If*

　　(i) $P(r)$ is a true statement;

and either

　　(ii) Whenever $k \ge r$ and $P(k)$ is true, then $P(k + 1)$ is true;

or

　　(ii') Whenever $P(j)$ is true for all j such that $r \le j < t$, then $P(t)$ is true,

then $P(n)$ is true for every $n \ge r$.

Proof Conditions (i) and (ii) are the analogue of Theorem C.1. Verify that the proof of Theorem C.1 carries over to the present case verbatim if 0 is replaced by r, 1 by $r + 1$, and $\mathbb{N}$ by the set $\mathbb{N}_r = \{n \mid n \in \mathbb{N} \text{ and } n \geq r\}$. Conditions (i) and (ii') are the analogue of Theorem C.2; its proof carries over similarly.   ◆

The final theorem to be proved here is not necessary in order to read the rest of the book. But it is a result that every serious mathematics student ought to know. It is also a good illustration of the fact that intuition can sometimes be misleading. Most people feel that the Well-Ordering Axiom is obvious, whereas the Principle of Complete Induction seems deeper and in need of some proof. But as we shall now see, these two statements are actually equivalent. Among other things, this suggests that the Well-Ordering Axiom is a good deal deeper than it first appears.

THEOREM C.4 *The following statements are equivalent:*

(1) *The Well-Ordering Axiom.*

(2) *The Principle of Mathematical Induction.*

(3) *The Principle of Complete Induction.*

Proof The proof of Theorem C.1 shows that $(1) \Rightarrow (2)$, and the proof of Theorem C.2 shows that $(2) \Rightarrow (3)$. To prove $(3) \Rightarrow (1)$, we assume the Principle of Complete Induction and let S be any subset of $\mathbb{N}$. To prove that the Well-Ordering Axiom holds, we must show

If S is nonempty, then S has a smallest element.

To do so, we shall prove the equivalent contrapositive statement

If S has no smallest element, then S is empty.

Assume S has no smallest element; to prove that S is empty we need only show that the following statement is true for every $n \in \mathbb{N}$:

(**) n is not an element of S.

Since 0 is the smallest element of $\mathbb{N}$, it is also the smallest element of any subset of $\mathbb{N}$ containing 0. Since S has no smallest element, 0 cannot be in S, and, hence, (**) is true when $n = 0$ (property (i) of Theorem C.2 holds). Suppose (**) is true for all j such that $0 \leq j < t$. Then none of the integers $0, 1, 2, \ldots, t - 1$ is in S, or equivalently, every element in S must be greater than or equal to t. If t were in S, then t would be the smallest element in S since $s \geq t$ for all $s \in S$. Since S has no smallest element, t is not in S. In other words, (**) is true when $n = t$. Thus the truth of (**) when $j < t$ implies its truth for t (property (ii) of Theorem C.2 holds). By the Principle of Complete Induction, (**) is true for all $n \in \mathbb{N}$. Therefore, S is empty, and the proof is complete.   ◆

◆ EXERCISES

A. **1.** Prove that the sum of the first n nonnegative integers is $n(n + 1)/2$. [*Hint:* Let $P(k)$ be the statement:

$$0 + 1 + 2 + \cdots + k = k(k + 1)/2.]$$

  2. Prove that for each nonnegative integer n, $2^n > n$.

  3. Prove that $2^{n-1} \le n!$ for every nonnegative integer n. [Recall that $0! = 1$ and for $n > 0$, $n! = 1 \cdot 2 \cdot 3 \cdots (n - 1)n$.]

  4. Let r be a real number, $r \ne 1$. Prove that for every integer $n \ge 1$,

$$1 + r + r^2 + r^3 + \cdots + r^{n-1} = \frac{r^n - 1}{r - 1}.$$

B. **5.** Prove that 4 is a factor of $7^n - 3^n$ for every positive integer n. [*Hint:* $7^{k+1} - 3^{k+1} = 7^{k+1} - 7 \cdot 3^k + 7 \cdot 3^k - 3^{k+1} = 7(7^k - 3^k) + (7 - 3)3^k$.]

  6. Prove that 3 is a factor of $4^n - 1$ for every positive integer n.

  7. Prove that 3 is a factor of $2^{2n+1} + 1$ for every positive integer n.

  8. Prove that 5 is a factor of $2^{4n-2} + 1$ for every positive integer n.

  9. Prove that 64 is a factor of $9^n - 8n - 1$ for every nonnegative integer n.

  10. Use the Principle of Complete Induction to show that every integer greater than 1 is a product of primes. [Recall that a positive integer p is prime provided that $p > 1$ and that the only positive integer factors of p are 1 and p.]

  11. Let B be a set of n elements. Prove that the number of different injective functions from B to B is $n!$. [$n!$ was defined in Exercise 3.]

  12. True or false: $n^2 - n + 11$ is prime for every nonnegative integer n. Justify your answer. [Primes were defined in Exercise 10.]

  13. Let B be a set of n elements.

   (a) If $n \ge 2$, prove that the number of two-element subsets of B is $n(n - 1)/2$.

   (b) If $n \ge 3$, prove that the number of three-element subsets of B is $n(n - 1)(n - 2)/3!$.

   (c) Make a conjecture as to the number of k-element subsets of B when $n \ge k$. Prove your conjecture.

  14. At a social bridge party every couple plays every other couple exactly once. Assume there are no ties.

   (a) If n couples participate, prove that there is a "best couple" in the following sense: A couple u is "best" provided that for every couple v, u beats v *or* u beats a couple that beats v.

   (b) Show by example that there may be more than one best couple.

15. What is wrong with the following "proof" that all roses are the same color. It suffices to prove the statement: In every set of n roses, all the roses in the set are the same color. If $n = 1$, the statement is certainly true. Assume the statement is true for $n = k$. Let S be a set of $k + 1$ roses. Remove one rose (call it rose A) from S; there are k roses remaining, and they must all be the same color by the induction hypothesis. Replace rose A and remove a different rose (call it rose B). Once again there are k roses remaining that must all be the same color by the induction hypothesis. Since the remaining roses include rose A, all the roses in S have the same color. This proves that the statement is true when $n = k + 1$. Therefore, the statement is true for all n by induction.

16. Let n be a positive integer. Suppose that there are three pegs and on one of them n rings are stacked, with each ring being smaller in diameter than the one below it, as shown here for $n = 5$:

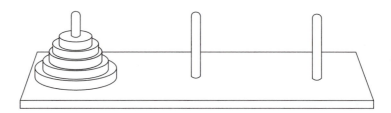

The game is to transfer all the rings to another peg according to these rules: (i) only one ring may be moved at a time; (ii) a ring may be moved to any peg but may never be placed on top of a smaller ring; (iii) the final order of the rings on the new peg must be the same as their original order on the first peg. Prove that the game can be completed in $2^n - 1$ moves and cannot be completed in fewer moves.

17. Let x be a real number greater than -1. Prove that for every positive integer $n, (1 + x)^n \geq 1 + nx$.

C. 18. Consider maps in the plane formed by drawing a finite number of straight lines (entire lines, not line segments). Use induction to prove that every such map may be colored with just two colors in such a way that any two regions with the same line segment as a common border have different colors. Two regions that have only a single point on their common border may have the same color. [This problem is a special case of the so-called Four-Color Theorem, which states that every map in the plane (with any continuous curves or segments of curves as boundaries) can be colored with at most four colors in such a way that any two regions that share a common border have different colors.]

APPENDIX **D**

Equivalence Relations

This appendix may be read anytime after you've finished Appendix B, but it is not needed in the text until Section 9.4. If you read it before that point, you should have no trouble with Examples 1–3 but may have to skip some of the unnumbered examples. Chapter 2 is a prerequisite for the examples labeled "integers," Chapter 6 for those labeled "rings," and Chapter 7 for those labeled "groups."

If A is a set, then any subset of $A \times A$ is called a **relation** on A. A relation T on A is called an **equivalence relation** provided that the subset T is

(i) **Reflexive:** $(a,a) \in T$ for every $a \in A$.

(ii) **Symmetric:** If $(a,b) \in T$, then $(b,a) \in T$.

(iii) **Transitive:** If $(a,b) \in T$ and $(b,c) \in T$, then $(a,c) \in T$.

If T is an equivalence relation on A and $(a,b) \in T$, we say that **a is equivalent to b** and write **$a \sim b$** instead of $(a,b) \in T$. In this notation, the conditions defining an equivalence relation become

(i) **Reflexive:** $a \sim a$ for every $a \in A$.

(ii) **Symmetric:** If $a \sim b$, then $b \sim a$.

(iii) **Transitive:** If $a \sim b$ and $b \sim c$, then $a \sim c$.

When this notation is used, the relation is usually defined without explicit reference to a subset of $A \times A$.

EXAMPLE 1 Let A be a set and define $a \sim b$ to mean $a = b$. In other words, the equivalence relation on A is the subset $T = \{(a,b) \mid a = b\}$ of $A \times A$. Then it is easy to see that $=$ is an equivalence relation.

EXAMPLE 2   The relation on the set $\mathbb{R}$ of real numbers defined by

$$r \sim s \text{ means } |r| = |s|$$

is an equivalence relation, as you can readily verify.

EXAMPLE 3*   Define a relation on the set $\mathbb{Z}$ of integers by

$$a \sim b \text{ means } a - b \text{ is a multiple of 3.}$$

For example, $17 \sim 5$ since $17 - 5 = 12$, a multiple of 3. Clearly $a \sim a$ for every a since $a - a = 0 = 3 \cdot 0$. To prove property (ii), suppose $a \sim b$. Then $a - b$ is a multiple of 3. Hence, $-(a - b)$ is also a multiple of 3. But $-(a - b) = b - a$. Therefore, $b \sim a$. To prove property (iii), suppose $a \sim b$ and $b \sim c$. Then $a - b$ and $b - c$ are multiples of 3 and so is their difference $(a - b) - (b - c) = a - c$, so that $a \sim c$. Thus $\sim$ is an equivalence relation (usually called congruence modulo 3 and denoted $a \equiv b \pmod{3}$).

EXAMPLE (INTEGERS)   If n is a fixed positive integer, the relation of congruence modulo n on the set $\mathbb{Z}$, defined by

$$a \equiv b \pmod{n} \text{ if and only if } a - b \text{ is a multiple of } n,$$

is an equivalence relation by Theorem 2.1.

EXAMPLE (RINGS)   If I is an ideal in the ring R, then the relation of congruence modulo I, defined by

$$a \equiv b \pmod{I} \text{ if and only if } a - b \in I,$$

is an equivalence relation on R by Theorem 6.4.

EXAMPLE (GROUPS)   If K is a subgroup of a group G, then the relation defined by

$$a \equiv b \text{ if and only if } ab^{-1} \in K$$

is an equivalence relation on G by Theorem 7.22.

Warning   It is quite possible to have a relation on a set that satisfies one or two, but not all three, of the properties that define an equivalence relation. For instance, the order relation $\leq$ on the set $\mathbb{R}$ of real numbers is reflexive and transitive but not symmetric; for other examples, see Exercises 8 and 9. Therefore, you must verify all three properties in order to prove that a particular relation is actually an equivalence relation.

* If you've already read Section 2.1, skip Example 3 here and below; it's just congruence modulo n when $n = 3$.

Let $\sim$ be an equivalence relation on a set A. If $a \in A$, then the **equivalence class** of a (denoted $[a]$) is the set of all elements in A that are equivalent to a, that is,

$$[a] = \{b \mid b \in A \text{ and } b \sim a\}.$$

In Example 2, for instance, the equivalence class $[9]$ of the number 9 consists of all real numbers b such that $b \sim 9$, that is, all numbers b such that $|b| = |9|$. Thus $[9] = \{9, -9\}$.

> **EXAMPLE (RINGS, GROUPS)**   If I is an ideal in a ring R, then an equivalence class under the relation of congruence modulo I is a coset $a + I = \{a + i \mid i \in I\}$. Similarly, if K is a subgroup of a group G, then an equivalence class of the relation congruence modulo K is a right coset $Ka = \{ka \mid k \in K\}$.

> **EXAMPLE 3**   (continued) The equivalence class of the integer 2 consists of all integers b such that $b \sim 2$, that is, all b such that $b - 2$ is a multiple of 3. But $b - 2$ is a multiple of 3 exactly when b is of the form $b = 2 + 3k$ for some integer k. Therefore,
>
> $$[2] = \{2 + 3k \mid k \in \mathbb{Z}\} = \{2 + 0, 2 \pm 3, 2 \pm 6, 2 \pm 9, \ldots\}$$
> $$= \{\ldots, -7, -4, -1, 2, 5, 8, 11, \ldots\}.$$
>
> A similar argument shows that the equivalence class $[8]$ consists of all integers of the form $8 + 3k$ ($k \in \mathbb{Z}$); consequently,
>
> $$[8] = \{\ldots, -7, -4, -1, 2, 5, 8, 11, 14, 17, \ldots\}.$$
>
> Thus $[2]$ and $[8]$ are the same set. Note that $2 \sim 8$. This is an example of

THEOREM D.1   *Let $\sim$ be an equivalence relation on a set A and $a, b \in A$. Then*

$$a \sim c \text{ if and only if } [a] = [c].$$

Proof*   Assume $a \sim c$. To prove that $[a] = [c]$, we first show that $[a] \subseteq [c]$. To do this, let $b \in [a]$. Then $b \sim a$ by definition. Since $a \sim c$, we have $b \sim c$ by transitivity. Therefore, $b \in [c]$ and $[a] \subseteq [c]$. Reversing the roles of a and c in this argument and using the fact that $c \sim a$ by symmetry, show that $[c] \subseteq [a]$. Therefore, $[a] = [c]$. Conversely, assume that $[a] = [c]$. Since $a \sim a$ by reflexivity, we have $a \in [a]$, and, hence, $a \in [c]$. The definition of $[c]$ shows that $a \sim c$. ◆

---

* If you've read Section 2.1, note that this proof and the proof of Corollary D.2 are virtually identical to the proofs of Theorem 2.3 and Corollary 2.4: Just replace $\equiv$ by $\sim$.

Generally when one has two sets, there are three possibilities: The sets are equal, the sets are disjoint, or the sets have some (but not all) elements in common. With equivalence classes, the third possibility cannot occur:

COROLLARY D.2 *Let ~ be an equivalence relation on a set A. Then any two equivalence classes are either disjoint or identical.*

Proof Let $[a]$ and $[c]$ be equivalence classes. If they are disjoint, then there is nothing to prove. If they are not disjoint, then $[a] \cap [c]$ is nonempty, and by definition there is an element b such that $b \in [a]$ and $b \in [c]$. By the definition of equivalence class, $b \sim a$ and $b \sim c$. Consequently, by transitivity and symmetry, $a \sim c$. Therefore, $[a] = [c]$ by Theorem D.1. ◆

A **partition** of a set A is a collection of nonempty, mutually disjoint* subsets of A whose union is A. Every equivalence relation $\sim$ on A leads to a partition as follows. Since $a \in [a]$ for each $a \in A$, every equivalence class is nonempty, and every element of A is in one. Distinct equivalence classes are disjoint by Corollary D.2. Therefore,

**The distinct equivalence classes of an equivalence
relation on a set A form a partition of A.**

Conversely, every partition of A leads to an equivalence relation whose equivalence classes are precisely the subsets of the partition (Exercise 21).

◆ EXERCISES

A.   **1.** Let P be a plane. If p, q are points in P, then $p \sim q$ means p and q are the same distance from the origin. Prove that $\sim$ is an equivalence relation on P.

2. Define a relation on the set $\mathbb{Q}$ of rational numbers by: $r \sim s$ if and only if $r - s \in \mathbb{Z}$. Prove that $\sim$ is an equivalence relation.

3. (a) Prove that the following relation on the set $\mathbb{R}$ of real numbers is an equivalence relation: $a \sim b$ if and only if $\cos a = \cos b$.

(b) Describe the equivalence class of 0 and the equivalence class of $\pi/2$.

4. If m and n are lines in a plane P, define $m \sim n$ to mean that m and n are parallel. Is $\sim$ an equivalence relation on P?

5. (a) Let $\sim$ be the relation on the ordinary coordinate plane defined by $(x,y) \sim (u,v)$ if and only if $x = u$. Prove that $\sim$ is an equivalence relation.

(b) Describe the equivalence classes of this relation.

* That is, any two of the subsets are disjoint.

6. Prove that the following relation on the coordinate plane is an equivalence relation: $(x,y) \sim (u,v)$ if and only if $x - u$ is an integer.

7. Let $f : A \to B$ be a function. Prove that the following relation is an equivalence relation of $A : u \sim v$ if and only if $f(u) = f(v)$.

8. Let $A = \{1, 2, 3\}$. Use the ordered-pair definition of a relation to exhibit a relation on A with the stated properties.

    (a) Reflexive, not symmetric, not transitive.

    (b) Symmetric, not reflexive, not transitive.

    (c) Transitive, not reflexive, not symmetric.

    (d) Reflexive and symmetric, not transitive.

    (e) Reflexive and transitive, not symmetric.

    (f) Symmetric and transitive, not reflexive.

9. Which of the properties (reflexive, symmetric, transitive) does the given relation have?

    (a) $a < b$ on the set $\mathbb{R}$ of real numbers.

    (b) $A \subseteq B$ on the set of all subsets of a set S.

    (c) $a \ne b$ on the set $\mathbb{R}$ of real numbers.

    (d) $(-1)^a = (-1)^b$ on the set $\mathbb{Z}$ of integers.

B. 10. If r is a real number, then $[\![r]\!]$ denotes the largest integer that is $\le r$; for instance $[\![\pi]\!] = 3$, $[\![7]\!] = 7$ and $[\![-1.5]\!] = -2$. Prove that the following relation is an equivalence relation on $\mathbb{R}$: $r \sim s$ if and only if $[\![r]\!] = [\![s]\!]$.

11. Let $\sim$ be defined on the set $\mathbb{R}^*$ of nonzero real numbers by: $a \sim b$ if and only if $a/b \in \mathbb{Q}$. Prove that $\sim$ is an equivalence relation.

12. Is the following relation an equivalence relation on $\mathbb{R}$: $a \sim b$ if and only if there exists $k \in \mathbb{Z}$ such that $a = 10^k b$.

13. In the set $\mathbb{R}[x]$ of all polynomials with real coefficients, define $f(x) \sim g(x)$ if and only if $f'(x) = g'(x)$, where $'$ denotes the derivative. Prove that $\sim$ is an equivalence relation on $\mathbb{R}[x]$.

14. Let T be the set of all continuous functions from $\mathbb{R}$ to $\mathbb{R}$ and define $f \sim g$ if and only if $f(2) = g(2)$. Prove that $\sim$ is an equivalence relation.

15. Prove that the relation on $\mathbb{Z}$ defined by $a \sim b$ if and only if $a^2 \equiv b^2 \pmod 6$ is an equivalence relation.

16. Let $S = \{(a,b) \mid a, b \in \mathbb{Z} \text{ and } b \ne 0\}$ and define $(a,b) \sim (c,d)$ if and only if $ad = bc$. Prove that $\sim$ is an equivalence relation on S.

17. Let $\sim$ be a symmetric and transitive relation on a set A. What is wrong with the following "proof" that $\sim$ is reflexive: $a \sim b$ implies $b \sim a$ by symmetry; then $a \sim b$ and $b \sim a$ imply $a \sim a$ by transitivity. [Also see Exercise 8(f).]

18. Let G be a group and define $a \sim b$ if and only if there exists $c \in G$ such that $b = c^{-1}ac$. Prove that $\sim$ is an equivalence relation on G.

19. (a) Let K be a subgroup of a group G and define $a \sim b$ if and only if $a^{-1}b \in K$. Prove that $\sim$ is an equivalence relation on G.

    (b) Give an example to show that the equivalence relation in part (a) need not be the same as the relation in the last example on page 528.

20. Let G be a subgroup of S_n. Define a relation on the set $\{1, 2, \ldots, n\}$ by $a \sim b$ if and only if $a = \sigma(b)$ for some σ in G. Prove that $\sim$ is an equivalence relation.

21. Let A be a set and $\{A_i \mid i \in I\}$ a partition of A. Define a relation on A by: $a \sim b$ if and only if a and b are in the same subset of the partition (that is, there exists $k \in I$ such that $a \in A_k$ and $b \in A_k$).

    (a) Prove that $\sim$ is an equivalence relation on A.

    (b) Prove that the equivalence classes of $\sim$ are precisely the subsets A_i of the partition.

APPENDIX **E**

◆

The Binomial Theorem

Appendix C and Section 3.2 are the prerequisites for this appendix. The material presented here is used in Section 10.6 and in occasional exercises elsewhere.

As we saw in the example on page 59,

$$(a + b)^2 = a^2 + 2ab + b^2$$

for any elements a, b in a *commutative* ring R. Similar calculations using distributivity and commutative multiplication show that

$$(a + b)^3 = a^3 + 3a^2b + 3ab^2 + b^3$$
$$(a + b)^4 = a^4 + 4a^3b + 6a^2b^2 + 4ab^3 + b^4.$$

There is a pattern emerging here, but it may not be obvious unless certain facts are pointed out first.

Recall that 0! is defined to be 1 and that for each positive integer n, the symbol $n!$ denotes the number $n(n - 1)(n - 2) \cdots 3 \cdot 2 \cdot 1$. For each k, with $0 \le k \le n$, the **binomial coefficient** $\binom{n}{k}$ is defined to be the number $\frac{n!}{k!(n - k)!}$. This number may appear to be a fraction, but every binomial coefficient is actually an integer (Exercise 6). For instance, $\binom{4}{1} = \frac{4!}{1!(4 - 1)!} = \frac{4 \cdot 3 \cdot 2 \cdot 1}{1 \cdot 3 \cdot 2 \cdot 1} = 4$, and similarly, $\binom{4}{2} = \frac{4!}{2!2!} = 6$. Note that these numbers appear as coefficients in the preceding expansion of $(a + b)^4$; in fact, you can readily verify that

$$(a + b)^4 = a^4 + \binom{4}{1}a^3b + \binom{4}{2}a^2b^2 + \binom{4}{3}ab^3 + b^4.$$

This is an example of

THEOREM E.1   (THE BINOMIAL THEOREM)   *Let R be a commutative ring and a, b ∈ R. Then for each positive integer n,*

$$(a + b)^n = a^n + \binom{n}{1}a^{n-1}b + \binom{n}{2}a^{n-2}b^2 + \cdots + \binom{n}{n-1}ab^{n-1} + b^n.$$

Proof   The proof is by induction on n. If $n = 1$, the theorem states that $(a + b)^1 = a^1 + b^1$, which is certainly true. Assume that the theorem is true when $n = k$, that is, that

$$(a + b)^k = a^k + \binom{k}{1}a^{k-1}b + \cdots + \binom{k}{r}a^{k-r}b^r + \cdots + \binom{k}{k-1}ab^{k-1} + b^k.$$

We must use this assumption to prove that the theorem is true when $n = k + 1$. By the definition of exponents $(a + b)^{k+1} = (a + b)(a + b)^k$. Applying the induction hypothesis to $(a + b)^k$ and using distributivity and commutative multiplication, we have

$$(a + b)^{k+1} = (a + b)(a + b)^k$$

$$= (a + b)\left[a^k + \binom{k}{1}a^{k-1}b + \cdots + \binom{k}{r}a^{k-r}b^r + \cdots \right.$$
$$\left. + \binom{k}{k-1}ab^{k-1} + b^k \right]$$

$$= a\left[a^k + \binom{k}{1}a^{k-1}b + \cdots + \binom{k}{r}a^{k-r}b^r + \cdots + \binom{k}{k-1}ab^{k-1} + b^k \right]$$
$$+ b\left[a^k + \binom{k}{1}a^{k-1}b + \cdots + \binom{k}{r}a^{k-r}b^r + \cdots + \binom{k}{k-1}ab^{k-1} + b^k \right]$$

$$= \left[a^{k+1} + \binom{k}{1}a^k b + \cdots + \binom{k}{r}a^{k-r+1}b^r + \cdots \right.$$
$$\left. + \binom{k}{k-1}a^2 b^{k-1} + ab^k \right]$$

$$+ \left[a^k b + \binom{k}{1}a^{k-1}b^2 + \cdots + \binom{k}{r}a^{k-r}b^{r+1} + \cdots \right.$$
$$\left. + \binom{k}{k-1}ab^k + b^{k+1} \right]$$

$$= a^{k+1} + \left[\binom{k}{1} + 1 \right]a^k b + \left[\binom{k}{2} + \binom{k}{1} \right]a^{k-1}b^2 + \cdots$$
$$+ \left[\binom{k}{r+1} + \binom{k}{r} \right]a^{k-r}b^{r+1} + \cdots + \left[1 + \binom{k}{k-1} \right]ab^k + b^{k+1}.$$

Exercise 5 (which you should do) shows that for $r = 0, 1, \ldots, k$

$$\binom{k}{r+1} + \binom{k}{r} = \binom{k+1}{r+1}.$$

Apply this fact to each of the coefficients in the last part of the equation above. For instance, $\binom{k}{1} + 1 = \binom{k}{1} + \binom{k}{0} = \binom{k+1}{1}$, and $\binom{k}{2} + \binom{k}{1} = \binom{k+1}{2}$, and so on. Then, from the first and last parts of the equation above we have

$$(a + b)^{k+1} = a^{k+1} + \binom{k+1}{1}a^k b + \binom{k+1}{2}a^{k-1}b^2 + \cdots$$
$$+ \binom{k+1}{r+1}a^{k-r}b^{r+1} + \cdots + \binom{k+1}{k}ab^k + b^{k+1}.$$

Therefore, the theorem is true when $n = k + 1$, and, hence, by induction it is true for every positive integer n.   ◆

◆ **EXERCISES**

A. 1. Let x and y be real numbers. Find the coefficient of $x^5 y^8$ in the expansion of $(2x - y^2)^9$. [*Hint:* Apply Theorem E.1 with $a = 2x$, $b = y^2$.]

2. If x and y are real numbers, what is the coefficient of $x^{12}y^6$ in the expansion of $(x^3 - 3y)^{10}$?

B. 3. Let r and n be integers with $0 < r < n$. Prove that $\binom{n}{r} = \binom{n}{n-r}$.

4. Prove that for any positive integer n, $2^n = \binom{n}{0} + \binom{n}{1} + \binom{n}{2} + \cdots + \binom{n}{n}$.
[*Hint:* $2^n = (1 + 1)^n$.]

5. Let r and k be integers such that $0 \leq r \leq k - 1$. Prove that $\binom{k}{r+1} + \binom{k}{r} = \binom{k+1}{r+1}$. [*Hint:* Use the fact that

$$(k - r)(k - (r + 1))! = (k - r)! = ((k + 1) - (r + 1))!]$$

to express each term on the left as a fraction with denominator $(k + 1)!(k - r)!$. Add the fractions, simplify the numerator, and compare the result with $\binom{k+1}{r+1}$.]

6. Let n be a positive integer. Use mathematical induction to prove this statement: For each integer r such that $0 \leq r \leq n$, $\binom{n}{r}$ is an integer. [*Hint:* For

$n = 1$ it is easy to calculate $\binom{1}{0} = 1 = \binom{1}{1}$; assume the statement is true for $n = k$ and use Exercise 5 to show that the statement is true for $n = k + 1$.]

7. Here are the first five rows of **Pascal's triangle:**

Row 0:					1				
Row 1:				1		1			
Row 2:			1		2		1		
Row 3:		1		3		3		1	
Row 4:	1		4		6		4		1

Note that each entry in a given row (except the 1's on the end) is the sum of the two numbers above it in the preceding row. For instance, the first 4 in row 4 is the sum of 1 and 3 in row 3; similarly, 6 in row 4 is the sum of the two 3's in row 3.

(a) Write out the next three rows of Pascal's triangle.

(b) Prove that the entries in row n of Pascal's triangle are precisely the coefficients in the expansion of $(a + b)^n$, that is, $\binom{n}{0}, \binom{n}{1}, \binom{n}{2}, \ldots, \binom{n}{n}$.

{*Hint:* Exercise 5 may be helpful.]

APPENDIX F

Matrix Algebra

This appendix may be read at any time after Section 3.1 but is needed only in Chapter 16. Throughout this appendix, *R is a ring with identity*.

Rings of 2×2 matrices with entries in $\mathbb{Z}$, $\mathbb{Q}$, $\mathbb{R}$, and $\mathbb{C}$ were introduced in Section 3.1. These matrices are special cases of this definition: An $n \times m$ **matrix** over R is an array of n horizontal rows and m vertical columns

$$\begin{pmatrix} r_{11} & r_{12} & r_{13} & \cdots & r_{1m} \\ r_{21} & r_{22} & r_{23} & \cdots & r_{2m} \\ r_{31} & r_{32} & r_{33} & \cdots & r_{3m} \\ \vdots & \vdots & \vdots & & \vdots \\ r_{n1} & r_{n2} & r_{n3} & \cdots & r_{nm} \end{pmatrix}$$

with each $r_{ij} \in R$. For example,

$$A = \begin{pmatrix} 7 & -6 & 4 & 10 & 0 \\ 1 & 0 & 5 & -2 & 1 \\ 3 & 3 & 4 & 12 & 9 \\ 0 & 5 & 2 & 0 & -8 \end{pmatrix} \quad B = \begin{pmatrix} 1 & 4 & 0 \\ 2 & 1 & 3 \\ 3 & 2 & 0 \end{pmatrix} \quad C = \begin{pmatrix} 1 & 0 & 1 & 0 \\ 0 & 1 & 1 & 1 \end{pmatrix}$$

$$4 \times 5 \text{ over } \mathbb{Z} \qquad\qquad 3 \times 3 \text{ over } \mathbb{Z}_5 \qquad 2 \times 4 \text{ over } \mathbb{Z}_2$$

Matrices are usually denoted by capital letters and their entries by lowercase letters with double subscripts indicating the row and column the entry appears in. For instance, in the matrix $A = (a_{ij})$ at the left, the entry in row 4 and column 2 is $a_{42} = 5$. In matrix C at the right, $c_{12} = 0$ and $c_{23} = 1$. Thus, for example, row i of an $n \times m$ matrix (r_{ij}) is

$$r_{i1}\ r_{i2}\ r_{i3}\ r_{i4} \cdots r_{im}.$$

The $n \times m$ **zero matrix** is the $n \times m$ matrix with 0_R in every entry. The **identity matrix I_n** is the $n \times n$ matrix with 1_R in positions 1-1, 2-2, 3-3, . . . , n-n, and 0_R in all other positions. For example, over the ring $\mathbb{R}$:

$$I_3 = \begin{pmatrix} 1 & 0 & 0 \\ 0 & 1 & 0 \\ 0 & 0 & 1 \end{pmatrix} \quad I_4 = \begin{pmatrix} 1 & 0 & 0 & 0 \\ 0 & 1 & 0 & 0 \\ 0 & 0 & 1 & 0 \\ 0 & 0 & 0 & 1 \end{pmatrix} \quad I_5 = \begin{pmatrix} 1 & 0 & 0 & 0 & 0 \\ 0 & 1 & 0 & 0 & 0 \\ 0 & 0 & 1 & 0 & 0 \\ 0 & 0 & 0 & 1 & 0 \\ 0 & 0 & 0 & 0 & 1 \end{pmatrix}.$$

The identity matrix I_n can be succinctly described by $I_n = (\delta_{ij})$, where δ_{ij} is the **Kronecker delta symbol,** defined by

$$\delta_{ij} = \begin{cases} 1_R \text{ if } i = j. \\ 0_R \text{ if } i \neq j. \end{cases}$$

It is sometimes convenient to think of a large matrix as being made up of two smaller ones. For example, if A is the 3×2 matrix

$$\begin{pmatrix} 4 & 2 \\ 1 & 0 \\ 3 & 5 \end{pmatrix}$$

over $\mathbb{Z}$, then $(I_3 \,|\, A)$ denotes the 3×5 matrix

$$\begin{pmatrix} 1 & 0 & 0 & 4 & 2 \\ 0 & 1 & 0 & 1 & 0 \\ 0 & 0 & 1 & 3 & 5 \end{pmatrix}.$$

Similarly, $\left(\dfrac{A}{I_2} \right)$ denotes the matrix $\begin{pmatrix} 2 & 3 \\ 4 & 6 \\ 1 & 0 \\ 0 & 1 \end{pmatrix}$, where $A = \begin{pmatrix} 2 & 3 \\ 4 & 6 \end{pmatrix}$.

If $A = (a_{ij})$ and $B = (b_{ij})$ are $n \times m$ matrices, then their **matrix sum** $A + B$ is the $n \times m$ matrix with $a_{ij} + b_{ij}$ in position i-j. In other words, just add the entries in corresponding positions, as in this example over $\mathbb{Z}_5$:

$$\begin{pmatrix} 1 & 3 & 4 \\ 0 & 2 & 1 \end{pmatrix} + \begin{pmatrix} 3 & 2 & 0 \\ 1 & 4 & 2 \end{pmatrix} = \begin{pmatrix} 4 & 0 & 4 \\ 1 & 1 & 3 \end{pmatrix}.$$

If A and B are of different sizes, their sum is not defined. But if A, B, C are $n \times m$ matrices, then Exercise 3 shows that *matrix addition is commutative* $[A + B = B + A]$ *and associative* $[A + (B + C) = (A + B) + C]$. The $n \times m$ zero matrix acts as an identity for addition (Exercise 4).

For reasons that are made clear in a linear algebra course, the product of matrices A and B is defined only when the number of *columns* of A is the same as the number of *rows* of B. The simplest case is the product of a $1 \times m$ matrix A consisting of a single row $(a_1 \ a_2 \ a_3 \ \cdots \ a_m)$ and an $m \times 1$ matrix B

consisting of a single column $\begin{pmatrix} b_1 \\ b_2 \\ \vdots \\ b_m \end{pmatrix}$.* The product is defined to be the 1×1 matrix whose single entry is the element

$$a_1b_1 + a_2b_2 + a_3b_3 + a_4b_4 + \cdots + a_mb_m.$$

For example, over $\mathbb{Z}$

(∗) $$(2 \quad 3 \quad 1)\begin{pmatrix} 4 \\ 0 \\ 2 \end{pmatrix} = 2 \cdot 4 + 3 \cdot 0 + 1 \cdot 2 = 10.$$

If A is an $n \times m$ matrix and B is an $m \times k$ matrix, then the **matrix product** AB is the $n \times k$ matrix (c_{ij}), where the entry in position i-j is the product of the ith row of A and the jth column of B:

$$c_{ij} = a_{i1}b_{1j} + a_{i2}b_{2j} + a_{i3}b_{3j} + a_{i4}b_{4j} + \cdots + a_{im}b_{mj} = \sum_{r=1}^{m} a_{ir}b_{rj}.$$

EXAMPLE The product of

$$A = \begin{pmatrix} 2 & 3 & 1 \\ 1 & 5 & 0 \end{pmatrix} \quad \text{and} \quad B = \begin{pmatrix} 4 & 2 & 6 & 3 \\ 0 & 1 & 2 & 1 \\ 2 & 6 & 0 & 2 \end{pmatrix}$$

is a 2×4 matrix whose entry in position 1-1 is 10 (the product of row 1 of A and column 1 of B as shown in (∗)). In position 2-3 the entry in AB is the product of row 2 of A and column 3 of B:

$$1 \cdot 6 + 5 \cdot 2 + 0 \cdot 0 = 16.$$

Similar calculations show that

$$AB = \begin{pmatrix} 2 & 3 & 1 \\ 1 & 5 & 0 \end{pmatrix} \begin{pmatrix} 4 & 2 & 6 & 3 \\ 0 & 1 & 2 & 1 \\ 2 & 6 & 0 & 2 \end{pmatrix} = \begin{pmatrix} 10 & 13 & 18 & 11 \\ 4 & 7 & 16 & 8 \end{pmatrix}.$$

The product BA is *not defined* because B has four columns but A has only two rows.

---

* A matrix with only one row is called a **row vector** and a matrix with only one column a **column vector.** Single subscripts are adequate to describe the entries of row and column vectors.

If A, B, C are matrices of appropriate sizes so that each of the products AB and BC is defined, then *matrix multiplication is associative:* $A(BC) = (AB)C$ (Exercise 7). Similarly, if E, F, G are matrices such that the products EG and FG are defined, then *the distributive law holds:* $(E + F)G = EG + FG$ (Exercise 5). The identity matrices act as *identity elements for multiplication* in this sense: If A is an $n \times m$ matrix, then $I_n \cdot A = A$ and $A \cdot I_m = A$ (Exercise 6). Even when both products AB and BA are defined, *matrix multiplication may not be commutative* (see the example on pages 44–45).

Let $M_n(R)$ denote the set of all $n \times n$ matrices over the ring R. Since all the matrices in $M_n(R)$ have the same number of columns and rows, both $A + B$ and AB and BA are defined for all $A, B \in M_n(R)$. The properties of matrix addition and multiplication listed above provide the proof of

THEOREM F.1 *If R is a ring with identity, then the set $M_n(R)$ of all $n \times n$ matrices over R is a noncommutative ring with identity I_n.*

◆ **EXERCISES**

NOTE: *Unless stated otherwise, all matrices are over a ring R with identity.*

A. 1. Assume A and B are matrices over $\mathbb{Z}$. Find $A + B$.

(a) $A = \begin{pmatrix} 1 & 2 & -2 & 0 \\ 3 & 5 & 7 & 11 \end{pmatrix} \quad B = \begin{pmatrix} 0 & -8 & 2 & 4 \\ 6 & 0 & 4 & 1 \end{pmatrix}$

(b) $A = \begin{pmatrix} 3 & 0 & 2 \\ 4 & 1 & 6 \\ 0 & 1 & 0 \\ 2 & -5 & 7 \end{pmatrix} \quad B = \begin{pmatrix} 1 & -2 & 0 \\ 3 & 0 & 4 \\ 0 & 7 & -6 \\ 1 & 6 & 0 \end{pmatrix}$

2. Assume A and B are matrices over $\mathbb{Z}_6$. Find AB and BA whenever the products are defined.

(a) $A = \begin{pmatrix} 2 & 4 \\ 1 & 5 \\ 3 & 0 \end{pmatrix} \quad B = \begin{pmatrix} 1 & 0 & 1 \\ 2 & 3 & 2 \end{pmatrix}$

(b) $A = \begin{pmatrix} 1 & 4 \\ 5 & 2 \end{pmatrix} \quad B = \begin{pmatrix} 0 & 5 & 3 \\ 1 & 0 & 2 \end{pmatrix}$

(c) $A = (3 \quad 2 \quad 1 \quad 0) \quad B = \begin{pmatrix} 1 & 0 & 0 \\ 0 & 1 & 0 \\ 0 & 0 & 1 \\ 1 & 1 & 1 \end{pmatrix}$

B. 3. Let $A = (a_{ij})$, $B = (b_{ij})$, and $C = (c_{ij})$ be $n \times m$ matrices. Prove that

(a) $A + B = B + A$ (b) $A + (B + C) = (A + B) + C$

4. If $A = (a_{ij})$ is an $n \times m$ matrix and Z is the $n \times m$ zero matrix, prove that $A + Z = A$.

5. (a) Let E and F be $1 \times m$ row vectors and $G = (g_{ij})$ an $m \times k$ matrix. Prove that $(E + F)G = EG + FG$.

   (b) Let $E = (e_{ij})$ and $F = (f_{ij})$ be $n \times m$ matrices and $G = (g_{ij})$ an $m \times k$ matrix. Prove that $(E + F)G = EG + FG$.

6. If A is an $n \times m$ matrix, prove that $I_n \cdot A = A$ and $A \cdot I_m = A$.

C. 7. Let $A = (a_{ij})$ be an $n \times m$ matrix, $B = (b_{ij})$ an $m \times k$ matrix, and $C = (c_{ij})$ a $k \times p$ matrix. Prove that $A(BC) = (AB)C$. [*Hint:* $BC = (d_{tj})$, where $d_{tj} = \sum_{r=1}^{k} b_{tr}c_{rj}$, and $AB = (e_{ir})$, where $e_{ir} = \sum_{t=1}^{m} a_{it}b_{tr}$. The *i-j* entry of $A(BC)$ is

$$\sum_{t=1}^{m} a_{it}d_{tj} = \sum_{t=1}^{m} a_{it}\left(\sum_{r=1}^{k} b_{tr}c_{rj}\right) = \sum_{t=1}^{m}\sum_{r=1}^{k} a_{it}b_{tr}c_{rj}.$$

Show that the *i-j* entry of $(AB)C$ is this same double sum.]

APPENDIX **G**

Polynomials

In high school there is some ambiguity about the "x" in polynomials. Sometimes x stands for a specific number (as in the equation $5x - 6 = 17$). Other times x doesn't seem to stand for any number—it's just a symbol that is algebraically manipulated (as in exercises such as $(x + 3)(x - 5) = x^2 - 2x - 15$).* Our goal here is to develop a rigorous definition of "polynomial" that removes this ambiguity. The prerequisites for this discussion are high-school algebra and Chapter 3.

As a prelude to the formal development, note that the polynomials from high school can be described without ever mentioning x. For instance, $5 + 6x - 2x^3$ is completely determined by its coefficients $(5, 6, 0, -2)$.** But $5 + 6x - 2x^3$ can also be written $5 + 6x - 2x^3 + 0x^4 + 0x^5 + 0x^6$. To allow for such additional "zero terms," we list the coefficients as an infinite sequence $(5, 6, 0, -2, 0, 0, 0, 0, \ldots)$ that ends in zeros.

Adding polynomials in this new notation is pretty much the same as before: Add the coefficients of corresponding powers of x, that is, add sequences coordinatewise:

$$
\begin{array}{ll}
5 + 6x \qquad\quad - 2x^3 & (5, \quad 6, 0, -2, 0, 0, 0, \ldots) \\
\underline{3 - 2x + 5x^2 - 4x^3} & \underline{(3, -2, 5, -4, 0, 0, 0, \ldots)} \\
8 + 4x + 5x^2 - 6x^3 & (8, \quad 4, 5, -6, 0, 0, 0, \ldots)
\end{array}
$$

Multiplication can also be described in terms of sequences, as we shall see. If you keep this model in mind, you will see clearly where the formal definitions and theorems come from.

Throughout the rest of this appendix, R is a ring with identity (not neces-

---

* Sometimes x is also used as a variable that can take infinitely many values (as in the function $f(x) = x^3 - x$). This usage is discussed in Section 4.4.

** 0 is the coefficient of x^2.

sarily commutative). A **polynomial with coefficients in the ring R** is defined to be an infinite sequence

$$(a_0, a_1, a_2, a_3, \ldots)$$

such that each $a_i \in R$ and only finitely many of the a_i are nonzero; that is, for some index k, $a_i = 0_R$ for all $i > k$. The elements $a_i \in R$ are called the **coefficients** of the polynomial.

The polynomials $(a_0, a_1, a_2, \ldots)$ and $(b_0, b_1, b_2, \ldots)$ are **equal** if they are equal as sequences, that is, if $a_0 = b_0$, $a_1 = b_1$, and in general, $a_i = b_i$ for every $i \geq 0$. **Addition of polynomials** is denoted by $\oplus$ and defined by the rule

$$(a_0, a_1, a_2, \ldots) \oplus (b_0, b_1, b_2, \ldots) =$$
$$(a_0 + b_0, a_1 + b_1, a_2 + b_2, \ldots, a_i + b_i, \ldots).$$

You should verify that the sequence on the right is actually a polynomial, that is, that after some point all its coordinates are zero (Exercise 2).

Multiplication of polynomials is denoted $\odot$ and defined by the rule*

$$(a_0, a_1, a_2, \ldots) \odot (b_0, b_1, b_2, \ldots) = (c_0, c_1, c_2, \ldots), \qquad \text{where}$$
$$c_0 = a_0 b_0$$
$$c_1 = a_0 b_1 + a_1 b_0$$
$$c_2 = a_0 b_2 + a_1 b_1 + a_2 b_0$$
$$\vdots$$
$$c_n = a_0 b_n + a_1 b_{n-1} + a_2 b_{n-2} + a_3 b_{n-3} + \cdots + a_{n-1} b_1 + a_n b_0$$
$$= \sum_{i=0}^{n} a_i b_{n-i}.$$

To show that the product defined here is actually a polynomial you must verify that after some point all the coordinates of $(c_0, c_1, \ldots)$ are zero (Exercise 2).

THEOREM G.1 *Let R be a ring with identity and P the set of polynomials with coefficients in R. Then P is a ring with identity. If R is commutative, then so is P.*

Proof Exercise 2 shows that P is closed under addition and multiplication. To show that addition in P is commutative, we note that $a_i + b_i = b_i + a_i$ for all $a_i, b_i \in R$ because R is a ring; therefore, in P

$$(a_0, a_1, a_2, \ldots) \oplus (b_0, b_1, b_2, \ldots)$$
$$= (a_0 + b_0, a_1 + b_1, \ldots) = (b_0 + a_0, b_1 + a_1, \ldots)$$
$$= (b_0, b_1, b_2, \ldots) \oplus (a_0, a_1, a_2, \ldots).$$

* To understand the formal definition, do the following multiplication problem and look at the coefficients of each power of x in the answer: $(a_0 + a_1 x + a_2 x^2)(b_0 + b_1 x + b_2 x^2)$.

Associativity of addition and the distributive laws are proved similarly. You can readily check that the multiplicative identity in P is the polynomial $(1_R, 0_R, 0_R, 0_R, \ldots)$, the zero element is the polynomial $(0_R, 0_R, 0_R, \ldots)$, and the solution of the equation $(a_0, a_1, a_2, \ldots) + X = (0_R, 0_R, 0_R, \ldots)$ is $X = (-a_0, -a_1, -a_2 \ldots)$.

To complete the proof that P is a ring with identity, we must show that multiplication is associative. Let $A, B, C \in P$, where

$$A = (a_0, a_1, a_2, \ldots) \qquad B = (b_0, b_1, b_2, \ldots) \qquad C = (c_0, c_1, c_2, \ldots).$$

Then the nth coordinate of $(A \odot B) \odot C$ is

$$(*) \qquad \sum_{i=0}^{n} (ab)_i c_{n-i} = \sum_{i=0}^{n} \left[\sum_{j=0}^{i} a_j b_{i-j} \right] c_{n-i} = \sum_{i=0}^{n} \sum_{j=0}^{i} a_j b_{i-j} c_{n-i}.$$

Exercise 6 shows that the last sum on the right is the same as

$$(**) \qquad \sum a_u b_v c_w,$$

where the sum is taken over all integers u, v, w such that $u + v + w = n$ and $u \geq 0, v \geq 0, w \geq 0$. On the other hand, the nth coordinate of $A \odot (B \odot C)$ is

$$(***) \qquad \sum_{r=0}^{n} a_r (bc)_{n-r} = \sum_{r=0}^{n} a_r \left[\sum_{s=0}^{n-r} b_s c_{n-r-s} \right] = \sum_{r=0}^{n} \sum_{s=0}^{n-r} a_r b_s c_{n-r-s}.$$

Exercise 6 shows that the last sum on the right is also equal to $(**)$. Since the nth coordinates of $(A \odot B) \odot C$ and $A \odot (B \odot C)$ are equal for each $n \geq 0$, $(A \odot B) \odot C = A \odot (B \odot C)$. The proof of the final statement of the theorem is left to the reader (Exercise 3). ◆

In the old notation, constant polynomials behave like ordinary numbers. In the new notation, constant polynomials are of the form $(r, 0, 0, 0, \ldots)$, and essentially the same thing is true:

THEOREM G.2 *Let P be the ring of polynomials with coefficients in the ring R. Let R^* be the set of all polynomials in P of the form $(r, 0_R, 0_R, 0_R, \ldots)$, with $r \in R$. Then R^* is a subring of P and is isomorphic to R.*

Proof Consider the function $f : R \to R^*$ given by

$$f(r) = (r, 0_R, 0_R, 0_R, \ldots).$$

You can readily verify that f is bijective. Furthermore,

$$f(r + s) = (r + s, 0_R, 0_R, 0_R, \ldots)$$
$$= (r, 0_R, 0_R, 0_R, \ldots) \oplus (s, 0_R, 0_R, 0_R) = f(r) + f(s)$$

and

$$f(rs) = (rs, 0_R, 0_R, 0_R, \ldots)$$
$$= (r, 0_R, 0_R, 0_R, \ldots) \odot (s, 0_R, 0_R, 0_R, \ldots) = f(r) \odot f(s).$$

Therefore, f is an isomorphism, and, hence, R^* is a subring. ◆

Now that the basic facts have been established, it's time to recover the "old" notation for polynomials. First, we want polynomials in R^* to look more like "constants" (elements of R), so

$(a, 0_R, 0_R, 0_R, \ldots)$ *will be denoted by the boldface letter* $\boldsymbol{a}$.

Next, reverting to the original source of our sequence notation,

$(0_R, 1_R, 0_R, 0_R, 0_R, \ldots)$ *will be denoted by* x.

There is no ambiguity about what x is here—it is a specific sequence in P; it is *not* an element of R or R^*, and it does *not* "stand for" any element of R or R^*.

This notation makes things look a bit more familiar. For instance,

$$(a, 0_R, 0_R, 0_R, \ldots) + (b, 0_R, 0_R, \ldots)(0_R, 1_R, 0_R, 0_R, \ldots)$$

becomes $\boldsymbol{a} + \boldsymbol{b}x$. Similarly, we would expect $\boldsymbol{c}x^3$ (the "constant" c times x^3) to be the sequence $(0_R, 0_R, 0_R, c, 0_R, 0_R, \ldots)$ with c in position 3.† But we can't just *assume* that everything works as it did in the old notation. The required proof is given in the next two results.

LEMMA G.3 *Let P be the ring of polynomials with coefficients in the ring R and x the polynomial $(0_R, 1_R, 0_R, 0_R, \ldots)$. Then for each element $a = (a, 0_R, 0_R, \ldots)$ of R^* and each integer $n \geq 1$:*

(1) $x^n = (0_R, 0_R, \ldots, 0_R, 1_R, 0_R, \ldots)$, where 1_R is in position n.
(2) $\boldsymbol{a}x^n = (0_R, 0_R, \ldots, 0_R, a, 0_R, \ldots)$, where a is in position n.

Proof The polynomial x can be described like this:

$$x = (e_0, e_1, e_2, \ldots), \quad \text{where } e_i = 0_R \text{ for all } i \neq 1, \text{ and } e_1 = 1_R.$$

Statement (1) will be proved by induction on n.‡ It is true for $n = 1$ by the definition of $x^1 = x$. Suppose that it is true for $n = k$, that is, suppose that

$$x^k = (d_0, d_1, d_2, \ldots), \quad \text{where } d_i = 0_R \text{ for } i \neq k, \text{ and } d_k = 1_R.$$

Then

$$x^{k+1} = x^k x = (d_0, d_1, d_2, \ldots)(e_0, e_1, e_2, \ldots) = (r_0, r_1, r_2, \ldots),$$

---

† Remember that in the polynomial $(r, s, t, \ldots)$ the element r is in position 0, s is in position 1, t is in position 2, etc.
‡ See Appendix C.

where for each $j \geq 0$,

$$r_j = \sum_{i=0}^{j} d_i e_{j-i}.$$

Since $e_i = 0_R$ _for_ $i \neq 1$ _and_ $d_i = 0_R$ _for_ $i \neq k$, _we have_

$$r_{k+1} = \underbrace{d_0 e_{k+1} + \cdots + d_{k-1} e_2}_{0} + d_k e_1 + \underbrace{d_{k+1} e_0}_{0} = d_k e_1 = 1_R 1_R = 1_R$$

and, for $j \neq k + 1$,

$$r_j = \underbrace{d_0 e_j + d_1 e_{j-1} + \cdots + d_{j-2} e_2}_{0} + d_{j-1} e_1 + \underbrace{d_j e_0}_{0}$$

$$= d_{j-1} e_1 = d_{j-1} 1_R = d_{j-1}.$$

But $j - 1 \neq k$ since $j \neq k + 1$. Therefore, $r_j = d_{j-1} = 0_R$ for all $j \neq k + 1$. Hence, $x^{k+1} = (r_0, r_1, r_2, \ldots) = (0_R, 0_R, \ldots, 0_R, 1_R, 0_R, \ldots)$, with 1_R in position $k + 1$. So (1) is true for $n = k + 1$ and, therefore, true for all n by induction.

A similar inductive argument proves (2); see Exercise 7. ◆

THEOREM G.4 _Let_ P _be the ring of polynomials with coefficients in the ring_ R. _Then_ P _contains an isomorphic copy_ R^* _of_ R _and an element_ x _such that_

   (1) $\boldsymbol{a}x = x\boldsymbol{a}$ _for every_ $a \in R^*$.
   (2) _Every element of_ P _can be written in the form_ $\boldsymbol{a_0} + \boldsymbol{a_1}x + \boldsymbol{a_2}x^2 + \cdots + \boldsymbol{a_n}x^n$.
   (3) _If_ $\boldsymbol{a_0} + \boldsymbol{a_1}x + \cdots + \boldsymbol{a_n}x^n = \boldsymbol{b_0} + \boldsymbol{b_1}x + \cdots + \boldsymbol{b_m}x^m$ _with_ $n \leq m$, _then_ $\boldsymbol{a_i} = \boldsymbol{b_i}$ _for_ $i \leq n$ _and_ $\boldsymbol{b_i} = \boldsymbol{0_R}$ _for_ $i > n$; _in particular,_
   (4) $\boldsymbol{a_0} + \boldsymbol{a_1}x + \boldsymbol{a_2}x^2 + \cdots + \boldsymbol{a_n}x^n = \boldsymbol{0_R}$ _if and only if_ $\boldsymbol{a_i} = \boldsymbol{0_R}$ _for every_ $i \geq 0$.

Proof Let x be as in Lemma G.3. The proof of (1) is left to the reader (Exercise 5).

(2) If $(a_0, a_1, a_2, \ldots) \in P$, then there is an index n such that $a_i = 0_R$ for all $i > n$. By Lemma G.3

$(a_0, a_1, a_2, \ldots, a_n, 0_R, 0_R, \ldots)$

$\quad = (a_0, 0_R, 0_R, \ldots) + (0_R, a_1, 0_R, \ldots) + (0_R, 0_R, a_2, 0_R, \ldots)$

$\qquad\qquad\qquad\qquad\qquad + \cdots + (0_R, \ldots, 0_R, a_n, 0_R, \ldots)$

$\quad = \boldsymbol{a_0} + \boldsymbol{a_1}x + \boldsymbol{a_2}x^2 + \cdots + \boldsymbol{a_n}x^n.$

(3) Reversing the argument in (2) shows that $\boldsymbol{a_0} + \boldsymbol{a_1}x + \cdots + \boldsymbol{a_n}x^n$ is the sequence $(a_0, a_1, a_2, \ldots, a_n, 0_R, 0_R, \ldots)$ and that $\boldsymbol{b_0} + \boldsymbol{b_1}x + \cdots +$

$\boldsymbol{b_m}x^m = (b_0, b_1, b_2, \ldots, b_m, 0_R, 0_R, \ldots)$. If these two sequences are equal, then we must have $a_i = b_i$ for $i \le n$ and $0_R = b_i$ for $n < i \le m$.

(4) is a special case of (3): Just let $\boldsymbol{b_i} = \boldsymbol{0_R}$. ◆

When polynomials are written in the form $\boldsymbol{a_0} + \boldsymbol{a_1}x + \cdots + \boldsymbol{a_n}x^n$, addition and multiplication look like they did in high school, except for the use of boldface print in certain symbols.

> **EXAMPLE**   In the ring of polynomials with real-number coefficients, the distributive laws and Theorems G.2 and G.4 show that
>
> $$\begin{aligned}
(\boldsymbol{3}x + \boldsymbol{1})(\boldsymbol{2}x + \boldsymbol{5}) &= (\boldsymbol{3}x + \boldsymbol{1})\boldsymbol{2}x + (\boldsymbol{3}x + \boldsymbol{1})\boldsymbol{5} \\
&= \boldsymbol{3}x\boldsymbol{2}x + \boldsymbol{1} \cdot \boldsymbol{2}x + \boldsymbol{3}x\boldsymbol{5} + \boldsymbol{1} \cdot \boldsymbol{5} \\
&= \boldsymbol{3} \cdot \boldsymbol{2}xx + \boldsymbol{1} \cdot \boldsymbol{2}x + \boldsymbol{3} \cdot \boldsymbol{5}x + \boldsymbol{1} \cdot \boldsymbol{5} \\
&= \boldsymbol{6}x^2 + \boldsymbol{17}x + \boldsymbol{5}.
\end{aligned}$$

In terms of elements, the distinction between boldface and regular print is important because $\boldsymbol{a}$ is a sequence, while a is an element of R. But in terms of algebraic structure, there is no need for distinction because R^* (consisting of all the boldface $\boldsymbol{a}$'s) is isomorphic to R (consisting of all the a's). Consequently, there is no harm in *identifying* R with its isomorphic copy R^* and writing the elements of $R \cong R^*$ in ordinary print.† Then polynomials look and behave as they did before. For this reason, the standard notation for the polynomial ring is $R[x]$, which we shall use hereafter instead of P.

We have now come full circle in terms of notation, with the added benefits of a rigorous justification of our past work with polynomials, a generalization of these concepts to rings, and a new viewpoint on polynomials. Beginning with a ring R with identity we have constructed an **extension ring** $R[x]$ of R (that is, a ring in which R is a subring). This extension ring contains an element x that commutes with every element of R. The element x is *not* in R and does *not* stand for an element of R. Every element of the extension ring can be written in an essentially unique way in terms of elements of R and powers of x. Because x has the property that $a_0 + a_1x + \cdots + a_nx^n = 0_R$ if and only if every $a_i = 0_R$, x is said to be **transcendental** over R or an **indeterminate** over R.‡

---

† You've been making this identification for years when, for example, you treat the constant polynomial 4 as if it were the real number 4. The identification question can be avoided by rewriting the definition of polynomial to say that a polynomial is either an element of R *or* a sequence $(a_1, a_2, \ldots)$ with at least one $a_i \ne 0_R$ for $i \ge 1$ and all a_i eventually zero. Then the polynomials actually contain R as a subset. The definitions of addition and multiplication, as well as the proofs of the theorems, then have to deal with several cases. Proceed in the obvious (but tiring) way until you have proved Theorem G.4 again.

‡ The latter terminology is a bit misleading since x is a well-defined element of $R[x]$.

Finally, what can be said about polynomials with coefficients in a ring R that does *not* have an identity? In this case, you need only find a ring S with identity that contains R as a subring. For example, the ring of even integers has no identity but is contained in the ring $\mathbb{Z}$ of all integers, which does have an identity; for the general case, see Exercise 37 in Section 3.3. Then form the ring $S[x]$ as before. Those polynomials whose coefficients are actually in R form a subring of $S[x]$; this is the ring $R[x]$. Note that when R does not have an identity, the polynomial x is *not* itself in $R[x]$.

◆ EXERCISES

A. **1.** Express each polynomial as a sequence and express each sequence as a polynomial.

    (a) $(0, 1, 0, 1, 0, 1, 0, 0, 0, \ldots)$

    (b) $(0, 1, 2, 3, 4, 5, 6, 6, 8, 9, 0, 0, 0, \ldots)$

    (c) $3x^6 - 5x^4 + 12x^3 - 3x^2 + 7.5x - 11$

    (d) $(x - 1)(x^3 - x^2 + 1)$

2. (a) If $(a_1, a_2, \ldots)$ and $(b_1, b_2, \ldots)$ are polynomials, show that their sum is a polynomial (that is, after some point all coordinates of the sum are zero).

    (b) Show that $(a_1, a_2, \ldots) \odot (b_1, b_2, \ldots)$ is a polynomial. [*Hint:* If $a_i = 0_R$ for $i > k$ and $b_i = 0_R$ for $i > t$, examine the ith coordinate of the product for $i > k + t$.]

3. Prove these parts of Theorem G.1:

    (a) addition in P is associative;

    (b) both distributive laws hold in P;

    (c) P is commutative if R is.

4. Complete the proof of Theorem G.2 by proving that

    (a) f is injective;   **(b)** f is surjective

5. Prove (1) in Theorem G.4.

B. **6. (a)** In the proof of Theorem G.1 (associative multiplication in P) show that

$$\sum_{i=0}^{n} \sum_{j=0}^{i} a_j b_{i-j} c_{n-i} = \sum a_u b_v c_w,$$ where the last sum is taken over all nonnegative integers u, v, w such that $u + v + w = n$. [*Hint:* Compare the two sums term by term; the sum of the subscripts of $a_j b_{i-j} c_{n-i}$ is n; to show that $a_u b_v c_w$ is in the other sum, let $j = u$ and $i = u + v$ and verify that $n - i = w$.]

(b) Show that $\sum\limits_{r=0}^{n} \sum\limits_{s=0}^{n-r} a_r b_s c_{n-r-s} = \sum a_u b_v c_w$ (last sum as in part (a)).

7. Prove (2) in Lemma G.3. [*Hint:* $\boldsymbol{a} = (a_0, a_1, a_2, \ldots)$, where $a_i = 0_R$ for $i > 1$, and by (1), $x^n = (d_0, d_1, d_2, \ldots)$, where $d_n = 1_R$ and $d_i = 0_R$ for $i \neq n$; use induction on n.]

8. Let R be an integral domain. Using sequence notation, prove that the polynomial ring $R[x]$ is also an integral domain.

9. Let R be a field. Using sequence notation, prove that the polynomial ring $R[x]$ is *not* a field. [*Hint:* Is $(0_R, 1_R, 0_R, 0_R, \ldots)$ a unit?]

C. 10. (a) Let $\mathbb{Q}[\pi]$ be the set of all real numbers of the form $r_0 + r_1\pi + r_2\pi^2 + \cdots + r_n\pi^n$, where $n \geq 0$ and each $r_i \in \mathbb{Q}$. Show that $\mathbb{Q}[\pi]$ is a subring of $\mathbb{R}$.

(b) Assume that $r_0 + r_1\pi + \cdots + r_n\pi^n = 0$ if and only if each $r_i = 0$. (This fact was first proved in 1882; the proof is beyond the scope of this book.) Prove that $\mathbb{Q}[\pi]$ is isomorphic to the polynomial ring $\mathbb{Q}[x]$.

Bibliography

This list contains all the books and articles referred to in the text, as well as a number of other books suitable for collateral reading, reference, and deeper study of particular topics. The list is far from complete. For the most part readability by students has been the chief selection criterion.

Except for [7], [8], and [15], the first 17 books can be read by anyone able to begin this book. Most of the remaining entries have parts of this text as prerequisites.

Abstract Algebra in General (Undergraduate Level)

These books contain approximately the same material as Chapters 1–11 of the text, but each of them provides a slightly different viewpoint and emphasis. Only [2] and [5] have a significant overlap with Chapters 12–16. Several of them also include linear algebra and a few other topics not covered in the text.

1. Beachy, J., and W. Blair, *Abstract Algebra,* 2nd edition. Prospect Heights, Ill.: Waveland Press, 1996.
2. Birkhoff, G., and S. Mac Lane, *A Survey of Modern Algebra,* 4th edition. New York: Macmillan, 1977.
3. Burton, D. M., *Abstract Algebra.* Dubuque, Iowa: Wm. C. Brown, 1988.
4. Fraleigh, J., *A First Course in Abstract Algebra,* 5th edition. Reading, Mass.: Addison-Wesley, 1994.
5. Gallian, J., *Contemporary Abstract Algebra,* 3rd edition. Lexington, Mass.: Heath, 1994.
6. Herstein, I. N., *Topics in Algebra,* 2nd edition. Lexington, Mass.: Xerox Publishing, 1975.

Abstract Algebra in General (Graduate Level)

These books have much deeper and more detailed coverage of the material in Chapters 1–11, as well as a large number of topics not discussed in the text.

7. Hungerford, T. W., *Algebra.* New York: Springer-Verlag, 1974.

8. Jacobson, N., *Basic Algebra I,* 2nd edition, and *Basic Algebra II,* 2nd edition. San Francisco: Freeman, 1985 and 1989.

Logic, Proof, and Set Theory

9. Galovich, S., *Doing Mathematics: An Introduction to Proofs and Problem Solving.* Philadelphia: Saunders College Publishing, 1993.
10. Halmos, P., *Naive Set Theory.* New York: Springer-Verlag, 1974.
11. Lucas, J., *Introduction to Abstract Mathematics,* 2nd edition. New York: Ardsley, 1990.
12. Smith, D., M. Eggen, and R. St. Andre, *A Transition to Advanced Mathematics,* 3rd edition. Monterey, Calif.: Brooks/Cole, 1990.
13. Solow, D., *How To Read and Do Proofs,* 2nd edition. New York: Wiley, 1990.

Number Theory

14. Burton, D. M., *Elementary Number Theory,* 3rd edition. Dubuque, Iowa: Wm. C. Brown, 1994.
15. Ireland, K., and M. Rosen, *A Classical Introduction to Modern Number Theory,* 2nd edition. New York: Springer-Verlag, 1993.
16. Rose, H. E., *A Course in Number Theory.* Oxford: Oxford University Press, 1988.
17. Rosen, K. H., *Elementary Number Theory and Its Applications,* 3rd edition. Reading, Mass.: Addison-Wesley, 1993.

Rings

18. Herstein, I. N., *Noncommutative Rings,* Carus Monograph 15. Washington, D.C.: Mathematical Association of America, 1968.
19. Kaplansky, I., *Commutative Rings,* revised edition. Chicago: University of Chicago Press, 1974.
20. McCoy, N., *Rings and Ideals,* Carus Monograph 8. Washington, D.C.: Mathematical Association of America, 1948.
21. Robinson, A., *Numbers and Ideals.* San Francisco: Holden-Day, 1965.
22. Stark, H., "A Complete Determination of Complex Quadratic Fields of Class Number One," *Michigan Mathematical Journal,* 14(1967), pp. 1–27.
23. Wilson, J. C., "A Principal Ideal Domain That Is Not a Euclidean Ring," *Mathematics Magazine,* 46(1973), pp. 34–38. A simplified version of part of this article is in Williams, K. S., "Note on Non-Euclidean Principal Ideal Domains", *Mathematics Magazine* 48(1975), pp. 176–177.

Groups

24. Gallian, J., "The Search for Finite Simple Groups," *Mathematics Magazine,* 49(1976), pp. 163–179.
25. Kaplansky, I., *Infinite Abelian Groups,* 2nd edition. Ann Arbor, Mich.: University of Michigan Press, 1969.
26. Rotman, J., *An Introduction to the Theory of Groups,* 4th edition. New York: Springer-Verlag, 1995.

27. Steen, L. A., "A Monstrous Piece of Research," *Science News,* 118(1980), pp. 204–206.

Fields and Galois Theory

28. Gaal, L., *Classical Galois Theory with Examples,* 4th edition. New York: Chelsea, 1988.
29. Hadlock, C. R., *Field Theory and Its Classical Problems,* Carus Monograph 19. Washington, D.C.: Mathematical Association of America, 1978.
30. Kaplansky, I., *Fields and Rings,* revised 2nd edition. Chicago: University of Chicago Press, 1972.
31. Niven, I., *Irrational Numbers,* Carus Monograph 11. Washington, D.C.: Mathematical Association of America, 1956.

Applied Algebra and Computer Science

32. Childs, L., *A Concrete Introduction to Higher Algebra,* 2nd edition. New York: Springer-Verlag, 1995.
33. Knuth, D. E., *The Art of Computer Programming: Seminumerical Algorithms,* Vol. 2, 2nd edition. Reading, Mass.: Addison-Wesley, 1981.
34. Lidl, R., and G. Pilz, *Applied Abstract Algebra.* New York: Springer-Verlag, 1984.
35. Mackiw, G., *Applications of Abstract Algebra.* New York: Wiley, 1985.

Cryptography

36. DeMillo, R. A., G. Davida, et al., *Applied Cryptology, Cryptographic Protocols, and Computer Security Models,* Proceedings of Symposia in Applied Mathematics, Vol. 29. Providence, R. I.: American Mathematical Society, 1983.
37. Diffie, W., and M. Hellman, "Privacy and Authentication: An Introduction to Cryptography," *Proc. of the IEEE,* 67(1979), pp. 397–427.
38. Rivest, R. L., A. Shamir, and L. Adleman, "A Method for Obtaining Digital Signatures and Public-Key Cryptosystems," *Communications of the A.C.M.,* 21(1978), pp. 120–126.
39. Simmons, G. J., "Cryptology: The Mathematics of Secure Communication," *Mathematical Intelligencer,* 1(1979), pp. 233–246.

Lattices and Boolean Algebras

40. Abbott, J. C., *Sets, Lattices, and Boolean Algebras.* Boston: Allyn and Bacon, 1969.
41. Birkhoff, G., *Lattice Theory.* Providence, R. I.: American Mathematical Society, 1967.
42. Hohn, F. E., *Applied Boolean Algebra,* 2nd edition. New York: Macmillan, 1966.
43. _____, "Some Mathematical Aspects of Switching," *American Mathematical Monthly,* 62(1955), pp. 75–90.
44. Stone, M. H., "The Theory of Representations of Boolean Algebras," *Transactions of the American Mathematical Society,* 40(1936), pp. 37–111.

Geometric Constructions

45. Dudley, U., *The Trisectors*. Washington, D.C.: Mathematical Association of America, 1994.

Algebraic Coding Theory

46. Hill, R., *A First Course in Coding Theory*. Oxford: Oxford University Press, 1990.
47. MacWilliams, F. J., and N. J. A. Sloane, *The Theory of Error-Correcting Codes*. Amsterdam: North-Holland, 1978.
48. Pless, V., *Introduction to the Theory of Error-Correcting Codes*. New York: Wiley, 1989.
49. Thompson, T. M., *From Error-Correcting Codes Through Sphere Packings to Simple Groups*. Washington, D.C.: Mathematical Association of America, 1984.

History

50. Boyer, C., *A History of Mathematics,* 2nd edition. New York: Wiley 1991.
51. Burton, D., *The History of Mathematics: An Introduction,* 3rd edition. Dubuque, Iowa: Wm. C. Brown, 1994.
52. Eves, H., *An Introduction to the History of Mathematics,* 6th edition. Philadelphia: Saunders College Publishing, 1990.
53. Van der Waerden, B. L., *A History of Algebra: From al-Khowarizmi to Emmy Noether*. New York: Springer-Verlag, 1985.

Answers and Suggestions for Selected Odd-Numbered Exercises

For exercises that ask for proofs, there may be a sketch of the full proof (you fill in minor details), a key part of the proof (you fill in the rest), or a comment that should enable you to find a proof.

Section 1.1, page 6

1. (a) quotient 15; remainder 17 (c) quotient 0; remainder 0 (e) quotient 117; remainder 12.
3. (a) quotient 6; remainder 19 (c) quotient 62,720; remainder 92 (e) quotient $-14{,}940$; remainder 335.
5. If $a = 3q + 1$, then $a^2 = (3q + 1)^2 = 9q^2 + 6q + 1 = 3(3q^2 + 2q) + 1$, which is of the form $3k + 1$ with $k = 3q^2 + 2q$. Use similar arguments when $a = 3q$ or $a = 3q + 2$.
9. By the Division Algorithm, every integer a is of the form $3q$ or $3q + 1$ or $3q + 2$. Compute a^3 in each case and proceed as in Exercise 5.

Section 1.2, page 12

1. (a) 8 (c) 1 (e) 9 (g) 592.
3. $a \mid b$ means $b = au$ for some integer u. Similarly, $b \mid c$ means $c = bv$ for some integer v. Combine these two equations to show that $c = a \cdot$ (something), which proves that $a \mid c$.
5. $a \mid b$ means $b = au$ for some integer u, and $b \mid a$ means $a = bv$ for some integer v. Combine the equations to show that $a = auv$, which implies that $1 = uv$. Since u and v are integers, what are the only possibilities?
11. (a) 1 or 2.
13. If $d \mid k$ and $d \mid a$, then $k = du$ and $a = dv$, so that $du = k = abc + 1 = dvbc + 1$. Use the first and last terms of the equations to show that $d \mid 1$.
15. Many possible answers, including (a) $8 = 56 \cdot 4 + 72(-3)$ (c) $1 = 143 \cdot 1943 + 227(-1224)$ (e) $592 = 4144 \cdot 2 + 7696(-1)$.

19. Suppose $d \mid a$ and $d \mid b$, so that $a = du$ and $b = dv$. Since $a \mid (b + c)$, $b + c = aw$. Hence, $c = aw - b = duw - dv = d(uw - v)$, so that $d \mid c$. Since $(b,c) = 1$, what can you conclude about d and (a,b)?

23. Every common divisor of a and (b,c) is also a common divisor of (a,b) and c. [*Proof:* If $d \mid (b,c)$, then $d \mid b$ and $d \mid c$ by the definition of (b,c). If $d \mid a$ also, then d is a common divisor of a and b, and, hence, $d \mid (a,b)$ by Corollary 1.4.] A similar argument shows that the common divisors of (a,b) and c are also common divisors of a and (b,c).

29. $d = cu + av$ for some u, v (Why?). Hence, $db = cbu + abv$. Use the fact that $ab = cw$ for some w (Why?) to show that $c \mid db$.

33. First show that every integer n is the sum of a multiple of 9 and the sum of its digits. [*Example:* $7842 = 7 \cdot 1000 + 8 \cdot 100 + 4 \cdot 10 + 2 = 7(999 + 1) + 8(99 + 1) + 4(9 + 1) + 2 = (7 \cdot 999 + 8 \cdot 99 + 4 \cdot 9) + (7 + 8 + 4 + 2) = 9(7 \cdot 111 + 8 \cdot 11 + 4) + (7 + 8 + 4 + 2)$.] Thus, every n is of the form $9k + r$, where r is the sum of the digits of n. Hence, n is divisible by 9 if and only if 9 divides r.

Section 1.3, page 18

1. (a) $5040 = 2^4 \cdot 3^2 \cdot 5 \cdot 7$ (c) $45{,}670 = 2 \cdot 5 \cdot 4567$

3. ($\Leftarrow$) Suppose p has the given property and let d be a divisor of p, say $p = dt$. By the property, $d = \pm 1$ (in which case $t = \pm p$) or $t = \pm 1$ (in which case $d = \pm p$). Thus the only divisors of p are ± 1 and $\pm p$, and p is prime.

7. (a) $3, 3^2, 3^3, \ldots, 3^s; 3 \cdot 5, 3^2 \cdot 5, 3^3 \cdot 5, \ldots, 3^s \cdot 5; 3 \cdot 5^2, 3^2 \cdot 5^2, 3^3 \cdot 5^2, \ldots, 3^s \cdot 5^2; 3 \cdot 5^3, \ldots; 3 \cdot 5^t, 3^2 \cdot 5^t, 3^3 \cdot 5^t, \ldots, 3^s \cdot 5^t; 5, 5^2, \ldots, 5^t.$

9. Every prime divisor of a^2 is also a divisor of a by Theorem 1.8, and similarly for b^2.

13. If c has prime decomposition $p_1 p_2 \cdots p_k$, then $ab = c^2 = p_1 p_1 p_2 p_2 \cdots p_k p_k$. Now p_1 must divide a or b by Theorem 1.8, say a. Since $(a,b) = 1$, p_1 cannot divide b. Hence, $(p_1)^2 \mid a$. By relabeling and reindexing if necessary, show that $a = p_1 p_1 p_2 p_2 \cdots p_j p_j = (p_1 p_2 \cdots p_j)^2$ and $b = p_{j+1} p_{j+1} \cdots p_k p_k = (p_{j+1} p_{j+2} \cdots p_k)^2$.

15. Every prime that divides a^3 must also divide a by Corollary 1.9, and similarly for b. Hence $p^3 \mid a^3$.

17. If $p > 3$ is prime, then $p = 6k + 1$ or $6k + 5$ (why can the other cases be eliminated?). If $p = 6k + 1$, then $p^2 + 2 = (6k + 1)^2 + 2 = 36k^2 + 12k + 3 = 3(12k^2 + 4k + 1)$. The other case is handled similarly.

Section 1.4, page 22

1. All of them.

5. Every three-digit integer is less than 1000, and $\sqrt{1000} \approx 31.62$. Use Theorem 1.13.

7. Verify that $x^n - 1 = (x - 1)(x^{n-1} + x^{n-2} + \cdots + x^2 + x + 1)$. Conclude that $y^{mn} - 1 = (y^m)^n - 1$ has $y^m - 1$ as a factor. Apply this fact with $y = 2$ and $p = mn$ to show that $2^p - 1$ is composite whenever p is.

13. Exercise 6 in Appendix E shows that $\binom{p}{k}$ is an integer. $\binom{p}{1} = p$, and for $k > 1$, the denominator of $\binom{p}{k}$ is the product of integers that are each strictly less than p.

Section 2.1, page 29

3. By Corollary 2.5, $a \equiv 0$ or $a \equiv 1$ or $a \equiv 2$ or $a \equiv 3 \pmod 4$. Hence, a^2 is congruent to 0^2 or 1^2 or 2^2 or $3^2 \pmod 4$ by Theorem 2.2.

9. (a) $(n - a)^2 = n^2 - 2na + a^2$. Hence, $(n - a)^2 - a^2$ is divisible by n.

11. (a) Every integer that is congruent to 4 (mod 5).   (c) Every integer that is congruent to 4 or 9 or 14 (mod 15).

13. (a) If $a \equiv b \pmod{2n}$, then $b = a + 2nk$ for some k (Why?). Hence, $b^2 = (a + 2nk)^2 = a^2 + 4nka + 4n^2k^2$, so that $b^2 - a^2 = 4n(ka + nk^2)$.

15. Let c and d be any integers. By Corollary 2.5, each is congruent (mod 4) to exactly one of 0, 1, 2, 3. If $c \equiv 2$ and $d \equiv 3$, then by Theorem 2.2, $c^2 + d^2 \equiv 2^2 + 3^2 \equiv 1 \pmod 4$. Since $a \equiv 3 \pmod 4$, we cannot have $a = c^2 + d^2$ in this case. The other cases are handled similarly.

17. ($\Rightarrow$) By the Division Algorithm, $a = qn + r$ and $b = pn + s$ with the remainders r and s satisfying $0 \leq r < n$ and $0 \leq s < n$. If $a \equiv b \pmod n$, then $a - b = kn$ (Why?), and, hence, $kn = (qn + r) - (pn + s)$, which implies that $r - s = (k - q + p)n$, that is, $n \mid (r - s)$. Since r and s are strictly less than n, this is impossible unless $r - s = 0$. To prove the converse, assume $r = s$ and show that $n \mid (a - b)$.

21. $a - b = nk$ for some k (Why?). Show that any common divisor of a and n also divides b, and that any common divisor of b and n also divides a. What does this say about (a,n) and (b,n)?

23. $10 \equiv 1 \pmod 9$; hence $10^n \equiv 1^n \equiv 1 \pmod 9$ by Theorem 2.2.

25. Note that $10 \equiv -1 \pmod{11}$ and use Theorem 2.2.

Section 2.2, page 35

1. (a)

+	0	1
0	0	1
1	1	0

·	0	1
0	0	0
1	0	1

(c)

+	0	1	2	3	4	5	6
0	0	1	2	3	4	5	6
1	1	2	3	4	5	6	0
2	2	3	4	5	6	0	1
3	3	4	5	6	0	1	2
4	4	5	6	0	1	2	3
5	5	6	0	1	2	3	4
6	6	0	1	2	3	4	5

·	0	1	2	3	4	5	6
0	0	0	0	0	0	0	0
1	0	1	2	3	4	5	6
2	0	2	4	6	1	3	5
3	0	3	6	2	5	1	4
4	0	4	1	5	2	6	3
5	0	5	3	1	6	4	2
6	0	6	5	4	3	2	1

3. (a) $a = 3$ or $a = 5$.
5. (a) $x = 0$ or 1 or 2 or 3 or 4   (c) $x = 0$ or 1 or 2 or 3.
9. (a) $a^5 + b^5$ (c) $a^2 + b^2$.

Section 2.3, page 39

1. (a) $a = 1, 2, 3, 4, 5,$ or 6   (c) $a = 1, 2, 4, 5, 7,$ or 8.
3. $ab = 0$ in $\mathbb{Z}_p$ means $p \mid ab$ in $\mathbb{Z}$. Apply Theorem 1.8 and translate the result into $\mathbb{Z}_p$.
7. (a) 16   (c) 21   (e) 6.
11. (a) 3, 9, 15.

Section 3.1, page 50

1. (a) Closure for addition.
5. (a) Subring without identity (every product is the zero matrix)   (c) Subring without identity   (e) Commutative subring with identity.

11.

+	0	S	A	B	C	D	E	F
0	0	S	A	B	C	D	E	F
S	S	0	F	E	D	C	B	A
A	A	F	0	D	E	B	C	S
B	B	E	D	0	F	A	S	C
C	C	D	E	F	0	S	A	B
D	D	C	B	A	S	0	F	E
E	E	B	C	S	A	F	0	D
F	F	A	S	C	B	E	D	0

·	0	S	A	B	C	D	E	F
0	0	0	0	0	0	0	0	0
S	0	S	A	B	C	D	E	F
A	0	A	A	0	0	A	A	0
B	0	B	0	B	0	B	0	B
C	0	C	0	0	C	0	C	C
D	0	D	A	B	0	D	A	B
E	0	E	A	0	C	A	E	C
F	0	F	0	B	C	B	C	F

13. (a)

+	(0,0)	(1,1)	(0,2)	(1,0)	(0,1)	(1,2)
(0,0)	(0,0)	(1,1)	(0,2)	(1,0)	(0,1)	(1,2)
(1,1)	(1,1)	(0,2)	(1,0)	(0,1)	(1,2)	(0,0)
(0,2)	(0,2)	(1,0)	(0,1)	(1,2)	(0,0)	(1,1)
(1,0)	(1,0)	(0,1)	(1,2)	(0,0)	(1,1)	(0,2)
(0,1)	(0,1)	(1,2)	(0,0)	(1,1)	(0,2)	(1,0)
(1,2)	(1,2)	(0,0)	(1,1)	(0,2)	(1,0)	(0,1)

·	(0,0)	(1,1)	(0,2)	(1,0)	(0,1)	(1,2)
(0,0)	(0,0)	(0,0)	(0,0)	(0,0)	(0,0)	(0,0)
(1,1)	(0,0)	(1,1)	(0,2)	(1,0)	(0,1)	(1,2)
(0,2)	(0,0)	(0,2)	(0,1)	(0,0)	(0,2)	(0,1)
(1,0)	(0,0)	(1,0)	(0,0)	(1,0)	(0,0)	(1,0)
(0,1)	(0,0)	(0,1)	(0,2)	(0,0)	(0,1)	(0,2)
(1,2)	(0,0)	(1,2)	(0,1)	(1,0)	(0,2)	(1,1)

17. The multiplicative identity is 6.

19. To prove that E is closed under $*$, you must verify that when a and b are even integers, so is $a * b = ab/2$. To prove that $*$ is associative, verify that $(a * b) * c = a * (b * c)$ as follows. By definition, $(a * b) * c = (ab/2) * c = \dfrac{(ab/2)c}{2}$. Express $a * (b * c)$ in terms of multiplication in $\mathbb{Z}$ and verify that the two expressions are equal. Commutativity of $*$ is proved similarly. To prove the distributive law, you must verify that $a * (b + c) = a * b + a * c$, that is, that $a(b + c)/2 = ab/2 + ac/2$. If there is a multiplicative identity e, then it must satisfy $e * a = a$ for every $a \in E$, which is equivalent to $ea/2 = a$ in $\mathbb{Z}$. But $ea/2 = a$ implies that $e = 2$.

29. Consider $R = \mathbb{Z}_2$, $S = \mathbb{Z}_3$ and examine the table in the answer to Exercise 13(a).

35. (b) Since H is contained in the ring $M(\mathbb{C})$, its addition is commutative and associative, its multiplication is associative, and the distributive law holds. So you need to verify only that H is closed under addition and multiplication, that the zero and identity matrices are in H, and that the negative of every matrix in H is also in H.

Section 3.2, page 62

1. (a) $a^2 - ab + ba - b^2$.

7. (a) yes.   (c) To show that S is closed under addition and multiplication, you must verify that $rb = 0_R$ and $sb = 0_R$ imply that $(r + s)b = 0_R$ and $(rs)b = 0_R$.

13. If $ub = 0_R$ and u is a unit with inverse v, multiply both sides by v. What is the definition of a zero divisor?

15. $ab = ac$ is equivalent to $a(b - c) = 0_R$.

25. (a) $(a + a)^2 = a + a$ because $x^2 = x$ for *every* x. But $(a + a)^2 = (a + a)(a + a) = a^2 + a^2 + a^2 + a^2 = a + a + a + a$.

Section 3.3, page 76

1. The tables for $\mathbb{Z}_2 \times \mathbb{Z}_3$ are in the answer to Exercise 13(a) of Section 3.1.

3. If $f(a) = f(b)$, then $(a,a) = (b,b)$, and, hence, $a = b$ by the equality rules for ordered pairs. Therefore, f is injective. $f(a + b) = (a + b,\, a + b) = (a,a) + (b,b) = f(a) + f(b)$. Complete the proof by showing that $f(ab) = f(a)f(b)$ and that f is surjective.

7. Use this fact to prove that f is injective: If $a - b\sqrt{2} = c - d\sqrt{2}$ (with $a, b, c, d \in \mathbb{Q}$), then $b = d$, which implies that $a = c$. [*Proof:* If $b \neq d$, then $\sqrt{2} = (c - a)/(d - b)$, contradicting the fact that $\sqrt{2}$ is an irrational real number.]

15. Use the same argument as in the answer to Exercise 7, with $\sqrt{2}$ replaced by i.

19. The multiplicative identity in $\mathbb{Z}^*$ is 0. If there is an isomorphism $f : \mathbb{Z} \to \mathbb{Z}^*$, Theorem 3.12 shows that f must satisfy $f(1) = 0$. Hence, $f(2) = f(1 + 1) = f(1) \oplus f(1) = 0 \oplus 0 = 0 + 0 - 1 = -1$. Similarly, $f(3) = f(1 + 2) = f(1) \oplus f(2) = 0 \oplus (-1) = 0 + (-1) - 1 = -2$. What is $f(4)$? $f(5)$? $f(-1)$? Find a formula for f. Then use this formula to show that f is injective, surjective, and a homomorphism.

25. (a) Because f and g are homomorphisms, $(f \circ g)(a + b) = f(g(a + b)) = f(g(a) + g(b)) = f(g(a)) + f(g(b)) = (f \circ g)(a) + (f \circ g)(b)$. A similar argument shows that $(f \circ g)(ab) = (f \circ g)(a) \cdot (f \circ g)(b)$.   (b) You must show two things: (1) if f and g are injective, so is $f \circ g$; and (2) if f and g are surjective, so is $f \circ g$. To prove (1), assume $(f \circ g)(a) = (f \circ g)(b)$, that is, $f(g(a)) = f(g(b))$. Then use the injectivity of f and g to show $a = b$.

33. (a) $\mathbb{Z}$ has an identity and E doesn't.   (c) The rings have different numbers of elements, and so no injective function is possible from $\mathbb{Z}_4 \times \mathbb{Z}_{14}$ to $\mathbb{Z}_{16}$.   (e) The equation $x + x = 0_R$ has a *nonzero* solution in $\mathbb{Z} \times \mathbb{Z}_2$ (What is it?) but not in $\mathbb{Z}$.

Section 4.1, page 88

1. (a) $3x^4 + x^3 + 2x^2 + 2$ (c) $x^5 - 1$.
3. (a) x^3; $x^3 + x^2$; $x^3 + x$; $x^3 + x^2 + x$; $x^3 + 1$; $x^3 + x^2 + 1$; $x^3 + x + 1$; $x^3 + x^2 + x + 1$.
5. (a) $q(x) = 3x^2 - 5x + 8$; $r(x) = -4x - 6$.
   (c) $q(x) = x^3 + 3x^2 + 2x + 3$; $r(x) = 4$.
9. Yes (read the definition of zero divisor and remember that R is a subset of $R[x]$).
15. If $0 \neq b \in R$, then $b \in R[x]$ and $1_R = bq(x) + r(x)$. Use the fact that $\deg b = 0$ to show that $r(x) = 0$ and $q(x) \in R$. Hence, every nonzero element of R has an inverse.

Section 4.2, page 93

1. If $0_F \neq c \in F$, then c has an inverse; hence, $f(x) = c(c^{-1}f(x))$.
5. (a) $x - 1$ (c) $x^2 - 1$ (e) $x - i$.
7. Since $f(x)$ divides every first-degree polynomial, it must have degree 1 or 0. If $f(x) = cx + d$ with $c \neq 0_F$, then $f(x) = c(x + dc^{-1})$, and, hence, $x + dc^{-1}$ also divides every nonconstant polynomial. Use Exercise 3 to reach a contradiction. Conclude that $\deg f(x) = 0$.
15. Every divisor of $h(x)$ is also a divisor of $f(x)$.

Section 4.3, page 98

1. (a) $x^3 + \frac{2}{3}x^2 + \frac{1}{3}x + \frac{5}{3}$ (c) $x^3 - ix + i$.
3. (a) $x^2 + x + 1$; $2x^2 + 2x + 2$; $3x^2 + 3x + 3$; $4x^2 + 4x + 4$.
7. ($\Rightarrow$) Suppose $f(x)$ is irreducible and $g(x) = cf(x)$, with $0_F \neq c \in F$. If $g(x) = r(x)s(x)$, then $f(x) = (c^{-1}r(x))s(x)$, and, hence, either $c^{-1}r(x)$ or $s(x)$ is a nonzero constant by Theorem 4.11. If $c^{-1}r(x)$ is a constant, show that $r(x)$ is also a constant. Hence, $g(x)$ is irreducible by Theorem 4.11.

9. (a) $x^2 + x + 1$ (c) $x^2 + 1$; $x^2 + x + 2$; $x^2 + 2x + 2$; $2x^2 + 2$; $2x^2 + x + 1$; $2x^2 + 2x + 1$.

11. If it were reducible, it would have a monic factor of degree 1 (Why?), that is, a factor of the form $x + a$ with $a \in \mathbb{Z}_7$. Verify that none of the seven possibilities is a factor.

15. $(x - 3)(x - 4)^3$.

21. (a) If $f(x) \in \mathbb{Z}_p[x]$ is a monic reducible quadratic, then it must factor as $f(x) = (cx + d)(c^{-1}x + e)$ for some $c, d, e \in \mathbb{Z}_p$ (Why?). Hence, $f(x) = c(x + dc^{-1})c^{-1}(x + ec) = (x + a)(x + b)$ with $a = dc^{-1}$ and $b = ec$. When counting the possible pairs of factors, remember that, for example, $(x + 2)(x + 3)$ is the same factorization as $(x + 3)(x + 2)$. Also consider factorizations such as $(x + 2)(x + 2)$.

23. (a) Proceed as in the answer to Exercise 11, with $\mathbb{Z}_5$ in place of $\mathbb{Z}_7$.

Section 4.4, page 104

1. (a) Many correct answers, including $f(x) = x^2 + x$.

3. (a) No; $f(-2) \neq 0$.   (c) Yes.

9. In $\mathbb{Z}_3[x]$: $x^2 + 1$; $x^2 + x + 2$; $x^2 + 2x + 2$.

13. (a) If $f(x) = cg(x)$ with $c \neq 0_F$, then $g(x) = c^{-1}f(x)$. Hence, $g(u) = 0_F$ implies $f(u) = 0_F$ and vice versa.

15. If $x^2 + 1$ is reducible, then $x^2 + 1 = (x + a)(x + b)$ for some $a, b \in \mathbb{Z}_p$ (see the answer to Exercise 21(a) of Section 4.3). Expand the right side.

19. (a) If $f(x) = (x - a)^k g(x)$ with $g(a) \neq 0$, then $f'(x) = k(x - a)^{k-1}g(x) + (x - a)^k g'(x)$. If a is a multiple root of $f(x)$, then $k \geq 2$ and $k - 1 \geq 1$. If a is a root of both $f(x)$ and $f'(x)$, show that $k \geq 2$.

Section 4.5, page 113

1. (a) $(-1)(x + 1)(x - 2)(x^2 + 1)$ (c) $xx(x + 2)(x - 1)(3x - 1)$
   (e) $(x + 3)(2x + 1)(x^2 + 1)$.

3. Use the Rational Root Test.

5. (a) Let $p = 2$.   (c) Let $p = 2$ or $p = 3$.

7. (a) Let $p = 5$ and use Corollary 4.18.

11. Apply Eisenstein's Criterion and Corollary 4.17.

17. A polynomial of degree k has $k + 1$ coefficients. There are n choices for each coefficient except the coefficient a_k of x^k. How many choices are there for a_k?

Section 4.6, page 117

1. (a) $1 - 2i$; $1 + 2i$; 3; -2 (c) $3 + 2i$; $3 - 2i$; $-1 + i$; $-1 - i$.

3. (a) $x^4 - 2$ in $\mathbb{Q}[x]$; $(x^2 + \sqrt{2})(x + \sqrt[4]{2})(x - \sqrt[4]{2})$ in $\mathbb{R}[x]$;
   $(x - \sqrt[4]{2}i)(x + \sqrt[4]{2}i)(x + \sqrt[4]{2})(x - \sqrt[4]{2})$ in $\mathbb{C}[x]$.   (c) $(x - 1)(x^2 - 5)$ in $\mathbb{Q}[x]$;
   $(x - 1)(x + \sqrt{5})(x - \sqrt{5})$ in $\mathbb{R}[x]$ and $\mathbb{C}[x]$.

5. Nonreal roots of $f(x)$ occur in pairs by Lemma 4.28.

Section 5.1, page 122

3. There are eight congruence classes.
5. Show that if $a \neq b$, then $x + a \not\equiv x + b \pmod{x^2 - 2}$. Hence, the equivalence classes of $x + a$ and $x + b$ are distinct.
9. See the answer to Exercise 17 of Section 2.1 with $f(x)$ and $g(x)$ in place of a and b.

Section 5.2, page 128

1.

+	[0]	[1]	[x]	[x + 1]	[x²]	[x² + 1]	[x² + x]	[x² + x + 1]
[0]	[0]	[1]	[x]	[x + 1]	[x²]	[x² + 1]	[x² + x]	[x² + x + 1]
[1]	[1]	[0]	[x + 1]	[x]	[x² + 1]	[x²]	[x² + x + 1]	[x² + x]
[x]	[x]	[x + 1]	[0]	[1]	[x² + x]	[x² + x + 1]	[x²]	[x² + 1]
[x + 1]	[x + 1]	[x]	[1]	[0]	[x² + x + 1]	[x² + x]	[x² + 1]	[x²]
[x²]	[x²]	[x² + 1]	[x² + x]	[x² + x + 1]	[0]	[1]	[x]	[x + 1]
[x² + 1]	[x² + 1]	[x²]	[x² + x + 1]	[x² + x]	[1]	[0]	[x + 1]	[x]
[x² + x]	[x² + x]	[x² + x + 1]	[x²]	[x² + 1]	[x]	[x + 1]	[0]	[1]
[x² + x + 1]	[x² + x + 1]	[x² + x]	[x² + 1]	[x²]	[x + 1]	[x]	[1]	[0]

·	[0]	[1]	[x]	[x + 1]	[x²]	[x² + 1]	[x² + x]	[x² + x + 1]
[0]	[0]	[0]	[0]	[0]	[0]	[0]	[0]	[0]
[1]	[0]	[1]	[x]	[x + 1]	[x²]	[x² + 1]	[x² + x]	[x² + x + 1]
[x]	[0]	[x]	[x²]	[x² + x]	[x + 1]	[1]	[x² + x + 1]	[x² + 1]
[x + 1]	[0]	[x + 1]	[x² + x]	[x² + 1]	[x² + x + 1]	[x²]	[1]	[x]
[x²]	[0]	[x²]	[x + 1]	[x² + x + 1]	[x² + x]	[x]	[x² + 1]	[1]
[x² + 1]	[0]	[x² + 1]	[1]	[x²]	[x]	[x² + x + 1]	[x + 1]	[x² + x]
[x² + x]	[0]	[x² + x]	[x² + x + 1]	[1]	[x² + 1]	[x + 1]	[x]	[x²]
[x² + x + 1]	[0]	[x² + x + 1]	[x² + 1]	[x]	[1]	[x² + x]	[x²]	[x + 1]

3.

+	[0]	[1]	[x]	[x + 1]
[0]	[0]	[1]	[x]	[x + 1]
[1]	[1]	[0]	[x + 1]	[x]
[x]	[x]	[x + 1]	[0]	[1]
[x + 1]	[x + 1]	[x]	[1]	[0]

·	[0]	[1]	[x]	[x + 1]
[0]	[0]	[0]	[0]	[0]
[1]	[0]	[1]	[x]	[x + 1]
[x]	[0]	[x]	[1]	[x + 1]
[x + 1]	[0]	[x + 1]	[x + 1]	[0]

7. $[ax + b] + [cx + d] = [(a + c)x + (b + d)]$; $[ax + b][cx + d] = [(ad + bc)x + (3ac + bd)]$.
11. Consider the product of $[x]$ with itself.

Section 5.3, page 132

3. By Corollary 5.5, the distinct elements of $F[x]/(x - a)$ are the classes of the form $[c]$ with $c \in F$. Use this to show that $F[x]/(x - a)$ is isomorphic to F.

5. (a) Verify that the multiplicative inverse of $r + s\sqrt{3}$ is $\dfrac{r}{t} - \dfrac{s}{t}\sqrt{3}$, where $t = r^2 - 3s^2$.

7. By Corollary 5.12, there is an extension field K of F that contains a root c_1 of $f(x)$. Hence, $f(x) = (x - c_1)g(x)$ in $K[x]$. Use Corollary 5.12 again to find an extension field L of K that contains a root c_2 of $g(x)$. Continue.

Section 6.1, page 141

1. To see that K is not an ideal, consider what happens when you multiply a constant polynomial by a polynomial of positive degree.

7. (a) $(0) = \{0\}$ and $(1) = (2) = (3) = (4) = \mathbb{Z}_5$ (c) $(0) = \{0\}$; $(1) = (5) = (7) = (11) = \mathbb{Z}_{12}$; $(2) = (6) = (10) = \{0, 2, 4, 6, 8, 10\}$; $(4) = (8) = \{0, 4, 8\}$; $(3) = (9) = \{0, 3, 6, 9\}$; $(6) = \{0, 6\}$.

13. (a) If $r \in R$ and $1_R \in I$, then $r = r \cdot 1_R \in I$. Hence, $R \subseteq I$ and thus $R = I$.

25. Use Theorem 6.1. K is nonempty because $f(0_R) = 0_S$ by Theorem 3.12, and, hence, $0_R \in K$. If $a, b \in K$, then $f(a) = 0_S$ and $f(b) = 0_S$ by the definition of K. To show that $a - b \in K$, you must prove that $f(a - b) = 0_S$. If $r \in R$, you must prove that $f(ra) = 0_S$ in order to show that $ra \in K$.

27. An element of $(m) \cap (n)$ is divisible by both m and n; hence, it is in (mn) (see Exercise 17 of Section 1.2).

29. ($\Rightarrow$) If $(a) = (b) = (0_R)$, show that $a = 0_R = b$ and, hence, $a = bu$ with $u = 1_R$. If $(a) = (b) \neq (0_R)$, then both a and b are nonzero and $a = a \cdot 1_R \in (a)$. Therefore, $a \in (b)$, so that $a = bu$ for some $u \in R$. Similarly, $b = av$ for some $v \in R$. Hence, $a = bu = avu$, which implies that $uv = 1_R$ (Theorem 3.10), so that u is a unit.

33. If $I \neq (3)$, show that I contains an element b such that $(3,b) = 1$. Use Theorem 1.3 to show that $1 \in I$ and, hence, by Exercise 13(a), $I = \mathbb{Z}$.

41. (b) If $f(x) \in \mathbb{Z}[x]$ has constant term c, then x divides $f(x) - c$, so that $f(x) \equiv c \pmod{J}$ by part (a). Hence, $f(x) + J = c + J$ by Theorem 6.6. If b, c are distinct integers, then $b - c$ cannot be divisible by x (Why?). Hence, $b - c \notin J$ and $b \not\equiv c \pmod{J}$. Therefore, $b + J \neq c + J$ by Theorem 6.6.

Section 6.2, page 151

3. By Exercise 14 in Section 6.1, the kernel of f is either (0_F) or F. Explain why it cannot be F. Hence, f is injective by Theorem 6.4 and, therefore, an isomorphism.

5. Consider the case when $R = \mathbb{Z}$ and I is the principal ideal (n). Then $\mathbb{Z}/I$ is just $\mathbb{Z}_n$. Is $\mathbb{Z}_n$ always an integral domain?

9. Apply the First Isomorphism Theorem to the identity map from R to R.

27. If $r + J$ is a nilpotent element of R/J, then for some n, we have $0_R + J = (r + J)^n = r^n + J$. Hence, $r^n \in J$ (Why?), which means that r^n is nilpotent in R. Hence, $(r^n)^m = 0_R$ for some m. But this says $r \in J$, and, hence, $r + J$ is the zero coset $0_R + J$.

Section 6.3, page 157

1. By the definition of composite, $n = cd$ with $1 < |c| < |n|$ and $1 < |d| < |n|$. Hence, c and d cannot be multiples of n. Thus $cd = n \in (n)$, but $c \notin (n)$ and $d \notin (n)$. Therefore, (n) is not a prime ideal.

3. (a) Use Theorem 2.8 to show that p is prime if and only if $\mathbb{Z}_p$ is a field. But $\mathbb{Z}_p = \mathbb{Z}/(p)$; apply Theorem 6.15.

5. The maximal ideals in $\mathbb{Z}_6$ are $\{0, 3\}$ and $\{0, 2, 4\}$.

7. If R is a field, use Exercise 14 of Section 6.1. If (0_R) is a maximal ideal, use Theorem 6.15 and Exercise 9 of Section 6.2.

9. If $p = cd$, then $cd \in (p)$. Since (p) is prime, either $c \in (p)$ or $d \in (p)$, say $c \in (p)$. Hence, $c = pv$ for some $v \in R$. Use this and the fact that $p = cd$ to show that d is a unit.

15. (b) M is not prime because, for example, $3 \cdot 7 = 0 \in M$, but $3 \notin M$ and $7 \notin M$.

17. I is an ideal by Exercise 24 of Section 6.2. Use the fact that $J \neq S$ (Why?) and surjectivity to show that $I \neq R$. If $rs \in I$, then $f(rs) \in J$. Hence, $f(r)f(s) \in J$ (Why?), so that $f(r) \in J$ or $f(s) \in J$ by primality. Therefore, $r \in I$ or $s \in I$, and, hence, I is prime.

19. ($\Rightarrow$) Suppose R has a unique maximal ideal M. Then $M \neq R$ by definition, and so M is contained in the set of nonunits by Exercise 13 of Section 6.1. If c is a nonunit, then the ideal $(c) \neq R$ (Why?). So (c) is contained in a maximal ideal by hypothesis. But M is the only maximal ideal. So $c \in (c) \subseteq M$. Since every nonunit is in M, the set of nonunits is the ideal M.

Section 7.1, page 171

1. $\begin{pmatrix} 1 & 2 & 3 \\ 2 & 3 & 1 \end{pmatrix}^{-1} = \begin{pmatrix} 1 & 2 & 3 \\ 3 & 1 & 2 \end{pmatrix}$ and $\begin{pmatrix} 1 & 2 & 3 \\ 3 & 1 & 2 \end{pmatrix}^{-1} = \begin{pmatrix} 1 & 2 & 3 \\ 2 & 3 & 1 \end{pmatrix}$. Each of the other permutations is its own inverse.

3. (a) 18   (c) 24   (e) 6.

7.

$\circ$	r_0	r_1	r_2	s	t	u
r_0	r_0	r_1	r_2	s	t	u
r_1	r_1	r_2	r_0	u	s	t
r_2	r_2	r_0	r_1	t	u	s
s	s	t	u	r_0	r_1	r_2
t	t	u	s	r_2	r_0	r_1
u	u	s	t	r_1	r_2	r_0

11. (a) $S_3 \times \mathbb{Z}_2$ is nonabelian of order 12 and $D_4 \times \mathbb{Z}_2$ is nonabelian of order 16.

15. (a) G is a group. *Closure:* If $a, b \in \mathbb{Q}$, then $a * b = a + b + 3 \in \mathbb{Q}$. *Associativity:* $(a * b) * c = (a + b + 3) * c = (a + b + 3) + c + 3 = a + b + c + 6 = a + (b + c + 3) + 3 = a * (b + c + 3) = a * (b * c)$. Verify that -3 is the identity element and that the inverse of a is $-6 - a$ because $a * (-6 - a) = a + (-6 - a) + 3 = -3$ and, similarly, $(-6 - a) * a = -3$.   (c) G is a group with identity 0. The inverse of a is $-a/(1 + a)$.

17. No; there is no identity e satisfying *both* $a * e = a$ and $e * a = a$ for *every* a.

25. If $ab = ac$, then $b = eb = (a^{-1}a)b = a^{-1}(ab) = a^{-1}(ac) = (a^{-1}a)c = ec = c$.

29. Let a, b, c be distinct elements of T. Let $\sigma \in A(T)$ be given by $\sigma(a) = b$, $\sigma(b) = a$, and $\sigma(t) = t$ for every other element of T. Let $\tau \in A(T)$ be given by $\tau(a) = b$, $\tau(b) = c$, $\tau(c) = a$, and $\tau(t) = t$ for every other element of T. Verify that $(\sigma \circ \tau)(a) = a$ and $(\tau \circ \sigma)(a) = c$; hence, $\sigma \circ \tau \neq \tau \circ \sigma$.

Section 7.2, page 178

1. $e = c^{-1}c = c^{-1}c^2 = (c^{-1}c)c = ec = c$.
5. If $f(a) = f(b)$, then $a^{-1} = b^{-1}$. Hence, $(a^{-1})^{-1} = (b^{-1})^{-1}$. Therefore, by Corollary 7.6, $a = (a^{-1})^{-1} = (b^{-1})^{-1} = b$. Thus f is injective. Corollary 7.6 can also be used to prove that f is surjective.
7. (a) 2   (c) 6.
9. (a) U_{10} has order 4; U_{24} has order 8.
11. If G is a finite group of order n and $a \in G$, then the $n + 1$ elements $a^0, a, a^2, a^3, \ldots,$ a^n cannot all be distinct. Hence, $a^i = a^j$ for some $i > j$. Apply Theorem 7.8.
15. (a) $x = a^{-1}b$ is a solution of $ax = b$ because $a(a^{-1}b) = (aa^{-1})b = eb = b$. If c is also a solution, then $ac = b = a(a^{-1}b)$. Hence, $c = a^{-1}b$ by Theorem 7.5(2).
25. If $a, b \in G$, then by hypothesis, $aa = e$, $bb = e$, and $abab = e$. Left multiply both sides of the last equation by ba and simplify.
27. Let $x = a^{-1}cb^{-1}$ and show that $axb = c$. To prove uniqueness, assume $ayb = c$ and show that $y = a^{-1}cb^{-1}$.
29. (b) In S_3, let $a = \begin{pmatrix} 1 & 2 & 3 \\ 2 & 1 & 3 \end{pmatrix}$ and $b = \begin{pmatrix} 1 & 2 & 3 \\ 1 & 3 & 2 \end{pmatrix}$. Verify that $|a| = 2$, $|b| = 2$, $ab = \begin{pmatrix} 1 & 2 & 3 \\ 2 & 3 & 1 \end{pmatrix}$, and $(ab)^4 = ab$.
31. Let $|a| = m$ and $|b| = n$, with $(m,n) = 1$. If $(ab)^k = e$ and $ab = ba$, then $a^k b^k = (ab)^k = e$, so that $a^k = b^{-k}$. Hence, $a^{kn} = (b^{-k})^n = (b^n)^{-k} = e$. Therefore, $m \mid kn$ by Theorem 7.8, and, hence, $m \mid k$ by Theorem 1.5. Similarly, $n \mid k$. So $mn \mid k$ (see Exercise 17 of Section 1.2).
33. $ab = b^4a \Rightarrow aba^{-1} = b^4 \Rightarrow (aba^{-1})^3 = (b^4)^3 \Rightarrow (aba^{-1})(aba^{-1})(aba^{-1}) = b^{12} \Rightarrow$ $ab^3a^{-1} = e$ (because $b^6 = e$) $\Rightarrow ab^3 = a \Rightarrow b^3 = e$. Therefore, $ab = b^4a = b^3ba = eba = ba$.

Section 7.3, page 187

1. Since e_H is the identity in H, $e_H e_H = e_H$. Apply Exercise 1 of Section 7.2 with $c = e_H$.
3. (a) If $a, b \in H \cap K$, then $a, b \in H$ and $a, b \in K$. Since H is a subgroup, $ab \in H$ and $a^{-1} \in H$. Similarly, $ab \in K$ and $a^{-1} \in K$. Hence, $ab \in H \cap K$ and $a^{-1} \in H \cap K$. Therefore, $H \cap K$ is a subgroup by Theorem 7.10.
11. (a) $\langle 1 \rangle = U_{15}$; $\langle 2 \rangle = \langle 8 \rangle = \{1, 2, 4, 8\}$; $\langle 4 \rangle = \{1, 4\}$; $\langle 7 \rangle = \langle 13 \rangle = \{1, 4, 7, 13\}$; $\langle 11 \rangle = \{1, 11\}$; $\langle 14 \rangle = \{1, 14\}$.
17. $1 = 2^4$; $2 = 2^1$; $4 = 2^2$; $7 = 13^3$; $8 = 2^3$; $11 = 2 \cdot 13$; $13 = 13^1$; $14 = 2^3 \cdot 13$.
21. (a) Using additive notation, we see that the group is cyclic with generator (1,1): $1(1,1) = (1,1)$; $2(1,1) = (0,2)$; $3(1,1) = (1,0)$; $4(1,1) = (0,1)$; $5(1,1) = (1,2)$; $6(1,1) = (0,0)$.

23. Since H is nonempty, there is some $c \in H$. By hypothesis, $e = cc^{-1} \in H$. If $d \in H$ then since $e \in H$, we have $d^{-1} = ed^{-1} \in H$. Use this and the fact that $d = (d^{-1})^{-1}$ to show that $c, d \in H$ implies $cd \in H$. Apply Theorem 7.10.

27. ($\Rightarrow$) If a is in the center of G, then $ag = ga$ for every $g \in G$. Hence, $C(a) = \{g \in G \mid ag = ga\} = G$.

31. If $x^{-1}ax$ and $x^{-1}bx \in x^{-1}Hx$ with $a, b \in H$, then $ab \in H$, and, hence, $(x^{-1}ax)(x^{-1}bx) = x^{-1}(ab)x \in x^{-1}Hx$. Show that $(x^{-1}ax)^{-1} = x^{-1}a^{-1}x \in x^{-1}Hx$. Apply Theorem 7.10.

35. If $a^n, b^n \in H$, then since G is abelian, $a^n b^n = (ab)^n \in H$. Also $(a^n)^{-1} = a^{-n} = (a^{-1})^n \in H$. Apply Theorem 7.10.

37. The subgroups of $\mathbb{Z}_{12}$ are $\{0\}$, $\{0, 6\}$, $\{0, 3, 6, 9\}$, $\{0, 4, 8\}$, $\{0, 2, 4, 6, 8, 10\}$, and $\mathbb{Z}_{12}$.

41. See Exercise 31 of Section 7.2.

45. If $(m,n) = d > 1$, then $m = dr$, $n = ds$, and $drs < mn$. If $(a,b) \in \mathbb{Z}_m \times \mathbb{Z}_n$, then $drs(ab) = (drsa, drsb) = (sma, rnb) = (0,0)$. Therefore, the order of (a,b) is a divisor of drs (by Theorem 7.8 in additive notation) and, hence, strictly less than mn. So (a,b) does not generate $\mathbb{Z}_m \times \mathbb{Z}_n$ (a group of order mn) by Theorem 7.14.

49. (a) Show that $U_{18} = \{1, 5, 7, 11, 13, 17\}$ is generated by 5.

Section 7.4, page 196

1. (a) *Homomorphism:* $f(x + y) = 3(x + y) = 3x + 3y = f(x) + f(y)$. *Surjective:* if $t \in \mathbb{R}$, then $f(t/3) = 3(t/3) = t$. *Injective:* if $f(x) = f(y)$, then $3x = 3y$, and, hence, $x = y$.

3. (a) $\begin{pmatrix} 1 & 0 \\ 0 & 1 \end{pmatrix}, \begin{pmatrix} 1 & 1 \\ 0 & 1 \end{pmatrix}, \begin{pmatrix} 0 & 1 \\ 1 & 0 \end{pmatrix}, \begin{pmatrix} 0 & 1 \\ 1 & 1 \end{pmatrix}, \begin{pmatrix} 1 & 1 \\ 1 & 0 \end{pmatrix}, \begin{pmatrix} 1 & 0 \\ 1 & 1 \end{pmatrix}$ (b) It is easier to compare tables if the corresponding elements are listed in the same order for each group. To determine a correspondence, remember that under an isomorphism, corresponding elements must have the same order.

5. Show that both groups are cyclic of order 4 and use Theorem 7.18.

11. $f(a^0) = f(e_G) = e_H = f(a)^0$. For positive integers, use induction: $f(a^1) = f(a) = f(a)^1$. If $f(a^k) = f(a)^k$, then $f(a^{k+1}) = f(a^k a^1) = f(a^k)f(a) = f(a)^k f(a) = f(a)^{k+1}$. Hence, $f(a^n) = f(a)^n$ for all $n \geq 0$. What about negative n?

15. ($\Rightarrow$) If G is abelian, then f is a homomorphism because $f(ab) = (ab)^{-1} = b^{-1}a^{-1} = a^{-1}b^{-1} = f(a)f(b)$. In this case, f is an isomorphism by Exercise 5 of Section 7.2.

21. If $a^n = e_G$, then by Exercise 11 and Theorem 7.19, $f(a)^n = f(a^n) = f(e_G) = e_H$. Similarly, if $f(a)^n = e_H$ then $f(a^n) = f(a)^n = e_H = f(e_G)$. Hence, $a^n = e_G$ since f is injective. So $a^n = e_G$ if and only if $f(a)^n = e_H$.

25. If $f, g \in \text{Inn } G$, then $f(a) = c^{-1}ac$ and $g(a) = d^{-1}ad$ for some c, d. Show that $(f \circ g)(a) = (dc)^{-1}a(dc)$ and, hence, $f \circ g \in \text{Inn } G$. Show that the inverse function h of f is given $h(a) = cac^{-1} = (c^{-1})^{-1}ac^{-1} \in \text{Inn } G$. Use Theorem 7.10.

29. (a) Verify that every nonidentity element of U_8 has order 2 but that this is not true for U_{10}. Hence, there is no isomorphism f by Exercise 21.   (c) Yes. Prove it by comparing operation tables.

35. (a) If $\theta_c(x) = \theta_c(y)$, then $xc^{-1} = yc^{-1}$. Hence, $x = y$ by Theorem 7.5. Therefore, θ_c is injective. If $x \in G$, then $xc \in G$ and $\theta_c(xc) = (xc)c^{-1} = x$. Hence, θ_c is surjective.

41. (a) Show that h and v both induce the same inner automorphism (that is, $h^{-1}ah = v^{-1}av$ for every $a \in D_4$). Do the same for r_0 and r_2, for r_1 and r_3, and for d and t. Then

show that the inner automorphisms induced by h, r_0, r_1, and d are all distinct (that is, no two of them have the same action on *every* element of D_4).

Section 7.5, page 206

1. ($\Rightarrow$) If $Ka = K$, then $a = ea \in Ka = K$. So $a \in K$.
3. (a) 4   (c) 1   (e) 6.
5. (a) 1, 2, 3, 4, 6, 8, 12, 24   (c) 1, 2, 4, 5, 8, 10, 16, 20, 40, 80.
7. 27, 720.
9. $H \cap K$ is a subgroup of H and of K, and so its order must divide p by Lagrange's Theorem. Hence, $|H \cap K|$ is either 1 (in which case $H \cap K = \langle e \rangle$) or p (in which case $H = H \cap K = K$).
11. If $e \neq a \in G$, then $\langle a \rangle$ is a nonidentity subgroup of G. Hence, $G = \langle a \rangle$. If $|G| = |a|$ has composite order, say $|a| = td$, then $\langle a^t \rangle$ is a subgroup of order d by Theorem 7.8. Use Theorem 7.28.
13. 2.
17. If $a^n = e_G$, then $f(a)^n = f(a^n) = f(e_G) = e_H$. So $|f(a)|$ divides n by Theorem 7.8.
19. A proper subgroup has order n, with $1 < n < pq$ and n a divisor of pq. Use Theorem 7.28.

Section 7.6, page 213

5. (b) If $\begin{pmatrix} 1 & c \\ 0 & 1 \end{pmatrix} \in N$ and $\begin{pmatrix} a & b \\ 0 & d \end{pmatrix} \in G$, then

$$\begin{pmatrix} a & b \\ 0 & d \end{pmatrix}^{-1}\begin{pmatrix} 1 & c \\ 0 & 1 \end{pmatrix}\begin{pmatrix} a & b \\ 0 & d \end{pmatrix} = \begin{pmatrix} 1/a & -b/ad \\ 0 & 1/d \end{pmatrix}\begin{pmatrix} 1 & c \\ 0 & 1 \end{pmatrix}\begin{pmatrix} a & b \\ 0 & d \end{pmatrix}$$

$$= \begin{pmatrix} 1/a & -b/ad \\ 0 & 1/d \end{pmatrix}\begin{pmatrix} a & b+cd \\ 0 & d \end{pmatrix} = \begin{pmatrix} 1 & cd/a \\ 0 & 1 \end{pmatrix} \in N.$$

7. $G^* = G \times \langle e \rangle$ is a subgroup by Exercise 5 of Section 7.3. It is normal by Theorem 7.34 since for any $(c,d) \in G \times H$ and $(a,e) \in G^*$, $(c,d)^{-1}(a,e)(c,d) = (c^{-1},d^{-1})(a,e)(c,d) = (c^{-1}ac, d^{-1}ed) = (c^{-1}ac, e) \in G^*$.
11. If $c \in G$, let f be the inner automorphism given by $f(x) = c^{-1}xc$ (see page 193). Since N is characteristic, $f(N) \subseteq N$, that is $c^{-1}Nc \subseteq N$. Hence, N is normal by Theorem 7.34.
17. Use Exercise 3 of Section 7.3 to show that $N \cap K$ is a subgroup of K. If $g \in K$ and $n \in N \cap K$, then $g \in G$, $n \in N$, and, hence, $g^{-1}ng \in N$ by the normality of N in G. But $n \in N \cap K$ implies that $n \in K$, and, hence, $g^{-1}ng \in K$ by closure in K. Therefore, $g^{-1}ng \in N \cap K$, so that $g^{-1}(N \cap K)g \subseteq N \cap K$. Hence, $N \cap K$ is normal in K by Theorem 7.34.
19. If $n \in N$ and $k \in K$, use normality to show that $k^{-1}(n^{-1}kn) = (k^{-1}n^{-1}k)n$ is in $K \cap N = \langle e \rangle$.
27. Let $N = \langle a \rangle$. Then $H = \langle a^k \rangle$ for some k by Theorem 7.16. If $g \in G$, then $g^{-1}ag \in N$ by normality; hence, $g^{-1}ag = a^t$ for some t. Consequently, for any $a^{ki} \in H$, $g^{-1}a^{ki}g = (g^{-1}ag)^{ki} = (a^t)^{ki} = (a^k)^{ti} \in H$.

33. N is a subgroup by Exercises 3 and 31 of Section 7.3. Show that N is normal in G.
35. By hypothesis, the cyclic group $\langle a \rangle$ is normal. Hence, $b^{-1}ab \in \langle a \rangle$, that is, $b^{-1}ab = a^k$ for some k.

Section 7.7, page 220

3. Show that $\mathbb{Z}_{18}/M$ is cyclic with generator $1 + M$; then show that $1 + M$ has order 6 in $\mathbb{Z}_{18}/M$.
5. $G/N \cong \mathbb{Z}_2$.
9. Since $ab = ba$ in G, $NaNb = Nab = Nba = NbNa$ in G/N.
13. $\mathbb{R}^*/\mathbb{R}^{**} \cong \mathbb{Z}_2$.
15. (a) If $m, n \in \mathbb{Z}$, then $n(m/n + \mathbb{Z}) = m + \mathbb{Z} = 0 + \mathbb{Z}$ in $\mathbb{Q}/\mathbb{Z}$.

Section 7.8, page 226

1. (a) *Homomorphism:* $f(x + y) = 3(x + y) = 3x + 3y = f(x) + f(y)$. Kernel $f = \{0, 4, 8\}$.   (c) Kernel $g = \{0, 2, 4, 6\}$.   (e) *Homomorphism:* $h([x]_{18} + [y]_{18}) = h([x + y]_{18}) = [2(x + y)]_3 = [2x + 2y]_3 = [2x]_3 + [2y]_3 = h([x]_{18}) + h([y]_{18})$.
5. (a) $\langle 0 \rangle$, $\mathbb{Z}_2$, $\mathbb{Z}_3$, $\mathbb{Z}_4$, $\mathbb{Z}_6$, $\mathbb{Z}_{12}$.
7. Kernel f is a normal subgroup of G, so what can it be?
9. The kernel is the subgroup of all multiples of $x - 3$ in $\mathbb{Z}[x]$.
11. Show that the map $f: \mathbb{R}^* \to \mathbb{R}^{**}$ given by $f(r) = |r|$ is a surjective homomorphism. What is its kernel?
19. Since $H \cong G/K$ by the First Isomorphism Theorem, it suffices to construct a bijection from the set S of all subgroups of G that contain K and the set T of all subgroups of G/K. If B is a subgroup of G that contains K, then B/K is a subgroup of G/K, so define $\theta: S \to T$ by $\theta(B) = B/K$. Then θ is surjective by Theorem 7.44. Show that θ is injective.

Section 7.9, page 235

1. (a) (173)   (c) (1476283).
3. (a) (12)(45)(679)   (c) (13)(254)(69)(78).
5. (a) odd   (c) even.
7. (a) 3   (c) 60.
11. $(a_1 a_2 \cdots a_k) = (a_1 a_k)(a_1 a_{k-1}) \cdots (a_1 a_4)(a_1 a_3)(a_1 a_2)$. There are $k - 1$ transpositions (one for each of $a_2, a_3, \ldots, a_k$). $k - 1$ is even if and only if k is odd.
15. Verify that $\tau\sigma = \sigma^{-1}\tau$; use this to show that any product of powers of σ and powers of τ is one of: σ, σ^2, σ^3, $\sigma^4 = (1)$, τ, $\sigma\tau$, $\sigma^2\tau$, or $\sigma^3\tau$.
23. Let $\tau = (ab)$ and express σ as a product of disjoint cycles. Since disjoint cycles commute by Exercise 12, all cycles in $\sigma\tau\sigma^{-1}$ not involving a or b will cancel and $\sigma\tau\sigma^{-1}$ will reduce to the form $\kappa(ab)\kappa^{-1}$, where κ has one of the following forms (in which a, b, x, y, u, v are distinct symbols): $(\cdots xaby \cdots)$; $(\cdots xbay \cdots)$; $(\cdots xay \cdots ubv \cdots)$; $(\cdots xay \cdots)$; $(\cdots ubv \cdots)$; or $(\cdots xay \cdots)(\cdots ubv \cdots)$. Verify that $\kappa(ab)\kappa^{-1}$ is a transposition in each case.

25. Adapt the proof of Theorem 7.51 with G in place of S_n.
29. There are three possible cases (where a, b, c, d are distinct symbols): $(ab)(ab)$, $(ab)(ac)$, and $(ab)(cd)$. But $(ab)(ab) = (1) = (abc)^3$; $(ab)(ac) = (acb)$; and $(ab)(cd) = (acb)(acd)$.

Section 7.10, page 241

1. (a) (123), (132), (124), (142), (134), (143), (234), (243).
3. (1).
5. Disjoint transpositions commute by Exercise 12 of Section 7.9.
7. If $N \neq (1)$, then N contains a nonidentity element σ. If $\tau \neq (1)$, then $\sigma\sigma = (1) = \sigma\tau$ implies that $\sigma = \tau$ by Theorem 7.5. Hence $N = \{(1), \sigma\}$; N is cyclic by Theorem 7.28.

Section 8.1, page 248

3. (a) $\{(0,0)\}$; $\{(0,0), (1,0)\}$; $\{(0,0), (0,1)\}$; $\{(0,0), (1,1)\}$; $\mathbb{Z}_2 \times \mathbb{Z}_2$.
5. $\mathbb{Z}_2 \times \mathbb{Z}_2$.
9. No.
13. (b) If D is normal, then for any a, $b \in G$, $(a,e,e)(b,b,b)(a,e,e)^{-1} \in D$. But $(a,e,e)(b,b,b)(a,e,e)^{-1} = (aba^{-1},b,b)$. Since this is in D, we must have $aba^{-1} = b$, which implies that $ab = ba$.
23. (a) Let $M = \langle(123)\rangle$ and $N = \langle(12)\rangle$ in S_3.
25. Use the homomorphism f in the proof of Theorem 8.1. If $f(a_1, \ldots ,a_k) = e$, then $a_i = (a_1 \cdots a_{i-1})^{-1}e(a_{i+1} \cdots a_k)^{-1}$. Use Lemma 8.2 and Corollary 7.6 repeatedly to show that $a_i \in N_i \cap N_1 \cdots N_{i-1}N_{i+1} \cdots N_k = \langle e\rangle$. Hence, f is injective by Theorem 7.40.
27. (a) What are the normal subgroups of S_3?

Section 8.2, page 260

1. If $p^n a = 0$ and $p^m b = 0$, then $p^n(-a) = -(p^n a) = 0$ and $p^{m+n}(a + b) = p^n p^m(a + b) = p^m(p^n a) + p^n(p^m b) = 0$. Hence, $a + b \in G(p)$ and $-a \in G(p)$. Use Theorem 7.10.
3. (a) $\mathbb{Z}_4 \oplus \mathbb{Z}_3$; $\mathbb{Z}_2 \oplus \mathbb{Z}_2 \oplus \mathbb{Z}_3$ (c) $\mathbb{Z}_2 \oplus \mathbb{Z}_3 \oplus \mathbb{Z}_5$ (e) $\mathbb{Z}_2 \oplus \mathbb{Z}_3 \oplus \mathbb{Z}_3 \oplus \mathbb{Z}_5$; $\mathbb{Z}_2 \oplus \mathbb{Z}_9 \oplus \mathbb{Z}_5$ (g) $\mathbb{Z}_2 \oplus \mathbb{Z}_2 \oplus \mathbb{Z}_2 \oplus \mathbb{Z}_3 \oplus \mathbb{Z}_5 \oplus \mathbb{Z}_5$; $\mathbb{Z}_2 \oplus \mathbb{Z}_4 \oplus \mathbb{Z}_3 \oplus \mathbb{Z}_5 \oplus \mathbb{Z}_5$; $\mathbb{Z}_8 \oplus \mathbb{Z}_3 \oplus \mathbb{Z}_5 \oplus \mathbb{Z}_5$; $\mathbb{Z}_2 \oplus \mathbb{Z}_2 \oplus \mathbb{Z}_2 \oplus \mathbb{Z}_3 \oplus \mathbb{Z}_{25}$; $\mathbb{Z}_2 \oplus \mathbb{Z}_4 \oplus \mathbb{Z}_3 \oplus \mathbb{Z}_{25}$; $\mathbb{Z}_8 \oplus \mathbb{Z}_3 \oplus \mathbb{Z}_{25}$.
5. (a) $2, 5^3$ (c) $2, 2, 2^2, 2^3, 3, 5, 5, 5, 5$.
7. (a) $2, 2$ and $2, 2$ (c) $2, 2^2$ and $2, 2^2$.
9. (a) G must contain an element of order p (Why?). If a has order p, then $pa = 0$.
13. If q is a prime other than p and if q divides $|G|$, use Exercise 12 to reach a contradiction.
19. (a) Exercise 1 is the special case when every element of finite order has order a power of p. Essentially the same proof works here.

Section 8.3, page 265

3. $\{(12)(34), (13)(24), (14)(23), (1)\}$ is the only Sylow 2-subgroup. The four Sylow 3-subgroups are $\langle(123)\rangle, \langle(124)\rangle, \langle(134)\rangle, \langle(234)\rangle$.
5. (a) 1 or 4.
7. (a) Show that G has a normal Sylow 7-subgroup.   (c) Show that G has a normal Sylow-11 subgroup.
9. If $a \in G$, then $(Na)^{p^n} = N$ in G/N, so that $a^{p^n} \in N$.
13. If $N_1, \ldots, N_k$ are the Sylow p-subgroups, use the injective homomorphism f in the proof of Exercise 25 of Section 8.1. Verify that $|G| = |N_1 \times \cdots \times N_k|$ and conclude that f is also surjective.
21. Show that there is a normal Sylow 3- or 5-subgroup. Note that if there are six Sylow 5-subgroups, G has 24 distinct elements of order 5 (Why?). Similarly, if there are ten Sylow 3-subgroups, G has 20 distinct elements of order 3.

Section 8.4, page 273

1. (a) $\{r_0\}, \{r_2\}, \{r_1, r_3\}, \{h, v\}, \{d, t\}$.
3. Look at $H = \{r_0, r_1, r_2, r_3\}$ in D_4.
5. $\langle(123)\rangle, \langle(124)\rangle, \langle(134)\rangle, \langle(234)\rangle$.
9. If C is the conjugacy class of $a \in G$, show that $f(C)$ is the conjugacy class of $f(a)$.
15. In the equation in Exercise 14 (c), verify that each $|C_i|$ is either 1 or a positive power of p. At least one $|C_i|$ is 1 because $\{e\}$ is a conjugacy class. Since $|N|$ is divisible by p, there must be more than one $|C_i| = 1$ and, hence, some nonidentity element of $Z(G)$ in N.
19. If $b \in N(N(K))$, then $b^{-1}N(K)b = N(K)$. Hence, $b^{-1}Kb \subseteq N(K)$, since $K \subseteq N(K)$. Verify that both K and $b^{-1}Kb$ are Sylow p-subgroups of $N(K)$ and, hence, conjugate in $N(K)$. But K is normal in $N(K)$, and so $b^{-1}Kb = K$. Hence, $b \in N(K)$.
21. If S is a Sylow p-subgroup containing H (Exercise 24), then every Sylow p-subgroup is of the form $a^{-1}Sa$ for some $a \in G$ and, therefore, contains $a^{-1}Ha$.

Section 8.5, page 281

1. First show that $p^2 \not\equiv 1 \pmod{q}$ [If $p^2 \equiv 1 \pmod{q}$, then q divides $p + 1$ or $p - 1$ (Why?). Use the facts that $p < q$ and $q \not\equiv 1 \pmod{p}$ to show that both possibilities lead to a contradiction.] Then use Theorem 8.30.

5. (a)

	e	a	a^2	a^3	b	ab	a^2b	a^3b
e	e	a	a^2	a^3	b	ab	a^2b	a^3b
a	a	a^2	a^3	e	ab	a^2b	a^3b	b
a^2	a^2	a^3	e	a	a^2b	a^3b	b	ab
a^3	a^3	e	a	a^2	a^3b	b	ab	a^2b
b	b	a^3b	a^2b	ab	a^2	a	e	a^3
ab	ab	b	a^3b	a^2b	a^3	a^2	a	e
a^2b	a^2b	ab	b	a^3b	e	a^3	a^2	a
a^3b	a^3b	a^2b	ab	b	a	e	a^3	a^2

7. Use Exercise 13 of Section 8.3 and Theorem 8.9.

11. $\{1, -1\}$.

15. How many Sylow p-subgroups does G have? Use Corollary 8.16.

Section 9.1, page 293

3. (a) True. *Proof:* $a \mid b$ means $b = au$ and $c \mid d$ means $d = cv$. Hence, $bd = aucv = ac(uv)$.

5. If a is an associate of b, then $a = bu$ for some unit u. Hence, $bu = a = bc$, and, therefore, $u = c$, a contradiction.

7. (a) Suppose $q = pu$, where p is irreducible and u is a unit. Suppose $q = rs$; then $rs = pu$, and, hence, $p = (pu)u^{-1} = (rs)u^{-1} = r(su^{-1})$. Since p is irreducible, r is a unit or su^{-1} is a unit by Theorem 9.1. But if su^{-1} is a unit, say $su^{-1}w = 1$, then s is a unit. Therefore, q is irreducible by Theorem 9.1.

15. (a) $\delta(ab) = \delta((su - tv) + (sv + tu)i) = (su - tv)^2 + (sv + tu)^2 = s^2u^2 - 2stuv + t^2v^2 + s^2v^2 + 2stuv + t^2u^2 = s^2u^2 + t^2v^2 + s^2v^2 + t^2u^2 = (s^2 + t^2)(u^2 + v^2) = \delta(a)\delta(b)$.

19. If $0_R \neq a \in R$, use Theorem 9.1 to show that a^2 can't be irreducible and, hence, must be a unit. Hence, a is a unit.

21. Suppose $p = rs$. Then $p \mid r$ or $p \mid s$. Show that r or s must be a unit and apply Theorem 9.1.

27. Assume that $\delta(a) = k$ for all nonzero $a \in R$. If $b \neq 0_R$, then there exist q, r such that $1_R = bq + r$, with $r = 0_R$ or $\delta(r) < \delta(b)$. The latter condition is impossible because $\delta(r) = k = \delta(b)$. Thus $r = 0_R$, and, hence, q is a multiplicative inverse of b.

Section 9.2, page 304

1. $(ab) \subseteq (b)$ since $b \mid ab$. If $(ab) = (b)$, then $ab \mid b$, say $abu = b$. Hence, $au = 1_R$, contradicting the fact that a is a nonunit.

5. See the second example on page 300.

11. If (a) is an ideal other than R, then a is not a unit (Why?) and, hence, must be divisible by an irreducible element p (Theorem 9.12). Hence, $(a) \subseteq (p)$, with (p) maximal by Exercise 10.

13. (b) Verify that $f: \mathbb{Z} \to \mathbb{Z}_6$, given by $f(a) = [a]$, is a surjective homomorphism.

15. By Theorem 9.8, $I = (b)$ for some nonzero b. If $a \in \mathbb{Z}[i]$, then $a = bq + r$ with $r = 0$ or $\delta(r) < \delta(b)$, and, hence, $a \equiv r \pmod{I}$. By Theorem 6.6, the number of distinct cosets of I (congruence classes mod I) is at most the number of possible r's under division by b. Show that there are only finitely many possible r's.

21. By Exercise 20, $d = au + bv$ for some $u, v \in R$. If $e \in S$ is a common divisor of a and b, then e necessarily divides d. Hence, d is a gcd of a and b in S.

29. For some d, $bc = ad$. If $a = r_1r_2 \cdots r_k$, $d = z_1z_2 \cdots z_n$, $b = p_1p_2 \cdots p_s$, and $c = q_1q_2 \cdots q_t$ with each p_i, q_i, r_i, z_i, irreducible, then $p_1p_2 \cdots p_sq_1q_2 \cdots q_t = r_1r_2 \cdots r_kz_1z_2 \cdots z_n$. So each r_i is an associate of p_j or q_j. But r_i cannot be an associate of any p_j (otherwise r_i would divide the gcd 1_R of a and b, which implies that the irreducible r_i is a unit).

Section 9.3, page 314

1. If $x = a, y = b, z = c$ is a solution of $x^n + y^n = z^n$ and $n = kt$, show that $x = a^t, y = b^t$, $z = c^t$ is a solution of $x^k + y^k = z^k$, contradicting the hypothesis.
3. $N(ab) = N((rm + snd) + (rn + sm)\sqrt{d}) = (rm + snd)^2 - d(rn + sm)^2 = r^2m^2 + 2mnrsd + s^2n^2d^2 - dr^2n^2 - 2mnrsd - ds^2m^2 = r^2m^2 + s^2n^2d^2 - dr^2n^2 - ds^2m^2 = (r^2 - ds^2)(m^2 - dn^2) = N(a)N(b).$
9. (a) Use Corollary 9.22.
17. ($\Rightarrow$) Let $a = u + v\sqrt{-5}$ and $b = w + z\sqrt{-5}$. If $r + s\sqrt{-5} \in P$, then $r + s\sqrt{-5} = 2a + (1 + \sqrt{-5})b = 2(u + v\sqrt{-5}) + (1 + \sqrt{-5})(w + z\sqrt{-5}) = (2u + w - 5z) + (2v + w + z)\sqrt{-5}$. Hence, $r - s = (2u + w - 5z) - (2v + w + z) = 2(u - v - 3z)$, so that $r \equiv s \pmod 2$.

Section 9.4, page 322

1. (2) $[a,b] = [ak,bk]$ because $a(bk) = b(ak)$.
3. $[a,1_R] + [b,1_R] = [a1_R + 1_Rb, 1_R1_R] = [a + b, 1_R] \in R^*$ and $[a,1_R][b,1_R] = [ab,1_R1_R] = [ab,1_R] \in R^*$; hence, R^* is closed under addition and multiplication. The zero element $[0_R,1_R]$ of F is in R^*. The negative of $[a,1_R]$ is $[-a,1_R] \in R^*$.
5. Verify that $f : F \to \{r + si \mid r, s \in \mathbb{Q}\}$ given by $f([a + bi, c + di]) = \left(\dfrac{ac + bd}{c^2 + d^2}\right) + \left(\dfrac{bc - ad}{c^2 + d^2}\right)i$ is an isomorphism.
11. $mu + nv = 1$ for some integers u and v by Theorem 1.3; u and v may be negative. Negative powers of a are defined in F and, hence, in F, $a = a^1 = a^{mu + nv} = a^{mu}a^{nv} = (a^m)^u(a^n)^v = (b^m)^u(b^n)^v = b^{mu + nv} = b^1 = b$.

Section 9.5, page 327

1. ($\Rightarrow$) If $f(x)$ is a unit in $R[x]$, then $f(x)g(x) = 1_R$ for some $g(x)$. By Theorem 4.2, $\deg f(x) + \deg g(x) = \deg 1_R = 0$. Hence, $\deg f(x) = 0 = \deg g(x)$, so that $f(x), g(x) \in R$. Hence, $f(x)$ is a unit in R.
3. ($\Rightarrow$) Assume p is irreducible in $R[x]$. If $p = rs$ in R, then either r or s is a unit in $R[x]$. Hence, r or s is a unit in R by Exercise 1. Therefore, p is irreducible in R by Theorem 9.1.
5. Since $c_1c_2 \cdots c_mf(x) = g(x)$, each c_i divides $g(x)$. Therefore, c_i is a unit in R because $g(x)$ is primitive.
9. First use the fact that $R[x]$ is a UFD to show that R is an integral domain. If c is a nonzero, nonunit element of R, then c is a nonzero, nonunit element of $R[x]$ by Exercise 1. Hence, $c = p_1p_2 \cdots p_k$, with each p_i irreducible in $R[x]$. Theorem 4.2 shows that each $p_i \in R$. Hence, p_i is irreducible in R by Exercise 3. Use the fact that $R[x]$ is a UFD to show that this factorization is unique up to order and associates in R.

Section 10.1, page 337

7. $a + bi = (b - 2a)i + a(1 + 2i) + 0(1 + 3i)$. Also, $a + bi = (-2a)i + (a - b)(1 + 2i) + b(1 + 3i)$.

9. Verify that $((-3/\sqrt{2}) - \sqrt{3})\sqrt{2} + \sqrt{3}(\sqrt{2} + i) + \sqrt{3}(\sqrt{3} - i) = 0$.

11. If the subset is $\{0_V, u_2, u_3, \ldots, u_n\}$, then $1_F 0_V + 0_F u_2 + 0_F u_3 + \cdots + 0_F u_n = 0_V$, with the first coefficient nonzero.

13. There exist $c_i \in F$, not all zero, such that $c_1 v_1 + \cdots + c_k v_k = 0_V$ since the v_i are linearly dependent. The set $\{v_1, \ldots, v_k, w_1, \ldots, w_t\}$ is linearly dependent because $c_1 v_1 + \cdots + c_k v_k + 0_F w_1 + \cdots + 0_F w_t = 0_V$ and not all the coefficients are zero.

15. For any $r + si \in \mathbb{C}$, $r + si = \left(\dfrac{r}{b} - \dfrac{cs}{bd}\right) b + \dfrac{s}{d}(c + di)$. Hence, $\{b, c + di\}$ spans $\mathbb{C}$ over $\mathbb{R}$. Prove that it is also linearly independent over $\mathbb{R}$.

23. (a) If $a + b\sqrt{2} + c\sqrt{3} = 0$, then $a + b\sqrt{2} = -c\sqrt{3}$. Squaring both sides and rearranging show that $2ab\sqrt{2} = 3c^2 - a^2 - 2b^2$. If $ab \neq 0$, then $\sqrt{2} = (3c^2 - a^2 - 2b^2)/2ab \in \mathbb{Q}$, which contradicts the fact that $\sqrt{2}$ is irrational. Hence, $a = 0$ or $b = 0$. If $a = 0$, then $b\sqrt{2} + c\sqrt{3} = 0$. Square both sides and make a similar argument to show that $bc = 0$. Hence, $b = 0$ or $c = 0$. But $a = 0$ and $b = 0$ imply that $c\sqrt{3} = 0$, whence, $c = 0$. Similarly, $a = 0$ and $c = 0$ imply that $b = 0$.

33. Suppose $c_1 u_1 + \cdots + c_t u_t + dw = 0_V$. If $d \neq 0_F$, then $w = -d^{-1}c_1 u_1 - d^{-1}c_2 u_2 - \cdots - d^{-1}c_t u_t$, a contradiction. Hence, $d = 0_F$. Then all the $c_i = 0_F$ because $\{u_1, \ldots, u_t\}$ is linearly independent.

37. ((i) $\Rightarrow$ (iii)) Suppose $S = \{v_1, \ldots, v_n\}$ spans K over F. Then some subset T of S is a basis of K over F by Exercise 32. Since $[K:F] = n$, T must have n elements, and, hence, $T = S$. Use Exercise 36 to prove (ii) $\Rightarrow$ (iii). (iii) implies (i) and (ii) by the definition of basis.

Section 10.2, page 345

3. Both $F(u + c)$ and $F(u)$ contain F by definition. Since $c \in F$ and $u \in F(u)$, $u + c \in F(u)$. Therefore, $F(u) \supseteq F(u + c)$, since $F(u + c)$ is the smallest subfield containing F and $u + c$. Conversely, $u = (u + c) - c \in F(u + c)$, so that $F(u) \subseteq F(u + c)$, since $F(u)$ is the smallest subfield containing F and u. Therefore, $F(u + c) = F(u)$.

5. (a) Verify that $3 + 5i$ is a root of $x^2 - 6x + 34$. (c) Verify that $1 + \sqrt[3]{2}$ is a root of $x^3 - 3x^2 + 3x - 3$.

7. By hypothesis, u is a root of some $p(x) \in F[x]$. But $F[x] \subseteq K[x]$, so that u is a root of $p(x) \in K[x]$.

9. $\sqrt{\pi}$ is a root of $x^2 - \pi \in \mathbb{Q}(\pi)[x]$.

11. 6.

15. By the Factor Theorem, $a + bi$ is a root of $f(x) = (x - (a + bi))(x - (a - bi))$. Verify that $f(x)$ has real coefficients.

17. (a) $x^4 - 2x^2 - 4$.

21. π is a root of $x^4 - \pi^4 \in \mathbb{Q}(\pi^4)[x]$ and, hence, is algebraic over $\mathbb{Q}(\pi^4)$. Therefore, $\{1, \pi, \pi^2, \pi^3\}$ is a basis by Theorem 10.7.

Section 10.3, page 350

3. Many correct answers, including (a) $\{1, \sqrt{5}, i, \sqrt{5}i\}$ (c) $\{1, \sqrt{2}, \sqrt{3}, \sqrt{5}, \sqrt{6}, \sqrt{10}, \sqrt{15}, \sqrt{30}\}$.

5. Use Corollary 4.18 to show that $x^2 + 1$ is irreducible over $\mathbb{Q}(\sqrt{3})$ and thus is the minimal polynomial of i over $\mathbb{Q}(\sqrt{3})$. Hence, $[\mathbb{Q}(\sqrt{3},i):\mathbb{Q}(\sqrt{3})] = 2$ and $[\mathbb{Q}(\sqrt{3},i):\mathbb{Q}] = [\mathbb{Q}(\sqrt{3},i):\mathbb{Q}(\sqrt{3})][\mathbb{Q}(\sqrt{3}):\mathbb{Q}] = 2 \cdot 2 = 4$.

7. $[K(u):F]$ is finite by Theorems 10.7 and 10.4. Hence, u is algebraic over F by Theorem 10.9. If $p(x) \in F[x]$ is the minimal polynomial of u over F and $q(x) \in K[x]$ is the minimal polynomial of u over K, then $q(x) \mid p(x)$ by Theorem 10.6. Hence, by Theorem 10.7, $[K(u):K] = \deg q(x) \le \deg p(x) = [F(u):F]$.

9. $[F(u):F]$ and $[K(u):F(u)]$ are finite by Theorems 10.4, 10.7, and 10.9 and Exercise 8. Apply Theorem 10.4 to $F \subseteq F(u) \subseteq K(u)$.

11. (a) Theorem 10.4 applied to $F \subseteq F(u) \subseteq F(u,v)$ shows that $m = \deg p(x) = [F(u):F]$ divides $[F(u,v):F]$. Similarly, $n \mid [F(u,v):F]$. Hence, $mn \mid [F(u,v):F]$ by Exercise 17 of Section 1.2. Use Theorem 10.4 and Exercise 7 to show that $[F(u,v):F] \le mn$. Therefore, $[F(u,v):F] = mn$.

13. Let $h(x) \in F(u)[x]$ be the minimal polynomial of v over $F(u)$; then $h(x) \mid q(x)$. By Exercise 11(a) and Theorems 10.4 and 10.7, $(\deg p(x))(\deg q(x)) = [F(u,v):F] = [F(u,v):F(u)][F(u):F] = (\deg h(x))(\deg p(x))$. Therefore, $\deg h(x) = \deg q(x)$, and, hence, $q(x) = kh(x)$ for some $k \in K$. Since $h(x)$ is irreducible over $F(u)$, so is $q(x)$.

15. If u is algebraic over E, then it is algebraic over F by Theorem 10.10 and Corollary 10.11.

Section 10.4, page 357

3. $\mathbb{Q}(\sqrt{5},i)$ is a splitting field; it has dimension 4 by Exercise 3 of Section 10.3.

7. The minimal polynomial $p(x)$ of u is irreducible in $F[x]$ and has a root in K. Therefore, $p(x)$ splits over $K = F(u)$.

11. The fourth roots of -1 are $(\pm \sqrt{2}/2) \pm (\sqrt{2}/2)i$, so that $\mathbb{Q}(\sqrt{2},i)$ is a splitting field.

15. $x^2 + 1$ is irreducible in $\mathbb{Z}_3[x]$ by Corollary 4.18. Hence, by Theorem 5.11, $\mathbb{Z}_3[x]/(x^2 + 1)$ is a field of nine elements that contains the roots $[x]$ and $[2x]$ of $x^2 + 1$.

21. If $p(x) \in K[x]$ is irreducible and u is a root of $p(x)$, then $K(u)$ is algebraic over K by Theorem 10.10. Therefore, u is algebraic over F by Corollary 10.11. Its minimal polynomial $q(x)$ over F splits over K and divides the irreducible $p(x)$ in $K[x]$ by Theorem 10.6. Show that $p(x)$ has degree 1 and apply Exercise 19.

Section 10.5, page 362

1. Every polynomial in $F[x]$ is also in $E[x]$.

7. (a) If $f(x) = a_n x^n + \cdots + a_0$ and $f'(x) = 0_F$, then for each $k > 0$, $(k1_F)a_k = ka_k = 0_F$. Since F has characteristic 0, $k1_F \ne 0_F$, and, hence, $a_k = 0$. Therefore, $f(x) = a_0$.

9. If $f(x)$ and $f'(x)$ are not relatively prime, then their gcd has a root u in some splitting field. Hence, u is a repeated root of $f(x)$ by Exercise 8, so that $f(x)$ is not separable.

13. Use the proof of Theorem 10.18, as in the example that follows the theorem.

Section 10.6, page 368

3. $na = a + a + \cdots + a = 1_R a + 1_R a + \cdots + 1_R a = (1_R + \cdots + 1_R)a = (n1_R)a = 0_R a = 0_R$.

5. Let $p = $ characteristic $F = $ characteristic K. F has order p^m, where $m = [F : \mathbb{Z}_p]$, by Theorem, 10.23, and, hence, $q = p^m$. Since $[K : \mathbb{Z}_p] = [K : F][F : \mathbb{Z}_p] = nm$, Theorem 10.23 shows that K has order $p^{mn} = q^n$.

13. Every element a of $\mathbb{Z}_p$ is a root of $x^p - x$ by the proof of Theorem 10.25. Hence, $a^p = a$ in $\mathbb{Z}_p$, which means that $a^p \equiv a \pmod{p}$ in $\mathbb{Z}$. If a is relatively prime to p in $\mathbb{Z}$, then a is a nonzero element of the field $\mathbb{Z}_p$ and, hence, has an inverse.

17. By Theorem 10.25, $E = \mathbb{Z}_p(u_1, \ldots, u_t) = F$, where the u_i are all the roots of $x^{p^n} - x$ in K.

Section 11.1, page 377

1. If $\sigma(c) = c$ for every $c \in F$, then $\sigma^{-1}(c) = \sigma^{-1}(\sigma(c)) = c$.

3. Use Theorem 10.7 to show that $\sigma(c) = c$ for all $c \in F(u)$.

5. Use Corollary 11.5 and Lagrange's Theorem 7.26.

9. (a) $p(x) = x^2 + x + 1$ (b) $\mathrm{Gal}_\mathbb{Q} \mathbb{Q}(\omega) \cong \mathbb{Z}_2$.

11. $\mathrm{Gal}_\mathbb{Q} \mathbb{Q}(\sqrt{2}, i) \cong \mathbb{Z}_2 \times \mathbb{Z}_2$.

Section 11.2, page 385

1. The number of intermediate fields is the same as the number of subgroups of $\mathrm{Gal}_F K$, which is finite by Theorem 11.11.

5. Four, of dimensions 10, 5, 2, and 1.

9. (a) Every subgroup of $\mathbb{Z}_n \cong \mathrm{Gal}_F K$ (in particular, $\mathrm{Gal}_E K$) is cyclic and normal by Theorem 7.16. By Theorem 11.11, $\mathrm{Gal}_F E \cong \mathrm{Gal}_F K / \mathrm{Gal}_E K$; apply Exercise 14 of Section 7.7.

11. (b) $[\mathbb{Q}(\sqrt[4]{2}) : \mathbb{Q}] = 4$ since $x^4 - 2$ is irreducible in $\mathbb{Q}[x]$ by Eisenstein's Criterion. $x^2 + 1$ is the minimal polynomial of i over $\mathbb{Q}(\sqrt[4]{2})$ by Corollary 4.18.

Section 11.3, page 395

1. (a) Many correct answers, including $\mathbb{Q} \subseteq \mathbb{Q}(\sqrt{5}) \subseteq \mathbb{Q}(\sqrt{5}, \sqrt{7}) \subseteq \mathbb{Q}(\sqrt{5}, \sqrt{7}, \sqrt[4]{2 + \sqrt{5}}) \subseteq \mathbb{Q}(\sqrt{5}, \sqrt{7}, \sqrt[4]{2 + \sqrt{5}}, \sqrt[4]{1 + \sqrt{7}})$.

5. (a) A_4 consists of the subgroup H and the eight 3-cycles (123), (132), (124), (142), (134), (143), (234), (243). Show that H is normal in A_4. Use the fact that all groups of

order ≤4 are abelian to show that the series $S_4 \supseteq A_4 \supseteq H \supseteq (1)$ satisfies the definition of solvability.

7. (a) ± 1 (c) $\pm 1, \pm i$ (e) $\pm 1,\ 1/2 \pm i\sqrt{3}/2,\ -1/2 \pm i\sqrt{3}/2$.

13. If K is the splitting field of a cubic polynomial, then $[K:F]$ is divisible by 3 (Why?) and ≤ 6 by Theorem 10.13. Hence, the Galois group is a subgroup of S_3 (Corollary 11.5) of order 3 or 6.

17. (a) $x^6 - 4x^3 + 4 = (x^3 - 2)^2$. $\mathbb{Q}(\sqrt[3]{2}, \omega)$ is a splitting field, where ω is a complex cube root of 1. $G \cong S_3$.   (c) $x^5 + 6x^3 + 9x = x(x^2 + 3)^2$. $\mathbb{Q}(i\sqrt{3})$ is a splitting field. $G \cong \mathbb{Z}_2$.   (e) $G \cong S_5$.

Chapter 12, page 405

1. If $ka \equiv 0 \pmod{p}$, then $p \mid ka$. But $(p,k) = 1$ (Why?). Hence, $p \mid a$ by Theorem 1.8, which is a contradiction.

3. (a) 0107   0512   2421   1479.

Section 13.1, page 410

3. If there is a solution, then 0, 1, or 2 is a solution by Exercise 2. Verify that this is not the case.

9. $x \equiv -30 \pmod{187}$.

11. $x \equiv -18 \pmod{210}$.

13. $x \equiv 204 \pmod{204{,}204}$.

19. ($\Leftarrow$) If $b - a = dk$ and $mu + nv = d$, then $muk + nvk = b - a$. Proceed as in the proof of Lemma 13.1.

Section 13.2, page 415

3. 7 is (1,2) and 8 is (2,3) in $\mathbb{Z}_3 \times \mathbb{Z}_5$. So the product is $(1 \cdot 2, 2 \cdot 3) = (2,1)$.

5. ($\Rightarrow$) If $f(r) = f(s)$, then *both* r and s are solutions of the system $x \equiv r \pmod{m_1}$, $x \equiv r \pmod{m_2}, \ldots, x \equiv r \pmod{m_r}$.

Section 13.3, page 419

1. (a) Repeated use of Corollary 13.6 shows that both are isomorphic to $\mathbb{Z}_3 \times \mathbb{Z}_4 \times \mathbb{Z}_5$ and, hence, to each other.

Section 14.1, page 427

1. (a) The proof given in Example 4 on page 422 also applies here.

3. It is not antisymmetric (consider 3 and -3).

7.

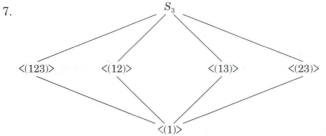

$\langle (123) \rangle$ $\langle (12) \rangle$ $\langle (13) \rangle$ $\langle (23) \rangle$

$\langle (1) \rangle$

15. In the set P of positive integers ordered by divisibility, the l.u.b. of a subset is the least common multiple of the integers in the set by Exercises 10 and 14. Consider the subset of primes.

17. (a) By the definition of g.l.b. and l.u.b., $a \wedge b \le a$ and $a \wedge b \le b \le b \vee c$, and, therefore, $a \wedge b \le a \wedge (b \vee c)$. Similarly, $a \wedge c \le a \wedge (b \vee c)$. Hence, $a \wedge (b \vee c)$ is an upper bound of $a \wedge b$ and $a \wedge c$, and, therefore, $(a \wedge b) \vee (a \wedge c) \le a \wedge (b \vee c)$.

21. $\mathbb{Z}$ is a principal ideal domain and $(a) \subseteq (c)$ if and only if $c \mid a$ (Lemma 9.9).

25. (a) 1 is the only minimal element of P. There is no maximal element since there is no positive integer that is divisible by *all* positive integers.   (c) Many correct answers, include the set S in Exercise 1.

Section 14.2, page 438

1. If I and J are both largest elements, then $I \le J$ since J is a largest element and $J \le I$ since I is a largest element. Hence, $I = J$.

3. c.

5. Show that if $mn = 105$, then n is the complement of m (that is, $[m,n] = 105$ and $(m,n) = 1$). Note that $105 = 3 \cdot 5 \cdot 7$.

9. Since $a = O \vee a$ for every a and $O \vee O' = I$ by definition, $O' = O \vee O' = I$.

13. By Theorem 14.4 you must show that for any positive integers a, b, c, $[a,(b,c)] = ([a,b],[a,c])$. By Exercise 11 in Section 1.3, it suffices to consider the case when p is prime and $a = p^r$, $b = p^s$, $c = p^t$. There are six cases ($r \le s \le t$, $s \le t \le r$, etc.). In the first case, for example, $[a,(b,c)] = \max(r, \min(s,t)) = s$ and $([a,b],[a,c]) = \min(\max(r,s), \max(r,t)) = s$.

17. By repeated use of De Morgan's laws and the distributive laws,
$((a \wedge b') \vee (a' \wedge b))' = (a \wedge b')' \wedge (a' \wedge b)' = (a' \vee b'') \wedge (a'' \vee b') =$
$(a' \vee b) \wedge (a \vee b') = ((a' \vee b) \wedge a) \vee ((a' \vee b) \wedge b') =$
$((a' \wedge a) \vee (b \wedge a)) \vee ((a' \wedge b') \vee (b \wedge b')) = (0 \vee (b \wedge a)) \vee ((a' \wedge b') \vee 0) =$
$(b \wedge a) \vee (a' \wedge b') = (a \wedge b) \vee (a' \wedge b')$.

21. ($\Rightarrow$) Since $O \le a \wedge b \le a$ and a is an atom, we must have $a \wedge b = O$ or $a \wedge b = a$.

23. ($\Leftarrow$) If $A = \{c\}$, then its only subsets are $\varnothing$ and A. Hence, $\varnothing \subsetneq B \subsetneq A$ is impossible, and A is an atom.

Section 14.3, page 448

1. (a) $p \wedge p'$.

3. (a) Use Theorem 14.6.

(c)

p	q	$p \Rightarrow q$	$p \wedge (p \Rightarrow q)$	$(p \wedge (p \Rightarrow q)) \Rightarrow q$
T	T	T	T	T
T	F	F	F	T
F	T	T	F	T
F	F	T	F	T

9.

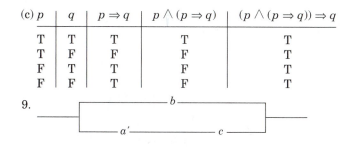

Chapter 15, page 460

3. (a) Begin as in the construction of the coordinate plane. Place the compass point on (1,0) and make a circle whose radius is the segment from (1,0) to (3,0). It intersects the vertical axis at Q. The right triangle with vertices (0,0), Q, (1,0) has hypotenuse of length 2 and one side of length 1. Hence the angle at Q (opposite the side of length 1) is a 30° angle, by a well-known theorem in plane geometry.   (c) Part (a) shows that a 90° angle can be trisected. Since a 30° angle can be *bisected,* a 45° angle can be trisected.

5. $\cos 3t = \cos(t + 2t) = \cos t \cos 2t - \sin t \sin 2t =$
$\cos t(2 \cos^2 t - 1) - \sin t(2 \sin t \cos t) = 2 \cos^3 t - \cos t - 2 \sin^2 t \cos t =$
$2 \cos^3 t - \cos t - 2(1 - \cos^2 t)\cos t = 4 \cos^3 t - 3 \cos t.$

7. No. To prove this, show that x must be the root of a cubic polynomial in $\mathbb{Q}[x]$ that has no rational roots.

9. No.

15. If $\sqrt{k} \in F$, then $F(\sqrt{k}) = F$. If $\sqrt{k} \notin F$, then the multiplicative inverse of a nonzero element $a + b\sqrt{k}$ of $F(\sqrt{k})$ is $c + d\sqrt{k}$, where $c = a/(a^2 - kb^2)$ and $d = -b/(a^2 - kb^2)$.

Section 16.1, page 472

1. Verify that C is closed under addition and, hence, is a subgroup by Theorem 7.11.

3. (a) 1   (c) 4.

5. (a) 0000, 1000, 0111, 1111   (c) 0000, 0010, 0101, 0111, 1001, 1011, 1100, 1110.

11. (c) If the ith coordinate is denoted by a subscript, then $(u + w)_i = u_i + w_i$ and $(v + w)_i = v_i + w_i$. Hence, $(u + v)_i = (v + w)_i$ if and only if $u_i = v_i$.

17. Many correct answers, including 00000, 11100, 00111, 11011.

21. $n = 5$.

25. Verify that an element of $B(n)$ has even Hamming weight if and only if it is the sum of an even number of elements of Hamming weight 1 (for instance, 110 = 100 + 010). Use this to show that the set of elements of even Hamming weight is closed under addition.

27. (a) .96059601   (c) .00058806   (e) .00000001.

Section 16.2, page 483

1. (a) $\begin{pmatrix} 0 & 0 \\ 1 & 1 \\ 1 & 0 \\ 0 & 1 \end{pmatrix}$ (c) $\begin{pmatrix} 1 \\ 1 \\ 0 \\ 1 \end{pmatrix}$

3. $\begin{pmatrix} 1 \\ 1 \\ 1 \\ 1 \\ 1 \\ 1 \end{pmatrix}$ 5. Several possible answers, including $\begin{pmatrix} 0 & 0 & 1 & 1 \\ 1 & 0 & 0 & 1 \\ 1 & 0 & 1 & 0 \\ 1 & 0 & 1 & 1 \\ 0 & 1 & 0 & 1 \\ 0 & 1 & 1 & 0 \\ 0 & 1 & 1 & 1 \\ 1 & 1 & 0 & 0 \\ 1 & 1 & 0 & 1 \\ 1 & 1 & 1 & 0 \\ 1 & 1 & 1 & 1 \\ 1 & 0 & 0 & 0 \\ 0 & 1 & 0 & 0 \\ 0 & 0 & 1 & 0 \\ 0 & 0 & 0 & 1 \end{pmatrix}$

13. An error is detected if and only if w is not a codeword. Note that $w = u + e$ and that the set of codewords is closed under addition.

Section 16.3, page 490

1. (a) If $f(x) = a_n x^n + \cdots + a_i x^i + \cdots + a_0$, then $f(x) + f(x) = (a_n + a_n)x^n + \cdots + (a_i + a_i)x^i + \cdots + (a_0 + a_0) = 0x^n + \cdots + 0x^i + \cdots + 0$ because $a_i + a_i = 0$ for every $a_i \in \mathbb{Z}_2$.

3. Verify that $1 + x + x^4$ has no roots in $\mathbb{Z}_2$ and, hence, no first- or third-degree factors. If there is a quadratic factor, it is either the product of two linear factors or irreducible. Use long division to show that the only irreducible quadratic (Exercise 2) is not a factor.

5. (a) Use the table to show that α^3 is a root of $f(x) = 1 + x + x^2 + x^3 + x^4$. It then suffices to show that $f(x)$ is irreducible. Use the method of Exercise 3.

7. (c) If $f([a_0 + a_1 x + \cdots + a_{n-1}x^{n-1}]) = (0,0, \ldots , 0)$, then $[a_0 + a_1 x + \cdots + a_{n-1}x^{n-1}] = [0]$, so that the kernel of f is the identity subgroup. Apply Theorem 7.40.

9. (a) $D(x) = x^2 + \alpha^4 x + \alpha$ has roots $1 = \alpha^0$ and $\alpha = \alpha^1$. Hence, the correct word is 000000000000000.   (c) $D(x) = x^2 + \alpha^{13}x + \alpha^4$ has roots α^9 and α^{10}. Hence, the correct word is 101010010110000.

Appendix B, page 514

1. (a) $\{-2, -1, 0, 1, 2, 3, 4, 5, 6, 7, 8\}$ (c) $\{1, 2\}$.
3. (a) Empty since $\sqrt{2}$ is irrational   (c) Empty.
7. $(a,0), (a,1)$ $(a,c), (b,0), (b,1), (b,c), (c,0), (c,1), (c,c)$.
11. (a) yes   (c) yes.
13. (a) Many correct answers, including the functions f, g, h, k given by $f(1) = a$, $f(2) = b$, $f(3) = c$, $f(4) = a$; $g(1) = c$, $g(2) = b$, $g(3) = a$, $g(4) = b$; $h(1) = b$, $h(2) = a$, $h(3) = c$, $h(4) = c$; $k(1) = c$, $k(2) = a$, $k(3) = a$, $k(4) = b$.   (c) There are six bijections from C to C.
19. If $(a,d) \epsilon A \times (B \cup C)$, then $a \epsilon A$ and $d \epsilon B$ or $d \epsilon C$. Therefore, $(a,d) \epsilon A \times B$ or $(a,d) \epsilon A \times C$, and, hence, $(a,d) \epsilon (A \times B) \cup (A \times C)$. Thus $A \times (B \cup C) \subseteq (A \times B) \cup (A \times C)$. Conversely, suppose $(r,s) \epsilon (A \times B) \cup (A \times C)$. Then $(r,s) \epsilon A \times B$ or $(r,s) \epsilon A \times C$. If $(r,s) \epsilon A \times B$, then $r \epsilon A$ and $s \epsilon B$ (and, hence, $s \epsilon B \cup C$), so that $(r,s) \epsilon A \times (B \cup C)$. Similarly, if $(r,s) \epsilon A \times C$, then $(r,s) \epsilon A \times (B \cup C)$. Therefore, $(A \times B) \cup (A \times C) \subseteq A \times (B \cup C)$, and, hence, the two sets are equal.
23. No; why not?
25. (a) If $f(a) = f(b)$, then $2a = 2b$. Dividing both sides by 2 shows that $a = b$. Therefore, f is injective.   (c) If $f(a) = f(b)$, then $a/7 = b/7$, which implies that $a = b$.
27. (a) If $(g \circ f)(a) = (g \circ f)(b)$, then $g(f(a)) = g(f(b))$. Since g is injective, $f(a) = f(b)$. This implies that $a = b$ because f is injective. Therefore, $g \circ f$ is injective.
29. (a) Let $d \epsilon D$. Since $g \circ f$ is surjective, there exists $b \epsilon B$ such that $(g \circ f)(b) = d$. Let $c = f(b) \epsilon C$. Then $g(c) = g(f(b)) = (g \circ f)(b) = d$. Hence, g is surjective.

Appendix C, page 525

1. $P(0)$ is true since $0 = 0(0 + 1)/2$. If $P(k)$ is true, then $1 + 2 + \cdots + k = k(k + 1)/2$. Add $k + 1$ to both sides and show that the right side is $(k + 1)(k + 2)/2$. This says that $P(k + 1)$ is true.
3. Let $P(n)$ be the statement $2^{n-1} \leq n!$. Verify that $P(0)$ and $P(1)$ are true. If $P(k)$ is true and $k \geq 1$, then $2^{k-1} \leq k!$ and $2 \leq k + 1$. Hence, $(2^{k-1})2 \leq k!(k + 1)$, that is, $2^k \leq (k + 1)!$. Thus $P(k + 1)$ is true.
7. Suppose the statement is true for k, that is, that 3 is a factor of $2^{2k+1} + 1$. Then $2^{2k+1} + 1 = 3t$, and, hence, $2^{2k+1} = 3t - 1$. To show that the statement is true for $k + 1$, note that $2^{2(k+1)+1} = 2^{2k+2+1} = 2^{2k+1}2^2 = (3t - 1)4 = 12t - 4 = 3(4t - 1) - 1$, and, hence, $2^{2(k+1)+1} + 1 = 3(4t - 1)$.
11. Let $B = \{b_1, b_2, \ldots, b_n\}$. In defining an injective function from B to B, there are n possible choices for the image of b_1, $n - 1$ choices for the image of b_2 (because b_2 can't have the same image as b_1), $n - 3$ choices for the image of b_3, and so on.
13. (a) Assume that a set of k elements has $k(k - 1)/2$ two-element subsets and that B has $k + 1$ elements. Choose $b \epsilon B$ and let $C = B - \{b\}$. Every two-element subset of B consists either of two elements of C or of b and one element of C. There are $k(k - 1)/2$ subsets of the first type by the induction hypothesis.

Appendix D, page 530

3. (a) $a \sim a$ since $\cos a = \cos a$. If $a \sim b$, then $\cos a = \cos b$ and, by the symmetric property of $=$, $\cos b = \cos a$; hence, $b \sim a$. If $a \sim b$ and $b \sim c$, then $\cos a = \cos b$ and $\cos b = \cos c$. Hence, $\cos a = \cos c$, and, therefore, $a \sim c$.
5. (b) The equivalence class of (r,s) is the vertical line through (r,s).
9. (a) Transitive   (c) Symmetric.
19. (b) Consider the subgroup $K = \{r_0, v\}$ of D_4.

Appendix E, page 535

1. 4032.
3. $\dbinom{n}{r} = \dfrac{n!}{r!(n-r)!} = \dfrac{n!}{(n-(n-r))!(n-r)!} = \dbinom{n}{n-r}.$

Appendix F, page 540

1. (a) $A + B = \begin{pmatrix} 1 & -6 & 0 & 4 \\ 9 & 5 & 11 & 12 \end{pmatrix}.$
3. (a) The entry in position i-j of $A + B$ is $a_{ij} + b_{ij}$. But $a_{ij} + b_{ij} = b_{ij} + a_{ij}$, which is the entry in position i-j of $B + A$. Hence, $A + B = B + A$.

Appendix G, page 548

1. (a) $x + x^3 + x^5$ (c) $(-11, 7.5, -3, 12, -5, 0, 3, 0, 0, 0, \ldots)$.
3. (a) $[(a_0, a_1, \ldots) \oplus (b_0, b_1, \ldots)] \oplus (c_0, c_1, \ldots)$

$= (a_0 + b_0, a_1 + b_1, \ldots) \oplus (c_0, c_1, \ldots)$

$= ((a_0 + b_0) + c_0, (a_1 + b_1) + c_1, \ldots)$

$= (a_0 + (b_0 + c_0), a_1 + (b_1 + c_1), \ldots)$

$= (a_0, a_1, \ldots) \oplus (b_0 + c_0, b_1 + c_1, \ldots)$

$= (a_0, a_1, \ldots) \oplus [(b_0, b_1, \ldots) \oplus (c_0, c_1, \ldots)].$

Index